THE OXFORD HANDBOOK OF

NEURONAL PROTEIN SYNTHESIS

THE OXFORD HANDBOOK OF

NEURONAL PROTEIN SYNTHESIS

Edited by

WAYNE S. SOSSIN

OXFORD
UNIVERSITY PRESS

OXFORD
UNIVERSITY PRESS

Oxford University Press is a department of the University of Oxford. It furthers the University's objective of excellence in research, scholarship, and education by publishing worldwide. Oxford is a registered trade mark of Oxford University Press in the UK and certain other countries.

Published in the United States of America by Oxford University Press
198 Madison Avenue, New York, NY 10016, United States of America.

Library of Congress Cataloging-in-Publication Data
Names: Sossin, Wayne S., editor.
Title: The Oxford handbook of neuronal protein synthesis / edited by Wayne S. Sossin.
Other titles: Oxford handbooks in neuroscience.
Description: New York : Oxford University Press, 2021. | Series: Oxford handbooks in neuroscience | Includes bibliographical references and index.
Identifiers: LCCN 2020047835 (print) | LCCN 2020047836 (ebook) |
ISBN 9780190686307 (hardback) | ISBN 9780190686321 (epub) |
ISBN 9780190686338 (ebook)
Subjects: MESH: Protein Biosynthesis | Neuronal Plasticity–physiology
Classification: LCC QH450.2 (print) | LCC QH450.2 (ebook) |
NLM WL 102 | DDC 572.8/845–dc23
LC record available at https://lccn.loc.gov/2020047835
LC ebook record available at https://lccn.loc.gov/2020047836

1 3 5 7 9 8 6 4 2

Printed by Marquis, Canada

Oxford Handbooks in Neuroscience

The *Oxford Handbooks in Neuroscience* series features authoritative, critical, and original perspectives on the state of the science, research trends, and future research needs in neuroscience, as well as on the relationship of research to applications in health and disease. Written by noted experts, these published review articles provide a critical survey of the current state of research on a topic and insightful discussions on the future directions of that research in this fast-moving field. *Oxford Handbooks in Neuroscience* is thus a premier source for sophisticated reviews on cutting-edge topics for experienced and new researchers, educators, leaders, technicians, and students. This series provides the kind of critical and original perspectives that readers need to advance the total scope of issues integral to, and associated with, neuroscience and its many related fields.

Gordon M. Shepherd
Editor in Chief
Oxford Handbooks in Neuroscience

Editorial Board

Contents

About the Editor xi
Contributors xiii
Preface xvii

PART I ROLE OF TRANSLATION FACTORS

1. Regulation of Protein Synthesis by eIF4E in the Brain 3
KLEANTHI CHALKIADAKI, STELLA KOULOULIA, CLIVE R. BRAMHAM, AND CHRISTOS G. GKOGKAS

2. Translational Control Through the eIF4E Binding Proteins in the Brain 23
ARGEL AGUILAR-VALLES, EDNA MATTA-CAMACHO, AND NAHUM SONENBERG

3. The Integrated Stress Response in Memory and Cognitive Disorders 43
JACQUNAE L. MAYS AND MAURO COSTA-MATTIOLI

4. The Role of the Eukaryotic Elongation Factor 2 (eEF2) Pathway in Neuronal Function 63
ELHAM TAHA AND KOBI ROSENBLUM

5. Sidekick No More: Neural Translation Control by p70 ribosomal S6 kinase 1 81
ADITI BHATTACHARYA

PART II ROLE OF SPECIFIC ELEMENTS IN THE MRNA

6. Dendritic Targeting and Regulatory RNA Control of Local Neuronal Translation 105
TAESUN EOM, ILHAM A. MUSLIMOV, ANNA IACOANGELI, AND HENRI TIEDGE

7. Internal Ribosome Entry Site-Mediated Translation in Neuronal Protein Synthesis 131
MARTIN HOLCIK

8. RNA Modifications in the Central Nervous System 153
DAN OHTAN WANG

9. Role of CPEB-Family Proteins in Memory 193
KAUSIK SI

10. FMRP and MicroRNAs in Neuronal Protein Synthesis 217
MONICA C. LANNOM AND STEPHANIE CEMAN

11. Focusing on mRNA Granules and Stalled Polysomes Amidst Diverse Mechanisms Underlying mRNA Transport, mRNA Storage, and Local Translation 239
MINA N. ANADOLU AND WAYNE S. SOSSIN

PART III PHYSIOLOGICAL ROLES OF TRANSLATION

12. Protein Synthesis and Synapse Specificity in Functional Plasticity 269
RADHA RAGHURAMAN, AMRITA BENOY, AND SREEDHARAN SAJIKUMAR

13. Regulation of Synaptic Homeostasis by Translational Mechanisms 297
MEGUMI MORI, JAY PENNEY, AND PEJMUN HAGHIGHI

14. Multiple Roles of RNA Regulatory Factors in Neuronal Development and Function in *C. elegans* 323
MATTHEW G. ANDRUSIAK AND YISHI JIN

15. Regulation of mRNA Translation in Axons 359
PRIYANKA PATEL, PABITRA K. SAHOO, AMAR N. KAR, AND JEFFERY L. TWISS

16. Protein Synthesis and Translational Control in Neural Stem Cell Development and Neurogenesis 397
LAMEES MOHAMMAD, JOSCELYN WISEMAN, SARAH ERICKSON, AND GUANG YANG

PART IV NEURONAL PROTEIN SYNTHESIS AND DISEASE

17. Translational Controls in Pain 427
Sarah Loerch, June Bryan De La Peña, Jane Song, Joseph J. Pancrazio, Theodore J. Price, and Zachary T. Campbell

18. Dysregulated Translation in Autism Spectrum Disorder 451
Emanuela Santini and Anders Borgkvist

19. Neuronal mRNA Translation in Addiction 487
Emma Puighermanal and Emmanuel Valjent

20. Dysregulated Protein Synthesis in Major Depressive Disorder 511
Chelcie F. Heaney and Kimberly F. Raab-Graham

21. Dysregulation of Neuronal Protein Synthesis in Alzheimer's Disease 533
Tao Ma

22. RNA-Binding Proteins and Translation in Neurodegenerative Disease 551
Kent E. Duncan

23. Role of Eukaryotic Initiation Factor eIF2B in Vanishing White Matter Disease 595
Truus E. M. Abbink, Lisanne E. Wisse, Xuemin Wang, and Christopher G. Proud

Index 619

About the Editor

Wayne S. Sossin received undergraduate degrees in biology and computer science from MIT in 1984 and a PhD from Stanford in biological sciences with Dr. Richard Scheller in 1989. He completed postdoctoral work with Dr. Schwartz at Columbia University in the Center for Neurobiology and Behavior before being appointed assistant professor at the Montreal Neurological Institute at McGill University in 1993, where he is now a James McGill Professor. He has been an EJLB Scholar, CIHR investigator, and FRSQ Checheur Nationaux. Dr. Sossin has published over 125 papers on the molecular and cellular processes underlying memory formation and maintenance, with a particular interest in the role of persistent protein kinases and the regulation of local translation.

Contributors

Truus E. M. Abbink, VU University Medical Center, The Netherlands

Argel Aguilar-Valles, McGill University, Canada

Mina N. Anadolu, McGill University, Canada

Matthew G. Andrusiak, University of California, San Diego, USA

Amrita Benoy, National University of Singapore, Singapore; Life Sciences Institute, Singapore

Aditi Bhattacharya, Institute for Stem Cell Biology and Regenerative Medicine, India

Anders Borgkvist, Karolinska Institute, Sweden

Clive R. Bramham, University of Bergen, Norway

Zachary T. Campbell, University of Texas at Dallas, USA

Stephanie Ceman, University of Illinois at Urbana Champaign, USA

Kleanthi Chalkiadaki, University of Edinburgh, UK

Mauro Costa-Mattioli, Baylor University, USA

June Bryan De La Peña, University of Texas at Dallas, USA

Kent E. Duncan, University Medical Center Hamburg-Eppendorf, Germany

Taesun Eom, State University of New York Downstate Medical Center, USA

Sarah Erickson, University of Calgary, Canada; Alberta Children's Hospital Research Institute, Canada

Christos G. Gkogkas, University of Edinburgh, UK

Pejmun Haghighi, Buck Institute for Research on Aging, USA; McGill University, Canada

Chelcie F. Heaney, Wake Forest University, USA

Martin Holcik, Carleton University, Canada

Anna Iacoangeli, State University of New York Downstate Medical Center, USA

Yishi Jin, University of California, San Diego, USA

Amar N. Kar, University of South Carolina, USA

Stella Kouloulia, University of Edinburgh, UK

Monica C. Lannom, University of Illinois at Urbana Champaign, USA

Sarah Loerch, Howard Hughes Medical Institute, USA

Tao Ma, Wake Forest University, USA

Edna Matta-Camacho, McGill University, Canada

Jacqunae L. Mays, Baylor University, USA

Lamees Mohammad, University of Calgary, Canada; Alberta Children's Hospital Research Institute, Canada

Megumi Mori, Buck Institute for Research on Aging, USA; University of Southern California, USA

Ilham A. Muslimov, State University of New York Downstate Medical Center, USA

Joseph J. Pancrazio, University of Texas at Dallas, USA

Priyanka Patel, University of South Carolina, USA

Jay Penney, McGill University, Canada

Theodore J. Price, University of Texas at Dallas, USA

Christopher G. Proud, South Australian Health and Medical Research Institute, Australia; University of Adelaide, Australia

Emma Puighermanal, Autonomous University of Barcelona, Spain

Kimberly F. Raab-Graham, Wake Forest University, USA

Radha Raghuraman, National University of Singapore, Singapore; Life Sciences Institute, Singapore

Kobi Rosenblum, University of Haifa, Israel

Pabitra K. Sahoo, University of South Carolina, USA

Sreedharan Sajikumar, National University of Singapore, Singapore; Life Sciences Institute, Singapore

Emanuela Santini, Karolinska Institute, Sweden

Kausik Si, Stowers Institute for Medical Research, USA; University of Kansas, USA

Nahum Sonenberg, McGill University, Canada

Jane Song, University of Texas at Dallas, USA

Wayne S. Sossin, Montreal Neurological Institute, Canada; McGill University, Canada

Elham Taha, University of Haifa, Israel

Henri Tiedge, State University of New York Downstate Medical Center, USA

Jeffery L. Twiss, University of South Carolina, USA

Emmanuel Valjent, University of Montpellier, France

Dan Ohtan Wang, Shenyang Pharmaceutical University, China; Kyoto University, Japan

Xuemin Wang, South Australian Health and Medical Research Institute, Australia; University of Adelaide, Australia

Joscelyn Wiseman, University of Calgary, Canada; Alberta Children's Hospital Research Institute, Canada

Lisanne E. Wisse, VU University Medical Center, The Netherlands

Guang Yang, University of Calgary, Canada; Alberta Children's Hospital Research Institute, Canada

Preface

WAYNE S. SOSSIN

TRANSLATIONAL control is the ability of cells to determine when, where, and how much protein will be produced from a particular mRNA. In concert with the regulated transcription of mRNAs and degradation of proteins, translational control contributes significantly to determining the proteome of cells. Our basic understanding of translational control has come from examining cells that continually replicate, such as cancer cell lines, bacteria, or yeast. To divide, cells need to duplicate their proteome; thus, much of translational control in these cells is focused on this important decision. Another area of focus for the field of translational control is in the contest between invaders of the cell (usually viruses) and the cell for control of the protein synthetic machinery, and here, the field has focused on the subset of translational regulators involved in this competition. Neurons are non-dividing cells, and the field of neuronal translational control is mainly devoted to the question of how neurons change their proteome in response to activity. In this, it shares some features of translational control during development, where the questions of where and when specific mRNAs are translated has dominated the field. However, developmental translational control represents a genetically encoded program whereby specific proteins are produced at specific times to drive proper development. In neurons, however, translational control is also used as a mechanism to modify neuronal properties and connections in response to ongoing input from the outside world. Indeed, protein synthesis has been extensively studied in the context of the synaptic plasticity that underlies memory formation. In addition, it is noteworthy that when protein synthesis is dysregulated, the nervous system is a sensitive marker of this dysregulation, often being more affected than other cellular systems. This may be due to the necessity of translational control for fine-tuning function in the nervous system more so than in other cellular systems.

This volume is devoted to the neuron-specific aspects of translational control, both in normal physiological function and in disease. Importantly, not only does the volume comprehensively review what is already known about translational control in the nervous system, but it addresses the important issues that remain to be resolved, as well as presenting new ways of thinking about these issues to invigorate future research. The volume is separated into four parts, the role of translation factors; the role of specific elements in the mRNA, physiological roles of translation, and neuronal protein synthesis and disease. Each section is previewed here.

Part I: Role of Translation Factors

The control of where, when, and how much proteins are produced is often mediated by the regulation of the protein factors that control the process of protein synthesis. Translation is canonically divided into three parts: initiation, where the mRNA is brought to the ribosome and the initiating methionine is chosen; elongation, where the genetic code is read and the protein is synthesized: and termination, where the protein is released and the ribosomes are recycled. This volume begins with one of the first steps of initiation. The eukaryotic initiation factor (eIF) 4E facilitates initiation through binding to the 7-methyl guanosine triphosphate cap found at the start of all mRNAs and to the ribosome-associated scaffold protein eIF4G. Two chapters cover the regulation of this interaction. First, the role of phosphorylation of eIF4E and its regulation of specific mRNAs to regulate synaptic plasticity is reviewed, and new proposed roles for eIF4E in the nucleus and in mRNA transport from the nucleus are discussed. Second, the eIF4E binding proteins (4EBPs) and their regulation of eI4E availability through phosphorylation of the 4EBPs by the mechanistic target of rapamycin (TOR) complex 1 (mTORC1) is comprehensively reviewed. The important role of this regulation in many aspects of neuronal development, plasticity, and disease is discussed. Another major regulator of translation initiation is through phosphorylation of eIF2α, which regulates the availability of the charged transfer RNA containing the initiating methionine; and this phosphorylation is a hallmark of the integrated stress response. Chapter 3 covers the important role of eIF2α phosphorylation regulation in determining whether or not a memory is formed through the ability of this regulation to control translation of critical factors involved in this decision. Interestingly, the idea that drugs that regulate eIF2α phosphorylation could be important treatments is presented. Eukaryotic elongation factors (eEFs), in particular eEF2, are also implicated in neuronal translational control, synaptic plasticity, and neurodegenerative diseases through regulating the elongation step of translation; and this area is reviewed in Chapter 4. As well as the 4EBPs, mTORC1 targets S6 kinase, and this is important for the regulation of both initiation and elongation. The role of S6 kinase in both physiological and pathological aspects of nervous system function is reviewed in Chapter 5.

Part II: Role of Specific Elements in the mRNA

While translation factors directly regulate translation, the particular mRNAs that are regulated are determined by properties of the mRNAs themselves, often through sequences recognized by RNA binding proteins (RBPs), which then interact with other factors, including the translation factors outlined in the first section, to determine translational control. The first chapter of this section (Chapter 6) focuses on the role of RBPs in mRNA transport to synaptic sites, reviewing the roles of a number of distinct RBPs as well as non-coding RNAs that play similar roles through binding to specific elements of mRNAs and coupling them to specific translation factors such as eIF4B. mRNAs with internal ribosome entry site (IRES) sequences in the 5′-untranslated region of mRNAs attract ribosomes independently of eIF4E and the 7-methyl cap. In Chapter 7, a number of important mRNAs that regulate aspects of nervous system function and have been identified as harboring an IRES are introduced, and the importance of this regulation in physiological and pathological regulation of the nervous system is discussed. The ability to post-transcriptionally modify mRNAs to regulate their translation greatly increases the complexity of translational control in the nervous system. In Chapter 8, the growing evidence that direct modification of the nucleotides in RNA plays a role in nervous system translational control is reviewed, as well as the technological hurdles that will need to be overcome to firmly establish the role of these modifications in nervous system function. Another post-transcriptional mechanism that regulates translation is controlling the length of the poly-A tail of mRNAs and, in particular, the cytoplasmic polyadenylation element binding protein (CPEB) that regulates this process. CPEBs play a key role in regulating mRNAs, whose translation is important for long-term memories; and the hypothesis that aggregation of CPEBs may result in long-lasting changes in neuronal translational control is reviewed in Chapter 9. Two highly studied regulators of nervous system translation, the RBP lost in fragile X syndrome (fragile X mental retardation protein) and microRNAs, are linked in Chapter 10, which thoroughly reviews both regulators and, importantly, the interactions between them. In the last chapter of this section (Chapter 11), the importance of macromolecular regulation of neuronal mRNAs, particularly the formation of non-vesicular mRNA granules, is reviewed; and the idea that these neuronal mRNA granules contain polysomes stalled at the elongation step is discussed.

Part III: Physiological Roles of Translation

While the first two sections of the volume focus on individual translation factors, specific mRNA modifications, and identified RBPs, the next two sections focus on physiological and pathological nervous system processes and review the importance of translational control in these processes. In the first chapter of this section (Chapter 12), the importance of protein synthesis in long-term plasticity is reviewed; and in particular, ideas on how protein synthesis is important for determining which of the thousands of synapses a neuron possesses are modified by plasticity are presented. Next, Chapter 13 reviews the evidence that neuronal homeostasis is mediated by protein synthesis mechanisms, and the importance of this homeostasis for neuronal function is discussed.

Caenorhabditis elegans is an important model for nervous system function because of the "elegant" genetics of this system, allowing unbiased screens to identify important regulators of nervous system function. Interestingly, using these unbiased screens, Chapter 14 describes how alternative splicing is a critical post-translational mRNA modification regulating many aspects of nervous system function. A unique feature of neurons are their axons, which can be extremely long, and thus present fundamental cell biological problems for somatic control. Chapter 15 describes how regulation of protein synthesis directly in axons can overcome this problem. Finally, while translational control has long been recognized as an important mediator of development, its role in neural stem cells has only recently been elucidated; and this important role is reviewed in Chapter 16.

Part IV: Neuronal Protein Synthesis and Disease

The nervous system is particularly sensitive to dysfunctions in protein synthesis, and many mutations in ribosomal proteins, translation factors, and RBPs lead to specific deficits in the nervous system; and the previous chapters in the volume highlight these specific deficits when discussing specific factors. In the last section of the volume, we examine specific nervous system–related diseases and the overall importance of translational control for this particular pathological state. The first chapter in this section, Chapter 17, examines the role of translation in pain, showing how noxious stimuli activate protein synthesis to lead to the persistent changes required for a long-lasting pain state and suggests that this state can be alleviated through alterations in translational control pathways. Next, Chapter 18 examines the role of translation in a neurodevelopmental context, discussing how many mutations that lead to autism spectrum disorder dysregulate translational control and the role of translation as a major hub for this disorder, with the possible consequence of using dysfunctional translation as a diagnostic for this disorder. Next, psychiatric disorders are examined. Chapter 19 outlines how addiction is mediated by long-term changes in neuronal circuits that require changes in translational control and emphasizes how identifying the translational changes in specific cell types will be key in understanding the mechanisms of addiction. Chapter 20 examines the role of translational control in depression, focusing on the important role modulation of the TOR pathway plays in depression and discussing if antidepressants based on this are feasible. The volume then moves on to studying neurodegenerative diseases. Chapter 21 reviews the evidence that specific translation factors are altered in Alzheimer's disease and discusses possible causative roles of these changes in the disease. Chapter 22 examines neurodegenerative diseases more generally and the important role that mutations of RBPs play in these diseases, through either aggregation or dominant negative effects of these mutations on many RNA-based processes. Chapter 23 examines probably the most potent example of how animal-wide changes in translation factors, in this case the eIF2B complex, cause a specific nervous system pathology, in this case vanishing white matter disease.

This volume should be an excellent resource for anyone in the neurosciences interested in the role of protein synthesis and anyone in the translational control field interested in the specialized applications of translational control in the nervous system. The intersection of these fields is an exciting place to be. I thank the *Oxford Handbooks in Neuroscience* editorial board for initiating the project and inviting me to be the editor of this volume. The quality of the book is due to the time and effort of the chapter authors, all of whom are experts in their fields. I also give special thanks to Ada Brunstein, editor in chief at Oxford University Press, for supporting the project, for her always enthusiastic attitude, and for ensuring that the volume stayed on schedule.

PART I

ROLE OF TRANSLATION FACTORS

CHAPTER 1

REGULATION OF PROTEIN SYNTHESIS BY EIF4E IN THE BRAIN

KLEANTHI CHALKIADAKI, STELLA KOULOULIA, CLIVE R. BRAMHAM, AND CHRISTOS G. GKOGKAS

INTRODUCTION

RIBOSOMES generate proteins via translation of cytoplasmic mRNA. Translation includes three main steps: initiation, elongation and termination/recycling (Hershey, Sonenberg, & Mathews, 2012). Nascent mRNAs, possessing the cap structure m7GpppN (where m is a methyl group and N is any nucleotide) at their 5′-end, are predominantly translated through cap-dependent translation (Raught & Gingras, 1999). In higher eukaryotes, the decoding from mRNA into amino acids involves several participants; the mRNA, the small-40S and large-60S ribosomal subunits, each one consisting of ribosomal RNA (rRNA) and proteins, the aminoacyl-tRNAs, ATP and various proteins termed eIFs (eukaryotic initiation factors; Hershey et al., 2012). Binding of eIF4E to the cap is indispensable for initiation, which is the rate-limiting step of cap-dependent translation (Marcotrigiano, Gingras, Sonenberg, & Burley, 1997; Sonenberg, Rupprechtt, Hechtt, & Shatkin, 1979).

Cap-dependent initiation commences with the formation of the 43S preinitiation complex. The Met-tRNA-eIF2-GTP ternary complex binds to the 40S ribosomal subunit (Hershey et al., 2012) and the 48S pre-initiation complex is now assembled, with the addition of eIF4E, eIF4G, eIF4A and eIF4B (Hershey et al., 2012). eIF4E, eIF4G and eIF4A form the eIF4F complex: eIF4E binds to the cap of mRNA and associates with eIF4G, a scaffolding protein, bridging the 40S ribosomal subunit and the mRNA by binding to eIF3 (Clemens, Elia, & Morley, 2013). eIF4A is a helicase that unwinds the secondary structure of mRNA (Svitkin et al., 2001). During initiation the eIF4F complex scans the 5′ UTR, and upon recognition of the initiation codon (AUG), the 60S ribosomal subunit is added in an eIF5-dependent step, followed by GTP hydrolysis (Hershey et al., 2012). Therefore, the 80S initiation complex is assembled and protein synthesis enters into its elongation phase (Clemens et al., 2013). The elaborate orchestration and regulation of initiation reflects an evolutionary conserved process, by which certain mRNAs are preferentially translated in different organisms and in different cell-types within the human body (Sonenberg & Hinnebusch, 2009).

mRNA Translation in the Brain

In the brain, formation of long-term memories (LTM) requires newly synthesized proteins. The very first evidence was found in 1963, when it was shown that puromycin, a protein-synthesis inhibitor, could block long-term aversive memory formation in mice (Flexner et al., 1963). Since then, a significant number of reports have supported an important role of translation in memory formation and in long-term synaptic plasticity. Early studies in Aplysia neurons and in isolated synaptic neuropil from the CA1 region of hippocampal slices have shown that long-term facilitation (Casadio et al., 1999; Martin et al., 1997) and synaptic plasticity (Kang & Schuman 1996; Huber et al., 2000) require local translational control mechanisms. By using protein synthesis inhibitors, Cracco showed that local protein synthesis in isolated synaptic neuropil from the CA1 region of hippocampal slices is necessary for long-term potentiation (LTP) induced by a brief double-tetanic stimulation (Cracco, 2005). Furthermore, early studies in the mammalian brain have revealed a differential effect of protein synthesis, between the two phases of long-term synaptic plasticity, suggesting that novel protein synthesis is only essential for the late-phase LTP (L-LTP) (Frey, Krug, Reymann, & Matthies, 1988; Stanton & Sarvey, 1984).

Subsequent work on the mechanisms of LTP maintenance has uncovered remarkable differences between hippocampal regions in the biochemical mechanisms and spatial-temporal control of translation, suggesting the existence of cell-type specific mechanism of translation (reviewed in Panja & Bramham, 2014. As discussed later in the chapter, major differences in the mechanisms of protein synthesis–dependent LTP exist between the perforant path input from entorhinal cortex to the dentate gyrus (DG) and the Schaffer collateral/commissural fiber input from CA3 to CA1. In addition, the mechanism of LTP is highly dependent on the pattern of synaptic stimulation used to induce LTP.

eIF4E-Related Translational Control Mechanisms in the Brain

Herein, we will review and discuss the current knowledge regarding the mechanisms of translational control mediated by the cap-binding protein eIF4E and how they may be linked to various aspects of brain function, such as synaptic plasticity, learning, memory, and disease.

eIF4E Regulation by eIF4E-Binding Proteins

Regulation of eIF4E activity, and therefore of cap-dependent protein synthesis, can occur through a family of inhibitory proteins, the eIF4E-binding proteins (4E-BPs). This family consists of three members, 4E-BP1, 4E-BP2 and 4E-BP3 (Raught & Gingras, 1999). 4E-BP1 is highly expressed in adipose tissue and muscle, whereas 4E-BP2 is the abundant brain paralog (J. L. Banko, 2005; Tsukiyama-Kohara et al., 2001). 4E-BPs inhibit eIF4E-dependent translation *in vitro* and *in vivo* by hindering the formation of the eIF4F complex, via competition with eIF4G for binding to eIF4E, which is essential for the recruitment of the ribosome to the mRNA (Haghighat, Mader, Pause, & Sonenberg, 1995; Pause, Methot, Svitkin, Merrick, & Sonenberg, 1994). The phosphorylation state of 4E-BPs determines their activity and subsequent effect on translation and is regulated by an evolutionarily conserved Ser/Thr kinase termed mTOR (mammalian/mechanistic Target of Rapamycin; Figure 1.1A). A more detailed description of the 4E-BP phosphorylation mechanism by mTOR can be found in the chapter by Aguilar-Valles, Matta-Camacho, and Sonenberg (this volume).

In the brain, several studies reveal a cardinal role for 4E-BPs in synaptic plasticity and memory. Phosphorylation of both eIF4E and 4E-BPs is substantially increased during LTP and LTM formation (Kelleher Iii, Govindarajan, Jung, Kang, & Tonegawa, 2004; Panja et al., 2009; Tang et al., 2002). The role of 4EBPs in synaptic transmission and memory formation is extensively discussed in chapter 2 by Aguilar-Valles, Matta-Camacho, and Sonenberg (this volume).

eIF4E Regulation by Phosphorylation

A second key mechanism of eIF4E regulation, is through mitogen-activated protein (MAP) kinase-interacting kinases (MNKs), MNK1 and MNK2. MNKs phosphorylate eIF4E in response to mitogens and cellular stress on Ser 209 (Joshi et al., 1995) through activation of either ERK or p38 pathway, stimulating cap-dependent translation (Gram, Knauf, & Tschopp, 2001; Ueda, Watanabe-Fukunaga, Fukuyama, Nagata, & Fukunaga, 2004). MNK1 phosphorylation by p38 or Erk1/2 MAPKs increases its interaction with eIF4G, thus facilitating eIF4E phosphorylation (Shveygert, Kaiser, Bradrick, & Gromeier, 2010; Figure 1.1A). The majority of the studies support a positive correlation of eIF4E phosphorylation and translation, reported by both in vitro and in vivo studies (Bramham, Jensen, & Proud, 2016; Lachance, Miron, Raught, Sonenberg, & Lasko, 2002; Panja et al., 2014; Pyronnet et al., 1999; Tuxworth, Saghir, Spruill, Menick, & McDermott, 2004). Interestingly, a selective effect of MNKs on translation of mRNAs containing both a cap-structure and a 5′- hairpin loop has been reported, in a cell-free

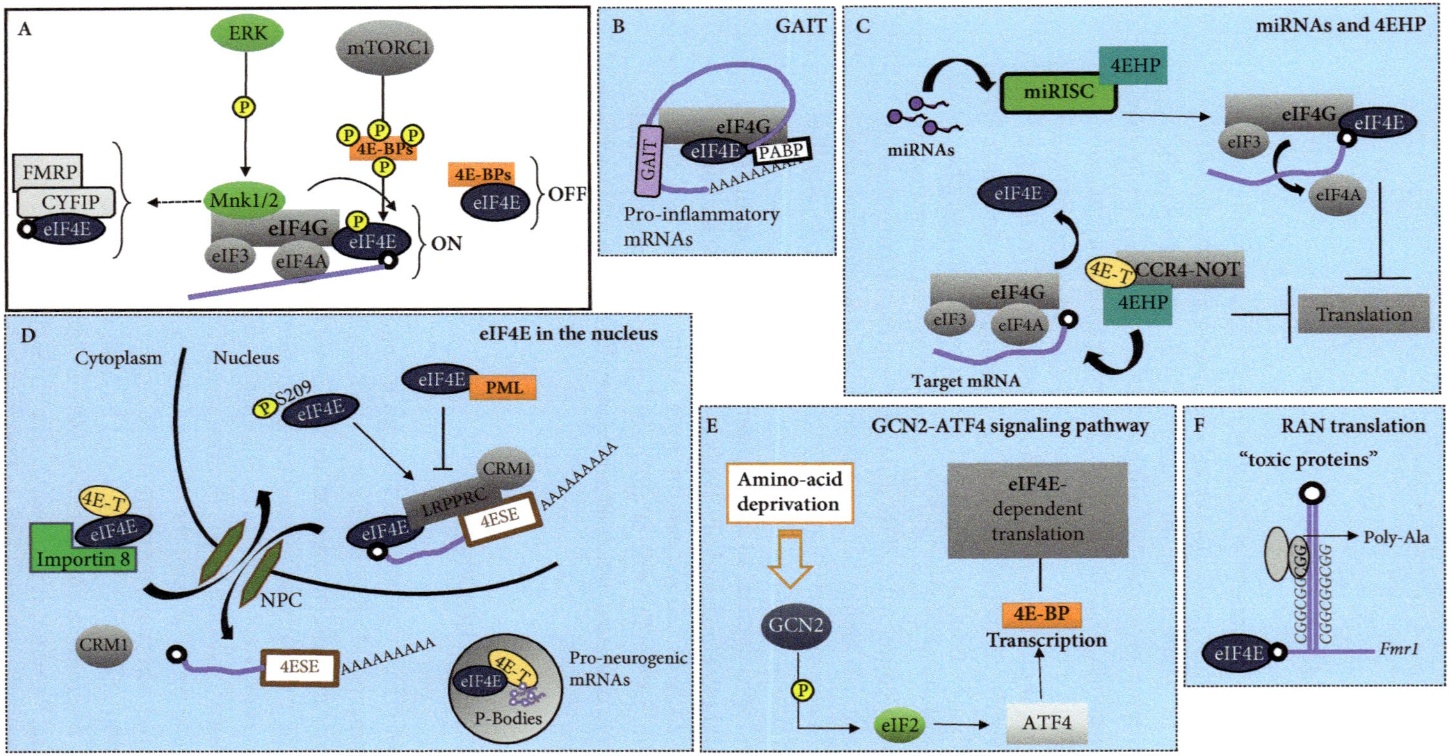

FIGURE 1.1. The role of eIF4E in mRNA translation and the regulation mechanisms. Schematic depiction of the signalling pathways and the interactions with eIF4E in the cytoplasm and the nucleus. A: eIF4E activity is controlled by mTORC1 and ERK signalling pathways. The diagram depicts the dual role of MNK1 in promoting protein synthesis through the dissociation of the CYFIP1/FMRP complex from the cap-structure the 4E-BP2 from the eIF4E. B: Certain phospho-eIF4E-sensitive mRNAs containing the 3′ UTR GAIT motif are preferentially translated. **C**: miRNAs and the homolog protein of eIF4E, 4E-HP, regulate mRNA translation, either by inhibiting the initiation step or by inducing mRNA degradation. D: eIF4E is implicated in the nuclear export of specific mRNAs that contain the 4ESE sequence, recruiting and interacting with several proteins. Importin-8, PML protein, and phosphorylation of eIF4E at Ser209 regulate the nuclear function of eIF4E. **E**: GCNT2 kinase and ATF4 binding protein regulate the transcription of 4E-BPs and, therefore, the mRNA translation. F: RAN translation is a non-canonical translation mechanism, identified in several neurodegenerative disorders, such as the Fragile X.

Abbreviations: 4E-BP2, eIF4E-binding protein 2; 4E-HP, eIF4E homologous protein; ATF4, activating-transcription factor 4; CYFIP1, cytoplasmic FMRP interacting protein; ERK, extracellular signal-regulated kinase; FMRP, fragile X mental retardation protein; GAIT, gamma interferon inhibitor of translation element; GCNT2 kinase, general control nonderepressible 2 kinase; MNK1, mitogen-activated protein (MAP) kinase-interacting serine/threonine-protein kinase 1; mTORC1, mammalian/mechanistic Target of Rapamycin complex 1; PML, promyelocytic leukemia protein; RAN translation, repeat-associated non-AUG translation; off, repression of translation; on, active translation.

study (Korneeva, Song, Gram, Edens, & Rhoads, 2016). Recently, a new phosphorylation site was identified on eIF4E. Mammalian site 20-like kinase (MST1) phosphorylation of eIF4E on T55 inhibits the synthesis of a subset of mRNAs, but paradoxically boosts non-canonical translation of long non-coding RNAs (such as *linc00689*; Min et al., 2017). *Linc00689* was previously shown to be implicated in autism spectrum disorders (ASD; Parikshak et al., 2016).

In the rodent brain, the N-methyl-D-aspartate subtype of glutamate receptor (NMDAR) holds a key a role in synaptic plasticity induction and in memory function (Bliss, T. V. P. & Collingridge, 1993; Morris, Anderson, Lynch, & Baudry, 1986). In the CA1 region of the hippocampus, NMDAR activation leads to MNK1 activation and eIF4E phosphorylation, in both dendritic and synaptic compartments (Jessica L. Banko, Hou, & Klann, 2004). NMDA-induced MNK1 activation is specifically PKA- and ERK-2 mediated (Jessica L. Banko et al., 2004). Additional evidence supporting the role of MNKs in brain function has been revealed with regards to mechanisms of LTP, in the dentate gyrus (DG). Panja and colleagues had initially shown that ERK-MNK, but not the mTORC1 signaling pathway, is essential for translation initiation and LTP consolidation in DG (Panja et al., 2009). It was later revealed that MNK1 has a dual role in translation initiation, through sustained activation of the brain-derived neurotrophic factor (BDNF) receptor, TrkB (Panja et al., 2014). In the early stage of LTP consolidation, MNK1 controls the translation repressor complex CYFIP1/FMRP, by triggering its dissociation from the mRNA cap structure, and thus promoting translation of FMRP-target mRNAs. In the late stage (2 hours after LTP induction), TrkB-MNK signaling triggers the removal of 4E-BP2 from eIF4E, and therefore promotes protein synthesis (Figure 1.1A). Interestingly, 4E-BP2 regulation occurs specifically in synaptic compartments of DG, as revealed by comparing whole DG lysates with DG-isolated synaptoneurosomes (Panja et al., 2014). Similar findings have been reported in cortical neurons (primary neuronal cultures). MNK1 but not MNK2 is crucial for BDNF-mediated protein synthesis regulation in cortical neurons, as proved by pharmacological and genetic approaches (Genheden et al., 2015). Given the strong link of eIF4E phosphorylation with the hippocampal DG and the recent findings that ablation of Ser209 eIF4E phosphorylation (Ser209Ala) does not affect CA1-dependent LTP or LTM (Amorim et al., 2018), it is highly plausible that eIF4E has a yet unidentified role in synaptic plasticity, learning and memory specifically via the DG.

Under normal conditions, eIF4E expression and phosphorylation increases in response to growth factors stimulation, such as BDNF (Genheden et al., 2015; Kanhema et al., 2006; Panja et al., 2014; Takei, Kawamura, Hara, Yonezawa, & Nawa, 2001) and NGF (Salehi & Mashayekhi, 2007), and thus, promotes new protein synthesis. However, eIF4E phosphorylation is also implicated in different pathological conditions, by mediating the translation of specific mRNAs. For example, a positive correlation between phospho-eIF4E and the hyperphosphorylated form of tau protein has been reported. Hyperphosphorylated tau is the core protein of neurofibrillary tangles, which constitute a characteristic marker of Alzheimer's disease, suggesting that increased eIF4E phosphorylation is involved in disease progression (X. Li et al., 2004). Another study demonstrated that increased eIF4E phosphorylation in a mouse model of fragile X syndrome (FXS) and in post-mortem brains from FXS patients is linked to increased metalloprotein 9 (MMP-9) levels. Interestingly, pharmacological or genetic inhibition of eIF4E phosphorylation reversed the FXS-like deficits, by regulating the aberrant translation of MMP-9 in FXS (Gantois et al., 2017; Gkogkas et al., 2014). Additional phospo-eIF4E dependent mRNAs have been identified, expanding our knowledge on the role of phospho-eIF4E in translational regulation in the brain (Amorim et al., 2018). Unbiased ribosome profiling in the forebrain of mice lacking phospho-Ser209 eIF4E revealed altered translation of mRNAs related to inflammation, extracellular matrix, pituitary hormones and serotonin pathway (Amorim et al., 2018). A significant proportion of phospho-eIF4E-sensitive mRNAs harbor a 3′ UTR sequence motif; the gamma interferon inhibitor of translation element (GAIT), which may explain their preferential translation (Amorim et al., 2018; Figure 1.1B).

In parallel, in the peripheral nervous system, it was shown that phosphorylation of eIF4E is required for BDNF mRNA translation, in dorsal root ganglion (DRG) neurons (Moy, Khoutorsky, Asiedu, Dussor, & Price, 2018). BDNF is implicated in pain plasticity and in conditions such as hyperalgesic priming (transition from acute to chronic pain; Melemedjian et al., 2013); therefore, phosphorylation of eIF4E seems to have a key role in regulating nociceptor plasticity (Muta et al., 2011).

eIF4E Regulation by Interaction with eIF4G

Translation initiation requires the interaction of eIF4E and eIF4G, which is endogenously regulated by 4E-BPs. Administration of the specific eIF4E-eIF4G interaction inhibitor, 4EGI-1, impaired LTM consolidation when infused in the lateral amygdala (LA) of rats (Hoeffer et al., 2011). However, no effects were found on memory formation or reconsolidation (Hoeffer et al., 2011). In the hippocampus, it was shown that 4EGI-1 disrupted L-LTP induction and blocked L-LTP-induced protein synthesis, without affecting basal synaptic transmission or presynaptic plasticity (Hoeffer et al., 2013). Based on these and previous reports (Artinian et al., 2008; Rodriguez-Ortiz & Bermudez-Rattoni, 2007), it has been proposed that memory consolidation requires cap-dependent translation, while reconsolidation is probably based on a cap-independent translation mechanisms. In parallel, at the molecular level it was shown that fear-conditioning training increased the frequency of polyribosomes counted on dendritic spines and on dendritic shafts, in the LA of rats (L. E. Ostroff, Cain, Bedont, Monfils, & LeDoux, 2010). Interestingly, electron microscopy imaging of LA revealed that blockade of eIF4E-eIF4G interaction, after fear-conditioning training affected only the polyribosomes located in the spine heads and dendritic shafts, without any impact on those located in the base or neck of the smallest spines (Linnaea E. Ostroff et al., 2017). Moreover, reducing eIF4E-eIF4G interactions with 4EGI-1 restored the balance between protein synthesis and actin dynamics in fragile X syndrome model mice (Santini et al., 2017).

eIF4E Regulation by MicroRNAs and 4EHP

Two main mechanisms of miRNA action in translational control are proposed: inhibition of translation initiation or induction of mRNA degradation (Oliveto, Mancino, Manfrini, & Biffo, 2017). In particular, miRNAs control translation by recruiting the miRNA-induced silencing complex (miRISC) on target mRNAs, which can then block translation initiation (Fukaya & Tomari, 2012; Mathonnet et al., 2007), followed by deadenylation and mRNA decay (Jonas & Izaurralde, 2015). Using in vitro studies, Fukao and colleagues have proposed that both eIF4AI and eIF4AII are important targets for the miRNA-mediated translation repression. They showed that miRISC induces the dissociation of eIF4AI and eIF4AII from the initiation complex, before eIF4E or eIF4G (Fukao et al., 2014), contributing to translation repression (Figure 1.1C). Furthermore, the homolog protein of eIF4E, eIF4E2 or 4EHP (4E-homologous protein), has been shown to directly interact with TNRC6 (tri-nucleotide-repeat containing proteins) of the miRISC. 4EHP competes with eIF4E for binding the cap-structure and unlike its homolog, it does not interact with eIF4G to facilitate translation (Chen & Gao, 2017). A parallel model proposes that 4EHP is recruited on the cap structure by direct interaction with the deadenylase complex CCR4-NOT and one of the eIF4E—binding proteins, 4E-T (4E transporter protein; Chapat et al., 2017). 4E-HP competes with eIF4E for binding to 4E-T protein and their interaction facilitates its binding to the cap-structure and thus, promoting translation repression (Figure 1.1C). However, in hypoxic conditions, a switch in 4E-HP function from translation-repressor to translation-activator has been identified (Uniacke et al., 2012). Interestingly, the 4E-HP-directed translation initiation, is used by cancer cells, in order to survive under hypoxic conditions of tumorigenesis (Melanson, Timpano, & Uniacke, 2017; Uniacke, Perera, Lachance, Francisco, & Lee, 2014). It has been further proposed that environmental oxygen levels determine the cap-dependent translation pathway that will be activated; eIF4E under high oxygen levels (physioxia/normoxia), both eIF4E and 4E-HP under mid-physioxia, or 4E-HP alone under hypoxic conditions (Timpano & Uniacke, 2016).

Recent advances in cancer research support a direct effect of miRNAs on eIF4E. For example, it has been found that miR-34c-3p and miR-141 suppress eIF4E expression, acting as a tumor suppressors in breast cancer and in non-small cell lung cancer (Liu et al., 2015; Wang, Ma, Ji, Xu, & Wei, 2017; Yao et al., 2015). Similarly, several other miRNAs (e.g.miRNA-497 and miR-455-3p) act as inhibitors of eIF4E mRNA expression, and suppress cell proliferation and survival, in various types of cancer (Jiang et al., 2014; Hassan, Zhao, Glover, Robinson, & Sidhu, 2019; Qi, Tian, Li, Wang, & Liu, 2019; Schröder et al., 2014; X. Yang et al., 2017; Zhao et al., 2017). In contrast, it has been shown that miRNA-558 has a positive correlation to eIF4E expression in neuroblastoma cancer, by facilitating its function in translation initiation (Qu et al., 2016).

miRNAs are implicated in brain synaptic plasticity. Their target mRNAs are implicated in several LTP- and memory-related cellular functions, such as in neurotransmitter release, transcription factors regulation and dendritogenesis. Rapid derepression of target mRNAs and regulation of synaptic transmission on the order of minutes has recently been shown to occur in the dentate gyrus in vivo (Berentsen et al., 2019). An extended discussion can been found in review by Ryan et al. (Ryan, Joilin, & Williams, 2015) and in chapter 10 by Lannom and Ceman (this volume).

Translation Repression by 4E-T

The eIF4E-binding protein, 4E-T, has been described as the shuttling protein that transfers eIF4E in the nucleus (Dostie, Ferraiuolo, Pause, Adam, & Sonenberg, 2000). Both of them are localized in the P-bodies, the cytoplasmic formations where mRNA decay takes place and has been shown to repress cap-dependent translation, facilitating the mRNA degradation pathway (Ferraiuolo et al., 2005; Figure 1.1D). Strikingly, human 4E-T was found to down-regulate translation of bound mRNAs, in an eIF4E-independent manner and without the presence of P-bodies (Kamenska et al., 2014). Furthermore, in the developing brain, 4E-T has been linked to cortical neurogenesis (G. Yang, Smibert, Kaplan, & Miller, 2014). It has been proposed that 4E-T/eIF4E complex represses neuronal differentiation in neural precursors, by binding mRNAs that promote neurogenesis, and thus inhibiting their translation (G. Yang et al., 2014). Enhancing these results, Zahr and colleagues have recently identified the mechanism that specifies the fate of the radial glial precursors (Zahr et al., 2018). 4E-T associates with the RNA-binding protein Pumilio2 and selectively suppresses the translation of some of the neuronal identity mRNAs (Zahr et al., 2018).

eIF4E in the Nucleus

The detection of eIF4E in the nucleus has been known for several years (Lejbkowicz et al., 1992). Since this observation, a new chapter in eIF4E has opened, expanding its function beyond the role in translation initiation. Nuclear eIF4E plays a significant role in the export of specific-mRNAs from the nucleus (Osborne & Borden, 2015), through direct association with their m7GTP binding-site (Borden, 2016). In the nucleus, eIF4E is found in spherical formations, nuclear bodies, where it co-localizes and interacts with other nuclear proteins (Lai & Borden, 2000). Common examples of eIF4E-dependent nuclear export involve the Cyclin D1 and Ornithine decarboxylase mRNAs, as shown by the increased transportation to the cytoplasm, under eIF4E-overexpressing conditions (Rousseau, Kaspar, Rosenwald, Gehrke, & Sonenberg, 1996). The mRNA molecules exported by eIF4E must be capped and contain a small (~50) nucleotide sequence in the 3′ UTR, the eIF4E sensitivity element (4ESE). This part of the sequence forms a secondary loop-shaped structure, which is required for the nuclear export (Culjkovic et al., 2008; Culjkovic, Topisirovic, Skrabanek, Ruiz-Gutierrez, & Borden, 2005, 2006). eIF4E recruits a specific RNA protein complex, the ribonucleoparticle (RNP) complex, which includes the leucine-rich pentatricopeptide repeat protein (LRPPRC; Ivan Topisirovic et al., 2009). The nuclear export of RNPs is conducted through the nuclear pore complex (NPC) and, unlike the bulk mRNAs, eIF4E-dependent mRNA export is mediated through the CRM1 receptor (Culjkovic et al., 2006; Ivan Topisirovic et al., 2009; Figure 1.1D). LRPPRC seems to play an important role in mRNP formation, and therefore in the eIF4E-mediated mRNA export mechanism. LRPPRC acts as a bridging factor, directly binding eIF4E, 4ESE and CRM1 receptor, contributing to mRNA export. Possibly, eIF4E exerts its nuclear role by altering the composition of the NPC (Culjkovic-Kraljacic, Baguet, Volpon, Amri, & Borden, 2012). Interestingly, the oncogenic properties of eIF4E have been attributed to its mRNA nuclear export activity. The first evidence was provided by Topisirovic et al. (I. Topisirovic et al., 2003) in cell cultures. Elevated eIF4E levels were reported in human leukemia specimens and it was shown that the nuclear function of eIF4E contributes to oncogenesis, by facilitating growth and by blocking cellular differentiation (I. Topisirovic et al., 2003).

One of the most recent roles identified for nuclear eIF4E is its implication in the 3′-end processing of specific mRNAs (Davis, Delaleau, & Borden, 2019), including cleavage and polyadenylation of the 3′-end of specific mRNAs. It is considered a maturation mechanism, which facilitates nuclear export of 4ESE-containing RNAs. eIF4E facilitates the production of the implicated proteins (cleavage and polyadenylation complex—CPA complex), and also physically interacts with the CPA machinery (Davis et al., 2019).

Regulation of Nuclear eIF4E

Nuclear translocation of eIF4E is assisted by other proteins. 4E-T has a 4E-binding site and a nuclear localization signal (NLS), which is required for the import through the importin aβ pathway (Dostie et al., 2000). Recently, another protein, importin 8, was identified as a regulator of nuclear localization of eIF4E. Interestingly, importin 8 specifically binds cap-free eIF4E and directly imports it in the nucleus (Volpon et al., 2016; Figure 1.1D).

The nuclear function of eIF4E is regulated by two additional proteins: promyelocytic leukemia protein (PML) interacts with eIF4E and decreases its affinity for the mRNA cap-structure (Cohen et al., 2001; Culjkovic et al., 2005) and proline-rich homeodomain (PRH) binds eIF4E and blocks its nuclear mRNA transport function and thus transformation, independently of PML (I. Topisirovic et al., 2003; Ivan Topisirovic et al., 2003). It has been also reported that phosphorylation of nuclear eIF4E at Ser209 can regulate eIF4E-mediated mRNA transport (Ivan Topisirovic, Ruiz-Gutierrez, & Borden, 2004; Figure 1.1D). Interestingly, phosphorylation at S209 facilitates the mRNA transport function, possibly by regulating the association of eIF4E with the binding proteins of RNP complex, and furthermore, it enhances its ability to transform cells (Ivan Topisirovic et al., 2004). Homeodomain protein HOXA9 is another modulator of nuclear eIF4E function. It competes with the inhibitory function of PRH protein, acting as a stimulator of eIF4E (I. Topisirovic et al., 2005). It was shown that HOXA9 binds eIF4E and colocalizes with nuclear bodies in a subset of myeloid leukemia specimens

(I. Topisirovic et al., 2005). Given the established role of eIF4E and phospho-eIF4E in synaptic plasticity, learning and memory, it would be topical to investigate how the nuclear role/localization of eIF4E may regulate those processes.

eIF4E Regulation by the GCN2-ATF4 Signaling Pathway

An alternative signaling pathway mediating 4E-BP activation has been identified in Drosophila larvae. It is active during normal development and under amino-acid deprivation conditions, demonstrating a significant role of 4E-BP in the lifespan of Drosophila (Kang et al., 2016). In particular, transcription of 4E-BP is accomplished by activation of kinase GCN2 and the ATF4 binding protein, and leads to the expression of specific proteins, upon eIF2A phosphorylation (Kang et al., 2016; Figure 1.1E). Similarly, it was shown that 4E-BP induced the immune response, upon bacterial infection (in Drosophila), by activation of the GCN2-ATF4 pathway (Vasudevan et al., 2017). Most recently, Li and colleagues (B. B. Li et al., 2018) developed a targeted assay for RNA translation, which allowed them to identify m-TOR independent regulators of ribosomal proteins (RP) translation, acting through the GCN2-eIF2a axis (B. B. Li et al., 2018).

eIF4E in RAN Translation

Repeat-associated non-AUG (RAN) translation is a non-canonical translation mechanism, which has been described in several neurodegenerative disorders, such as myotonic dystrophy, Huntington's disease and Fragile x-associated tremor/ataxia(Rodriguez & Todd, 2019; Zu et al., 2011). RAN translation is 30–40 percent as efficient as the canonical translation (Kearse et al., 2016) and it occurs in pathological conditions associated with nucleotide repeat expansions and expression of toxic proteins (Figure 1.1F). In the case of the fragile X gene, FMR1, a CGG repeat expansion normally exists. Larger expansions lead to loss of FMR protein and to the fragile X phenotype (Nelson, Orr, & Warren, 2013). Kearse and colleagues studied the mechanism of CGG RAN translation. Importantly, they revealed that CGG RAN translation has variable initiation points (depending on the reading frame and the repeat length), while it shares common features with the canonical translation; it is cap-dependent and requires eIF4E and eIF4A to initiate (Kearse et al., 2016). However, the precise implication of eIF4E regulation in RAN translation remains to be elucidated.

Future Directions

eIF4E binds to all capped mRNAs, yet it preferentially stimulates the translation of certain mRNAs without affecting global translation. This paradox highlights the evolutionary conserved function of eIF4E, which is yet to be fully elucidated in the brain. Cancer cells and viruses preferentially target eIF4F to bias translational control of their mRNAs (viral or oncogenes), without shutting down global protein synthesis (Truitt & Ruggero, 2016). A predominant model in the field posits that via 4E-BPs and eIF4G, eIF4E can selectively dissociate from the cap of eIF4E-sensitive mRNAs, thus ensuring cap-dependent translation of all other cellular mRNAs. Indeed, only half of total eIF4E is adequate for normal development (Truitt et al., 2015). Another plausible explanation is that knockdown of eIF4E does not affect global translation due to homeostatic ubiquitination and proteasomal degradation of hypo-phosphorylated 4E-BPs (Yanagiya et al., 2012).

Cap-dependent translation is required for synthesizing new proteins for LTP and LTM—yet ablation of phospho-eIF4E in CA1 hippocampus does not affect LTP or LTM. Understanding the role of eIF4E phosphorylation downstream of MNK kinases in the hippocampus will elucidate a key avenue of eIF4E regulation.

Which mRNAs does eIF4E regulate in different tissues and cell types or for different neurodevelopmental disorders endophenotypes? To decipher how eIF4E regulates protein synthesis in the brain, explain the pleiotropic effects of monogenic neurodevelopmental disorders (such as FXS and tuberous sclerosis) it will be essential to study cell-type specific cap-dependent translation using the plethora of available "omics" techniques (Alvarez-Castelao, Schanzenbächer, Langer, & Schuman, 2019; Dedic, Chen, & Deussing, 2018; Dieterich et al., 2007; Heiman, Kulicke, Fenster, Greengard, & Heintz, 2014; Thomas et al., 2012). Linking cell-type specific translational control to NDD endophenotypes will further our understanding of the pathophysiology of NDDs and bolster the discovery of novel therapeutics.

What is the function of nuclear eIF4E in neurons? Nuclear eIF4E has multiple role in oncogenesis. It will be important to determine whether eIF4E regulation of 3′-end processing and nuclear export of mRNAs plays a role in neuronal development and plasticity.

Imaging eIF4E activity. Current understanding of eIF4E regulation and function is largely based on biochemical methods using bulk affinity-purification techniques such as pull-downs. To elucidate how eIF4E functions are coordinated in neurons, from nucleus to dendritic spines, there is a growing need for live-cell imaging approaches such as single-particle tracking of eIF4E and FRET-based measurement of eIF4E protein-protein interactions.

References

Alvarez-Castelao, B., Schanzenbächer, C. T., Langer, J. D., & Schuman, E. M. (2019). Cell-type-specific metabolic labeling, detection and identification of nascent proteomes in vivo. *Nature Protocols*, *14*(2), 556–575. https://doi.org/10.1038/s41596-018-0106-6

Amorim, I. S., Kedia, S., Kouloulia, S., Simbriger, K., Gantois, I., Jafarnejad, S. M., . . . Gkogkas, C. G. (2018). Loss of eIF4E phosphorylation engenders depression-like behaviors via selective mRNA translation. *Journal of Neuroscience*, *38*(8), 2118–2133.

Artinian, J., McGauran, A.-M. T., De Jaeger, X., Mouledous, L., Frances, B., & Roullet, P. (2008). Protein degradation, as with protein synthesis, is required during not only long-term spatial memory consolidation but also reconsolidation. *European Journal of Neuroscience, 27*(11), 3009–3019.

Banko, J. L. (2005). The translation repressor 4E-BP2 is critical for eIF4F complex formation, synaptic plasticity, and memory in the hippocampus. *Journal of Neuroscience, 25*(42), 9581–9590.

Banko, Jessica L., Hou, L., & Klann, E. (2004). NMDA receptor activation results in PKA- and ERK-dependent Mnk1 activation and increased eIF4E phosphorylation in hippocampal area CA1. *Journal of Neurochemistry, 91*(2), 462–470.

Berentsen, B., Patil, S., Rønnestad, K., Goff, K. M., Pajak, M., Simpson, T. I.,... Bramham, C. R. (2019). MicroRNA-34a acutely regulates synaptic efficacy in the adult dentate gyrus in vivo. *Molecular Neurobiology*. https://doi.org/10.1007/s12035-019-01816-1

Bliss, T. V. P. & Collingridge, G. L. (1993). A synaptic model of memory: LTP in the hippocampus. *Nature, 361*, 31–39.

Borden, K. L. B. (2016). The eukaryotic translation initiation factor eIF4E wears a "cap" for many occasions. *Translation, 4*(2), e1220899. https://doi.org/10.1080/21690731.2016.1220899

Bramham, C. R., Jensen, K. B., & Proud, C. G. (2016). Tuning specific translation in cancer metastasis and synaptic memory: Control at the MNK–eIF4E axis. *Trends in Biochemical Sciences, 41*(10), 847–858. https://doi.org/10.1016/j.tibs.2016.07.008

Casadio, A., Martin, K. C., Giustetto, M., Zhu, H., Chen, M., Bartsch, D.,... Kandel, E. R. (1999). A transient, neuron-wide form of CREB-mediated long-term facilitation can be stabilized at specific synapses by local protein synthesis. *Cell, 99*(2), 221–237. https://doi.org/10.1016/S0092-8674(00)81653-0

Chapat, C., Jafarnejad, S. M., Matta-Camacho, E., Hesketh, G. G., Gelbart, I. A., Attig, J.,... Sonenberg, N. (2017). Cap-binding protein 4EHP effects translation silencing by microRNAs. *Proceedings of the National Academy of Sciences of the United States of America, 114*(21), 5425–5430. https://doi.org/10.1073/pnas.1701488114

Chen, S., & Gao, G. (2017). MicroRNAs recruit eIF4E2 to repress translation of target mRNAs. *Protein and Cell, 8*(10), 750–761. https://doi.org/10.1007/s13238-017-0444-0

Clemens, M. J., Elia, A., & Morley, S. J. (2013). Requirement for the eIF4E binding proteins for the synergistic down-regulation of protein synthesis by hypertonic conditions and mTOR inhibition, *8*(8). https://doi.org/10.1371/journal.pone.0071138

Cohen, N., Sharma, M., Kentsis, A., Perez, J. M., Strudwick, S., & Borden, K. L. B. (2001). PML RING suppresses oncogenic transformation by reducing the affinity of eIF4E for mRNA. *EMBO Journal, 20*(16), 4547–4559. https://doi.org/10.1093/emboj/20.16.4547

Cracco, J. B., Serrano, P., Moskowitz, S. I., Bergold, P. J., & Sacktor, T. C. (2005). Protein synthesis-dependent LTP in isolated dendrites of CA1 pyramidal cells. *Hippocampus, 15*(5), 551–556. https://doi.org/10.1002/hipo.20078

Culjkovic, B., Tan, K., Orolicki, S., Amri, A., Meloche, S., & Borden, K. L. B. (2008). The eIF4E RNA regulon promotes the Akt signaling pathway. *Journal of Cell Biology, 181*(1), 51–63. https://doi.org/10.1083/jcb.200707018

Culjkovic, B., Topisirovic, I., Skrabanek, L., Ruiz-Gutierrez, M., & Borden, K. L. B. (2005). eIF4E promotes nuclear export of cyclin D1 mRNAs via an element in the 3'UTR. *Journal of Cell Biology, 169*(2), 245–256. https://doi.org/10.1083/jcb.200501019

Culjkovic, B., Topisirovic, I., Skrabanek, L., Ruiz-Gutierrez, M., & Borden, K. L. B. (2006). eIF4E is a central node of an RNA regulon that governs cellular proliferation. *Journal of Cell Biology, 175*(3), 415–426. https://doi.org/10.1083/jcb.200607020

Culjkovic-Kraljacic, B., Baguet, A., Volpon, L., Amri, A., & Borden, K. L. B. (2012). The oncogene eIF4E reprograms the nuclear pore complex to promote mRNA export and oncogenic transformation. *Cell Reports*, *2*(2), 207–215. https://doi.org/10.1016/j.celrep.2012.07.007

Davis, M. R., Delaleau, M., & Borden, K. L. B. (2019). Nuclear eIF4E stimulates 3'-end cleavage of target RNAs. *Cell Reports*, *27*(5), 1397–1408.e4. https://doi.org/10.1016/j.celrep.2019.04.008

Dedic, N., Chen, A., & Deussing, J. M. (2018). The CRF family of neuropeptides and their receptors—mediators of the central stress response. *Current Molecular Pharmacology*, *11*(1). https://doi.org/10.2174/1874467210666170302104053

Dieterich, D. C., Lee, J. J., Link, A. J., Graumann, J., Tirrell, D. A., & Schuman, E. M. (2007). Labeling, detection and identification of newly synthesized proteomes with bioorthogonal non-canonical amino-acid tagging. *Nature Protocols*, *2*(3), 532–540. https://doi.org/10.1038/nprot.2007.52

Dostie, J., Ferraiuolo, M., Pause, A., Adam, S. A., & Sonenberg, N. (2000). A novel shuttling protein, 4E-T, mediates the nuclear import of the mRNA 5' cap-binding protein, eIF4E. *The EMBO Journal*, *19*(12), 3142–3156. https://doi.org/10.1093/emboj/19.12.3142

Ferraiuolo, M. A., Basak, S., Dostie, J., Murray, E. L., Schoenberg, D. R., & Sonenberg, N. (2005). A role for the eIF4E-binding protein 4E-T in P-body formation and mRNA decay. *Journal of Cell Biology*, *170*(6), 913–924. https://doi.org/10.1083/jcb.200504039

Flexner, J. B., Flexner, L. B., & Stellar, E. (1963). Memory in mice as affected by intracerebral puromycin. *Science*, *141*(3575), 57 LP–59. https://doi.org/10.1126/science.141.3575.57

Frey, U., Krug, M., Reymann, K. G., & Matthies, H. (1988). Anisomycin, an inhibitor of protein synthesis, blocks late phases of LTP phenomena in the hippocampal CA1 region in vitro. *Brain Research*, *452*(1), 57–65. https://doi.org/https://doi.org/10.1016/0006-8993(88)90008-X

Fukao, A., Mishima, Y., Takizawa, N., Oka, S., Imataka, H., Pelletier, J., . . . Fujiwara, T. (2014). MicroRNAs trigger dissociation of eIF4AI and eIF4AII from target mRNAs in humans. *Molecular Cell*, *56*(1), 79–89. https://doi.org/10.1016/j.molcel.2014.09.005

Fukaya, T., & Tomari, Y. (2012). MicroRNAs mediate gene silencing via multiple different pathways in Drosophila. *Molecular Cell*, *48*(6), 825–836. https://doi.org/10.1016/j.molcel.2012.09.024

Gantois, I., Khoutorsky, A., Popic, J., Aguilar-Valles, A., Freemantle, E., Cao, R., . . . Sonenberg, N. (2017). Metformin ameliorates core deficits in a mouse model of fragile X syndrome. *Nature Medicine*, *23*(6), 674–677. https://doi.org/10.1038/nm.4335

Genheden, M., Kenney, J. W., Johnston, H. E., Manousopoulou, A., Garbis, S. D., & Proud, C. G. (2015). BDNF stimulation of protein synthesis in cortical neurons requires the MAP kinase-interacting kinase MNK1. *Journal of Neuroscience*, *35*(3), 972–984.

Gkogkas, C. G., Khoutorsky, A., Cao, R., Jafarnejad, S. M., Prager-Khoutorsky, M., Giannakas, N., . . . Sonenberg, N. (2014). Pharmacogenetic inhibition of eIF4E-dependent Mmp9 mRNA translation reverses Fragile X Syndrome-like Phenotypes. *Cell Reports*, *9*(5), 1742–1755. https://doi.org/10.1016/j.celrep.2014.10.064

Gram, H., Knauf, U., & Tschopp, C. (2001). Negative regulation of protein translation by mitogen-activated protein negative regulation of protein translation by mitogen-activated protein kinase-interacting kinases 1 and 2. *Molecular and Cellular Biology*, *21*(16), 5500–5511. https://doi.org/10.1128/MCB.21.16.5500

Haghighat, A., Mader, S., Pause, A., & Sonenberg, N. (1995). Repression of cap-dependent translation by 4E-binding protein 1: competition with p220 for binding to eukaryotic initiation factor-4E, *14*(22), 5701–5709.

Hassan, N., Zhao, J. T., Glover, A., Robinson, B. G., & Sidhu, S. B. (2019). Reciprocal interplay of miR-497 and MALAT1 promotes tumourigenesis of adrenocortical cancer. *Endocrine-Related Cancer, 26*(7), 677–688. https://doi.org/10.1530/ERC-19-0036

Heiman, M., Kulicke, R., Fenster, R. J., Greengard, P., & Heintz, N. (2014). Cell type-specific mRNA purification by translating ribosome affinity purification (TRAP). *Nature Protocols, 9*(6), 1282–1291. https://doi.org/10.1038/nprot.2014.085

Hershey, J. W. B., Sonenberg, N., & Mathews, M. B. (2012). Principles of translational control: An overview. *Cold Spring Harbor Perspectives in Biology, 4*(12). https://doi.org/10.1101/cshperspect.a009829

Hoeffer, C. A., Cowansage, K. K., Arnold, E. C., Banko, J. L., Moerke, N. J., Rodriguez, R.,... Klann, E. (2011). Inhibition of the interactions between eukaryotic initiation factors 4E and 4G impairs long-term associative memory consolidation but not reconsolidation. *Proceedings of the National Academy of Sciences of the United States of America, 108*(8), 3383–3388. https://doi.org/10.1073/pnas.1013063108

Hoeffer, C. A., Santini, E., Ma, T., Arnold, E. C., Whelan, A. M., Arnold, E. C.,... Klann, E. (2013). Multiple components of eIF4F are required for protein synthesis-dependent hippocampal long-term potentiation. *Journal of Neurophysiology, 109*(1), 68–76. https://doi.org/10.1152/jn.00342.2012

Huber, K. M., Kayser, M. S., & Bear, M. F. (2000). Role for rapid dendritic protein synthesis in hippocampal mGluR-dependent long-term depression. *Science, 288*(5469), 1254 LP–1256. https://doi.org/10.1126/science.288.5469.1254

Jiang, C. C., Croft, A., Tseng, H. Y., Guo, S. T., Jin, L., Hersey, P., & Zhang, X. D. (2014). Repression of microRNA-768-3p by MEK/ERK signalling contributes to enhanced mRNA translation in human melanoma. *Oncogene, 33*(20), 2577–2588. https://doi.org/10.1038/onc.2013.237

Jonas, S., & Izaurralde, E. (2015). Towards a molecular understanding of microRNA-mediated gene silencing. *Nature Reviews Genetics, 16*(7), 421–433. https://doi.org/10.1038/nrg3965

Joshi, B., Cai, A. L., Keiper, B. D., Minich, W. B., Mendez, R., Beach, C. M.,... Rhoads, R. E. (1995). Phosphorylation of eukaryotic protein synthesis initiation factor 4E at Ser-209. *The Journal of Biological Chemistry, 270*(24), 14597–14603. https://doi.org/10.1074/jbc.270.24.14597

Kamenska, A., Lu, W. T., Kubacka, D., Broomhead, H., Minshall, N., Bushell, M., & Standart, N. (2014). Human 4E-T represses translation of bound mRNAs and enhances microRNA-mediated silencing. *Nucleic Acids Research, 42*(5), 3298–3313. https://doi.org/10.1093/nar/gkt1265

Kang, H., & Schuman, E. M. (1996). A requirement for local protein synthesis in neurotrophin-induced hippocampal synaptic plasticity. *Science, 273*(5280), 1402 LP–1406. https://doi.org/10.1126/science.273.5280.1402

Kang, M.-J., Vasudevan, D., Kang, K., Kim, K., Park, J.-E., Zhang, N.,... Don Ryoo, H. (2016). 4E-BP is a target of the GCN2–ATF4 pathway during Drosophila development and aging. *Journal of Cell Biology, 216*(1).

Kanhema, T., Dagestad, G., Panja, D., Tiron, A., Messaoudi, E., Håvik, B.,... Bramham, C. R. (2006). Dual regulation of translation initiation and peptide chain elongation during BDNF-induced LTP in vivo: Evidence for compartment-specific translation control. *Journal of Neurochemistry, 99*(5), 1328–1337.

Kearse, M. G., Green, K. M., Krans, A., Rodriguez, C. M., Linsalata, A. E., Goldstrohm, A. C., & Todd, P. K. (2016). CGG repeat-associated non-AUG translation utilizes a cap-dependent scanning mechanism of initiation to produce toxic proteins. *Molecular Cell*, *62*(2), 314–322. https://doi.org/10.1016/j.molcel.2016.02.034

Kelleher 3rd, R. J., Govindarajan, A., Jung, H.-Y., Kang, H., & Tonegawa, S. (2004). Translational control by MAPK signaling in long-term synaptic plasticity and memory. *Cell*, *116*, 467–479. Retrieved from http://www.cell.com/cgi/content/full/116/3/467/DC1

Korneeva, N. L., Song, A., Gram, H., Edens, M. A., & Rhoads, R. E. (2016). Inhibition of mitogen-activated protein kinase (MAPK)-interacting kinase (MNK) preferentially affects translation of mRNAs containing both a 5'-terminal cap and hairpin. *Journal of Biological Chemistry*, *291*(7), 3455–3467. https://doi.org/10.1074/jbc.M115.694190

Lachance, P. E. D., Miron, M., Raught, B., Sonenberg, N., & Lasko, P. (2002). Phosphorylation of eukaryotic translation initiation factor 4E is critical for growth. *Molecular and Cellular Biology*, *22*(6), 1656 LP–1663.

Lai, H. K., & Borden, K. L. B. (2000). The promyelocytic leukemia (PML) protein suppresses cyclin D1 protein production by altering the nuclear cytoplasmic distribution of cyclin D1 mRNA. *Oncogene*, *19*(13), 1623–1634. https://doi.org/10.1038/sj.onc.1203473

Lejbkowicz, F., Goyer, C., Darveau, A., Neron, S., Lemieux, R., & Sonenberg, N. (1992). A fraction of the mRNA 5' cap-binding protein, eukaryotic initiation factor 4E, localizes to the nucleus. *Proceedings of the National Academy of Sciences of the United States of America*, *89*(20), 9612–9616. https://doi.org/10.1073/pnas.89.20.9612

Li, B. B., Qian, C., Gameiro, P. A., Liu, C.-C., Jiang, T., Roberts, T. M., . . . Zhao, J. J. (2018). Targeted profiling of RNA translation reveals mTOR-4EBP1/2-independent translation regulation of mRNAs encoding ribosomal proteins. *Proceedings of the National Academy of Sciences*, *115*(40), E9325–E9332. https://doi.org/10.1073/pnas.1805782115

Li, X., An, W. L., Alafuzoff, I., Soininen, H., Winblad, B., & Pei, J. J. (2004). Phosphorylated eukaryotic translation factor 4E is elevated in Alzheimer brain. *NeuroReport*, *15*(14), 2237–2240. https://doi.org/10.1097/00001756-200410050-00019

Liu, F., Wang, X., Li, J., Gu, K., Lv, L., Zhang, S., . . . Yu, Y. (2015). miR-34c-3p functions as a tumour suppressor by inhibiting eIF4E expression in non-small cell lung cancer. *Cell Proliferation*, *48*(5), 582–592. https://doi.org/10.1111/cpr.12201

Marcotrigiano, J., Gingras, A. C., Sonenberg, N., & Burley, S. K. (1997). Cocrystal structure of the messenger RNA 5' cap-binding protein (eIF4E) bound to 7-methyl-GDP. *Cell*, *89*(6), 951–961. https://doi.org/10.1016/s0092-8674(00)80280-9

Martin, K. C., Casadio, A., Zhu, H., Yaping, E., Rose, J. C., Chen, M., . . . Kandel, E. R. (1997). Synapse-specific, long-term facilitation of aplysia sensory to motor synapses: A function for local protein synthesis in memory storage. *Cell*, *91*(7), 927–938. https://doi.org/10.1016/s0092-8674(00)80484-5

Mathonnet, G., Fabian, M. R., Svitkin, Y. V., Parsyan, A., Huck, L., Murata, T., . . . Sonenberg, N. (2007). MicroRNA inhibition of translation initiation in vitro by targeting the cap-binding complex eIF4F. *Science*, *317*(5845), 1764–1767. https://doi.org/10.1126/science.1146067

Melanson, G., Timpano, S., & Uniacke, J. (2017). The eIF4E2-directed hypoxic cap-dependent translation machinery reveals novel therapeutic potential for cancer treatment. *Oxidative Medicine and Cellular Longevity*, *2017*. https://doi.org/10.1155/2017/6098107

Melemedjian, O. K., Tillu, D. V., Asiedu, M. N., Mandell, E. K., Moy, J. K., Blute, V. M., . . . Price, T. J. (2013). BDNF regulates atypical PKC at spinal synapses to initiate and maintain a centralized chronic pain state. *Molecular Pain*, *9*(1), 1–14. https://doi.org/10.1186/1744-8069-9-12

Min, K.-W., Davila, S., Zealy, R. W., Lloyd, L. T., Lee, I. Y., Lee, R., . . . Yoon, J.-H. (2017). eIF4E phosphorylation by MST1 reduces translation of a subset of mRNAs, but increases lncRNA translation. *Biochimica et Biophysica Acta (BBA)—Gene Regulatory Mechanisms, 1860*(7), 761–772. https://doi.org/10.1016/j.bbagrm.2017.05.002

Morris, R. G., Anderson, E., Lynch, G. S., & Baudry, M. (1986). Selective impairment of learning and blockade of long-term potentiation by an N-methyl-D-aspartate receptor antagonist, AP5. *Nature, 319*(6056), 774–776. https://doi.org/10.1038/319774a0

Moy, J. K., Khoutorsky, A., Asiedu, M. N., Dussor, G., & Price, T. J. (2018). eIF4E phosphorylation influences Bdnf mRNA translation in mouse dorsal root ganglion neurons. *Frontiers in Cellular Neuroscience, 12*(February), 1–10. https://doi.org/10.3389/fncel.2018.00029

Muta, D., Makino, K., Nakamura, H., Yano, S., Kudo, M., & Kuratsu, J. I. (2011). Inhibition of eIF4E phosphorylation reduces cell growth and proliferation in primary central nervous system lymphoma cells. *Journal of Neuro-Oncology, 101*(1), 33–39. https://doi.org/10.1007/s11060-010-0233-6

Nelson, D. L., Orr, H. T., & Warren, S. T. (2013). The unstable repeats: Three evolving faces of neurological disease. *Neuron, 77*(5), 825–843. https://doi.org/10.1016/j.neuron.2013.02.022

Oliveto, S., Mancino, M., Manfrini, N., & Biffo, S. (2017). Role of microRNAs in translation regulation and cancer. *World Journal of Biological Chemistry, 8*(1), 45. https://doi.org/10.4331/wjbc.v8.i1.45

Osborne, M. J., & Borden, K. L. B. (2015). The eukaryotic translation initiation factor eIF4E in the nucleus: Taking the road less traveled. *Immunological Reviews, 263*(1), 210–223. https://doi.org/10.1111/imr.12240

Ostroff, L. E., Cain, C. K., Bedont, J., Monfils, M. H., & LeDoux, J. E. (2010). Fear and safety learning differentially affect synapse size and dendritic translation in the lateral amygdala. *Proceedings of the National Academy of Sciences, 107*(20), 9418–9423. https://doi.org/10.1073/pnas.0913384107

Ostroff, E., Botsford, B., Gindina, S., Cowansage, K. K., LeDoux, J. E., Klann, E., & Hoeffer, C. (2017). Accumulation of polyribosomes in dendritic spine heads, but not bases and necks, during memory consolidation depends on cap-dependent translation initiation. *Journal of Neuroscience, 37*(7), 1862–1872.

Panja, D., & Bramham, C. R. (2014). BDNF mechanisms in late LTP formation: A synthesis and breakdown. *Neuropharmacology, 76*(PART C), 664–676. https://doi.org/10.1016/j.neuropharm.2013.06.024

Panja, D., Dagyte, G., Bidinosti, M., Wibrand, K., Kristiansen, Å. M., Sonenberg, N., & Bramham, C. R. (2009). Novel translational control in arc-dependent long term potentiation consolidation in vivo. *Journal of Biological Chemistry, 284*(46), 31498–31511. https://doi.org/10.1074/jbc.M109.056077

Panja, D., Kenney, J. W., D'Andrea, L., Zalfa, F., Vedeler, A., Wibrand, K., . . . Bramham, C. R. (2014). Two-stage translational control of dentate gyrus LTP consolidation is mediated by sustained BDNF-TrkB signaling to MNK. *Cell Reports, 9*(4), 1430–1445. https://doi.org/10.1016/j.celrep.2014.10.016

Parikshak, N. N., Swarup, V., Belgard, T. G., Irimia, M., Ramaswami, G., Gandal, M. J., . . . Geschwind, D. H. (2016). Genome-wide changes in lncRNA, splicing, and regional gene expression patterns in autism. *Nature, 540*, 423. Retrieved from https://doi.org/10.1038/nature20612

Pause, A., Methot, N., Svitkin, Y., Merrick, W. C., & Sonenberg, N. (1994). Dominant negative mutants of mammalian translation initiation factor elF-4A define a critical role for elF-4F in cap-dependent and cap-independent initiation of translation. *EMBO Journal, 13*(5), 1205–1215.

Pyronnet, S., Imataka, H., Gingras, A. C., Fukunaga, R., Hunter, T., & Sonenberg, N. (1999). Human eukaryotic translation initiation factor 4G (eIF4G) recruits Mnk1 to phosphorylate eIF4E. *EMBO Journal*, *18*(1), 270–279. https://doi.org/10.1093/emboj/18.1.270

Qi, N. N., Tian, S., Li, X., Wang, F. L., & Liu, B. (2019). Up-regulation of microRNA-496 suppresses proliferation, invasion, migration and in vivo tumorigenicity of human osteosarcoma cells by targeting eIF4E. *Biochimie*, *163*, 1–11. https://doi.org/10.1016/j.biochi.2019.04.017

Qu, H., Zheng, L., Song, H., Jiao, W., Li, D., Fang, E., ... Tong, Q. (2016). microRNA-558 facilitates the expression of hypoxia-inducible factor 2 alpha through binding to 5';-untranslated region in neuroblastoma. *Oncotarget*, *7*(26). https://doi.org/10.18632/oncotarget.9813

Raught, B., & Gingras, A. (1999). eIF4E activity is regulated at multiple levels. *The International Journal of Biochemistry & Cell Biology*. *31*(1), 43–57. https://doi.org/10.1016/S1357-2725(98)00131-9

Rodriguez, C. M., & Todd, P. K. (2019). New pathologic mechanisms in nucleotide repeat expansion disorders. *Neurobiology of Disease*, *130*, 104515. https://doi.org/10.1016/j.nbd.2019.104515

Rodriguez-Ortiz, C. J., & Bermudez-Rattoni, F. (2007). Memory reconsolidation or updating consolidation? In F. Bermudez-Rattoni (Ed.), *Neural Plasticity and Memory: From Genes to Brain Imaging*. Boca Raton, FL: CRC Press/Taylor/Taylor & Francis. Chapter 11. Available from: https://www.ncbi.nlm.nih.gov/books/NBK3905/

Rousseau, D., Kaspar, R., Rosenwald, I., Gehrke, L., & Sonenberg, N. (1996). Translation initiation of ornithine decarboxylase and nucleocytoplasmic transport of cyclin D1 mRNA are increased in cells overexpressing eukaryotic initiation factor 4E. *Proceedings of the National Academy of Sciences of the United States of America*, *93*(3), 1065–1070. https://doi.org/10.1073/pnas.93.3.1065

Ryan, B., Joilin, G., & Williams, J. M. (2015). Plasticity-related microRNA and their potential contribution to the maintenance of long-term potentiation. *Frontiers in Molecular Neuroscience*, *8*(Feb.), 1–17. https://doi.org/10.3389/fnmol.2015.00004

Salehi, Z., & Mashayekhi, F. (2007). Eukaryotic translation initiation factor 4E (eIF4E) expression in the brain tissue is induced by infusion of nerve growth factor into the mouse cisterna magnum: An in vivo study. *Molecular and Cellular Biochemistry*, *304*(1–2), 249–253. https://doi.org/10.1007/s11010-007-9507-5

Santini, E., Huynh, T. N., Longo, F., Koo, S. Y., Mojica, E., D'Andrea, L., ... Klann, E. (2017). Reducing eIF4E-eIF4G interactions restores the balance between protein synthesis and actin dynamics in fragile X syndrome model mice. *Science Signaling*, *10*(504). https://doi.org/10.1126/scisignal.aan0665

Schröder, J., Ansaloni, S., Schilling, M., Liu, T., Radke, J., Jaedicke, M., ... Bertram, L. (2014). MicroRNA-138 is a potential regulator of memory performance in humans. *Frontiers in Human Neuroscience*, *8*(July), 1–10. https://doi.org/10.3389/fnhum.2014.00501

Shveygert, M., Kaiser, C., Bradrick, S. S., & Gromeier, M. (2010). Regulation of eukaryotic initiation factor 4E (eIF4E) phosphorylation by mitogen-activated protein kinase occurs through modulation of Mnk1-eIF4G interaction. *Molecular and Cellular Biology*, *30*(21), 5160–5167.

Sonenberg, N., & Hinnebusch, A. G. (2009). Regulation of translation initiation in eukaryotes: Mechanisms and biological targets. *Cell*, *136*(4), 731–745. https://doi.org/10.1016/j.cell.2009.01.042

Sonenberg, N., Rupprechtt, K. M., Hechtt, S. M., & Shatkin, A. J. (1979). Eukaryotic mRNA cap binding protein: Purification by affinity chromatography on Sepharose-coupled m7GDP (7-methylguanosine/eukaryotic initiation factors/protein synthesis). *Biochemistry, 76*(9), 4345–4349. Retrieved from https://www.ncbi.nlm.nih.gov/pmc/articles/PMC411571/pdf/pnas00009-0193.pdf

Stanton, P. K., & Sarvey, J. M. (1984). Blockade of long-term potentiation in rat hippocampal CA1 region by inhibitors of protein synthesis. *The Journal of Neuroscience?: The Official Journal of the Society for Neuroscience, 4*(12), 3080–3088. Retrieved from http://www.ncbi.nlm.nih.gov/pubmed/6502226

Svitkin, Y. V., Pause, A., Haghighat, A., Pyronnet, S., Witherell, G., Belsham, G. J., & Sonenberg, N. (2001). The requirement for eukaryotic initiation factor 4A (eIF4A) in translation is in direct proportion to the degree of mRNA 5 9 secondary structure. *RNA, 7*(3), 382–394.

Takei, N., Kawamura, M., Hara, K., Yonezawa, K., & Nawa, H. (2001). Brain-derived neurotrophic factor enhances neuronal translation by activating multiple initiation processes: Comparison with the effects of insulin. *Journal of Biological Chemistry, 276*(46), 42818–42825. https://doi.org/10.1074/jbc.M103237200

Tang, S. J., Reis, G., Kang, H., Gingras, A.-C., Sonenberg, N., & Schuman, E. M. (2002). A rapamycin-sensitive signaling pathway contributes to long-term synaptic plasticity in the hippocampus. *Proceedings of the National Academy of Sciences of the United States of America, 99*(1), 467–472. https://doi.org/10.1073/pnas.012605299

Thomas, A., Lee, P. J., Dalton, J. E., Nomie, K. J., Stoica, L., Costa-Mattioli, M., . . . Dierick, H. A. (2012). A versatile method for cell-specific profiling of translated mrnas in Drosophila. *PLoS ONE, 7*(7), 1–8. https://doi.org/10.1371/journal.pone.0040276

Timpano, S., & Uniacke, J. (2016). Human cells cultured under physiological oxygen utilize two cap-binding proteins to recruit distinct mRNAs for translation. *The Journal of Biological Chemistry, 291*(20), 10772–10782. https://doi.org/10.1074/jbc.M116.717363

Topisirovic, I., Guzman, M. L., McConnell, M. J., Licht, J. D., Culjkovic, B., Neering, S. J., . . . Borden, K. L. B. (2003). Aberrant eukaryotic translation initiation factor 4E-dependent mRNA transport impedes hematopoietic differentiation and contributes to leukemogenesis. *Molecular and Cellular Biology, 23*(24), 8992–9002.

Topisirovic, I., Kentsis, A., Perez, J. M., Guzman, M. L., Jordan, C. T., & Borden, K. L. B. (2005). Eukaryotic translation initiation factor 4E activity is modulated by HOXA9 at multiple levels. *Molecular and Cellular Biology, 25*(3), 1100–1112.

Topisirovic, Ivan, Culjkovic, B., Cohen, N., Perez, J. M., Skrabanek, L., & Borden, K. L. B. (2003). The proline-rich homeodomain protein, PRH, is a tissue-specific inhibitor of eiF4E-dependent cyclin D1 mRNA transport and growth. *EMBO Journal, 22*(3), 689–703. https://doi.org/10.1093/emboj/cdg069

Topisirovic, Ivan, Ruiz-Gutierrez, M., & Borden, K. L. B. (2004). Phosphorylation of the eukaryotic translation initiation factor eIF4E contributes to its transformation and mRNA transport activities. *Cancer Research, 64*(23), 8639–8642. https://doi.org/10.1158/0008-5472.CAN-04-2677

Topisirovic, Ivan, Siddiqui, N., Lapointe, V. L., Trost, M., Thibault, P., Bangeranye, C., . . . Borden, K. L. B. (2009). Molecular dissection of the eukaryotic initiation factor 4E (eIF4E) export-competent RNP. *EMBO Journal, 28*(8), 1087–1098. https://doi.org/10.1038/emboj.2009.53

Truitt, M. L., Conn, C. S., Shi, Z., Pang, X., Tokuyasu, T., Coady, A. M., . . . Ruggero, D. (2015). Differential requirements for eIF4E dose in normal development and cancer. *Cell, 162*(1), 59–71. https://doi.org/10.1016/j.cell.2015.05.049

Truitt, M. L., & Ruggero, D. (2016). New frontiers in translational control of the cancer genome. *Nature Reviews Cancer*, *16*(5), 288–304. https://doi.org/10.1038/nrc.2016.27

Tsukiyama-Kohara, K., Poulin, F., Kohara, M., DeMaria, C. T., Cheng, A., Wu, Z., . . . Sonenberg, N. (2001). Adipose tissue reduction in mice lacking the translational inhibitor 4E-BP1. *Nature Medicine*, *7*(10), 1128–1132. https://doi.org/10.1038/nm1001-1128

Tuxworth, W. J., Saghir, A. N., Spruill, L. S., Menick, D. R., & McDermott, P. J. (2004). Regulation of protein synthesis by eIF4E phosphorylation in adult cardiocytes: The consequence of secondary structure in the 5'- untranslated region of mRNA. *Biochemical Journal*, *378*(1), 73–82. https://doi.org/10.1042/BJ20031027

Ueda, T., Watanabe-Fukunaga, R., Fukuyama, H., Nagata, S., & Fukunaga, R. (2004). Mnk2 and Mnk1 are essential for constitutive and inducible phosphorylation of eukaryotic initiation factor 4E but not for cell growth or development. *Molecular and Cellular Biology*, *24*(15), 6539–6549.

Uniacke, J., Holterman, C. E., Lachance, G., Franovic, A., Jacob, M. D., Fabian, M. R., . . . Lee, S. (2012). An oxygen-regulated switch in the protein synthesis machinery. *Nature*, *486*(7401), 126–129. https://doi.org/10.1038/nature11055

Uniacke, J., Perera, J. K., Lachance, G., Francisco, C. B., & Lee, S. (2014). Cancer cells exploit eIF4E2-directed synthesis of hypoxia response proteins to drive tumor progression. *Cancer Research*, *74*(5), 1379–1389. https://doi.org/10.1158/0008-5472.CAN-13-2278

Vasudevan, D., Clark, N. K., Sam, J., Cotham, V. C., Ueberheide, B., Marr, M. T., & Ryoo, H. D. (2017). The GCN2-ATF4 signaling pathway induces 4E-BP to bias translation and boost antimicrobial peptide synthesis in response to bacterial infection. *Cell Reports*, *21*(8), 2039–2047. https://doi.org/10.1016/j.celrep.2017.10.096

Volpon, L., Culjkovic-Kraljacic, B., Osborne, M. J., Ramteke, A., Sun, Q., Niesman, A., . . . Borden, K. L. B. (2016). Importin 8 mediates m 7 G cap-sensitive nuclear import of the eukaryotic translation initiation factor eIF4E. *Proceedings of the National Academy of Sciences of the United States of America*, *113*(19), 5263–5268. https://doi.org/10.1073/pnas.1524291113

Wang, D., Ma, J., Ji, X., Xu, F., & Wei, Y. (2017). miR-141 regulation of EIF4E expression affects docetaxel chemoresistance of non-small cell lung cancer. *Oncology Reports*, *37*(1), 608–616. https://doi.org/10.3892/or.2016.5214

Yanagiya, A., Suyama, E., Adachi, H., Svitkin, Y. V, Aza-Blanc, P., Imataka, H., . . . Sonenberg, N. (2012). Translational homeostasis via the mRNA cap-binding protein, eIF4E. *Molecular Cell*, *46*(6), 847–858. https://doi.org/10.1016/j.molcel.2012.04.004

Yang, G., Smibert, C. A., Kaplan, D. R., & Miller, F. D. (2014). An eIF4E1/4E-T complex determines the genesis of neurons from precursors by translationally repressing a proneurogenic transcription program. *Neuron*, *84*(4), 723–739. https://doi.org/10.1016/j.neuron.2014.10.022

Yang, X., Zang, J., Pan, X., Yin, J., Xiang, Q., Yu, J., . . . Lei, X. (2017). MiR-503 inhibits proliferation making human hepatocellular carcinoma cells susceptible to 5-fluorouracil by targeting EIF4E. *Oncology Reports*, *37*(1), 563–570. https://doi.org/10.3892/or.2016.5220

Yao, Y. S., Qiu, W. S., Yao, R. Y., Zhang, Q., Zhuang, L. K., Zhou, F., . . . Yue, L. (2015). MiR-141 confers docetaxel chemoresistance of breast cancer cells via regulation of EIF4E expression. *Oncology Reports*, *33*(5), 2504–2512. https://doi.org/10.3892/or.2015.3866

Zahr, S. K., Yang, G., Kazan, H., Borrett, M. J., Yuzwa, S. A., Voronova, A., . . . Miller, F. D. (2018). A translational repression complex in developing mammalian neural stem cells that regulates neuronal specification. *Neuron*, *97*(3), 520–537. e6. https://doi.org/10.1016/j.neuron.2017.12.045

Zhao, Y., Yan, M., Yun, Y., Zhang, J., Zhang, R., Li, Y.,…Jiang, H. (2017). MicroRNA-455-3p functions as a tumor suppressor by targeting eIF4E in prostate cancer. *Oncology Reports*, *37*(4), 2449–2458. https://doi.org/10.3892/or.2017.5502

Zu, T., Gibbens, B., Doty, N. S., Gomes-Pereira, M., Huguet, A., Stone, M. D.,…Ranum, L. P. W. (2011). Non-ATG-initiated translation directed by microsatellite expansions. *Proceedings of the National Academy of Sciences of the United States of America*, *108*(1), 260–265. https://doi.org/10.1073/pnas.1013343108

CHAPTER 2

TRANSLATIONAL CONTROL THROUGH THE EIF4E BINDING PROTEINS IN THE BRAIN

ARGEL AGUILAR-VALLES, EDNA MATTA-CAMACHO, AND NAHUM SONENBERG

EUKARYOTIC TRANSLATION INITIATION FACTOR 4E BINDING PROTEINS: BASIC BIOCHEMICAL CONSIDERATIONS

EUKARYOTIC translation initiation factor 4E (eIF4E) is central to mRNA translation initiation as a component of the eIF4F translation initiation complex, along with eIF4G and eIF4A. eIF4E recognizes the 5′-cap structure, enabling the recruitment of the small ribosomal subunit to mRNA through eIF4G and eIF3 (Sonenberg, Morgan, Merrick, & Shatkin, 1978).

eIF4E cap-binding cavity specifically interacts with the $m^7G(5')ppp(5')N$ (where N is typically G or A) cap of an mRNA (reviewed in (Rhoads, 2009; Topisirovic, Svitkin, Sonenberg, & Shatkin, 2011). eIF4E also interacts with eIF4G, on the convex side of the cap-binding cavity, by binding to a $YX_4L\phi$ motif on eIF4G (where ϕ denotes hydrophobic residue, usually leucine, methionine, or phenylalanine). The interaction of eIF4E with the mRNA 5′-cap structure is dramatically enhanced by eIF4G (Haghighat, Mader, Pause, & Sonenberg, 1995; Ptushkina et al., 1998).

A *bona fide* mechanism for activity regulation of eIF4E is through its interaction with the eIF4E-binding proteins (4E-BPs). 4E-BPs are small, acidic, heat-stable proteins first described in vertebrates as a three-member protein family—4E-BP1, 4E-BP2 and 4E-BP3—which share a 55% amino acid sequence identity.

The 4E-BPs bind eIF4E and prevent its interaction with eIF4G. 4E-BPs use the same $YX_4L\phi$ sequence to competitively bind eIF4E (Hershey et al., 1999; Marcotrigiano, Gingras, Sonenberg, & Burley, 1999), preventing eIF4E-eIF4G interaction and inhibiting cap-dependent translation by blocking the assembly of the eIF4F complex (Martineau, Azar, Bousquet, & Pyronnet, 2013; Rhoads, 2009; Richter & Sonenberg, 2005). Deletion of this sequence in the 4E-BPs or mutation of either the tyrosine or the leucine to alanine abrogates their eIF4E binding (Mader, Lee, Pause, & Sonenberg, 1995; Poulin, Gingras, Olsen, Chevalier, & Sonenberg, 1998).

The association of 4E-BPs with eIF4E is reversible and regulated by phosphorylation. Mammalian 4E-BPs are targets of mTOR (mammalian/mechanistic target of rapamycin) kinase, a key sensor of nutrient status. Increased phosphorylation of 4E-BPs promotes dissociation of eIF4E and thus increases translation activity (Gingras et al., 1999).

4E-BP Phosphorylation

The 4E-BPs contain three functional domains: the eIF4E-binding domain (Mader et al., 1995), a C-terminal TOR (target of rapamycin) signaling motif (TOS; Schalm & Blenis, 2002), and an N-terminal RAIP (Arg13, Ala14, Ile15, Pro16) motif (Tee & Proud, 2002). The TOS and RAIP motifs contribute to the binding of Raptor (Eguchi et al., 2006; Lee, Healy, Fonseca, Hayashi, & Proud, 2008), a scaffold protein of mTOR complex 1 (mTORC1) that bridges mTOR and the 4E-BPs and is necessary for efficient phosphorylation of the 4E-BPs (Hara et al., 2002; Schalm, Fingar, Sabatini, & Blenis, 2003).

Phosphorylated 4E-BPs have been generally accepted as a marker of activated mTORC1 signaling. Seven phosphorylation sites have been identified in 4E-BP1: Tyr 37, Thr 46, Ser 65, Thr 70, Ser 83, Ser 101, and Ser 112 (numbering based on human 4E-BP1; Martineau et al., 2013). The first five phosphorylation sites are phylogenetically conserved among eukaryotes; mTORC1-dependent phosphorylation of 4E-BP1 proceeds in a hierarchical way: initial phosphorylation of Thr 37 and Thr 46 is followed by Thr 70 and Ser 65 (Gingras et al., 1999; Gingras, Raught, & Sonenberg, 2001). Phosphorylation at Ser 101 is required for efficient phosphorylation at Ser 65 (Wang, Li, Parra, Beugnet, & Proud, 2003), while phosphorylation at Ser 112, directly affects binding of 4E-BP1 to eIF4E, without influencing phosphorylation of other sites (Wang et al., 2003). 4E-BP2 and 4E-BP3 lack residues corresponding to both Ser 101 and Ser 112 (Wang et al., 2003).

Recently, several other kinases have been shown to phosphorylate 4E-BP1 (Qin, Jiang, & Zhang, 2016). For example, GSK3β phosphorylates 4E-BP1 at Thr 37 and Thr 46 in some cancer cells (Shin et al., 2014); P38-dependet phosphorylation of 4E-BP1 has also been shown in vitro in response to UV irradiation (G. Liu, Zhang, Bode, Ma, & Dong, 2002); cyclin-dependent kinase, cdc2/CDK1, can phosphorylate 4EBP1 at Thr 70 in HeLa cells (Heesom, Gampel, Mellor, & Denton, 2001) and in a breast cancer cell line (Greenberg & Zimmer, 2005). Finally, LRRK2 (leucine-rich repeat kinase 2) is one of the physiological kinases for 4E-BP at the Thr 37/46 sites, and it is the only one, other than mTOR, that that has been shown to phosphorylate 4E-BP in the nervous system of Drosophila (Imai et al., 2008). It is yet to be determined whether any of these kinases contributed to the regulation of the 4E-BPs in the mammalian brain.

Structural studies in solution using nuclear magnetic resonance show that multisite phosphorylation induces folding of the intrinsically disordered 4E-BP2, the major neural isoform of the family of the three mammalian proteins (Bah et al., 2015). Bah and coworkers demonstrated that phosphorylation of Thr 37 and Thr 46 induces folding of residues Pro 18-Arg 62 of 4E-BP2 into a four-stranded β-domain that sequesters the helical $YX_4L\Phi$ motif inside a partially buried β-strand, blocking its accessibility to eIF4E (Bah et al., 2015). The structurally rigid state of phosphorylated Thr 37-Thr 46 of 4E-BP2 presented a decreased affinity for eIF4E by a factor of approximately 4000 (Bah et al., 2015).

Distribution of 4E-BPs in the Brain

Although there is no comprehensive mapping of the 4E-BPs in the mammalian brain, initial studies investigating the distribution of the 4E-BPs across different tissues found that 4E-BP2 is the most abundant isoform in the brain, while low but detectable levels of 4E-BP1 were also detected (Banko et al., 2005; Tsukiyama-Kohara et al., 2001). In contrast, there are no detectable levels of 4E-BP3 in the brain at baseline conditions (Banko et al., 2005; Tsukiyama-Kohara et al., 2001).

In contrast to the widespread distribution of 4E-BP2 (Banko et al., 2005), 4E-BP1 appears to exhibit a restricted expression pattern in the brain, with higher levels in the suprachiasmatic nucleus, while sparse 4E-BP1 immunoreactivity is also detected in the hippocampus and cortex (Cao et al., 2013).

Deamidation of 4E-BP2 in the Adult Mouse Brain

4E-BP2 is widely expressed in the adult mouse brain (Banko et al., 2005; Tsukiyama-Kohara et al., 2001); however, its phosphorylation is not readily detectable in the adult CNS in unstimulated conditions (Bidinosti, Ran, et al., 2010). Although the reason for the lack of 4E-BP2 phosphorylation in CNS is still unclear, research on this topic led to the discovery of an additional posttranslational modification for 4E-BP2.

4E-BP2 undergoes a pH-dependent non-enzymatic deamidation in the adult mouse brain, which results in greater affinity for raptor and lower affinity for eIF4E (Bidinosti, Ran, et al., 2010). The deamidation can occur in two arginine residues (N99 and N102) that are not conserved in the murine or human 4E-BP1, therefore it is hypothesized that only 4E-BP2 undergoes this modification (Bidinosti, Ran, et al., 2010). Importantly, 4E-BP2 deamidation correlates with an age-dependent decrease in activation, at baseline, of the PI3K-Akt-mTOR signaling pathway in whole brain extracts (Bidinosti, Ran, et al., 2010). This decrease in upstream signaling did not correlate with a decrease in eIF4F complex; given the lower affinity of deamidated 4E-BP2 for eIF4E in adult brain, it was proposed that the progressive increases in 4E-BP2 deamidation contributed to keeping a steady level of eIF4F throughout development (Bidinosti, Ran, et al., 2010). Once deamidated, 4E-BP2 becomes a substrate for protein L-isoaspartyl methyltransferase (PIMT)-mediated repair of isoaspartyl residues (Bidinosti, Martineau, Frank, & Sonenberg, 2010).

Genes Regulated by 4E-BP1 and 4E-BP2 in the Brain

A landmark study in mouse embryonic fibroblasts (MEFs) used ribosome profiling (Ingolia, Ghaemmaghami, Newman, & Weissman, 2009) to determine which mRNAs were sensitive to mTOR inhibition (Thoreen et al., 2012). This analysis determined that the most sensitive mRNAs to the mTOR inhibitor Torin-1 were TOP or TOP-like mRNAs, which are defined by a cytidine immediately after the 5′ cap, followed by an uninterrupted stretch of 4–14 pyrimidines (Jefferies, Reinhard, Kozma, & Thomas, 1994; Meyuhas, 2000), and tend to encode proteins associated with translation (Iadevaia, Caldarola, Tino, Amaldi, & Loreni, 2008; Meyuhas, 2000). Most of the Torin-1 sensitive mRNAs were regulated through the 4E-BPs (Thoreen et al., 2012). Furthermore, contrary to the hypothesis that RNAs with long and complex 5′ untranslated regions (UTRs) are regulated through a 4E-BP-dependent mechanism (Hay & Sonenberg, 2004), this study found no evidence that 5′ UTR length or complexity correlated positively with sensitivity to mTOR inhibition (Thoreen et al., 2012).

In the brain, there is experimental evidence that mTORC1 signaling is necessary for the translation of several mRNAs that are not involved in the protein synthesis process, that is, Ca^{2+}/calmodulin-dependent kinase II alpha (CamKIIa), AMPA receptor subunits, microtubule-associated protein (MAP2), postsynaptic density protein 95 (PSD-95), and glutamate receptor interaction protein (Hou & Klann, 2004; Mameli, Balland, Lujan, & Luscher, 2007; Schratt, Nigh, Chen, Hu, & Greenberg, 2004; Slipczuk et al., 2009). Presumably, the 4E-BPs contribute to the translational regulation of a number of these and other genes, but an exhaustive unbiased approach has yet to be applied to identify the targets of their regulation.

Intriguingly, absence of *Eif4ebp2* did not result in changes in total protein synthesis in the adult brain (Gkogkas et al., 2013), although several mRNA targets for 4EBP1 and 4E-BP2 have been already confirmed/identified. For example, *Gria1* and *Gria2* mRNAs, encoding for the GluA1 and GluA2 AMPA receptor subunits (Ran et al., 2013), were identified as sensitive to 4E-BP2 and contributing to the synaptic effects observed in *Eif4ebp2* KO mice (Ran et al., 2013). In the study of autism-related genes, it was demonstrated that translation of neuroligins 1, 2, and 3 is increased in the hippocampi of *Eif4ebp2* KO and *Eif4e* transgenic mice (Gkogkas et al., 2013). It is unclear whether 4E-BP1 also contributes to the translational regulation of these genes in the brain; however, at least one mRNA has been shown to be regulated by 4E-BP1 in the brain: the *Vip* mRNA, encoding the vasoactive intestinal peptide (VIP; Cao et al., 2013).

The presence of several secondary structures in the 5′ UTR in the neuroligin genes was proposed to be involved in the sensitivity to 4E-BP2 (Gkogkas et al., 2013). The latter suggests that some structural elements in the 5′UTR of the mRNAs can contribute to the sensitivity toward the 4E-BPs (Gkogkas et al., 2013). The reason behind the apparent discrepancy between these findings and those in MEFs is unclear. One potential explanation is the contrast between acute effects of mTOR inhibition (as those studied in MEFs; Thoreen et al., 2012) and the long-term adaptations that occur with the chronic absence of the 4E-BPs as studied in the *Eif4ebp1 −/−* and *Eif4ebp2 −/−* mice. In addition, there is evidence that in adult mice, the genes related to the translational machinery itself are translationally suppressed in the hippocampus (Cho et al., 2015) through an unknown mechanism, adding another layer of complexity in the understanding of the translational control in the brain.

4E-BP2 in Brain Development and Neurodevelopmental Disorders

Overactive mTORC1 signaling is a signature of many disorders with cortical malformations, ranging from tuberous sclerosis complex with focal dysplasias to hemimegalencephaly with more diffuse, hemispheric aberrations (Crino, 2011). Focal increased activity of mTORC1 in the anterior cingulate cortex (ACC) at E15.5 leads to increased neuronal soma size, increased complexity of dendritic arbors, and mislamination, whereby neurons with greater mTORC1 are ectopically retained in deeper cortical layers (i.e., neurons fail to migrate out toward the external layers 2/3 of ACC; Lin, Hsieh, Kimura, Malone, & Bordey, 2016). These effects, along with increased cap-dependent translation, are dependent on the 4E-BPs, since they can be normalized when a phosphorylation-resistant mutant form of 4E-BP1 is overexpressed. In support of this, knock-down of endogenous 4E-BP2, leading to enhanced cap-dependent translation, is sufficient to induce ectopic localization of neurons to deeper layers of the cortex, indicating that the 4E-BPs are central in regulating cortical lamination and neuronal morphology (Lin et al., 2016).

Intriguingly, markers of decreased autophagy and increased endoplasmic reticulum stress (both induced by increased mTORC1 in neuronal progenitors) appear normalized after overexpression of phosphorylation-resistant 4E-BP1 and induced by 4E-BP2 knock down (Lin et al., 2016). These results suggest the possibility that cap-dependent translational control may influence the autophagy and ER stress response, in addition to the direct effect of mTORC1 on these cellular functions. The extent to which translation-regulated changes to autophagy and ER stress signaling contributes, if at all, to mTORC1-induced mislamination is yet to be determined (Lin et al., 2016).

In addition, 4E-BP2 is important for mTORC1-mediated control of axon elongation in ACC neurons in vivo. In this regard, hyperactive mTORC1 at E15.5 in ACC neurons led to increased axonal growth projecting contralaterally into the corpus callosum and into the contralateral cortices (Gong et al., 2015). This effect was reversed by co-transfecting a constitutively active 4E-BP1 (Gong et al., 2015). Neither axonal branching, direction, or growth appeared altered by the manipulation of mTORC1 or the 4E-BPs (Gong et al., 2015). Intriguingly, 4E-BP1 phosphorylation stimulated β-actin translation in *Xenopus laevis* retinal growth cones, which controlled directional turning of the growth cone (Leung et al., 2006). It is unclear whether the involvement of 4E-BP1 in growth cone response to guidance cues also extends to the mammalian brain.

In early postnatal development, mTORC1 activity in the subventricular zone (SVZ) controls neural stem cell (NSC) self-renewal, and generation of transit amplifying cells (TACs) and, therefore, regulates the production of newborn neurons (Hartman et al., 2013). In this regard, decreased and increased mTORC1 function decreases and increases, respectively, generation of TACs at the expense of NSC self-renewal. In other words, mTORC1 activity does not induce NSCs to enter the cell cycle but, rather, is a regulator of their differentiation into their daughter cells: Mash1+ TACs in the neonatal SVZ. Importantly, this effect was mediated by 4E-BP2 and cap-dependent translation, and not S6K1/2, as constitutively active 4E-BP1 mimicked the effects of decreased mTORC1, while knock-down of 4E-BP2 mimicked those of overactive mTORC1 (Hartman et al., 2013).

Overall, during embryonic development the 4E-BP2 control cortical migration, it appears to contribute to autophagy and ER stress, dendritic arborization, and axonal growth. Furthermore, 4E-BP2 is an important regulator of postnatal neurogenesis, by controlling the differentiation of neuronal stem cells in the SVZ. These findings are central, given the overactivation of mTORC1 in several disorders with high rates of autism, including tuberous sclerosis (Tsai et al., 2014), PTEN, and fragile X syndrome (Sharma et al., 2010).

Role of 4E-BP2 in Autism and Synaptic Transmission

Supporting the importance of 4E-BPs and cap-dependent translation in autism spectrum disorders (ASD), absence of 4E-BP2 in *Eif4ebp2* KO mice leads to decreased social interaction, decrease ultrasonic vocalizations, and increased repetitive behaviors (Gkogkas et al., 2013). The social deficits could be rescued by pharmacological inhibition of eIF4F. In this regard, while ASD is a neurodevelopmental disorder, countless studies have demonstrated adult reversal of behavioral and synaptic phenotypes in ASD mouse models (e.g., Aguilar-Valles et al., 2015; Gkogkas et al., 2013). In this regard, acute administration of group I metabotropic glutamate receptor antagonists rescued the social deficits and increased repetitive behaviors (Aguilar-Valles et al., 2015), suggesting that like *Fmr1* KO mice, a model of fragile X syndrome, increased cap-dependent mRNA translation in the brain results in increased activity of mGluR1 and mGluR5, which can be targeted to normalize autism-like behaviors.

Intriguingly, no major structural alterations are evident in the brain of *Eif4ebp2* KO mice (Banko et al., 2005), although studies in this mouse strain may have not been detailed enough to reveal them. Thus, the relationship between the proposed developmental roles of 4E-BP2 (Gong et al., 2015; Hartman et al., 2013; Lin et al., 2016; described in the previous section) and the behavioral deficits observed in the full body knockout mice (Aguilar-Valles et al., 2015; Banko, Hou, Poulin, Sonenberg, & Klann, 2006; Banko et al., 2005, 2007; Gkogkas et al., 2013) remain unexplored and deserve more attention.

In addition to the behavioral deficits, *Eif4ebp2* KO mice present increased hippocampal synaptic activity, measured in CA1 pyramidal cells (Bidinosti, Martineau, et al., 2010; Gkogkas et al., 2013). CA1 miniature excitatory postsynaptic currents are increased in amplitude, frequency, and total charge (Bidinosti, Ran, et al., 2010; Gkogkas et al., 2013). In addition, miniature inhibitory postsynaptic currents are also increased in amplitude and total charge, but not in frequency, thus creating and excitatory/inhibitory (E/I) imbalance that is thought to contribute to the autistic-like behavioral deficits (Gkogkas et al., 2013). Importantly, overexpressed neuroligin 1, but not neuroligin 2, was involved in the induction of increased E/I ratio and decreased social interaction in the *Eif4ebp2* KO mice (Gkogkas et al., 2013).

Intriguingly, attempted rescue of increased mEPSC in hippocampal neurons with constitutively deamidated 4E-BP2 results in normalized mEPSC frequency and amplitude; yet, some aspects of the mEPSC remained altered with deamidated 4E-BP2 transfection. Specifically, there was an increase in total charge of mESPC and a slower rise and decay kinetics compared to neurons expressing the wildtype non-deamidated 4E-BP2 (Bidinosti, Martineau, et al., 2010). These findings suggest that in postnatal development, 4E-BP2 deamidation contributes to an increase in mEPSC charge and slower kinetics in the hippocampus, although this remains to be directly demonstrated (Bidinosti, Martineau, et al., 2010).

The effects of 4E-BP2 in synaptic transmission appear to be region specific, as enhanced mTORC1 activity in the ACC leads to reduced EPSC frequency in pyramidal neurons, which can be rescued by overexpressing a constitutively active form of 4E-BP1 (Lin et al., 2016). In addition, enhanced mTORC1 in ACC pyramidal neurons results in increased the membrane resting potential, which was also dependent on 4E-BP control of translation.

4E-BPs in the Control of Circadian Rhythms

Circadian clocks have evolved to adaptively align internal biological processes to daily environmental changes (Lim & Allada, 2013). Circadian rhythms are self-sustaining, persisting even in the absence of external time cues. However, the period of these clocks only approximates 24 h. These self-sustaining clocks are reset by oscillating inputs such as light, temperature, or feeding to synchronize with the 24 h environment (Lim & Allada, 2013). The mechanism for intracellular generation of circadian rhythms is based in part on a transcription-translation feedback loop in which CLOCK and BMAL1 proteins induce the expression of the *Period* (*Per1, Per2* and *Per3*) and *Cryptochrome* genes (*Cry1* and *Cry2*), whose protein products interact and inhibit CLOCK/BMAL1 transcriptional activity (Lim & Allada, 2013).

mRNA translation is also a critical event for light-entrainment of the clock. Along these lines, work performed in a wide range of clock model systems has shown that the application of translation inhibitors suppresses light entrainment (Johnson & Nakashima, 1990; Murakami, Nishi, Katayama, & Nasu, 1995; Raju, Yeung, & Eskin, 1990; Zhang, Takahashi, & Turek, 1996). Furthermore, rhythmic activation of signaling pathways that regulate translation initiation, in particular the mTORC1 pathway, results in circadian time-dependent phosphorylation of translation factors in suprachiasmatic nucleus (SCN) clock neurons, as well as peripheral clock tissues (Cao, Anderson, Jung, Dziema, & Obrietan, 2011; Jouffe et al., 2013).

Regarding the specific involvement of the 4E-BPs in mammals, 4E-BP1 is enriched in the central clock pacemaker, the suprachiasmatic nucleus (SCN; Cao et al., 2013). In the SCN, 4E-BP1 phosphorylation varies with the circadian rhythm in mice kept in constant darkness (Cao et al., 2013), and this phosphorylation can be increased by light stimulation, downstream of mTORC1 activation (Cao, Lee, Cho, Saklayen, & Obrietan, 2008). One of the main phenotypes of the *Eif4ebp1* KO mice is the accelerated re-entrainment of the circadian wheel-running behavior in "jet-lag" models, where there the light schedule is either delayed or advanced abruptly (Cao et al., 2013). This effect is reflected by concurrent changes in PER1 and PER2 levels, that recover faster in *Eif4ebp1* KO compared to wildtype mice (Cao et al., 2013). *Eif4ebp1* KO mice are also more resilient to the disruptive effects of constant light on circadian wheel-running compared to wildtype mice (Cao et al., 2013). These phenotypes indicate the importance of the mTORC1-4E-BP1 activation in light-induced resetting of the circadian clock. Furthermore, as 4E-BP1 controls the translation of the *Vip* mRNA (Cao et al., 2013), which is required for coordination of circadian rhythms both in the SCN and in behavior (Aton, Colwell, Harmar, Waschek, & Herzog, 2005), and its elevated translation in the brain of 4E-BP1 may be responsible for the observed circadian phenotypes.

Roles in Synaptic Plasticity and Memory Formation

Several studies have demonstrated activation and requirement of the mTORC1 for memory formation (Costa-Mattioli, Sossin, Klann, & Sonenberg, 2009; Hoeffer & Klann, 2010; Qi, Mizuno, Yonezawa, Nawa, & Takei, 2010), including phosphorylation of mTOR, and the mTORC1 substrates S6K1 and 4E-BP1/2 (Kelleher, Govindarajan, Jung, Kang, & Tonegawa, 2004; Saraf, Luo, Morris, & Storm, 2014; Tang et al., 2002). Furthermore, downstream of 4E-BP phosphorylation, there is a dose-dependent increase in eIF4F formation following LTP-inducing electrical and LTD-inducing chemical stimulation in the hippocampus (Banko et al., 2005). However, the role of mRNA translation in memory consolidation is much more nuanced, involving not only activating (such as mTORC1), but also translationally repressive pathways (Cho et al., 2015; Costa-Mattioli et al., 2005, 2007).

As 4E-BP2 appears to be more ubiquitously expressed 4E-BP in the brain (Tsukiyama-Kohara et al., 2001), these mutant mice have been further investigated for effects on memory formation, although it is possible that the more anatomically restricted, 4E-BP1, may also have a role in this process. Since mTORC1 is important for memory formation (Stoica et al., 2011), it may thus follow that a model that mimics constitutive mTORC1 activation, at least in one of its downstream signaling branches (such as the *Eif4ebp2* KO mice), may result in enhanced plasticity and memory formation. Indeed, early-long term potentiation (E-LTP) in the CA1 region of the hippocampus, which is relevant for memory encoding, is exacerbated in *Eif4ebp2* KO mice, and this is sensitive to protein synthesis and transcription inhibitors (Banko et al., 2005). However, late-LTP (L-LTP), induced by stronger stimulation (either 4 HFS or 4 TBS trains), is absent/occluded in *Eif4ebp2* KO mice. Moreover, *Eif4ebp2* KO mice are impaired in long-term contextual fear memory test and Morris water maze, although their memory for cued-fear conditioning is intact (Banko et al., 2005). *Eif4ebp2* KO mice also show impaired hippocampal-dependent working memory in the T-maze spontaneous alternation (Banko et al., 2007). *Eif4ebp2* KO mice do not have any exploratory/ambulatory alteration in the open field but have impaired motor learning and compromised motor coordination and balance (Banko et al., 2007). They do not show any impairment in step-through passive avoidance, or anxiety in the elevated plus maze (Banko et al., 2007). These results indicate that the absence of this translation repressor is necessary for proper memory encoding/retrieval. These effects on cognition may also be the result of a faulty developmental process due to the continuous absence of 4E-BP2 in *Eif4ebp2* KO mice.

Importantly, *Eif4ebp2* KO mice also present exacerbated chemically induced hippocampal LTD (Banko et al., 2006). This form of plasticity, induced by the type I mGluR agonist, (S)-3,5-Dihydroxyphenylglycine (DHPG), normally induces 4E-BP phosphorylation and eIF4F complex activation (Banko et al., 2006). Further characterization of this phenotype revealed that anisomycin, a protein synthesis inhibitor, can normalize but not block LTD in 4E-BP2 KO mice (Aguilar-Valles et al., 2015). Intriguingly, synaptically induced LTD, which also depends on protein translation, was completely absent in 4E-BP2 KO mice (Aguilar-Valles et al., 2015). The reason behind these mechanistic differences between chemically and electrically induced hippocampal LTD remains unexplored. Furthermore, it is still unclear how these effects on LTD may impact hippocampal-dependent cognitive processing. These findings are also partly divergent from those in the *Fmr1* y/- mice, where chemically induced hippocampal LTD is also exacerbated (Huber, Kayser, & Bear, 2000) but is protein synthesis independent (i.e., insensitive to anisomycin; Hou et al., 2006). Adding to the confusion, a recent study showed that hippocampal mGluR-LTD (both chemically and synaptically induced) is normal in mice in which *Raptor* was knocked out in excitatory cells (Zhu, Chen, Mays,

Stoica, & Costa-Mattioli, 2018). It still remains to be investigated whether the mTORC1-4E-BP pathway in inhibitory interneurons plays any role in the induction of mGluR-LTD.

Outside the hippocampus, 4E-BP2 appears to have a different role in memory formation; indeed, *Eif4ebp2* KO mice demonstrated an enhanced performance in a conditioned taste aversion task by avoiding the saccharin and NaCl solutions to a higher degree than wild-types following a one trial pairing of saccharin or NaCl with LiCl (Banko et al., 2007). This type of conditioned learning involves the amygladar complex, the insular cortex, and the reward circuitry (Yamamoto & Yasoshima, 2007), suggesting that 4E-BP2 in other brain circuits can be acting as a physiological brake in the formation of learning.

Roles of 4E-BPs in Psychiatric Disorders

Substance abuse models. Most studies analyzing the role of mTORC1 signaling on the behavioral effects of drugs of abuse have focused on determining the activation (phosphorylation) of elements of this signaling pathway, as well as the effect of rapamycin. To date, there is no study on the direct effects of the 4E-BPs on addiction pathophysiology.

For example, several studies have shown that the psychostimulant cocaine regulates mTOR signaling in the reward circuitry, particularly the nucleus accumbens (Bailey, Ma, & Szumlinski, 2012; Sutton & Caron, 2015; Wu, McCallum, Glick, & Huang, 2011), and that this is important for some of the behavioral effects of this psychostimulant. Furthermore, activation of dopamine receptor 1 (D1R) is involved in the induction of mTORC1 activation and 4E-BP phosphorylation in the NAc (Sutton & Caron, 2015). Systemic injection of rapamycin blocked the expression of cocaine-induced locomotion (and sensitization) and conditional place preference (Bailey et al., 2012; James et al., 2014; Wu et al., 2011), and knockout of *Mtor* or *Raptor* in D1R expressing neurons reduced the locomotor response to acute cocaine treatment (Sutton & Caron, 2015). Infusions of rapamycin in the NAc have been shown to block the development of methamphetamine-induced conditional place preference (Narita et al., 2005).

In addition to psychostimulants, alcohol administration has also been shown to activate mTORC1-mediated signaling in the NAc of mice, including phosphorylation of 4E-BPs, and that rapamycin treatment decreases expression of alcohol-induced locomotor sensitization and place preference, as well as excessive alcohol intake and seeking (Neasta, Ben Hamida, Yowell, Carnicella, & Ron, 2010). Morphine also activates mTORC1 in the VTA (Mazei-Robison et al., 2011), but it remains undetermined whether 4E-BP-dependent translational control of gene expression plays any role in the synaptic and behavioral neuroadaptations to this drug of abuse.

In contrast, amphetamine administration does not affect mTORC1 signaling (Biever et al., 2016); thus, the activation of mTORC1 signaling may not be a common mechanism shared by all psychostimulants or in general drugs of abuse. However, many studies have successfully used rapamycin to temper the effects of drugs of abuse, suggesting a key role for mTORC1 in reward and relapse behaviors.

Depressive disorders. Decreases in mTORC1 activation are observed in several models of depressive-like behaviors induced by chronic stress, including inescapable shock exposure (Li et al., 2010) or three weeks of chronic-unpredictable stress (Li et al., 2011). In addition, REDD1 (regulated in development and DNA damage response 1), an inhibitor of mTORC1 signaling (Katiyar et al., 2009), is increased by stress, and its overexpression in the mPFC is sufficient to produce synapse loss and depressive-like behavior (Ota et al., 2014). REDD1 is also found to be increased in postmortem tissue from individuals with depression, consistent with the possibility that REDD1 could contribute to neuronal atrophy and depressive behaviors in patients (Ota et al., 2014).

Consistently, several anti-depressant treatments produce an increase in mTORC1 activation and 4E-BP phosphorylation, including serotonin reuptake inhibitors (X. L. Liu et al., 2015; Park et al., 2014) and NMDA blocker such as ketamine (Li et al., 2010; Paul et al., 2014) and MK-801 (Yoon et al., 2008). mTOR activation is required for ketamine's behavioral antidepressant actions, as intracerebroventricular pre-treatment with rapamycin blocks ketamine induced synaptic molecular changes and antidepressant actions in mice (Li et al., 2010) and rats (Holubova et al., 2016).

Intriguingly, rapamycin has antidepressant effects on its own (Cleary et al., 2008; Zhou et al., 2013), suggesting that the role of mTORC1 and its downstream targets in depression pathophysiology and antidepressant treatment may be more complex than anticipated. In this regard, the mechanistic contribution of 4E-BPs in the development of stress-induced synaptic alterations, depression-like behaviors and the antidepressant effects of SSRIs and NMDA blockers remains to be elucidated.

Anti-psychotic treatments. A potential role for 4E-BPs in the mechanism of action of antipsychotic treatments exists, as 4E-BP phosphorylation is increased by the antipsychotic drug haloperidol, a dopamine receptor type 2 (D2R) antagonist, in primary striatal D2R-positive neurons (Bowling et al., 2014). Furthermore, haloperidol-induced increase in dendritic morphological complexity in D2R- positive neurons was 4E-BP dependent (Bowling et al., 2014), which coincided with marked changes in the pattern of protein synthesis, including increased abundance of cytoskeletal proteins and proteins associated with translation machinery (Bowling et al., 2014).

As for depression, the role of 4E-BPs in schizophrenia pathophysiology and treatment may be more nuanced, since models of schizophrenia induced by chronic treatment with the NMDA antagonist, MK-801, also result in increased phosphorylation of 4E-BP in the rat prefrontal cortex (Yoon et al., 2008), and rapamycin treatment reversed cognitive and affective deficits caused by *Disc1* knockdown (Zhou et al., 2013). Gene mutations in *DISC1* (disrupted-in-schizophrenia 1) are linked to psychiatric illness, including bipolar disorder, and depression (St. Clair et al., 1990).

4E-BPs in Neurodegenerative Diseases

Mutations in *PINK1* and *PARK2* cause autosomal recessive parkinsonism (Farrer, 2006), a neurodegenerative disorder that is characterized by the loss of dopaminergic neurons.

In Drosophila, overexpression of the translation inhibitor Thor (4E-BP) can suppress all the pathologic phenotypes of mutants in *park* and *Pink1*, including degeneration of dopaminergic neurons (Tain et al., 2009). Consistently, 4E-BP was shown to be hyperphosphorylated by the most common cause of parkinsonism, dominant mutations in LRRK2 (Imai et al., 2008). These results remain to be verified in mammalian animal models and in tissue from patients bearing the mutations linked to Parkinson's disease.

Concluding Remarks

4E-BPs are a central mechanism in the control of cap-dependent translation in the brain. They are major players in the control of brain development, affecting cortical and hippocampal synaptic transmission, neuronal morphology and migration of neurons. A testament of their importance in neurodevelopment is the finding that the absence of 4E-BP2 results in autistic-like alterations, as well as major deficits in learning and memory and hippocampal plasticity. These effects occur through the regulation of several key mRNAs, such as the AMPA subunits GluA1 and GluA2, and the neuroligins 1-3. Furthermore, 4E-BPs are central in circadian rhythm regulation, and have potential roles in the pathophysiology and treatment of psychiatric (depressive disorders, schizophrenia) and neurodegenerative disorders (Parkinson's). As such, their pharmacological targeting (either enhancement or inhibition) is a promising avenue for the treatment of psychiatric and neurological disorders.

References

Aguilar-Valles, A., Matta-Camacho, E., Khoutorsky, A., Gkogkas, C., Nader, K., Lacaille, J. C., & Sonenberg, N. (2015). Inhibition of group I metabotropic glutamate receptors reverses autistic-like phenotypes caused by deficiency of the translation repressor eIF4E binding protein 2. *Journal of Neuroscience, 35*(31), 11125–11132. https://doi.org/10.1523/JNEUROSCI.4615-14.2015

Aton, S. J., Colwell, C. S., Harmar, A. J., Waschek, J., & Herzog, E. D. (2005). Vasoactive intestinal polypeptide mediates circadian rhythmicity and synchrony in mammalian clock neurons. *Nature Neuroscience, 8*(4), 476–483. https://doi.org/10.1038/nn1419

Bah, A., Vernon, R. M., Siddiqui, Z., Krzeminski, M., Muhandiram, R., Zhao, C., . . . Forman-Kay, J. D. (2015). Folding of an intrinsically disordered protein by phosphorylation as a regulatory switch. *Nature, 519*(7541), 106–109. https://doi.org/10.1038/nature13999

Bailey, J., Ma, D., & Szumlinski, K. K. (2012). Rapamycin attenuates the expression of cocaine-induced place preference and behavioral sensitization. *Addiction Biology, 17*(2), 248–258. https://doi.org/10.1111/j.1369-1600.2010.00311.x

Banko, J. L., Hou, L., Poulin, F., Sonenberg, N., & Klann, E. (2006). Regulation of eukaryotic initiation factor 4E by converging signaling pathways during metabotropic glutamate receptor-dependent long-term depression. *Journal of Neuroscience, 26*(8), 2167–2173. https://doi.org/10.1523/JNEUROSCI.5196-05.2006

Banko, J. L., Merhav, M., Stern, E., Sonenberg, N., Rosenblum, K., & Klann, E. (2007). Behavioral alterations in mice lacking the translation repressor 4E-BP2. *Neurobiology of Learning and Memory, 87*(2), 248–256. https://doi.org/10.1016/j.nlm.2006.08.012

Banko, J. L., Poulin, F., Hou, L., DeMaria, C. T., Sonenberg, N., & Klann, E. (2005). The translation repressor 4E-BP2 is critical for eIF4F complex formation, synaptic plasticity, and memory in the hippocampus. *Journal of Neuroscience, 25*(42), 9581–9590. https://doi.org/10.1523/JNEUROSCI.2423-05.2005

Bidinosti, M., Martineau, Y., Frank, F., & Sonenberg, N. (2010). Repair of isoaspartate formation modulates the interaction of deamidated 4E-BP2 with mTORC1 in brain. *Journal of Biological Chemistry, 285*(25), 19402–19408. https://doi.org/10.1074/jbc.M110.120774

Bidinosti, M., Ran, I., Sanchez-Carbente, M. R., Martineau, Y., Gingras, A. C., Gkogkas, C., ... Sonenberg, N. (2010). Postnatal deamidation of 4E-BP2 in brain enhances its association with raptor and alters kinetics of excitatory synaptic transmission. *Molecular Cell, 37*(6), 797–808. https://doi.org/10.1016/j.molcel.2010.02.022

Biever, A., Boubaker-Vitre, J., Cutando, L., Gracia-Rubio, I., Costa-Mattioli, M., Puighermanal, E., & Valjent, E. (2016). Repeated exposure to D-amphetamine decreases global protein synthesis and regulates the translation of a subset of mRNAs in the striatum. *Frontiers in Molecular Neuroscience, 9*, 165. https://doi.org/10.3389/fnmol.2016.00165

Bowling, H., Zhang, G., Bhattacharya, A., Perez-Cuesta, L. M., Deinhardt, K., Hoeffer, C. A., ... Chao, M. V. (2014). Antipsychotics activate mTORC1-dependent translation to enhance neuronal morphological complexity. *Science Signaling, 7*(308), ra4. https://doi.org/10.1126/scisignal.2004331

Cao, R., Anderson, F. E., Jung, Y. J., Dziema, H., & Obrietan, K. (2011). Circadian regulation of mammalian target of rapamycin signaling in the mouse suprachiasmatic nucleus. *Neuroscience, 181*, 79–88. https://doi.org/10.1016/j.neuroscience.2011.03.005

Cao, R., Lee, B., Cho, H. Y., Saklayen, S., & Obrietan, K. (2008). Photic regulation of the mTOR signaling pathway in the suprachiasmatic circadian clock. *Molecular and Cellular Neuroscience, 38*(3), 312–324. https://doi.org/10.1016/j.mcn.2008.03.005

Cao, R., Robinson, B., Xu, H., Gkogkas, C., Khoutorsky, A., Alain, T., ... Sonenberg, N. (2013). Translational control of entrainment and synchrony of the suprachiasmatic circadian clock by mTOR/4E-BP1 signaling. *Neuron, 79*(4), 712–724. https://doi.org/10.1016/j.neuron.2013.06.026

Cho, J., Yu, N. K., Choi, J. H., Sim, S. E., Kang, S. J., Kwak, C., ... Kaang, B. K. (2015). Multiple repressive mechanisms in the hippocampus during memory formation. *Science, 350*(6256), 82–87. https://doi.org/10.1126/science.aac7368

Cleary, C., Linde, J. A., Hiscock, K. M., Hadas, I., Belmaker, R. H., Agam, G., ... Einat, H. (2008). Antidepressive-like effects of rapamycin in animal models: Implications for mTOR inhibition as a new target for treatment of affective disorders. *Brain Research Bulletin, 76*(5), 469–473. https://doi.org/10.1016/j.brainresbull.2008.03.005

Costa-Mattioli, M., Gobert, D., Harding, H., Herdy, B., Azzi, M., Bruno, M., ... Sonenberg, N. (2005). Translational control of hippocampal synaptic plasticity and memory by the eIF2alpha kinase GCN2. *Nature, 436*(7054), 1166–1173. https://doi.org/10.1038/nature03897

Costa-Mattioli, M., Gobert, D., Stern, E., Gamache, K., Colina, R., Cuello, C., ... Sonenberg, N. (2007). eIF2alpha phosphorylation bidirectionally regulates the switch from short- to long-term synaptic plasticity and memory. *Cell, 129*(1), 195–206. https://doi.org/10.1016/j.cell.2007.01.050

Costa-Mattioli, M., Sossin, W. S., Klann, E., & Sonenberg, N. (2009). Translational control of long-lasting synaptic plasticity and memory. *Neuron, 61*(1), 10–26. https://doi.org/10.1016/j.neuron.2008.10.055

Crino, P. B. (2011). mTOR: A pathogenic signaling pathway in developmental brain malformations. *Trends in Molecular Medicine, 17*(12), 734–742. https://doi.org/10.1016/j.molmed.2011.07.008

Eguchi, S., Tokunaga, C., Hidayat, S., Oshiro, N., Yoshino, K., Kikkawa, U., & Yonezawa, K. (2006). Different roles for the TOS and RAIP motifs of the translational regulator protein 4E-BP1 in the association with raptor and phosphorylation by mTOR in the regulation of cell size. *Genes Cells, 11*(7), 757–766. https://doi.org/10.1111/j.1365-2443.2006.00977.x

Farrer, M. J. (2006). Genetics of Parkinson disease: Paradigm shifts and future prospects. *Nature Reviews Genetics, 7*(4), 306–318. https://doi.org/10.1038/nrg1831

Gingras, A. C., Gygi, S. P., Raught, B., Polakiewicz, R. D., Abraham, R. T., Hoekstra, M. F., ... Sonenberg, N. (1999). Regulation of 4E-BP1 phosphorylation: A novel two-step mechanism. *Genes & Development, 13*(11), 1422–1437.

Gingras, A. C., Raught, B., & Sonenberg, N. (2001). Regulation of translation initiation by FRAP/mTOR. *Genes & Development, 15*(7), 807–826. https://doi.org/10.1101/gad.887201

Gkogkas, C. G., Khoutorsky, A., Ran, I., Rampakakis, E., Nevarko, T., Weatherill, D. B., ... Sonenberg, N. (2013). Autism-related deficits via dysregulated eIF4E-dependent translational control. *Nature, 493*(7432), 371–377. https://doi.org/10.1038/nature11628

Gong, X., Zhang, L., Huang, T., Lin, T. V., Miyares, L., Wen, J., ... Bordey, A. (2015). Activating the translational repressor 4E-BP or reducing S6K-GSK3beta activity prevents accelerated axon growth induced by hyperactive mTOR in vivo. *Human Molecular Genetics, 24*(20), 5746–5758. https://doi.org/10.1093/hmg/ddv295

Greenberg, V. L., & Zimmer, S. G. (2005). Paclitaxel induces the phosphorylation of the eukaryotic translation initiation factor 4E-binding protein 1 through a Cdk1-dependent mechanism. *Oncogene, 24*(30), 4851–4860. https://doi.org/10.1038/sj.onc.1208624

Haghighat, A., Mader, S., Pause, A., & Sonenberg, N. (1995). Repression of cap-dependent translation by 4E-binding protein 1: Competition with p220 for binding to eukaryotic initiation factor-4E. *EMBO Journal, 14*(22), 5701–5709.

Hara, K., Maruki, Y., Long, X., Yoshino, K., Oshiro, N., Hidayat, S., ... Yonezawa, K. (2002). Raptor, a binding partner of target of rapamycin (TOR), mediates TOR action. *Cell, 110*(2), 177–189.

Hartman, N. W., Lin, T. V., Zhang, L., Paquelet, G. E., Feliciano, D. M., & Bordey, A. (2013). mTORC1 targets the translational repressor 4E-BP2, but not S6 kinase 1/2, to regulate neural stem cell self-renewal in vivo. *Cell Reports, 5*(2), 433–444. https://doi.org/10.1016/j.celrep.2013.09.017

Hay, N., & Sonenberg, N. (2004). Upstream and downstream of mTOR. *Genes & Development, 18*(16), 1926–1945. https://doi.org/10.1101/gad.1212704

Heesom, K. J., Gampel, A., Mellor, H., & Denton, R. M. (2001). Cell cycle-dependent phosphorylation of the translational repressor eIF-4E binding protein-1 (4E-BP1). *Current Biology, 11*(17), 1374–1379.

Hershey, P. E., McWhirter, S. M., Gross, J. D., Wagner, G., Alber, T., & Sachs, A. B. (1999). The Cap-binding protein eIF4E promotes folding of a functional domain of yeast translation initiation factor eIF4G1. *Journal of Biological Chemistry, 274*(30), 21297–21304.

Hoeffer, C. A., & Klann, E. (2010). mTOR signaling: At the crossroads of plasticity, memory and disease. *Trends Neuroscience, 33*(2), 67–75. https://doi.org/10.1016/j.tins.2009.11.003

Holubova, K., Kleteckova, L., Skurlova, M., Ricny, J., Stuchlik, A., & Vales, K. (2016). Rapamycin blocks the antidepressant effect of ketamine in task-dependent manner. *Psychopharmacology (Berlin), 233*(11), 2077–2097. https://doi.org/10.1007/s00213-016-4256-3

Hou, L., Antion, M. D., Hu, D., Spencer, C. M., Paylor, R., & Klann, E. (2006). Dynamic translational and proteasomal regulation of fragile X mental retardation protein controls mGluR-dependent long-term depression. *Neuron, 51*, 441–454.

Hou, L., & Klann, E. (2004). Activation of the phosphoinositide 3-kinase-Akt-mammalian target of rapamycin signaling pathway is required for metabotropic glutamate receptor-dependent long-term depression. *Journal of Neuroscience, 24*(28), 6352–6361. https://doi.org/10.1523/JNEUROSCI.0995-04.2004

Huber, K. M., Kayser, M. S., & Bear, M. F. (2000). Role for rapid dendritic protein synthesis in hippocampal mGluR-dependent long-term depression. *Science, 288*(5469), 1254–1257. https://doi.org/10.1126/science.288.5469.1254

Iadevaia, V., Caldarola, S., Tino, E., Amaldi, F., & Loreni, F. (2008). All translation elongation factors and the e, f, and h subunits of translation initiation factor 3 are encoded by 5′-terminal oligopyrimidine (TOP) mRNAs. *RNA, 14*(9), 1730–1736. https://doi.org/10.1261/rna.1037108

Imai, Y., Gehrke, S., Wang, H. Q., Takahashi, R., Hasegawa, K., Oota, E., & Lu, B. (2008). Phosphorylation of 4E-BP by LRRK2 affects the maintenance of dopaminergic neurons in Drosophila. *EMBO Journal, 27*(18), 2432–2443. https://doi.org/10.1038/emboj.2008.163

Ingolia, N. T., Ghaemmaghami, S., Newman, J. R., & Weissman, J. S. (2009). Genome-wide analysis in vivo of translation with nucleotide resolution using ribosome profiling. *Science, 324*(5924), 218–223. https://doi.org/10.1126/science.1168978

James, M. H., Quinn, R. K., Ong, L. K., Levi, E. M., Charnley, J. L., Smith, D. W., . . . Dayas, C. V. (2014). mTORC1 inhibition in the nucleus accumbens "protects" against the expression of drug seeking and "relapse" and is associated with reductions in GluA1 AMPAR and CAMKIIalpha levels. *Neuropsychopharmacology, 39*(7), 1694–1702. https://doi.org/10.1038/npp.2014.16

Jefferies, H. B., Reinhard, C., Kozma, S. C., & Thomas, G. (1994). Rapamycin selectively represses translation of the "polypyrimidine tract" mRNA family. *Proceedings of the National Academy of Sciences USA, 91*(10), 4441–4445.

Johnson, C. H., & Nakashima, H. (1990). Cycloheximide inhibits light-induced phase shifting of the circadian clock in Neurospora. *Journal of Biological Rhythms, 5*(2), 159–167. https://doi.org/10.1177/074873049000500207

Jouffe, C., Cretenet, G., Symul, L., Martin, E., Atger, F., Naef, F., & Gachon, F. (2013). The circadian clock coordinates ribosome biogenesis. *PLoS Biology, 11*(1), e1001455. https://doi.org/10.1371/journal.pbio.1001455

Katiyar, S., Liu, E., Knutzen, C. A., Lang, E. S., Lombardo, C. R., Sankar, S., . . . Chiang, G. G. (2009). REDD1, an inhibitor of mTOR signalling, is regulated by the CUL4A-DDB1 ubiquitin ligase. *EMBO Reports, 10*(8), 866–872. https://doi.org/10.1038/embor.2009.93

Kelleher, R. J., 3rd, Govindarajan, A., Jung, H. Y., Kang, H., & Tonegawa, S. (2004). Translational control by MAPK signaling in long-term synaptic plasticity and memory. *Cell, 116*(3), 467–479. https://doi.org/10.1016/S0092-8674(04)00115-1

Lee, V. H., Healy, T., Fonseca, B. D., Hayashi, A., & Proud, C. G. (2008). Analysis of the regulatory motifs in eukaryotic initiation factor 4E-binding protein 1. *FEBS Journal, 275*(9), 2185–2199. https://doi.org/10.1111/j.1742-4658.2008.06372.x

Leung, K. M., van Horck, F. P., Lin, A. C., Allison, R., Standart, N., & Holt, C. E. (2006). Asymmetrical beta-actin mRNA translation in growth cones mediates attractive turning to netrin-1. *Nature Neuroscience, 9*(10), 1247–1256. https://doi.org/10.1038/nn1775

Li, N., Lee, B., Liu, R. J., Banasr, M., Dwyer, J. M., Iwata, M.,... Duman, R. S. (2010). mTOR-dependent synapse formation underlies the rapid antidepressant effects of NMDA antagonists. *Science, 329*(5994), 959–964. https://doi.org/10.1126/science.1190287

Li, N., Liu, R. J., Dwyer, J. M., Banasr, M., Lee, B., Son, H.,... Duman, R. S. (2011). Glutamate N-methyl-D-aspartate receptor antagonists rapidly reverse behavioral and synaptic deficits caused by chronic stress exposure. *Biological Psychiatry, 69*(8), 754–761. https://doi.org/10.1016/j.biopsych.2010.12.015

Lim, C., & Allada, R. (2013). Emerging roles for post-transcriptional regulation in circadian clocks. *Nature Neuroscience, 16*(11), 1544–1550. https://doi.org/10.1038/nn.3543

Lin, T. V., Hsieh, L., Kimura, T., Malone, T. J., & Bordey, A. (2016). Normalizing translation through 4E-BP prevents mTOR-driven cortical mislamination and ameliorates aberrant neuron integration. *Proceedings of the National Academy of Sciences USA, 113*(40), 11330–11335. https://doi.org/10.1073/pnas.1605740113

Liu, G., Zhang, Y., Bode, A. M., Ma, W. Y., & Dong, Z. (2002). Phosphorylation of 4E-BP1 is mediated by the p38/MSK1 pathway in response to UVB irradiation. *Journal of Biological Chemistry, 277*(11), 8810–8816. https://doi.org/10.1074/jbc.M110477200

Liu, X. L., Luo, L., Mu, R. H., Liu, B. B., Geng, D., Liu, Q., & Yi, L. T. (2015). Fluoxetine regulates mTOR signalling in a region-dependent manner in depression-like mice. *Science Reports, 5*, 16024. https://doi.org/10.1038/srep16024

Mader, S., Lee, H., Pause, A., & Sonenberg, N. (1995). The translation initiation factor eIF-4E binds to a common motif shared by the translation factor eIF-4 gamma and the translational repressors 4E-binding proteins. *Molecular and Cellular Biology, 15*(9), 4990–4997.

Mameli, M., Balland, B., Lujan, R., & Luscher, C. (2007). Rapid synthesis and synaptic insertion of GluR2 for mGluR-LTD in the ventral tegmental area. *Science, 317*(5837), 530–533. https://doi.org/10.1126/science.1142365

Marcotrigiano, J., Gingras, A. C., Sonenberg, N., & Burley, S. K. (1999). Cap-dependent translation initiation in eukaryotes is regulated by a molecular mimic of eIF4G. *Molecular Cell, 3*(6), 707–716. https://doi.org/10.1016/S1097-2765(01)80003-4

Martineau, Y., Azar, R., Bousquet, C., & Pyronnet, S. (2013). Anti-oncogenic potential of the eIF4E-binding proteins. *Oncogene, 32*(6), 671–677. https://doi.org/10.1038/onc.2012.116

Mazei-Robison, M. S., Koo, J. W., Friedman, A. K., Lansink, C. S., Robison, A. J., Vinish, M.,... Nestler, E. J. (2011). Role for mTOR signaling and neuronal activity in morphine-induced adaptations in ventral tegmental area dopamine neurons. *Neuron, 72*(6), 977–990. https://doi.org/10.1016/j.neuron.2011.10.012

Meyuhas, O. (2000). Synthesis of the translational apparatus is regulated at the translational level. *European Journal of Biochemistry, 267*(21), 6321–6330.

Murakami, N., Nishi, R., Katayama, T., & Nasu, T. (1995). Inhibitor of protein synthesis phase-shifts the circadian oscillator and inhibits the light induced-phase shift of the melatonin rhythm in pigeon pineal cells. *Brain Research, 693*(1–2), 1–7.

Narita, M., Akai, H., Kita, T., Nagumo, Y., Narita, M., Sunagawa, N.,... Suzuki, T. (2005). Involvement of mitogen-stimulated p70-S6 kinase in the development of sensitization to the methamphetamine-induced rewarding effect in rats. *Neuroscience, 132*(3), 553–560. https://doi.org/10.1016/j.neuroscience.2004.12.050

Neasta, J., Ben Hamida, S., Yowell, Q., Carnicella, S., & Ron, D. (2010). Role for mammalian target of rapamycin complex 1 signaling in neuroadaptations underlying alcohol-related disorders. *Proceedings of the National Academy of Sciences USA, 107*(46), 20093–20098. https://doi.org/10.1073/pnas.1005554107

Ota, K. T., Liu, R. J., Voleti, B., Maldonado-Aviles, J. G., Duric, V., Iwata, M.,...Duman, R. S. (2014). REDD1 is essential for stress-induced synaptic loss and depressive behavior. *Nature Medicine, 20*(5), 531–535. https://doi.org/10.1038/nm.3513

Park, S. W., Lee, J. G., Seo, M. K., Lee, C. H., Cho, H. Y., Lee, B. J.,...Kim, Y. H. (2014). Differential effects of antidepressant drugs on mTOR signalling in rat hippocampal neurons. *International Journal of Neuropsychopharmacology, 17*(11), 1831–1846. https://doi.org/10.1017/S1461145714000534

Paul, R. K., Singh, N. S., Khadeer, M., Moaddel, R., Sanghvi, M., Green, C. E.,...Wainer, I. W. (2014). (R,S)-Ketamine metabolites (R,S)-norketamine and (2S,6S)-hydroxynorketamine increase the mammalian target of rapamycin function. *Anesthesiology, 121*(1), 149–159. https://doi.org/10.1097/ALN.0000000000000285

Poulin, F., Gingras, A. C., Olsen, H., Chevalier, S., & Sonenberg, N. (1998). 4E-BP3, a new member of the eukaryotic initiation factor 4E-binding protein family. *Journal of Biological Chemistry, 273*(22), 14002–14007.

Ptushkina, M., von der Haar, T., Vasilescu, S., Frank, R., Birkenhager, R., & McCarthy, J. E. (1998). Cooperative modulation by eIF4G of eIF4E-binding to the mRNA 5' cap in yeast involves a site partially shared by p20. *EMBO Journal, 17*(16), 4798–4808. https://doi.org/10.1093/emboj/17.16.4798

Qi, S., Mizuno, M., Yonezawa, K., Nawa, H., & Takei, N. (2010). Activation of mammalian target of rapamycin signaling in spatial learning. *Neuroscience Research, 68*(2), 88–93. https://doi.org/10.1016/j.neures.2010.06.008

Qin, X., Jiang, B., & Zhang, Y. (2016). 4E-BP1, a multifactor regulated multifunctional protein. *Cell Cycle, 15*(6), 781–786. https://doi.org/10.1080/15384101.2016.1151581

Raju, U., Yeung, S. J., & Eskin, A. (1990). Involvement of proteins in light resetting ocular circadian oscillators of Aplysia. *American Journal of Physiology, 258*(1 Pt 2), R256–262.

Ran, I., Gkogkas, C. G., Vasuta, C., Tartas, M., Khoutorsky, A., Laplante, I.,...Lacaille, J. C. (2013). Selective regulation of GluA subunit synthesis and AMPA receptor-mediated synaptic function and plasticity by the translation repressor 4E-BP2 in hippocampal pyramidal cells. *Journal of Neuroscience, 33*(5), 1872–1886. https://doi.org/10.1523/JNEUROSCI.3264-12.2013

Rhoads, R. E. (2009). eIF4E: New family members, new binding partners, new roles. *Journal of Biological Chemistry, 284*(25), 16711–16715. https://doi.org/10.1074/jbc.R900002200

Richter, J. D., & Sonenberg, N. (2005). Regulation of cap-dependent translation by eIF4E inhibitory proteins. *Nature, 433*(7025), 477–480. https://doi.org/10.1038/nature03205

Saraf, A., Luo, J., Morris, D. R., & Storm, D. R. (2014). Phosphorylation of eukaryotic translation initiation factor 4E and eukaryotic translation initiation factor 4E-binding protein (4EBP) and their upstream signaling components undergo diurnal oscillation in the mouse hippocampus: Implications for memory persistence. *Journal of Biological Chemistry, 289*(29), 20129–20138. https://doi.org/10.1074/jbc.M114.552638

Schalm, S. S., & Blenis, J. (2002). Identification of a conserved motif required for mTOR signaling. *Current Biology, 12*(8), 632–639.

Schalm, S. S., Fingar, D. C., Sabatini, D. M., & Blenis, J. (2003). TOS motif-mediated raptor binding regulates 4E-BP1 multisite phosphorylation and function. *Current Biology, 13*(10), 797–806.

Schratt, G. M., Nigh, E. A., Chen, W. G., Hu, L., & Greenberg, M. E. (2004). BDNF regulates the translation of a select group of mRNAs by a mammalian target of rapamycin-phosphatidylinositol 3-kinase-dependent pathway during neuronal development. *Journal of Neuroscience, 24*(33), 7366–7377. https://doi.org/10.1523/JNEUROSCI.1739-04.2004

Sharma, A., Hoeffer, C. A., Takayasu, Y., Miyawaki, T., McBride, S. M., & Klann, E. (2010). Dysregulation of mTOR signaling in fragile X syndrome. *Journal of Neuroscience, 30*, 694–702.

Shin, S., Wolgamott, L., Tcherkezian, J., Vallabhapurapu, S., Yu, Y., Roux, P. P., & Yoon, S. O. (2014). Glycogen synthase kinase-3beta positively regulates protein synthesis and cell proliferation through the regulation of translation initiation factor 4E-binding protein 1. *Oncogene, 33*(13), 1690–1699. https://doi.org/10.1038/onc.2013.113

Slipczuk, L., Bekinschtein, P., Katche, C., Cammarota, M., Izquierdo, I., & Medina, J. H. (2009). BDNF activates mTOR to regulate GluR1 expression required for memory formation. *PLoS One, 4*(6), e6007. https://doi.org/10.1371/journal.pone.0006007

Sonenberg, N., Morgan, M. A., Merrick, W. C., & Shatkin, A. J. (1978). A polypeptide in eukaryotic initiation factors that crosslinks specifically to the 5'-terminal cap in mRNA. *Proceedings of the National Academy of Sciences USA, 75*(10), 4843–4847.

St. Clair, D., Blackwood, D., Muir, W., Walker, M., St Clair, D., Muir, W., . . . Evans, H. J. (1990) Association within a family of a balanced autosomal translocation with major mental illness. *The Lancet, 336*(8706), 13–16. https://doi.org/10.1016/0140-6736(90)91520-K

Stoica, L., Zhu, P. J., Huang, W., Zhou, H., Kozma, S. C., & Costa-Mattioli, M. (2011). Selective pharmacogenetic inhibition of mammalian target of Rapamycin complex I (mTORC1) blocks long-term synaptic plasticity and memory storage. *Proceedings of the National Academy of Sciences USA, 108*(9), 3791–3796. https://doi.org/10.1073/pnas.1014715108

Sutton, L. P., & Caron, M. G. (2015). Essential role of D1R in the regulation of mTOR complex1 signaling induced by cocaine. *Neuropharmacology, 99*, 610–619. https://doi.org/10.1016/j.neuropharm.2015.08.024

Tain, L. S., Mortiboys, H., Tao, R. N., Ziviani, E., Bandmann, O., & Whitworth, A. J. (2009). Rapamycin activation of 4E-BP prevents parkinsonian dopaminergic neuron loss. *Nature Neuroscience, 12*(9), 1129–1135. https://doi.org/10.1038/nn.2372

Tang, S. J., Reis, G., Kang, H., Gingras, A. C., Sonenberg, N., & Schuman, E. M. (2002). A rapamycin-sensitive signaling pathway contributes to long-term synaptic plasticity in the hippocampus. *Proceedings of the National Academy of Sciences USA, 99*(1), 467–472. https://doi.org/10.1073/pnas.012605299

Tee, A. R., & Proud, C. G. (2002). Caspase cleavage of initiation factor 4E-binding protein 1 yields a dominant inhibitor of cap-dependent translation and reveals a novel regulatory motif. *Molecular and Cellular Biology, 22*(6), 1674–1683.

Thoreen, C. C., Chantranupong, L., Keys, H. R., Wang, T., Gray, N. S., & Sabatini, D. M. (2012). A unifying model for mTORC1-mediated regulation of mRNA translation. *Nature, 485*(7396), 109–113. https://doi.org/10.1038/nature11083

Topisirovic, I., Svitkin, Y. V., Sonenberg, N., & Shatkin, A. J. (2011). Cap and cap-binding proteins in the control of gene expression. *Wiley Interdisciplinary Reviews: RNA, 2*(2), 277–298. https://doi.org/10.1002/wrna.52

Tsai, V., Parker, W. E., Orlova, K. A., Baybis, M., Chi, A. W., Berg, B. D., . . . Crino, P. B. (2014). Fetal brain mTOR signaling activation in tuberous sclerosis complex. *Cerebral Cortex, 24*(2), 315–327. https://doi.org/10.1093/cercor/bhs310

Tsukiyama-Kohara, K., Poulin, F., Kohara, M., DeMaria, C. T., Cheng, A., Wu, Z., . . . Sonenberg, N. (2001). Adipose tissue reduction in mice lacking the translational inhibitor 4E-BP1. *Nature Medicine, 7*(10), 1128–1132. https://doi.org/10.1038/nm1001-1128

Wang, X., Li, W., Parra, J. L., Beugnet, A., & Proud, C. G. (2003). The C terminus of initiation factor 4E-binding protein 1 contains multiple regulatory features that influence its function and phosphorylation. *Molecular and Cellular Biology, 23*(5), 1546–1557.

Wu, J., McCallum, S. E., Glick, S. D., & Huang, Y. (2011). Inhibition of the mammalian target of rapamycin pathway by rapamycin blocks cocaine-induced locomotor sensitization. *Neuroscience, 172*, 104–109. https://doi.org/10.1016/j.neuroscience.2010.10.041

Yamamoto, T., & Yasoshima, Y. (2007). Electrophysiological representation of taste memory. In F. Bermudez-Rattoni (Ed.), *Neural plasticity and memory: From genes to brain imaging.* Boca Raton (FL).

Yoon, S. C., Seo, M. S., Kim, S. H., Jeon, W. J., Ahn, Y. M., Kang, U. G., & Kim, Y. S. (2008). The effect of MK-801 on mTOR/p70S6K and translation-related proteins in rat frontal cortex. *Neuroscience Letters, 434*(1), 23–28. https://doi.org/10.1016/j.neulet.2008.01.020

Zhang, Y., Takahashi, J. S., & Turek, F. W. (1996). Critical period for cycloheximide blockade of light-induced phase advances of the circadian locomotor activity rhythm in golden hamsters. *Brain Research, 740*(1–2), 285–290.

Zhou, M., Li, W., Huang, S., Song, J., Kim, J. Y., Tian, X., . . . Silva, A. J. (2013). mTOR inhibition ameliorates cognitive and affective deficits caused by Disc1 knockdown in adult-born dentate granule neurons. *Neuron, 77*(4), 647–654. https://doi.org/10.1016/j.neuron.2012.12.033

Zhu, P. J., Chen, C. J., Mays, J., Stoica, L., & Costa-Mattioli, M. (2018). mTORC2, but not mTORC1, is required for hippocampal mGluR-LTD and associated behaviors. *Nature Neuroscience, 21*(6), 799–802. https://doi.org/10.1038/s41593-018-0156-7

CHAPTER 3

THE INTEGRATED STRESS RESPONSE IN MEMORY AND COGNITIVE DISORDERS

JACQUNAE L. MAYS AND MAURO COSTA-MATTIOLI

INTRODUCTION

THE ISR is a central protein quality control mechanism that helps the brain and the organism to respond to different stimuli and maintain health. Different stresses, including protein homeostasis (proteostasis) defects, nutrient deprivation, viral infection, and oxidative stress are sensed by four specialized kinases (GCN2, PERK, PKR, and HRI), whose activities converge on phosphorylation of a single serine on the alpha subunit of the translation initiation factor eIF2. In addition, two dedicated phosphatase complexes antagonize the phosphorylation of eIF2. Both phosphatases contain a common catalytic core subunit, the protein phosphatase 1 (PP1), and a regulatory subunit (GADD34 or CReP), thus rendering the phosphatase specific to eIF2. The phosphorylation of eIF2 blocks eIF2's dedicated guanine nucleotide exchange factor (GEF) eIF2B, which results in a reduction in general translation. Paradoxically, phosphorylation of eIF2 also triggers the translation of specific mRNAs, including the transcription factor *ATF4*.

Behavioral and experimental observations in humans and animals have shown that memory storage is constituted by two temporally distinct forms: short-term memory, lasting minutes to hours, and long-term memory, lasting days, weeks and even a lifetime (McGaugh, 2000; Kandel, 2001). Memory consolidation is a process by which protein synthesis is required for the formation of a stable long-term memory (Davis & Squire, 1984; Kandel, 2001). Ever since Agranoff & Klinger (Agranoff & Klinger, 1964) and Barondes & Cohen (Barondes & Cohen, 1966) showed that protein synthesis determines whether a long-term memory is formed, hundreds of studies have been published attempting to identify the key factors that regulate this formation. Several transcription factors, including the cAMP response element binding protein (CREB), have originally been proposed to behave as long-term memory switches. However, drugs or compounds that promote their activity and enhance long-term memory formation remain elusive.

Recent convergent genetic and pharmacological discoveries in several animal models and across different levels of analysis from multiple laboratories support the idea that the ISR, a conserved signaling network that responds to a variety of stress conditions to restore protein homeostasis by tuning protein synthesis rates, is a universal regulator of long-term memory formation (Costa-Mattioli & Walter, 2020). In addition, the ISR bidirectionally regulates long-lasting physical changes in synaptic function that are widely believed to encode long-term memory formation, including long-term potentiation (LTP) and long-term depression (LTD; Costa-Mattioli & Walter, 2020).

Importantly, the ISR is a causative mechanism underlying the cognitive deficits and neurodegeneration in a broad range of brain disorders (Costa-Mattioli & Walter, 2020). The studies on the ISR have not only impacted the neurobiology of memory formation and cognitive disorders, but also industryefforts to develop drugs that reverse cognitive problems and prevent neurodegeneration. Today, three major pharmaceutical companies (Calico Life Sciences, Sanofi, and Denali Therapeutics) are targeting the ISR with the aim of reestablishing brain health.

In this chapter, we discuss the evidence that translational reprogramming via the ISR is a central molecular switch for long-term memory formation. In addition, we address how inhibition of the ISR emerges as a promising therapeutic avenue for the treatment of cognitive disorders and neurodegenerative disorders.

Translational Reprogramming by the ISR

Neurons harbor a wide diversity of mRNAs encoding proteins with various functions. Translation of mRNAs occurs in three steps: initiation, elongation, and termination, with initiation being the most tightly regulated step. Regulation of mRNA translation is important especially under conditions that compromise the health of neurons. The ISR is one such system in place for stress-induced translational control. eIF2 is the central hub of the ISR heterotrimeric G-protein switch that is activated by GTP binding. Eukaryotic initiation factor 2B (eIF2B) acts as a guanine exchange factor (GEF) for eIF2 and exchanges its bound guanosine diphosphate (GDP) for guanosine triphosphate (GTP; Figure 3.1). eIF2•GTP recruits the initiator methionyl-transfer RNA (Met-$tRNA_i^{Met}$) to the preinitiation complex consisting of the small 40S ribosomal subunit and other supporting initiation factors (Figure 3.1). The preinitiation complex is brought to the 5′-7-methylguanosine cap of the mRNA transcript where it scans in the 5′- to 3′- direction in search of the start

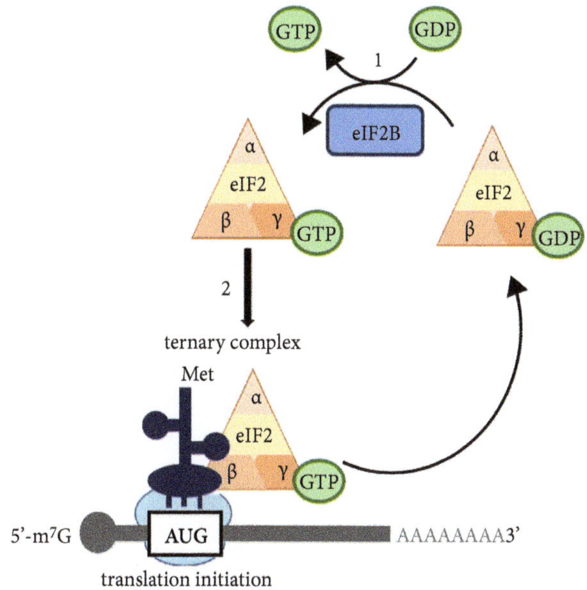

FIGURE 3.1. The ISR hub eIF2 regulates translation initiation. Under normal conditions, the central hub of the ISR, eIF2, is activated by GTP binding. eIF2B serves as a GEF for eIF2, exchanging its GDP for GTP (1). Upon activation, eIF2 recruits the Met-tRNA$_i^{Met}$ to the preinitiation ternary complex, consisting of the small ribosomal subunit and other supporting initiation factors (2). At the 5′ cap of the mRNA transcript, the ternary complex moves in the 3′ direction to identify the AUG start codon. Upon identification, the GTP bound to eIF2 is hydrolyzed, inactivating eIF2. GDP-bound eIF2 disassociates from the ternary complex and mRNA transcript (3), and can again be activated by eIF2B to repeat another round of translation initiation.

codon (AUG; Figure 3.1). Once the complex identifies and binds to the start codon, GTP is hydrolyzed (eIF2•GTP to eIF2•GDP) and eIF2•GDP is released from the preinitiation complex and mRNA transcript (Figure 3.1). The larger, 60S ribosomal subunit now joins the complex and translation elongation resumes. Once released, eIF2•GDP is free to be recycled to eIF2•GTP by eIF2B once again for another round of translation initiation.

In response to stress, the ISR exerts translational control through phosphorylation of the alpha subunit of eIF2 at serine-51, thus repressing translation initiation. eIF2 is phosphorylated by four known kinases: (1) general control nonderepressible 2 (GCN2), which detects amino acid deficiency; (2) double-stranded RNA-activated protein kinase (PKR), which detects double-stranded RNA; (3) the PKR-like ER kinase (PERK), which detects endoplasmic reticulum (ER) stress due to unfolded proteins; and (4) the heme-regulated kinase (HRI), which detects the redox state of the cell and heme availability in reticulocytes. eIF2B is a decameric complex for which eIF2 can serve as both a substrate and an inhibitor. Once phosphorylated, eIF2•GDP tightly binds via its alpha and gamma subunits and sequesters eIF2B such that eIF2B is unable to exchange eIF2•GDP for eIF2•GTP (Gordiyenko et al., 2019; Figure 3.2). Recently, cryo-EM studies revealed

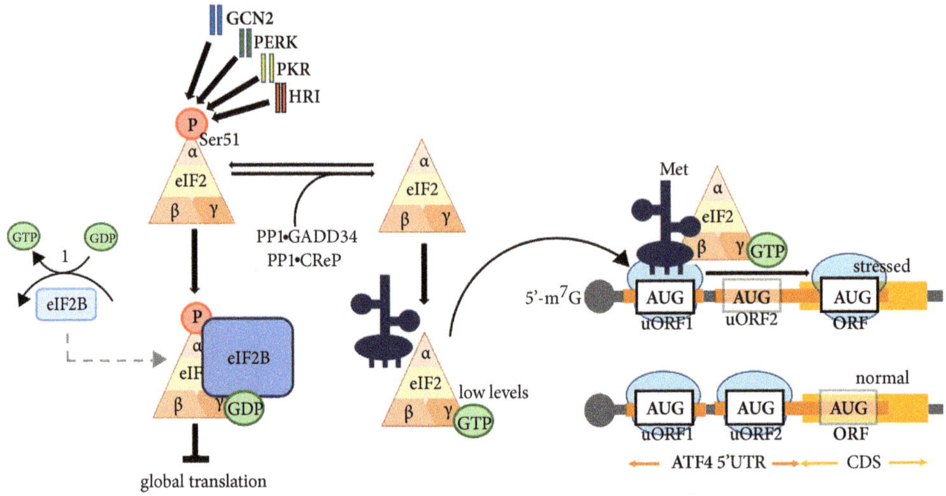

FIGURE 3.2. The ISR reduces global translation and promotes translation of specific mRNAs by regulating ternary complex availability. In response to stress, eIF2 is phosphorylated on its alpha-subunit at Serine 51 by four known kinases: GCN2, PERK, PKR, and HRI. When eIF2 is phosphorylated, it binds eIF2B via its alpha and gamma subunits, sequestering available eIF2B and preventing its GEF activity towards eIF2. Thus, phosphorylated eIF2 remains GDP bound, reducing the pool of ternary complex for translation initiation resulting in reduced general translation. eIF2 is dephosphorylated by one of two phosphatases: GADD34:PP1 or CReP:PP1. Under stress conditions, there are very low levels of dephosphorylated eIF2 available for translation initiation. Insufficient ternary complex availability promotes the translation of specific mRNAs with uORFs, such as *ATF4*. Under normal conditions, sufficient ternary complexes tend to initiate translation at uORF2, which overlaps with the coding sequence of *ATF4*, preventing translation of *ATF4*. Under stress conditions, reduced ternary complexes now bypass uORF2 and initiate translation at the AUG within the coding sequence of *ATF4*, promoting *ATF4* translation.

eIF2B to exist in one of two structures depending on the state of eIF2: the more stable "productive" complex, in which eIF2B has GEF activity towards unphosphorylated eIF2, and the "nonproductive" complex, in which GEF activity is restricted towards phosphorylated eIF2, due to repulsion of eIF2-P at the productive interface (Kashiwagi et al., 2019; Kenner et al., 2019). Thus, when eIF2 is phosphorylated, active eIF2•GTP levels are reduced, decreasing the formation and pool of ternary complex, and inhibiting general translation initiation (Figure 3.2).

Phosphorylation of eIF2 not only regulates global translation initiation rates, but also regulates translation of specific mRNAs. For mRNAs containing one or more upstream open reading frames (uORFs) in their 5′ untranslated region (UTR), eIF2 selectively translates and expresses specific genes from their transcripts. Activating transcription factor 4 (*ATF4*) is one such mRNA transcript whose selective translation is widely known to be upregulated following phosphorylation of eIF2 (Harding et al., 2000;

Vattem & Wek, 2004). The mRNA encoding ATF4 contains two uORFs in its 5′UTR. Under normal conditions there is a high concentration of available ternary complex. Sufficient ternary complexes initiate at uORF1 and the ribosome dissociates when it reaches the stop codon for uORF1; uORF1 thereby prevents the ribosome from ever reaching the downstream uORFs (Figure 3.2; Harding et al., 2000). If the eIF2-bound ribosome escapes uORF1 under normal conditions, initiation will likely occur at uORF2, which overlaps with the coding region of *ATF4*, providing a second layer of regulation of the *ATF4* coding region. When cells are stressed, the pool of ternary complex is greatly reduced. Thus, the scanning 40S ribosome now scans past the start codons for uORF1 and 2 without initiating, permitting recognition of the AUG of the main coding sequence (CDS) of *ATF4* (Figure 3.2).

Phosphorylation of eIF2 (eIF2-P) is a tightly regulated process. In addition to the four dedicated kinases, two phosphatase complexes reverse eIF2-P. eIF2 dephosphorylation is catalyzed by the catalytic subunit protein phosphatase 1 (PP1) bound to one of two regulatory subunits: stress-inducible GADD34 (PP1•GADD34), which is induced in response to ISR activation or the constitutively expressed CReP (PP1•CReP; Figure 3.2; Novoa et al., 2001; Jousse et al., 2003). Thus, the ISR responds to stress such that energy-consuming translation is halted, selective translation of genes that help reestablish cell homeostasis is promoted, and general translation is resumed once cell stress is resolved. If the stress cannot be mitigated, the ISR triggers apoptosis to eliminate the damaged cell.

The ISR Regulates Long-Term Memory Formation

While changes in gene expression (e.g., transcriptional or epigenetic modifications) are required for long-term memory storage (Alberini & Kandel, 2014), translation is the ultimate step in the control of the functional output of a gene. In addition, in response to activity neurons can rapidly regulate protein synthesis locally at dendrites without altering mRNA synthesis and/or transport (Sutton & Schuman, 2006). The synthesized proteins are believed to be required to bring about the modulation of the strength of synaptic connections during learning and memory. While other translational control mechanisms have been shown to regulate memory formation (Richter & Klann, 2009; Buffington et al., 2014), the discovery that the phosphorylation of eIF2 modulates synaptic plasticity and bidirectionally regulates long-term memory in several animal models provides the strongest evidence that the translational program controlled by the ISR underlies, at least in part, the consolidation of long-term memory. Here we will discuss the evidence supporting the role of the ISR in memory storage.

Neuronal Activity Leading to Long-lasting Increases in Synaptic Function Inactivates the ISR

Activity-dependent modulation of synaptic strength, and more specifically, long-term potentiation (LTP), is believed to be one of the key mechanisms underlying memory storage (Neves et al., 2008). In slices, LTP induced by high frequency (tetanic) stimulation of an afferent pathway in the hippocampus, a region crucially required for memory formation (Scoville & Milner, 1957), can last for several hours. By contrast, in anesthetized animals hippocampal LTP can last for days, or even weeks (Bliss & Lomo, 1973). Interestingly, LTP-stimulation in hippocampal slices or *in vivo* reduces the phosphorylation of eIF2 (Costa-Mattioli et al., 2005; Panja et al., 2009; Trinh et al., 2014), a process that is mediated by activation of N-methyl-D-aspartate (NMDA) receptors (NMDARs). In addition, chemical agents that promote LTP, such as forskolin or the brain-derived neurotrophic factor (BDNF), also reduce eIF2 phosphorylation (Takei et al., 2001; Svitkin et al., 2005). Consistent with the idea that LTP underlies memory storage, the decrease in eIF2-P in the hippocampus is seen immediately after cognitive training (Costa-Mattioli et al., 2007). Thus, the ISR integrates a diverse set of both intracellular and extracellular stimuli to promote long-lasting protein synthesis-dependent changes in synaptic function underlying long-term memory storage.

Genetic and Pharmacological Inhibition of the ISR Enhances Long-term Memory

The first evidence that the ISR was causally involved in long-term memory formation came from studies using mice with reduced eIF2-P (Costa-Mattioli et al., 2005, 2007). Briefly, mice lacking the eIF2 kinase *GCN2* (*GCN2−/−* mice) or *Eif2s1*$^{S/A}$ heterozygous knock-in mice (where the single phosphorylatable serine 51 is replaced by alanine) show enhanced long-term memory formation. The Morris water maze is a spatial memory task in which mice swim in a pool of opaque water and use visual cues (placed on the wall of the testing room) to remember the location of a hidden platform. Mice are trained for several days for up to four sessions per day, after which control mice remember the location of the hidden platform. However, if a more demanding training protocol (only one training session per day) is used, only mice with reduced eIF2 phosphorylation (not normal mice) recall the location of the platform. Consistent with these findings, mice with reduced eIF2 phosphorylation show enhanced memory when trained in classic conditioning tasks, including Pavlovian fear conditioning, where animals learn that an auditory or visual stimulus predicts a foot shock, and conditioned taste aversion, where animals learn to associate a novel taste with nausea.

The memory-enhancing abilities associated with inhibition of the ISR were further confirmed by studies genetically targeting the eIF2 kinase PKR. Genetic deletion of PKR

(*Pkr–/–*mice) enhances long-term spatial and fear memories (P. J. Zhu et al., 2011). Remarkably, pharmacological inhibition of PKR with an inhibitor also improved long-term memory performance (P. J. Zhu et al., 2011), thus providing the first evidence that the ISR can be targeted with drugs to enhance long-term memory. Subsequent studies show that inhibition of the ISR by either genetic or pharmacological inhibition of PKR also enhances long-term taste memory (Stern et al., 2013). In addition, local inhibition of PERK in the insular cortex with an inhibitor (GSK2606414) or an shRNA against PERK enhances long-term taste memory (Sharma et al., 2018). Contrastingly, conditional deletion of PERK in the forebrain leads to impaired working memory (Trinh et al., 2012), which is likely due to an eIF2-P-independent process and the regulation of calcium dynamics by PERK (S. Zhu, Henninger, et al., 2016; S. Zhu, McGrath, et al., 2016). Notably, a new small molecule inhibitor of the ISR (ISRIB) developed by Dr. Peter Walter and his group enhances long-term memory formation by pharmacologically mimicking the effects of eIF2 dephosphorylation (Sidrauski et al., 2013). ISRIB is an activator of eIF2B that reverses the translational effects induced by eIF2-P.

Inhibition of the ISR not only promotes long-term memory in rodents (mice and rats), but also in birds. In a very elegant study, Batista and colleagues (Batista et al., 2016) have recently found that pharmacological inhibition of the ISR with either a PKR inhibitor or ISRIB enhances imprinting in birds, a behavior that is both innate and learned. Interestingly, inhibition of the ISR allows animals to form imprinted auditory memory, even outside of the critical (learning) period (Batista et al., 2016). Given that inhibition of the ISR enhances behavioral imprinting, it would be interesting to determine whether the ISR is involved in language acquisition, which takes place during an early critical developmental period and requires experience-dependent plasticity.

Thus, translational reprogramming by the ISR emerges as a primary way to promote memory formation.

Activation of the ISR Prevents Long-term Memory Formation in Humans and Animals

The first evidence that activation of the ISR impairs long-term spatial and fear memory arises from studies using Sal003, an inhibitor of the eIF2 phosphatase complexes (Boyce et al., 2005; Robert et al., 2006), which increases eIF2-P upon local injection into the hippocampus (Costa-Mattioli et al., 2007). Accordingly, injection of Sal003 blocks imprinted memories during the critical period in chicks (Batista et al., 2016). Moreover, using a very clever and elegant pharmacogenetic approach to activate PKR in a subregion of the hippocampus (the CA1 region), Nakazawa and colleagues (Jiang et al., 2010) demonstrated that local PKR-mediated increase in eIF2-P impairs long-term memory formation. Similar results were obtained by a recently developed cell-type specific drug-inducible method (Shrestha et al., 2020). Finally, the importance of the ISR in long-term memory formation is underscored by the identification of mutations in humans in key ISR components (Borck et al., 2012; Moortgat et al., 2016; Skopkova et al., 2017) that lead to activation of the ISR.

Collectively, these gain- and loss-of-function experiments demonstrate that the ISR is a universal regulator of long-term memory formation.

The ISR Controls the Two Major Forms of Synaptic Plasticity in the Mammalian Brain

Synaptic plasticity, the ability of neurons to regulate the strength of their synaptic connections in response to activity, is thought to be fundamental for information and long-term memory storage. It is widely believed that a given experience could be stored in the brain by modifying the strength of synapses, e.g., by promoting the strength of some key pathways within a circuit and weakening others (Malenka & Bear, 2004). The selective strengthening and weakening of excitatory synapses in the mammalian brain are manifested by two cellular models of synaptic plasticity, LTP and long-term depression (LTD), respectively. While several molecular processes are able to modulate LTP and LTD, the addition and removal of synaptic AMPA-type glutamate receptors constitute a common mechanism that is often required to express these two forms of synaptic plasticity, respectively (Kelleher et al., 2004; Kessels & Malinow, 2009).

Intriguingly, both LTP and LTD require new protein synthesis (Buffington et al., 2014). How could two cellular processes (LTP and LTD) leading to two different outcomes (synaptic strengthening and weakening, respectively) share the same molecular mechanism (protein synthesis)? Like memory storage, which can be separated into short- and long-term components, LTP occurs in two temporally distinct phases. Early LTP (E-LTP) is typically induced by a single train of tetanic stimulation to hippocampal fibers and lasts only 1-2 hours. In contrast, late LTP (L-LTP), which is generally induced by several tetanic trains (typically four), lasts for many hours. While E-LTP is independent of new protein synthesis, L-LTP requires new protein synthesis (Kandel, 2001; Kelleher et al., 2004). This and other studies support the notion that the molecular mechanism underlying long-term memory storage can be studied at the cellular level by studying LTP and its different phases. Indeed, for nearly 40 years since its discovery, LTP has become the most studied model of synaptic plasticity in the mammalian brain. Interestingly, L-LTP inducing protocols reduced eIF2-P (Costa-Mattioli et al., 2005; Panja et al., 2009; Trinh et al., 2014) in the hippocampus. Moreover, in mice in which the ISR is genetically inhibited (*Gcn2−/−*, *Eif2s1*$^{S/A}$, *and Pkr−/−*), all of which exhibit reduced eIF2-P in the hippocampus, a short-lasting E-LTP-inducing protocol elicits a protein synthesis-dependent L-LTP (Costa-Mattioli et al., 2005, 2007; P. J. Zhu et al., 2011). These studies reveal that the ISR represses the conversion of E-LTP into L-LTP. Accordingly, activation of the ISR by genetic (Costa-Mattioli et al., 2005) or pharmacological means

(Costa-Mattioli et al., 2007), blocks protein synthesis-dependent L-LTP induced by multiple tetanic trains. Interestingly, the bidirectional regulation of LTP by the ISR is not restricted to the hippocampus since cocaine-induced LTP in the ventral tegmental area (VTA) is also controlled by the ISR (Huang et al., 2016; Placzek et al., 2016). Thus, the ISR is a broad mechanism that regulates protein synthesis-dependent synaptic function underlying long-term memory formation.

LTD is the other major form of synaptic plasticity in the brain (Sajikumar et al., 2005). Glutamate receptors play an important function in learning and memory. NMDA-receptor dependent LTD (NMDAR-LTD), a type of glutamate receptor-dependent LTD, was originally reported to be protein synthesis independent (Huber et al., 2000). However, the long-term maintenance of a type of NMDAR-dependent LTD studied by the Frey lab requires protein synthesis (Sajikumar & Frey, 2004). Synaptic cross-tagging, which refers to L-LTP/LTD in one input inducing L-LTP/LTD in a separate input, and vice versa, requires de novo translation of plasticity-related mRNAs, and is known to be a mechanism by which NMDAR-LTD is promoted (Sajikumar & Frey, 2004; Sajikumar et al., 2005). It is currently unknown whether the ISR is involved in this type of NMDAR-LTD or L-LTP/LTD cross tagging.

Another type of glutamate receptor-dependent LTD, metabotropic glutamate receptor-mediated LTD (mGluR-LTD), also requires protein synthesis (Huber et al., 2000; Lüscher & Huber, 2010). Furthermore, we and others have found that the mGluR agonist DHPG (3,5-dihydroxyphenylglycine; R. C. Malenka, 1994; Palmer et al., 1997) activates the ISR, as determined by increased eIF2-P (Di Prisco et al., 2014; Trinh et al., 2014). Using paired single-cell recordings where a given neuron is genetically modified and its neighbor serves as a control, Di Prisco and colleagues (Di Prisco et al., 2014) found that genetically or pharmacologically reducing the ISR prevented mGluR-LTD. Conversely, direct activation of the ISR promoted LTD. These state-of-the-art, genetic, single cell manipulations show that the ISR is both necessary and sufficient for mGluR-LTD. Taken together, these findings support the notion that the state of activation of the ISR determines whether LTP or LTD is induced. Briefly, if the ISR is inhibited L-LTP is facilitated, but mGluR-LTD is impaired. By contrast, if the ISR is activated, L-LTP is impaired, but mGluR-LTD is enhanced. To our knowledge, no other pathway operates in this manner, bidirectionally regulating protein synthesis-dependent LTP and LTD. The ISR may be the main mechanism by which mGluR-LTD prevents LTP in several regions of the brain (Lüscher & Malenka, 2011).

The ISR's Mechanisms of Action May Differ in Function Based on Cell Type

How does ISR-mediated protein synthesis reprogramming contribute to the formation of long-term memory? Inhibition of the ISR leads to an increase in general translation and reduction in translation of specific mRNAs, including the typical ISR reporter *ATF4*. ATF4 has been shown to inhibit CREB (cAMP response element-binding protein)-mediated gene expression, which is required for long-term memory formation (Kida et al., 2002). PKR activation in hippocampal CA1 pyramidal neurons impairs contextual long-term memory without reducing general translation (Jiang et al., 2010). Using a similar cell type specific approach, Klann and colleagues show that activation of ISR sensor PKR impairs memory formation by reducing translation rates (Shrestha et al., 2020). However, this method is unable to dissociate the inhibition of general translation with the increase in specific translation of *ATF4*, since *ATF4* levels are also increased in this model (Shrestha et al., 2020). Moreover, blocking ATF4 function in forebrain neurons leads to enhanced spatial long-term memory (Chen et al., 2003), but see recent results showing that reduction of *ATF4* by injecting an shRNA in the hippocampus indeed impairs memory (Pasini et al., 2015).

During mGluR-LTD, the ISR regulates the synthesis of the X-linked mental retardation protein oligophrenin-1 (OPHN1) (Nadif Kasri et al., 2011; Placzek et al., 2016). Like *ATF4* mRNA, *OPHN1* mRNA contains 5′-uORFs and its translation is increased by activation of the ISR. Since OPHN1 interacts with endophilin A2/3 and promotes the endocytosis of AMPAR, this remains an interesting mechanism by which the ISR regulates synaptic plasticity (Placzek et al., 2016).

Finally, mice in which the ISR's sensor kinase PKR is either genetically or pharmacologically inhibited exhibit reduced inhibitory synaptic transmission (P. J. Zhu et al., 2011). How the ISR mechanistically impacts inhibitory synaptic transmission during long-term memory formation remains to be determined.

The ISR in Cognitive and Neurodegenerative Disorders

The ability of the ISR to integrate neuronal signaling and different stresses, regulate both general protein synthesis and gene specific translation, and bidirectionally regulate the two major forms of synaptic plasticity in the brain may explain why the ISR is so central to cognitive and neurodegenerative disorders (Figure 3.3), including Down syndrome (DS), the most common form of intellectual disability. The ISR is activated in the brains of Ts65Dn mice (a model system for DS) and individuals with DS, as well as DS patient-derived induced pluripotent stem cells (iPSCs; P. J. Zhu et al., 2019). More importantly, inhibition of the ISR either upstream at the level of the ISR's sensor kinase PKR or downstream by manipulating the central ISR signaling hub (eIF2/eIF2B) reverses the cognitive decline and alterations in synaptic plasticity in DS mice. Given that DS is characterized by a high incidence of Alzheimer's disease (AD), tuning of the ISR also reverses the cognitive decline in AD. Indeed, in the brain of AD individuals and a mouse model of AD (APP-PS1 mice) that expresses AD-related mutant amyloid precursor

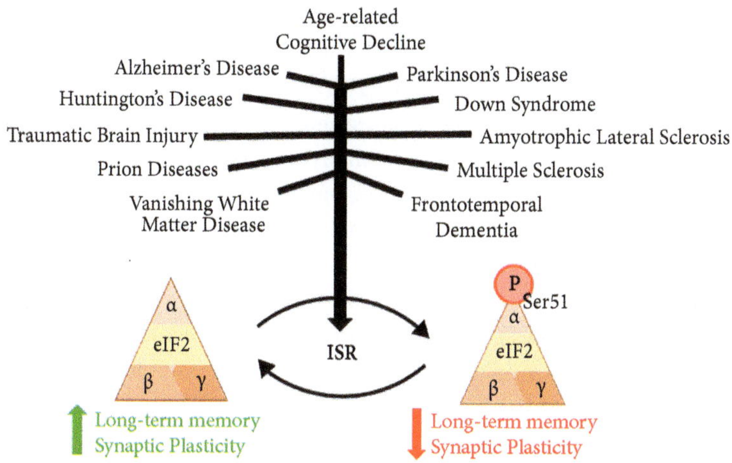

FIGURE 3.3. The ISR is a converging molecular target of a variety of brain disorders. Aberrant ISR activation is a causative mechanism implicated in multiple neurological disorders related to memory and cognition.

protein (*APP*) and presenilin 1 (*PSEN1*) the ISR is activated, as determined by increased eIF2-P (Ma et al., 2013). More importantly, genetic inhibition of the ISR's sensors GCN2 and PERK improved the deficits in long-term memory (Ma et al., 2013). Accordingly, genetic or pharmacological inhibition of the PKR branch of the ISR reverses the cognitive decline in other mouse models of AD, including the 5xFAD transgenic mice, $A\beta_{1\text{-}42}$-injected mice, and ApoE4 mice (Ma et al., 2013). Thus, inhibition of the ISR reverses the cognitive decline associated with DS and AD.

In traumatic brain injury (TBI), the PERK branch of the ISR is activated (Sen et al., 2017) and inhibition of the ISR with ISRIB rescues the long-term memory and synaptic plasticity abnormalities in a mouse model of TBI (Chou et al., 2017). Surprisingly, inhibition of the eIF2 phosphatase with the compound guanabenz, resulting in increased eIF2-P, also reverses memory deficits in TBI (Hood et al., 2018). Given that eIF2 activation relieves ER stress, further increase of eIF2 activation also proves to be neuroprotective in TBI (Hood et al., 2018).

The ISR in Neurodegeneration

The ISR is also activated in prion disease, a fatal neurodegenerative disorder characterized by misfolded protein aggregates (Moreno et al., 2012), and amyotrophic lateral sclerosis (ALS), a neurodegenerative disease affecting motor neurons (Kim et al., 2014). Suppression of the ISR by overexpressing GADD34 in the brain, treatment with ISRIB, PERK inhibitor, or trazodone, prevented neuronal loss and survival in prion and ALS disease mouse models (Moreno et al., 2013; Kim et al., 2014; M. Halliday et al., 2015;

M. Halliday et al., 2017). Accordingly, an inhibitor of PERK was neuroprotective in Drosophila models of early-onset Parkinson disease (PD; Celardo et al., 2016), and in a mouse model of frontotemporal dementia (FTD; Radford et al., 2015). Similarly, inhibition of the PERK branch of the ISR rescues neurodegeneration in a mouse model of Alzheimer's disease (Devi & Ohno, 2014). Finally, suppression of the ISR's sensor kinase PKR by using genetics or pharmacology protected neurons from Aβ accumulation (Carret-Rebillat et al., 2015) as well as excitotoxicity (Tronel et al., 2014). Thus, inhibition of ISR emerges as a central mechanism for the treatment of neurodegenerative disorders

The ISR is also implicated in Huntington's Disease (HD), a familial neurodegenerative disease that is caused by misfolded proteins and a mutated form of the huntingtin gene (Krzyzosiak et al., 2018). Interestingly, unlike the neurodegenerative diseases mentioned earlier, promotion of Raphin1, an inhibitor of CReP•PP1, reduced protein synthesis and huntingtin inclusions in the brain of a mouse model for HD (Krzyzosiak et al., 2018). The authors posit that increased eIF2-P and subsequent reduced translation allows for proteostasis to be attained through increased chaperone capacity for misfolded proteins (Krzyzosiak et al., 2018). Thus, at least in some conditions, targeting eIF2 phosphatases also improves eIF2-mediated neurodegeneration.

Mutations in ISR Components in Human Disease Resulting in Activation of the ISR

Mutations in central components of the ISR are also implicated in brain-related disorders. More than 170 mutations in eIF2B result in vanishing white matter disease (VWMD), an autosomal recessive leukoencephalopathy characterized by chronic demyelination (Wong et al., 2018). These mutations destabilize decameric eIF2B, impairing its catalytic GEF activity towards eIF2α (Wong et al., 2018). ISRIB is able to stabilize eIF2B and inhibit the persistent activation of the ISR associated with VWMD; ISRIB reduces ATF4 levels and increases global protein synthesis in HEK293T cells with mutated eIF2B (Wong et al., 2018). Thus, ISRIB or ISRIB analogs could potentially be used for VWMD patients in which eIF2B assembly and activity is impaired. Similarly, mice harboring a human eIF2B mutation exhibit increased ATF4, motor impairment, and myelin loss which are all rescued using 2BAct, a small molecule eIF2B activator (Wong et al., 2019). This suggests that activators of eIF2B such as ISRIB may specifically improve symptoms of patients with VWMD.

Mutations in *EIF2S3*, which encodes the gamma subunit of eIF2, result in MEHMO syndrome, an X-linked intellectual disability disorder (Young-Baird et al., 2019, 2020). MEHMO by name is characterized by mental impairment, epilepsy, hypogonadism, microcephaly, and obesity (Young-Baird et al., 2020). Currently, there are five identified mutations in *EIF2S3* that result in MEHMO syndrome (Young-Baird et al., 2020). Human iPSCs carrying the eIF2γ-I465Sfs*4 (eIF2γ-fs) mutation exhibit activation of the ISR, as determined by reduced global translation and increased *ATF4* expression (Young-Baird et al., 2020). Interestingly, in these cells, eIF2-P is not altered, but the ISR is activated; eIF2α cannot bind to the Met-tRNA$_i^{Met}$, reducing ternary complex formation, which results in reduced translation rates and increased *ATF4* translation. Further, ISRIB restored global translation and suppressed *ATF4* expression in eIF2γ-fs iPSCs (Young-Baird et al., 2020).

Other mutations that result in activation of the ISR also cause intellectual disability, highlighting the role of the ISR in memory formation. Loss-of-function mutations in the gene encoding CReP are associated with intellectual disability, short stature, and diabetes (Abdulkarim et al., 2015; Kernohan et al., 2015). These mutations prevent the formation of the CReP•PP1 phosphatase complex, thus resulting in increased eIF2-P. In future experiments, it would be interesting to generate mouse models carrying the human mutations in CReP and those resulting in MEHMO syndrome, and determine whether these mice exhibit cognitive problems.

Future Directions

Translational control by the ISR is crucial for cognitive function. Importantly, when the ISR is activated through eIF2-P, long-term memory formation and synaptic plasticity are impaired. Conversely, inactivation of the ISR promotes memory. Thus, the ISR serves as a molecular switch for memory, making it an excellent therapeutic target for memory-related disorders, such as Down syndrome and Alzheimer's disease. While great strides have been made to understand the role of the ISR in the brain, there are still questions that remain unanswered. Little is known about the specific mechanism by which the ISR regulates protein synthesis and resultant changes in memory. An important feature of the ISR is the selective translation of specific mRNAs with uORFs. One of the most widely studied mRNAs that is selectively translated following eIF2-P is *ATF4*, which serves as a transcriptional target for numerous genes. While much focus has been given to *ATF4*, over 50 percent of mRNAs contain uORFs. Future studies are desirable to identify other translational targets of the ISR and to understand how mRNAs are selectively translated over others despite so many mRNAs containing uORFs. In addition, given that the ISR is implicated in various neurological diseases, it will be interesting to identify the exact translational program triggered by aberrant eIF2 under different conditions. Moreover, recent studies suggest separate roles for the ISR in different cell types and cellular locations; it is unknown whether the ISR is differentially activated or translates different targets between excitatory and inhibitory neurons, at the soma versus the dendrites, or at the pre- versus the post-synapse. As the eIF2 kinases respond to different types of stress, it is also possible that activation of distinct kinases underlies this differential localization of the ISR response. Given that the ISR is emerging as a therapeutic target for multiple neurological disorders, future research is warranted for identification of therapeutic activators and inhibitors of ISR-pathway components that are specifically dysregulated in these disorders. Finally, to better model these diseases and test potential pharmaceutical drugs targeting the ISR, the generation of novel mouse models in which the ISR is persistently activated will be of great use.

References

Abdulkarim, B., Nicolino, M., Igoillo-Esteve, M., Daures, M., Romero, S., Philippi, A., . . . Julier, C. (2015). A missense mutation in PPP1R15B causes a syndrome including diabetes, short stature, and microcephaly. *Diabetes*, *64*(11), 3951–3962. https://doi.org/10.2337/db15-0477

Agranoff, B. W., & Klinger, P. D. (1964). Puromycin effect on memory fixation in the goldfish. *Science*, *146*(3646), 952–953. https://doi.org/10.1126/science.146.3646.952

Alberini, C. M., & Kandel, E. R. (2014). The regulation of transcription in memory consolidation. *Cold Spring Harbor Perspectives in Biology*, *7*(1), a021741. https://doi.org/10.1101/cshperspect.a021741

Barondes, S. H., & Cohen, H. D. (1966). Puromycin effect on successive phases of memory storage. *Science*, *151*(3710), 594–595. https://doi.org/10.1126/science.151.3710.594

Batista, G., Johnson, J. L., Dominguez, E., Costa-Mattioli, M., & Pena, J. L. (2016). Translational control of auditory imprinting and structural plasticity by eIF2α. *ELife*, *5*. https://doi.org/10.7554/eLife.17197

Bliss, T. V., & Lomo, T. (1973). Long-lasting potentiation of synaptic transmission in the dentate area of the anaesthetized rabbit following stimulation of the perforant path. *Journal of Physiology*, *232*(2), 331–356. https://doi.org/10.1113/jphysiol.1973.sp010273

Borck, G., Shin, B.-S., Stiller, B., Mimouni-Bloch, A., Thiele, H., Kim, J.-R., . . . Basel-Vanagaite, L. (2012). EIF2γ mutation that disrupts eIF2 complex integrity links intellectual disability to impaired translation initiation. *Molecular Cell*, *48*(4), 641–646. https://doi.org/10.1016/j.molcel.2012.09.005

Boyce, M., Bryant, K. F., Jousse, C., Long, K., Harding, H. P., Scheuner, D., . . . Yuan, J. (2005). A selective inhibitor of eIF2alpha dephosphorylation protects cells from ER stress. *Science*, *307*(5711), 935–939. https://doi.org/10.1126/science.1101902

Buffington, S. A., Huang, W., & Costa-Mattioli, M. (2014). Translational control in synaptic plasticity and cognitive dysfunction. *Annual Review of Neuroscience*, *37*(1), 17–38. https://doi.org/10.1146/annurev-neuro-071013-014100

Carret-Rebillat, A.-S., Pace, C., Gourmaud, S., Ravasi, L., Montagne-Stora, S., Longueville, S., . . . Hugon, J. (2015). Neuroinflammation and Aβ accumulation linked to systemic inflammation are decreased by genetic PKR down-regulation. *Scientific Reports*, *5*, 8489. https://doi.org/10.1038/srep08489

Celardo, I., Costa, A. C., Lehmann, S., Jones, C., Wood, N., Mencacci, N. E., . . . Martins, L. M. (2016). Mitofusin-mediated ER stress triggers neurodegeneration in pink1/parkin models of Parkinson's disease. *Cell Death & Disease*, *7*(6), e2271. https://doi.org/10.1038/cddis.2016.173

Chen, A., Muzzio, I. A., Malleret, G., Bartsch, D., Verbitsky, M., Pavlidis, P., . . . Kandel, E. R. (2003). Inducible enhancement of memory storage and synaptic plasticity in transgenic mice expressing an inhibitor of ATF4 (CREB-2) and C/EBP proteins. *Neuron*, *39*(4), 655–669. https://doi.org/10.1016/s0896-6273(03)00501-4

Chou, A., Krukowski, K., Jopson, T., Zhu, P. J., Costa-Mattioli, M., Walter, P., & Rosi, S. (2017). Inhibition of the integrated stress response reverses cognitive deficits after traumatic brain injury. *Proceedings of the National Academy of Sciences of the United States of America, 114*(31), E6420–E6426. https://doi.org/10.1073/pnas.1707661114

Costa-Mattioli, M., Gobert, D., Harding, H., Herdy, B., Azzi, M., Bruno, M., . . . Sonenberg, N. (2005). Translational control of hippocampal synaptic plasticity and memory by the eIF2alpha kinase GCN2. *Nature, 436*(7054), 1166–1173. https://doi.org/10.1038/nature03897

Costa-Mattioli, M., Gobert, D., Stern, E., Gamache, K., Colina, R., Cuello, C., . . . Sonenberg, N. (2007). EIF2α Phosphorylation bidirectionally regulates the switch from short- to long-term synaptic plasticity and memory. *Cell, 129*(1), 195–206. https://doi.org/10.1016/j.cell.2007.01.050

Costa-Mattioli, M., & Walter, P. (2020). The integrated stress response: From mechanism to disease. *Science, 368*(6489). https://doi.org/10.1126/science.aat5314

Davis, H. P., & Squire, L. R. (1984). Protein synthesis and memory: A review. *Psychological Bulletin, 96*(3), 518–559.

Devi, L., & Ohno, M. (2014). PERK mediates eIF2α phosphorylation responsible for BACE1 elevation, CREB dysfunction and neurodegeneration in a mouse model of Alzheimer's disease. *Neurobiology of Aging, 35*(10), 2272–2281. https://doi.org/10.1016/j.neurobiolaging.2014.04.031

Di Prisco, G. V., Huang, W., Buffington, S. A., Hsu, C.-C., Bonnen, P. E., Placzek, . . . Costa-Mattioli, M. (2014). Translational control of mGluR-dependent long-term depression and object-place learning by eIF2α. *Nature Neuroscience, 17*(8), 1073–1082. https://doi.org/10.1038/nn.3754

Gordiyenko, Y., Llácer, J. L., & Ramakrishnan, V. (2019). Structural basis for the inhibition of translation through eIF2α phosphorylation. *Nature Communications, 10*(1), 2640. https://doi.org/10.1038/s41467-019-10606-1

Halliday, M., Radford, H., Sekine, Y., Moreno, J., Verity, N., le Quesne, J., . . . Mallucci, G. R. (2015). Partial restoration of protein synthesis rates by the small molecule ISRIB prevents neurodegeneration without pancreatic toxicity. *Cell Death & Disease, 6*, e1672. https://doi.org/10.1038/cddis.2015.49

Halliday, Mark, Radford, H., Zents, K. A. M., Molloy, C., Moreno, J. A., Verity, N. C., . . . Mallucci, G. R. (2017). Repurposed drugs targeting eIF2α-P-mediated translational repression prevent neurodegeneration in mice. *Brain: A Journal of Neurology, 140*(6), 1768–1783. https://doi.org/10.1093/brain/awx074

Harding, H. P., Novoa, I., Zhang, Y., Zeng, H., Wek, R., Schapira, M., & Ron, D. (2000). Regulated translation initiation controls stress-induced gene expression in mammalian cells. *Molecular Cell, 6*(5), 1099–1108. https://doi.org/10.1016/s1097-2765(00)00108-8

Hood, K. N., Zhao, J., Redell, J. B., Hylin, M. J., Harris, B., Perez, A., . . . Dash, P. K. (2018). Endoplasmic reticulum stress contributes to the loss of newborn hippocampal neurons after traumatic brain injury. *Journal of Neuroscience: The Official Journal of the Society for Neuroscience, 38*(9), 2372–2384. https://doi.org/10.1523/JNEUROSCI.1756-17.2018

Huang, W., Placzek, A. N., Viana Di Prisco, G., Khatiwada, S., Sidrauski, C., Krnjević, K., . . . Costa-Mattioli, M. (2016). Translational control by eIF2α phosphorylation regulates vulnerability to the synaptic and behavioral effects of cocaine. *ELife, 5*. https://doi.org/10.7554/eLife.12052

Huber, K. M., Kayser, M. S., & Bear, M. F. (2000). Role for rapid dendritic protein synthesis in hippocampal mGluR-dependent long-term depression. *Science, 288*(5469), 1254–1257. https://doi.org/10.1126/science.288.5469.1254

Jiang, Z., Belforte, J. E., Lu, Y., Yabe, Y., Pickel, J., Smith, C. B., . . . Nakazawa, K. (2010). EIF2alpha Phosphorylation-dependent translation in CA1 pyramidal cells impairs hippocampal memory consolidation without affecting general translation. *Journal of Neuroscience: The Official Journal of the Society for Neuroscience*, *30*(7), 2582–2594. https://doi.org/10.1523/JNEUROSCI.3971-09.2010

Jousse, C., Oyadomari, S., Novoa, I., Lu, P., Zhang, Y., Harding, H. P., & Ron, D. (2003). Inhibition of a constitutive translation initiation factor 2alpha phosphatase, CReP, promotes survival of stressed cells. *Journal of Cell Biology*, *163*(4), 767–775. https://doi.org/10.1083/jcb.200308075

Kandel, E. R. (2001). The molecular biology of memory storage: A dialogue between genes and synapses. *Science*, *294*(5544), 1030–1038. https://doi.org/10.1126/science.1067020

Kashiwagi, K., Yokoyama, T., Nishimoto, M., Takahashi, M., Sakamoto, A., Yonemochi, M., . . . Ito, T. (2019). Structural basis for eIF2B inhibition in integrated stress response. *Science*, *364*(6439), 495–499. https://doi.org/10.1126/science.aaw4104

Kelleher, R. J., Govindarajan, A., & Tonegawa, S. (2004). Translational regulatory mechanisms in persistent forms of synaptic plasticity. *Neuron*, *44*(1), 59–73. https://doi.org/10.1016/j.neuron.2004.09.013

Kenner, L. R., Anand, A. A., Nguyen, H. C., Myasnikov, A. G., Klose, C. J., McGeever, L. A., . . . Frost, A. (2019). EIF2B-catalyzed nucleotide exchange and phosphoregulation by the integrated stress response. *Science*, *364*(6439), 491–495. https://doi.org/10.1126/science.aaw2922

Kernohan, K. D., Tétreault, M., Liwak-Muir, U., Geraghty, M. T., Qin, W., Venkateswaran, S., . . . Boycott, K. M. (2015). Homozygous mutation in the eukaryotic translation initiation factor 2alpha phosphatase gene, PPP1R15B, is associated with severe microcephaly, short stature and intellectual disability. *Human Molecular Genetics*, *24*(22), 6293–6300. https://doi.org/10.1093/hmg/ddv337

Kessels, H. W., & Malinow, R. (2009). Synaptic AMPA receptor plasticity and behavior. *Neuron*, *61*(3), 340–350. https://doi.org/10.1016/j.neuron.2009.01.015

Kida, S., Josselyn, S. A., Peña de Ortiz, S., Kogan, J. H., Chevere, I., Masushige, S., & Silva, A. J. (2002). CREB required for the stability of new and reactivated fear memories. *Nature Neuroscience*, *5*(4), 348–355. https://doi.org/10.1038/nn819

Kim, H.-J., Raphael, A. R., LaDow, E. S., McGurk, L., Weber, R. A., Trojanowski, J. Q., . . . Bonini, N. M. (2014). Therapeutic modulation of eIF2α phosphorylation rescues TDP-43 toxicity in amyotrophic lateral sclerosis disease models. *Nature Genetics*, *46*(2), 152–160. https://doi.org/10.1038/ng.2853

Krzyzosiak, A., Sigurdardottir, A., Luh, L., Carrara, M., Das, I., Schneider, K., & Bertolotti, A. (2018). Target-based discovery of an inhibitor of the regulatory phosphatase PPP1R15B. *Cell*, *174*(5), 1216–1228.e19. https://doi.org/10.1016/j.cell.2018.06.030

Lüscher, C., & Huber, K. M. (2010). Group 1 mGluR-dependent synaptic long-term depression: Mechanisms and implications for circuitry and disease. *Neuron*, *65*(4), 445–459. https://doi.org/10.1016/j.neuron.2010.01.016

Lüscher, C., & Malenka, R. C. (2011). Drug-evoked synaptic plasticity in addiction: From molecular changes to circuit remodeling. *Neuron*, *69*(4), 650–663. https://doi.org/10.1016/j.neuron.2011.01.017

Ma, T., Trinh, M. A., Wexler, A. J., Bourbon, C., Gatti, E., Pierre, P., . . . Klann, E. (2013). Suppression of eIF2α kinases alleviates Alzheimer's disease-related plasticity and memory deficits. *Nature Neuroscience*, *16*(9), 1299–1305. https://doi.org/10.1038/nn.3486

Malenka, R. C. (1994). Synaptic plasticity in the hippocampus: LTP and LTD. *Cell, 78*(4), 535–538. https://doi.org/10.1016/0092-8674(94)90517-7

Malenka, Robert C., & Bear, M. F. (2004). LTP and LTD: An embarrassment of riches. *Neuron, 44*(1), 5–21. https://doi.org/10.1016/j.neuron.2004.09.012

McGaugh, J. L. (2000). Memory—A century of consolidation. *Science, 287*(5451), 248–251. https://doi.org/10.1126/science.287.5451.248

Moortgat, S., Désir, J., Benoit, V., Boulanger, S., Pendeville, H., Nassogne, M.-C., . . . & Maystadt, I. (2016). Two novel EIF2S3 mutations associated with syndromic intellectual disability with severe microcephaly, growth retardation, and epilepsy. *American Journal of Medical Genetics. Part A, 170*(11), 2927–2933. https://doi.org/10.1002/ajmg.a.37792

Moreno, J. A., Halliday, M., Molloy, C., Radford, H., Verity, N., Axten, J. M., . . . Mallucci, G. R. (2013). Oral treatment targeting the unfolded protein response prevents neurodegeneration and clinical disease in prion-infected mice. *Science Translational Medicine, 5*(206), 206ra138. https://doi.org/10.1126/scitranslmed.3006767

Moreno, J. A., Radford, H., Peretti, D., Steinert, J. R., Verity, N., Martin, M. G., . . . Mallucci, G. R. (2012). Sustained translational repression by eIF2α-P mediates prion neurodegeneration. *Nature, 485*(7399), 507–511. https://doi.org/10.1038/nature11058

Nadif Kasri, N., Nakano-Kobayashi, A., & Van Aelst, L. (2011). Rapid synthesis of the X-linked mental retardation protein OPHN1 mediates mGluR-dependent LTD through interaction with the endocytic machinery. *Neuron, 72*(2), 300–315. https://doi.org/10.1016/j.neuron.2011.09.001

Neves, G., Cooke, S. F., & Bliss, T. V. P. (2008). Synaptic plasticity, memory and the hippocampus: A neural network approach to causality. *Nature Reviews. Neuroscience, 9*(1), 65–75. https://doi.org/10.1038/nrn2303

Novoa, I., Zeng, H., Harding, H. P., & Ron, D. (2001). Feedback inhibition of the unfolded protein response by GADD34-mediated dephosphorylation of eIF2alpha. *Journal of Cell Biology, 153*(5), 1011–1022. https://doi.org/10.1083/jcb.153.5.1011

Palmer, M. J., Irving, A. J., Seabrook, G. R., Jane, D. E., & Collingridge, G. L. (1997). The group I mGlu receptor agonist DHPG induces a novel form of LTD in the CA1 region of the hippocampus. *Neuropharmacology, 36*(11–12), 1517–1532. https://doi.org/10.1016/s0028-3908(97)00181-0

Panja, D., Dagyte, G., Bidinosti, M., Wibrand, K., Kristiansen, A.-M., Sonenberg, N., & Bramham, C. R. (2009). Novel translational control in Arc-dependent long term potentiation consolidation in vivo. *Journal of Biological Chemistry, 284*(46), 31498–31511. https://doi.org/10.1074/jbc.M109.056077

Pasini, S., Corona, C., Liu, J., Greene, L. A., & Shelanski, M. L. (2015). Specific downregulation of hippocampal ATF4 reveals a necessary role in synaptic plasticity and memory. *Cell Reports, 11*(2), 183–191. https://doi.org/10.1016/j.celrep.2015.03.025

Placzek, A. N., Prisco, G. V. D., Khatiwada, S., Sgritta, M., Huang, W., Krnjević, . . . Costa-Mattioli, M. (2016). EIF2α-mediated translational control regulates the persistence of cocaine-induced LTP in midbrain dopamine neurons. *ELife, 5*. https://doi.org/10.7554/eLife.17517

Radford, H., Moreno, J. A., Verity, N., Halliday, M., & Mallucci, G. R. (2015). PERK inhibition prevents tau-mediated neurodegeneration in a mouse model of frontotemporal dementia. *Acta Neuropathologica, 130*(5), 633–642. https://doi.org/10.1007/s00401-015-1487-z

Richter, J. D., & Klann, E. (2009). Making synaptic plasticity and memory last: Mechanisms of translational regulation. *Genes & Development, 23*(1), 1–11. https://doi.org/10.1101/gad.1735809

Robert, F., Kapp, L. D., Khan, S. N., Acker, M. G., Kolitz, S., Kazemi, S., . . . Pelletier, J. (2006). Initiation of protein synthesis by hepatitis C virus is refractory to reduced eIF2. GTP.Met-tRNA(i)(Met) ternary complex availability. *Molecular Biology of the Cell*, *17*(11), 4632–4644. https://doi.org/10.1091/mbc.e06-06-0478

Sajikumar, S., & Frey, J. U. (2004). Late-associativity, synaptic tagging, and the role of dopamine during LTP and LTD. *Neurobiology of Learning and Memory*, *82*(1), 12–25. https://doi.org/10.1016/j.nlm.2004.03.003

Sajikumar, S., Navakkode, S., Sacktor, T. C., & Frey, J. U. (2005). Synaptic tagging and cross-tagging: The role of protein kinase Mzeta in maintaining long-term potentiation but not long-term depression. *Journal of Neuroscience: The Official Journal of the Society for Neuroscience*, *25*(24), 5750–5756. https://doi.org/10.1523/JNEUROSCI.1104-05.2005

Scoville, W. B., & Milner, B. (1957). Loss of recent memory after bilateral hippocampal lesions. *Journal of Neurology, Neurosurgery, and Psychiatry*, *20*(1), 11–21. https://doi.org/10.1136/jnnp.20.1.11

Sen, T., Gupta, R., Kaiser, H., & Sen, N. (2017). Activation of PERK elicits memory Impairment through Inactivation of CREB and downregulation of PSD95 after traumatic brain injury. *Journal of Neuroscience: The Official Journal of the Society for Neuroscience*, *37*(24), 5900–5911. https://doi.org/10.1523/JNEUROSCI.2343-16.2017

Sharma, V., Ounallah-Saad, H., Chakraborty, D., Hleihil, M., Sood, R., Barrera, I., . . . Rosenblum, K. (2018). Local inhibition of PERK enhances memory and reverses age-related deterioration of cognitive and neuronal properties. *Journal of Neuroscience: The Official Journal of the Society for Neuroscience*, *38*(3), 648–658. https://doi.org/10.1523/JNEUROSCI.0628-17.2017

Shrestha, P., Ayata, P., Herrero-Vidal, P., Longo, F., Gastone, A., LeDoux, J. E., . . . Klann, E. (2020). Cell-type-specific drug-inducible protein synthesis inhibition demonstrates that memory consolidation requires rapid neuronal translation. *Nature Neuroscience*, *23*(2), 281–292. https://doi.org/10.1038/s41593-019-0568-z

Sidrauski, C., Acosta-Alvear, D., Khoutorsky, A., Vedantham, P., Hearn, B. R., Li, H., Walter, P. (2013). Pharmacological brake-release of mRNA translation enhances cognitive memory. *ELife*, *2*, e00498. https://doi.org/10.7554/eLife.00498

Skopkova, M., Hennig, F., Shin, B.-S., Turner, C. E., Stanikova, D., Brennerova, K., . . . Kalscheuer, V. M. (2017). EIF2S3 mutations associated with severe X-Linked intellectual disability syndrome MEHMO. *Human Mutation*, *38*(4), 409–425. https://doi.org/10.1002/humu.23170

Stern, E., Chinnakkaruppan, A., David, O., Sonenberg, N., & Rosenblum, K. (2013). Blocking the eIF2α kinase (PKR) enhances positive and negative forms of cortex-dependent taste memory. *The Journal of Neuroscience: The Official Journal of the Society for Neuroscience*, *33*(6), 2517–2525. https://doi.org/10.1523/JNEUROSCI.2322-12.2013

Sutton, M. A., & Schuman, E. M. (2006). Dendritic protein synthesis, synaptic plasticity, and memory. *Cell*, *127*(1), 49–58. https://doi.org/10.1016/j.cell.2006.09.014

Svitkin, Y. V., Herdy, B., Costa-Mattioli, M., Gingras, A.-C., Raught, B., & Sonenberg, N. (2005). Eukaryotic translation initiation factor 4E availability controls the switch between cap-dependent and internal ribosomal entry site-mediated translation. *Molecular and Cellular Biology*, *25*(23), 10556–10565. https://doi.org/10.1128/MCB.25.23.10556-10565.2005

Takei, N., Kawamura, M., Hara, K., Yonezawa, K., & Nawa, H. (2001). Brain-derived neurotrophic factor enhances neuronal translation by activating multiple initiation processes: Comparison with the effects of insulin. *The Journal of Biological Chemistry*, *276*(46), 42818–42825. https://doi.org/10.1074/jbc.M103237200

Trinh, M. A., Kaphzan, H., Wek, R. C., Pierre, P., Cavener, D. R., & Klann, E. (2012). Brain-specific disruption of the eIF2α kinase PERK decreases ATF4 expression and impairs behavioral flexibility. *Cell Reports, 1*(6), 676–688. https://doi.org/10.1016/j.celrep.2012.04.010

Trinh, M. A., Ma, T., Kaphzan, H., Bhattacharya, A., Antion, M. D., Cavener, D. R.,... Klann, E. (2014). The eIF2α kinase PERK limits the expression of hippocampal metabotropic glutamate receptor-dependent long-term depression. *Learning & Memory, 21*(5), 298–304. https://doi.org/10.1101/lm.032219.113

Tronel, C., Page, G., Bodard, S., Chalon, S., & Antier, D. (2014). The specific PKR inhibitor C16 prevents apoptosis and IL-1β production in an acute excitotoxic rat model with a neuroinflammatory component. *Neurochemistry International, 64*, 73–83. https://doi.org/10.1016/j.neuint.2013.10.012

Vattem, K. M., & Wek, R. C. (2004). Reinitiation involving upstream ORFs regulates ATF4 mRNA translation in mammalian cells. *Proceedings of the National Academy of Sciences of the United States of America, 101*(31), 11269–11274. https://doi.org/10.1073/pnas.0400541101

Wong, Y. L., LeBon, L., Basso, A. M., Kohlhaas, K. L., Nikkel, A. L., Robb, H. M.,... Sidrauski, C. (2019). EIF2B activator prevents neurological defects caused by a chronic integrated stress response. *ELife, 8*. https://doi.org/10.7554/eLife.42940

Wong, Y. L., LeBon, L., Edalji, R., Lim, H. B., Sun, C., & Sidrauski, C. (2018). The small molecule ISRIB rescues the stability and activity of vanishing white matter disease eIF2B mutant complexes. *ELife, 7*. https://doi.org/10.7554/eLife.32733

Young-Baird, S. K., Lourenço, M. B., Elder, M. K., Klann, E., Liebau, S., & Dever, T. E. (2020). Suppression of MEHMO syndrome mutation in eIF2 by small molecule ISRIB. *Molecular Cell, 77*(4), 875–886.e7. https://doi.org/10.1016/j.molcel.2019.11.008

Young-Baird, S. K., Shin, B.-S., & Dever, T. E. (2019). MEHMO syndrome mutation EIF2S3-I259M impairs initiator Met-tRNAiMet binding to eukaryotic translation initiation factor eIF2. *Nucleic Acids Research, 47*(2), 855–867. https://doi.org/10.1093/nar/gky1213

Zhu, P. J., Huang, W., Kalikulov, D., Yoo, J. W., Placzek, A. N., Stoica, L.,... Costa-Mattioli, M. (2011). Suppression of PKR promotes network excitability and enhanced cognition by interferon-γ-mediated disinhibition. *Cell, 147*(6), 1384–1396. https://doi.org/10.1016/j.cell.2011.11.029

Zhu, P. J., Khatiwada, S., Cui, Y., Reineke, L. C., Dooling, S. W., Kim, J. J.,... Costa-Mattioli, M. (2019). Activation of the ISR mediates the behavioral and neurophysiological abnormalities in Down syndrome. *Science, 366*(6467), 843–849. https://doi.org/10.1126/science.aaw5185

Zhu, S., Henninger, K., McGrath, B. C., & Cavener, D. R. (2016). PERK regulates working memory and protein synthesis-dependent memory flexibility. *PloS One, 11*(9), e0162766. https://doi.org/10.1371/journal.pone.0162766

Zhu, S., McGrath, B. C., Bai, Y., Tang, X., & Cavener, D. R. (2016). PERK regulates Gq protein-coupled intracellular $Ca2^{+}$ dynamics in primary cortical neurons. *Molecular Brain, 9*(1), 87. https://doi.org/10.1186/s13041-016-0268-5

CHAPTER 4

THE ROLE OF THE EUKARYOTIC ELONGATION FACTOR 2 (EEF2) PATHWAY IN NEURONAL FUNCTION

ELHAM TAHA AND KOBI ROSENBLUM

INTRODUCTION

THE dynamic process of mRNA translation consists of three phases: initiation, elongation, and termination, where each phase requires specific translation factors (Merrick, 2010). The initiation phase of mRNA translation, the most regulated phase of the three, involves interaction between the cap structure located at the 5′ end of the eukaryotic mRNA and eukaryotic initiation factor 4E (eIF4E; Costa-Mattioli, Sossin, Klann, & Sonenberg, 2009; Gingras, Raught, & Sonenberg, 1999). This binding is highly regulated and leads to a cascade of protein phosphorylation, one of which is the phosphorylation of 4E-BP, which in turn regulates the binding of 4E-BP to eIF4E. The phosphorylation of 4E-BP occurs downstream of the mTOR pathway and releases the inhibition of eIF4E and facilitates the initiation phase of translation (Banko & Klann, 2008). In addition, the initiation phase is regulated by the phosphorylation of the α-subunit of eIF2 at serine 51 (ser51), which converts eIF2 to a competitive inhibitor of eIF2B, a guanine-nucleotide exchange factor that is required to generate the GTP-eIF2, a necessary factor for a new translation cycle (Sonenberg & Hinnebusch, 2009). Using behavioral genetics, biochemistry, and electrophysiological measurements, it has been demonstrated that eIF2α phosphorylation on ser51 plays a crucial role in regulating long-term memory and synaptic plasticity (Costa-Mattioli et al., 2007; Costa-Mattioli &

Sonenberg, 2008; Ounallah-Saad, Sharma, Edry, & Rosenblum, 2014; Sharma et al., 2018; Stern, Chinnakkaruppan, David, Sonenberg, & Rosenblum, 2013). However, the elongation phase of translation has been less studied in relation to learning, memory, and brain function and dysfunction.

The elongation phase is the second step of mRNA translation, during which amino acids are added to the growing peptide chain. The main regulation at the elongation phase is via the phosphorylation of the eukaryotic elongation factor 2 (eEF2) on Thr-56 by its only known kinase, eukaryotic elongation factor 2 kinase (eEF2K), resulting in inactivation of eEF2 by preventing it from binding to the ribosome and inhibition of the mRNA translation elongation phase (Carlberg, Nilsson, & Nygård, 1990; Justin W. Kenney, Moore, Wang, & Proud, 2014; Taha, Gildish, Gal-Ben-Ari, & Rosenblum, 2013). Dephosphorylation of eEF2 on Thr-56 increases general protein synthesis but also can differentially regulates the translation of plasticity related mRNA's such as BDNF, α-CaMKII, and Arc according to their unique 5′ and/or 3′ untranslated regions. (Adaikkan, Taha, Barrera, David, & Rosenblum, 2018; Autry et al., 2011; Taha et al., 2013; Verpelli et al., 2010). In the present review, we focus on the function of the eEF2/eEF2K pathway in the brain.

eEF2/eEF2K Pathway Function in Neurons

eEF2 is a monomeric protein of 95 kDa, containing 857 amino acid residues, which is a member of the G-protein superfamily and is bound to GTP at the N-terminus (Bartish & Nygård, 2008). During the elongation phase of protein synthesis, aminoacyl-tRNA attached to a newly recruited amino acid enters the A-site of the ribosome via the eukaryotic elongation factors 1A and 1B (eEF1A and eEF1B). eEF2 mediates the translocation of peptidyl-tRNA from the ribosomal A-site to its P-site by GTP hydrolysis, following which eEF2 leaves the ribosome in its inactive form as eEF2-GDP (Browne & Proud, 2002; Jobe, Liu, Gutierrez-Vargas, & Frank, 2019; Ling & Ermolenko, 2016; Montanaro, Sperti, Testoni, & Mattioli, 1976).

eEF2 has a single kinase known to phosphorylate it, eEF2 kinase (eEF2K), which phosphorylates eEF2 on Thr-56, thereby inhibiting protein synthesis translation elongation. The Thr-56 residue is highly conserved and even identical in mammals and yeast. However, eEF2K homologues are absent in other eukaryotes (fungi, plants). In nematodes and yeast, homologues of eEF2 can be phosphorylated by a different kinase, Rck2, on the homologuesThr-56 (Teige, Scheikl, Reiser, Ruis, & Ammerer, 2001). In addition, eEF2K is not expressed in insects, particularly Drosophila, a major research model of nervous system function. Although Thr-56 is conserved, there does not appear to be any kinase in this system, resulting in constantly unphosphorylated Thr-56 (Manning, Plowman, Hunter, & Sudarsanam, 2002). Functionally, eEF2K is a unique kinase, which belongs to a small group termed α-kinases and is the only α-kinase whose activity is dependent on Ca^{2+}-ions, known as Ca^{2+}/CaM-kinase III (CaMKIII). The family of α-kinases represents atypical protein kinases that show no sequence similarity to classical protein kinases (Nairn, Bhagat, & Palfrey, 1985; A. G. Ryazanov, Pavur, & Dorovkov, 1999; A.G. Ryazanov, Natapov, Shestakova, Severin, & Spirin, 1988; Yamaguchi, Matsushita, Nairn, & Kuriyan, 2001). While the single enzyme/single substrate relationship between eEF2 and eEF2K makes the eEF2/eEF2K pathway unique and relatively easy to study using genetic tools, pharmacological research remains limited, since currently there are limited efficient small molecule inhibitors.

As mentioned above, eEF2 is part of the protein synthesis process, which occurs locally in dendritic, axonal, and post-synaptic locations, where it is known to be required for memory consolidation and synaptic plasticity. Local protein synthesis requires a local reservoir of ribosomes, translation factors including eEF2, and a subset of mRNAs (Chotiner, Khorasani, Nairn, O'Dell, & Watson, 2003; Gal-Ben-Ari et al., 2012; Miyashiro, Dichter, & Eberwine, 1994; Smith & Schuman, n.d.; Steward & Levy, 1982; Taha et al., 2013). The phosphorylation of eEF2 by eEF2K activity is subject to complex regulation by various stimuli in different cells including neurons. eEF2K, in contrast to eEF2, is controlled by many proteins and serves as a junction point for integrating intracellular information, which in turn affects the rate of mRNA translation. For example, eEF2K is negatively regulated by the mammalian target of rapamycin complex 1 (mTORC1), which is in turn activated by hormones, growth factors, and the availability of amino acids (Efeyan, Zoncu, & Sabatini, 2012; Proud, 2013). In addition, eEF2K is positively regulated by cAMP and AMP-activated protein kinase, AMPK, a major sensor of low cellular energy status (Hardie, Ross, & Hawley, 2012; Johanns et al., 2017). eEF2K is also regulated in response to stress-stimulated MAP kinase (MAPK) cascades, either directly by MAPKs or by their downstream effectors (Knebel, Morrice, & Cohen, 2001; see Figure 4.1).

In the central nervous system, eEF2 activity is regulated by glutamate, the primary excitatory neurotransmitter, which stimulates eEF2 phosphorylation in both neurons and isolated preparations of synaptic compartments, mainly via Ca^{2+} influx in an N-methyl-D-aspartate receptor (NMDAR)-dependent manner. In addition, glutamatergic signaling also activates the mTORC1/S6K1 and MEK/ERK pathways, which negatively regulate eEF2K (Lenz & Avruch, 2005; Marin et al., 1997; Scheetz, Nairn, & Constantine-Paton, 2000; see Figure 4.1). The interplay between these inputs determines temporal dynamics of eEF2 phosphorylation, which in turn affect mRNA translation elongation rates. In cortical primary neurons, elevating synaptic activity results in Ca^{2+}-coordinated changes manifested in a rapid and transient increase in eEF2 phosphorylation that depends on AMPA and NMDA-type glutamate receptors (GluRs), in parallel to a decrease in eEF2 phosphorylation that depends on the MEK/ERK and mTORC1 pathways (J. W. Kenney et al., 2015). In addition to the regulation of eEF2K through AMPA and NMDA-type glutamate receptors, eEF2K activity is controlled through mGluR receptor signaling. It was shown that eEF2K and FMRP control the de novo translation of Arc, which is required for mGluR-LTD (Park et al., 2008).

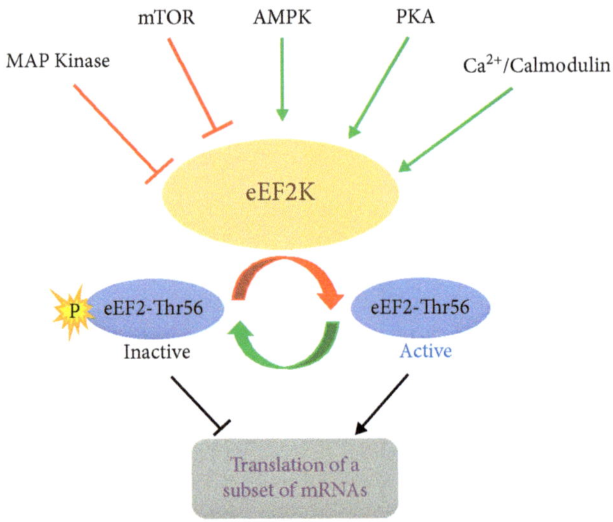

FIGURE 4.1. Regulation of eEF2K by AMPK, Ca^{2+}/calmodulin, PKA, mTOR, and MAP kinase pathways. eEF2K is positively regulated by the AMPK, PKA, and Ca^{2+}/calmodulin pathways, and negatively regulated by the mTOR and MAP kinase pathways. Activation of eEF2K leads to phosphorylation of eEF2 on Thr-56 and later to protein synthesis inhibition. Deactivation of eEF2K leads to dephosphorylation of eEF2 on Thr-56 and increased translation of a subset of mRNAs.

Little is currently known regarding the neurotransmitters that regulate eEF2 phosphorylation. David and colleagues (2020) showed that activation of the D1 dopamine receptor but not D2 receptor leads to eEF2 dephosphorylation at Thr-56 in cortical neuronal culture and in cortical and hippocampal lysates. In addition, D1 dopamine receptor activation resulted in decreased eEF2 phosphorylation levels in dendrites together with a correlative increase in local mRNA translation. Moreover, NMDA receptor and the mTOR and ERK pathways are upstream of the D1 dopamine receptor-dependent eEF2 dephosphorylation (David et al.,2020).

Translation elongation plays essential roles in multiple aspects of protein biogenesis (Chaney & Clark, 2015). In particular, ribosome stalling during the elongation phase, which is manifested in the local accumulation of ribosomes at specific codon positions of mRNAs, can lead to various biological consequences, such as regulation of protein expression and pathological conditions (Ishimura et al., 2014; Tuller et al., 2010). In addition, mGluR activation reduces the number of stalled polyribosomes and induces initiation-independent peptide synthesis in neurites. Such reactivation of neuronal translation stalled at the elongation phase is an elegant mechanism that allows spatial and temporal regulation of synaptic protein synthesis upon synaptic signaling (Graber et al., 2013). In the nervous system, neuronal proteostasis is maintained by the dynamic integration of different processes that regulate the synthesis, degradation, folding, and localization of proteins (Rosenberg et al., 2014; Stein & Frydman, 2019).

What is the end-result of manipulating eEF2 phosphorylation levels in terms of proteostasis? The answer is complex, as expression levels of protein populations can decrease or increase via unknown mechanisms or because of the structure of specific mRNAs (Justin W. Kenney et al., 2016). Although eEF2K activation is a negative regulator of general protein synthesis, it was shown that it specifically facilitates the synthesis of certain proteins, such as BDNF, α-CaMKII, and Arc. In this context, early studies provided evidence that impairing elongation could promote the translation of specific mRNAs. It is suggested that decreased elongation rates may relieve competition between mRNAs at the initiation stage, thus allowing mRNAs that are poor initiators, as defined by their structure, to be recruited into active polyribosomes (K. Belelovsky, Elkobi, Kaphzan, Nairn, & Rosenblum, 2005; Katya Belelovsky, Kaphzan, Elkobi, & Rosenblum, 2009; Park et al., 2008; Scheetz et al., 2000; Walden & Thach, 1986). In line with this work, low doses of cycloheximide, an inhibitor of translation elongation, increase rather than decrease the synthesis rates of a few proteins such as Arc and α-CaMKII, and thus rescue deficits in mGluR-dependent LTD in eEF2K-KO mice (Park et al., 2008). It is possible that different concentrations of the inhibitors can differentially affect proteostasis. Overall, the above data suggest that the regulation of eEF2K in neurons is complex, and additional research is necessary to understand how and why it is differentially regulated under different conditions and in different neuronal preparations.

eEF2/eEF2K Pathway Regulation in Memory and Synaptic Plasticity

The regulation of mRNA translation elongation has been found to play an important role in memory and synaptic plasticity (Taha et al., 2013). Following novel taste learning, eEF2 phosphorylation increases within 20 min in the gustatory cortex (GC) with no change in the total amount of eEF2 protein. In addition, eEF2 phosphorylation is not altered following exposure to an insufficient amount of novel taste required to induce aversion toward it (K. Belelovsky et al., 2005; Gildish et al., 2012). Genetically engineered eEF2K-KI mice containing a point mutation in the catalytic domain of eEF2K, which express residual activity of eEF2K and low levels of eEF2 phosphorylation, are impaired in conditioned taste aversion (CTA) learning, taste learning that requires the association between a novel taste (conditioned stimulus) and malaise (the unconditioned stimulus), resulting in novel taste aversion (Gal-Ben-Ari. & Rosenblum, 2012; Gildish et al., 2012; Rosenblum, Meiri, & Dudai, 1993; Yiannakas & Rosenblum, 2017). In contrast to the impaired CTA phenotype in eEF2K-KI mice, these mice showed normal novel taste learning. This suggests that eEF2 phosphorylation is critical for associative taste memory formation. To better understand the function of the eEF2 pathway in CTA learning, brain activity as measured by manganese-enhanced magnetic resonance imaging (MEMRI) was examined. CTA training induced accumulation of Mn^{2+} in brains of both eEF2K-KI and wild type mice, however, Mn^{2+} was localized in specific brain areas in the wild type group, whereas a less localized pattern was observed in the eEF2K-KI mice. Indeed, eEF2K-KI mice had abnormal brain activity during CTA learning compared to wild type mice, but no differences were detected under basal conditions. These results suggest that abnormality of memory processing on the system level underlies impaired associative taste memory detected in eEF2K-KI mice (Gildish et al., 2012).

In addition to GC- and amygdala-dependent taste learning, eEF2 phosphorylation state is modulated following hippocampal-dependent contextual fear conditioning and hippocampal-independent cued fear conditioning (Im et al., 2009). Following fear conditioning training, eEF2 was dephosphorylated within 0.5-2 h in the hippocampus and the amygdala of trained mice. Furthermore, hippocampus-specific eEF2K transgenic mice (hip-eEF2K-tg mice) that exhibit overexpression of eEF2K were used in this study. In order to generate the hip-eEF2K-tg mice, two independent transgenic mouse strains, eEF2K transgenic mice and Cre transgenic mice were crossed. The eEF2K transgenic mice were characterized by Cre recombination-dependent deletion of a transcriptional stop signal located upstream of the eEF2K transgene so that the transgene would be expressed only in a Cre-dependent cell- and region-specific manner. In the Cre-eEF2K double transgenic mice, Cre expression was limited to neurons in most parts of the hippocampus, and these mice exhibited hippocampus-specific eEF2K overexpression, which was confirmed by in situ hybridization (Im et al., 2009). In these mice, the dephosphorylation of eEF2 was prevented in the hippocampus following fear conditioning training, leading to impairment in long term contextual fear memory, but not long term cued fear memory. Moreover, hip-eEF2K-tg mice were impaired in long-term hippocampus-dependent (spatial) memory. Overall, these results suggest that the eEF2 pathway differentially contributes to cortical-dependent taste learning and hippocampal-dependent memory in the brain.

While the studies described thus far used rodent models (mice and rats), a large body of other studies in the field used the *Aplysia* model. The most obvious advantage of using *Aplysia* rather than rodents in neuroscience research is the large neurons that facilitate physiological, biochemical, and genomic studies at the level of single cells. In addition, the simplicity of the *Aplysia* nervous system provides an opportunity to easily understand the distinct neural circuits underlying behavior. Moreover, *Aplysia* is considered as an ideal model to explore the effect of serotonin (5-HT) at the sensory-to-motor neuron synapse that induces synaptic facilitation, an increase in synaptic strength that underlies sensitization (Kandel, 2001).

In addition, the ability to generate synaptic facilitation seen in vivo in isolated neurons is a unique advantage of the *Aplysia* model (Kandel, 2001). In *Aplysia*, it was shown that long-term facilitation (LTF), a process underlying learning, requires translational control via the eEF2/eEF2K pathway (Weatherill et al., 2011). Moreover, phosphorylation of eEF2 and eEF2K was found to be regulated differentially in the soma and neurites of sensory neurons in *Aplysia* through the mTORC1 pathway (Weatherill et al., 2011).

Spaced (non-continuous stimulation with intervals) or massed (one continuous stimulation without intervals) exposure to 5-HT induces a form of intermediate-term facilitation (ITF) that requires new protein synthesis (Kandel, 2012; Mauelshagen, Sherff, & Carew, 1998). Interestingly, McCamphill and colleagues (2015) showed that phosphorylation of eEF2 is increased in *Aplysia* sensory neurons by massed application of serotonin (5-HT) but decreased by spaced applications (McCamphill, Farah, Anadolu, Hoque, & Sossin, 2015). Moreover, the same group showed that eEF2 dephosphorylation is necessary for mTORC1-dependent translation and plasticity. Furthermore, LTF-induced decrease in eEF2 phosphorylation was blocked by expression of an eEF2K modified to be resistant to mTORC1 regulation, and blocking eEF2 dephosphorylation also blocked the increase in local sensorin synthesis, a neuropeptide whose local translation and release is required for LTF (McCamphill, Ferguson, & Sossin, 2017). These findings suggest that, in *Aplysia*, eEF2 acts as a biochemical sensor that is able to decode distinct neuronal activity patterns, resulting in differential protein synthesis and synaptic plasticity. Furthermore, the findings identify eEF2 dephosphorylation as a critical target for mTORC1 long-term plasticity and sensorin expression.

An earlier study by Sutton and colleagues (2007) elegantly demonstrated that eEF2 is a key biochemical sensor that couples neuronal transmission to local protein synthesis. It was demonstrated that intrinsic action potential-mediated network activity in cultured hippocampal neurons leads to eEF2 dephosphorylation, whereas spontaneous neurotransmitter release (i.e., miniature neurotransmission) strongly promotes eEF2 phosphorylation (Sutton, Taylor, Ito, Pham, & Schuman, 2007). Furthermore, Verpelli and colleagues (2010) showed that mGluR/eEF2K-induced eEF2 phosphorylation promotes spine stability by controlling dendritic BDNF synthesis (Verpelli et al., 2010). Taken together, the above results indicate that, in neurons, eEF2 serves as a biochemical sensor that is tuned to synaptic activity.

Despite the intensive research into the neurobiological function of the eEF2/eEF2K pathway, a complete understanding of the role of the eEF2K/eEF2 pathway in synapses and neural networks is not clear. To that aim, we and the Sala group conducted a series of biochemical, electrophysiological, and behavioral experiments in eEF2K-KO mice (Heise et al., 2017; Alexey G. Ryazanov, 2002). First, we identified the function of eEF2K in excitatory and inhibitory neurotransmission. We used whole-cell patch-clamp recordings to measure miniature excitatory postsynaptic currents (mPSCs) in dentate gyrus (DG) granule cells, and the analysis was carried out employing an in vitro eEF2K gain-of-function design (using primary neuronal cultures) and an in vivo loss-of-function design (eEF2K-KO mice). We found that mEPSC parameters were not different in eEF2K-KO mice compared with WT littermates. However, we observed a strong increase in the miniature inhibitory postsynaptic current (mIPSC) frequency and amplitude in hippocampal granule cells of eEF2K-KO mice. In addition, LTP induced by high frequency stimulation (HFS) or theta-burst stimulation (TBS) in the CA1 region of the hippocampus was normal in eEF2K-KO.

Next, since there was no effect on excitatory synaptic transmission, we explored the function of eEF2K in inhibitory synaptic transmission. We studied how eEF2K activity specifically affects the GABAergic system in vivo by first measuring the level of tonic inhibition in hippocampal slices in eEF2K-KO mice and WT littermates. Indeed, loss of eEF2K in eEF2K-KO mice increases GABAergic synaptic transmission in the DG as identified by increased tonic inhibition. At the molecular level, to explore which synaptic proteins are controlled by eEF2K activity and mediate the modulatory activity on the excitation/inhibition balance, we utilized stable isotope labeling in cell culture (SILAC). Interestingly, we found that the presynaptic protein synapsin 2b and α5 containing $GABA_A$ receptors were upregulated as a result of in vitro and in vivo eEF2K-KO and thus interferes with the excitation/inhibition balance.

At the behavioral level, the eEF2K-KO mice showed a mild hippocampal-dependent phenotype with no cortical dependent taste learning phenotype. These data show that eEF2K-KO mice exhibit normal excitatory synaptic transmission while inhibitory synaptic transmission is clearly increased at the presynaptic and postsynaptic sites. At the same time, the behavioral phenotype is very mild, however, it is more pronounced in DG-dependent behaviors. Finally, we found that the eEF2K-KO mice were resistant to epileptic seizures when challenged with epilepsy inducing drugs. Strikingly, using pharmacological and genetic approaches to reduce eEF2K activity, we were able to rescue the epileptic phenotype in SYN1-KO mice, a mouse model of human epilepsy. These results point toward the possibility of targeting eEF2K activity in the context of epilepsy, raising the possibility to pharmacologically reduce seizure susceptibility and duration in patients.

eEF2/eEF2K Pathway in Neurological Diseases

Global dysregulation of protein synthesis resulting from abnormal activation in the translation machinery, especially dysregulation of eEF2 phosphorylation may contribute to pathogenesis in a subset of neurodegenerative and neuropsychiatric disorders (Beckelman et al., 2019; Jan et al., 2018; Monteggia, Gideons, & Kavalali, 2013; see Figure 4.2).

Neuropsychiatric Disorders

Intensive research regarding the cellular and molecular mechanisms of ketamine, an antidepressant drug, has demonstrated that ketamine administration reduces eEF2 phosphorylation on Thr-56 in both the cortex and the hippocampus of wild-type C57BL/6 mice (Autry et al., 2011; Monteggia et al., 2013). This reduced phosphorylation is correlated with increased expression of BDNF and the GluA1 subunit of the AMPAR (Kavalali & Monteggia, 2015). Upregulated BDNF and GluA1 have been shown to trigger evoked synaptic transmission to mediate the long-lasting effects of ketamine

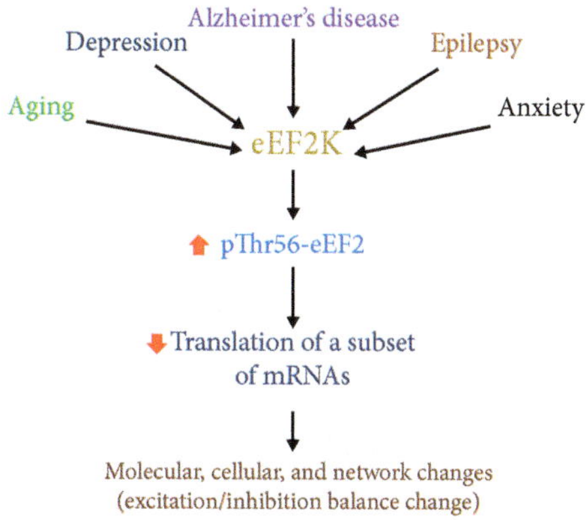

FIGURE 4.2. Dysregulation of eEF2 phosphorylation may contribute to pathogenesis in Alzheimer's disease, epilepsy, anxiety, depression, and aging. peEF2 levels in the brain are increased in Alzheimer's disease, epilepsy, anxiety, depression, and aging, which leads to reduced levels of protein synthesis. This chronic activation of eEF2K can induce positive feedback loop at the different levels (i.e., molecular, cellular, and network). Thus, the eEF2K pathway may serve as a potential target for brain related diseases.

(Autry et al., 2011; Kavalali & Monteggia, 2015; Zhou et al., 2014). This mechanistic insight is further supported by the findings that eEF2K inhibitors induce fast antidepressant effects in mice (Autry et al., 2011). In agreement with previous findings, we demonstrated that the CaMKII and eEF2K pathways mediate the antidepressant action of ketamine. In addition, ketamine administration leads to increased protein synthesis and differential regulation of CaMKII function, manifested as autoinhibition (pT305 phosphorylation) followed by autoactivation (pT286 phosphorylation) of CaMKIIα in the hippocampus and cortex. The inhibition phase of CaMKII, which lasted 10 to 20 minutes after administration of ketamine, occurred concurrently with eEF2K-dependent increased protein synthesis. Moreover, ketamine administration–dependent delayed induction of GluA1 (24 hours) was regulated by the activation of CaMKII. The eEF2K-KO mice were resistant to ketamine treatment, suggesting that synaptic activity-driven eEF2K dynamics is necessary for the action of ketamine. These findings suggest that eEF2K and CaMKII are major molecular substrates mediating the rapid antidepressant effect of ketamine (Adaikkan et al., 2018).

In addition to the critical role the eEF2 pathway plays in the antidepressant action of ketamine, it was recently shown that the eEF2 pathway is necessary for the anxiolytic effects of oxytocin (Martinetz et al., 2019). Interestingly, the authors showed that oxytocin application leads to eEF2 dephosphorylation in a hypothalamic cell line and in vivo within the paraventricular nucleus of the hypothalamus (PVN) in rats. In addition, eEF2 activation following oxytocin stimulation is mediated through the PKC and MEK1/2 signaling pathways. Next, the authors confirmed that oxytocin promotes de novo protein synthesis in hypothalamic cells, while inhibition of protein synthesis using anisomycin within the PVN prevents the anxiolytic effect of oxytocin in rats. Moreover, direct activation of eEF2 within the PVN by locally inhibiting eEF2K is sufficient to reduce anxiety-like behavior, indicating that the anxiolytic effect of oxytocin is mediated specifically by eEF2. Finally, the authors determined the mRNAs and proteins affected by intra-PVN oxytocin administration using an Affymetrix microarray. One of the proteins found to be upregulated by oxytocin was the neuropeptide Y receptor 5, which is known to be involved in anxiety-like behavior (Heilig, 2004). Importantly, infusion of a specific neuropeptide Y receptor 5 agonist into the PVN led to decreased anxiety-related behavior, while pretreatment with a neuropeptide Y receptor 5 antagonist prevented the anxiolytic effect of oxytocin. Taken together, this study identified eEF2 as a novel effector of PVN oxytocin-induced anxiety, and demonstrated that increasing protein synthesis through the eEF2 pathway may serve as a potential target for anxiety related disorders (Martinetz et al., 2019).

Normal Aging

With aging, there is decline in certain aspects of cognitive abilities such as learning and memory (Klencklen, Després, & Dufour, 2012). Interestingly, it was shown recently that memory decline and behavioral inflexibility in aged mice are correlated with dysregulation of protein synthesis capacity. In addition, dysregulation of the mTOR pathway as well as hyper-phosphorylation of eEF2 and its upstream regulator AMPK were found in aged mice, indicating repression of general protein synthesis. Furthermore, old mice, compared to young mice, displayed impairments in spatial memory, working memory, and behavioral flexibility. Moreover, an early form of LTP in hippocampal slices was inhibited in old mice. These findings give insight into the molecular mechanisms underlying age related cognitive decline at the mRNA translation level (Yang, Zhou, & Ma, 2019).

Alzheimer's Disease

Alzheimer's disease (AD) is the most common form of age-related dementia, characterized by cognitive impairment and neurodegeneration. Impaired translational capacity and ribosomal function have been observed in the brains of human AD patients. Abnormal hyper-phosphorylation of eEF2 was observed in postmortem brain tissue from AD patients and in the brains of AD mouse models (Li, Alafuzoff, Soininen, Winblad, & Pei, 2005; Ma et al., 2014). Interestingly, Beckelman and colleagues (2019) showed that genetic reduction of eEF2K in AD mouse models suppressed AD-associated eEF2 hyperphosphorylation and improved memory deficits and hippocampal LTP impairments. Furthermore, eEF2K reduction alleviated AD-associated defects in dendritic spine morphology, postsynaptic density formation, and de novo protein synthesis (Beckelman et al., 2019).

Addiction

Cocaine craving involves neuroadaptations in the circuitry underlying reward and motivation (Wolf, 2016). The nucleus accumbens brain area plays a central role in cocaine craving circuitry. It was shown that the synaptic response to cocaine is protein synthesis-dependent using brain slices prepared from rats incubated with anisomycin, a protein synthesis inhibitor. In addition, it was shown that, at the molecular level, mTOR-dependent signaling as well as eEF2 (Thr56) and eIF2α (ser51) dephosphorylation are required for expression of incubated cocaine seeking. These findings suggest that translation regulation at the initiation and elongation levels may play a critical role in cocaine treatment strategies (Werner, Stefanik, Milovanovic, Caccamise, & Wolf, 2018).

Conclusions and Future Directions

Intensive molecular, behavioral, and cellular research work points to the regulation of mRNA translation as a key pathway underlying neuronal and circuit homeostasis in the brain (Cajigas, Will, & Schuman, 2010; Heise et al., 2017; Kapur, Monaghan, & Ackerman, 2017; Ramocki & Zoghbi, 2008; Richter, Bassell, & Klann, 2015). The eEF2K-KI mice serve as an important genetic model to study the role of the eEF2 pathway in learning and memory. However, the fact that these mice display residual activity of the kinase and low levels of eEF2 phosphorylation complicates interpretation of the findings. Using the new eEF2K-KO mice, we and others were able to better understand the function of the eEF2/eEF2K pathway in epilepsy, aging, and AD. However, due to the fact that the KO mutation is expressed all over the brain and during development, a new conditional eEF2K-KO mouse model is required to answer basic questions about the function of the eEF2/eEF2K pathway and protein synthesis in excitatory versus inhibitory neurons in specific brain regions. Moreover, future studies in the field will have to address questions of circuit specificity and which cells in specific regions participate in memory formation or retention under genetic manipulation of eEF2K. To date, one eEF2K-KI mouse model and two eEF2K-KO mouse models have been used by different labs to study the role of the eEF2 pathway in brain function. Importantly, the conditional eEF2K-KO mice would serve as an ideal tool to better understand the role of eEF2K in health and disease conditions.

One significant issue that has not been adequately addressed is how translational control at the elongation phase regulates the translation of a subset of mRNAs. Thus, additional bioinformatics and structural analyses of specific mRNAs that are more sensitive to elongation inhibition are needed. This will enhance our understanding of possible targeting of memory related synaptic proteins. Another unresolved issue is the lack of potent eEF2K small molecule inhibitors to be used in brain research. Much work is needed in order to design and validate potent and specific eEF2K inhibitors that can be used in pharmacological brain studies, which would also provide possible treatments for epilepsy, aging, anxiety, and AD. Finally, it would be valuable to learn about the function of mRNA translation under specific manipulation of eEF2K in live neurons using new tools, such as the innovative tRNA-FRET method for labeling newly synthesized proteins using fluorescently labeled tRNAs (Koltun et al.,2020), or specifically in presynaptic or postsynaptic compartments of a specific neuronal subtype (Heumüller, Glock, Rangaraju, Biever, & Schuman, 2019). These new tools combined with conditional eEF2K-KO mice can help provide insight into the significance of mRNA translation at the elongation phase in excitatory and inhibitory neuronal subtypes in a temporal manner as well as at the circuit level, by inhibiting neurons that project to specific areas in order to probe the importance of different cell populations in learning and memory.

Acknowledgments

This research was supported by a grant from the Israel Science Foundation (ISF 946/17) and TransNeuro ERANET JPND supported by the Israel Ministry of Health Grant 3-14616 to K. R.

E. T. was supported by the Israeli Planning and Budgeting Committee Program Fellowships, the Ministry of Science and Technology Program Fellowships, and Israel Society for Neuroscience for outstanding PhD students.

We thank members of K. R.'s laboratory for suggestions and discussion, specifically Dr. Shunit Gal-Ben-Ari for reading and editing the manuscript and Ms. Iness Gildish, Mr. Mohammad Khamaisy, and Ms. Neta Sa'ar, who have taken part in the research project on eEF2.

References

Adaikkan, C., Taha, E., Barrera, I., David, O., & Rosenblum, K. (2018). Calcium/calmodulin-dependent protein kinase ii and eukaryotic elongation factor 2 kinase pathways mediate the antidepressant action of ketamine. *Biological Psychiatry*, *84*(1), 65–75. https://doi.org/10.1016/j.biopsych.2017.11.028

Autry, A. E., Adachi, M., Nosyreva, E., Na, E. S., Los, M. F., Cheng, P., ... Monteggia, L. M. (2011). NMDA receptor blockade at res t triggers rapid behavioural antidepressant responses. *Nature*, *475*(7354), 91–95. https://doi.org/10.1038/nature10130

Banko, J. L., & Klann, E. (2008). Chapter 4, Cap-dependent translation initiation and memory. In *Progress in Brain Research*, *169*. https://doi.org/10.1016/S0079-6123(07)00004-0

Bartish, G., & Nygård, O. (2008). Importance of individual amino acids in the Switch I region in eEF2 studied by functional complementation in S. cerevisiae. *Biochimie*, *90*(5), 736–748. https://doi.org/10.1016/J.BIOCHI.2008.01.006

Beckelman, B. C., Yang, W., Kasica, N. P., Zimmermann, H. R., Zhou, X., Keene, C. D., ... Ma, T. (2019). Genetic reduction of eEF2 kinase alleviates pathophysiology in Alzheimer's disease model mice. *Journal of Clinical Investigation*, *129*(2), 820–833. https://doi.org/10.1172/JCI122954

Belelovsky, K., Elkobi, A., Kaphzan, H., Nairn, A. C., & Rosenblum, K. (2005). A molecular switch for translational control in taste memory consolidation. *European Journal of Neuroscience*, 22(10), 2560–2568. https://doi.org/10.1111/j.1460–9568.2005.04428.x

Belelovsky, Katya, Kaphzan, H., Elkobi, A., & Rosenblum, K. (2009). Biphasic activation of the mTOR pathway in the gustatory cortex is correlated with and necessary for taste learning. *Journal of Neuroscience*, *29*(23), 7424–7431. https://doi.org/10.1523/JNEUROSCI.3809-08.2009

Browne, G. J., & Proud, C. G. (2002). Regulation of peptide-chain elongation in mammalian cells. *European Journal of Biochemistry*, *269*(22), 5360–5368. https://doi.org/10.1046/j.1432-1033.2002.03290.x

Cajigas, I. J., Will, T., & Schuman, E. M. (2010). Protein homeostasis and synaptic plasticity. *EMBO Journal*, *29*(16), 2746–2752. https://doi.org/10.1038/emboj.2010.173

Carlberg, U., Nilsson, A., & Nygård, O. (1990). Functional properties of phosphorylated elongation factor 2. *European Journal of Biochemistry*, *191*(3), 639–645. https://doi.org/10.1111/j.1432-1033.1990.tb19169.x

Chaney, J. L., & Clark, P. L. (2015). Roles for synonymous codon usage in protein biogenesis. *Annual Review of Biophysics*, *44*(1), 143–166. https://doi.org/10.1146/annurev-biophys-060414-034333

Chotiner, J., Khorasani, H., Nairn, A., O'Dell, T., & Watson, J. (2003). Adenylyl cyclase-dependent form of chemical long-term potentiation triggers translational regulation at the elongation step. *Neuroscience*, *116*(3), 743–752. https://doi.org/10.1016/S0306-4522(02)00797-2

Costa-Mattioli, M., Gobert, D., Stern, E., Gamache, K., Colina, R., Cuello, C., . . . Sonenberg, N. (2007). eIF2α Phosphorylation bidirectionally regulates the switch from short- to long-term synaptic plasticity and memory. *Cell*, *129*(1), 195–206. https://doi.org/10.1016/j.cell.2007.01.050

Costa-Mattioli, M., & Sonenberg, N. (2008). Chapter 5, Translational control of gene expression: A molecular switch for memory storage. In *Progress in Brain Research*, *169*. https://doi.org/10.1016/S0079-6123(07)00005-2

Costa-Mattioli, M., Sossin, W. S., Klann, E., & Sonenberg, N. (2009). Translational control of long-lasting synaptic plasticity and memory. *Neuron*, *61*(1), 10–26. https://doi.org/10.1016/j.neuron.2008.10.055

David, O., Barrera, I., Gould, N., Gal-Ben-Ari, S., & Rosenblum, K. (2020). D1 dopamine receptor activation induces neuronal eEF2 pathway-dependent protein synthesis. *Frontiers in Molecular Neuroscience*, *13*, 67. https://doi.org/10.3389/fnmol.2020.00067

Efeyan, A., Zoncu, R., & Sabatini, D. M. (2012). Amino acids and mTORC1: from lysosomes to disease. *Trends in Molecular Medicine*, *18*(9), 524–533. https://doi.org/10.1016/J.MOLMED.2012.05.007

Gal-Ben-Ari, S., Kenney, J. W., Ounalla-Saad, H., Taha, E., David, O., Levitan, D., . . . Rosenblum, K. (2012). Consolidation and translation regulation. *Learning and Memory*, *19*(9), 410–422. https://doi.org/10.1101/lm.026849.112

Gal-Ben-Ari., S., & Rosenblum, K. (2012). Molecular mechanisms underlying memory consolidation of taste information in the cortex. *Frontiers in Behavioral Neuroscience*, 5(January), 1–15. https://doi.org/10.3389/fnbeh.2011.00087

Gildish, I., Manor, D., David, O., Sharma, V., Williams, D., Agarwala, U., . . . Rosenblum, K. (2012). Impaired associative taste learning and abnormal brain activation in kinase-defective eEF2K mice. *Learning and Memory*, *19*(3), 116–125. https://doi.org/10.1101/lm.023937.111

Gingras, A.-C., Raught, B., & Sonenberg, N. (1999). eIF4 Initiation factors: Effectors of mRNA recruitment to ribosomes and regulators of translation. *Annual Review of Biochemistry*, *68*(1), 913–963. https://doi.org/10.1146/annurev.biochem.68.1.913

Graber, T. E., Hébert-Seropian, S., Khoutorsky, A., David, A., Yewdell, J. W., Lacaille, J. C., & Sossin, W. S. (2013). Reactivation of stalled polyribosomes in synaptic plasticity. *Proceedings of the National Academy of Sciences of the United States of America, 110*(40), 16205–16210. https://doi.org/10.1073/pnas.1307747110

Hardie, D. G., Ross, F. A., & Hawley, S. A. (2012). AMPK: A nutrient and energy sensor that maintains energy homeostasis. *Nature Reviews Molecular Cell Biology, 13*(4), 251–262. https://doi.org/10.1038/nrm3311

Heilig, M. (2004). The NPY system in stress, anxiety and depression. *Neuropeptides, 38*(4), 213–224. https://doi.org/10.1016/J.NPEP.2004.05.002

Heise, C., Taha, E., Murru, L., Ponzoni, L., Cattaneo, A., Guarnieri, F. C., . . . Sala, C. (2017). EEF2K/eEF2 pathway controls the excitation/inhibition balance and susceptibility to epileptic seizures. *Cerebral Cortex, 27*(3), 2226–2248. https://doi.org/10.1093/cercor/bhw075

Heumüller, M., Glock, C., Rangaraju, V., Biever, A., & Schuman, E. M. (2019). A genetically encodable cell-type-specific protein synthesis inhibitor. *Nature Methods, 16*(8), 699–702. https://doi.org/10.1038/s41592-019-0468-x

Im, H. I., Nakajima, A., Gong, B., Xiong, X., Mamiya, T., Gershon, E. S., . . . Tang, Y. P. (2009). Post-training dephosphorylation of eEF-2 promotes protein synthesis for memory consolidation. *PLoS ONE, 4*(10). https://doi.org/10.1371/journal.pone.0007424

Ishimura, R., Nagy, G., Dotu, I., Zhou, H., Yang, X.-L., Schimmel, P., . . . Ackerman, S. L. (2014). Ribosome stalling induced by mutation of a CNS-specific tRNA causes neurodegeneration. *Science, 345*(6195), 455 LP–459. https://doi.org/10.1126/science.1249749

Jan, A., Jansonius, B., Delaidelli, A., Bhanshali, F., An, Y. A., Ferreira, N., . . . Sorensen, P. H. (2018). Activity of translation regulator eukaryotic elongation factor-2 kinase is increased in Parkinson disease brain and its inhibition reduces alpha synuclein toxicity. *Acta Neuropathologica Communications, 6*(1), 54. https://doi.org/10.1186/s40478-018-0554-9

Jobe, A., Liu, Z., Gutierrez-Vargas, C., & Frank, J. (2019). New insights into ribosome structure and function. *Cold Spring Harbor Perspectives in Biology, 11*(1), 1–18. https://doi.org/10.1101/cshperspect.a032615

Johanns, M., Pyr dit Ruys, S., Houddane, A., Vertommen, D., Herinckx, G., Hue, L., . . . Rider, M. H. (2017). Direct and indirect activation of eukaryotic elongation factor 2 kinase by AMP-activated protein kinase. *Cellular Signalling, 36*, 212–221. https://doi.org/10.1016/J.CELLSIG.2017.05.010

Kandel, E. R. (2001). The molecular biology of memory storage: A dialogue between genes and synapses. *Science, 294*(5544), 1030 LP–1038. https://doi.org/10.1126/science.1067020

Kandel, E. R. (2012). The molecular biology of memory: CAMP, PKA, CRE, CREB-1, CREB-2, and CPEB. *Molecular Brain, 5*(1), 1. https://doi.org/10.1186/1756-6606-5-14

Kapur, M., Monaghan, C. E., & Ackerman, S. L. (2017). Regulation of mRNA translation in neurons—: A matter of life and death. *Neuron, 96*(3), 616–637. https://doi.org/10.1016/j.neuron.2017.09.057

Kavalali, E. T., & Monteggia, L. M. (2015). How does ketamine elicit a rapid antidepressant response? *Current Opinion in Pharmacology, 20*, 35–39. https://doi.org/10.1016/J.COPH.2014.11.005

Kenney, J. W., Genheden, M., Moon, K. M., Wang, X., Foster, L. J., & Proud, C. G. (2016). Eukaryotic elongation factor 2 kinase regulates the synthesis of microtubule-related proteins in neurons. *Journal of Neurochemistry, 136*(2), 276–284. https://doi.org/10.1111/jnc.13407

Kenney, J. W., Moore, C. E., Wang, X., & Proud, C. G. (2014). Eukaryotic elongation factor 2 kinase, an unusual enzyme with multiple roles. *Advances in Biological Regulation, 55*, 15–27. https://doi.org/10.1016/j.jbior.2014.04.003

Kenney, J. W., Sorokina, O., Genheden, M., Sorokin, A., Armstrong, J. D., & Proud, C. G. (2015). Dynamics of elongation factor 2 kinase regulation in cortical neurons in response to synaptic activity. *Journal of Neuroscience*, *35*(7), 3034–3047. https://doi.org/10.1523/jneurosci.2866-14.2015

Klencklen, G., Després, O., & Dufour, A. (2012). What do we know about aging and spatial cognition? Reviews and perspectives. *Ageing Research Reviews*, *11*(1), 123–135. https://doi.org/10.1016/J.ARR.2011.10.001

Knebel, A., Morrice, N., & Cohen, P. (2001). A novel method to identify protein kinase substrates: eEF2 kinase is phosphorylated and inhibited by SAPK4/p38δ. *EMBO Journal*, *20*(16), 4360–4369. https://doi.org/10.1093/emboj/20.16.4360

Koltun, B., Ironi, S., Gershoni-Emek, N., Barrera, I., Hleihil, M., Nanguneri, S., . . . & Rosenblum, K. (2020). Measuring mRNA translation in neuronal processes and somata by tRNA-FRET. *Nucleic Acids Research*, *48*(6), e32. https://doi.org/10.1093/nar/gkaa042.

Lenz, G., & Avruch, J. (2005). Glutamatergic regulation of the p70S6 kinase in primary mouse neurons. *Journal of Biological Chemistry*, *280*(46), 38121–38124. https://doi.org/10.1074/jbc.C500363200

Li, X., Alafuzoff, I., Soininen, H., Winblad, B., & Pei, J. J. (2005). Levels of mTOR and its downstream targets 4E-BP1, eEF2, and eEF2 kinase in relationships with tau in Alzheimer's disease brain. *FEBS Journal*, *272*(16), 4211–4220. https://doi.org/10.1111/j.1742-4658.2005.04833.x

Ling, C., & Ermolenko, D. N. (2016). Structural insights into ribosome translocation. *Wiley Interdisciplinary Reviews: RNA*, *7*(5), 620–636. https://doi.org/10.1002/wrna.1354

Ma, T., Chen, Y., Vingtdeux, V., Zhao, H., Viollet, B., Marambaud, P., & Klann, E. (2014). Inhibition of AMP-activated protein kinase signaling alleviates impairments in hippocampal synaptic plasticity induced by amyloid β. *Journal of Neuroscience*, *34*(36), 12230–12238. https://doi.org/10.1523/JNEUROSCI.1694-14.2014

Manning, G., Plowman, G. D., Hunter, T., & Sudarsanam, S. (2002). Evolution of protein kinase signaling from yeast to man. *Trends in Biochemical Sciences*, *27*(10), 514–520. https://doi.org/10.1016/S0968-0004(02)02179-5

Marin, P., Nastiuk, K. L., Daniel, N., Girault, J. A., Czernik, A. J., Glowinski, J., . . . Prémont, J. (1997). Glutamate-dependent phosphorylation of elongation factor-2 and inhibition of protein synthesis in neurons. *Journal of Neuroscience*, *17*(10), 3445–3454. https://doi.org/10.1523/jneurosci.17-10-03445.1997

Martinetz, S., Meinung, C. P., Jurek, B., von Schack, D., van den Burg, E. H., Slattery, D. A., & Neumann, I. D. (2019). De novo protein synthesis mediated by the eukaryotic elongation factor 2 is required for the anxiolytic effect of oxytocin. *Biological Psychiatry*, *85*(10), 802–811. https://doi.org/10.1016/j.biopsych.2019.01.010

Mauelshagen, J., Sherff, C. M., & Carew, T. J. (1998). Differential induction of long-term synaptic facilitation by spaced and massed applications of serotonin at sensory neuron synapses of Aplysia californica. *Learning and Memory*, *5*(3), 246–256. https://doi.org/10.1101/lm.5.3.246

McCamphill, P. K., Farah, C. A., Anadolu, M. N., Hoque, S., & Sossin, W. S. (2015). Bidirectional regulation of eEF2 phosphorylation controls synaptic plasticity by decoding neuronal activity patterns. *Journal of Neuroscience*, *35*(10), 4403–4417. https://doi.org/10.1523/JNEUROSCI.2376-14.2015

McCamphill, P. K., Ferguson, L., & Sossin, W. S. (2017). A decrease in eukaryotic elongation factor 2 phosphorylation is required for local translation of sensorin and long-term facilitation in Aplysia. *Journal of Neurochemistry*, *142*(2), 246–259. https://doi.org/10.1111/jnc.14030

Merrick, W. C. (2010). Eukaryotic protein synthesis: Still a mystery. *Journal of Biological Chemistry, 285*(28), 21197–21201. https://doi.org/10.1074/jbc.R110.111476

Miyashiro, K., Dichter, M., & Eberwine, J. (1994). On the nature and differential distribution of mRNAs in hippocampal neurites: Implications for neuronal functioning. *Proceedings of the National Academy of Sciences of the United States of America, 91*(23), 10800–10804. https://doi.org/10.1073/pnas.91.23.10800

Montanaro, L., Sperti, S., Testoni, G., & Mattioli, A. (1976). Effect of elongation factor 2 and of adenosine diphosphate ribosylated elongation factor 2 on translocation. *Biochemical Journal, 156*(1), 15–23. https://doi.org/10.1042/bj1560015

Monteggia, L. M., Gideons, E., & Kavalali, E. T. (2013). The role of eukaryotic elongation factor 2 kinase in rapid antidepressant action of ketamine. *Biological Psychiatry, 73*(12), 1199–1203. https://doi.org/10.1016/j.biopsych.2012.09.006

Nairn, A. C., Bhagat, B., & Palfrey, H. C. (1985). Identification of calmodulin-dependent protein kinase III and its major M(r) 100,000 substrate in mammalian tissues. *Proceedings of the National Academy of Sciences of the United States of America, 82*(23), 7939–7943. https://doi.org/10.1073/pnas.82.23.7939

Ounallah-Saad, H., Sharma, V., Edry, E., & Rosenblum, K. (2014). Genetic or pharmacological reduction of PERK enhances cortical-dependent taste learning. *Journal of Neuroscience, 34*(44), 14624–14632. https://doi.org/10.1523/JNEUROSCI.2117-14.2014

Park, S., Park, J. M., Kim, S., Kim, J. A., Shepherd, J. D., Smith-Hicks, C. L., . . . Worley, P. F. (2008). Elongation factor 2 and fragile x mental retardation protein control the dynamic translation of Arc/Arg3.1 essential for mGluR-LTD. *Neuron, 59*(1), 70–83. https://doi.org/10.1016/j.neuron.2008.05.023

Proud, C. G. (2013). MTORC1 regulates the efficiency and cellular capacity for protein synthesis. *Biochemical Society Transactions, 41*(4), 923–926. https://doi.org/10.1042/BST20130036

Ramocki, M. B., & Zoghbi, H. Y. (2008). Failure of neuronal homeostasis results in common neuropsychiatric phenotypes. *Nature, 455*(7215), 912–918. https://doi.org/10.1038/nature07457

Richter, J. D., Bassell, G. J., & Klann, E. (2015). Dysregulation and restoration of translational homeostasis in fragile X syndrome. *Nature Reviews Neuroscience, 16*, 595. Retrieved from https://doi.org/10.1038/nrn4001

Rosenberg, T., Gal-Ben-Ari, S., Dieterich, D. C., Kreutz, M. R., Ziv, N. E., Gundelfinger, E. D., & Rosenblum, K. (2014). The roles of protein expression in synaptic plasticity and memory consolidation. *Frontiers in Molecular Neuroscience, 7*(November), 1–14. https://doi.org/10.3389/fnmol.2014.00086

Rosenblum, K., Meiri, N., & Dudai, Y. (1993). Taste memory: The role of protein synthesis in gustatory cortex. *Behavioral and Neural Biology, 59*(1), 49–56. https://doi.org/10.1016/0163-1047(93)91145-D

Ryazanov, A. G. (2002). Elongation factor-2 kinase and its newly discovered relatives. *FEBS Letters, 514*(1), 26–29. https://doi.org/10.1016/S0014-5793(02)02299-8

Ryazanov, A. G., Natapov, P. G., Shestakova, E. A., Severin, F. F., & Spirin, A. S. (1988). Phosphorylation of the elongation factor 2: The fifth Ca2+/calmodulin-dependent system of protein phosphorylation. *Biochimie, 70*(5), 619–626. https://doi.org/10.1016/0300-9084(88)90245-3

Ryazanov, A. G., Pavur, K. S., & Dorovkov, M. V. (1999). Alpha-kinases: A new class of protein kinases with a novel catalytic domain. *Current Biology, 9*(2), 43–45. https://doi.org/10.1016/s0960-9822(99)80006-2

Scheetz, A. J., Nairn, A. C., & Constantine-Paton, M. (2000). NMDA receptor-mediated control of protein synthesis at developing synapses. *Nature Neuroscience*, *3*(3), 211–216. https://doi.org/10.1038/72915

Sharma, V., Ounallah-Saad, H., Chakraborty, D., Hleihil, M., Sood, R., Barrera, I., . . . Rosenblum, K. (2018). Local inhibition of PERK enhances memory and reverses age-related deterioration of cognitive and neuronal properties. *Journal of Neuroscience*, *38*(3), 648–658. https://doi.org/10.1523/JNEUROSCI.0628-17.2017

Smith, W. B., Aakalu, G., & Schuman, E. M. (2001). Local protein synthesis in neurons. *Current Biology*, *11*(22), 901–903.

Sonenberg, N., & Hinnebusch, A. G. (2009). Regulation of translation initiation in eukaryotes: Mechanisms and biological targets. *Cell*, *136*(4), 731–745. https://doi.org/10.1016/j.cell.2009.01.042

Stein, K. C., & Frydman, J. (2019). The stop-and-go traffic regulating protein biogenesis: How translation kinetics controls proteostasis. *Journal of Biological Chemistry*, *294*(6), 2076–2084. https://doi.org/10.1074/jbc.REV118.002814

Stern, E., Chinnakkaruppan, A., David, O., Sonenberg, N., & Rosenblum, K. (2013). Blocking the eIF2α kinase (PKR) enhances positive and negative forms of cortex-dependent taste memory. *Journal of Neuroscience*, *33*(6), 2517–2525. https://doi.org/10.1523/JNEUROSCI.2322-12.2013

Steward, O., & Levy, W. B. (1982). Preferential localization of polyribosomes under the base of dendritic spines in granule cells of the dentate gyrus. *Journal of Neuroscience*, *2*(3), 284–291. https://doi.org/10.1523/jneurosci.02-03-00284.1982

Sutton, M. A., Taylor, A. M., Ito, H. T., Pham, A., & Schuman, E. M. (2007). Postsynaptic decoding of neural activity: eEF2 as a biochemical sensor coupling miniature synaptic transmission to local protein synthesis. *Neuron*, *55*(4), 648–661. https://doi.org/10.1016/j.neuron.2007.07.030

Taha, E., Gildish, I., Gal-Ben-Ari, S., & Rosenblum, K. (2013). The role of eEF2 pathway in learning and synaptic plasticity. *Neurobiology of Learning and Memory*, *105*, 100–106. https://doi.org/10.1016/j.nlm.2013.04.015

Teige, M., Scheikl, E., Reiser, V., Ruis, H., & Ammerer, G. (2001). Rck2, a member of the calmodulin-protein kinase family, links protein synthesis to high osmolarity MAP kinase signaling in budding yeast. *Proceedings of the National Academy of Sciences of the United States of America*, *98*(10), 5625–5630. https://doi.org/10.1073/pnas.091610798

Tuller, T., Carmi, A., Vestsigian, K., Navon, S., Dorfan, Y., Zaborske, J., . . . Pilpel, Y. (2010). An evolutionarily conserved mechanism for controlling the efficiency of protein translation. *Cell*, *141*(2), 344–354. https://doi.org/10.1016/j.cell.2010.03.031

Verpelli, C., Piccoli, G., Zibetti, C., Zanchi, A., Gardoni, F., Huang, K., . . . Sala, C. (2010). Synaptic activity controls dendritic spine morphology by modulating eEF2-dependent BDNF synthesis. *Journal of Neuroscience*, *30*(17), 5830–5842. https://doi.org/10.1523/JNEUROSCI.0119-10.2010

Walden, W. E., & Thach, R. E. (1986). Translational control of gene expression in a normal fibroblast. Characterization of a subclass of mRNAs with unusual kinetic properties. *Biochemistry*, *25*(8), 2033–2041. https://doi.org/10.1021/bi00356a030

Weatherill, D. B., McCamphill, P. K., Pethoukov, E., Dunn, T. W., Fan, X., & Sossin, W. S. (2011). Compartment-specific, differential regulation of eukaryotic elongation factor 2 and its kinase within Aplysia sensory neurons. *Journal of Neurochemistry*, *117*(5), 841–855. https://doi.org/10.1111/j.1471-4159.2011.07251.x

Werner, C. T., Stefanik, M. T., Milovanovic, M., Caccamise, A., & Wolf, M. E. (2018). Protein translation in the nucleus accumbens is dysregulated during cocaine withdrawal and required for expression of incubation of cocaine craving. *Journal of Neuroscience*, *38*(11), 2683–2697. https://doi.org/10.1523/JNEUROSCI.2412-17.2018

Wolf, M. E. (2016). Synaptic mechanisms underlying persistent cocaine craving. *Nature Reviews Neuroscience*, *17*, 351. Retrieved from https://doi.org/10.1038/nrn.2016.39

Yamaguchi, H., Matsushita, M., Nairn, A. C., & Kuriyan, J. (2001). Crystal structure of the atypical protein kinase domain of a TRP channel with phosphotransferase activity. *Molecular Cell*, *7*(5), 1047–1057. https://doi.org/10.1016/S1097-2765(01)00256-8

Yang, W., Zhou, X., & Ma, T. (2019). Memory decline and behavioral inflexibility in aged mice are correlated with dysregulation of protein synthesis capacity. *Frontiers in Aging Neuroscience*, *11*(September), 1–9. https://doi.org/10.3389/fnagi.2019.00246

Yiannakas, A., & Rosenblum, K. (2017). The insula and taste learning. *Frontiers in Molecular Neuroscience*, *10*(November), 1–24. https://doi.org/10.3389/fnmol.2017.00335

Zhou, W., Wang, N., Yang, C., Li, X.-M., Zhou, Z.-Q., & Yang, J.-J. (2014). Ketamine-induced antidepressant effects are associated with AMPA receptors-mediated upregulation of mTOR and BDNF in rat hippocampus and prefrontal cortex. *European Psychiatry*, *29*(7), 419–423. https://doi.org/10.1016/j.eurpsy.2013.10.005

CHAPTER 5

SIDEKICK NO MORE

Neural Translation Control by p70 ribosomal S6 kinase 1

ADITI BHATTACHARYA

A Historical Timeline of S6K1 Research in Neurobiology

Ribosomal S6 kinase B1 (gene ID: 6198, EC 2.7.11.1), or p70 ribosomal S6 kinase 1 (S6K1), is a well-known serine-threonine kinase belonging to the AGC kinase superfamily (Pearce, Kommander, & Alessi, 2010). It is a member of a family of related kinases that share the nominal ability to phosphorylate S6 ribosomal protein but have differing roles in cell biology. The S6 kinase A group is commonly referred to as the p90 RSK or p90^{rsk} kinases. The kinases of the group have four isoforms, containing an extra kinase domain, and are almost exclusively activated by mitogen-activated kinases and mainly influence transcription (Frödin & Gammeltoft, 1999). The S6 kinase B group includes two members, namely S6K1 and S6K2, which can also be referred to as p70^{rsk}. The S6K1 and 2 split likely resulted from a gene duplication and translocation event, placing the two isoforms on different chromosomes with highly a conserved kinase domain but with divergent sequences at the N- and C-termini. This high sequence conservation at the business area of the kinases is often cited as a reason for studying S6K1 more than S6K2. However, there is a growing body of compelling evidence, stemming largely from cancer studies, that show that S6K1 and S6K2 have different cellular localization, interactomes, and disease involvement (see Pardo & Seckl, 2013 for an in-depth review). While there is an inherent distinction in the roles played by the two S6Ks, individual knock down of S6K1 or S6K2 induce compensatory upregulation of the other kinase (Nardella et al., 2011), while deletion of both causes perinatal lethality in mice (Pende et al., 2004). Due to the presence of two nuclear localization sequences, S6K2 is largely thought to regulate transcription and cell fate determination programs. There are probably a handful of studies on the exclusive role of S6K2 in the brain. Therefore, this review will discuss studies on the involvement of S6K1 in brain function in health and disease.

Contrary to the popular notion of S6K1 being an acolyte of the mechanistic target of rapamycin complex 1 (mTORC1) and protein kinase B (Akt), its existence was indicated, prior to the formal discovery mTOR (as early as 1980s), by multiple studies pertaining to growth factor signaling (Roberts & Morelos, 1982; Wettenhall, Chesterman, Walker, & Morgan, 1983; Novak-Hofer & Thomas, 1984). The kinase was subsequently cloned from rat liver extracts in 1990 (Banerjee et al., 1990; Kozma et al., 1990). In the interim, studies in neuroscience indicate that S6K1 was implicated in insulin and fibroblast growth factor signaling in astrocytes (Pierre, Toru-Delbauffe, Gavaret, Pomerance, & Jacquemin, 1986; Gavaret et al., 1989) as the only bonafide kinase to phosphorylate S6 protein that is a constituent of the 40S ribosomal unit. The first report involving S6K1 in neurons identified this kinase to be important in insulin signaling in cultured fetal neurons as well as distinct from protein kinase-C and cAMP-dependent protein kinase (Heidenreich & Toledo, 1989). In an oft-cited paper, Jefferies et al. (1997) provided a critical input that S6K1 activity was rapamycin-sensitive and hence activated by mTOR, which was then corroborated across multiple cell lines and model systems. Another finding of this study was that S6K1/2 controls translation of 5′ TOP mRNA, which subsequently has been disproved by evidence from cells that lack S6K1 and S6K2 but still show robust 5′TOP mRNA translation.

In the1990s, tremendous strides were made in parallel in the fields of molecular mechanisms of learning and memory (see Kandel, 2001; Malenka & Nicoll, 1999), identifying the translation control components and their sequence of action for proper translation (see Costa-Mattioli, Sossin, Klann, & Sonenberg, 2009; Kozak, 1992) in non-neuronal systems. Pioneering work was also done in uncovering the role of translation in learning and memory (Kang & Schuman, 1996) implicating signaling cascades that eventually were shown to require S6K1 activation. Therefore, by early 2000s, S6K1 was accepted widely as an influencer in synaptic protein synthesis. With the development of the S6K1, S6K2 knock-out mouse (Shima et al., 1998; Pende et al., 2004), researchers gained a tool to directly examine the role of S6K1 or S6K1 or both at a systems level. This facilitated a spate of studies that dissected the role of this kinase in many neurobiological phenomena that I discuss later in the chapter. This genetic deletion has also been used to dissect the role of S6K1 in multiple disease models of autism, Parkinson's disease, Alzheimer's disease, and schizophrenia (Bhattacharya et al., 2012; Caccamo et al., 2015; Bowling et al., 2014; Liu & Lu, 2010). Another major enabler to S6K1 research arrived in 2010 with Pfizer introducing the first small molecule inhibitor, PF 4708671 (Pearce et al., 2010b). This was followed by a handful of other reports of new blockers identified by using rational drug design approaches (discussed later in the chapter). These reagents offer methods to dissect S6K1 function in a larger array of contexts and may yield candidates for translation into clinical arenas.

As compared to the massive body of work that has focused on mTORC1 and ERK in neurobiology, S6K1 remains understudied and is largely considered one of the conduits of these critical hub kinases. Of the approximately 500 papers published involving S6K1 actions in the brain, only a handful dissect the role of this kinase alone. In contrast, there are more than 5000 papers on S6K1/2 that deal with longevity, caloric balance, muscle maintenance and memory etc., all outside the nervous system. It is perhaps time we started to closely examine the role of S6K1 in neural maintenance and function. This review is focused mainly on dissecting S6K1 mediated control of neural translation, its effects on learning, memory, behavior and its dysregulation in disease. In doing so, the review will attempt to highlight confounds, current limitations and possible modes of investigation for the future.

S6K1 Signaling and Synaptic Plasticity

Like most AGC kinases, resting S6K1 has a bilobed fold arrangement in which the N- and C-termini interact with each other, hiding the catalytic domain within (Sunami et al., 2010). The beginning of the C-terminal harbors an activation T-loop, which includes the serine threonine residues that are targets of upstream kinases and mediate the conformation change that renders the kinase domain and active site accessible. Hence S6K1 requires a carefully orchestrated set of activating signals that lead to phosphorylation of its multiple serine-threonine residues (see Magnuson, Ekim, & Fingar, 2012 for extensive explanation). This allows S6K1 to serve as a signal aggregator of many signals and stimuli initiated by a plethora of cell surface receptors. In neuronal cells, S6K1 activation has been primarily studied downstream of Trk-B- BDNF, Insulin-like growth factor (IGF1/2), metabotropic glutamate 1/5 receptor (mGluR5), and dopamine D1/2 receptors (Becker, Ibrahim, Cui, Lee, & Yee, 2011; Bhattacharya et al., 2012; Bowling et al., 2014; Zhou, Lin, Zheng, Sutton, & Wang, 2010).

Neurotrophin and growth factor signaling stimulation canonically activates phosphatidyl inositol 3 kinase (PI-3K) and protein kinase B (or Akt) and PDK1. As receptor tyrosine kinases, growth factor signaling also activates the Ras-Raf pathway that induces mitogen-activated kinase (MAK kinase) signaling. Akt inhibits the TSC1-2 complex that then relieves the brake on mTORC1, causing phosphorylation of Thr 389, which strongly cooperates with the Thr229 site phosphorylation mediated by PDK1 (Magnuson et al., 2012). MAP kinase activation of S6K1 has been mainly studied in context of extracellular—signal regulated kinase (ERK1/2) which phosphorylates Ser 421 on S6K1 (Lehman, Calvo, & Gomez-Cambronero, 2003). ERK mediated phosphorylation may occur through influencing TSC1-2 activity differently as reported by Winter, Jefferson, & Kimball (2011). Apart from mTORC1 and ERK1/2, S6K1 may also be activated by Rho-GTP-cdc 42 signaling and atypical protein kinase C isoforms λ and ζ, though the precise phosphorylation sites have yet to be explored in great detail (Fang et al., 2003; Chuo and Blenis, 1996; Romanelli, Martin, Toker, & Blenis, 1999). S6K1 is also reported to be activated at Thr 371 by glycogen synthase kinase 3β (GSK-3 β) facilitated by mTORC1 (Shin, Wolgamott, Yub, Blenis, & Yoona, 2011). S6K1 activity is regulated by dephosphorylation by PP2A in which the kinase forms a complex (Hahn, Miranda, Francis, Vendrell, Zorzano, & Teleman., 2010). Acetylation by p300/CBP (cAMP responsive element binding protein binding protein) and PCAF (p300/CBP associated factor) at the C-terminal domain which antagonizes Thr 389 phosphorylation and kinase activity has been reported to occur in response to Sirtuin signaling (Hong, Zhao, Lombard, Fingar, & Inoki, 2014). S6K1 also undergoes ubiquitination via ROC1 action in response to mitogenic activity, which may serve as a negative feedback to S6K1 signaling (Wang et al., 2008). These latter mechanisms have not been investigated in any detail in a neurobiological context.

Once active, S6K1 phosphorylates an array of targets which channel pro-anabolic signals downstream to translation, mRNA quality control, transcription, ribosome biogenesis leading to cell growth, motility, differentiation, synaptic plasticity and changes in neuronal morphology. Relevant to mRNA processing and translation are eukaryotic initiation factor 4B (eIF4B), eukaryotic elongation factor 2 (eEF2) via suppressing eEF2 kinase, PDCD4, SKAR and the traditional ribosomal protein S6 (Magnuson et al., 2012). S6K1 was also purported to phosphorylate Fragile X Mental Retardation Protein (FMRP), a key translation regulatory protein (Narayanan et al., 2008). It was believed to be the only known kinase for FMRP before work by Bartley, O'Keefe, & Bordey (2014) disproved this assertion. In this study using two knockout of Tsc1, the authors elevated mTORC1-S6K1 signaling, which did not affect Ser499 phosphorylation of FMRP. They also showed that this phosphorylation remains intact in *S6K1* KO mice. Concurrent phosphorylation of eIF4B (activating) and PDCD4 (inhibiting) increase the helicase activity of eIF4A, which then promotes translation initiation. At the same time, phosphorylation of eEF2K by S6K1 inactivates it, which then relieves the dampening of eEF2 which in its dephosphorylated form mediates translation of ribosome of the loaded mRNA. SKAR serves as a conduit to recruiting S6K1 to exon junction complexes of newly processed mRNA for serving as a quality controller for pioneer round of translation (Ma, Yoon, Richardson, Julich, & Blenis, 2008). Ribosomal S6 protein is phosphorylated by S6K1, but is also a target of RSK sister family of kinases. Feedback signaling of S6K1 attenuates growth factor signaling by way of IRS-1, mTORC1 by acting on Raptor and GSK-3 and on mTORC2 by phosphorylating Rictor (Harrington et al., 2004; Julien, Carriere, Moreau, & Roux, 2010).

The role of mTORC1 and ERK 1/2 in early and long-term plasticity has been exhaustively studied (see Graber, McCamphill, & Sossin, 2013; Kelleher, Govindarajan, & Tonegawa, 2004). This was largely made possible by the availability of specific small molecule inhibitors (like rapamycin and SL-327) that are blood-brain-barrier permeable. In such conditions of mTORC1 and ERK blockade, S6K1 has been mostly used as a readout examining levels of phosphorylated Thr 389 or 421 with the assumption of almost linear signal transmission. This has led to a skewed informational database on S6K1 activity, which suggests its involvement in many synaptic plasticity paradigms *ipso facto* mTOR and ERK activation. Only recently have investigators commenced using S6K1 specific blocker PF-4708671, so most previous work has been with the *S6K1* KO mouse. Antion, Hou, Wong, Hoeffer, & Klann, (2008a) investigated traditional long-term potentiation (LTP) paradigms in this *S6K1* KO mice to find that the deletion of S6K1 did not impair L-LTP evoked by 4 high frequency trains stimuli (4x HFS), 2x HFS and theta-burst stimulation. Intriguingly, the authors found that S6K1 deletion impaired E-LTP, which is believed to be protein-synthesis independent. This may indirectly be contributed by the abrogation of negative feedback to IRS-1, leading to a pervasive increase of Akt phosphorylation. Recently, abrogation of S6K1 target- Serine 366 site on eEF2K in Aplysia has shown that it impairs long-term facilitation (LTF, McCamphill, Ferguson, & Sossin, 2017).

The protein synthesis dependence of LTD in the hippocampus is well documented (Huber, Kayser, & Bear, 2000; Graber et al., 2013). While there exists much debate on whether translation initiation or elongation is the greater contributor to LTD, evidence for the contribution of S6K1 to the process is conflicting, at best. Reports show that rapamycin and W13 (inhibiting CamKII) incubation of WT hippocampal slices does abrogate LTD and decreased S6K1 activation (Sethna et al., 2016). Additionally, mice with δRGD deletion of TSC2 (Chevere-Torres et al., 2012) showed no appreciable S6K1 activation but had enhanced ERK1/2 signaling, yet impaired LTD. However. induction remains intact with an increase in S6rp phosphorylation and translation of elongation factor 1 alpha in S6K1 KO LTD (eIF1a; Antion et al., 2008a). Subsequently, S6K1 knock down was shown to have no effect on manifestation of LTD induced by mGluR5 activation while also rescuing exaggerated LTD in Fmr 1 KO model of Fragile X syndrome (Bhattacharya et al., 2012). Therefore, it appears that in conditions where S6K1 is developmentally ablated, LTD expression is possibly routed through sister kinases like S6K2, Rsk etc. In conditions where S6K1 exists and is acutely perturbed by targeting signaling partners, LTD and perhaps LTP is impaired. This may arise from S6K1 being present in multiple signaling complexes that change according to stimulus and localization. An interesting example of such dynamics is provided by Bernard et al. (2013), where early life seizure induced enhancement in mGluR-LTD is S6K1 and protein synthesis dependent. Activated S6K1 levels in this model are not enhanced immediately after the insult, but accumulate slowly, showing a significant increase by postnatal day 60. S6K1 was found to be in a dynamic complex with FMRP and protein-phosphatase 2A (PP2A) which changed localization from cytosolic to synaptic compartments on the induction of LTD.

Though S6K1 is ubiquitously expressed, important in a variety of tissues, the available commercial antibodies do not work well in immunohistochemical experiments and have largely been used for western blot analyses. In the mouse brain, in situ hybridization using antisense probe to *S6K1* reveals high expression in the hippocampal formation and sparse expression across the rest of the areas (http://mouse.brain-map.org/gene/show/48349; Lein et al., 2007), which is perplexing given notable levels of S6K1 detected in striatum, cortex, and cerebellum. Within a neuron again, epitope tagged S6K1 have been expressed that show localization in the cell body and neurites (Cammalleri et al., 2003). A very recent study shows that application of PF-4708671 can reduce S6 ribosomal protein phosphorylation across the post synaptic cell (Pirbhoy, Farris, Steward, 2017). Hence data of direct S6K1 visualization is limited and is an important area of research given the extensive nature of decentralized function in neuron and glia (Jung, Gkogkas, Sonenberg & Holt, 2014). In terms of translation, and in response to a specific stimulus, the amount of somatic and dendritic protein synthesis that is controlled by localized pools of S6K1 is still very limited. An associated question is, do more somatically localized S6K1 bear greater contribution to transcription over translation? Finally, it is unclear if only certain neurons are recruited in response to a specific stimuli (like in fear conditioning) and hence tracking S6K1 activity at whole tissue level would again dilute effects. In summary, though much is purported to be known about S6K1 signaling and synaptic plasticity, many confounds and contradictions still persist.

S6K1, Translation and Structural Plasticity of Neurons

On the heels of the finding that mTORC1 and ERK1/2 underwrite many critical aspects of LTP and LTD came reports of their implication in structural plasticity of spines and dendritic arbors (Wu, Deisseroth, & Tsein, 2001, Jaworski & Sheng, 2006). Subsequently, several studies linked local and general protein synthesis to spine changes and axon regeneration (Piper et al., 2006; Verma et al., 2005). In 2005, Jaworski, Spangler, Seeburg, Hoogenraad, & Sheng demonstrated how over activation of Akt, or depression of mTORC1 and S6K1 modulated dendritic arbors and neuron morphology in cultured cells. Jointly, S6K1 had already been implicated in cell size and longevity (Pende et al., 2004; Selman et al., 2009). A sufficient number of studies show clear implication of S6K1 in structural morphology in neurons, some independent of the involvement of protein synthesis.

Evidence of a translation-driven role for S6K1 in structural plasticity is manifold. Just overexpressing a constitutively active (T389E) species of S6K1 can increase neuronal complexity (Dwyer, Maldonado-Aviles, Lepack, DiLeone, & Duman, 2015). S6K1 inhibition by either genetic deletion or pharmacological inhibition, normalized both enhanced protein synthesis and increased filopodial spines in the mouse model of fragile X syndrome (FXS; Bhattacharya et al., 2012; Bhattacharya et al., 2016). Surprisingly, in *S6K1* KO, no appreciable size difference was seen in neurons or in spine density from littermate wild-type (Bhattacharya et al., 2012). But this was only one brain area and done at one specific age. In the case of wild-type CA1 neurons though, S6K1 inhibition did cause an increase in spine density (Bhattacharya et al., 2016). Bowling et al. (2014), demonstrated the Haloperidol, a typical antipsychotic, increased protein synthesis in striatal neurons and dendritic morphology that was substantially reduced upon expression of an shRNA against S6K1. This was also found to decrease Haloperidol-induced increases in translation. A proteomic survey done in this study showed proteins that regulated cytoskeletal rearrangements and scaffold proteins that hold up spines are maximally activated upon Haloperidol stimulation, implicating a role of S6K1 in this process. In Angelman Syndrome model mice, harboring a deletion of UBE3A proteosomal unit, hippocampal slices treated with PF-4708671 (specific S6K1 inhibitor) normalized actin polymerization, LTP, similar to rapamycin (Sun et al., 2016) in indicating that normalizing proteostasis leads to changes in spines as well. Ancillary to these spine related effects, a report by Yang et al., (2014) demonstrated that S6K1 and 4E-binding protein (4E-BP) play different roles in axon regeneration in the murine optic nerve crush model. S6K1 activation in this model promotes regeneration while a complimentary study (Gong et al., 2015) showed that activating 4E-BP or repressing S6K1-GSK3β impedes the process by interfering with mTORC1 signaling and translation. Finally, S6K1 dependent translation of myelin basic protein was reported by Michel, Zhao, Karl, Lewis, Fyffe-Maricich (2015), shown to be more ERK-dependent than mTORC1. These data are summarized in Figure 5.1 showing protein synthesis dependent interactions in violet arrows.

In contrast, there also exist studies that showcase S6K1's role in structural plasticity independent of translation control (shown in Figure 5.1 with green arrows). As early as 1998, Burnett et al., using a yeast-two hybrid screen had identified Neurabin (Neural tissue specific binding protein) as a binding partner for S6K1. Neurabin was subsequently found to be a neuronal spinophilin, which interact with last five amino acids of S6K1 and increases its phosphotransferase activity. Subsequently, Buchsbaum, Connolly, & Feig, 2003, verified that spinophilin binding brings S6K1 in proximity of Rac, activating this pathway that is critical to changes in spine architecture. S6K1 has been shown to phosphorylate Rictor in the mTORC2 complex (known to regulate actin remodeling) in its feedback to Akt in non-neuronal systems (Julien et al., 2010). Contradicting this is the report that in the Rictor knock out mice there are spine changes with no concomitant change in S6K1 activation levels (Huang et al., 2013), leaving the room for further scrutiny of this signaling loop in the future. Further support for an mTOR independent pathway to S6K1 that impinges in spinogenesis was provided by reports like Lai, Liang, Fei, Huang, & Ip (2015) where BDNF- induced spine changes were shown to involve cyclin dependent kinase 5 (Cdk 5) which phosphorylated S6K1 at Ser 411, leaving the Thr 389 site untouched. In this pathway S6rp seemed to be the downstream S6K1 target rather than eEF2K. It remains to be verified how S6rp causes no changes in ribosomal processivity in this case. A very recent report (Al-Ali et al., 2017) of spinal injury in mice reported that S6K1 inhibition promoted regeneration and is a potential "druggable" target for spinal injury conditions. The authors suggest that PF-4708671 application or siRNA knock down of S6K1 in cultured hippocampal cells increased neurite outgrowth. This blockade of S6K1 activates the feedback loop of S6K1 to PI3K which in turn activates mTORC1 and 2, though it abrogates any translation dependent effects downstream of S6K1.

Taken together, it is clear that S6K1 intimately influences neuronal architecture and structural plasticity. Given the neuro-centric thrust of most studies, it is likely that the

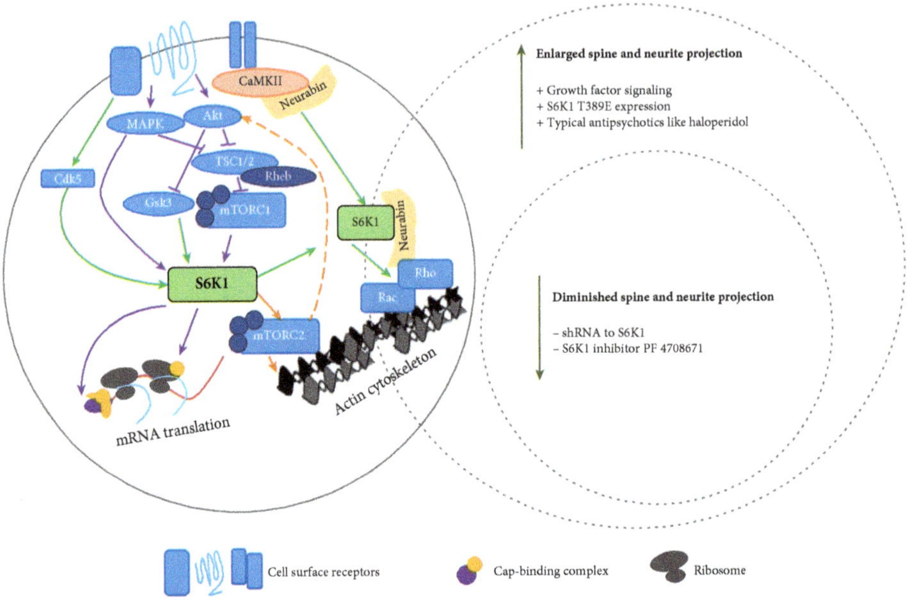

FIGURE 5.1. Multi-modal influences of S6K1 on the structural plasticity of neuronal systems. S6K1 mediates spine and dendritic morphology via translation-dependent and independent mechanisms. Downstream of MAPK, Akt, and mTORC1, S6K1 influences the synthesis of proteins involved in synaptic scaffolding, actin remodeling, and energetic regulation (violet lines). S6K1 also phosphorylates the mTORC2 complex and feedback to Akt in some systems that again impacts actin remodeling (orange lines). Finally, S6K1 interacts with a spinophilin, Neurabin, integrating signals via GSK-3 or Cdk5, to Rac-Rho cascades known to be responsible for structural plasticity (green lines). Evidence for increased spine and neurite projection involving S6K1 includes growth factor signaling (e.g., BDNF), expression of constitutively active S6K1, and action of Dopamine D2 receptor antagonists. Conversely shRNA or active site blockers of S6K1 decrease spine and neurite projection and spine density.

same effect should show up in other neural cells as well. S6K1 can modulate this process both by translation and translation-independent modes, though the downstream effects through both mechanism remain to be worked out.

S6K1 in Behavior

Behavioral studies have provided the keenest insight into signaling distinctly mediated by S6K1. In this section, behavioral results of modulating S6K1 in wild type animals will be discussed, while the following section addresses the effects of S6K1 in rodent disease models. A key but less highlighted feature of most studies discussed here: the effect of genetic versus pharmacological removal of S6K1 is not always concurrent.

The first behavioral study contrasted the effects of knocking out S6K1 versus S6K2 on mouse behavior (Antion et al., 2008b). Ablation of S6K1 affected early onset of fear memory, conditioned taste aversion (CTA) and impaired acquisition of platform location in Morris Water Maze (MWM) test of spatial memory. Another important phenotype was hypoactivity in S6K1 KO mice, which has been employed later to titrate optimum dosing of S6K1 inhibitors (Huynh, Santini, & Klann, 2014; and Bhattacharya et al., 2016). In contrast, deletion of S6K2 KO reported decreased contextual fear memory, a reduction in latent inhibition of CTA, and intact spatial learning in MWM (Antion et al., 2008b). An important aspect to note here is the mixed background of knock out mice that have been speculated to contribute to the behavioral phenotypes observed. Another set of behavioral test battery was carried out in Bhattacharya et al., (2012) to test the efficacy of genetically deleting S6K1 to rescue phenotypes of fragile X syndrome (FXS). In this study, background was a mix of 129/SvJ and C57/Bl6, with a greater contribution of the latter due to more frequent backcrosses. S6K1 KO alone continued to remain hypoactive with a diminished performance in the rotarod test as well. S6K1 KO mice showed indifferent preference to social approach test (mice vs. object), with a distinct preference to the novel over the familiar mice in the novelty phase of the test. The mice also showed no impairment in behavioral flexibility as tested by the Y-maze, but buried more marbles in the eponymous test which may correlate to increased preservative/anxiogenic behavior. Interestingly, S6K1 abrogation in FXS model mice rescued a wide range of phenotypes, to be discussed further. In 2016, pharmacologic blockade of S6K1 in WT mice using two small molecule inhibitors showed no effect on social preference while slightly increasing reversal times in the Y-maze, though both showed enhanced marble burying (Bhattacharya et al., 2016).

A separate approach to model depression and the implication of S6K1 therein was adopted by Dwyer et al., (2015) using viral microinjection in the prefrontal cortex (PFC). Expression of constitutively active form of S6K1 (T389A δ CT) had an anti-depressant effect on rats, with increased locomotion, decreased times in forced swim test and abrogated the decreased sucrose preference after chronic stress. Suppression S6K1 signaling by expressing a kinase inactive form (K100R) showed opposite effects and blocked the anti-depressant effects of ketamine. This study established a strong connection between the state of S6K1 activity and depressive phenotypes. It remains to be validated by a pharmacologic approach.

S6K1 has also been implicated to mediate gustatory learning (Belelovsky, Kaphzan, Elkobi, & Rosenblum, 2009) using the CTA paradigm. However, S6K1 does not contribute to feeding behavior directly, though it regulates glucose homeostasis. Shown in a report using a conditionally driven S6K1 in AgrP and POMC neurons of hypothalamus, the deletion of kinase changed intrinsic excitability of these neurons (Smith et al., 2015). This effect of S6K1 is thought to be intimately related to its role in regulating metabolism.

The role of S6K1 in fear memory has been better dissected as fallout of interest in the contribution of protein synthesis in various phases of this learning paradigm. S6K1 signaling also seems to be region-specific in fear, a point that is not well appreciated. For instance, Gafford, Parsons, & Helmstetter (2013) measured phospho-S6K1 levels 1, 10 and 36 days after contextual fear conditioning in the dorsal hippocampus (DH) and anterior cingulate cortex (ACC). 1 day after phospho-S6K1 levels were elevated in the DH with an increase only after 36 days in the ACC, alluding to the temporally spaced translational phases of memory. S6K1 seems to mediate not memory acquisition, but rather reconsolidation and extinction. Hyunh et al. (2014) examined the dependence of cued fear memory retrieval/reconsolidation on S6K1 activity. They reported that the effect of rapamycin inhibition on retrieval could only be recapitulated if both eIF4E and S6K1 were blocked simultaneously. S6K1 did underwrite the persistence of reconsolidated memory 10 days later. This seemed to impinge via ERK signaling to Ser 421 site S6K1. In a follow-up paper in 2017, the same team showed that this ERK-S6K1 signaling is key to within-session extinction learning in the BLA and involves phosphorylation of AMPA receptor sub unit GluA1 (Huynh et al., 2017).

Overall, current data suggests that in most behaviors that have previously been shown to require mTORC1 activation, there exists a reliance on S6K1 signaling. What is implicit in the reports is the region specificity of this effect, though no concerted effort has been devoted in closely examining this issue. There is also significant divergence in behavioral phenotypes when S6K1 is genetically ablated versus shorter time suppression by blockers or viral expression of mutant kinase. It is now possible to undertake circuit limited experiments where S6K1 would be perturbed in some cells and intact in others to see how the kinase modulates behavioral outcomes at this level.

S6K1 in Neurological Disease

There is considerable ongoing effort in studying the role of S6K1 in etiology of diabetes, obesity, liver function and cancer (Um, D'Alessio, & Thomas, 2004; Savinska et al., 2004; Ben-Hur et al., 2013; Tavares et al., 2015). Alongside, the Akt-ERK-mTOR signaling in neurology has implications in neurodevelopmental, neurodegenerative and traumatic injury conditions. S6K1 plays a substantial role in the pathology of disorders caused by this signaling nexus but also in implicated in the pathology of certain conditions in a stand-alone capacity.

Given the central role that mTOR plays in anabolism and development, dysregulated mTORC1 signaling in neurodevelopment disorders (NDD) is expected. The best studied model in this subset of conditions is FXS. S6K1 was shown to be elevated as a part of dysregulated mTORC1 signaling in the *Fmr1* KO mice hippocampus (Sharma et al., 2010) and, in patient-derived fibroblasts, lymphocytes and post-mortem brain tissue (Kumari et al., 2014; Hoeffer et al., 2012). Therefore S6K1 seemed to be intimately associated with the eccentric protein synthesis seen in FXS and a potential target for intervention. Genetic deletion of S6K1 in the background of *Fmr1* loss (Bhattacharya et al., 2012) resulted in restored proteostasis, mTORC1 signaling, synthesis of FMRP target proteins relevant to synaptic plasticity, DHPG-evoked LTD and abnormal filopodial spines. In terms of behavioral phenotypes, S6K1 ablation rescued behavioral inflexibility and inappropriate social interaction, but not marble burying repetitive behavior. Given *S6K1* KO are small (Shima et al., 1998), rescue of weight gain was an expected fallout. Additionally, macroorchidism in FXS has been shown to be caused by abnormal proliferation of Sertoli cells (Slegtenhorst-Eegdeman et al., 1998)—a process known to require S6K1 activity downstream of follicle stimulating hormone (FSH; Lécureuil et al. 2005). Therefore, rescue of macroorchidism by *S6K1* genetic deletion was again expected. A key issue unresolved in this study was whether the effects were mediated by S6K1 inhibition on translation initiation or elongation.

However, the effects of pharmacologically targeting S6K1 in FXS, in neurodevelopmentally complete adult mice, were published recently (Bhattacharya et al., 2016). Head-to-head comparison of two S6K1 blockers, PF-4708671 and FS-115 was done in *Fmr1* and wild-type littermates. FS-115 is more brain-penetrant and remains in the CNS for a longer time than PF-4708671. While both agents reduced translation in hippocampal and cortical brain preparations, they preferentially affected eEF2 and S6 phosphorylation rather than eIF4B, implying an elongation bias in FXS. Additionally, increased availability of FS115 in the brain of treated FXS mice normalized marble burying behavior, in part due to continued S6K1 blockade or effects on other AGC kinases. PF-4708671 was only able to correct macro-orchidism, while FS-115 impacted both phenotypes arguing for a CNS-driven control of weight and metabolism in FXS. Both agents successfully countered abnormal social interactions and behavioral inflexibility, with higher dosing acutely inducing a hypo-locomotor effect. Through these studies, it is apparent that S6K1 mediated protein synthesis plays a key role in FXS phenotypes, but requires more pre-clinical validation before it can be considered for clinical trials. For example, the effect of short-term S6K1 inhibition on LTD remains to be clarified as does studying S6K1 depression in other brain areas like the amygdala that underpins the anxiety phenotypes in seen with this syndrome.

An allied effect of S6K1 in NDD was reported recently by Huang, Chen, & Page (2016) in a germline, heterozygous *Pten* knock down model. The authors found a transient spike in mTORC1-S6K1 activity at post-natal day 14 (P14) in *Pten* +/– mice pups that caused increased axonal branching and connectivity between the medial PFC to the basolateral amygdala (BLA). This hyper-connectivity was coincident with impaired social behavioral and attendant anxiety in these mice. Interestingly, S6K1 blockade between P4 and P14 using PF-4708671 was successful in normalizing this phenotype, but a similar dose regime in adulthood did not have any effect. The authors also report that a similar spike in mTORC1 activity is also seen in young FXS mice, which argues for a conserved etiology of regional alterations in signaling and connections in syndromic models of ASD. Additionally, Angelman syndrome is reported to have imbalances in proteostasis, which are addressable by manipulating S6K1 signaling. Sun et al (2016) reported that improved LTP and spine maintenance could be achieved by administering PF-4708671, rapamycin and mTORC2 activator A-443654. S6K1 inhibition can counteract the phenotypes of STRADA deletion that cause rare neurodevelopmental disorders called PMSE (polyhydramnios, megacephaly, symptomatic epilepsy; Parker et al., 2013). STRADA, or STE20-related kinase adaptor alpha, is a pseudo-kinase that binds LKB1, a kinase upstream of AMP-activated kinases, and monitors energy homeostasis, cell size and senescence in cells.

In contrast, the relationship in aging between protein synthesis and S6K1 signaling appear to be complex. While it is well established that protein synthesis declines in the aging brain (Schimanski & Barnes, 2010), equally strong evidence exists for the beneficial effects of suppressing S6K1 signaling to improve longevity across multiple models (Selman et al., 2009; Kennedy & Kaeberlein, 2009). At the outset, these two phenomena seem contradictory, and the defining quality could well be the timing of the intervention. That S6K1 signaling should be lower in the aging brain can be reconciled in the light of increased oxidative stress, autophagy and eIF2alpha stress signaling (Keller et al., 2004; Ma et al., 2013; Massaad & Klann, 2011). However, S6K1 hyperactivity contributes to increased insulin resistance in neurodegenerative conditions (Cholerton Baker, Craft 2013), which can trigger aging. In support of the latter model, Caccamo et al., (2015) presented work that shows that reducing S6K1 expression improved spatial memory and synaptic deficits in the 3xAD mouse model for Alzheimer's disease by suppressing the translation of Tau and BACE1. This study solely focuses the effect of having reduced S6K1 expression a priori before onset of AD on disease pathology. Though not implicating any upstream signaling effects, the authors squarely demonstrate that the effect involves translation control and reinforces the fact that suppressing the translation of specific pathologic proteins in AD is of therapeutic value. However, if intervention is to be attempted after the advent of amyloid β in the system, Zare, Motamedi, Digaleh, Khodagholi, and Maghsoudi (2015) suggest that activating S6K1 can help prevent cell death. Similarly, suppression of S6K1 signaling leads to neuronal death in Parkinson disease models (Xu et al., 2014). In the current state, it is unclear what the biggest contributor to S6K1 signaling is in a degenerating brain and if it is entirely neuronal. A plausible hypothesis is that S6K1 signaling changes in response to peripheral metabolism and at some point uncouples from mTORC1 in aging brain. It is clear that much work will be devoted to this solving this conundrum in the future.

Finally, efforts of all pre-clinical studies converge upon a search for a translatable therapeutic approach. Most therapeutic efforts in the industry are geared to controlling S6K1 in the context of obesity, cancer and diabetes. The key breakthroughs have come with the advent of a slew of small molecule inhibitors, from Pfizer (PF-4708671; Pearce et al., 2010), Axon MedChem (LY2584702; Tolcher et al., 2014), Sentinel Oncology (FS-115; Bhattacharya et al., 2016), and a battery of candidates from rational drug design screens (Qin et al., 2015). Currently. there are no clinical trials ongoing targeting S6K1 for any disease. S6K1 can also be inhibited by multi-specificity AGC kinase inhibitors like H89 2HCl and AT 13148 (Xi et al., 2016; Liu et al., 2016). Efforts to target S6K1 for neurological disorders using these molecules are at a relatively early stage, with the key challenges being to find blood-brain-barrier-permeable agents and titrating the ideal dosage for increased effectiveness while minimizing side effects such as decreased locomotion.

Outstanding Questions, Experimental Limitations, and Looking Forward

Dedicated research by multiple groups across the world is slowly bringing S6K1 out of the shadow of mTORC1 and ERK1/2 signaling. That S6K1 is a strong regulatory member of cellular signaling is well accepted outside the central nervous system, but this notion has been slow to spread for neurobiology, in part due to the absence of reliable small molecule inhibitors. However, some serious experimental limitations still exist in the field that require acknowledgement prior to addressing them. Additionally, there seems to be a paucity of clinical information about S6K1 that will impede future translational work. Next I attempt to collate the many pitfalls that plague holistic S6K1 research now, which must be investigated, interpreted and overcome in the future.

Better Reagents Fuel Better Research

Neuronal translation is equal parts somatic and decentralized in processes (dendritic and axonal). While this notion is strongly entrenched in neurobiology, another form of translational localization is becoming more apparent. For neurons to work effectively, astroglial and oligodendritic cells need to also perform their functions effectively. Our view of translation has been constrained to only consider the neuronal aspects; it is time to embrace that it is very likely that there are orchestrated changes in translation across neurons and glial cells to respond to any stimulus. This becomes relevant when one considers that S6K1 is ubiquitously expressed in all cells and likely will regulate translation differently downstream of cell-specific or cell autonomous cues. This should be an easily achievable goal with immunohistochemistry.

However, in reality, experimenters routinely note the intractability of most commercially available phospho-S6K1 antibodies for use in brain sections. Therefore, usually phosphorylated S6 ribosomal protein is used as a proxy, which is influenced by activation status of Rsk kinases and actually provides a measure of the level of translational activation. A quick look at the product sheets of these antibodies shows that almost all validation is done in non-neuronal cells and hardly any in tissue lysates or fixed section. Probing *S6K1* KO mouse brain lysates with these antibodies is currently the best way to determine specificity, which becomes tedious for researchers without access to such control samples. Since a large number of these primary antibodies are rabbit polyclonals, lot to lot variations is a major concern. This may be alleviated with more monoclonal and recombinant choices appearing in the market. A reliable and consistent antibody source for S6K1 that works in multiple applications will address a slew of questions about relative localization of activated and non-activated forms of this kinase in polarized neural cells.

There is also a lack of genetic reagents to perturb S6K1 function that are readily deployable for neuroscience. Conditional knockdowns of S6K1 exist, but have been used only sparingly for neurobiological studies (Smith et al., 2015). It is unclear how much upregulation of S6K2 occurs in global deletion of S6K1, since no reliable antibody exists for this isoform. Bowling et al., employed an effective shRNA for S6K1 in neuronal cultures which paradoxically hasn't been used often. With the advent of CRISPR/Cas9 technology, it should become easier to develop guides that knockdown S6K1 in targeted fashion. AAV2 constructs with S6K1 are available (Dwyer et al., 2015), and we should see more studies with these reagents being published in the future, however conditional Tet On/Off systems have yet to be made.

Natural Variation to Discover Phenotype Stratification

The Val633Met variation in BDNF is perhaps one of the most intensely studied examples of a genetic change in neuropsychiatry. Usually naturally occurring variants that are not immediately disease causing but prodromal or correlate with early onset or symptom severity have been less studied in neurobiology as compared to cancer. The translation control pathway is usually under tremendous purifying pressure to maintain them error-free and seldom come up in large genetic studies. That said, duplications and truncations that enhance S6K1 activity have been found in breast cancers and other solid tumors. Additionally, ExAC (http://exac.broadinstitute.org) and other mutation aggregator websites provide a list of common, rare and ultra-rare S6K1 exonic variants that exist in humans across the world. These are likely to cause subtle changes in enzyme activity that first need to be cataloged and then cross-referenced in existing exome databases for their presence or absence in specific patient populations. In the light of small molecules being developed for S6K1, knowing how these would react with variant S6K1 proteins will be critically important in determining candidate stratification and treatment outcome measures in trials feature S6K1 inhibition. With increased cognizance given to the role that altered metabolomics plays in the expression of neurological disease, studying S6K1 variants gathers greater significance.

Steady State versus Activity-Driven Translation

In almost all studies of S6K1, and, to an extent, translation control in neural systems, no distinction is made between steady-state and stimulus-driven states. This may be in unconscious bias since any brain function is a combinatorial outcome of these two phases, but this becomes important when considering pharmacological interventions. For example, unpublished data from multiple sources allude to that fact that perturbing S6K1 activity in slices incubated in ACSF alone with no stimulation will not affect translation as profoundly as rapamycin would. However, the same perturbation in the wake of agents that induce chemical LTP/LTD have a more dramatic effect. The reduced role of S6K1 in steady state translation is also alluded to by the relatively benign effects of complete deletion in knock out animals. It is very likely that S6K1 is actively involved in stimulus-driven state changes in neural systems, which is exemplified by its critical role in glial transformation (Nakamura, Garcia, & Pieper, 2008). More refined experiments are required to clearly delineate this mechanism.

What, Where, When and How Much?

It is well documented that behavior and related neurobiological phenomena are related in many instances to translation by S6K1. This translation control not only has a temporal order, but likely happens in the different cells making up the tri-partite synapse at different rates, producing different proteins. Thus far, most studies have focused on the "when," and future studies must look to answer the "where," "what" and "how much." The will become critical as it is increasingly being noted that different brain areas and circuits may regulate their proteomic flux differently in the face of the same stimulus. Hence a system administration of a drug (like rapamycin or PF-4708671) will impact all areas equally and lead to off-target results in neural circuits that don't increase translation in response to a signal. Knowing in what translation case S6K1 is downstream of mTORC1 and in which S6K1 acts independently will help sort out which ways to deliver the drug, what drug combinations will work etc. Neurobiology as a whole is shifting to acknowledge that region and sub-type specificity in circuits underlie much of what the brain expresses. It is time that the proteomic and molecular map is also put in place to track these changes and help inform future drug development for disease conditions.

Acknowledgments

A. B. is supported by funds from the Department of Biotechnology, Government of India and the FRAXA Research Foundation. I apologize for not including the work of all authors in the interest of presenting a focused view of the existing literature and highlighting areas that require greater attention in S6K1 biology.

References

Al-Ali, H., Ding, Y., Slepak, T., Wu, W., Sun, Y., Martinez, Y., . . . Bixby, J. L. (2017). The mTOR Substrate S6 Kinase 1 (S6K1) is a negative regulator of axon regeneration and a potential drug target for central nervous system injury. *Journal of Neuroscience*, *37*(30), 7079–7095.

Antion, M. D., Hou, L., Wong, H., Hoeffer, C. A., & Klann, E. (2008a). mGluR-dependent long-term depression is associated with increased phosphorylation of S6 and synthesis of elongation factor 1A but remains expressed in S6K-deficient mice. *Molecular Cellular Biology*, *28*(9), 2996–3007.

Antion, M. D., Merhav, M., Hoeffer, C. A., Reis, G., Kozma, S. C., Thomas, G.,... Klann, E. (2008b). Removal of S6K1 and S6K2 leads to divergent alterations in learning, memory, and synaptic plasticity. *Learning & Memory, 15*(1), 29–38.

Banerjee, P., Ahmad, M. F., Grove, J. R., Kozlosky, C., Price, D. J., & Avruch, J. (1990). Molecular structure of a major insulin/mitogen-activated 70-kDa S6 protein kinase. *Proceedings of National Academy Science of USA, 87*(21), 8550–8554.

Bartley, C. M., O'Keefe, R. A., & Bordey, A. (2014). FMRP S499 Is Phosphorylated Independent of mTORC1-S6K1 activity. *PLoS ONE, 9*(5), e96956.

Becker, M. A., Ibrahim, Y. H., Cui, X., Lee, A. V., & Yee, D. (2011). The IGF pathway regulates erα through a S6k1-dependent mechanism in breast cancer cells. *Molecular Endocrinology, 25*(3), 516–528.

Belelovsky, K., Kaphzan, H., Elkobi, A., & Rosenblum, K. (2009). Biphasic activation of the mTOR pathway in the gustatory cortex is correlated with and necessary for taste learning. *Journal of Neuroscience, 29*(23), 7424–7431.

Ben-Hur, V., Denichenko, P., Siegfried, Z., Maimon, A., Krainer, A., Davidson, B., & Karni, R. (2013). S6K1 alternative splicing modulates its oncogenic activity and regulates mTORC1. *Cell Reports, 3*(1), 103–115.

Bernard, P. B., Castano, A. M., O'Leary, H., Simpson, K., Browning, M. D., & Benke, T. A. (2013). Phosphorylation of FMRP and alterations of FMRP complex underlie enhanced mLTD in adult rats triggered by early life seizures. *Neurobiology of Disease, 59*, 1–17.

Bhattacharya, A., Kaphzan, H., Alvarez-Dieppa, A. C., Murphy, J. P., Pierre, P., & Klann E. (2012). Genetic removal of p70 S6 kinase 1 corrects molecular, synaptic, and behavioral phenotypes in fragile X syndrome mice. *Neuron, 76*(2), 325–337.

Bhattacharya, A., Mamcarz, M., Mullins, C., Choudhury, A., Boyle, R. G., Smith, D. G.,... Klann, E. (2016). Targeting translation control with p70 S6 kinase 1 inhibitors to reverse phenotypes in fragile x syndrome mice. *Neuropsychopharmacology, 41*(8), 1991–2000.

Bowling, H., Zhang, G., Bhattacharya, A., Pérez-Cuesta, L. M., Deinhardt, K., Hoeffer, C. A., ... Chao M. V. (2014). Antipsychotics activate mTORC1-dependent translation to enhance neuronal morphological complexity. *Science Signaling, 7*(308), ra4.

Buchsbaum, R. J., Connolly, B. A., & Feig, L. A. (2003). Regulation of p70 S6 kinase by complex formation between the Rac guanine nucleotide exchange factor (Rac-GEF) Tiam1 and the scaffold spinophilin. *Journal of Biological Chemistry, 278*(21), 18833–18841.

Burnett, P. E., Blackshaw, S., Lai, M. M., Qureshi, I. A, Burnett, A. F., Sabatini, D. M., & Snyder, S. H. (1998). Neurabin is a synaptic protein linking p70 S6 kinase and the neuronal cytoskeleton. *Proceedings of National Academy of Science of USA, 95*(14), 8351–8356.

Caccamo, A., Branca, C., Talboom, J. S., Shaw, D. M., Turner, D., Ma, L.,... Oddo, S. (2015). Reducing ribosomal protein s6 kinase 1 expression improves spatial memory and synaptic plasticity in a mouse model of Alzheimer's disease. *Journal of Neuroscience, 35*(41), 14042–14056.

Cammalleri, M., Lütjens, R., Berton, F., King, A. R., Simpson, C., Francesconi, W., & Sanna, P. P. (2003). Time-restricted role for dendritic activation of the mTOR-p70S6K pathway in the induction of late-phase long-term potentiation in the CA1. *Proceedings of National Academy of Science of USA, 100*(24), 14368–14373.

Chévere-Torres, I., Kaphzan, H., Bhattacharya, A., Kang, A., Maki, J. M., Gambello, M. J.,... Klann, E. (2012). Metabotropic glutamate receptor-dependent long-term depression is impaired due to elevated ERK signaling in the ΔRG mouse model of tuberous sclerosis complex. *Neurobiology of Disease, 45*(3), 1101–1110.

Cholerton, B., Baker, L. D., & Craft, S. (2013). Insulin, cognition, and dementia. *European Journal of Pharmacology, 719*(1–3), 170–179.

Chou, M. M., & Blenis, J. (1996). The 70Kd S6 kinase complexes with and is activated by the Rho family G proteins Cdc42 and Rac1. *Cell, 85*, 573–583.

Costa-Mattioli, M., Sossin, W. S., Klann, E., & Sonenberg, N. (2009). Translational control of long-lasting synaptic plasticity and memory. *Neuron, 61*(1), 10–26.

Dwyer, J. M., Maldonado-Avilés, J. G., Lepack, A. E., DiLeone, R. J., & Duman, R. S. (2015). Ribosomal protein S6 kinase 1 signaling in prefrontal cortex controls depressive behavior. *Proceedings of National Academy of Science of USA, 112*(19), 6188–6193.

Fang, Y., Park, I. H., Wu, A. L., Du, G., Huang, P., Frohman, M. A.,... Chen, J. (2003). PLD1 regulates mTOR signaling and mediates Cdc42 activation of S6K1. *Journal of Biological Chemistry, 13*(23), 2037–2044.

Frödin, M., & Gammeltoft, S. (1999). Role and regulation of 90 kDa ribosomal S6 kinase (RSK) in signal transduction. *Molecular and Cellular Endocrinology, 151*(1–2), 65–77.

Gafford, G. M., Parsons, R. G., & Helmstetter, F. J. (2013). Memory accuracy predicts hippocampal mTOR pathway activation following retrieval of contextual fear memory. *Hippocampus, 23*(9), 842–847.

Gavaret, J. M., Matricon, C., Pomerance, M., Jacquemin, C., Toru-Delbauffe, D., & Pierre, M. (1989). Activation of S6 kinase in astroglial cells by FGFa and FGFb. *Brain Research Dev Brain Research, 45*(1), 77–82.

Gong, X., Zhang, L., Huang, T., Lin, T. V., Miyares, L., Wen, J., Hsieh, L., & Bordey, A. (2015). Activating the translational repressor 4E-BP or reducing S6K-GSK3β activity prevents accelerated axon growth induced by hyperactive mTOR in vivo. *Human Molecular Genetics, 24*(20), 5746–5758.

Graber, T. E., McCamphill, P. K., & Sossin, W. S. (2013). A recollection of mTOR signaling in learning and memory. *Learning & Memory, 20*(10), 518–530.

Hahn, K., Miranda, M., Francis, V. A., Vendrell, J., Zorzano, A., & Teleman, A. A. (2010). PP2A regulatory subunit PP2A-B′ counteracts S6K phosphorylation. *Cell Metabolism, 11*(5), 438–444.

Harrington, L. S., Findlay, G. M., Gray, A., Tolkacheva, T., Wigfield, S., Rebholz, H.,... Lamb R. F. (2004). The TSC1–2 tumor suppressor controls insulin-PI3K signaling via regulation of IRS proteins. *Journal of Cell Biology, 166*(2), 213–223.

Heidenreich, K. A., & Toledo, S. P. (1989). Insulin receptors mediate growth effects in cultured fetal neurons. II. Activation of a protein kinase that phosphorylates ribosomal protein S6. *Endocrinology, 125*(3), 1458–1463.

Hoeffer, C. A., Sanchez, E., Hagerman, R. J., Mu, Y., Nguyen, D. V., Wong, H.,... Tassone, F. (2012). Altered mTOR signaling and enhanced CYFIP2 expression levels in subjects with fragile X syndrome. *Genes Brain Behavior, 11*(3), 332–341.

Hong, S., Zhao, B., Lombard, D. B., Fingar, D. C., & Inoki, K. (2014). Cross-talk between sirtuin and mammalian target of rapamycin complex 1 (mTORC1) signaling in the regulation of S6 kinase 1 (S6K1) phosphorylation. *Journal of Biological Chemistry, 289*(19), 13132–13141. doi: 10.1074/jbc.M113.520734.

Huang, W., Zhu, P. J., Zhang, S., Zhou, H., Stoica, L., Galiano, M.,... Costa-Mattioli, M. (2013). mTORC2 controls actin polymerization required for consolidation of long-term memory. *Nature Neuroscience, 16*(4), 441–448.

Huang, W. C., Chen, Y., & Page, D. T. (2016). Hyperconnectivity of prefrontal cortex to amygdala projections in a mouse model of macrocephaly/autism syndrome. *Nature Communications, 7*, 13421.

Huber, K. M., Kayser, M. S., & Bear, M. F. (2000). Role for rapid dendritic protein synthesis in hippocampal mGluR-dependent long-term depression. *Science, 288*(5469), 1254–1257.

Huynh, T. N., Santini, E., & Klann, E. (2014). Requirement of mammalian target of rapamycin complex 1 downstream effectors in cued fear memory reconsolidation and its persistence. *Journal of Neuroscience, 34*(27), 9034–9039.

Huynh, T. N., Santini, E., Mojica, E., Fink, A. E., Hall, B. S., Fetcho, R. N.,... Klann, E. (2017). Activation of a novel p70 S6 kinase 1-dependent intracellular cascade in the basolateral nucleus of the amygdala is required for the acquisition of extinction memory. *Molecular Psychiatry*. doi: 10.1038/mp.2017.99.

Jaworski, J., & Sheng, M. (2006). The growing role of mTOR in neuronal development and plasticity. *Molecular Neurobiology, 34*(3), 205–219.

Jaworski, J., Spangler, S., Seeburg, D. P., Hoogenraad, C. C., & Sheng, M.(2005). Control of dendritic arborization by the phosphoinositide-3′-kinase-Akt-mammalian target of rapamycin pathway. *Journal of Neuroscience, 25*(49), 11300–11312.

Jefferies, H. B., Fumagalli, S., Dennis, P. B., Reinhard, C., Pearson, R. B., & Thomas, G. (1997). Rapamycin suppresses 5′TOP mRNA translation through inhibition of p70s6k. *EMBO Journal, 16*(12), 3693–3704.

Julien, L. A., Carriere, A., Moreau, J., & Roux, P. P. (2010). mTORC1-activated S6K1 phosphorylates Rictor on threonine 1135 and regulates mTORC2 signaling. *Molecular & Cellular Biology, 30*(4), 908–921.

Jung, H., Gkogkas, C. G., Sonenberg, N., & Holt, C. E. (2014). Remote control of gene function by local translation. *Cell, 157*(1), 26–40.

Kandel, E. R. (2001). The molecular biology of memory storage: a dialogue between genes and synapses. *Science, 294*(5544), 1030–1038.

Kang, H., & Schuman, E. M. (1996). A requirement for local protein synthesis in neurotrophin-induced hippocampal synaptic plasticity. *Science, 273*(5280), 1402–1406.

Kelleher, R. J. 3rd, Govindarajan, A., & Tonegawa, S. (2004). Translational regulatory mechanisms in persistent forms of synaptic plasticity. *Neuron, 44*(1), 59–73.

Keller, J. N., Dimayuga, E., Chen, Q., Thorpe, J., Gee, J., & Ding, Q. (2004). Autophagy, proteasomes, lipofuscin, and oxidative stress in the aging brain. *International Journal of Biochemistry Cell Biology, 36*(12), 2376–2391.

Kennedy, B. K., & Kaeberlein, M. (2009). Hot topics in aging research: Protein translation. *Aging Cell, 8*(6), 617–623.

Kozak, M. (1992). Regulation of translation in eukaryotic systems. *Annual Review of Cell Biology, 8*, 197–225.

Kozma, S. C., Ferrari, S., Bassand, P., Siegmann, M., Totty, N., & Thomas, G. (1990). Cloning of the mitogen-activated S6 kinase from rat liver reveals an enzyme of the second messenger subfamily. *Proceedings of National Academy of Science of USA, 87*(19), 7365–7369.

Kumari, D., Bhattacharya, A., Nadel, J., Moulton, K., Zeak, N. M., Glicksman, A.,... Usdin, K. (2014). Identification of fragile X syndrome specific molecular markers in human fibroblasts: A useful model to test the efficacy of therapeutic drugs. *Human Mutation, 35*(12), 1485–1494.

Lai, K. O., Liang, Z., Fei, E., Huang, H., & Ip, N. Y. (2015). Cyclin-dependent kinase 5 (Cdk5)-dependent phosphorylation of p70 ribosomal s6 kinase 1 (S6K) is required for dendritic spine morphogenesis. *Journal of Biological Chemistry, 290*(23), 14637–14646.

Lécureuil, C., Tesseraud, S., Kara, E., Martinat, N., Sow, A., Fontaine, I.,... Crépieux, P. (2005). Follicle-stimulating hormone activates p70 ribosomal protein S6 kinase by protein kinase

A-mediated dephosphorylation of Thr 421/Ser 424 in primary Sertoli cells. *Molecular Endocrinology*, *19*, 1812–1820.

Lehman, J. A., Calvo, V., & Gomez-Cambronero, J. (2003). Mechanism of ribosomal p70S6 kinase activation by granulocyte macrophage colony-stimulating factor in neutrophils: Cooperation of a MEK-related, THR421/SER424 kinase and a rapamycin-sensitive, m-TOR-related THR389 kinase. *Journal of Biological Chemistry*, *278*(30), 28130–28138.

Lein, E. S., Hawrylycz, M. J., Ao, N., Ayres, M., Bensinger, A., Bernard, A.,... Jones, A. R. (2007). Genome-wide atlas of gene expression in the adult mouse brain. *Nature*, *445*, 168–176.

Liu, S., & Lu, B. (2010). Reduction of protein translation and activation of autophagy protect against PINK1 pathogenesis in Drosophila melanogaster. *PLoS Genetics*, *6*(12), e1001237.

Liu, X., Müller, F., Wayne, A. S., & Pastan, I. (2016). Protein kinase inhibitor H89 enhances the activity of pseudomonas exotoxin a-based immunotoxins. *Molecular Cancer & Therapeutics*, *5*, 1053–1062.

Ma, X. M., Yoon, S. O, Richardson, C. J., Jülich, K., & Blenis, J. (2008). SKAR links pre-mRNA splicing to mTOR/S6K1-mediated enhanced translation efficiency of spliced mRNAs. *Cell*, *133*(2), 303–313.

Ma, T., Trinh, M. A., Wexler, A., Bourbon, C., Gatti, E., Pierre, P., Cavener, D. R., Klann, E. (2013). Suppression of eIF2α kinases alleviates Alzheimer's disease-related plasticity and memory deficits. *Nature Neuroscience*, *16*(9), 1299–1305.

Magnuson, B., Ekim, B., & Fingar, D. C. (2012). Regulation and function of ribosomal protein S6 kinase (S6K) within mTOR signalling networks. *Biochemical Journal*, *441*(1), 1–21.

Malenka, R. C., & Nicoll, R. A. (1999). Long-term potentiation—a decade of progress? *Science*, *285*(5435), 1870–1874.

Massaad, C. A., & Klann, E. (2011). Reactive oxygen species in the regulation of synaptic plasticity and memory. *Antioxidant and Redox Signal*, *14*(10), 2013–2054.

McCamphill, P. K., Ferguson, L., & Sossin, W. S. (2017). A decrease in eukaryotic elongation factor 2 phosphorylation is required for local translation of sensorin and long-term facilitation in Aplysia. *Journal of Neurochemistry*, *142*(2), 246–259.

Michel, K., Zhao, T., Karl, M., Lewis, K., & Fyffe-Maricich, S. L. (2015). Translational control of myelin basic protein expression by ERK2 MAP kinase regulates timely remyelination in the adult brain. *Journal of Neuroscience*, *35*(20), 7850–7865.

Nakamura, J. L., Garcia, E., & Pieper, R. O. (2008). S6K1 plays a key role in glial transformation. *Cancer Research* *68*(16), 6516–6523.

Narayanan, U., Nalavadi, V., Nakamoto, M., Thomas, G., Ceman, S., Bassell, G. J., & Warren S. T. (2008). S6K1 phosphorylates and regulates fragile X mental retardation protein (FMRP) with the neuronal protein synthesis-dependent mammalian target of rapamycin (mTOR) signaling cascade. *Journal of Biological Chemistry*, *283*(27), 18478–18482.

Nardella, C., Lunardi, A., Fedele, G., Clohessy, J. G., Alimonti, A., Kozma, S. C.,... Pandolfi, P. P. (2011). Differential expression of S6K2 dictates tissue-specific requirement for S6K1 in mediating aberrant mTORC1 signaling and tumorigenesis. *Cancer Research*, *71*(10), 3669–3675.

Novak-Hofer, I., & Thomas, G. (1984). An activated S6 kinase in extracts from serum- and epidermal growth factor-stimulated Swiss 3T3 cells. *Journal of Biological Chemistry*, *259*(9), 5995–6000.

Pardo, O. E., & Seckl, M. J. (2013). S6K2: The neglected S6 kinase family member. *Frontiers of Oncology*, *3*, 191.

Parker, W. E., Orlova, K. A., Parker, W. H., Birnbaum, J. F., Krymskaya, V. P., Goncharov, D. A.,...Crino, P. B. (2013). Rapamycin prevents seizures after depletion of STRADA in a rare neurodevelopmental disorder. *Science Translational Medicine, 5*(182), 182ra53.

Pearce, L. R., Alton, G. R., Richter, D. T., Kath, J. C., Lingardo, L., Chapman, J.,...Alessi D. R. (2010b). Characterization of PF-4708671, a novel and highly specific inhibitor of p70 ribosomal S6 kinase (S6K1). *Biochemical Journal, 431*(2), 245–255.

Pearce, L. R., Komander, D., Alessi, D. R. (2010). The nuts and bolts of AGC protein kinases. *Nature Reviews Molecular Cell Biology, 11*, 9–22. doi: 10.1038/nrm2822.

Pende, M., Um, S. H., Mieulet, V., Sticker, M., Goss, V. L., Mestan, J.,...Thomas, G. (2004). S6K1(-/-)/S6K2(-/-) mice exhibit perinatal lethality and rapamycin-sensitive 5′-terminal oligopyrimidine mRNA translation and reveal a mitogen-activated protein kinase-dependent S6 kinase pathway. *Molecular & Cellular Biology, 24*(8), 3112–3124.

Pierre, M., Toru-Delbauffe, D., Gavaret, J. M., Pomerance, M., & Jacquemin, C. (1986). Activation of S6 kinase activity in astrocytes by insulin, somatomedin C and TPA. *FEBS Letters, 206*(1), 162–166.

Piper, M., Anderson, R., Dwivedy, A., Weinl, C., van Horck, F., Leung, K. M.,...Holt, C. (2006). Signaling mechanisms underlying Slit2-induced collapse of Xenopus retinal growth cones. *Neuron, 49*(2), 215–228.

Pirbhoy, P. S., Farris, S., & Steward, O. (2017). Synaptically driven phosphorylation of ribosomal protein S6 is differentially regulated at active synapses versus dendrites and cell bodies by MAPK and PI3K/mTOR signaling pathways. *Learning & Memory, 24*(8), 341–357.

Qin, J., Rajaratnam, R., Feng, L., Salami, J., Barber-Rotenberg, J. S., Domsic, J.,...Marmorstein, R. (2015). Development of organometallic S6K1 inhibitors. *Journal of Medicinal Chemistry, 58*(1), 305–314.

Roberts, S., & Morelos, B. S. (1982). Inhibition of cerebral protein kinase activity and cyclic AMP-dependent ribosomal-protein phosphorylation in experimental hyperphenylalaninaemia. *Biochemical Journal, 202*(2), 343–351.

Romanelli, A., Martin, K. A., Toker, A., & Blenis, J. (1999). p70 S6K is regulated by protein kinase C eta and participants in a phosphoinositide 3 kinase-regulated signaling complex. *Molecular & Cellular Biology, 19*, 2921–2928.

Savinska, L. O., Lyzogubov, V. V., Usenko, V. S., Ovcharenko, G. V., Gorbenko, O. N., Rodnin, M. V.,...Filonenko, V. V. (2004). Immunohistochemical analysis of S6K1 and S6K2 expression in human breast tumors. *Eksp Onkol, 26*(1), 24–30.

Schimanski, L. A., & Barnes, C. A. (2010). Neural protein synthesis during aging: Effects on plasticity and memory. *Frontiers in Aging Neuroscience*, 2, article 26. doi: 10.3389/fnagi.2010.00026. eCollection 2010.

Selman, C., Tullet, J. M., Wieser, D., Irvine, E., Lingard, S. J., Choudhury, A. I.,...Withers, D. J. (2009). Ribosomal protein S6 kinase 1 signaling regulates mammalian life span. *Science, 326*(5949), 140–144.

Sethna, F., Zhang, M., Kaphzan, H., Klann, E., Autio, D., Cox, C. L., & Wang, H. (2016). Calmodulin activity regulates group I metabotropic glutamate receptor-mediated signal transduction and synaptic depression. *Journal of Neuroscience Research, 94*(5), 401–408.

Sharma, A., Hoeffer, C. A., Takayasu, Y., Miyawaki, T., McBride, S. M., Klann, E., & Zukin, R. S. (2010). Dysregulation of mTOR signaling in fragile X syndrome. *Journal of Neuroscience*, *30*(2), 694–702.

Shima, H., Pende, M., Chen, Y., Fumagalli, S., Thomas, G., & Kozma, S. C. (1998). Disruption of the p70(s6k)/p85(s6k) gene reveals a small mouse phenotype and a new functional S6 kinase. *EMBO Journal*, *17*(22), 6649–6659.

Shin, S., Wolgamott, L., Yub, Y., Blenis, J., & Yoona, S. O. (2011). Glycogen synthase kinase (GSK)-3 promotes p70 ribosomal protein S6 kinase (p70S6K) activity and cell proliferation. *Proceedings of National Academy of Sciences of USA*, *108*(47), E1204–E1213.

Slegtenhorst-Eegdeman, K. E., de Rooij, D. G., Verhoef-Post, M., van de Kant, H. J., Bakker, C. E., Oostra, B. A., . . . Themmen, A. P. (1998). Macroorchidism in FMR1 knockout mice is caused by increased Sertoli cell proliferation during testicular development. *Endocrinology*, *139*(1), 156–162.

Smith, M. A., Katsouri, L., Irvine, E. E., Hankir, M. K., Pedroni, S. M., Voshol, P. J., . . . Withers, D. J. (2015). Ribosomal S6K1 in POMC and AgRP neurons regulates glucose homeostasis but not feeding behavior in mice. *Cell Reports*, *11*(3), 335–343.

Sun, J., Liu, Y., Tran, J., O'Neal, P., Baudry, M., & Bi, X. (2016). mTORC1-S6K1 inhibition or mTORC2 activation improves hippocampal synaptic plasticity and learning in Angelman syndrome mice. *Cell Molecular Life Science*, *73*(22), 4303–4314.

Sunami, T., Byrne, N., Diehl, R. E., Funabashi, K., Hall, D. L., Ikuta, M., . . . Sharma, S. (2010). Structural basis of human p70 ribosomal S6 kinase-1 regulation by activation loop phosphorylation. *Journal of Biological Chemistry*, *285*(7), 4587–4594.

Tavares, M. R., Pavan, I. C., Amaral, C. L., Meneguello, L., Luchessi, A. D., & Simabuco, F. M. (2015). The S6K protein family in health and disease. *Life Science*, *131*, 1–10.

Tolcher, A., Goldman, J., Patnaik, A., Papadopoulos, K. P., Westwood, P., Kelly, C. S., . . . Rosen, L. S. (2014). A phase I trial of LY2584702 tosylate, a p70 S6 kinase inhibitor, in patients with advanced solid tumours. *European Journal of Cancer*, *50*(5), 867–875.

Um, S. H., D'Alessio, D., Thomas, G. (2004). Nutrient overload, insulin resistance, and the ribosomal protein S6 Kinase 1, S6K1. *Cell Metabolism*, *6*, 393–402.

Verma, P., Chierzi, S., Codd, A. M., Campbell, D. S., Meyer, R. L., Holt, C. E., & Fawcett, J. W. (2005). Axonal protein synthesis and degradation are necessary for efficient growth cone regeneration. *Journal of Neuroscience*, *25*(2), 331–342.

Wang, M. L., Panasyuk, G., Gwalter, J., Nemazanyy, I., Fenton, T., Filonenko, V., & Gout, I. (2008). Regulation of ribosomal protein S6 kinases by ubiquitination. *Biochemical Biophysical Research Communications*, *369*, 382–387.

Wettenhall, R. E., Chesterman, C. N., Walker, T., & Morgan, F. J. (1983). Phosphorylation sites for ribosomal S6 protein kinases in mouse 3T3 fibroblasts stimulated with platelet-derived growth factor. *FEBS Letters*, *162*(1), 171–176.

Winter, J. N., Jefferson, L. S., & Kimball, S. R. (2011). ERK and Akt signaling pathways function through parallel mechanisms to promote mTORC1 signaling. *American Journal of Physiology*, *300*(5), C1172–C1180.

Wu, G. Y., Deisseroth, K., & Tsien, R. W. (2001). Spaced stimuli stabilize MAPK pathway activation and its effects on dendritic morphology. *Nature Neuroscience*, *4*(2), 151–158.

Xi, Y., Niu, J., Shen, Y., Li, D., Peng, X., & Wu, X. (2016). AT13148, a first-in-class multi-AGC kinase inhibitor, potently inhibits gastric cancer cells both in vitro and in vivo. *Biochemical Biophysical Research Communications, 478*(1), 330–336.

Xu, Y., Liu, C., Chen, S., Ye, Y., Guo, M., Ren, Q., ... Chen, L. (2014). Activation of AMPK and inactivation of Akt result in suppression of mTOR-mediated S6K1 and 4E-BP1 pathways leading to neuronal cell death in in vitro models of Parkinson's disease. *Cell Signaling, 26*(8), 1680–1689.

Yang, L., Miao, L., Liang, F., Huang, H., Teng, X., Li, S., ... Hu, Y. (2014). The mTORC1 effectors S6K1 and 4E-BP play different roles in CNS axon regeneration. *Nature Communications, 5*, 5416. doi: 10.1038/ncomms6416.

Zare, N., Motamedi, F., Digaleh, H., Khodagholi, F., & Maghsoudi, N. (2015). Collaboration of geldanamycin-activated P70S6K and Hsp70 against beta-amyloid-induced hippocampal apoptosis: An approach to long-term memory and learning. *Cell Stress Chaperones, 20*(2), 309–319.

Zhou, X., Lin, D. S, Zheng, F., Sutton, M. A., & Wang, H. (2010). Intracellular calcium and calmodulin link brain-derived neurotrophic factor to p70S6 kinase phosphorylation and dendritic protein synthesis. *Journal of Neuroscience Research, 88*(7), 1420–1432.

PART II

ROLE OF SPECIFIC ELEMENTS IN THE MRNA

CHAPTER 6

DENDRITIC TARGETING AND REGULATORY RNA CONTROL OF LOCAL NEURONAL TRANSLATION

TAESUN EOM, ILHAM A. MUSLIMOV, ANNA IACOANGELI, AND HENRI TIEDGE

CONTROL of gene expression is essential for the temporal and spatial modulation of cellular form and function (Mathews, Sonenberg, & Hershey, 2007). Gene expression is regulated at the levels of transcription, RNA processing, translation, and post-translational modification. Translation of mRNAs is an important mechanism in the control of gene expression as regulation of protein synthesis directly and rapidly impacts cellular function (Van Der Kelen, Beyaert, Inze, & De Veylder, 2009). Translation proceeds in the three phases of initiation, elongation, and termination, and various translation factors cooperate with mRNAs, tRNAs, and ribosomal subunits in these phases. Translational initiation, which typically is rate-limiting, is often a site of regulation in the translation pathway.

MECHANISM OF TRANSLATION INITIATION

Translation initiation unfolds in three steps: (i) formation of the 43S preinitiation complex (PIC), (ii) formation of the 48S initiation complex, and (iii) assembly of the 80S ribosome. In eukaryotes, translation initiation requires participation of 12 or more eIFs in the pathway that concludes with the formation of 80S ribosomes (Bourgeois, Mortreux, & Auboeuf, 2016; Costa-Mattioli, Sossin, Klann, & Sonenberg, 2009; Hinnebusch, 2017).

A ternary complex, consisting of eIF2, Met-tRNA$_i^{Met}$, and GTP, initially associates with the small 40S ribosomal subunit, together with eIFs 1, 1A, and 3, resulting in the formation of the 43S PIC (Jackson, Hellen, & Pestova, 2010). Higher-order structures in the 5′ untranslated region (UTR) of an mRNA are unwound by eIF4A, a factor that is in part integrated in eIF4F and operates in conjunction with eIF4B. A heterotrimeric complex, eIF4F consists of eIF4E (a cap-binding protein), eIF4A (a DEAD-box RNA helicase), and eIF4G (a scaffold protein interacting with eIF4E, eIF4A, eIF3, and poly(A)-binding protein (PABP)). eIF4A is an ATP-dependent RNA helicase which, at times assisted by RNA helicases Ddx3 and Dhx29, unwinds mRNA higher-order structures between the 5′ end and the initiator codon (Hinnebusch, 2017). eIF4B plays a dual role: it (i) stimulates the helicase activity of eIF4A and (ii) mediates recruitment of the 43S PIC to the mRNA. The latter is achieved by interactions with eIF3 and with 18S rRNA, the latter a component of the 40S ribosomal subunit. eIF4B contains two RNA-binding domains, an RNA recognition motif (RRM) located in the N-terminal region and two arginine-rich motifs (ARMs) located in the C-terminal region (Méthot, Pause, Hershey, & Sonenberg, 1994; Méthot, Pickett, Keene, & Sonenberg, 1996). eIF4B unspecifically binds mRNAs through the ARMs and specifically binds stem-loop structures in 18S rRNA through the RRM (Méthot et al., 1996). Thus, eIF4B plays a role as a bridge between the 5′ end of the mRNA and the 43S PIC (Méthot et al., 1996). In addition, PABP stimulates translation initiation by binding to the mRNA 3' poly(A) tail and by simultaneously interacting with eIF4G, thus effectively circularizing the mRNA (Pestova, Lorsch, & Hellen, 2007).

Once the 43S PIC has attached to the mRNA, it translocates ("scans") through the 5′ UTR until it recognizes a competent start codon, resulting in 48S initiation complex formation. Subsequently, the 80S ribosome is formed following factor release and joining of the 60S large ribosomal subunit.

RNA Helicases in Translation Initiation

Various RNA helicases participate in splicing, nuclear export, translation, cytoplasmic mRNA decay, miRNA-induced gene silencing, and RNA transport (Bourgeois et al., 2016). Here we will focus on RNA helicases associated with translation initiation.

The paramount RNA helicase operating in translation initiation is eIF4A. The eIF4A family consists of eIFs 4A1, 4A2, and 4A3 of which eIFs 4A1 and 4A2 share 95% amino acid identity whereas eIF4A3 shares 65% identity with the other two isoforms (Bourgeois et al., 2016; Yoder-Hill, Pause, Sonenberg, & Merrick, 1993). Structured 5′ UTRs interfere with the ability of the 43S PIC to bind to and move along the mRNA, and eIFs 4A1 and 4A2 promote 5′ UTR unwinding during 43S PIC recruitment and scanning (Pestova et al., 2007). While eIF4A3 is a component of exon junction complex, it also participates in the unwinding of structured 5′ UTRs (Choe et al., 2014).

Translation in Neurons

A CNS principal neuron typically entertains more than 1,000 synaptic connections with other neurons (Stiles & Jernigan, 2010). Synapses are regulated independently in an input-specific manner (Bartley, Huang, Huber, & Gibson, 2008). The term synaptic plasticity refers to the ability of synapses to strengthen or weaken in response to changes in the input they receive (Hughes, 1958). Long-lasting synaptic plasticity and long-term memory require new protein synthesis (Dahm, Kiebler, & Macchi, 2007; Kang & Schuman, 1996; Kindler, Wang, Richter, & Tiedge, 2005). Locally controlled translation allows select proteins to be synthesized at the synapse upon demand (Job & Eberwine, 2001). Regulators are therefore essential to control local translation at the synapse. The fragile X mental retardation protein (FMRP) (Bhakar, Dolen, & Bear, 2012), miRNAs (Fabian, Sonenberg, & Filipowicz, 2010) and brain cytoplasmic (BC) RNAs (Iacoangeli & Tiedge, 2013) have been identified as translational repressors in neurons in the basal default state. Mechanisms of translational control by FMRP and BC RNAs have been studied in depth and will be discussed in the following.

Fragile X syndrome (FXS), resulting from functional absence of FMRP, is a common cause of intellectual disability and autism (Bassell & Warren, 2008). Loss of FMRP is typically the consequence of transcriptional silencing induced by 200 or more CGG triplet repeats in the 5′ UTR of the FMR1 gene. Lack of FMRP results in impaired neuronal translational control, i.e. in inadequate translational repression of FMRP target mRNAs (Weiler et al., 1997). FMRP inhibits translation at the level of elongation by stalling translocation of ribosomes along the mRNA (Darnell et al., 2011). In addition, FMRP is associated with Dicer and Argonaute 2 (Ago2), components of the RNA-induced silencing complex (Chen & Joseph, 2015).

In Fmr1 KO animals, various proteins are expressed at elevated levels, including the α subunit of the Ca^{2+}/calmodulin-dependent protein kinase II (CaMKIIα), striatal-enriched protein tyrosine phosphatase (STEP), potassium channels Kv3.1 and Kv4.2, microtubule-associated protein 1B, and PI3K enhancer (Goebel-Goody et al., 2012; Lee et al., 2011; Lu et al., 2004;Hou et al., 2006; Strumbos, Brown, Kronengold, Polley, & Kaczmarek, 2010). HITS-CLIP analysis (Darnell et al., 2011) identified 842 FMRP target mRNAs, including a subset encoding pre- and post-synaptic proteins. The data indicate that translation of selective mRNAs is regulated by FMRP to control expression of proteins associated with synaptic functions (Darnell & Klann, 2013). Lack of FMRP in Fmr1 KO mice results in increased susceptibility to audiogenic (sound-induced) seizures (Musumeci et al., 2000).

BC RNAs, constituting a subtype of non-protein-coding, small cytoplasmic RNAs (scRNAs), are neuronal regulatory RNAs that function as repressors of translation initiation in the basal default state (Eom, Berardi, Zhong, Risuleo, & Tiedge, 2011; Eom et al., 2014; Wang et al., 2002; Wang et al., 2005). BC RNAs are predominantly expressed in neurons (DeChiara & Brosius, 1987; Tiedge, Chen, & Brosius, 1993) as non-neuronal expression appears restricted to cells of the germ line (Muslimov et al., 2002) and subsets of cancer cells (Chen, Böcker, Brosius, & Tiedge, 1997; Iacoangeli et al., 2004).

Regulatory BC RNAs include rodent BC1 RNA and primate BC200 RNA. BC1 and BC200 RNAs are not orthologs as their evolutionary pedigrees are different. They are phylogenetically distinct (Martignetti & Brosius, 1993a): a common ancestor does not exist, with BC200 RNA being restricted to primates, BC1 RNA to rodents (Iacoangeli & Tiedge, 2013; Kim, Kass, & Deininger, 1995; Skryabin et al., 1998; Tiedge et al., 1993). Orthologs of either BC1 RNA or BC200 RNA have not been detected in other mammalian orders (Martignetti & Brosius, 1993a; 1993b). Despite the fact that BC1 and BC200 RNAs are evolutionary unrelated, the two RNAs are considered functional analogs as in all tests applied, their modes of action were found equivalent (Eom et al., 2011; Iacoangeli & Tiedge, 2013). BC1 and BC200 RNAs appear to have been independently but convergently recruited into their neuronal functional roles with the beginning of the mammalian radiation about 65 million years ago (Brosius, 2005; Iacoangeli & Tiedge, 2013). BC RNAs, in terms of their convergent evolutionary history, are thus comparable to animal wings. Wings are important for most birds, bats, and butterflies, but there is no common wing ancestor: animal wings are functionally analogous, not phylogenetically orthologous. We will further discuss RNA evolutionary mechanisms in the section "Activity-Dependent Transport."

BC1 and BC200 RNAs are of limited sequence similarity but feature identical 3' C-loop motifs that are critical for binding to eIF4B (Eom et al., 2011). BC RNAs repress protein synthesis by interacting with eIFs 4A and 4B and with PABP (Eom et al., 2011, 2014; Lin, Pestova, Hellen, & Tiedge, 2008; Wang et al., 2002; Wang et al., 2005). BC RNA translational repression is mediated mainly through interactions with eIFs 4A and 4B whereas interactions with PABP account for at most 20% of the repression mechanism (Eom et al., 2011; Lin et al., 2008). BC RNAs bind to eIF4A through their central A-rich regions and the 3' stem-loop interfaces, causing inhibition of eIF4A helicase activity (Eom et al., 2011; Lin et al., 2008). They bind to eIF4B through their 3′ C-loop motifs in a high-affinity interaction that competitively inhibits binding of 18S rRNA to the factor and, as a consequence, prevents recruitment of the 43S PIC to the mRNA (Eom et al., 2011).

Phosphorylation Status in Translational Regulation

Phosphorylation and dephosphorylation of proteins are known to be among the most frequent posttranslational modification events and play important roles as molecular switches to regulate protein function. In response to various extracellular stimuli and intracellular messengers, signaling cascades regulate translation of mRNAs by modulating phosphorylation of components of the translational machinery (Roux & Topisirovic, 2012).

FMRP Phosphorylation Status

Phosphorylation of FMRP occurs primarily at serine 499 (S499), and FMRP thus phosphorylated associates with stalled ribosomes (Ceman et al., 2003), that is, is linked to reduced translation elongation. In addition, phosphorylation of FMRP promotes formation of Ago2-miRNA inhibitory complexes on target mRNAs, causing translational repression (Muddashetty et al., 2011). S499-phosphorylated FMRP thus acts as a negative translational regulator in the basal state. Upon stimulation of group I metabotropic glutamate receptors (mGluRs), protein phosphatase 2A (PP2A) is rapidly activated to dephosphorylate FMRP at S499. This dephosphorylation enhances translation of FMRP target mRNAs as dephosphorylated FMRP does not stall polyribosomes and does not promote formation of miRNA inhibitory complexes (Muddashetty et al., 2011; Narayanan et al., 2007; 2008; Niere, Wilkerson, & Huber, 2012). Thus, phosphorylation and dephosphorylation of FMRP are key regulatory events in the activity-dependent control of neuronal protein synthesis (Figure 6.1).

eIF4B Phosphorylation Status

Serine 406 (S406) and serine 422 (S422) are known phosphorylation sites on eIF4B (Raught et al., 2004; van Gorp et al., 2009). Phosphorylation of eIF4B at S422 enhances its affinity for the eIF3 complex and stimulates the helicase activity of eIF4A (Shahbazian et al., 2006). Several kinases phosphorylate eIF4B at S422, including S6 kinase (S6K), a downstream target of PI3K/mTOR signaling (Raught et al., 2004), p90 ribosomal protein S6 kinase (RSK), a downstream target of ERK1/2 MAPK signaling (Shahbazian et al., 2006), and protein kinase B (PKB), a component of the PI3K-mTOR pathway (van Gorp et al., 2009). In contrast, the significance of S406 eIF4B phosphorylation in translational control has remained poorly understood until recently.

Phosphorylation of eIF4B at S406 significantly increases the factor's affinity for BC RNAs (whereas phosphorylation at S422 does not), with the result that it is higher for BC RNAs than it is for 18S rRNA (Eom et al., 2014). S406 dephosphorylation reduces the factor's affinity for BC RNAs by almost two orders of magnitude. In neurons, such dephosphorylation occurs rapidly, within 1 min, following stimulation of group I mGluRs and activation of PP2A. As a result, the eIF4B-BC RNA complex dissociates, enabling recruitment of the 43S PIC and thus translation initiation. PP2A is subsequently deactivated (beginning after 2 min) and eIF4B is rephosphorylated at S406 (within 10 min of initial mGluR activation) (Eom et al., 2014; Narayanan et al., 2007). As a result, BC RNA translational control switches back from permissive to repressive (Figure 6.1). This mechanism establishes an activity-dependent window of opportunity for the synaptic synthesis of select proteins, effectively controlling the amount of protein that can be produced following one round of receptor activation. Tight control of local protein synthesis is essential, as excessive production may give rise to adverse phenotypic consequences (see the next section).

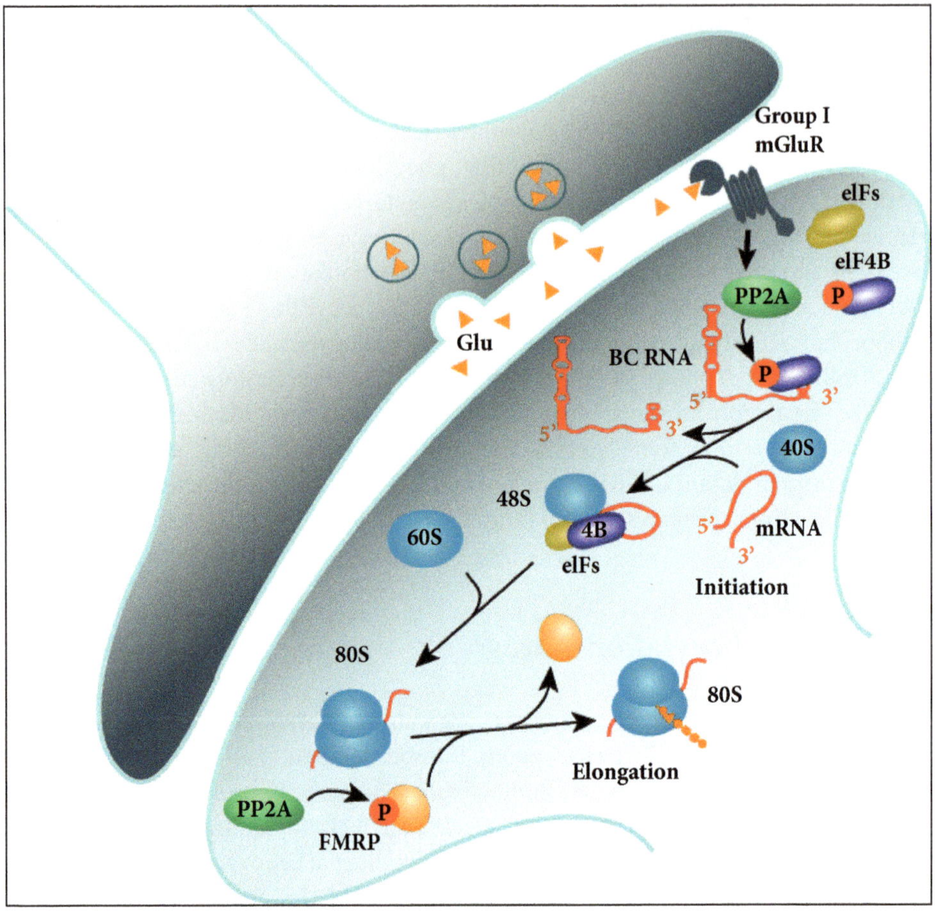

FIGURE 6.1. Phosphorylation status in translational control. BC RNAs and FMRP repress translation of their target mRNAs at the levels of initiation and elongation, respectively. During initiation, binding of BC RNAs to eIF4B competitively inhibits binding of 18S rRNA (and, consequently, the 43S PIC) to the factor, thus repressing translation initiation. FMRP stalls translocation of ribosomes along mRNAs in the elongation phase of translation. As a result, the two translational control systems (BC RNAs and FMRP) repress translation of a subset of mRNAs at the synapse in the basal state. Upon neuronal stimulation via group I mGluR signaling, PP2A is rapidly activated to dephosphorylate eIF4B at S406 and FMRP at S499. Dephosphorylation of eIF4B at S406 decreases its binding affinity for BC RNAs, releasing BC RNAs from the eIF4B-BC RNA complex (Eom et al., 2014). eIF4B engages with the 40S small ribosomal subunit in the 43S PIC to initiate translation. Dephosphorylation of FMRP ends polyribosome stalling, thus allowing elongation to resume.

Which kinase is responsible for the rephosphorylation of eIF4B at S406? Maternal and embryonic leucine zipper kinase (MELK) phosphorylates eIF4B at S406 (but not at S422) in cancer cells (Wang et al., 2016). Synthesis of myeloid cell leukemia 1 protein, an anti-apoptotic protein important for cancer cell survival during cell division, is associated with phosphorylation of eIF4B by MELK (Wang et al., 2016). The eIF4B phosphorylation status may differentially impact translation in different cell types, owing to cell-type specific availability of additional interacting factors.

The dual phosphorylation status of eIF4B, at S406 and S422, is a determinant of the reversible, activity-dependent switch between repressive and permissive states of translation initiation in neurons (Eom et al., 2014). The available evidence is in support of the following scenario. In the repressive state, S406 is phosphorylated, allowing high-affinity binding of BC RNAs and thus disallowing interactions with the 43S PIC, as a consequence repressing translation initiation. At the same time, S422 is dephosphorylated, and binding to eIF3 is not enabled. Conversely, in the permissive state, S406 is PP2A-dephosphorylated, enabling eIF4B interactions with the 43S PIC, whereas S422 is phosphorylated, enabling interactions with eIF3. As a result, translation is initiated.

Translational Dysregulation: Phenotypic Consequences

Absence of regulatory BC1 RNA in the BC1 KO animal model results in exaggerated protein synthesis that is dependent on group I mGluR activation and MEK-ERK signaling (Zhong et al., 2009). Thus, in the basal neuronal default state, BC1 RNA negatively regulates neuronal translation, acting as a break downstream of translational stimulation mediated by activation of group I mGluRs (Chuang et al., 2005; Huber, Kayser, & Bear, 2000; Zhong et al., 2009). Absence of this break gives rise to excessive cortical gamma oscillations, prolonged epileptiform discharges in hippocampal CA3 pyramidal cells, and convulsive seizures upon auditory stimulation in vivo (Zhong et al., 2009).

As cortical gamma oscillations have been implicated in prefrontal cortical mechanisms of perception and cognition (Cho, Konecky, & Carter, 2006; Cho et al., 2015), dysregulation of such oscillations in BC1 KO animals raised the question whether these animals are cognitively impaired. Recent work (Iacoangeli, Dosunmu, Eom, Stefanov, & Tiedge, 2017) indicates that this is indeed the case. BC1 KO animals excessively self-groom, an activity associated with autism-like behavior in rodents (Kalueff et al., 2016; McFarlane et al., 2008; Silverman, Tolu, Barkan, & Crawley, 2010). Furthermore, behavioral analysis using a modified Attentional Set Shift Task assay has shown that while learning and memory per se appear normal, BC1 KO animals are impaired in the cognitive control of acquired memories. When presented with a novel situation that conflicts with previously stored information, BC1 KO animals (but not corresponding WT animals) continue to operate in the framework of the older but now superseded memory. The cognitive errors are of the regressive type (Iacoangeli et al., 2017): a BC1 KO animal will, even after having made a serendipitously correct choice and having been rewarded for it, revert to making a series of consecutive incorrect decisions. The previously stored but now irrelevant information has remained dominant over subsequently acquired relevant but conflicting information (Figure 6.2). Thus, the animals display lack of behavioral flexibility as supervisory cognitive control is defective (Iacoangeli et al., 2017).

In addition, Briz et al. (2017) recently reported that BC1 KO animals exhibited impaired texture recognition and social interactions, similar to what has been observed with animal models of FXS and autism spectrum disorder (ASD) (Pasciuto et al., 2015; Santos, Kanellopoulos, & Bagni, 2014). The novel object recognition test (NORT), using visual NORT (vNORT) and texture NORT (tNORT), was applied to examine whether visual or sensory experience is associated with learning. BC1 KO animals preferred a novel object over a familiar one in the vNORT, but had no preference for a novel object in the tNORT (Briz et al., 2017). In a three-chamber test, BC1 KO animals showed deficits in sociability but not in social memory (Briz et al., 2017).

The question arises how BC1 RNA impacts animal behavior. BC1 RNA represses synthesis of a subset of synaptic proteins (including PSD-95, NR2B and mGluR5) in postsynaptic microdomains (Briz et al., 2017; Zhong et al., 2009). In BC1 KO mice, elevated levels of such proteins lead to increased spine density and decreased dendritic complexity, similar to what has been observed in Fmr1 KO mice, an FXS animal model (Restivo et al., 2005; De Rubeis, Fernandez, Buzzi, Di Marino, & Bagni, 2012)). Elevated PSD-95 levels cause increased postsynaptic clustering and stabilization of spines, and increase the number and size of spines along dendrites (De Roo, Klauser, Mendez, Poglia, & Muller, 2008; El-Husseini, Schnell, Chetkovich, Nicoll, & Bredt, 2000). In BC1 KO animals, significantly elevated levels of PSD-95 are associated with spine abnormalities, including enlarged spine heads and PSDs (Briz et al., 2017; Zhong et al., 2009). Such abnormalities may skew synaptic excitation-repression equilibria, causing neuronal hyperexcitability which, in BC1 KO animals, has been observed in the form of an increased propensity for prolonged epileptiform discharges and a susceptibility to audiogenic seizures (Zhong et al., 2009; Iacoangeli & Tiedge, 2013; Zhong et al., 2009).

In addition, excessive cortical gamma frequency oscillations are a hallmark of BC1 KO mice (Zhong et al., 2009). Gamma frequency band activity has been implicated in cortical mechanisms of cognitive processing (Fries, 2009), specifically as it relates to prefrontal cortical network functionality and cognitive flexibility (Cho et al., 2006; 2015). Impaired gamma synchrony may cause discoordination of network connectivity and reduced cognitive flexibility, thus contributing to autism spectrum disorder (ASD) phenotypes (D'Cruz et al., 2013; Uhlhaas & Singer, 2012. In summary, recent work (Briz et al., 2017; Iacoangeli et al., 2017) is in support of the notion that impaired neuronal translational control may give rise to cognitive phenotypes that resemble ASD manifestations (Aguilar-Valles et al., 2015; Gkogkas et al., 2013; Santini et al., 2013).

Strain type may constitute a confound in the phenotypic analysis of mice. Iacoangeli et al (2017) found that cognitive competence in the C57BL/6J strain is generally lower than that in a C57BL/6J and 129X1/SvJ mixed-background strain. Translational control may potentially be impacted in C57BL/6J mice as a result of a mutation in a CNS-specific tRNA gene which, in the presence of a second, interacting mutation, may cause ribosome stalling (Ishimura et al., 2014). Possible deleterious consequences of such an epistatic tRNA gene mutation may thus affect the readout of other mutations (Ishimura et al., 2014).

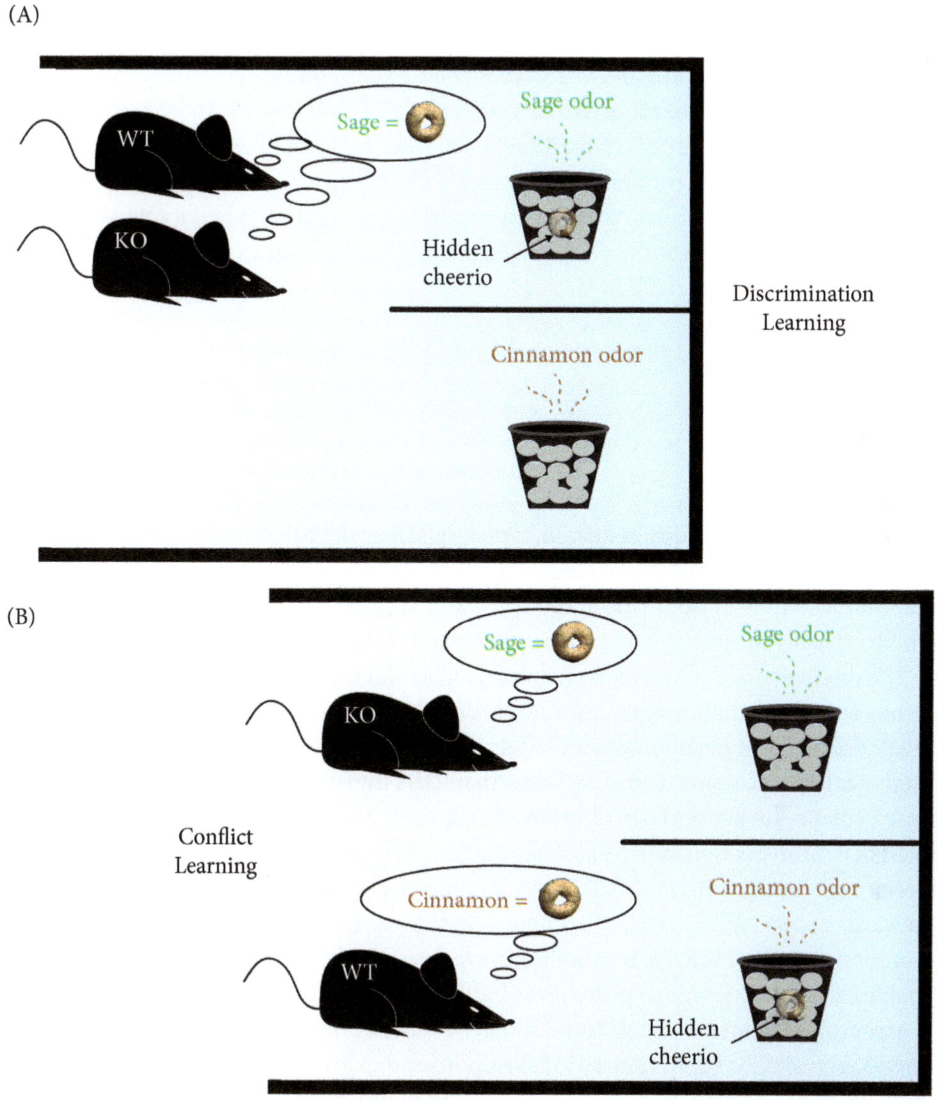

FIGURE 6.2. BC RNA control in animal cognition. (**A**) BC1 KO and WT animals perform equally well in discrimination learning. They learn and memorize that a scent of sage, but not of cinnamon, is predicting a reward (a cheerio). (**B**) BC1 KO animals are significantly impaired in cognitive flexibility. In conflict learning sessions, WT animals rapidly adjust their retrieval strategy as they learn and memorize that the reward-predicting odor has been switched from sage to cinnamon. BC1 KO animals, in contrast, operate on the basis of the earlier acquired information and continue to search for a reward in the sage-scented bowl even when confronted with repeated negative feedback. Thus, in the absence of BC RNA translational control, cognitive flexibility is impaired as animal behavior is dominated by previously established but now outdated outcome expectancies (Iacoangeli et al., 2017).

Adapted from Iacoangeli et al. (2017).

This is directly relevant in cases where such other mutations impact neuronal translational control pathways, in particular considering the possibility that tRNA mutation-induced potential ribosome stalling may be exacerbated under conditions when rapid or local synthesis of proteins is required in the brain (Darnell, 2014). In conclusion, any mutation impacting neuronal translation in a mouse-model on the C57BL/6J background may potentially face phenotypic interactions with that strain's tRNA mutation. Caution is therefore advisable when performing translational control phenotypic work with mutant model mice on the C57BL/6J background (Darnell, 2014).

Dendritic RNA Targeting

Dendritic RNAs

Once a neuron is fully differentiated, distal dendritic domains can be several hundred micrometers away from the nucleus. Dendritic arborizations of mature neurons may feature thousands of dendritic spines (Loew & Hell, 2013), micro-protrusions each of which typically forms a synapse with a presynaptic terminal.

The discovery of RNAs, ribosomes and translation factors in dendrites and dendritic spines suggested that synapse form and function can be modulated directly through regulation of local protein synthesis (Steward & Levy, 1982; Sutton & Schuman, 2006; Tiedge & Brosius, 1996). Today we know that RNA transport serves as an important cellular protein-sorting and distribution mechanism, and that mRNAs are locally translated in neurons as well as in other eukaryotic cell types (Grossman, Aldridge, Weiler, & Greenough, 2006; Martin & Zukin, 2006; Pfeiffer & Huber, 2006; Schuman, Dynes, & Steward, 2006; Wells, 2006). Through this mechanism, neurons acquire spatial and temporal control of mRNA translation in synapto-dendritic domains. Neurons are thus enabled to rapidly produce proteins at individual dendritic branches or even at single synapses in response to local stimulation (Besse & Ephrussi, 2008). Thus, RNA transport is requisite for rapid translational responses that are independent of transcription in the nucleus or translation in the cell body (Shav-Tal & Singer, 2005).

Local translation occurs during development in processes of immature neurons as well as in dendrites of fully differentiated neurons (Kennedy & Ehlers, 2006). The first step for such regulation is the selective transport of a specific set of RNAs (including mRNAs, tRNAs, and regulatory RNAs) to dendritic destination sites. Well-established dendritically localized mRNAs include those encoding microtubule-associated protein 2 (MAP2) (Garner, Tucker, & Matus, 1988), the α subunit of Ca^{2+}/calmodulin-dependent protein kinase II (CaMKIIα) (Burgin et al., 1990), inositol 1,4,5-triphosphate receptor type 1 (Furuichi et al., 1993), neurogranin (Landry, Watson, Kashima, & Campagnoni, 1994), activity-regulated cytoskeleton-associated protein (Arc, arg3.1) (Link et al., 1995; Lyford et al., 1995), dendrin (Herb et al., 1997), cAMP response element binding protein (CREB) (Crino et al., 1998), β-actin (Eom, Antar, Singer, & Bassell, 2003), and protein kinase M-zeta (PKMζ) (Muslimov et al., 2004). In addition, the synapto-dendritic presence of rRNAs, tRNAs and regulatory BC RNAs has been documented (Kleiman, Banker, & Steward, 1993; Tiedge et al., 1993; Tiedge & Brosius, 1996; Tiedge, Fremeau, Weinstock, Arancio, & Brosius, 1991). These discoveries raised questions concerning mechanisms of RNA transport and the nature of spatial determinants that specify RNA delivery to dendritic destination sites.

Cis-Acting Targeting Elements and Trans-Acting Transport Factors

Neuronal RNAs located in synapto-dendritic domains contain cis-acting dendritic targeting elements (DTEs). Deletion of a DTE impairs dendritic localization of the RNA (Blichenberg et al., 1999; Mori, Imaizumi, Katayama, Yoneda, & Tohyama, 2000; Muslimov et al., 1997; Muslimov et al., 2004). In dendritic mRNAs, DTEs are found within UTRs as well as in protein-coding regions (Andreassi & Riccio, 2009; Moore, 2005; Muslimov et al., 2004). It is likely that localized neuronal RNAs use more than one type of DTE as combinations of different DTEs can mediate distinct steps in transport and localization (Blichenberg et al., 1999; Muslimov et al., 2004; Muslimov, Iacoangeli, Brosius, & Tiedge, 2006; Subramanian et al., 2011). Localization of BC RNAs and CaMKIIα mRNA appears to depend on multiple DTEs (Gao, Tatavarty, Korza, Levin, & Carson, 2008; Mori et al., 2000; Muslimov et al., 2004; 2006; Subramanian et al., 2011).

RNA transport is mediated by trans-acting RNA-binding proteins (RBPs) which recognize DTEs and specify RNA delivery (Gao et al., 2008; Huang, Jung, Sarkissian, & Richter, 2002; Kwon et al., 2002; Muslimov et al., 2006). RNAs with multiple DTEs can be recognized by different RBPs, further increasing the complexity of neuronal RNA targeting mechanisms (Bassell & Kelic, 2004; Bullock, Ringel, Ish-Horowicz, & Lukavsky, 2010). Dendritically localized RNAs and their cognate transport factors associate to form ribonucleoprotein (RNP) complexes. These complexes engage, directly or indirectly, with motor proteins that move them along the dendritic cytoskeleton (Doyle & Kiebler, 2011; Kanai, Dohmae, & Hirokawa, 2004).

Transport RBPs are mediators of subcellular RNA delivery. Such RBPs include zipcode binding protein (ZBP), Staufen, and heterogeneous nuclear ribonucleoprotein (hnRNP) A2 (Figure 6.3; Gao et al., 2008; Muslimov, Patel, Rose, & Tiedge, 2011; Shan, Munro, Barbarese, Carson, & Smith, 2003; Tang, Meulemans, Vazquez, Colaco, & Schuman, 2001; Zhang et al., 2001). Several neuronal mRNAs, including those encoding CaMKIIα, Arc, MAP2, and neurogranin, are delivered to dendrites via the hnRNP A2 pathway (which also mediates targeting of myelin basic protein mRNA in oligodendrocytes; Gao et al., 2008; Hoek, Kidd, Carson, & Smith, 1998; Shan et al., 2003; Tübing et al., 2010).

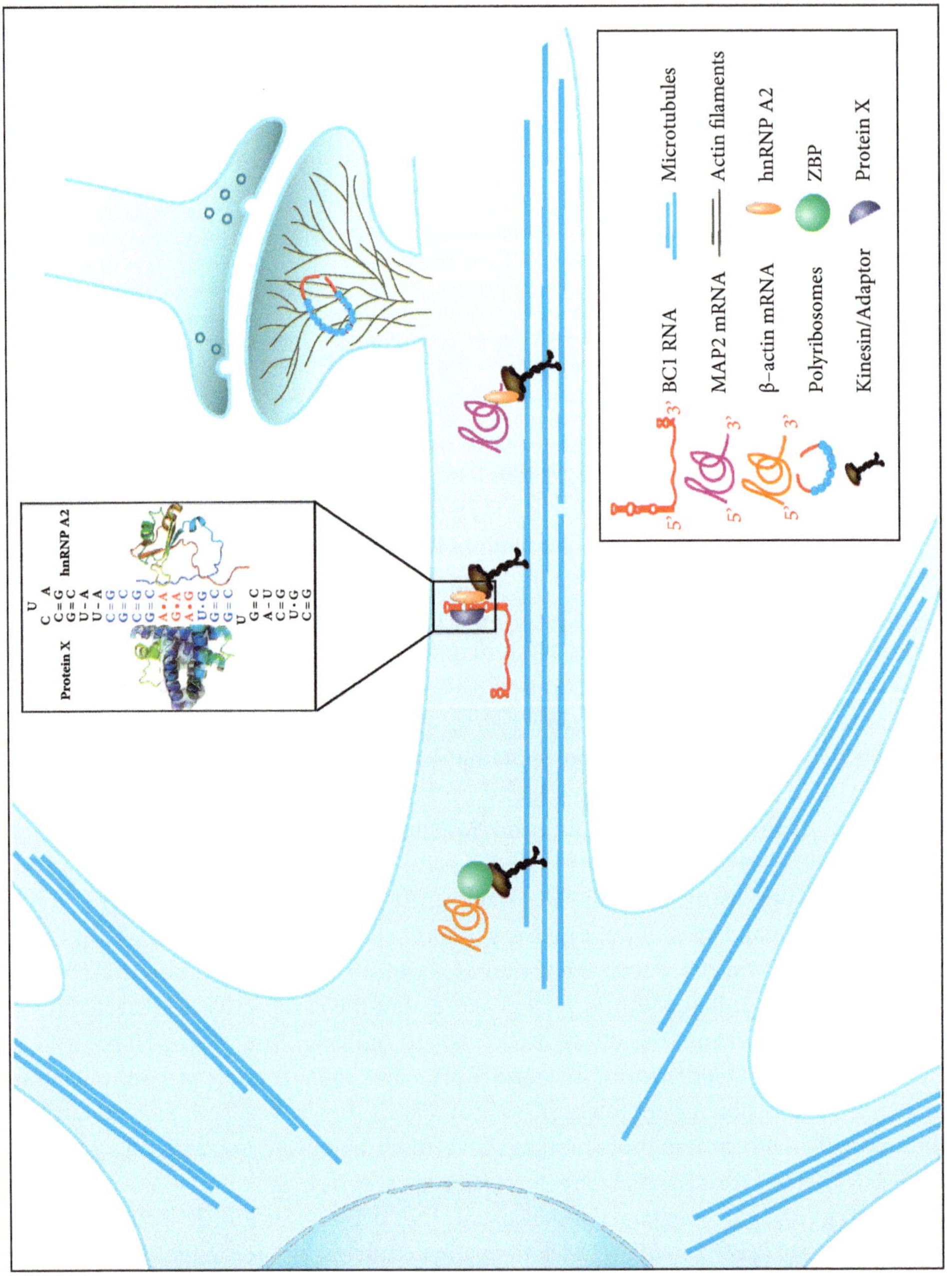

FIGURE 6.3. RNA transport in dendrites. The hnRNPA2 and ZBP pathways are two major RNA targeting mechanisms in neurons. RNA DTEs are recognized by transport RBPs to form RNPs. RNP interactions with molecular motors may require participation of additional adapter proteins. ZBP interacts with the β-actin mRNA zipcode while hnRNP A2 interacts with various RNA DTEs, including those in BC RNAs and MAP 2 mRNA. Inset: hnRNP A2 interacts with the GA-motif DTE in the 5′ stem-loop domain of BC1 RNA. A second transport factor, which remains currently unidentified (Protein X, represented in a hypothetical rendering), is likely participating in dendritic BC RNA transport via DTE interactions (Muslimov et al., 2011).

Dendritic Localization of Regulatory BC RNAs

The double-stranded 5′ stem-loop domains of BC1 and BC200 RNAs contain similarly structured DTEs (Iacoangeli & Tiedge, 2013; Muslimov et al., 2006, 2011). BC RNA DTEs feature architectural motif components that underlie their dendritic targeting competence. In rodent BC1 RNA, the 5′ stem-loop domain harbors an apical GA motif (Muslimov et al., 2006). GA motifs are constructed around noncanonical (non-Watson-Crick) purine•purine nucleotide pairings, including in particular two G•A/A•G pairs arranged in tandem (Iacoangeli & Tiedge, 2013; Tiedge, 2006). These tandem pairs engage in the trans-Hoogsteen/sugar edge format of hydrogen bonding (Iacoangeli & Tiedge, 2013; Lescoute, Leontis, Massire, & Westhof, 2005). They, together with an additional non-Watson-Crick A•A pair, constitute the noncanonical core of the BC1 GA motif. This core is flanked on both sides by several G = C standard Watson-Crick base pairs, affording the motif with high structural stability (Muslimov et al., 2006).

The BC1 GA motif interacts with transport factor hnRNP A2 which is required for the delivery of BC1 RNA to synapto-dendritic domains (Muslimov et al., 2011). The motif is indispensable for dendritic targeting as conversion of its noncanonical purine•purine core to standard WC pairings abolishes hnRNP A2 binding and distal dendritic delivery (Muslimov et al., 2006, 2011). In addition, a GA motif-adjacent unpaired U residue at position 22 (U22) is requisite for dendritic targeting although it is not interacting with hnRNP A2 (Muslimov et al., 2006, 2011). Thus, other transport factors, in addition to hnRNP A2, may participate in BC RNA dendritic delivery.

Primate BC200 RNA, interacting with hnRNP A2 and being delivered to dendrites (Muslimov et al., 2011), copies the BC1 RNA DTE structural arrangements. A double-stranded 5′ stem-loop domain harbors a bipartite GA motif with noncanonical cores featuring A•A and tandem G•A/A•G pairings (Skryabin et al., 1998). As in BC1 RNA, these cores are flanked by standard Watson-Crick G = C pairs and a motif-adjacent unpaired U residue. The architectural attributes of BC1 and BC200 5′ motif structures are thus remarkably similar, and future work will provide further insight into the question of how interactions with RNA transport factors and dendritic delivery competence are encoded by BC RNA architectural motifs.

Recent work with transgenic mouse lines expressing mutant BC1 RNAs has confirmed a 5′ DTE requirement for dendritic localization (Robeck, Skryabin, Rozhdestvensky, Skryabin, & Brosius, 2016). However, apparent methodological insufficiencies are cause for concern (see accompanying technical comment, Sci. Rep. 6, 28300, 2016, published online). For example, the use of ^{35}S-labeled probes with thick (30 μm) tissue sections is problematic as 50% of the emitted beta particles decompose within 25 μm (Bicknese, Shahrokh, Shohet, & Verkman, 1992), causing arbitrary variations in the spatial distribution of labeling intensities as a result of differential-absorption quenching artifacts (Smolen & Beaston-Wimmer, 1990).

Similar to BC RNA GA motifs, CGG-repeat RNA forms stable stem-loops that interact with hnRNP A2 (Sofola et al., 2007; Swanson & Orr, 2007). GA-motif BC RNA DTEs and CGG-repeat stem-loops both feature noncanonical purine•purine base pairing (G•G in the latter case) (Muslimov et al., 2006; Napierala, Michalowski, de Mezer, & Krzyzosiak, 2005; Sobczak et al., 2010; Tiedge, 2006; Zumwalt, Ludwig, Hagerman, & Dieckmann, 2007). GA targeting motifs and CGG-repeat stem-loops compete with each other for binding to hnRNP A2, causing reduced dendritic delivery of BC1 RNA in the presence of expanded CGG-repeat RNA (Muslimov et al., 2011). These observations are relevant with respect to the fragile X premutation disorder which is caused by CGG repeat expansion in the 5′ UTR of the FMR1 gene (Hagerman, 2013).

Activity-Dependent Transport

BC RNA genes originated through retroposition (Kim, Martignetti, Shen, Brosius, & Deininger, 1994; Martignetti & Brosius, 1993a). Gene duplication and retroposition are two major but distinct mechanisms of genomic diversification in eukaryotes. Whereas genes encoding mRNAs (and thus proteins) are frequently dispersed in the genome by gene duplication, genes encoding non-protein-coding (including regulatory) RNAs are often dispersed by retroposition (Brosius, 1991, 2005; Cordaux & Batzer, 2009; Herbert, 2004; Kazazian, 2004). In the retroposition mechanism, cellular transcripts are reverse transcribed and inserted into the genome as retroposons, often in high copy numbers (Cordaux & Batzer, 2009; Iacoangeli & Tiedge, 2013; Kazazian, 2004;. Non-protein-coding RNA genes are frequently transcribed by RNA polymerase III (Pol III). Because such genes—in contrast to Pol II transcribed protein-coding genes—carry promoters within their RNA-coding regions, a retroposon derived form a Pol III transcript may acquire the status of a transcriptionally competent new gene (Iacoangeli & Tiedge, 2013).

The retroposition mechanism has been an innovative force in the shaping and remodeling of eukaryotic genomes. About 45% of the human genome has been retroposition-generated (Herbert, 2004), and over two thirds of the genomic content has been generated by this and other RNA-to-DNA conversion mechanisms (Brosius, 1999; De Koning, Gu, Castoe, Batzer, & Pollock, 2011). In stark contrast, protein-coding genes make up little more than 1% of the human genome (Taft, Pheasant, & Mattick, 2007).

About three quarters of the human genome are transcribed into RNAs (Djebali et al., 2012), most of them likely with regulatory functions. The need for RNA regulators has apparently increased significantly during mammalian phylogenetic development as a result of growing organismal complexities (Mattick, 2003; Taft et al., 2007). The evolution of such RNAs is rapid and ongoing, and evolutionary constraints are often such that it is more likely higher-order than primary structure that is under selective pressure in phylogenetic RNA development (Noller, 2005; Leontis, Lescoute, & Westhof, 2006; Pang, Frith, & Mattick, 2006) Such constraints may include motif recognition by RBPs which is often mediated by RNA 3D structure rather than sequence (Grandin, 2010; Lescoute et al., 2005; Noller, 2005).

Also by retroposition, BC1 RNA has served as a master gene for ID element amplification and dissemination in rodent genomes (Kim et al., 1994). ID elements are identical or similar to the 5′ BC1 domain, including the cis-acting DTE (Kim et al., 1994; Muslimov et al., 1997). Thus, an ID element, retroposed into a host mRNA gene and transcribed in neurons, may confer dendritic targeting competence to that mRNA (Buckley et al., 2011). The retroposition mechanism has thus helped disseminate elements with DTE potential in rodent genomes during phylogenetic development (Iacoangeli & Tiedge, 2013; Muslimov et al., 2014).

ID elements have been identified in introns of cytoplasmic transcripts to which they confer dendritic targeting competence (Buckley et al., 2011). They are also contained in a number of neuronal mRNAs that are targeted to dendrites in an activity-dependent manner (Muslimov et al., 2014). This mechanism relies on high-affinity interactions of noncanonical RNA motif structures with hnRNP A2. The ID element 4 subtype (ID4) interacts with hnRNP A2 in a Ca^{2+} concentration range of 50 nM to 5 μM, and the binding affinity peaks with a K_D = 200 pM at 500 nM Ca^{2+}. mRNAs that carry ID-type noncanonical motif DTEs are delivered to dendrites upon β-adrenergic receptor activation which, in turn, causes influx of Ca^{2+} through voltage-dependent calcium channels (Muslimov et al., 2014). Intracellular Ca^{2+} waves can be triggered by transient rises in postsynaptic $[Ca^{2+}]_i$, reaching dendritic $[Ca^{2+}]_i$ amplitudes of 1 μM or more (Berridge, Lipp, & Bootman, 2000; Grienberger & Konnerth, 2012; Ross, 2012). Once such a Ca^{2+} transient has reached a local threshold of 500 nM, it can cause a switch to the high-affinity conformation of the ID targeting element, enabling binding of hnRNP A2 and dendritic delivery (Muslimov et al., 2014). The activity-dependent ID dendritic delivery mechanism allows neurons to supply synapto-dendritic domains with select RNAs on demand.

In conclusion, dendritic RNA targeting is an essential prerequisite for the locally controlled, activity-dependent synthesis of proteins at the synapse. Impaired neuronal RNA transport mechanisms, increasing evidence indicates, may be contributing to physiological and cognitive dysfunction in neurological disorders.

Outlook

In-depth understanding of molecular-cellular mechanisms that regulate dendritic RNA transport and local protein synthesis is critically important in dissecting how dysregulation of such mechanisms can cause disease. We anticipate that future work will elucidate mechanisms of (i) synapse-to-soma or synapse-to-nucleus signaling, (ii) proximal vs. distal dendritic RNA delivery, (iii) local transition from microtubule-based dendritic delivery to actin-based distribution and docking in postsynaptic microdomains, (iv) local regulation of protein synthesis in response to stimulation or injury, during development or in disease, and (v) local translational control in synaptic plasticity, learning, memory, and cognition.

Recent work has shown that translational dysregulation may underlie human cognitive and behavioral dysfunction (Darnell, 2011). A properly maintained excitation-inhibition balance is a hallmark of neuronal function as any skewing of such balance beyond physiological ranges may cause hypo- or hyperexcitability. Dysregulation of translation initiation causes ASD-like phenotypes in mouse models (Aguilar-Valles et al., 2015; Gkogkas et al., 2013; Santini et al., 2013). Recent genome-wide analyses reveal aberrant expression of non-protein-coding RNAs in ASD patients (Parikshak et al., 2016; Ziats & Rennert, 2013). BC1 KO mice exhibit ASD-like behavioral manifestations (Iacoangeli et al., 2017). Undoubtedly, future advances in the area of translational regulation in neurons will improve our understanding of how translational dysregulation contributes to neurological and cognitive disorders.

Acknowledgments

We thank the members of the Robert F. Furchgott Center for advice and discussion. Work in the authors' laboratory was in part supported by NIH grants NS046769 and DA026110 (to H.T.).

References

Aguilar-Valles, A., Matta-Camacho, E., Khoutorsky, A., Gkogkas, C., Nader, K., Lacaille, J. C., & Sonenberg, N. (2015). Inhibition of group i metabotropic glutamate receptors reverses autistic-like phenotypes caused by deficiency of the translation repressor eif4e binding protein 2. *Journal of Neuroscience*, *35*(31), 11125–11132.

Andreassi, C., & Riccio, A. (2009). To localize or not to localize: mRNA fate is in 3′ UTR ends. *Trends in Cell Biology*, *19*(9), 465–474.

Bartley, A. F., Huang, Z. J., Huber, K. M., & Gibson, J. R. (2008). Differential activity-dependent, homeostatic plasticity of two neocortical inhibitory circuits. *Journal of Neurophysiology*, *100*(4), 1983–1994.

Bassell, G. J., & Kelic, S. (2004). Binding proteins for mRNA localization and local translation, and their dysfunction in genetic neurological disease. *Current Opinion in Neurobiology*, *14*(5), 574–581.

Bassell, G. J., & Warren, S.T. (2008). Fragile X syndrome: Loss of local mRNA regulation alters synaptic development and function. *Neuron*, *60*(2), 201–214.

Berridge, M. J., Lipp, P., & Bootman, M. D. (2000). The versatility and universality of calcium signalling. *Nature Reviews Molecular Cell Biology*, *1*(1), 11–21.

Besse, F., & Ephrussi, A. (2008). Translational control of localized mRNAs: Restricting protein synthesis in space and time. *Nature Reviews Molecular Cell Biology*, *9*(12), 971–980.

Bhakar, A. L., Dolen, G., & Bear, M. F. (2012). The pathophysiology of fragile X (and what it teaches us about synapses). *Annual Review of Neuroscience*, *35*, 417–443.

Bicknese, S., Shahrokh, Z., Shohet, S. B., & Verkman, A. S. (1992). Single photon radioluminescence. I. Theory and spectroscopic properties. *Biophysical Journal*, *63*(5), 1256–1266.

Blichenberg, A., Schwanke, B., Rehbein, M., Garner, C.C., Richter, D., & Kindler, S. (1999). Identification of a cis-acting dendritic targeting element in MAP2 mRNAs. *Journal of Neuroscience*, *19*(20), 8818–8829.

Bourgeois, C. F., Mortreux, F., & Auboeuf, D. (2016). The multiple functions of RNA helicases as drivers and regulators of gene expression. *Nature Reviews Molecular Cell Biology*, *17*(7), 426–438.

Briz, V., Restivo, L., Pasciuto, E., Juczewski, K., Mercaldo, V., Lo, A. C.,... Bagni, C. (2017). The non-coding RNA BC1 regulates experience-dependent structural plasticity and learning. *Nature Communications*, *8*(1), 293.

Brosius, J. (1991). Retroposons—seeds of evolution. *Science*, *251*(4995), 753.

Brosius, J. (1999). Genomes were forged by massive bombardments with retroelements and retrosequences. *Genetica*, *107*(1–3), 209–238.

Brosius, J. (2005). Echoes from the past—are we still in an RNP world? *Cytogenetic and Genome Research*, *110*(1–4), 8–24.

Buckley, P. T., Lee, M. T., Sul, J. Y., Miyashiro, K. Y., Bell, T. J., Fisher, S. A.,... Eberwine, J. (2011). Cytoplasmic intron sequence-retaining transcripts can be dendritically targeted via ID element retrotransposons. *Neuron*, *69*(5), 877–884.

Bullock, S. L., Ringel, I., Ish-Horowicz, D., & Lukavsky, P. J. (2010). A′-form RNA helices are required for cytoplasmic mRNA transport in Drosophila. *Nature Structural & Molecular Biology*, *17*(6), 703–709.

Burgin, K. E., Waxham, M. N., Rickling, S., Westgate, S.A., Mobley, W. C., & Kelly, P. T. (1990). In situ hybridization histochemistry of Ca^{2+}/calmodulin-dependent protein kinase in developing rat brain. *Journal of Neuroscience*, *10*(6), 1788–1798.

Ceman, S., O'Donnell, W. T., Reed, M., Patton, S., Pohl, J., & Warren, S. T. (2003). Phosphorylation influences the translation state of FMRP-associated polyribosomes. *Human Molecular Genetics*, *12*(24), 3295–3305.

Chen, E., & Joseph, S. (2015). Fragile X mental retardation protein: A paradigm for translational control by RNA-binding proteins. *Biochimie*, *114*, 147–154.

Chen, W., Böcker, W., Brosius, J., & Tiedge, H. (1997). Expression of neural BC200 RNA in human tumours. *Journal of Pathology*, *183*(3), 345–351.

Cho, K. K., Hoch, R., Lee, A.T., Patel, T., Rubenstein, J.L., & Sohal, V. S. (2015). Gamma rhythms link prefrontal interneuron dysfunction with cognitive inflexibility in Dlx5/6(+/−) mice. *Neuron*, *85*(6), 1332–1343.

Cho, R. Y., Konecky, R. O., & Carter, C. S. (2006). Impairments in frontal cortical gamma synchrony and cognitive control in schizophrenia. *Proceedings of the National Academy of Sciences of the United States of America*, *103*(52), 19878–19883.

Choe, J., Ryu, I., Park, O. H., Park, J., Cho, H., Yoo, J. S.,... Kim, Y. K. (2014). eIF4AIII enhances translation of nuclear cap-binding complex-bound mRNAs by promoting disruption of secondary structures in 5′ UTR. *Proceedings of the National Academy of Sciences of the United States of America*, *111*(43), E4577–E4586.

Chuang, S. C., Zhao, W., Bauchwitz, R., Yan, Q., Bianchi, R., & Wong, R. K. (2005). Prolonged epileptiform discharges induced by altered group I metabotropic glutamate receptor-mediated synaptic responses in hippocampal slices of a fragile X mouse model. *Journal of Neuroscience*, *25*(35), 8048–8055.

Cordaux, R., & Batzer, M. A. (2009). The impact of retrotransposons on human genome evolution. *Nature Reviews Genetics*, *10*(10), 691–703.

Costa-Mattioli, M., Sossin, W.S., Klann, E., & Sonenberg, N. (2009). Translational control of long-lasting synaptic plasticity and memory. *Neuron*, *61*(1), 10–26.

Crino, P., Khodakhah, K., Becker, K., Ginsberg, S., Hemby, S., & Eberwine, J. (1998). Presence and phosphorylation of transcription factors in developing dendrites. *Proceedings of the National Academy of Sciences of the United States of America*, *95*(5), 2313–2318.

D'Cruz, A. M., Ragozzino, M. E., Mosconi, M. W., Shrestha, S., Cook, E. H., & Sweeney, J. A. (2013). Reduced behavioral flexibility in autism spectrum disorders. *Neuropsychology*, *27*(2), 152–160.

Dahm, R., Kiebler, M., & Macchi, P. (2007). RNA localisation in the nervous system. *Seminars in Cell & Developmental Biology*, *18*(2), 216–223.

Darnell, J. C. (2011). Defects in translational regulation contributing to human cognitive and behavioral disease. *Current Opinion in Genetics & Development*, *21*(4), 465–473.

Darnell, J. C., Van Driesche, S. J., Zhang, C., Hung, K. Y., Mele, A., Fraser, C. E., . . . Darnell, R. B. (2011). FMRP stalls ribosomal translocation on mRNAs linked to synaptic function and autism. *Cell*, *146*(2), 247–261.

Darnell, J. C., & Klann, E. (2013). The translation of translational control by FMRP: Therapeutic targets for FXS. *Nature Neuroscience*, *16*(11), 1530–1536.

Darnell, J. C. (2014). Molecular biology. Ribosome rescue and neurodegeneration. *Science*, 345(6195), 378–379.

De Koning, A. P., Gu, W., Castoe, T. A., Batzer, M. A., & Pollock, D. D. (2011). Repetitive elements may comprise over two-thirds of the human genome. *PLOS Genetics*, *7*(12), e1002384.

De Roo, M., Klauser, P., Mendez, P., Poglia, L., & Muller, D. (2008). Activity-dependent PSD formation and stabilization of newly formed spines in hippocampal slice cultures. *Cerebral Cortex*, *18*(1), 151–161.

De Rubeis, S., Fernández, E., Buzzi, A., Di Marino, D., & Bagni, C. (2012) Molecular and cellular aspects of mental retardation in the Fragile X syndrome: From gene mutation/s to spine dysmorphogenesis. In M. Kreutz & C. Sala (Eds.), *Synaptic plasticity. Advances in experimental medicine and biology* (pp. 517–551). Vienna, Austria: Springer.

DeChiara, T. M., & Brosius, J. (1987). Neural BC1 RNA: cDNA clones reveal nonrepetitive sequence content. *Proceedings of the National Academy of Sciences of the United States of America*, *84*(9), 2624–2628.

Djebali, S., Davis, C. A., Merkel, A., Dobin, A., Lassmann, T., Mortazavi, A., . . . Gingeras, T. R. (2012). Landscape of transcription in human cells. *Nature*, *489*(7414), 101–108.

Doyle, M., & Kiebler, M. A. (2011). Mechanisms of dendritic mRNA transport and its role in synaptic tagging. *EMBO Journal*, *30*(17), 3540–3552.

El-Husseini, A. E., Schnell, E., Chetkovich, D. M., Nicoll, R. A., & Bredt, D. S. (2000). PSD-95 involvement in maturation of excitatory synapses. *Science*, *290*(5495), 1364–1368.

Eom, T., Antar, L. N., Singer, R. H., & Bassell, G. J. (2003). Localization of a β-actin messenger ribonucleoprotein complex with zipcode-binding protein modulates the density of dendritic filopodia and filopodial synapses. *Journal of Neuroscience*, 23(32), 10433–10444.

Eom, T., Berardi, V., Zhong, J., Risuleo, G., & Tiedge, H. (2011). Dual nature of translational control by regulatory BC RNAs. *Molecular and Cellular Biology*, *31*(22), 4538–4549.

Eom, T., Muslimov, I. A., Tsokas, P., Berardi, V., Zhong, J., Sacktor, T. C., & Tiedge, H. (2014). Neuronal BC RNAs cooperate with eIF4B to mediate activity-dependent translational control. *Journal of Cell Biology*, *207*(2), 237–252.

Fabian, M. R., Sonenberg, N., & Filipowicz, W. (2010). Regulation of mRNA translation and stability by microRNAs. *Annual Review of Biochemistry*, *79*, 351–379.

Fries, P. (2009). Neuronal gamma-band synchronization as a fundamental process in cortical computation. *Annual Review of Neuroscience*, *32*, 209–224.

Furuichi, T., Simon-Chazottes, D., Fujino, I., Yamada, N., Hasegawa, M., Miyawaki, A., . . . Mikoshiba, K. (1993). Widespread expression of inositol 1,4,5-trisphosphate receptor type 1 gene (Insp3r1) in the mouse central nervous system. *Receptors Channels*, *1*(1), 11–24.

Gao, Y., Tatavarty, V., Korza, G., Levin, M.K., & Carson, J.H. (2008). Multiplexed dendritic targeting of alpha calcium calmodulin-dependent protein kinase II, neurogranin, and activity-regulated cytoskeleton-associated protein RNAs by the A2 pathway. *Molecular Biology of the Cell*, *19*(5), 2311–2327.

Garner, C. C., Tucker, R. P., & Matus, A. (1988). Selective localization of messenger RNA for cytoskeletal protein MAP2 in dendrites. *Nature, 336*(6200), 674–677.

Gkogkas, C. G., Khoutorsky, A., Ran, I., Rampakakis, E., Nevarko, T., Weatherill, D. B.,... Sonenberg, N. (2013). Autism-related deficits via dysregulated eIF4E-dependent translational control. *Nature, 493*(7432), 371–377.

Goebel-Goody, S. M., Wilson-Wallis, E. D., Royston, S., Tagliatela, S. M., Naegele, J. R., & Lombroso, P. J. (2012). Genetic manipulation of STEP reverses behavioral abnormalities in a fragile X syndrome mouse model. *Genes, Brain and Behavior, 11*(5), 586–600.

Grandin, K. (Ed.) (2010). *The Nobel Prizes 2009*. Stockholm, Sweden: Nobel Foundation.

Grienberger, C., & Konnerth, A. (2012). Imaging calcium in neurons. *Neuron, 73*(5), 862–885.

Grossman, A. W., Aldridge, G. M., Weiler, I. J., & Greenough, W. T. (2006). Local protein synthesis and spine morphogenesis: Fragile X syndrome and beyond. *Journal of Neuroscience, 26*(27), 7151–7155.

Hagerman, R. J. (2013). Epilepsy drives autism in neurodevelopmental disorders. *Developmental Medicine & Child Neurology, 55*(2), 101–102.

Herb, A., Wisden, W., Catania, M. V., Marechal, D., Dresse, A., & Seeburg, P. H. (1997). Prominent dendritic localization in forebrain neurons of a novel mRNA and its product, dendrin. *Molecular and Cellular Neuroscience, 8*(5), 367–374.

Herbert, A. (2004). The four Rs of RNA-directed evolution. *Nature Genetics, 36*(1), 19–25.

Hinnebusch, A. G. (2017). Structural insights into the mechanism of scanning and start codon recognition in eukaryotic translation initiation. *Trends in Biochemical Sciences, 42*(8), 589–611.

Hoek, K. S., Kidd, G. J., Carson, J. H., & Smith, R. (1998). hnRNP A2 selectively binds the cytoplasmic transport sequence of myelin basic protein mRNA. *Biochemistry, 37*(19), 7021–7029.

Hou, L., Antion, M. D., Hu, D., Spencer, C. M., Paylor, R., & Klann, E. (2006). Dynamic translational and proteasomal regulation of fragile X mental retardation protein controls mGluR-dependent long-term depression. *Neuron, 51*(4), 441–454.

Huang, Y. S., Jung, M. Y., Sarkissian, M., & Richter, J. D. (2002). N-methyl-D-aspartate receptor signaling results in Aurora kinase-catalyzed CPEB phosphorylation and alpha CaMKII mRNA polyadenylation at synapses. *EMBO Journal, 21*(9), 2139–2148.

Huber, K. M., Kayser, M. S., & Bear, M. F. (2000). Role for rapid dendritic protein synthesis in hippocampal mGluR-dependent long-term depression. *Science, 288*(5469), 1254–1257.

Hughes, J. R. 1958. Post-tetanic potentiation. *Physiological Reviews 38*(1), 91–113.

Iacoangeli, A., Lin, Y., Morley, E. J., Muslimov, I. A., Bianchi, R., Reilly, J.,... Tiedge, H.(2004). BC200 RNA in invasive and preinvasive breast cancer. *Carcinogenesis, 25*(11), 2125–2133.

Iacoangeli, A., & Tiedge, H. (2013). Translational control at the synapse: Role of RNA regulators. *Trends in Biochemical Sciences, 38*(1), 47–55.

Iacoangeli, A., Dosunmu, A., Eom, T., Stefanov, D. G., & Tiedge, H. (2017). Regulatory BC1 RNA in cognitive control. *Learning & Memory, 24*(7), 267–277.

Ishimura, R., Nagy, G., Dotu, I., Zhou, H., Yang, X.L., Schimmel, P.,... Ackerman, S. L. (2014). RNA function. Ribosome stalling induced by mutation of a CNS-specific tRNA causes neurodegeneration. *Science, 345*(6195), 455–459.

Jackson, R. J., Hellen, C. U., & Pestova, T. V. (2010). The mechanism of eukaryotic translation initiation and principles of its regulation. *Nature Reviews Molecular Cell Biology, 11*(2), 113–127.

Job, C., & Eberwine, J. (2001). Localization and translation of mRNA in dendrites and axons. *Nature Reviews Neuroscience, 2*(12), 889–898.

Kalueff, A. V., Stewart, A. M., Song, C., Berridge, K. C., Graybiel, A. M., & Fentress, J. C. (2016). Neurobiology of rodent self-grooming and its value for translational neuroscience. *Nature Reviews Neuroscience, 17*(1), 45–59.

Kanai, Y., Dohmae, N., & Hirokawa, N. (2004). Kinesin transports RNA: Isolation and characterization of an RNA-transporting granule. *Neuron, 43*(4), 513–525.

Kang, H., & Schuman, E. M. (1996). A requirement for local protein synthesis in neurotrophin-induced hippocampal synaptic plasticity. *Science, 273*(5280), 1402–1406.

Kazazian, H. H., Jr. (2004). Mobile elements: Drivers of genome evolution. *Science, 303*(5664), 1626–1632.

Kennedy, M. J., & Ehlers, M. D. (2006). Organelles and trafficking machinery for postsynaptic plasticity. *Annual Review of Neuroscience, 29*, 325–362.

Kim, J., Martignetti, J. A., Shen, M. R., Brosius, J., & Deininger, P. (1994). Rodent BC1 RNA gene as a master gene for ID element amplification. *Proceedings of the National Academy of Sciences of the United States of America, 91*(9), 3607–3611.

Kim, J., Kass, D. H., & Deininger, P. L. (1995). Transcription and processing of the rodent ID repeat family in germline and somatic cells. *Nucleic Acids Research, 23*(12), 2245–2251.

Kindler, S., Wang, H., Richter, D., & Tiedge H. (2005). RNA transport and local control of translation. *Annual Review of Cell and Developmental Biology, 21*, 223–245.

Kleiman, R., Banker, G., & Steward, O. (1993). Subcellular distribution of rRNA and poly(A) RNA in hippocampal neurons in culture. *Molecular Brain Research, 20*(4), 305–312.

Kwon, S., Abramson, T., Munro, T. P., John, C.M., Kohrmann, M., & Schnapp, B. J. (2002). UUCAC- and vera-dependent localization of VegT RNA in Xenopus oocytes. *Current Biology, 12*(7), 558–564.

Landry, C. F., Watson, J. B., Kashima, T., & Campagnoni, A. T. (1994). Cellular influences on RNA sorting in neurons and glia: An in situ hybridization histochemical study. *Molecular Brain Research, 27*(1), 1–11.

Lee, H. Y., Ge, W. P., Huang, W., He, Y., Wang, G. X., Rowson-Baldwin, A., . . . Jan, L. Y. (2011). Bidirectional regulation of dendritic voltage-gated potassium channels by the fragile X mental retardation protein. *Neuron, 72*(4), 630–642.

Leontis, N. B., Lescoute, A., & Westhof, E. (2006). The building blocks and motifs of RNA architecture. *Current Opinion in Structural Biology, 16*(3), 279–287.

Lescoute, A., Leontis, N. B., Massire, C., & Westhof, E. (2005). Recurrent structural RNA motifs, isostericity matrices and sequence alignments. *Nucleic Acids Research, 33*(8), 2395–2409.

Lin, D., Pestova, T. V., Hellen, C. U., & Tiedge H. (2008). Translational control by a small RNA: dendritic BC1 RNA targets the eukaryotic initiation factor 4A helicase mechanism. *Molecular and Cellular Biology, 28*(9), 3008–3019.

Link, W., Konietzko, U., Kauselmann, G., Krug, M., Schwanke, B., Frey, U., & Kuhl, D. (1995). Somatodendritic expression of an immediate early gene is regulated by synaptic activity. *Proceedings of the National Academy of Sciences of the United States of America, 92*(12), 5734–5738.

Loew, L. M. & Hell, S. W. (2013). Superresolving dendritic spines. *Biophysical Journal, 104*(4), 741–743.

Lu, R., Wang, H., Liang, Z., Ku, L., O'Donnell W. T., Li, W., . . . Feng, Y. (2004). The fragile X protein controls microtubule-associated protein 1B translation and microtubule stability in brain neuron development. *Proceedings of the National Academy of Sciences of the United States of America, 101*(42), 15201–15206.

Lyford, G. L., Yamagata, K., Kaufmann, W. E., Barnes, C.A., Sanders, L. K., Copeland, N. G.,...Worley, P. F. (1995). Arc, a growth factor and activity-regulated gene, encodes a novel cytoskeleton-associated protein that is enriched in neuronal dendrites. *Neuron, 14*(2), 433–445.

Martignetti, J. A., & Brosius, J. (1993a). BC200 RNA: A neural RNA polymerase III product encoded by a monomeric Alu element. *Proceedings of the National Academy of Sciences of the United States of America, 90*(24), 11563–11567.

Martignetti, J. A., & Brosius, J. (1993b). Neural BC1 RNA as an evolutionary marker: Guinea pig remains a rodent. *Proceedings of the National Academy of Sciences of the United States of America, 90*(20), 9698–9702.

Martin, K. C. & Zukin, R. S. (2006). RNA trafficking and local protein synthesis in dendrites: An overview. *Journal of Neuroscience, 26*(27), 7131–7134.

Mathews, M. B., Sonenberg, N., & Hershey, J. W. B. (Eds.) (2007). *Translational control in biology and medicine.* Cold Spring Harbor, NY: Cold Spring Harbor Laboratory Press.

Mattick, J. S. (2003). Challenging the dogma: The hidden layer of non-protein-coding RNAs in complex organisms. *Bioessays, 25*(10), 930–939.

McFarlane, H. G., Kusek, G. K., Yang, M., Phoenix, J. L., Bolivar, V. J., & Crawley, J. N. (2008). Autism-like behavioral phenotypes in BTBR T+tf/J mice. *Genes, Brain and Behavior, 7*(2), 152–163.

Méthot, N., Pause, A., Hershey, J. W., & Sonenberg, N. (1994). The translation initiation factor eIF-4B contains an RNA-binding region that is distinct and independent from its ribonucleoprotein consensus sequence. *Molecular and Cellular Biology, 14*(4), 2307–2316.

Méthot, N., Pickett, G., Keene, J. D., & Sonenberg, N. (1996). In vitro RNA selection identifies RNA ligands that specifically bind to eukaryotic translation initiation factor 4B: The role of the RNA recognition motif. *RNA, 2*(1), 38–50.

Moore, M. J. (2005). From birth to death: The complex lives of eukaryotic mRNAs. *Science, 309*(5740), 1514–1518.

Mori, Y., Imaizumi, K., Katayama, T., Yoneda, T., & Tohyama, M. (2000). Two cis-acting elements in the 3′ untranslated region of alpha-CaMKII regulate its dendritic targeting. *Nature Neuroscience, 3*(11), 1079–1084.

Muddashetty, R. S., Nalavadi, V. C., Gross, C., Yao, X., Xing, L., Laur, O.,...Bassell, G. J. (2011). Reversible inhibition of PSD-95 mRNA translation by miR-125a, FMRP phosphorylation, and mGluR signaling. *Molecular Cell, 42*(5), 673–688.

Muslimov, I. A., Santi, E., Homel, P., Perini, S., Higgins, D., & Tiedge H. (1997). RNA transport in dendrites: A cis-acting targeting element is contained within neuronal BC1 RNA. *Journal of Neuroscience, 17*(12), 4722–4733.

Muslimov, I. A., Lin, Y., Heller, M., Brosius, J., Zakeri, Z., & Tiedge H. (2002). A small RNA in testis and brain: Implications for male germ cell development. *Journal of Cell Science, 115*(6), 1243–1250.

Muslimov, I. A., Nimmrich, V., Hernandez, A. I., Tcherepanov, A., Sacktor, T. C., & Tiedge H. (2004). Dendritic transport and localization of protein kinase Mzeta mRNA: Implications for molecular memory consolidation. *Journal of Biological Chemistry, 279*(50), 52613–52622.

Muslimov, I. A., Iacoangeli, A., Brosius, J., & Tiedge H. (2006). Spatial codes in dendritic BC1 RNA. *Journal of Cell Biology, 175*(3), 427–439.

Muslimov, I. A., Patel, M. V., Rose, A., & Tiedge, H. (2011). Spatial code recognition in neuronal RNA targeting: Role of RNA-hnRNP A2 interactions. *Journal of Cell Biology, 194*(3), 441–457.

Muslimov, I. A., Tuzhilin, A., Tang, T. H., Wong, R. K., Bianchi, R., & Tiedge H. (2014). Interactions of noncanonical motifs with hnRNP A2 promote activity-dependent RNA transport in neurons. *Journal of Cell Biology, 205*(4), 493–510.

Musumeci, S. A., Bosco, P., Calabrese, G., Bakker, C., De Sarro, G. B., Elia, M., . . . Oostra, B. A. (2000). Audiogenic seizures susceptibility in transgenic mice with fragile X syndrome. *Epilepsia, 41*(1), 19–23.

Napierala, M., Michalowski, D., de Mezer, M., & Krzyzosiak, W. J. (2005). Facile FMR1 mRNA structure regulation by interruptions in CGG repeats. *Nucleic Acids Research, 33*(2), 451–463.

Narayanan, U., Nalavadi, V., Nakamoto, M., Pallas, D. C., Ceman, S., Bassell, G. J., & Warren, S. T. (2007). FMRP phosphorylation reveals an immediate-early signaling pathway triggered by group I mGluR and mediated by PP2A. *Journal of Neuroscience, 27*(52), 14349–14357.

Narayanan, U., Nalavadi, V., Nakamoto, M., Thomas, G., Ceman, S., Bassell, G. J., & Warren, S. T. (2008). S6K1 phosphorylates and regulates fragile X mental retardation protein (FMRP) with the neuronal protein synthesis-dependent mammalian target of rapamycin (mTOR) signaling cascade. *Journal of Biological Chemistry, 283*(27), 18478–18482.

Niere, F., Wilkerson, J. R., & Huber, K. M. (2012). Evidence for a fragile X mental retardation protein-mediated translational switch in metabotropic glutamate receptor-triggered Arc translation and long-term depression. *Journal of Neuroscience, 32*(17), 5924–5936.

Noller, H. F. (2005). RNA structure: Reading the ribosome. *Science, 309*(5740), 1508–1514.

Pang, K. C., Frith, M. C., & Mattick, J. S. (2006). Rapid evolution of noncoding RNAs: Lack of conservation does not mean lack of function. *Trends in Genetics, 22*(1), 1–5.

Parikshak, N. N., Swarup, V., Belgard, T. G., Irimia, M., Ramaswami, G., Gandal, M. J., . . . Geschwind, D. H. (2016). Genome-wide changes in lncRNA, splicing, and regional gene expression patterns in autism. *Nature, 540*(7633), 423–427.

Pasciuto, E., Borrie, S. C., Kanellopoulos, A. K., Santos, A. R., Cappuyns, E., D'Andrea, L., . . . Bagni, C. (2015). Autism spectrum disorders: Translating human deficits into mouse behavior. *Neurobiology of Learning and Memory, 124*, 71–87.

Pestova, T., Lorsch, J. R., & Hellen, C. U. (2007). The mechanism of translation initiation in eukaryotes. In M. B. Mathews, N. Sonenberg, & J. W. B. Hershey (Eds.), *Translational control in biology and medicine* (pp. 87–128). Cold Spring Harbor, NY: Cold Spring Harbor Laboratory Press.

Pfeiffer, B. E., & Huber, K. M. (2006). Current advances in local protein synthesis and synaptic plasticity. *Journal of Neuroscience, 26*(27), 7147–7150.

Raught, B., Peiretti, F., Gingras, A. C., Livingstone, M., Shahbazian, D., Mayeur, G. L., . . . Hershey, J. W. (2004). Phosphorylation of eucaryotic translation initiation factor 4B Ser422 is modulated by S6 kinases. *EMBO Journal, 23*(8), 1761–1769.

Restivo, L., Ferrari, F., Passino, E., Sgobio, C., Bock, J., Oostra, B. A., . . . Ammassari-Teule, M. (2005). Enriched environment promotes behavioral and morphological recovery in a mouse model for the fragile X syndrome. *Proceedings of the National Academy of Sciences of the United States of America, 102*(32), 11557–11562.

Robeck, T., Skryabin, B. V., Rozhdestvensky, T. S., Skryabin, A. B., & Brosius, J. (2016). BC1 RNA motifs required for dendritic transport in vivo. *Scientific Reports, 6*, 28300.

Ross, W. N. (2012). Understanding calcium waves and sparks in central neurons. *Nature Reviews Neuroscience, 13*(3), 157–168.

Roux, P. P. & Topisirovic, I. (2012). Regulation of mRNA translation by signaling pathways. *Cold Spring Harbor Perspectives in Biology, 4*(11), a012252.

Santini, E., Huynh, T. N., MacAskill, A. F., Carter, A. G., Pierre, P., Ruggero, D., ... Klann, E. (2013). Exaggerated translation causes synaptic and behavioural aberrations associated with autism. *Nature*, *493*(7432), 411–415.

Santos, A. R., Kanellopoulos, A. K., & Bagni, C. (2014). Learning and behavioral deficits associated with the absence of the fragile X mental retardation protein: What a fly and mouse model can teach us. *Learning & Memory*, *21*(10), 543–555.

Schuman, E. M., Dynes, J. L., & Steward, O. (2006). Synaptic regulation of translation of dendritic mRNAs. *Journal of Neuroscience*, *26*(27), 7143–7146.

Shahbazian, D., Roux, P. P., Mieulet, V., Cohen, M. S., Raught, B., Taunton, J., ... Sonenberg, N. (2006). The mTOR/PI3K and MAPK pathways converge on eIF4B to control its phosphorylation and activity. *EMBO Journal*, *25*(12), 2781–2791.

Shan, J., Munro, T. P., Barbarese, E., Carson, J. H., & Smith, R. (2003). A molecular mechanism for mRNA trafficking in neuronal dendrites. *Journal of Neuroscience*, *23*(26), 8859–8866.

Shav-Tal, Y., & Singer, R. H. (2005). RNA localization. *Journal of Cell Science*, *118*(18), 4077–4081.

Silverman, J. L., Tolu, S. S., Barkan, C. L., & Crawley, J. N. (2010). Repetitive self-grooming behavior in the BTBR mouse model of autism is blocked by the mGluR5 antagonist MPEP. *Neuropsychopharmacology*, *35*(4), 976–989.

Skryabin, B. V., Kremerskothen, J., Vassilacopoulou, D., Disotell, T. R., Kapitonov, V. V., Jurka, J., & Brosius, J. (1998). The BC200 RNA gene and its neural expression are conserved in Anthropoidea (Primates). *Journal of Molecular Evolution*, *47*(6), 677–685.

Smolen, A. J., & Beaston-Wimmer, P. (1990). Quantitative analysis of in situ hybridization using image analysis. In M.-F. Chesselt (Ed.), *In situ hybridization histochemistry* (pp. 175–188). Boca Raton: CRC Press.

Sobczak, K., Michlewski, G., de Mezer, M., Kierzek, E., Krol, J., Olejniczak, M., ... Krzyzosiak, W. J. (2010). Structural diversity of triplet repeat RNAs. *Journal of Biological Chemistry*, *285*(17), 12755–12764.

Sofola, O. A., Jin, P., Qin, Y., Duan, R., Liu, H., de Haro, M., ... Botas, J. (2007). RNA-binding proteins hnRNP A2/B1 and CUGBP1 suppress fragile X CGG premutation repeat-induced neurodegeneration in a Drosophila model of FXTAS. *Neuron*, *55*(4), 565–571.

Steward, O., & Levy, W. B. (1982). Preferential localization of polyribosomes under the base of dendritic spines in granule cells of the dentate gyrus. *Journal of Neuroscience*, *2*(3), 284–291.

Stiles, J., & Jernigan, T. L. (2010). The basics of brain development. *Neuropsychology Review*, *20*(4), 327–348.

Strumbos, J. G., Brown, M. R., Kronengold, J., Polley, D. B., & Kaczmarek, L. K. (2010). Fragile X mental retardation protein is required for rapid experience-dependent regulation of the potassium channel Kv3.1b. *Journal of Neuroscience*, *30*(31), 10263–10271.

Subramanian, M., Rage, F., Tabet, R., Flatter, E., Mandel, J. L., & Moine, H. (2011). G-quadruplex RNA structure as a signal for neurite mRNA targeting. *EMBO Report*, *12*(7), 697–704.

Sutton, M. A., & Schuman, E. M. (2006). Dendritic protein synthesis, synaptic plasticity, and memory. *Cell*, *127*(1), 49–58.

Swanson, M. S. & Orr, H. T. (2007). Fragile X tremor/ataxia syndrome: Blame the messenger! *Neuron* *55*(4), 535–537.

Taft, R. J., Pheasant, M., & Mattick, J. S. (2007). The relationship between non-protein-coding DNA and eukaryotic complexity. *Bioessays*, *29*(3), 288–299.

Tang, S. J., Meulemans, D., Vazquez, L., Colaco, N., & Schuman, E. (2001). A role for a rat homolog of staufen in the transport of RNA to neuronal dendrites. *Neuron*, *32*(3), 463–475.

Tiedge, H., Fremeau, R. T. Jr., Weinstock, P. H., Arancio, O., & Brosius J. (1991). Dendritic location of neural BC1 RNA. *Proceedings of the National Academy of Sciences of the United States of America*, *88*(6), 2093–2097.

Tiedge, H., Chen, W., & Brosius J. (1993). Primary structure, neural-specific expression, and dendritic location of human BC200 RNA. *Journal of Neuroscience*, *13*(6), 2382–2390.

Tiedge, H., & Brosius J. (1996). Translational machinery in dendrites of hippocampal neurons in culture. *Journal of Neuroscience*, *16*(22), 7171–7181.

Tiedge, H. (2006). K-turn motifs in spatial RNA coding. *RNA Biology*, *3*(4), 133–139.

Tübing, F., Vendra, G., Mikl, M., Macchi, P., Thomas, S., & Kiebler, M. A. (2010). Dendritically localized transcripts are sorted into distinct ribonucleoprotein particles that display fast directional motility along dendrites of hippocampal neurons. *Journal of Neuroscience*, *30*(11), 4160–4170.

Uhlhaas, P. J. & Singer, W. (2012). Neuronal dynamics and neuropsychiatric disorders: Toward a translational paradigm for dysfunctional large-scale networks. *Neuron*, *75*(6), 963–980.

Van Der Kelen, K., Beyaert, R., Inze, D., & De Veylder, L. (2009). Translational control of eukaryotic gene expression. *Critical Reviews in Biochemistry and Molecular Biology*, *44*(4), 143–168.

van Gorp, A. G., van der Vos, K. E., Brenkman, A. B., Bremer, A., van den Broek, N., Zwartkruis, F., . . . Coffer, P. J. (2009). AGC kinases regulate phosphorylation and activation of eukaryotic translation initiation factor 4B. *Oncogene*, *28*(1), 95–106.

Wang, H., Iacoangeli, A., Popp, S., Muslimov, I. A., Imataka, H., Sonenberg, N., . . . Tiedge ,H. (2002). Dendritic BC1 RNA: Functional role in regulation of translation initiation. *Journal of Neuroscience*, *22*(23), 10232–10241.

Wang, H., Iacoangeli, A., Lin, D., Williams, K., Denman, R. B., Hellen, C. U., & Tiedge H. (2005). Dendritic BC1 RNA in translational control mechanisms. *Journal of Cell Biology*, *171*(5), 811–821.

Wang, Y., Begley, M., Li, Q., Huang, H. T., Lako, A., Eck, M. J., . . . Zhao, J. J. (2016). Mitotic MELK-eIF4B signaling controls protein synthesis and tumor cell survival. *Proceedings of the National Academy of Sciences of the United States of America*, *113*(35), 9810–9815.

Weiler, I. J., Irwin, S. A., Klintsova, A. Y., Spencer, C. M., Brazelton, A. D., Miyashiro, K., . . . Greenough, W. T. (1997). Fragile X mental retardation protein is translated near synapses in response to neurotransmitter activation. *Proceedings of the National Academy of Sciences of the United States of America*, *94*(10), 5395–5400.

Wells, D. G. (2006). RNA-binding proteins: A lesson in repression. *Journal of Neuroscience*, *26*(27), 7135–7138.

Yoder-Hill, J., Pause, A., Sonenberg, N., & Merrick, W. C. (1993). The p46 subunit of eukaryotic initiation factor (eIF)-4F exchanges with eIF-4A. *Journal of Biological Chemistry*, *268*(8), 5566–5573.

Zhang, H. L., Eom, T., Oleynikov, Y., Shenoy, S. M., Liebelt, D. A., Dictenberg, J. B., . . . Bassell, G. J. (2001). Neurotrophin-induced transport of a β-actin mRNP complex increases β-actin levels and stimulates growth cone motility. *Neuron*, *31*(2), 261–275.

Zhong, J., Chuang, S.C., Bianchi, R., Zhao, W., Lee, H., Fenton, A.A., . . . Tiedge H. (2009). BC1 regulation of metabotropic glutamate receptor-mediated neuronal excitability. *Journal of Neuroscience*, *29*(32), 9977–9986.

Ziats, M. N. & Rennert, O. M. (2013). Aberrant expression of long noncoding RNAs in autistic brain. *Journal of Molecular Neuroscience*, 49(3), 589–593.

Zumwalt, M., Ludwig, A., Hagerman, P. J., & Dieckmann, T. (2007). Secondary structure and dynamics of the r(CGG) repeat in the mRNA of the fragile X mental retardation 1 (FMR1) gene. *RNA Biology*, 4(2), 93–100.

CHAPTER 7

INTERNAL RIBOSOME ENTRY SITE-MEDIATED TRANSLATION IN NEURONAL PROTEIN SYNTHESIS

MARTIN HOLCIK

REGULATION of protein synthesis (translation) is a key cellular process that underpins cellular survival. Traditionally, the transcription and translation of the genome was considered a highly correlated phenomenon. This notion has been challenged by the demonstration that mRNA transcriptional outputs correlate with only about 40% of the total protein content in a cell (Schwanhausser et al., 2011; Tebaldi et al., 2012). This disconnect is further accentuated upon epidermal growth factor stimulation, for example, where up to 90% of transcripts exhibited uncoupled translation from transcription (Tebaldi et al., 2012). Interestingly, these studies also showed that highly expressed transcripts are sometimes poorly translated and, vice versa, poorly transcribed genes can be translated efficiently. This suggests that transcription and translation are largely independent of each other, further strengthening the notion that translational control has a major impact on regulating the proteome under certain conditions. It is therefore not surprising that many aberrant cellular processes require modification of the translation machinery and translation output (i.e., proteome). This is often the case in various stress responses and diseases. In addition, it is also particularly important in functioning of highly specialized cells, such as those of the central nervous system.

Neurons make hundreds if not thousands of synaptic connections that require independent maintenance and regulation. In addition, these synaptic connections have to specifically and appropriately respond to physiological stimuli that are unique to each synapse (reviewed in Iacoangeli & Tiedge, 2013). Localized translation is ideally suited to support this function, as it allows for precisely tailored, mosaic repertoire of proteins in individual dendrites and axons. In addition, localized and mRNA-specific translation can support experience-dependent and site-specific modulation of protein complement at each synapse, which is believed to be the basis for synaptic function, plasticity, and learning (for reviews see additional chapters in this volume).

General Control of Translation

Several recently written outstanding reviews describe in detail the key regulatory checkpoints of translation (e.g., Hershey, Sonenberg, & Mathews, 2012; Hinnebusch & Lorsch, 2012; Roux & Topisirovic, 2012). Therefore, only a summary is given here, with an emphasis on those translation factors and regulatory steps that intersect canonical and non-canonical modes of translation.

Translation begins with the recognition of mRNA by cellular translation apparatus and ends with the production of a protein. This entire process is for didactic purposes divided into four steps—initiation, elongation, termination, and ribosome recycling—but it is important to keep in mind that translation is a continuous process. All four steps are regulated; however, the majority of the control seems to center on the initiation step, which is thought to be rate limiting. This is probably because it is more effective to regulate (e.g., stop) the commencement of translation than to deal with unfinished, partly finished, or unwanted proteins that could arise at the wrong time or place and thus be detrimental to cells' homeostasis, function, or survival.

The chief method of translation initiation occurs by means of the so-called cap-dependent scanning mode, which is the main source of de novo synthesized cellular proteins under normal growth conditions. Following synthesis in the nucleus, virtually all eukaryotic mRNAs are modified at both the 5′ and 3′ termini by addition of m^7G cap and poly(A) tail, respectively. The purpose of these modifications is to protect the mRNA against degradation and to promote engagement of the mRNA by the translational apparatus. The recognition of mRNA by the translation apparatus occurs *via* the m^7G cap and a specialized cap-binding protein, eIF4E (but see exceptions, to be discussed). In turn, eIF4E is bound by eIF4G (a scaffolding protein) and an RNA helicase eIF4A (this three-protein complex is termed eIF4F), and subsequently a cohort of additional initiation factors, including eIF3E, which recruit the small ribosomal submit to the 5′ end of the mRNA (Figure 7.1). It is believed that the mRNA-bound 43S ribosomal complex then repositions (scans) along the mRNA until the appropriate initiation codon (most often AUG) in the proper context is recognized and the polypeptide chain synthesis begins. It should be noted that the poly(A) tail and its associated poly(A)-binding protein PABP play a role of a translational enhancer, since the recruitment of 43S to the mRNA is greatly enhanced by eIF4G-PABP interaction (Hentze, Gebauer, & Preiss, 2007). This type of translation is referred to as cap-dependent, implying that the ability of translation machinery to recognize mRNA is absolutely dependent on the presence of m^7G cap at the 5′ end of the mRNA.

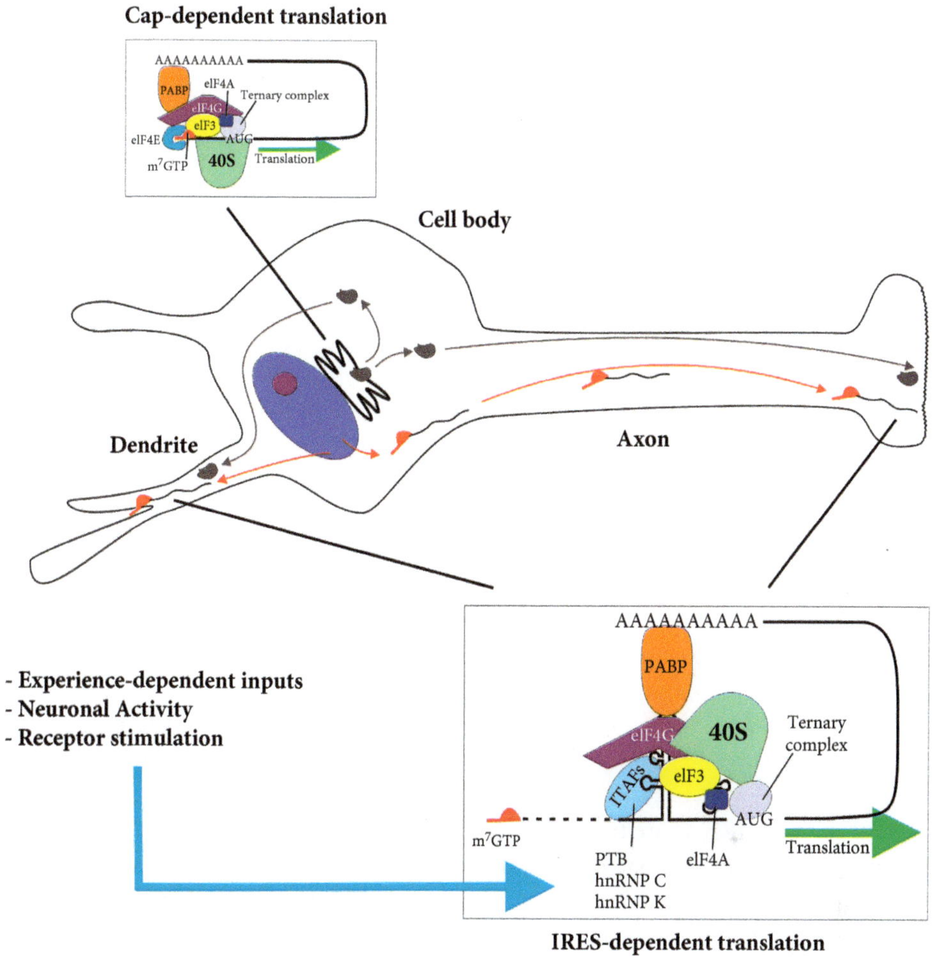

FIGURE 7.1. Cap-dependent and IRES-dependent modes of translation coexist in a single cell but are responsible for production of proteins under distinct circumstances. The bulk of the protein synthesis in cell body proceeds by a conventional, cap-dependent mechanism, schematically shown in the upper insert. For simplicity, only eukaryotic initiation factors (eIF) pertinent to this article are indicated. The 5′ m7GpppX cap structure (shown in red) is recognized and bound by the cap-binding protein eIF4E (light blue) and is then bridged with the 40S ribosomal subunit (green) by an adapter molecule eIF4G (purple). The binding of eIF4G to the 40S subunit is facilitated by eIF3 (yellow). eIF4A (dark blue) is an RNA-dependent ATPase and RNA helicase involved in the unwinding of the secondary structure of the 5′ UTR. Circularization of the mRNA is facilitated by PABP (orange) that interacts with the poly(A) tail and eIF4G. Delivery of the initiator Met-tRNA is brought about by the ternary complex (grey). Both proteins (brown) and mRNA (red) are transported to distal parts of the cell where they engage in a spatially and temporary regulated translation of select cohorts of mRNA by the IRES-dependent mechanism, schematically shown in the bottom insert. Some canonical factors such as eIF3, eIF4A, eIF4G and PABP are likely required by IRES and are shared with the cap-dependent mechanism. The rate of the specialized translation, however, is modulated by neuronal activity, experience-dependent inputs, receptor simulation, and other environmental, physiological, and pathological inputs. This layer of regulation is likely executed through IRES-specific ITAFs, such as PTB, hnRNP C, and hnRNP K that bind to distinct domains of IRES. Individual components of the translation machinery are not drawn to scale. The comprehensive and detailed description of eukaryotic translation and the roles of individual translation factors can be found in (Hershey et al., 2012).

In addition to this main mode of translation initiation, an existence of some other, alternative way(s) to initiate translation was also hypothesized, since it was observed that under conditions in which cap-dependent translation is severely compromised (e.g., viral infection or nutrient deprivation), a subset of mRNAs was shown to be still efficiently translated (e.g., Bushell et al., 2006; Johannes, Carter, Eisen, Brown, & Sarnow, 1999; Morley, Coldwell, & Clemens, 2005). In addition, it was observed that the proper cellular stress response (which is invariably accompanied by reduction in global protein synthesis) requires a reprograming of the cellular proteome, and this can be facilitated by the alternative initiation of translation (Spriggs, Bushell, & Willis, 2010; Spriggs, Stoneley, Bushell, & Willis, 2008). It was therefore proposed that selective translation that occurs by the alternative mode of translation initiation (most commonly by the Internal Ribosome Site Entry [IRES] mechanism) is a key mechanism that is required for cellular adaptation to stress and is used by cells to fine-tune their stress response (Holcik & Sonenberg, 2005; Holcik, Sonenberg, & Korneluk, 2000; Liwak, Faye, & Holcik, 2012; Ruggero, 2013; Silvera, Formenti, & Schneider, 2010).

Internal Ribosome Entry Site-Dependent Mode of Translation Initiation

Several cellular mRNAs are translated by a mechanism that does not appear to rely on the m7G cap and its associated factors. Cap-independent initiation of translation was first observed with RNA viruses (in particular from the *picornaviridae* family), whose RNA is naturally uncapped and yet efficiently translated by the host translation machinery (Pelletier & Sonenberg, 1988). These viruses also encode proteases that cleave several critical eukaryotic initiation factors to block translation of host proteins. For example, upon infection of cells with polio virus, the virus-encoded protease 2A cleaves eIF4G to inactivate the eIF4F cap-binding complex and consequently prevents the ribosome recruitment to capped mRNAs (K. A. Lee & Sonenberg, 1982). The consequences of this cleavage are twofold: (1) it prevents the synthesis of host proteins that might interfere with viral propagation (therefore functioning as a suppressor of innate immunity), and (2) it ensures that the host cell's translational machinery is now fully available for virus protein translation. This mechanism of alternative translation initiation was termed *internal initiation* (Pelletier & Sonenberg, 1988). Instead of cap, distinct functional RNA elements, termed IRES (Internal Ribosome Entry Site) elements, are found in the non-translated region of viral RNA and guide the recruitment of the 40S ribosome directly to or near the initiation codon (Figure 7.1). The primary sequence and secondary structures of these elements are very characteristic and are used to group similar IRES elements into distinct families of related elements. Importantly, the primary sequence and secondary structure also determines whether the IRES requires binding of additional protein(s) for its proper function. Thus, ribosomal recruitment can occur in the absence of any protein factors (as with dicistrovirus intergenic IRES, which binds directly to the 40S and 80S ribosomal subunits by mimicking the shape of tRNA (Kerr, Ma, Jang, Thompson, & Jan, 2016)), or with the aid of various combinations of canonical initiation factors (such as eIF3E, eIF5, and eIF5B) and auxiliary proteins (termed ITAFs—IRES Trans-Acting Factors; to be discussed) in case of other viral IRES, and as dictated by the IRES sequence and structure (for a comprehensive review of factor requirements on viral IRES see Jackson, 2013). These observations spurred study of cellular mRNAs to determine whether a similar mechanism(s) exists. Indeed, it was observed that a subset of cellular mRNAs (up to 10%) was efficiently translated in cells infected with poliovirus (which inhibits cap-dependent translation; Johannes et al., 1999). Subsequent studies described functional IRES elements in a variety of cellular mRNAs. Interestingly, many of these mRNAs encode proteins involved in processes such as cell proliferation and apoptosis, and are critical in determining the survival of a cell under physiological and pathophysiological stress conditions (Holcik & Sonenberg, 2005). This is, perhaps, not surprising, given that these cellular processes require strict control of gene expression. IRES-mediated translation thus provides a means for escaping the global decline in protein synthesis and allows the selective translation of specific mRNAs that are required under given condition. Thus, we and others have proposed that the selective regulation of IRES-mediated translation is important for the regulation of cell death and survival (Holcik & Sonenberg, 2005; Holcik, Sonenberg, et al., 2000; Lewis & Holcik, 2005; Sachs, Sarnow, & Hentze, 1997; Silvera et al., 2010). Indeed, the experimental data from many laboratories have now validated this hypothesis in many models (reviewed in Elroy-Stein & Merrick, 2007; Silvera et al., 2010). The in vivo evidence supporting this concept is particularly striking. For example, the selective impairment of IRES-mediated translation but not cap-dependent translation results in enhanced apoptosis of hematopoietic progenitors and stem cells, leading to the fatal progressive bone marrow failure syndrome known as *dyskeratosis congenita* and its associated tumorigenesis (Bellodi, Kopmar, & Ruggero, 2010; Bellodi, Krasnykh, et al., 2010; Yoon et al., 2006). A switch to IRES-dependent translation has been implicated in breast cancer growth and angiogenic potential in vivo (Braunstein et al., 2007) and in the acquired resistance of cancer cells to radiation-induced apoptosis (Gu et al., 2009). Similarly, HPV-induced transformation of human keratinocytes is accompanied by a switch from cap-dependent to IRES-dependent translation (Mizrachy-Schwartz, Kravchenko-Balasha, Ben-Bassat, Klein, & Levitzki, 2007).

It needs to be stressed, however, that several of the cellular IRESs were subsequently shown to be artefacts of poorly controlled experiments. Many, if not most of the early studies to describe and characterize cellular IRES relied almost exclusively on a use of artificial bi-cistronic DNA constructs. In these constructs, the cap-dependent translation is monitored through the expression of the 5′ cistron, which is followed by a short linker preceding the second cistron. The suspected IRES is then inserted into the linker region of the bi-cistronic construct and translation of the second cistron is taken to indicate IRES activity. However, translation of the second cistron could also be a result of a cryptic promoter within the suspected IRES (giving rise to an independent, capped mRNA transcript harboring a second cistron), or alternative splicing (that would result in excision of the first cistron; Holcik et al., 2005). While utilization of stringent controls such as RNA transfections (to eliminate cryptic promoter activity), qRT-PCR, Northern blots, or siRNAs targeting specific cistrons (to monitor integrity of the mRNA) can discern between these alternate explanation, not all published experiments were carefully designed (Jackson, 2013). Justifiably, this resulted in questioning of the validity of cellular IRES mechanism in general and is still a hotly debated topic (Jackson, 2013; Komar, Mazumder, & Merrick, 2012). Nevertheless, rigorous experiments, both in vitro and in vivo, confirmed several cellular IRESs to be *bona fide* regulatory elements that during cellular stress drive translation of their respective proteins in a manner that seems independent of eIF4E and/or cap. Further evidence for the existence of this mechanism is now emerging from large, genome-wide studies. For example, a recent high-throughput bicistronic assay identified hundreds of novel IRES elements in the human transcriptome (Weingarten-Gabbay et al., 2016). Subsequent characterization of these elements confirmed some previously identified elements (such as secondary structure, short sequence motifs, and base pairing with 18S rRNA; Baird, Turcotte, Korneluk, & Holcik, 2006), but also added new potential mechanisms (IRES within the 3′UTR) of how cellular IRESs operate (Weingarten-Gabbay et al., 2016; Weingarten-Gabbay & Segal, 2016).

Unlike viral IRES, the mechanism of cellular IRESs is still poorly understood. Complicating the issue is the fact that cellular IRESs do not share sequence of structure similarities (Baird, Lewis, Turcotte, & Holcik, 2007; Baird et al., 2006) and they function in a different manner from most viral IRESs. That is, most cellular IRES elements require binding of some of the canonical initiation factors, as well as auxiliary ITAFs that modulate the IRES activity (Holcik & Sonenberg, 2005). Most of the ITAFs identified thus far are RNA-binding proteins that fulfill a variety of functions, including involvement in mRNA splicing, export, and stress granule formation, as well as important roles in cap-dependent translation initiation (Komar & Hatzoglou, 2011). The binding of ITAFs can either enhance or repress IRES activity; it is thought that the positive regulators act either as RNA chaperons that aid in the formation of the proper IRES structure (e.g., Mitchell, Spriggs, Coldwell, Jackson, & Willis, 2003) or by directly recruiting the ribosome to the mRNA (e.g., Thakor et al., 2017). The precise mechanism of how the repressive ITAFs function is not clear, however. Interestingly, many ITAFs shuttle between the nucleus and cytoplasm, and this shuttling is regulated by posttranslational modifications such as phosphorylation in response to a variety of triggers (e.g., Courteau et al., 2015). Therefore, the cytoplasmic availability of positive or negative regulators can also determine the strength of IRES translation (Cammas, Lewis, Vagner, & Holcik, 2008).

IRES Translation in CNS

Although the IRES-dependent translation initiation has been characterized most thoroughly in the context of cellular stress response, and in particular during tumorigenesis and cancer formation, recent data strongly suggest that it plays a critical role in the function of CNS, both under physiological but also pathological conditions. The first observation of IRES-dependent translation was made in dendrites. Because the concentration of components of the translation apparatus is thought to be low in dendrites, Pinkstaff et al. searched for alternative means of translation of dendritically localized mRNAs (Pinkstaff, Chappell, Mauro, Edelman, & Krushel, 2001). They identified five mRNAs whose 5′ UTRs support IRES-dependent translation and are less sensitive to the inhibition of eIF4E. These mRNAs encode for activity regulated cytoskeletal protein (ARC), the α subunit of calcium-calmodulin-dependent kinase II (αCaM Kinase II), dendrin, the microtubule-associated protein 2 (MAP2), and neurogranin (RC3). In addition to supporting translation in bi-cistronic constructs in established cell lines, the authors further demonstrated the ability of RC3 IRES to support translation in dendrites of primary hippocampal neurons. Notably, the IRES-dependent translation was relatively more efficient in the dendrite than in the cell body, suggesting that this mechanism might be responsible for the synthesis of proteins required for the strengthening of active synapse (Pinkstaff et al., 2001).

The physiological role of IRES-mediated translation in neurons was demonstrated in sensory neurons of sea slug *Aplysia californica*. Induction of ovulation and egg-laying behavior in *Aplysia* is triggered by an egg-laying hormone (ELH) that is produced by bag cell neurons (W. Lee & Wayne, 1998). Early observations determined that the electrical afterdischarge (AD), which leads to depletion of ELH, also rapidly stimulates rate of translation of ELH from an existing, stable ELH mRNA (W. Lee & Wayne, 1998). This increase in ELH translation is mediated by an IRES within the 5′ UTR of the ELH mRNA whose activity is stimulated by AD (Dyer et al., 2003). Importantly, the AD leads to dephosphorylation of eIF4E, which is thought to attenuate cap-dependent translation. A similar mechanism has been described during mitosis in which the decreased phosphorylation of eIF4E correlates with induction of IRES-mediated translation (Pyronnet, Dostie, & Sonenberg, 2001). Indeed, forced dephosphorylation of eIF4E in bag cell neurons was sufficient to cause the switch to IRES-dependent translation (Dyer et al., 2003).

Neurotrophin receptors affect multiple cellular functions, in particular during the development and maintenance of the nervous system (Huang & Reichardt, 2003). TrkB is one member of this family of receptors. TrkB is activated by BDNF and promotes local protein synthesis, glutamate receptor phosphorylation, synaptic efficacy, and enhanced cell survival (Dobson, Minic, Nielsen, Amiott, & Krushel, 2005). Although TrkB expression is regulated at the level of transcription, mRNA stability, and protein half-life, regulation at the level of protein synthesis would be expected in neurons, in particular since TrkB mRNA is transported to dendrites for local translation (Righi, Tongiorgi, & Cattaneo, 2000). The 5′ UTR of human TrkB mRNA was thus examined for its ability to selectively control translation of TrkB. It was found that it harbors an IRES that supports translation of a reporter mRNA in neuronal cells even when cap-dependent translation is inhibited by overexpression of 4E-BP (Dobson et al., 2005). A similar situation was observed with mouse TrkB, which was shown to contain two distinct IRES elements in two alternatively transcribed mRNA that, however, encode the same TrkB open reading frame (Timmerman, Pfingsten, Kieft, & Krushel, 2007). These two IRES are controlled differently, based on the differentiation state of the SH-SY5Y neuronal cell line used; while the IRES found in exon 1 is constitutively active, the exon 2 IRES is active only in retinoic acid differentiated SH-SY5Y cells. Mechanistically, this could be accomplished through an RNA binding protein PTB. Although PTB binds both exon 1 and exon2 IRESs, only exon 2 IRES requires PTB for its activity (Timmerman et al., 2007). Although neither study examined the translation of the endogenous TrkB, it has been noted that in the young rat brains the TrkB mRNA is synthesized during ischemic injury and activation of the TrkB receptor promotes cell survival (Narumiya, Ohno, Tanaka, Yamano, & Shimada, 1998). Since in other model systems the IRES-dependent translation has been linked to enhanced cell survival (Holcik & Sonenberg, 2005), these observations suggest that it might be the IRES-dependent translation mechanism that mediates enhance expression of TrkB under these conditions.

Fibroblast growth factor 2 (FGF-2) is one of the key growth factors that shapes both embryonal and adult CNS development (Woodbury & Ikezu, 2014). Owing to its crucial developmental role, FGF-2 expression is regulated both spatially and temporally, coinciding with development of specific brain regions (Woodbury & Ikezu, 2014). In particular, the FGF-2 mRNA generates five protein isoforms through the use of multiple alternative translation initiation sites (Arnaud et al., 1999; Florkiewicz & Sommer, 1989) and an IRES (Vagner et al., 1995). The activity of the FGF-2 IRES was measured using a transgenic reporter mouse and was shown to mimic expression of the endogenous FGF-2, with particularly high expression in the brain (Creancier, Morello, Mercier, & Prats, 2000). Further dissection of the FGF-2 IRES activity pattern showed strict spatiotemporal regulation during embryogenesis and into adulthood with marked peak in postnatal day 7, which is coincidental with neuronal maturation, and it was particularly enriched in synaptoneurosomes (Audigier et al., 2008). In isolated cortical neurons the FGF-2 IRES activity was stimulated by co-culturing neurons with astrocytes, and this could be recapitulated by using astrocyte-conditioned medium, suggesting a release of stimulatory diffusible factor from astrocytes (Audigier et al., 2008). Interestingly, FGF-2 itself was identified as one such stimulatory factor (Audigier et al., 2008). In addition, blockage of spontaneous electrical activity in cultured neurons by tetrodotoxin attenuated FGF-2 IRES activity, suggesting that spontaneous electrical activity contributes to the regulation of FGF-2 IRES during development (Audigier et al., 2008).

IRES Translation Is Linked to Neurodegenerative Disease

Alzheimer's Disease

Expression of the amyloid precursor protein (APP) is a critical factor in the development of Alzheimer's disease since it serves as a precursor of β-amyloid peptide in the brain (Farlow, 1998). While investigating the translation of mRNAs during the global translational repression through mitosis Qin and Sarnow noted that APP mRNA remained associated with actively translating ribosome in mitotically arrested HeLa cells (Qin & Sarnow, 2004). Subcloning of the APP 5′ UTR into a bi-cistronic DNA vector confirmed the presence of an IRES element. Subsequent work from the Krushel group confirmed the IRES-mediated translation of APP in neuronal cells and further elucidated the mechanisms controlling translation of APP (Beaudoin, Poirel, & Krushel, 2008). Expression of APP increased in rat neural C6 or human SH-SY5Y neuroblastoma cells in which cap-dependent translation was inhibited with an mTOR inhibitor rapamycin, or in cells in which eIF4E has been depleted by an siRNA. The validity of a *bona fide* IRES was confirmed using DNA and RNA constructs both in cells and in vitro. Most importantly, however, this paper connected the IRES mechanism of APP translation to physiologically relevant conditions. The evidence from postmortem brains from individuals with Alzheimer's disease linked two conditions—increased intracellular iron concentration and acute ischemic injury—with the presence of β-amyloid peptide plaques in the brain (Connor, Menzies, St Martin, & Mufson, 1992; Richardson, 2004). Simulation of these conditions in C6 or SH-SY5Y cells led to an increase in APP mRNA translation as well as APP IRES activity (Beaudoin et al., 2008). This work suggests that IRES-dependent translation of APP is critical for APP expression. Although the role of this mechanism in normal, physiological conditions of the brain is unclear, it suggests that specific targeting of this mechanism could be exploited therapeutically in Alzheimer's disease.

Another characteristic of the Alzheimer's disease is the occurrence of neurofibrillary tangles of hyperphosphorylated protein tau (Harada et al., 1994). Using a variety of stringent methods the 5′ UTR of tau mRNA was shown to harbor an IRES element that was active in human neuroblastoma cell line SK-N-SH or in vitro (Veo & Krushel, 2009). In addition, in cells with siRNA-mediated knock-down of eIF4E the synthesis of endogenous tau protein increased, providing further support for the importance of IRES in the translation of tau. Further biochemical analysis and sequence mapping of the tau IRES disclosed extensive secondary structure and domain architecture that were more similar to viral than to cellular IRESs (Veo & Krushel, 2012). Interestingly, two naturally occurring single nucleotide polymorphisms (SNP) within the tau IRES that are linked to the development of Parkinson disease resulted in a complete loss of IRES activity (Veo & Krushel, 2012). This observation is both intriguing and puzzling; the formation of neurofibrillary tangles has been seen in patients with Parkinson disease (Arima et al., 1999), yet the Parkinson disease-associated SNPs in tau 5′ UTR suppresses tau IRES activity, which should lead to decreased tau expression and diminish tangles. Since the effect of these SNPs on the expression of endogenous tau have not been examined it is possible that the tau IRES behaves differently when found in the context of the endogenous mRNA.

Parkinson Disease

A pathological signature of Parkinson disease are fibrillar aggregates known as Lewy bodies. Although Lewy bodies contain over 70 different molecules, the predominant constituent is α-synuclein, SNCA (Wakabayashi, Tanji, Mori, & Takahashi, 2007). The 5′ UTR of human *SNCA* mRNA was examined for bearing translation regulatory elements and was shown to substantially enhance translation of a reporter construct and to exhibit IRES-activity in both the DNA and RNA-based reporter systems (Koukouraki & Doxakis, 2016). Rapamycin, an mTOR inhibitor, can be used to block cap-dependent translation by activating eIF4E inhibiting proteins, 4E-BPs (Beretta, Gingras, Svitkin, Hall, & Sonenberg, 1996) while the IRES-dependent translation remains mainly undisturbed (Shi, Sharma, Wu, Lichtenstein, & Gera, 2005). Interestingly, the 5′ UTR of *SNCA* could support rapamycin-independent translation in Neuro-2a cells. Furthermore, translation of the endogenous α-synuclein was observed in rapamycin treated HEK-293 cells, suggesting the use of an IRES. Since SNCA expression is deregulated in Parkinson disease the authors also tested if various stress inducers, particularly those associated with ageing brain, alter the activity of SNCA IRES. Indeed, depolarization with KCl, iron accumulation, serum deprivation and oxidative stress all induced both the SNCA IRES activity as well as endogenous SNCA protein levels, indicating that SNCA IRES contributes to accumulation of SNCA under stress conditions (Koukouraki & Doxakis, 2016). Targeting SNCA IRES could therefore represent additional strategy to combat α-synuclein toxicity.

BiP, also known as GRP78, is a molecular chaperone involved in the resolution of the unfolded protein response activated by an endoplasmic reticulum stress in the cells (Casas, 2017). BiP is gaining interest in the field of neurodegeneration since many, if not all age-related neurodegenerative disorders are commonly associated with the accumulation of misfolded or aggregated proteins. Indeed, the levels of BiP are altered in brains of Alzheimer's and Parkinson's disease patients and, conversely, are decreased during aging (Casas, 2017). The expression of BiP was shown to increase during ischemic preconditioning and regeneration, while decrease during neurodegenerative processes (Penas, Casas, Robert, Fores, & Navarro, 2009; Penas et al., 2011; Zhang et al., 2015). In addition, a mutation in BiP disrupts proper development of the thalamocortical exon projections and other forebrain axon tracks (Favero et al., 2013).

BiP function is controlled primarily at the level of its interaction with and dissociation from the ER membrane-anchored sensor proteins PERK, IRE1 and ATF6. In addition, BiP expression is regulated at the level of translation initiation by an IRES, in particular in response to cellular stress (Johannes & Sarnow, 1998; Kim & Jang, 2002; Macejak & Sarnow, 1991). Although the BiP IRES and its interacting proteins were characterized before (Johannes & Sarnow, 1998; Kim, Back, Rho, Lee, & Jang, 2001; Kim, Hahm, & Jang, 2000; Kim & Jang, 2002; Macejak & Sarnow, 1991; Thoma, Bergamini, Galy, Hundsdoerfer, & Hentze, 2004), it was shown only recently that BiP IRES is active in sensory axons (Pacheco & Twiss, 2012). Using axonally targeted fluorescent bicistronic reporter Pacheco and Twiss detected robust rat BiP IRES activity in both the cell body and axons in cultured rat DRG neurons. The axonal activity occurred even after photobleaching and was sensitive to inhibitors of translation, suggesting that the adult rodent sensory neurons have the capacity to support BiP IRES-mediated translation (Pacheco & Twiss, 2012). Although this study did not examine translation of the endogenous BiP in axons, it was shown previously that BiP mRNA is transported to and subsequently locally translated in sensory axons (Willis et al., 2005).

Stimulation of IRES Translation by Opioids

Opioids, such as morphine, signal through specific receptors that are expressed primarily in the central nervous system (Tao & Auerbach, 2002). The mu-opioid receptor (MOR) is the key mediator of the analgesic effect and its expression is regulated by both transcriptional and posttranscriptional mechanisms (P. T. Lee et al., 2014; Song, Choi, Law, Wei, & Loh, 2017). One of the proteins that activates transcription of MOR in neuronal cells is an RNA binding protein hnRNP K (Choi et al., 2008). Notably, injection of morphine into mice elicited robust neuron-specific increase in hnRNP K protein levels, and similar increases were seen in rat primary cortical neurons, or in HEK 293 cells expressing MOR that were treated with morphine (P. T. Lee et al., 2014). Further dissection of the mechanism responsible for morphine-mediated induction of hnRNP K identified an IRES element in the 5′ UTR of hnRNP K mRNA whose activity can be enhanced by morphine (P. T. Lee et al., 2014). In addition, morphine treatment results in a cytoplasmic accumulation of hnRNP K. It has been described for several ITAFs, most notably hnRNP A1, that cytoplasmic accumulation of these proteins (in particular during cellular stress conditions) is needed for their ITAF activity (Cammas et al., 2008). Interestingly, hnRNP K was shown to interact with its own 5′ UTR and act as an ITAF, and this interaction is enhanced by morphine, suggesting an existence of a positive feedback loop: morphine treatment results in enhanced translation of hnRNP K which in turn drives both the transcription of MOR receptor (nuclear hnRNP K) as well as strengthening hnRNP K interaction with its own IRES (cytoplasmic hnRNP K), thus increasing expression levels of hnRNP K. In mice, opioid receptor activation is required for hnRNP K expression since the treatment of animals with naloxone, an opioid antagonist prior to morphine injection prevented the morphine-dependent stimulation of hnRNP K expression (P. T. Lee et al., 2014). Finally, siRNA-mediated reduction of hnRNP K in mice produced approximately 30% inhibition of basal nociceptive sensitivity (measured by tail-flick latency) and significantly attenuated morphine-mediated analgesic effect (P. T. Lee et al., 2014). These data thus provide a first mechanistic glimpse of the translational control of opioid-based pain control.

IRES Control of Neuronal Apoptosis

IRES-dependent mechanism of translation initiation is particularly well suited to maintain or enhance expression of survival proteins under conditions of cellular stress (Holcik & Sonenberg, 2005). This has been nicely demonstrated, both in vitro and in vivo for a number of pro-survival proteins, particularly in the context of tumorigenesis (Bellodi, Kopmar, et al., 2010; Bellodi, Krasnykh, et al., 2010; Braunstein et al., 2007;

Faye et al., 2015; Gu et al., 2009; Yoon et al., 2006). One such anti-apoptotic protein is XIAP, the prototypical member of the Inhibitor of Apoptosis (IAP) protein family. The IAP proteins are critical regulators of apoptosis; a subset of IAPs binds to and inhibits key caspases involved both in the initiation and the execution steps (Holcik & Korneluk, 2001; Salvesen & Duckett, 2002). Cellular levels of XIAP protein are regulated by several mechanisms including protein degradation and changes in the protein synthesis. However, a mechanism that predominates during cellular stress is the selective XIAP translation via an IRES element (Bevilacqua et al., 2010; Durie et al., 2011; Gu et al., 2009; Holcik, Lefebvre, Yeh, Chow, & Korneluk, 1999; Holcik, Yeh, Korneluk, & Chow, 2000; Lewis et al., 2007; Muaddi et al., 2010; Nevins, Harder, Korneluk, & Holcik, 2003; Riley, Jordan, & Holcik, 2010; Yamagiwa, Marienfeld, Meng, Holcik, & Patel, 2004). XIAP is encoded by two mRNAs with distinct 5′ UTRs; the major, shorter 5′ UTR promotes a basal level of XIAP expression under normal growth conditions, while the less abundant, longer 5′UTR contains an IRES and supports cap-independent translation during stress (Riley et al., 2010). Importantly, the IRES-driven upregulation of XIAP in response to irradiation, serum deprivation, glucose deficiency, IL-6 treatment or cellular transformation as well as various other experimental settings, in situation when normal cap-dependent translation fails, has been shown to enhance cell survival, suggesting a central role for IRES-dependent translation of XIAP in mediating cellular fate, both in cells and in vivo (Aird et al., 2008; Aird, Ghanayem, Peplinski, Lyerly, & Devi, 2010; Blais et al., 2006; Durie et al., 2011; Gu et al., 2009; Holcik, Yeh, et al., 2000; Mizrachy-Schwartz et al., 2007; Muaddi et al., 2010; Nevins et al., 2003; Riley et al., 2010; Thakor & Holcik, 2012; Ungureanu et al., 2006; Warnakulasuriyarachchi, Cerquozzi, Cheung, & Holcik, 2004; Yamagiwa et al., 2004; Yoon et al., 2006). Clearly, in these cases, and we suspect more generally, the role IRES-mediated translation plays for the inhibition of apoptosis is biologically important. In fact, specific targeting of XIAP IRES translation was recently used to develop cancer therapeutics (Gu et al., 2016).

Cerebral ischemic injury activates apoptotic cell death in neurons, and is accompanied by a transient increase in XIAP levels in rescued cells in rats and mice (Siegelin, Kossatz, Winckler, & Rami, 2005; Spahn, Blondeau, Heurteaux, Dehghani, & Rami, 2008). Similarly, hippocampal HT22 cells treated with staurosporine exhibit an increase in XIAP expression (Spahn et al., 2008). In both models of neuronal cell death the expression of the RNA binding protein hnRNP C1 has been observed to coincide with XIAP expression (Spahn et al., 2008). Although direct evidence for the regulation of XIAP by hnRNP C1 during ischemic injury is not available, previous work has identified hnRNP C1 as one of the ITAFs that enhances IRES-dependent translation of the XIAP IRES (Holcik, Gordon, & Korneluk, 2003) suggesting that IRES-mediated control of XIAP contributes to the regulation of neuronal apoptosis.

N-myc is an oncogenic transcription factor of the myc family that is expressed primarily in neuronal tissues during early development. Similarly to other orthologs of the Myc family of proteins, misregulated expression of N-myc is linked to cancer, in particular neuroblastoma (Ruiz-Perez, Henley, & Arsenian-Henriksson, 2017). N-myc is essential for the proper fetal development as N-myc null mice die between 10.5 and 12.5 days of gestation with significant defects in organ development, most notably heart and the cranial and spinal ganglia. In the context of neuroblastomas, N-myc has been shown to have a dual role in the regulation of apoptosis—it is involved in the upregulation of pro-apoptotic phorbol-12-myristate-13-acetate-induced protein 1 (NOXA) and it sensitizes cells to cytotoxic drugs (Ham et al., 2016). On the other hand, through the regulation of p53 and H-Twist N-myc can increase apoptotic resistance as well (Valsesia-Wittmann et al., 2004; Yogev et al., 2016).

All members of the Myc family contain an IRES in their mRNA (Cobbold et al., 2008). However, these IRESs exhibit distinct patterns of expression and canonical factors requirement (Cobbold et al., 2008; Jopling & Willis, 2001; Spriggs et al., 2009). Notably, the N-myc IRES activity decreases during neuronal differentiation (Jopling & Willis, 2001) thus recapitulating, at least partially, decreasing expression of endogenous N-myc during development (Ruiz-Perez et al., 2017). In addition, the N-myc IRES has a lower requirement for the ternary complex (Spriggs et al., 2009), and would therefore be expected to operate efficiently under conditions of increased eIF2α phosphorylation, such as during hypoxia. Although the N-myc IRES requires eIF3 for its function, the mode of recruitment of eIF3 to the N-myc IRES appears to be independent of eIF4F (Spriggs et al., 2009). Such recruitment mode has been described for the XIAP IRES, in which eIF3 binds directly to the XIAP IRES RNA in a structure-dependent manner, followed by recruitment of PABP and the 40S ribosome (Thakor et al., 2017). Whether the N-myc IRES recruits eIF3 in the same way remains to be shown.

Conclusion

IRES-mediated translation has evolved to allow cells to fine-tune their proteome under various conditions. It is also emerging as a critical mechanism that allows for localized and precisely timed expression of distinct proteins in cellular extensions, and in response to environmental cues (Figure 7.1). We are only beginning to understand the full extent by which IRES translation shapes the function of CNS, and by extension how dysregulated IRES translation (or mutations in key parts of the IRES system) contributes to disease. Nevertheless, several small molecule inhibitors of IRES-mediated translations were reported in recent years (Berry et al., 2011; Didiot et al., 2013; Vaklavas et al., 2015; Venkatesan & Dasgupta, 2001; Zhou, Rynearson, Ding, Brunn, & Hermann, 2013). Although none of these approaches specifically looked for inhibitors of neuronal IRES (such as APP or SNCA) they established a methodology framework that could be extended to those mRNAs that utilize IRES translation specifically in CNS.

Future Directions

Despite a considerable progress in elucidating the role of IRES-dependent translation in the function of CNS, there remain many unanswered questions. For example, are there multiple mechanisms of alternative translation initiation that are used preferentially in response to various cues? Is the IRES mediated translation used to coordinate expression of distinct cohorts of mRNAs to allow for neuronal plasticity? Are there genetic links between diseases of CNS and translation machinery players? And could these be specifically targeted for the development of therapies?

References

Aird, K. M., Ding, X., Baras, A., Wei, J., Morse, M. A., Clay, T., . . . Devi, G. R. (2008). Trastuzumab signaling in ErbB2-overexpressing inflammatory breast cancer correlates with X-linked inhibitor of apoptosis protein expression. *Molecular Cancer Therapy*, *7*(1), 38–47. https://doi.org/10.1158/1535-7163.mct-07-0370

Aird, K. M., Ghanayem, R. B., Peplinski, S., Lyerly, H. K., & Devi, G. R. (2010). X-linked inhibitor of apoptosis protein inhibits apoptosis in inflammatory breast cancer cells with acquired resistance to an ErbB1/2 tyrosine kinase inhibitor. *Molecular Cancer Therapy*, *9*(5), 1432–1442. https://doi.org/10.1158/1535-7163.mct-10-0160

Arima, K., Hirai, S., Sunohara, N., Aoto, K., Izumiyama, Y., Ueda, K., . . . Kawai, M. (1999). Cellular co-localization of phosphorylated tau- and NACP/alpha-synuclein-epitopes in Lewy bodies in sporadic Parkinson's disease and in dementia with Lewy bodies. *Brain Research*, *843*(1–2), 53–61.

Arnaud, E., Touriol, C., Boutonnet, C., Gensac, M. C., Vagner, S., Prats, H., & Prats, A. C. (1999). A new 34-kilodalton isoform of human fibroblast growth factor 2 is cap dependently synthesized by using a non-AUG start codon and behaves as a survival factor. *Molecular and Cellular Biology*, *19*(1), 505–514.

Audigier, S., Guiramand, J., Prado-Lourenco, L., Conte, C., Gonzalez-Herrera, I. G., Cohen-Solal, C., . . . Prats, A. C. (2008). Potent activation of FGF-2 IRES-dependent mechanism of translation during brain development. *RNA*, *14*(9), 1852–1864. https://doi.org/1810.1261/rna.790608. Epub 792008 Aug 790601.

Baird, S. D., Lewis, S. M., Turcotte, M., & Holcik, M. (2007). A search for structurally similar cellular internal ribosome entry sites. *Nucleic Acids Research*, *35*(14), 4664–4677. https://doi.org/10.1093/nar/gkm483

Baird, S. D., Turcotte, M., Korneluk, R. G., & Holcik, M. (2006). Searching for IRES. *RNA*, *12*(10), 1755–1785. https://doi.org/10.1261/rna.157806

Beaudoin, M. E., Poirel, V. J., & Krushel, L. A. (2008). Regulating amyloid precursor protein synthesis through an internal ribosomal entry site. *Nucleic Acids Research*, *36*(21), 6835–6847. https://doi.org/6810.1093/nar/gkn6792. Epub 2008 Oct 6825.

Bellodi, C., Kopmar, N., & Ruggero, D. (2010). Deregulation of oncogene-induced senescence and p53 translational control in X-linked dyskeratosis congenita. *EMBO Journal*, *29*(11), 1865–1876.

Bellodi, C., Krasnykh, O., Haynes, N., Theodoropoulou, M., Peng, G., Montanaro, L., & Ruggero, D. (2010). Loss of function of the tumor suppressor DKC1 perturbs p27 translation control and contributes to pituitary tumorigenesis. *Cancer Research*, *70*(14), 6026–6035.

Beretta, L., Gingras, A. C., Svitkin, Y. V., Hall, M. N., & Sonenberg, N. (1996). Rapamycin blocks the phosphorylation of 4E-BP1 and inhibits cap-dependent initiation of translation. *EMBO Journal*, *15*(3), 658–664.

Berry, K. E., Peng, B., Koditek, D., Beeman, D., Pagratis, N., Perry, J. K., ... Shih, I. H. (2011). Optimized high-throughput screen for hepatitis C virus translation inhibitors. *Journal of Biomolecular Screening*, *16*(2), 211–220. https://doi.org/210.1177/1087057110391665.

Bevilacqua, E., Wang, X., Majumder, M., Gaccioli, F., Yuan, C. L., Wang, C., ... Hatzoglou, M. (2010). eIF2alpha phosphorylation tips the balance to apoptosis during osmotic stress. *Journal of Biological Chemistry*, *285*(22), 17098–17111. https://doi.org/10.1074/jbc.M110.109439

Blais, J. D., Addison, C. L., Edge, R., Falls, T., Zhao, H., Wary, K., ... Bell, J. C. (2006). Perk-dependent translational regulation promotes tumor cell adaptation and angiogenesis in response to hypoxic stress. *Molecular and Cellular Biology*, *26*(24), 9517–9532. Epub 2006 Oct 9519.

Braunstein, S., Karpisheva, K., Pola, C., Goldberg, J., Hochman, T., Yee, H., ... Schneider, R. J. (2007). A hypoxia-controlled cap-dependent to cap-independent translation switch in breast cancer. *Molecular and Cellular Biology*, *28*(3), 501–512.

Bushell, M., Stoneley, M., Kong, Y. W., Hamilton, T. L., Spriggs, K. A., Dobbyn, H. C., ... Willis, A. E. (2006). Polypyrimidine tract binding protein regulates IRES-mediated gene expression during apoptosis. *Molecular Cell*, *23*(3), 401–412.

Cammas, A., Lewis, S. M., Vagner, S., & Holcik, M. (2008). Post-transcriptional control of gene expression through subcellular relocalization of mRNA binding proteins. *Biochemical Pharmacology*, *76*(11), 1395–1403. https://doi.org/10.1016/j.bcp.2008.05.022

Casas, C. (2017). GRP78 at the centre of the stage in cancer and neuroprotection. *Frontiers in Neuroscience*, *11:177*. (doi), 10.3389/fnins.2017.00177. eCollection 02017.

Choi, H. S., Song, K. Y., Hwang, C. K., Kim, C. S., Law, P. Y., Wei, L. N., & Loh, H. H. (2008). A proteomics approach for identification of single strand DNA-binding proteins involved in transcriptional regulation of mouse mu opioid receptor gene. *Molecular and Cellular Proteomics*, *7*(8), 1517–1529. Https://doi.org/1510.1074/mcp.M800052-MCP800200. Epub 802008 May 800052.

Cobbold, L. C., Spriggs, K. A., Haines, S. J., Dobbyn, H. C., Hayes, C., de Moor, C. H., ... Willis, A. E. (2008). Identification of internal ribosome entry segment (IRES)-trans-acting factors for the Myc family of IRESs. *Molecular and Cellular Biology*, *28*(1), 40–49. Epub 2007 Oct 2029.

Connor, J. R., Menzies, S. L., St Martin, S. M., & Mufson, E. J. (1992). A histochemical study of iron, transferrin, and ferritin in Alzheimer's diseased brains. *Journal of Neuroscience Research*, *31*(1), 75–83.

Courteau, L., Crasto, J., Hassanzadeh, G., Baird, S. D., Hodgins, J., Liwak-Muir, U., ... Holcik, M. (2015). Hexokinase 2 controls cellular stress response through localization of an RNA-binding protein. *Cell Death and Disease*, *6*, e1837. https://doi.org/10.1038/cddis.2015.209

Creancier, L., Morello, D., Mercier, P., & Prats, A. C. (2000). Fibroblast growth factor 2 Internal ribosome entry site (IRES) activity ex vivo and in transgenic mice reveals a stringent tissue-specific regulation. *Journal of Cell Biology*, *150*(1), 275–281.

Didiot, M. C., Hewett, J., Varin, T., Freuler, F., Selinger, D., Nick, H., ... Parker, C. N. (2013). Identification of cardiac glycoside molecules as inhibitors of c-Myc IRES-mediated translation. *Journal of Biomolecular Screening*, *18*(4), 407–419. https://doi.org/10.1177/1087057112466698

Dobson, T., Minic, A., Nielsen, K., Amiott, E., & Krushel, L. (2005). Internal initiation of translation of the TrkB mRNA is mediated by multiple regions within the 5′ leader. *Nucleic Acids Research*, *33*(9), 2929–2941.

Durie, D., Lewis, S. M., Liwak, U., Kisilewicz, M., Gorospe, M., & Holcik, M. (2011). RNA-binding protein HuR mediates cytoprotection through stimulation of XIAP translation. *Oncogene, 30*(12), 1460–1469.

Dyer, J. R., Michel, S., Lee, W., Castellucci, V. F., Wayne, N. L., & Sossin, W. S. (2003). An activity-dependent switch to cap-independent translation triggered by eIF4E dephosphorylation. *Nature Neuroscience, 6*(3), 219–220.

Elroy-Stein, O., & Merrick, W. (2007). Translation initiation via cellular internal ribosome entry sites. In M. B. Mathews, N. Sonenberg, & J. Hershey (Eds.), *Translational Control in Biology and Medicine* (pp. 155–172). Cold Spring Harbor, NY: Cold Spring Harbor Laboratory Press.

Farlow, M. R. (1998). Etiology and pathogenesis of Alzheimer's disease. *American Journal of Health-System Pharmacy.*, 55(Suppl 2), S5–10.

Favero, C. B., Henshaw, R. N., Grimsley-Myers, C. M., Shrestha, A., Beier, D. R., & Dwyer, N. D. (2013). Mutation of the BiP/GRP78 gene causes axon outgrowth and fasciculation defects in the thalamocortical connections of the mammalian forebrain. *The Journal of Comparative Neurology, 521*(3), 677–696. https://doi.org/610.1002/cne.23199.

Faye, M. D., Beug, S. T., Graber, T. E., Earl, N., Xiang, X., Wild, B., ... Holcik, M. (2015). IGF2BP1 controls cell death and drug resistance in rhabdomyosarcomas by regulating translation of cIAP1. *Oncogene, 34*(12), 1532–1541. https://doi.org/10.1038/onc.2014.90

Florkiewicz, R. Z., & Sommer, A. (1989). Human basic fibroblast growth factor gene encodes four polypeptides: Three initiate translation from non-AUG codons. *Proceedings of the National Academy of Sciences USA, 86*(11), 3978–3981.

Gu, L., Zhang, H., Liu, T., Zhou, S., Du, Y., Xiong, J., ... Zhou, M. (2016). Discovery of dual inhibitors of MDM2 and XIAP for cancer treatment. *Cancer Cell, 17*(16), 30400–30407.

Gu, L., Zhu, N., Zhang, H., Durden, D. L., Feng, Y., & Zhou, M. (2009). Regulation of XIAP translation and induction by MDM2 following irradiation. *Cancer Cell, 15*(5), 363–375.

Ham, J., Costa, C., Sano, R., Lochmann, T. L., Sennott, E. M., Patel, N. U., ... Faber, A. C. (2016). Exploitation of the apoptosis-primed state of MYCN-amplified neuroblastoma to develop a potent and specific targeted therapy combination. *Cancer Cell*, 29(2), 159–172. https://doi.org/110.1016/j.ccell.2016.1001.1002.

Harada, A., Oguchi, K., Okabe, S., Kuno, J., Terada, S., Ohshima, T., ... Hirokawa, N. (1994). Altered microtubule organization in small-calibre axons of mice lacking tau protein. *Nature, 369*(6480), 488–491.

Hentze, M. W., Gebauer, F., & Preiss, T. (2007). cis-regulatory sequences and trans-acting factors in translational control. In M. B. Mathews, N. Sonenberg, & J. W. B. Hershey (Eds.), *Translational control in biology and medicine* (pp. 269–298). New York, NY: Cold Spring Harbor Laboratory Press.

Hershey, J. W., Sonenberg, N., & Mathews, M. B. (2012). Principles of translational control: An overview. *Cold Spring Harbor Perspectives in Biology, 4*(12). https://doi.org/10.1101/cshperspect.a011528

Hinnebusch, A. G., & Lorsch, J. R. (2012). The mechanism of eukaryotic translation initiation: New insights and challenges. *Cold Spring Harbor Perspectives in Biology, 4*(10). https://doi.org/10.1101/cshperspect.a011544

Holcik, M., Gordon, B. W., & Korneluk, R. G. (2003). The internal ribosome entry site-mediated translation of antiapoptotic protein XIAP is modulated by the heterogeneous nuclear ribonucleoproteins C1 and C2. *Molecular and Cellular Biology*, 23(1), 280–288.

Holcik, M., Graber, T., Lewis, S. M., Lefebvre, C. A., Lacasse, E., & Baird, S. (2005). Spurious splicing within the XIAP 5′ UTR occurs in the Rluc/Fluc but not the betagal/CAT bicistronic reporter system. *RNA*, *11*(11), 1605–1609. https://doi.org/10.1261/rna.2158605

Holcik, M., & Korneluk, R. G. (2001). XIAP, the guardian angel. *Nature Reviews Molecular Cell Biology*, *2*(7), 550–556. https://doi.org/10.1038/35080103

Holcik, M., Lefebvre, C., Yeh, C., Chow, T., & Korneluk, R. G. (1999). A new internal-ribosome-entry-site motif potentiates XIAP-mediated cytoprotection. *Nature Cell Biology*, *1*(3), 190–192. https://doi.org/10.1038/11109

Holcik, M., & Sonenberg, N. (2005). Translational control in stress and apoptosis. *Nature Reviews Molecular Cell Biology*, *6*(4), 318–327. https://doi.org/10.1038/nrm1618

Holcik, M., Sonenberg, N., & Korneluk, R. G. (2000). Internal ribosome initiation of translation and the control of cell death. *Trends in Genetics*, *16*(10), 469–473. https://doi.org/10.1016/s0168-9525(00)02106-5

Holcik, M., Yeh, C., Korneluk, R. G., & Chow, T. (2000). Translational upregulation of X-linked inhibitor of apoptosis (XIAP) increases resistance to radiation induced cell death. *Oncogene*, *19*(36), 4174–4177. https://doi.org/10.1038/sj.onc.1203765

Huang, E. J., & Reichardt, L. F. (2003). Trk receptors: Roles in neuronal signal transduction. *Annual Review of Biochemestry*, *72*, 609–642. Epub 2003 Mar 2027.

Iacoangeli, A., & Tiedge, H. (2013). Translational control at the synapse: Role of RNA regulators. *Trends in Biochemical Sciences*, *38*(1), 47–55. https://doi.org/10.1016/j.tibs.2012.1011.1001. Epub 2012 Dec 1014.

Jackson, R. J. (2013). The current status of vertebrate cellular mRNA IRESs. *Cold Spring Harbor Perspectives in Biology*, *5*(2). https://doi.org/10.1101/cshperspect.a011569

Johannes, G., Carter, M. S., Eisen, M. B., Brown, P. O., & Sarnow, P. (1999). Identification of eukaryotic mRNAs that are translated at reduced cap binding complex eIF4F concentrations using a cDNA microarray. *Proceedings of the National Academy of Sciences USA*, *96*(23), 13118–13123.

Johannes, G., & Sarnow, P. (1998). Cap-independent polysomal association of natural mRNAs encoding c-myc, BiP, and eIF4G conferred by internal ribosome entry sites. *RNA*, *4*(12), 1500–1513.

Jopling, C. L., & Willis, A. E. (2001). N-myc translation is initiated via an internal ribosome entry segment that displays enhanced activity in neuronal cells. *Oncogene*, *20*(21), 2664–2670.

Kerr, C. H., Ma, Z. W., Jang, C. J., Thompson, S. R., & Jan, E. (2016). Molecular analysis of the factorless internal ribosome entry site in Cricket Paralysis virus infection. *Scientific Reports*, *6*, 37319. https://doi.org/10.1038/srep37319

Kim, Y. K., Back, S. H., Rho, J., Lee, S. H., & Jang, S. K. (2001). La autoantigen enhances translation of BiP mRNA. *Nucleic Acids Research*, *29*(24), 5009–5016.

Kim, Y. K., Hahm, B., & Jang, S. K. (2000). Polypyrimidine Tract-binding protein inhibits translation of Bip mRNA. *Journal of Molecular Biology*, *304*(2), 119–133.

Kim, Y. K., & Jang, S. K. (2002). Continuous heat shock enhances translational initiation directed by internal ribosomal entry site. *Biochemical and Biophysical Research Communications*, *297*(2), 224–231.

Komar, A. A., & Hatzoglou, M. (2011). Cellular IRES-mediated translation: The war of ITAFs in pathophysiological states. *Cell Cycle*, *10*(2), 229–240.

Komar, A. A., Mazumder, B., & Merrick, W. C. (2012). A new framework for understanding IRES-mediated translation. *Gene*, *502*(2), 75–86. https://doi.org/10.1016/j.gene.2012.04.039

Koukouraki, P., & Doxakis, E. (2016). Constitutive translation of human alpha-synuclein is mediated by the 5′-untranslated region. *Open Biology, 6*(4), 160022. https://doi.org/160010.161098/rsob.160022. Epub 162016 Apr 160020.

Lee, K. A., & Sonenberg, N. (1982). Inactivation of cap-binding proteins accompanies the shut-off of host protein synthesis by poliovirus. *Proceedings of the National Academy of Sciences USA, 79*(11), 3447–3451.

Lee, P. T., Chao, P. K., Ou, L. C., Chuang, J. Y., Lin, Y. C., Chen, S. C.,...Yeh, S. H. (2014). Morphine drives internal ribosome entry site-mediated hnRNP K translation in neurons through opioid receptor-dependent signaling. *Nucleic Acids Research, 42*(21), 13012–13025. https://doi.org/13010.11093/nar/gku11016. Epub 12014 Oct 13031.

Lee, W., & Wayne, N. L. (1998). The roles of transcription and translation in mediating the effect of electrical afterdischarge on neurohormone synthesis in Aplysia bag cell neurons. *Endocrinology, 139*(12), 5109–5115.

Lewis, S. M., & Holcik, M. (2005). IRES in distress: Translational regulation of the inhibitor of apoptosis proteins XIAP and HIAP2 during cell stress. *Cell Death Differ, 12*(6), 547–553. https://doi.org/10.1038/sj.cdd.4401602

Lewis, S. M., Veyrier, A., Hosszu Ungureanu, N., Bonnal, S., Vagner, S., & Holcik, M. (2007). Subcellular relocalization of a trans-acting factor regulates XIAP IRES-dependent translation. *Molecular Biology of the Cell, 18*(4), 1302–1311.

Liwak, U., Faye, M. D., & Holcik, M. (2012). Translation control in apoptosis. *Experimental Oncology, 34*(3), 218–230.

Macejak, D. G., & Sarnow, P. (1991). Internal initiation of translation mediated by the 5′ leader of a cellular mRNA. *Nature, 353*(6339), 90–94.

Mitchell, S. A., Spriggs, K. A., Coldwell, M. J., Jackson, R. J., & Willis, A. E. (2003). The Apaf-1 internal ribosome entry segment attains the correct structural conformation for function via interactions with PTB and unr. *Molecular Cell, 11*(3), 757–771.

Mizrachy-Schwartz, S., Kravchenko-Balasha, N., Ben-Bassat, H., Klein, S., & Levitzki, A. (2007). Optimization of energy-consuming pathways towards rapid growth in HPV-transformed cells. *PLoS One*, 2(7), e628. https://doi.org/https://doi.org/10.1371/journal.pone.0000628

Morley, S. J., Coldwell, M. J., & Clemens, M. J. (2005). Initiation factor modifications in the preapoptotic phase. *Cell Death and Differentiation, 12*(6), 571–584.

Muaddi, H., Majumder, M., Peidis, P., Papadakis, A. I., Holcik, M., Scheuner, D.,...Koromilas, A. E. (2010). Phosphorylation of eIF2alpha at serine 51 is an important determinant of cell survival and adaptation to glucose deficiency. *Molecular Biology of the Cell, 21*(18), 3220–3231. https://doi.org/10.1091/mbc.E10-01-0023

Narumiya, S., Ohno, M., Tanaka, N., Yamano, T., & Shimada, M. (1998). Enhanced expression of full-length TrkB receptors in young rat brain with hypoxic/ischemic injury. *Brain Research, 797*(2), 278–286.

Nevins, T. A., Harder, Z. M., Korneluk, R. G., & Holcik, M. (2003). Distinct regulation of internal ribosome entry site-mediated translation following cellular stress is mediated by apoptotic fragments of eIF4G translation initiation factor family members eIF4GI and p97/DAP5/NAT1. *Journal of Biological Chemistry, 278*(6), 3572–3579.

Pacheco, A., & Twiss, J. L. (2012). Localized IRES-dependent translation of ER chaperone protein mRNA in sensory axons. *PLoS ONE, 7*(7), e40788. https://doi.org/40710.41371/journal.pone.0040788. Epub 0042012 Jul 0040724.

Pelletier, J., & Sonenberg, N. (1988). Internal initiation of translation of eukaryotic mRNA directed by a sequence derived from poliovirus RNA. *Nature*, *334*(6180), 320–325.

Penas, C., Casas, C., Robert, I., Fores, J., & Navarro, X. (2009). Cytoskeletal and activity-related changes in spinal motoneurons after root avulsion. *Journal of Neurotrauma*, *26*(5), 763–779. https://doi.org/710.1089/neu.2008-0661.

Penas, C., Font-Nieves, M., Fores, J., Petegnief, V., Planas, A., Navarro, X., & Casas, C. (2011). Autophagy, and BiP level decrease are early key events in retrograde degeneration of motoneurons. *Cell Death and Differentiation*, *18*(10), 1617–1627. https://doi.org/1610.1038/cdd.2011.1624. Epub 2011 Mar 1625.

Pinkstaff, J. K., Chappell, S. A., Mauro, V. P., Edelman, G. M., & Krushel, L. A. (2001). Internal initiation of translation of five dendritically localized neuronal mRNAs. *Proceedings of the National Academy of Sciences*, *98*(5), 2770–2775.

Pyronnet, S., Dostie, J., & Sonenberg, N. (2001). Suppression of cap-dependent translation in mitosis. *Genes and Development*, *15*(16), 2083–2093.

Qin, X., & Sarnow, P. (2004). Preferential translation of internal ribosome entry site-containing mRNAs during the mitotic cycle in mammalian cells. *Journal of Biological Chemistry*, *279*(14), 13721–13728. Epub 12004 Jan 13721.

Richardson, D. R. (2004). Novel chelators for central nervous system disorders that involve alterations in the metabolism of iron and other metal ions. *Annals of the New York Academy of Sciences*, *1012*, 326–341.

Righi, M., Tongiorgi, E., & Cattaneo, A. (2000). Brain-derived neurotrophic factor (BDNF) induces dendritic targeting of BDNF and tyrosine kinase B mRNAs in hippocampal neurons through a phosphatidylinositol-3 kinase-dependent pathway. *Journal of Neuroscience*, *20*(9), 3165–3174.

Riley, A., Jordan, L. E., & Holcik, M. (2010). Distinct 5′ UTRs regulate XIAP expression under normal growth conditions and during cellular stress. *Nucleic Acids Research*, *38*(14), 4665–4674. https://doi.org/10.1093/nar/gkq241

Roux, P. P., & Topisirovic, I. (2012). Regulation of mRNA translation by signaling pathways. *Cold Spring Harbor Perspectives in Biology*, *4*(11). https://doi.org/10.1101/cshperspect.a012252

Ruggero, D. (2013). Translational control in cancer etiology. *Cold Spring Harbor Perspectives in Biology*, *5*(2). https://doi.org/10.1101/cshperspect.a012336

Ruiz-Perez, M. V., Henley, A. B., & Arsenian-Henriksson, M. (2017). The MYCN protein in health and disease. *Genes (Basel)*, *8*(4).(pii), E113. https://doi.org/110.3390/genes8040113.

Sachs, A. B., Sarnow, P., & Hentze, M. W. (1997). Starting at the beginning, middle, and end: translation initiation in eukaryotes. *Cell*, *89*(6), 831–838.

Salvesen, G. S., & Duckett, C. S. (2002). Apoptosis: IAP proteins: blocking the road to death's door. *Nature Reviews Molecular Cell Biology*, *3*(6), 401–410.

Schwanhausser, B., Busse, D., Li, N., Dittmar, G., Schuchhardt, J., Wolf, J., . . . Selbach, M. (2011). Global quantification of mammalian gene expression control. *Nature*, *473*(7347), 337–342.

Shi, Y., Sharma, A., Wu, H., Lichtenstein, A., & Gera, J. (2005). Cyclin D1 and c-Myc internal ribosome entry site (IRES)-dependent translation is regulated by AKT activity and enhanced by rapamycin through a p38 MAPK and ERK-dependent pathway. *Journal of Biological Chemistry*, 280 (12), 10964–10973.

Siegelin, M. D., Kossatz, L. S., Winckler, J., & Rami, A. (2005). Regulation of XIAP and Smac/DIABLO in the rat hippocampus following transient forebrain ischemia. *Neurochemistry International*, *46*(1), 41–51.

Silvera, D., Formenti, S. C., & Schneider, R. J. (2010). Translational control in cancer. *Nature Reviews Cancer, 10*(4), 254–266.

Song, K. Y., Choi, H. S., Law, P. Y., Wei, L. N., & Loh, H. H. (2017). Post-transcriptional regulation of the human mu-opioid receptor (MOR) by morphine-induced RNA binding proteins hnRNP K and PCBP1. *Journal of Cell Physiology, 232*(3), 576–584. https://doi.org/510.1002/jcp.25455. Epub 22016 Jun 25424.

Spahn, A., Blondeau, N., Heurteaux, C., Dehghani, F., & Rami, A. (2008). Concomitant transitory up-regulation of X-linked inhibitor of apoptosis protein (XIAP) and the heterogeneous nuclear ribonucleoprotein C1-C2 in surviving cells during neuronal apoptosis. *Neurochemistry Research, 33*(9), 1859–1868. https://doi.org/1810.1007/s11064-11008-19658-11060. Epub 12008 Mar 11025.

Spriggs, K. A., Bushell, M., & Willis, A. E. (2010). Translational regulation of gene expression during conditions of cell stress. *Molecular Cell, 40*(2), 228–237.

Spriggs, K. A., Cobbold, L. C., Jopling, C. L., Cooper, R. E., Wilson, L. A., Stoneley, M., . . . Willis, A. E. (2009). Canonical initiation factor requirements of the Myc family of internal ribosome entry segments. *Molecular and Cellular Biology, 29*(6), 1565–1574. https://doi.org/1510.1128/MCB.01283-01208. Epub 02009 Jan 01285.

Spriggs, K. A., Stoneley, M., Bushell, M., & Willis, A. E. (2008). Re-programming of translation following cell stress allows IRES-mediated translation to predominate. *Biology of the Cell, 100*(1), 27–38.

Tao, R., & Auerbach, S. B. (2002). Opioid receptor subtypes differentially modulate serotonin efflux in the rat central nervous system. *Journal of Pharmacology and Experimental Therapeutics, 303*(2), 549–556.

Tebaldi, T., Re, A., Viero, G., Pegoretti, I., Passerini, A., Blanzieri, E., & Quattrone, A. (2012). Widespread uncoupling between transcriptome and translatome variations after a stimulus in mammalian cells. *BMC Genomics, 13*, 220. https://doi.org/10.1186/1471-2164-13-220

Thakor, N., & Holcik, M. (2012). IRES-mediated translation of cellular messenger RNA operates in eIF2alpha-independent manner during stress. *Nucleic Acids Research, 40*(2), 541–552. https://doi.org/10.1093/nar/gkr701

Thakor, N., Smith, M. D., Roberts, L., Faye, M. D., Patel, H., Wieden, H. J., . . . Holcik, M. (2017). Cellular mRNA recruits the ribosome via eIF3-PABP bridge to initiate internal translation. *RNA Biology, 14*(5), 553–567. https://doi.org/10.1080/15476286.2015.1137419

Thoma, C., Bergamini, G., Galy, B., Hundsdoerfer, P., & Hentze, M. W. (2004). Enhancement of IRES-Mediated Translation of the c-myc and BiP mRNAs by the Poly(A) Tail Is Independent of Intact eIF4G and PABP. *Molecular Cell, 15*(6), 925–935.

Timmerman, S. L., Pfingsten, J. S., Kieft, J. S., & Krushel, L. A. (2007). The 5′ leader of the mRNA encoding the mouse neurotrophin receptor TrkB contains two internal ribosomal entry sites that are differentially regulated. *PLoS One, 3*(9), e3242. https://doi.org/3210.1371/journal.pone.0003242.

Ungureanu, N. H., Cloutier, M., Lewis, S. M., de Silva, N., Blais, J. D., Bell, J. C., & Holcik, M. (2006). Internal ribosome entry site-mediated translation of Apaf-1, but not XIAP, is regulated during UV-induced cell death. *Journal of Biological Chemistry, 281*(22), 15155–15163. Epub 12006 Apr 15154.

Vagner, S., Gensac, M. C., Maret, A., Bayard, F., Amalric, F., Prats, H., & Prats, A. C. (1995). Alternative translation of human fibroblast growth factor 2 mRNA occurs by internal entry of ribosomes. *Molecular and Cellular Biology, 15*(1), 35–44.

Vaklavas, C., Meng, Z., Choi, H., Grizzle, W. E., Zinn, K. R., & Blume, S. W. (2015). Small molecule inhibitors of IRES-mediated translation. *Cancer Biology and Therapy*, *16*(10), 1471–1485. https://doi.org/1410.1080/15384047.15382015.11071729. Epub 15382015 Jul 15384015.

Valsesia-Wittmann, S., Magdeleine, M., Dupasquier, S., Garin, E., Jallas, A. C., Combaret, V., . . . Puisieux, A. (2004). Oncogenic cooperation between H-Twist and N-Myc overrides failsafe programs in cancer cells. *Cancer Cell*, *6*(6), 625–630.

Venkatesan, A., & Dasgupta, A. (2001). Novel fluorescence-based screen to identify small synthetic internal ribosome entry site elements. *Molecular and Cellular Biology*, *21*(8), 2826–2837.

Veo, B. L., & Krushel, L. A. (2009). Translation initiation of the human tau mRNA through an internal ribosomal entry site. *Journal of Alzheimers Disease*, *16*(2), 271–275. https://doi.org/210.3233/JAD-2009-0978.

Veo, B. L., & Krushel, L. A. (2012). Secondary RNA structure and nucleotide specificity contribute to internal initiation mediated by the human tau 5′ leader. *RNA Biology*, *9*(11), 1344–1360. https://doi.org/1310.4161/rna.22181. Epub 22012 Sep 22120.

Wakabayashi, K., Tanji, K., Mori, F., & Takahashi, H. (2007). The Lewy body in Parkinson's disease: molecules implicated in the formation and degradation of alpha-synuclein aggregates. *Neuropathology*, *27*(5), 494–506.

Warnakulasuriyarachchi, D., Cerquozzi, S., Cheung, H. H., & Holcik, M. (2004). Translational induction of the inhibitor of apoptosis protein HIAP2 during endoplasmic reticulum stress attenuates cell death and is mediated via an inducible internal ribosome entry site element. *Journal of Biological Chemistry*, *279*(17), 17148–17157. https://doi.org/10.1074/jbc.M308737200

Weingarten-Gabbay, S., Elias-Kirma, S., Nir, R., Gritsenko, A. A., Stern-Ginossar, N., Yakhini, Z., . . . Segal, E. (2016). Comparative genetics. Systematic discovery of cap-independent translation sequences in human and viral genomes. *Science*, *351(6270)*.(pii), aad4939. https://doi.org/https://doi.org/4910.1126/science.aad4939. Epub 2016 Jan 4914.

Weingarten-Gabbay, S., & Segal, E. (2016). Toward a systematic understanding of translational regulatory elements in human and viruses. *RNA Biology*, *13*(10), 927–933. Epub 2016 Jul 2021.

Willis, D., Li, K. W., Zheng, J. Q., Chang, J. H., Smit, A. B., Kelly, T., . . . Twiss, J. L. (2005). Differential transport and local translation of cytoskeletal, injury-response, and neurodegeneration protein mRNAs in axons. *Journal of Neuroscience*, *25*(4), 778–791.

Woodbury, M. E., & Ikezu, T. (2014). Fibroblast growth factor-2 signaling in neurogenesis and neurodegeneration. *Journal of Neuroimmune Pharmacology*, *9*(2), 92–101. https://doi.org/https://doi.org/110.1007/s11481-11013-19501-11485. Epub 12013 Sep 11421.

Yamagiwa, Y., Marienfeld, C., Meng, F., Holcik, M., & Patel, T. (2004). Translational regulation of x-linked inhibitor of apoptosis protein by interleukin-6: A novel mechanism of tumor cell survival. *Cancer Research*, *64*(4), 1293–1298.

Yogev, O., Barker, K., Sikka, A., Almeida, G. S., Hallsworth, A., Smith, L. M., . . . Chesler, L. (2016). p53 Loss in MYC-driven neuroblastoma leads to metabolic adaptations supporting radioresistance. *Cancer Research*, *76*(10), 3025–3035. https://doi.org/https://doi.org/3010.1158/0008-5472.CAN-3015-1939. Epub 2016 Mar 3029.

Yoon, A., Peng, G., Brandenburg, Y., Zollo, O., Xu, W., Rego, E., & Ruggero, D. (2006). Impaired control of IRES-mediated translation in X-linked dyskeratosis congenita. *Science*, *312*(5775), 902–906.

Zhang, X. Y., Zhang, T. T., Song, D. D., Zhou, J., Han, R., Qin, Z. H., & Sheng, R. (2015). Endoplasmic reticulum chaperone GRP78 is involved in autophagy activation induced by ischemic preconditioning in neural cells. *Molecular Brain*, *8*:20. https://doi.org/10.1186/s13041-13015-10112-13043.

Zhou, S., Rynearson, K. D., Ding, K., Brunn, N. D., & Hermann, T. (2013). Screening for inhibitors of the hepatitis C virus internal ribosome entry site RNA. *Bioorganic & Medicinal Chemistry*, *21*(20), 6139–6144. https://doi.org/6110.1016/j.bmc.2013.6103.6054. Epub 2013 Apr 6132.

CHAPTER 8

RNA MODIFICATIONS IN THE CENTRAL NERVOUS SYSTEM

DAN OHTAN WANG

INTRODUCTION

MOST if not all ribonucleic acids (RNAs) in nature are chemically modified. The chemically modified nucleosides are derived from the four standard nucleosides, adenosine, guanosine, cytidine, and uridine, but add distinct chemical and structural properties to the modified nucleosides. If we define RNA modification as RNA variance independent of genomic sequence decoded by the RNA polymerase, there are more than 170 modified RNA nucleosides identified today, distributed commonly or uniquely to the three life domains: bacteria, archea, and eukaryotes (Boccaletto et al., 2018). Functional characterizations of these modifications have shown that they concern the functionally important and most conserved regions and render the modified RNA to function more efficiently and accurately. By the same categorizing definition, RNA modification also includes post-transcriptional insertion, deletion, and substitution/editing. They also occur to natural RNA sequences in a nonrandom manner, sometimes guided by small nucleolar RNAs (snoRNAs) or specific ribonucleoprotein complexes (RNPs). For example, an endonuclease, a 3′-uridylyl transferase or TUTase (trypanosome mitochondrial 3′ terminal uridylyl transferase), an RNA ligase, and a trans-acting guide RNA (gRNA) can provide sufficient information for uridine insertion/deletion to specific RNA positions (Stuart & Panigrahi, 2002).

More than six decades have passed since the first RNA modification 5′-ribosyluracil, later named pseudouridine, was discovered thus marking the beginning of the field (Davis and Allen 1957; Cohn 1960). But it is not until recently that the genome-wide characterizations of RNA modifications on RNA of low abundance, and the discoveries of enzymes that can reverse the modification process have started to illustrate a highly prevalent, dynamic, and potent regulatory pathway vitally important to life and health (Dominissini et al., 2012a; Jia et al., 2011a; Meyer et al., 2012c). Currently termed as "RNA epigenetics" or "Epitranscriptomics," a new field to investigate this "expanded genetic vocabulary" is burgeoning and rapidly integrating with cancer biology, stem-cell biology, tissue regeneration, developmental biology, neuroscience, and many other life science areas. Fueled by the advanced sequencing technology, RNA purification and cloning routines, streamlined recombinant protein production, quantitative analysis, system approaches, and many other new technologies that have just become available now, the field is rapidly evolving, quickly synthesizing new hypotheses and results. In this chapter, I will introduce major RNA modifications in the central nervous system (CNS), and postulate answers to two questions. (1) Can RNA modifications react to environmental or physiological changes in CNS? (2) Can changes in RNA modification, either in stoichiometry, nucleotide positions, or decoding rules, regulate brain cell function? Although limited, the current data suggest positive answers to both questions and a highly dynamic regulatory layer of gene expression.

Finally, I will highlight the ongoing efforts for decoding the "RNA epigenetic code," or mapping the modifications to single nucleotides and quantifying the amount of modification. This information will generate an essential body of knowledge for understanding the function and mechanism of RNA modification in the CNS and will allow the eventual manipulation of RNA modification pathways for the benefit of brain health.

CNS, a Unique Location for Gene Expression and RNA Modifications

RNA modifications have been identified in all types of RNAs and in all organisms. CNS is no exception but may rather have unique demands for RNA modification to fine-tune gene expression. Epitranscriptomic mechanisms may be of particular importance in the CNS. First, brain contains more types and subtypes of cells than any other organs, thus expresses more diverse genetic programs under each cell identity. Second, cognitive development of individuals requires postnatally experience-driven and activity-dependent gene expression to precisely assemble functional brain circuits (Z. Shi et al., 2018). Third, the extensive and elaborate structures of neurons and other brain cells require sophisticated RNA transport mechanisms and regulated local translation for spatially controlled protein synthesis (Eberwine, Belt, Kacharmina, & Miyashiro, 2002; Misra et al., 2016). Post-transcriptional regulation such as RNA modification may be of particular importance for forming activity-dependent circuit connectivity at high resolution. Finally, the thousands of synaptic compartments shared by a single neuron, the far distances of the synapses from the soma where genetic materials are housed, the extraordinary diversity and complexity of these synapses, may demand neurons to use exquisite location- and context-sensitive regulatiory mechanisms of protein synthesis that can be mediated by RNA modification. Indeed, a recent investigation combining RNA-seq, ribo-seq, and proteomics at different stages of neuron differentiation has revealed an increasing degree of uncoupling between transcripts and protein products, with a larger number of specific transcripts undergoing enhanced or suppressed translation (Baser et al., 2019).

This result is corroborated by other findings that neuronal RNAs harbor longer 3′UTRs (untranslated regions) that contain more microRNA targeting sites (Tushev et al., 2018). Two important trans-acting RNA regulators: miRNA and long noncoding RNA (lncRNA) species have more diverse species in CNS than in other tissues; about 40 percent of known lncRNAs are specifically expressed in the brain, referring to 4000 to 20,000 lncRNA genes whose diversity increases along evolution (Derrien et al., 2012). RNA modification not only diversifies the output of gene expression, but also provides finer spatiotemporal regulation, thus enhancing the diversity, flexibility, and responsiveness of gene expression programs. In support of this role, later appearance of A-to-I editing enzyme in vertebrate adenosine deaminase (ADAR3) has its expression largely restricted to brain, and inosine is more abundant in brain messenger RNA (mRNA) with an expansion in human brain (Paul 1998a; Chen et al. 2000; Kim 2004). In addition, NSUN (NOP2/Sun) methyltransferase family members responsible for RNA methylation including 5-methylcytosine (m5C) synthesis are enriched in the developing brain, consistent with the role of this RNA modification pathway in the evolution of cognitive function and neurocognitive development (Chi & Delgado-Olguín, 2013). The most abundant internal mRNA modification N6-methyl-adenosine (m6A) has higher expression levels and there is a wider spread of m6A-containing RNA species in the brain (Livneh, Moshitch-Moshkovitz, Amariglio, Rechavi, & Dominissini, 2020). An albeit controversial m6A reader protein, fragile X mental retardation protein (FMRP), is linked to Fragile-X syndrome, a genetic mental retardation disorder despite the universal expression pattern of FMRP protein (Jin, 2000a). Furthermore, modifications to specific nucleotides of transfer RNAs (tRNAs) have been reported in numerous neurological diseases despite their universal expression in the body (see later section on tRNA modification). Such studies indicate that CNS may be particularly sensitive to a dysfunctional RNA modification regulatory pathway.

RNA Modification in tRNA, rRNA, UsnRNA, and snoRNA in CNS

RNA molecules in the brain exist in various classes with rRNA, tRNA, and small RNA molecules in the nucleus (small nuclear RNA, snRNA; and small nucleolar RNA, snoRNA) being the most abundant. tRNAs and rRNAs, which play catalytic roles in protein synthesis, uridylate-rich small nuclear RNAs (UsnRNAs) in regulating RNA splicing, and snoRNAs in regulating chemical modifications of rRNA are all highly modified with a variety of chemical moieties that facilitate their structural stability and function (Massenet et al., 1999; Sloan et al., 2017).

tRNAs

tRNAs are highly structured molecules described as a "clover leaf" and contain a plethora of modifications from simple as methylation to complex as 5-methoxycarbonylmethyl thiouridine (mcm5s2U; Frye, Harada, Behm, & He, 2018). These tRNA modifications play important roles in protecting tRNAs from degradation, maintaining their tertiary structures and facilitating accuracy and efficiency during translation. Of particular significance are the conserved locations with astonishing diverse modifications in the anticodon stem loop (including the wobble position) and junction positions between key secondary structural elements of tRNA. Next we describe how dysregulation of tRNA modifications are linked to a spectrum of neurological and neurodevelopmental disorders, attesting to the health relevance of tRNA modifications and their particular importance in the developing CNS.

In eukaryotes, the formation of an inosine through A-to-I editing (inosine forms base pairs with U, C, or A, I:C > I:U > I:A, thus increases flexibility in base-pairing potential) at the wobble position of tRNA is catalyzed by a heterodimeric adenosine deaminase complex (ADAT: adenosine deaminase tRNA specific) consisting of ADAT2 and ADAT3 (Gerber, 1999). A missense mutation in human ADAT3 is the most common cause of autosomal-recessive intellectual disability in Arabian countries (Alazami et al. 2013). In addition, a frameshift mutation in the ADAT3 gene has been identified in patients exhibiting intellectual disability and microcephaly (Salehi Chaleshtori et al., 2018). One can hypothesize that the pathological pathway is triggered by disrupted function of wobble inosine and compromised translation fidelity, efficiency, and codon usage. However, cellular and molecular consequences of these mutations are yet to be clarified, for example how changes in wobble inosine affect gene expression, to further establish the link between tRNA modification to cognitive function. It is also not clear why this particular loss of a tRNA modification selectively regulates brain development.

Uridine at the wobble position of tRNA undergoes multi-steps modification catalyzed by Elongator complex, Trm9-Trm112 complex, and Ncs2-Ncs3 complex (Huang 2005; Karlsborn et al. 2014; Bourgeois et al. 2017). Mutations that disrupt the activity of any protein subunits of these complexes therefore affect synthesis of the wobble uridine modifications. Mutations causing mis-splicing of ELP1 (ELP: elongator acetyltransferase complex subunit) reduces its expression level and mcm5s2U in patients exhibiting familial dysautonomia, a disorder of autonomic nervous system, which controls involuntary actions such as digestion, breathing, tear production, and the regulation of blood pressure and body temperature (Anderson et al., 2001). Missense mutations in ELP2 subunit also exhibit neurological disorders characterized by movement disorders along with severe intellectual disability and behavioral abnormalities (Cohen et al., 2015; Najmabadi et al., 2011). Allelic variants of ELP3 have been associated with motor neuron degeneration through tRNA modification, further highlighting the link between the Elongator complex and human neuropathologies (Bento-Abreu et al., 2018; Simpson et al., 2009). The profound effect of modification of this wobble uridine lies in its function of ensuring efficient tRNA interactions within the ribosome while preventing errors of reading frame, as suggested by biochemical analysis and structural studies. Studies using ribosome profiling to measure ribosome occupancy in eukaryotic cells and in mouse forebrains with "Elongator-related mutants" have revealed stalling of ribosomes at A-ending codons such as "AGA" and "CUA" (Chou, Donnard, Gustafsson, Garber, & Rando, 2017; Laguesse et al., 2015). Such alterations can trigger accumulation of unfolded proteins which subsequently lead to (hyper)activation of ubiquitin/proteasome system (UPS), unfolded protein response (UPR), and endoplasmic reticulum (ER) stress pathways. Of note, UPR has been shown to affect cortical neurogenesis by impairing the balance between direct and indirect neurogenesis pathway, thus cortex development (Laguesse et al., 2015).

Another prominent tRNA chemical transformation in the anti-codon stem is the isomerization of uridine at position 38 and 39 catalyzed by the pseudouridylate synthase PUS3. A human truncation mutation or null variant in this gene results in severe neurodevelopmental delay, intellectual disability, and reduced life span (Abdelrahman, Al-Shamsi, Ali, & Al-Gazali, 2018; Shaheen et al., 2016), thus supporting a key developmental requirement for PUS3.

Not far from Ψ38 and Ψ39, C32 and N34 are methylated at the 2′-hydroxyl group thus Cm(32) and Nm(34), which is required for another modification wybutosine at position 37 of phenylalanine tRNA (tRNA-Phe; Guy et al., 2012). Notably, the human enzyme responsible for Cm(32) and Nm(34), FTSJ1 (putative ribosomal RNA methyltransferase), has been identified as the cause of non-syndromic X-linked intellectual disability (NSXLID), with mild to moderate cognitive impairment without further clinical features (Perche-Letuvée, Molle, Forouhar, Mulliez, & Atta, 2014), suggesting a functional link specifically between tRNA-Phe and CNS development.

N6-threonyl-carbamoyl-adenosine (t6A) modification is present at position 37 of nearly every tRNA for "ANN" codons, catalyzed by endopeptidase and other proteins of small size (KEOPS) complex. Autosomal-recessive mutations in the protein subunits of this complex have been identified in Galloway-Mowat syndrome, a complex disorder exhibiting neurological abnormalities and developmental delay (Edvardson et al., 2017). Since t6A in the anti-codon stem is important for enhancing and stabilizing anticodon-codon interactions and preventing aberrant codon recognition, a deficiency in this modification may lead to UPR and other stress responses related to proteostasis. The relevance of t6A 37 of tRNA to CNS function has yet to be discovered.

More modifications are found in tRNA outside of the anti-codon stem loop and junctions, including 1-methylguanosine (m1G) at position 9 catalyzed by tRNA methyltransferase 10 (TRMT10) and removed by AlkB homolog 1, histone H2A dioxygenase (ALKBH1), N2,N2-dimethylguanosine (m2,2G) modification catalyzed by Trm1, clusters of 5-methylcytosine (m5C) catalyzed by DNMT2, NSUN2, and NSUN6 at positions 48/49/50, and 7-methylguanosine (m7G) at position 46 by methyltransferase like 1 (METTL1)/WDR4 complex. Genome-wide association studies have associated mutations in TRMT10A to diabetes and microcephaly, TRMT1 to autosomal-recessive intellectual disability, facial dysmorphisms, and developmental delay (Davarniya et al., 2015). Individuals with homozygous mutations in NSUN2 exhibit a variety of neurodevelopmental defects, including mild brain abnormalities and facial dysmorphism (Abbasi-Moheb et al., 2012; Fahiminiya et al., 2014; Khan et al., 2012; Martinez et al., 2012). Consistent with a role for NSUN2-catalyzed m5C modifications in neurodevelopment, NSUN2 is highly expressed during mouse embryogenesis and is specifically enriched in the brain (Blanco et al., 2011). Loss of NSUN2 in either flies or mice leads to perturbations in memory, cognition and behavior (Blanco et al., 2014). Cellular tRNAs lacking m5C due to NSUN2 deficiency are targeted for cleavage, generating 5′ tRNA fragments during the process that repress global cap-dependent translation and impacting cell motility and adhesion (Blanco et al., 2014; Tuorto et al., 2012a). NSUN2 has also been shown to modify mRNAs as well as the non-coding RNA vault, an abundant non-coding RNA localized to dendritic shafts (Hussain et al., 2013). Autosomal-recessive mutations in WDR4 cause a distinct form of microcephalic primordial dwarfism with growth delay along with intellectual disability (Shaheen et al., 2015; Trimouille et al., 2018). A severe reduction in m7G in tRNA-Phe may result in instability of tRNA through de-aminoacylation and the rapid tRNA decay pathway (RTD).

In summary, the versatile modifications at strategic positions, each adding distinct physiochemical properties (e.g., charges, size, hydrophobicity, etc.) to the modified tRNA reveal fundamental roles of RNA modifications: (1) stabilizing the modified RNA and preventing them from degradation; (2) maintaining or creating structural changes; iii) promoting fidelity and flexibility of tRNA in decoding mRNA, biasing codon usage, translation speed and co-translational folding, etc. A common pathological pathway related to dysregulation or hypomodification of tRNA modification could be to disturb proteostasis which triggers GAAC pathway (general amino acid control), RTD pathway, UPS and UPR pathways in the cells. When other factors that compromise these responding pathways co-exist, a diseased state leading to cell malfunction may be triggered. Neural tissues may be more vulnerable, that failing of growth factor, activity-dependent protein synthesis may result in delayed development and inappropriate wiring properties.

Due to the length limit, I will not elaborate on the thermodynamic regulation, structural alterations, physicochemical properties, and mechanistic aspects of tRNA modifications. Future studies are required to determine the mechanisms underlying vulnerability of CNS to dysregulated proteostasis through tRNA modifications, especially in the context of neurodegenerative diseases.

rRNA

Ribosomes are universally conserved translation molecular machines to mediate translation. Human ribosomes are composed by 28S, 18S, 5S, 5.8S rRNAs together with other >80 proteins with the rRNAs being the catalytical core for translation (Ban, 2000). A recent comprehensive characterization of human rRNAs using mass spectrometry has identified 14 types of modifications at 228 internal sites (4 in 5.8S, 91 in 18S, and 133 in 28S rRNA) with the vast majority being pseudouridylation and 2′O-methylation (Taoka et al., 2018). Note that modifications at some sites remain controversial among multiple systematic mapping methods. These modifications are positioned to highly conserved sites such as the decoding and tRNA binding sites, the peptidyltransferase center, and the inter-subunit interface, suggesting their role to stabilize the ribosomes and regulate translation (Sloan et al., 2017; Taoka et al., 2018). The majority of these sites are fully modified with stoichiometry of >=85 percent with little polymorphism, but some 2′O-Methyl and ψsites are partially modified and show heterogeneity in ribosomes in an environmental and developmental-responsive manner (Taoka et al., 2015). However, how such heterogeneity is achieved and how ribosome function is affected has yet to be investigated.

Two enzymatic modification pathways, snoRNA-guided and protein-only pathways, are responsible for the site-specific modifications of rRNA during ribosome maturation. A point mutation in the EMG1 gene in Bowen-Conradi syndrome and a chromosomal deletion encompassing the WBSCR22 and WBSCR20/NSUN5 genes in Williams-Beuren syndrome have linked rRNA modification dysregulation to severe developmental delay in human (Doll & Grzeschik, 2001; Warda et al., 2016). However the molecular and cellular mechanisms underlying the pathogenicity are yet to be identified and whether the lack of any specific rRNA modification is disease causing is unclear (Sloan et al., 2017).

Compared to tRNA, much less is known about the function of rRNA modification in CNS. Nonetheless, considering the central role of rRNA in ribosome and translation regulation, how this pathway influences CNS development and function warrants future investigations.

UsnRNA and snoRNA

While both uridylate-rich small nuclear RNA (UsnRNA) and small nucleolar RNA (snoRNA) are on average less than 200nt and function in the nucleus, they are two distinct classes transcribed by different polymerases and carry out distinct functions.

In eukaryotes, UsnRNA (U1, U2, U4, U5, and U6) are critical components of a massive RNA processing machine, the spliceosome, responsible for splicing out introns of pre-mRNAs. Remarkably, multiple rounds of RNA modifications occur before Uridylate-rich snRNPs (UsnRNPs) mature and become active. The first round involves the 5′ m7G cap modification occurring to all transcripts made by type II polymerase, which functions to prevent the RNA from degradation and signals nuclear export. The second round occurs in the cytoplasm to convert m7G cap to a trimethyl-2,2,7-G cap (m3G), which serves as a signal for reimporting UsnRNAs into the nucleus. Finally, a third round occurs in the nucleus that the UsnRNAs undergo 2′-O-methylation, pseudouridylation, and N6-methylation. All five spliceosomal UsnRNAs are extensively modified in regions known important for splicing, suggesting functional roles of the modifications. Biochemical and biophysical evidences on the structural effects of these modifications support their role for both the biogenesis and for the splicing activities of spliceosome (Y. Lin & Kielkopf, 2008).

For the synthesis of these modifications, specific nuclear loci such as Cajal bodies have been identified where small noncoding BoxH/ACA and BoxC/D RNAs (or scaRNAs for small Cajal body-specific RNAs) guide 2′-O-methylation and pseudouridylation to specific nucleotide positions of UsnRNA. For N6-methylation, methyltransferase like 16 (METTL16) has been identified as the m6A methyltransferase for U6 UsnRNA, and meanwhile targets the methionine adenosyltransferase 2A (MAT2A) mRNA encoding S-adenosylmethionine (SAM) synthetase for splicing. Thus reducing SAM in cells can trigger splicing and production of MAT2A to up-regulate SAM synthesis to maintain cellular SAM homeostasis (Pendleton et al., 2017).

In CNS, neuron- and brain-enriched splicing factors such as Nova, CELF, Fox, Hu/ELAV, and STAR/GSG families have been identified, indicating neuron-specific splicing programs. Whether UsnRNA modifications may play a role in generating or accommodating neuron-specific spliceosomes (e.g., autism-related alternative splicing regulatory pathways) is currently unknown. Unlike tRNA modifications, genetic linkage and biological studies have not linked snRNA modifications to CNS diseases.

Small nucleolar RNAs (snoRNAs) are located to nucleoli, a particular subnuclear location joint by Cajal bodies for ribosome assembly and maturation (Verheggen, 2001). The localization of snoRNAs to these nuclear bodies are consistent with their major role to guide position-specific 2′O-methylation and pseudouridylation to rRNA as part of the biosynthesis pathway of the ribosome. Interestingly, BoxH/ACA and BoxC/D scaRNAs at Cajal bodies are also responsible for UsnRNA modification (but not for tRNA modifications) thus are important for the biosynthesis of spliceosomes.

A subgroup of snoRNAs are predominantly expressed in the brain (b-snoRNAs). One evolutionally conserved snoRNA HBII-52 appears to repress ADAR2-mediated A-to-I editing in 5-HT(2C) receptor subunit encoding RNA, indicating a mechanism of brain-specific snoRNA-mediated RNA modification in generating functional diversity of serotonin signaling pathways of high prevalence to neuropsychiatric diseases (Vitali et al., 2005). In rodents, hippocampal expression of MBII-52 is temporarily upregulated in 90 minutes after contextual conditioning, opposite to another b-snoRNAs MBII-48, which is downregulated (Rogelj, Hartmann, Yeo, Hunt, & Giese, 2003). Whether this regulation plays a role in regulating 5HT(2C) receptor expression and function of serotonin pathway in association to learning and memory consolidation is yet to be tested. Another b-snoRNA, HBII-85 (SNORD [small nucleolar RNA, C/D Box] 116), has also been shown to have reduced expression level in Prader-Willi syndrome patients (Gallagher, Pils, Albalwi, & Francke, 2002). These studies strongly support neurological function of snoRNAs and brain-specific snoRNAs.

Although snoRNAs are known to be modified and play a role in rRNA and UsnRNA modification, whether these biological functions serve the etiology of brain disorders such as Prader-Willi syndrome is unclear. It is noteworthy that b-snoRNAs appear not to be involved in modification of rRNAs and snRNAs due to lacking of sequence complementarity (Cavaille et al., 2000; Huttenhofer, 2001; Rogelj & Giese, 2004). snoRNA also functions in miRNA-mediated silencing and alternative splicing, thus a dysregulated and pathological gene expression network may be triggered by excessive or insufficient expression of snoRNAs.

RNA Modification in mRNA, miRNA, and lncRNA in CNS

Compared to rRNA, tRNA, snRNA and snoRNA biotypes, mRNA, miRNA, and lncRNAs are expressed with much lower abundance, resulting in delayed discoveries of their chemical modifications. For example, m6A, the most abundant internal mRNA modification, was discovered in 1974 (17 years later than the discovery of pseudouridine), only when concentrating mRNAs using oligo d(T) became available.

mRNA

For messenger RNAs (mRNA), the nucleotides at the 5′-end are highly modified. In fact, every new mRNA and UsnRNA transcribed by Pol II starts with m7Gppp (m7G), or the "cap," which serves to protect the new transcripts from 5′-exoribonulease activity and to recruit and anchor cap-binding complex. In the nucleus, CBC (cap-binding complex) recruited to the cap structure is required for efficient splicing and nuclear export; later in cytoplasm, CBC is replaced by eukaryotic translation initiation factor 4E (eIF4E) for bulk translation. Recently, several studies showed that unlike other eukaryote cells where CBC is quickly replaced by eIF4E after nuclear export, presumably after the first round of translation, in neurons CBC travels with specific mRNAs with exon junction components along microtubules far into the distal dendrites, suggesting transportation of translationally silent mRNAs in neurons (Vessey et al., 2010; Dan O. Wang et al., 2017). In addition to the cap structure, the 2′-O-methylnucleosides following m7G are also widely distributed and well established for their role in stabilizing the transcripts and promoting splicing and export. A di-methyl nucleotide N6,2-O-dimethyladenosine (m6Am) has been recently identified as an abundant cap-dependent modification at the

5′ ends of mRNAs and function to antagonize eIF4E thus negatively impact translation (Akichika et al., 2019; Sendinc et al., 2019).

Thus the modifications at the 5′ ends are a common feature of mRNAs, regulating their stability, transportation, and translation. In internal sequences of mRNA, inosine (I), N6-methyladenosine (m6A), 5-methylcytosine (m5C), 5-hydroxymethylcytosine (hm5C), pseudouridine (ψ), N1-methyadenosine (m1A), and 2′-O-methylnucleotides (Nm) are the major modifications.

m6A

In 1974, using high-speed liquid chromatograph, m6A was detected as the most abundant chemical modification on mammalian polyadenylated RNAs (Desrosiers, Friderici, & Rottman, 1974; Perry & Kelley, 1976). This discovery was made possible by a then new poly(A) RNA purification technique using oligo d(T) to purify polyadenylated RNAs through hybridization, thus this relatively low abundant RNA species could be concentrated (Harley, 1987). Annotation of m6A to specific mRNA was not possible since sequence information was lost when the mRNAs were digested into single nucleosides before detection. Until 1984, only two m6A sites were mapped, one to a simian virus 40 RNA, the other to the 3′UTR of bovine prolactin mRNA in 20 percent of the mRNA transcripts (Canaani, Kahana, Lavi, & Groner, 1979; Horowitz, Horowitz, Nilsen, Munns, & Rottman, 1984). In 2012, transcriptome-wide m6A profiles were generated in two independent studies using antibody-capture and massive parallel sequencing. The two studies revealed thousands of modified genes at conserved sites (Dominissini et al., 2012b; Meyer et al., 2012a). However, issues on limited mapping specificity, resolution, and coverage remain to be addressed. The new technologies allowing transcriptome-wide mapping of m6A and other chemical modifications at single-nucleotide resolution are paving the way to grant us a clearer view on these potent gene expression elements. This area of development will be discussed in a later section.

m6A perhaps has the richest literature in neuroscience disproportionally to any other types of modifications. For this reason, I will devote a whole chapter to m6A regulation and its known function in the CNS. I also refer the readers to a recently published comprehensive review on m6A in the nervous system (Livneh et al., 2020).

Inosine (I)

Inosine (I) is generated by hydrolytic deamination of adenosine (A), catalyzed by adenosine deaminase acting RNA (ADAR). Inosine exists in mRNA in a tissue-specific manner and is most abundant in brain mRNA, but not restricted to mRNAs. Since inosine is recognized as guanosine by the translational machinery, its decoding results in amino acid replacement, which has been shown to significantly change the properties of the encoded proteins such as brain-specific receptors. Due to the excellent literature on RNA editing and their regulatory pathways in the CNS, I will not further elaborate on this important RNA modification/editing in this chapter but refer the readers to an excellent review (Behm & Öhman, 2016).

5-Methylcytosine (5mC) and 5-Hydroxymethylcytosine (5hmC)

5-methylcytosine (5mC) is found in multiple RNA species such as tRNA, rRNA, mRNA and lncRNA, mediated by different corresponding enzymes. In mRNA, 5mC modification is mediated by a NOP2/Sun domain RNA methyltransferase family member NSUN2, whose loss of function in mouse and human led to motor, neurodevelopmental and cognitive defects, suggesting the neurological function of 5mC in mRNA (Abbasi-Moheb et al., 2012; Blanco et al., 2014; Flores et al., 2017; Khan et al., 2012; Khoddami & Cairns, 2013; Komara, Al-Shamsi, Ben-Salem, Ali, & Al-Gazali, 2015; Martinez et al., 2012; Squires et al., 2012; Tuorto et al., 2012a; X. Yang et al., 2017). Bisulfite conversion-based genome-wide 5mC detection (bisulfite converts unmethylated cytidines to "U" in later sequencing while methylated cytidines remains as "C") revealed 5mC in mRNAs at CG dinucleotides around translation initiation sites, which is recognized and bound by Aly/REF export factor (ALYREF) for nuclear export (Delatte et al., 2016a; Hussain et al., 2013; Khoddami & Cairns, 2013; Squires et al., 2012; X. Yang et al., 2017). Not only immediately after translation initiation sites, m5C sites are also enriched in 3′UTRs, especially in the neighborhood of Argonaute I–IV binding sites, suggesting a role of m5C in gene silencing (Squires et al., 2012). Recognized by ALYREF for export and bound by Argonaute to the vicinity provide a mechanistic illustration of 5mC function in regulating RNA export and translation in cells, both extremely important for mRNA trafficking and local translation regulation in neurons.

An oxidized form of 5mC generated by ten-eleven translocator family TET, 5-hydroxymethylcytosine (5hmC), has also been found in mRNAs and with the highest levels in the brain (Delatte et al., 2016a; Fu et al., 2014; Huber, von Watzdorf, & Marx, 2016; X. Yang et al., 2017). In flies, TET knock-down led to brain malformations in the larva and lethality at pupal stage, however the functional attribution to 5hmC-RNA is obscured by the fact that TET enzymes target both DNA and RNA 5mC. With the readers unidentified, whether cells read out 5hmC as independent signals from 5mC are unknown.

MicroRNA

microRNAs (miRNAs) are small (21–23nt) but powerful players in RNA silencing pathways inducing mRNA degradation, translational repression, and transcriptional silencing. They contain a rich repertoire of modified nucleotides such as inosine, m6A, and 2′-O-Me. 2′-O-Me protects miRNAs from small RNA decay pathways that are triggered by 3′-uridylation (Y. Zhao et al., 2012). For inosine, specific primary miRNA (pri-miRNA) transcripts can contain dsRNA structures recognizable and targeted by ADARs. In case of miR-142, two edited sites to inosine increases resistance of pre-miR-142 to cleavage by Drosha, thus preventing pri- to pre-miRNA processing and reducing mature miR-142 in the cell (W. Yang et al., 2006). Another example of A-to-I editing in regulating microRNA biogenesis occurs at -1 and +3 positions of miR-151 which prevents pre-miR-151 from being cleaved by Dicer, resulting in an accumulation of pre-miR-151 specifically in human cerebral cortex, amygdala and mouse central nervous system (Kawahara et al., 2007). Finally, editing of miR-376 cluster RNAs occurs within the "seed sequence" in the mature miRNA sequence. This edited inosine does not affect processing for the expression of mature miR-376 RNAs, however redirects the miRNAs to target a new set of target mRNAs and suppress their translation. One target of the edited miR-376 is phosphoribosyl pyrophosphate synthetase 1, an enzyme involved in the uric-acid synthesis pathway (Y. Kawahara et al., 2007). Thus A-to-I editing in pri- and pre-miRs not only influence microRNA expression but also function, in a tissue and cell-specific manner.

Another RNA modification pathway regulating microRNA biosynthesis is m6A. m6A methyltransferase METTL3 targets both mRNA and pri-miRNAs in the nucleus and m6A marks pri-miRNAs for DGCR8 recognition and thus cleavage by DROSHA-DGCR8 complex. Depletion of METTL3 from MDA-MB-231 cells reduces the binding of DGCR8 to pri-miRNAs and results in a global reduction of mature miRNAs and concomitant accumulation of unprocessed pri-miRNAs (Alarcón, Lee, Goodarzi, Halberg, & Tavazoie, 2015). Since the majority of miRNAs require DGCR8 for processing and maturation, and miRNAs play a central role in regulating neuron survival and function, the requirement of m6A for the maturation of miRNAs strongly suggests essential m6A function in neuron (De Pietri Tonelli et al., 2008).

Currently, the biological function of most endogenous chemical modifications on miRNAs remain largely unknown, however the feasibility of synthesizing siRNAs with pre-designed chemical modifications allows their function to be explored both in vitro and in vivo. In the clinical sector and the field of translational medicine, chemical modifications are being used to improve the performance of RNA-based therapeutic reagents, such as antisense oligonucleotide (ASO), small interfering RNAs (siRNAs), miRs and anti-miRs. The conserved regulation mechanisms between human and other species provide experimental evidence-based rationale designing strategy for RNA therapeutic agents. Various types and extensive degrees of chemical modifications have been exploited to prolong half-life of the therapeutic siRNA in human body, increase target specificity and decrease undesired immune response and cellular toxicity (Lennox & Behlke, 2011). Such modifications include replacing phosphate backbone with phosphorothioate, adding chemical moieties to the vulnerable 2′-hydroxy group at the ribose with 2′-O-methyl group or unnatural 2′-fluoro, 2′-O-methoxyethyl, 2′-O-4′-C-methylene, modifying 5′ ends with 5′-(E)-vinylphosphoate, 5′-methylphosphate and 5′-phosphorothioate etc. (Khvorova & Watts, 2017). These synthetic RNA mimics have been proved not only important for drug development, but also for understanding how cells respond to different modifications.

lncRNA

Long noncoding RNAs (lncRNAs) refer to a class of noncoding RNAs >200nt in length. Its rather "non-functional" naming indicates diverse and largely unexplored functions. But recently several examples of lncRNAs and circular RNAs (circRNAs) have shown their importance in epigenetic regulations and association with neuropsychiatric conditions in humans such as schizophrenia and depression (Issler et al., 2020; Zimmerman et al., 2020), warranting new investigations in the biological function of this class of RNA. A variety of modifications including m5C, m6A, ψ, and A-to-I editing have been documented in lncRNAs.

m5C sites are found in numerous nuclear lncRNA such as the catalytic ncRNA subunit of RNase P RPPH1 and epigenetic silencing factors XIST (X inactive specific transcript) and HOTAIR (HOX transcript antisense RNA). RNase P is involved in the processing of tRNA precursors, and may also play a role in RNA polymerase III transcription (Jarrous & Reiner, 2007). The identified m5C site resides in the P12 domain of RPPH1 that may be important for RPPH1 function in transcription of 7SL RNA, 5S rRNA and several other targets. m6A is identified in metastasis associated lung adenocarcinoma transcript 1 (Malat1) in a hairpin stem loop structures. The presence of m6A increases recognition and binding of a U5-tract by heterogeneous nuclear ribonucleoprotein C (HNRNPC) through thermodynamically induced structural changes (N. Liu et al., 2015). Malat1 is upregulated in mature neurons in an activity-dependent manner and in turn positively influence synaptogenesis through regulating splicing of synapse proteins (Bernard et al., 2010). Meanwhile, Malat1 is also dysregulated in various cancer cells. How m6A-mediated dynamic interaction with HNRNPC contribute to its function remains a mystery. Another lncRNA X-linked X-inactive-specific transcript (XIST) is extensively modified with m6A (at least 78 sites in murine embryonic stem cells) mediated by RBM15 (RNA binding motif protein 15) and METTL3 methyltransferase complex (Patil et al., 2016). m6A-modified XIST can be recognized by a nuclear m6A reader YTHDC1 (YTH domain containing 1), which is necessary for XIST-mediated gene silencing of X chromosome inactivation in the embryonic stem cells (Ha et al., 2018; Patil et al., 2016). These studies indicate that RNA modification crosstalks with epigenetic mechanisms and chromatin regulation to control transcription. Deposition of m6A on other lncRNA can be identified in sequencing datasets however their function is yet to be discovered.

m6A in CNS from Development to Regeneration

As the most abundant mRNA internal modification, m6A is the current front runner among all mRNA chemical modifications with emerging roles in CNS. m6A is electrically neutral, neither changes coding of the modified adenosine nor its base-pairing to uridine. Instead, m6A regulates kinetics of every step of RNA metabolism, such as RNA processing (Ke et al., 2017; Xiao et al., 2016; X. Zhao et al., 2014), nuclear export (Zheng et al., 2013), RNA degradation (Ke et al., 2017; X. Wang et al., 2014), and translation (Coots et al., 2017; Meyer et al., 2015; X. Wang et al., 2015; Zhou et al., 2015). The widely spread m6A sites on mRNAs and their intimate involvement in RNA metabolism together present a potent mechanism to regulate gene expression in a dynamically responsive manner.

With the dynamic regulators, the writers, erasers, and readers identified for m6A modification, a variety of loss- and gain-of-function cell lines and animal models have been generated which have facilitated our understanding of cellular dynamics and physiological function of this modification in brain.

m6A "Writer" Complex

The vast majority of m6A modifications are deposited co-transcriptionally to the RNA polymerase II-derived transcripts in the nucleus, mediated by a multimeric molecular complex consisting of METTL3, the catalytic enzyme, METTL14, inactive as transferase but active as allosteric partner for METTL3, together with several other auxiliary components such as VIRMA (vir like m6A methyltransferase associated), ZC3H13 (zinc finger CCCH-type containing 13), RBM15/15B, and CBLL1 that render not only selectivity of m6A sites, but possibly dynamic regulation in response to environmental cues, stress, and intracellular conditions. The selectivity may be mediated by the interaction of the "writer" complex with specific histone markers (e.g., H3K36), transcription factors, and RNA-binding protein (e.g., RBM15 and RBM15B; Bertero et al., 2018; Huang et al., 2019; Slobodin et al., 2017), but whether all mRNAs follow one or more of these rules is unclear.

Current investigations on m6A function largely rely on genetic manipulating the writers, such as to ectopically augment or knockout Mettl3 or Mettl14. In human pluripotent cells, constitutive knock-out of Mettl3 impedes stem cell differentiation and cell-fate specification (Geula et al., 2015). In mouse models, conventional Mettl3-KO mice are embryonic lethal at E7.5 (Geula et al., 2015), thus multiple groups have used conditional KO mice spatiotemporally controlled by specific promoters. In a neural progenitor-specific knockout mouse model of Mettl3 or Mettl14, cortical radial glial cells (RGCs) have prolonged cell cycles, resulting in delayed neurogenesis and gliogenesis

(Yoon et al., 2017). Contradictory to this study, an independent study reported conditional knockout of Mettl14 in RGCs resulting premature differentiation of the progenitor cells (Y. Wang et al., 2018). Thus the spatially and temporally coordinated cortical neurogenesis are disordered, in concordance with the expression of several transcriptomes representing multiple developmental stages mixed up in the same cell. It is noteworthy that the alteration of gene expression may be partially mediated by epigenetic mechanisms that loss of m6A triggers such as increases in H3 trimetylation of lysine 4 and lysine 27, and H3 triacetylation at lysine 27 (Y. Wang et al., 2018). A similar phenomenon was recapitulated in human iPS cell-derived forebrain organoids, indicting an evolutionarily conserved role of m6A in regulation of cortical neurogenesis (Yoon et al., 2017). In cerebellum, a high level of m6A expression is observed and Mettl3 depletion results in cerebellar hypoplasia due to the increased apoptosis of the newborn cerebellar granule cells (C.-X. Wang et al., 2018). To study the role of Mettl3 in mature neurons, Zhang et al used mice expressing conditional alleles of Mettl3 in a CaMK2-Cre background where Mettl3 is deleted in the postnatal forebrain in excitatory neurons, including cortex and hippocampus. These mice show deficits in consolidating weak stimuli-induced memories (e.g., only one foot shock is given during fear conditioning). Importantly reintroducing Mettl3 to the dorsal hippocampus was sufficient for rescuing the phenotype, thus m6A modification of mRNAs upon learning is necessary and sufficient for enhancing the strength of memories (Zeyu Zhang et al., 2018). In this study, lack of degeneration or morphological changes of hippocampal neurons is intriguing since Mettl3 deletion has been shown to induce cell death in cultured hippocampal neurons (Merkurjev et al., 2018), degeneration of Purkinje neurons and apoptosis in cerebellar granule neurons (C.-X. Wang et al., 2018). In a separate study, adult striatum-specific Mettl14 knockout mice failed in motor learning and locomotor response to dopamine signaling, having altered neuronal excitability and reduced spike frequency adaptation in striatal neurons (Koranda et al., 2018). In a sciatic nerve lesion model, loss of METTL14 or YTHDF1 (YTH N6-methyladenosine RNA binding protein 1; m6A reader protein) function attenuates protein synthesis in response to the injury thus functional recovery. The same study demonstrates that both regeneration of peripheral nerves (dorsal root ganglion) and central nerves (retinal ganglion neurons) require m6A signaling pathway (Y.-L. Weng et al., 2018).

m6A "Erasers"

Two demethylases, FTO (fat mass and obesity associated) and ALKBH5 (AlkB homolog 5, histone H2A dioxygenase), have been identified to oxidatively remove m6A from mRNAs in Fe^{2+} and α-ketoglutarate-dependent manner (Jia et al., 2011b; Zheng et al., 2013). The two enzymes are co-expressed in many cell types and tissues, but each is sufficient for conducting demethylation activity, thus may be able to compensate each other's demethylation activity when one is absent. However, current literature suggests that FTO and ALKBH5 target specific mRNAs as substrates and perform independent functions. For example, in conventional knockout or dopamine neuron-specific FTO knockout mice, only a subset of transcripts showed upregulated methylation instead of global upregulation (Hess et al., 2013). In addition, FTO has especially higher expression level in the brain and while the loss-of-function of ALKBH5 yields mouse infertility (Zheng et al., 2013), FTO knockout mice showed leaner mass, impaired adult neurogenesis, smaller brain, impaired spatial learning and memory, and altered reward response (Gao et al., 2010; Hess et al., 2013; L. Li et al., 2017; Walters et al., 2017). In addition, knocking down FTO expression specifically in the medial prefrontal cortex (mPFC) region enhances consolidation of cued fear memory in rodents, indicating that m6A plays a facilitating role in memory formation (Widagdo et al., 2016). The substrate specificity of FTO however, remains controversial. in vitro studies have indicated that m1A, m6A, m6Am can all be oxidized by FTO but m6A could be the main substrate in vivo (Mauer et al., 2017, 2019; Mauer & Jaffrey, 2018; J. Wei et al., 2018; X. Zhang et al., 2019). Multi-substrates catalyzed by a single enzyme and/or a single targeted by multiple enzymes are typical and challenging issues for the epitranscriptomics field to address because the physiological function and dysregulation can be obscured by other types of modifications affected by similar enzymes. In cerebellum, deletion of Alkbh5 gene causes abnormal shapes of cerebellum and altered response to hypobaric hypoxia (Ma et al., 2018).

m6A "Readers/Anti-readers"

In contrast to a single m6A "writer" complex, and two "eraser" proteins, at least a dozen m6A readers have been identified, indicating diverse and flexible output from a single type of RNA modification (H. Shi, Wei, & He, 2019). Based on their actions toward the modified nucleotide, the reader proteins can be further divided into "readers" and "anti-readers" that the readers are attracted to and the anti-readers are repelled by m6A (Edupuganti et al., 2017). Currently more readers are known than anti-readers. Five members of YT521-B homology (YTH) family (a family with more than 170 family members sharing a highly conserved RNA-binding domain YTH) are known as m6A readers (Stoilov, Rafalska, & Stamm, 2002). YTHDF1, YTHDF2, and YTHDF3 are cytoplasmic m6A readers and thus regulate mature mRNAs and their translation and degradation (H. Shi et al., 2017). YTHDC1 is a nuclear m6A reader that regulates mRNA splicing and nuclear export (Xiang, et al., 2017; Xiao et al., 2016). YTHDC2 localizes to both cytoplasm and nucleus, and facilitates degradation of specific modified mRNAs (Kretschmer et al., 2018). Other readers using non-YTH domains to recognize m6A include IGF2 mRNA-binding proteins (IGF2BPs) and translation initiation factor 3 (eIF3). The recruitment of eIF3 to m6A sites in the 5′UTR of mRNAs or circular RNAs (circRNAs) has been proposed as a cap-independent translation activation mechanism (Meyer et al., 2015; Y. Yang et al., 2017; Zhou et al., 2015). Known anti-readers are Ras GTPase-activating protein binding protein 1 (G3BP1) and 2 (G3BP2), important for initiating stress granule formation also for neuron development and plasticity (Arguello, DeLiberto, & Kleiner, 2017; Edupuganti et al., 2017; Martin et al., 2013; Sahoo et al., 2018; Tourrière et al., 2003). Except for the YTH domain, the binding mode and structural basis for m6A recognition for many readers are yet to be delineated. A third class of proteins that bind to m6A-mRNAs do not directly recognize the modified nucleotide, but rely on the m6A-induced local unfolding to generate the recognition motif, such as HNRNPC and HNRPNG (N. Liu et al., 2015, 2017). Thus m6A-dependent signal transduction by readers is mediated by different mechanisms to exert various effects. Loss of YTHDF2 results in delayed cortical neural development and late embryonic lethality, due to the delay in degrading developmental transcripts (M. Li et al., 2018). Loss of YTHDF1 in mice attenuates axon regeneration, causes axon misguidance, also impairs learning and memory (H. Shi et al., 2018; Y. Weng et al., 2018; Zhuang et al., 2019). The readers can function cooperatively or antagonistically. One dynamic regulation mechanism of m6A function has been shown to be through competition for m6A sites by readers. m6A reader proline rich coiled-coil 2A (PRRC2A) functionally antagonizes YTHDF2 to stabilize a subset of m6A transcripts including oligodendrocyte transcription factor 2 (Olig2), a key determination transcription factor for oligodendrocytes. Thus loss of PRRC2A results in hypomyelination and consequent locomotive and cognitive deficits without affecting neurogenesis (Wu et al., 2019).

m6A in Regulation of Local Translation in Neurons

Activity-induced persistent structural and functional plasticity is a key property of neurons that requires corresponding global and local changes in proteome (Heo et al., 2018; Dan Ohtan Wang, Martin, & Zukin, 2010). De novo protein synthesis in the subcellular compartments of neurons such as axons, growth cones, synapses is coupled to activity-dependent signaling pathways in a stimulus-, transcript-specific and spatially restricted manner (Holt, Martin, & Schuman, 2019; D. O. Wang et al., 2009). This remarkable specificity has been shown to be mediated by a rich molecular toolbox of trans-, and cis-regulatory elements and factors such as miRNAs, RNA-binding proteins, poly(A) tail extension, nonsense-mediated decay, RNA granule structures, eIF4E binding proteins, and local structures. But the number of known transcripts whose translation is regulated through one or more of these mechanisms is still limited. m6A modification is prevalent, dynamic, reversible, and regulates both RNA stability and translation rates, thus is well poised to regulate translation of specific mRNAs at individual transcript level in response to activity. With mRNAs trafficked into distal subcellular departments of neurons, m6A may travel together with the mRNAs (or even play an instructive role for targeting) and serve as a local epigenetic mark for activity-dependent local translation (Roy, Shiina, & Wang, 2020). However the full complement of synaptic epitranscriptome and their local regulation mechanisms are yet to be discovered.

To identify RNA species with m6A modification at synapses, we purified synaptosomes from healthy adult mouse forebrains, immunoprecipitated RNA fragments with anti-m6A antibody and performed massively parallel sequencing. Synaptosomes represent "tripartite synapses," containing pre- and post-synaptic compartments, and the surrounding glial terminals, thus the synaptsomal m6A-epitranscriptome we identified, a composite of 4,469 m6A sites distributed to 2,921 genes, belongs to all three cellular compartments of the tripartite synapses. Many genes in this list are implicated in neurodevelopment and neuropsychiatric disease pathways, especially 1,266 synaptically hypermethylated genes. The synaptically hypermethylated genes are enriched for synaptic functions such as synapse assembly, maturation, organization, and modulation of transmission, whereas the synaptically hypomethylated genes are enriched for metabolic functions. This functional partitioning between the synaptically hyper- and hypomethylated transcripts indicates that m6A can mark groups of functionally related transcripts at synapses that can be locally translated upon neuronal activity (Merkurjev et al., 2018). One of highly methylated, autism-associated synaptic mRNA *Apc*, showed diminished protein expression upon YTHDF1 knock-down in cultured rodent hippocampal neurons. Smaller spines and dampened synaptic transmission were measured in concordance with reduced glutamate receptor expression from the plasma membrane (Merkurjev et al., 2018). These in vitro results are consistent with the cellular and behavioral phenotypes in Ythdf1-KO mice which show deficits in hippocampus-dependent memory consolidation (H. Shi et al., 2018). How stimuli or neuron activities activate YTHDF1-mediated local translation however remains unknown. In addition to YTHDF1, other regulators such as METTL14, FTO, YTHDF1, YTHDF2, and YTHDF3 are also localized to distal neurites in vivo and in vitro, supporting dynamic and cooperative regulation. However, the function of these proteins in dendrites remains unknown.

One controversial reader protein of m6A, fragile X mental retardation protein (FMRP), is also known as an activity-dependent RNA-binding protein and local translational repressor/activator in dendrites and at synapses (Chang et al., 2017; Edupuganti et al., 2017). Loss of function of FMRP in humans causes fragile X syndrome, the most common form of inherited intellectual disability, indicating that dysregulated translation of m6A-RNA at synapses may be particularly important for cognitive development (Jin, 2000b). Evidence that FMRP affects dynamic behavior of the m6A-mRNA such as nuclear export and translation is emerging but whether FMRP is a direct m6A reader is under active debate (Edens et al., 2019; Hsu et al., 2019). Inconsistent behavioral phenotypes in loss-of-function rodent models of FMRP have been reported, indicating a complex role of this regulatory pathway (Roy et al., 2020).

In axons, local translation can be induced by neurotrophic factors, guidance cues, synaptic partners, and injury-triggered signals. Similarly to dendrites, the specific signaling pathways regulating activity-specific translational response remains poorly understood (Cioni, Koppers, & Holt, 2018). FTO mRNA is locally translated in axons and in response to nerve growth factor (NGF) that promotes axon growth. The de novo synthesized FTO triggers demethylation of *Gap43* mRNA and activates its axonal translation, leading to axon elongation (Yu et al., 2018). In crossing axons of spinal commissural neurons, the m6A reader YTHDF1 binds to and facilitates translation of m6A-modified mRNA. Deletion of Ythdf1 specifically in this neuron type results in misprojection of the pre-crossing axons into motor columns, indicating a critical role of m6A and YTHDF1 in axon guidance (Zhuang et al., 2019).

Together m6A and its dynamic regulation of local translation in both dendritic and axonal compartments indicates an important role of this modification for building, maintaining, and modulating circuit connectivity of CNS. Future investigations of m6A and other RNA modifications in regulating local translation, circuits development and plasticity are clearly warranted. What are the operational mechanisms? Mostly intriguingly, transcripts with multiple m6A modification sites have been shown to enhance phase separation of the RNAs and their binding proteins, a potential mechanism for neuronal granule formation responsible for RNA transportation (Ries et al., 2019). Details on what neuronal granules may bear m6A and how this modification may change the physiochemical properties of the granules are yet to be elucidated.

RNA Modifications Are Inducible and Erasable

The cellular and functional dynamics of RNA modification can be regulated through the "writers" (enzymes responsible for adding the modification to the designated RNA nucleotides), "erasers" (enzymes responsible for chemically converting the modified nucleotides back to A, G, C, or U), and "readers" (proteins that distinguish the modification status and direct the modified RNA into different metabolic and functional pathway). RNA degradation is another route to "erase" the modification. In yeast, m6A modification pathway is specifically associated with meiosis and no m6A demethylase has been identified. Thus massive degradation of the meiosis-related and m6A-modified transcripts effectively erase this modification from yeast cells (Bodi, Button, Grierson, & Fray, 2010; Gasch et al., 2000).

An initial hurdle in determining the function of a single RNA modification was that many RNA modifications are non-essential in the sense that organisms can survive without a certain RNA modification. This "inconvenient" feature and the partial modification status of many RNAs indicate a regulatory role of RNA modifications in specific circumstances such when cells are stressed, viral infected, deprived of nutrition, or damaged as was found in "non-essential" post-translational protein modifications. Indeed, excellent studies have demonstrated dynamic responses in different types of modifications on multiple biotypes of RNAs, to UV-light, DNA repair, and cancer malignancy. Here we focus specifically on the relevant events in CNS, where the extent of specific modifications varies depending on the cellular environment and physiological/pathological states. For other types and context of inducible and dynamic RNA modification regulation, I suggest excellent reviews on this matter (Yi & Pan, 2011).

One of the earliest studies suggesting the functional significance of reversible RNA modification in CNS came from a knockout mouse study of FTO, the first identified RNA demethylase to catalyze the oxidative demethylation of m6A (Jia et al., 2011b). Later m1a and m6Am were also added to the substrate list of FTO (J. Wei et al., 2018). In Fto knockout mice, basal locomotor was increased and animal's sensitivity to the locomotor- and reward- stimulatory actions of cocaine was enhanced (Hess et al., 2013). At molecular level, a subset of mRNAs important for dopamine signaling pathways revealed increased adenosine methylation and altered expression in the midbrain and striatum of Fto-deficient mice, indicating the behavioral abnormity is mediated by the altered dopamine transmission, an important signaling pathway for psychomotor activity, drug seeking behavior, and learning and memory (Hess et al., 2013).

m6A

Upregulation of m6A has been shown in different cellular contexts, some particularly relevant to CNS during development. Although an increased brain m6A level in mature brain was initially published (Meyer et al., 2012b), a later study showed decreased cerebellar expression level of m6A along with METTL3, METTL14, and WTAP (WT1 associated protein) between P7 and P60 (Ma et al., 2018). Another study reported decreased m6A expression in pallium and dorsal sub ventricular zone along with METTL3 and METTL14 between E15.5 and P2, in parallel to the restricted neurogenic capacity at P2 (Donega et al., 2018), thus m6A dynamics can be regulated in a region- and cell type-specific manner.

Several studies have addressed varied extent of m6A after fear conditioning (H. Shi et al., 2018; Walters et al., 2017; Widagdo et al., 2016; Zeyu Zhang et al., 2018) stress (Bai et al., 2018; Engel et al., 2018; Xiang, Laurent, Hsu, Nachtergaele, et al., 2017), stroke (Chokkalla et al., 2019), and axon regeneration (Y.-L. Weng et al., 2018). However, the detailed regulation mechanisms and cellular pathways are yet awaiting discovery. In these studies, the induced m6A modifications are commonly found in the "response-specific" transcripts such as those involved in synaptic plasticity, transcription factors, axon regeneration, suggesting a role of this pathway in generating a temporary "emergency" mode of gene regulation in response to the stimuli. Due to the role of m6A to prioritize translation of the modified RNA, and increase RNA turnover by targeting them to degradation, responsive m6A modification ensures a timely and less lasting response so that the energy expenses associated with the emergency mode of gene regulation can be minimized. This phenomenon is consistent with the role of m6A in cell differentiation and lineage determination where it allows smooth shifts of genetic programs from one development stage to the next. Downregulation of m6A in specific transcripts can be caused by degradation of the modified transcripts or through the demethylation activity of the enzymes. In animals missing METTL3, learning becomes less efficient but remains possible, given that excessive training is provided to the animals, corroborating the modulatory role of this pathway (Zeyu Zhang et al., 2018). These results demonstrate cellular dynamics of m6A is associated with functional and behavioral adaptation of CNS.

How is m6A induced under these conditions? m6A "writer" complexes METTL3-METTL14 and regulatory subunits responsible for mRNA m6A modification are expressed universally in CNS. Acute stress in rodent downregulates Mettl3 mRNA but upregulates Fto, Alkbh5, and Ythdc1 (Engel et al., 2018). In cancer cells, SUMOylation of the METTL3 has been shown to repress its methylation activity without affecting its expression level (Du et al., 2018). One study has suggested relocation of FTO to "leave" synapse upon fear conditioning (Walters et al., 2017). Such spatial regulation may induce a temporary increased level of m6A at synapse, which in turn facilitates learning-induced local translation. Compliment to this result, when the accumulation of m6A is amplified following knockdown of the RNA demethylase FTO, memory is enhanced, indicating FTO limits the fear-induced memory (Widagdo et al., 2016). However, the upstream pathways in CNS that potentially couple their expression to neuronal activity and environmental stimuli remain largely unexplored. It is worth mentioning that demethylation activity of FTO is strictly dependent on the cofactors Fe^{2+} and 2′ ketoglutarate thus changing local concentrations of these soluble factors can also influence demethylation activity (Jia et al., 2011b).

Inosine

A-to-I RNA editing has important roles in the evolution of cognitive function with specific enzymes evolved in primate brains and huge expansion of presence in human brain (Paul, 1998b). Many important neuronally expressed genes in neuromodulation and excitatory signal pathways contain A-to-I conversions, which influence splice-site choice, miRNA targeting capacity, and coding thus this RNA modification pathway is a requisite for neuronal function. Developmentally regulated A-to-I conversions, in the 5-HT2C subunit of serotonin receptors generate variants with distinct G-protein coupling properties; in the GluR2 subunit of AMPA receptors change Ca2+ permeability of the channels; in the voltage-gated potassium channels alter their trafficking properties and in GABA receptors alter their kinetics, subunit assembly and cell surface expression. These changes collectively alter the intrinsic excitability, signal transduction, and transmission of neurons (Burns et al., 1997; Higuchi et al., 2000; Hoopengardner, 2003; Niswender, Copeland, Herrick-Davis, Emeson, & Sanders-Bush, 1999; Sommer, Köhler, Sprengel, & Seeburg, 1991).

Interestingly, editing in Cyfip2 (cytoplasmic FMR1 interacting protein 2), a dendritically localized mRNA that leads to a lysine to glutamate (K/E) amino acid replacement, is dynamically regulated during brain development and in response to neuronal activity (Sanjana, Levanon, Hueske, Ambrose, & Li, 2012; Wahlstedt, Daniel, Enstero, & Ohman, 2009). However, whether the edited status of Cyfip2 is associated with localized Cyfip2 mRNA and its local translation, is currently unknown. The spatial and temporal information of RNA editing is yet to be gained to understand the correct context and function for their dynamic regulation.

m1A

m1A can be found in eukaryotic tRNA, rRNA, mRNA, and mitochondrial RNA. Unlike tRNA and rRNA, the prevalence of m1A in mRNA remains largely controversial, ranging from 0.015 percent and up to 0.16 percent of adenosine in mammalian cells and tissues, from a handful to thousands modified (Dominissini et al., 2016; Dominissini & Rechavi, 2017; X. Li et al., 2016, 2017), reviewed in (C. Zhang & Jia, 2018). These sites can be added to RNA enzymatically by TRMT6/61A, TRMT61B, and TRMT10C (Safra et al., 2017), and erased by demethylases ALKBH1 and ALKBH3 (X. Li et al., 2016; F. Liu et al., 2016). Since m1A negatively impacts translation, its dynamic regulation can serve as a "releasable brake" on protein synthesis. Many questions remain to be answered regarding the cellular dynamics of m1A, such as what and where are the cellular "readers" of m1A? how are the methylation and demethylation activities regulated? What is the function of the translational inhibition and disinhibition to CNS cells?

Pseudouridine (ψ)

Many potentially functional ψ sites in coding and noncoding transcripts enriched in the brain are recently identified yet their function remains unknown (X. Li et al., 2015a). Given the isomerized uridine improves RNA stacking and interaction with RNA-binding proteins, this modification may influence the way the modified RNA is recognized in cells (Davis, 1995). As discussed in previous chapters, pseudouridines have critical structural roles in translation function by ribosomes and tRNAs, therefore its dynamic regulation may have global effects on translation activity. In HeLa cells, many ψ sites in mRNA respond to environmental stress such as serum starvation, suggesting functional dynamics (Carlile et al., 2014). However, whether this induction occurs in the brain in relation to physiological or pathological functions is yet to be tested.

To Decode the "RNA Epigenetic Code"

After the first modified nucleosides were discovered, a big challenge for RNA modification field was to "map" them to individual genes. As the modifications were detected in the RNA hydrolysates, or a soup of enzyme-digested RNA nucleotides, sequence information to be compared to the genome reference was lost. Later on, biochemical analysis using a combination of ribonuclease digestion and chromatography, primer extension assays were used to identify motifs and specific modifications for a handful of genes. Without amplification technology, earlier studies were conducted in highly abundant rRNA and tRNAs, whose single species could be isolated where multiple abundant RNA modifications such as inosine and pseudouridine were mapped, annotated, and functionally demonstrated. But modifications in RNA of low abundance such as mRNA was only mapped genome-widely to mouse and human genomes in 2012 with a newly developed technology now known as "m6A-seq," which uses an antibody to enrich m6A-containing RNA fragments followed by massively parallel sequencing (Dominissini et al., 2012b; Meyer et al., 2012a).

Currently the newly established and emerging techniques are focused on "global-scaled modification mapping with precision" often dubbed as "modification"-seq, which is to detect the modification status of any given position/residue/nucleotide in any given transcript for the entire transcriptome. Such mapping techniques rely on RNA-seq in combination with specific preparation methods to generate detectable "signatures" in RNA-seq. The preparation process can be classified into three categories: enzyme-based, antibody-based, and reaction-based. Enzyme-based methods typically use one of the three enzymes: 1. reverse transcriptase that generates a signature in the first-strand cDNA synthesis (Z. Wang, Gerstein, & Snyder, 2009); 2. enzymes to generate specific conversions such as C to U in the proximity of the modification (Rosenberg, Hamilton, Mwangi, Dewell, & Papavasiliou, 2011); 3. enzymes that remove a specific modification which would have impeded first strand cDNA synthesis, resulting in detectable "truncation" or "mutation" signatures. Antibody-based methods as the name indicated use modification-specific antibodies for affinity-enrichment of transcripts fragments, and further combined with UV-crosslinking to generate "truncations" or "mutations" depending on the property of the antibodies. Finally, reaction-based techniques that rely on specific chemical reactions to introduce detectable changes to the modified nucleotides in the current sequencing technology. I provide a small table (Table 8.1) here to share a flavor on the variety of modifications amenable to a variety of deep sequencing

Table 8.1. New and Emerging Methods to Detect RNA Modifications

m6A	MeRIP-Seq or m6A-seq[1-2], Low input m6A-seq[3-5], miCLIP-m6A-seq[6,7], PA-m6A-Seq[8], LAIC-seq[9-10], Maz-seq[11-12], DART-seq[13], SMART-seq[14], Nanopore RNA-seq[15]
hm5Ch	MeRIP-seq[16]
m7G	TRAC-seq[17], AlkAniline-Seq[18]
m1A	ARM-seq[19], DM-tRNA-Seq[20]
m5C	Bis-seq[21-28]
m3C	HAMR-seq[29,30]
ψ	Pseudo-seq[31-33]
Nm	2′O-Me-seq[34,35], RTL-P-seq[36,37], RiboMethSeq[38-40], Nm(RibOxi)-Seq[41,42]
I	ICE-seq[43]

Sources: [1] Dominissini et al. 2012b; [2] Meyer et al. 2012b; [3] Merkurjev et al. 2018; [4] Weng et al. 2018; [5] Zeng et al. 2018; [6] Linder et al. 2015; [7] Grozhik et al. 2017; [8] Chen, Luo and He 2015; [9] Molinie et al. 2016; [10] Grozhik et al. 2017; [11] Garcia-Campos et al. 2019; [12] Zhang et al. 2019; [13] Meyer 2019; [14] Vilfan et al. 2013; [15] Liu et al. 2019; [16] Delatte et al. 2016b; [17] Lin et al. 2018; [18] Marchand et al. 2018; [19] Cozen et al. 2015; [20] Zheng et al. 2015; [21] Schaefer et al. 2008; [22] Tuorto et al. 2012b; [23] Edelheit et al. 2013; [24] Bourgeois et al. 2015; [25] Müller et al. 2015; [26] Amort et al. 2017; [27] Legrand et al. 2017; [28] Z. Wei et al. 2018; [29] Ryvkin et al. 2013; [30] Kuksa et al. 2017; [31] Carlile et al. 2014; [32] Schwartz et al. 2014; [33] Li et al. 2015b; [34] Maden 2001; [35] Incarnato et al. 2017; [36] Dong et al. 2012; [37] Aschenbrenner and Marx 2016; [38] Birkedal et al. 2014; [39] Gumienny et al. 2016; [40] Marchand et al. 2016; [41] Dai et al. 2017; [42] Zhu, Pirnie and Carmichael 2017; [43] Suzuki et al. 2015.

approaches, but refer to the readers a more comprehensive survey on current and near future "modification"-seqs by Yuri Motorin and Mark Helm (Motorin & Helm, 2019).

Powered by these revolutionary new methods, our understanding of the epitranscriptome will quickly evolve. Some development may be especially exciting for neuroscientists such as to sequence RNA modifications in specific circuits, cell populations, and at specific time points. Previously, molecular profiling of neurons based on projection/connectivity has been reported that can theoretically be adapted to study m6A and other modifications in specific cell types under specific conditions (Knight et al., 2012).

Nonetheless, new methods are yet to be developed that are sufficiently advanced to allow us to do genome-wide detection in a quantitative and sensitive manner so that small but meaningful changes can be resolved by neurobiologists in their studies.

Some Other Interesting Phenomena and Emerging Ideas

Queuosine, a quanosine analogue often found in the first position of anticodon of tRNAs that biases codon preference and prevent read-through (Meier, Suter, Grosjean, Keith, & Kubli, 1985), which has been found to be incororparted into the brain tRNA, can not be synthesized by eukaryotes, thus has to be salvaged from diet or from intestinal microflora (Vinayak & Pathak, 2010). Such studies are revealing new exciting possibilities of microflora-brain interaction. Where else can the modification come from? New studies are underway to show that metabolites from food, drugs, and spices can generate reactive intermediates that add adducts to intracellular RNA in hepatocytes. The downstream impact such as removal and repair of these RNA adducts is largely unexplored such as whether the RNA can gain or lose function because of these adducts. Can such "RNA adducts" become an experimenting playground for the RNA regulation pathway to evolve? These are exciting possibilities that warrant future investigation.

Another class of epigenetic regulator, piwi-interacting RNA (piRNA) are known to have 2′O-methylation at 3′end in plants, flies, and rodents (Houwing et al., 2007; Kirino & Mourelatos, 2007; Saito et al., 2007). Given the role of piRNAs in silencing transposable elements that invade the genome by inserting themselves into DNA elements, the methylation of piRNA may play important role in the biogenesis of this class of RNA and function in genome surveillance during germline development.

Conclusions and Future Directions

The brain is an intricate and complex organ that underlies our cognition and behavior. Although we are living in an era of privilege where we can read our own genetic makeups, and may soon become capable of editing them at will, we remain woefully ignorant of how our genetic information is transformed into the functional brain circuits that make us walking, thinking, and talking human beings. The function of the nervous system relies heavily on protein synthesis and is more susceptible than other organs to imbalanced proteostasis, which is a major driver of 51 neurodevelopmental and neurodegenerative disease (Sossin & Costa-Mattioli, 2019). RNA modifications and other post-transcriptional regulations by nature are poised to generate coding diversity, to influence translation efficiency and fidelity, to generate timely and appropriate amount of response to electrical and chemical activities, to couple gene expression to cellular metabolism and circadian rhythm, thus are deeply tied to the stability and dynamic regulation of proteome and metabolism. Identifying the full complement of the "neuro-epitranscriptome" together with the cellular regulatory mechanisms in CNS is therefore a big step toward understanding "neuro-epitranscriptomics."

In a most comprehensive book published in 2005 on RNA modification, "Fine-Tuning of RNA Functions by Modification and Editing" edited by Henri Grosjean, not a single chapter was dedicated to their function in nervous system. Research in the past fifteen years has revealed that RNA modifications are involved in all major functions of CNS, including but not limited to cortical neurogenesis, circadian rhythm, reward, addiction, stress, learning and memory, stroke, spinal injury, etc. Nonetheless, the function of most of RNA chemical modifications and their cellular regulation remain to be unveiled. With advent of analytical tools, future studies will lead the way toward harvesting on this potent cellular pathway for maintaining a healthy and functional CNS.

Acknowledgments

This work was supported by KAKENHI 17H03546, 19H04907, 19H05212, and AMED 18dm 0307023h 0001, NSFC 31971335, Xingliao Talents Program XLYC1802007, Department of Education of Liaoning Province 1911520092, and Hirose research grant to DOW.

References

Abbasi-Moheb, L., Mertel, S., Gonsior, M., Nouri-Vahid, L., Kahrizi, K., Cirak, S.,... Kuss, A. W. (2012). Mutations in NSUN2 cause autosomal-recessive intellectual disability. *American Journal of Human Genetics*, 90(5), 847–855. https://doi.org/10.1016/j.ajhg.2012.03.021

Abdelrahman, H. A., Al-Shamsi, A. M., Ali, B. R., & Al-Gazali, L. (2018). A null variant in PUS3 confirms its involvement in intellectual disability and further delineates the associated neurodevelopmental disease. *Clinical Genetics*, 94(6), 586–587. https://doi.org/10.1111/cge.13443

Akichika, S., Hirano, S., Shichino, Y., Suzuki, T., Nishimasu, H., Ishitani, R.,... Suzuki, T. (2019). Cap-specific terminal N 6 -methylation of RNA by an RNA polymerase II–associated methyltransferase. *Science*, 363(6423), 1–13. https://doi.org/10.1126/science.aav0080

Alarcón, C. R., Lee, H., Goodarzi, H., Halberg, N., & Tavazoie, S. F. (2015). N6-methyladenosine marks primary microRNAs for processing. *Nature*, 519(7544), 482–485. https://doi.org/10.1038/nature14281

Alazami, A. M., Hijazi, H., Al-Dosari, M. S., Shaheen, R., Hashem, A., Aldahmesh, M. A., . . . Alkuraya, F. S. (2013). Mutation in ADAT3, encoding adenosine deaminase acting on transfer RNA, causes intellectual disability and strabismus. *Journal of Medical Genetics*, 50(7), 425–430. https://doi.org/10.1136/jmedgenet-2012-101378

Amort, T., Rieder, D., Wille, A., Khokhlova-Cubberley, D., Riml, C., Trixl, L., . . . Lusser, A. (2017). Distinct 5-methylcytosine profiles in poly(A) RNA from mouse embryonic stem cells and brain. *Genome Biology*, 18(1), 1. https://doi.org/10.1186/s13059-016-1139-1

Anderson, S. L., Coli, R., Daly, I. W., Kichula, E. A., Rork, M. J., Volpi, S. A., . . . Rubin, B. Y. (2001). Familial dysautonomia is caused by mutations of the IKAP gene. *The American Journal of Human Genetics*, 68(3), 753–758. https://doi.org/10.1086/318808

Arguello, A. E., DeLiberto, A. N., & Kleiner, R. E. (2017). RNA chemical proteomics reveals the N 6-methyladenosine (m 6 A)-regulated protein–RNA interactome. *Journal of the American Chemical Society*, 139(48), 17249–17252. https://doi.org/10.1021/jacs.7b09213

Aschenbrenner, J., & Marx, A. (2016). Direct and site-specific quantification of RNA 2′-O-methylation by PCR with an engineered DNA polymerase. *Nucleic Acids Research*, 44(8), 3495–3502. https://doi.org/10.1093/nar/gkw200

Bai, L., Tang, Q., Zou, Z., Meng, P., Tu, B., Xia, Y., . . . Chen, C. (2018). m6A Demethylase FTO regulates dopaminergic neurotransmission deficits caused by arsenite. *Toxicological Sciences*, 165(2), 431–446. https://doi.org/10.1093/toxsci/kfy172

Ban, N. (2000). The complete atomic structure of the large ribosomal subunit at 2.4 A resolution. *Science*, 289(5481), 905–920. https://doi.org/10.1126/science.289.5481.905

Baser, A., Skabkin, M., Kleber, S., Dang, Y., Gülcüler Balta, G. S., Kalamakis, G., . . . Martin-Villalba, A. (2019). Onset of differentiation is post-transcriptionally controlled in adult neural stem cells. *Nature*, 566(7742), 100–104. https://doi.org/10.1038/s41586-019-0888-x

Behm, M., & Öhman, M. (2016). RNA Editing: A contributor to neuronal dynamics in the mammalian brain. *Trends in Genetics*, 32(3), 165–175. https://doi.org/10.1016/j.tig.2015.12.005

Bento-Abreu, A., Jager, G., Swinnen, B., Rué, L., Hendrickx, S., Jones, A., . . . Robberecht, W. (2018). Elongator subunit 3 (ELP3) modifies ALS through tRNA modification. *Human Molecular Genetics*, 27(7), 1276–1289. https://doi.org/10.1093/hmg/ddy043

Bernard, D., Prasanth, K. V, Tripathi, V., Colasse, S., Nakamura, T., Xuan, Z., . . . Bessis, A. (2010). A long nuclear-retained non-coding RNA regulates synaptogenesis by modulating gene expression. *EMBO Journal*, 29(18), 3082–3093. https://doi.org/10.1038/emboj.2010.199

Bertero, A., Brown, S., Madrigal, P., Osnato, A., Ortmann, D., Yiangou, L., . . . Vallier, L. (2018). The SMAD2/3 interactome reveals that TGFβ controls m6A mRNA methylation in pluripotency. *Nature*, 555(7695), 256–259. https://doi.org/10.1038/nature25784

Birkedal, U., Christensen-Dalsgaard, M., Krogh, N., Sabarinathan, R., Gorodkin, J., & Nielsen, H. (2014). Profiling of ribose methylations in RNA by high-throughput sequencing. *Angewandte Chemie International Edition*, n/a-n/a. https://doi.org/10.1002/anie.201408362

Blanco, S., Dietmann, S., Flores, J. V, Hussain, S., Kutter, C., Humphreys, P., . . . Frye, M. (2014). Aberrant methylation of tRNAs links cellular stress to neuro-developmental disorders. *EMBO Journal*, 33(18), 2020–2039. https://doi.org/10.15252/embj.201489282

Blanco, S., Dietmann, S., Flores, J. V, Hussain, S., Kutter, C., Humphreys, P.,... Frye, M. (2014). Aberrant methylation of tRNAs links cellular stress to neuro-developmental disorders. *EMBO Journal*, 33(18), 2020–2039. https://doi.org/10.15252/embj.201489282

Blanco, S., Kurowski, A., Nichols, J., Watt, F. M., Benitah, S. A., & Frye, M. (2011). The RNA–methyltransferase Misu (NSun2) poises epidermal stem cells to differentiate. *PLoS Genetics*, *7*(12), e1002403. https://doi.org/10.1371/journal.pgen.1002403

Boccaletto, P., Machnicka, M. A., Purta, E., Piątkowski, P., Bagiński, B., Wirecki, T. K.,... Bujnicki, J. M. (2018). MODOMICS: a database of RNA modification pathways. 2017 update. *Nucleic Acids Research*, 46(D1), D303–D307. https://doi.org/10.1093/nar/gkx1030

Bodi, Z., Button, J. D., Grierson, D., & Fray, R. G. (2010). Yeast targets for mRNA methylation. *Nucleic Acids Research*, 38(16), 5327–5335. https://doi.org/10.1093/nar/gkq266

Bourgeois, G., Létoquart, J., van Tran, N., & Graille, M. (2017). Trm112, a protein activator of methyltransferases modifying actors of the eukaryotic translational apparatus. *Biomolecules*, 7(4), 7. https://doi.org/10.3390/biom7010007

Bourgeois, G., Ney, M., Gaspar, I., Aigueperse, C., Schaefer, M., Kellner, S.,... Motorin, Y. (2015). Eukaryotic rRNA modification by yeast 5-methylcytosine-methyltransferases and human proliferation-associated antigen p120. *PLOS One*, 10(7), e0133321. https://doi.org/10.1371/journal.pone.0133321

Burns, C. M., Chu, H., Rueter, S. M., Hutchinson, L. K., Canton, H., Sanders-Bush, E., & Emeson, R. B. (1997). Regulation of serotonin-2C receptor G-protein coupling by RNA editing. *Nature*, *387*(6630), 303–308. https://doi.org/10.1038/387303a0

Canaani, D., Kahana, C., Lavi, S., & Groner, Y. (1979). Identification and mapping of N 6 methyladenosine containing sequences in Simian Virus 40 RNA. *Nucleic Acids Research*, *6*(8), 2879–2899. https://doi.org/10.1093/nar/6.8.2879

Carlile, T. M., Rojas-Duran, M. F., Zinshteyn, B., Shin, H., Bartoli, K. M., & Gilbert, W. V. (2014). Pseudouridine profiling reveals regulated mRNA pseudouridylation in yeast and human cells. *Nature*, *515*(7525), 143–146. https://doi.org/10.1038/nature13802

Cavaille, J., Buiting, K., Kiefmann, M., Lalande, M., Brannan, C. I., Horsthemke, B.,... Huttenhofer, A. (2000). Identification of brain-specific and imprinted small nucleolar RNA genes exhibiting an unusual genomic organization. *Proceedings of the National Academy of Sciences USA*, 97(26), 14311–14316. https://doi.org/10.1073/pnas.250426397

Chang, M., Lv, H., Zhang, W., Ma, C., He, X., Zhao, S.,... Tong, W.-M. (2017). Region-specific RNA m 6 A methylation represents a new layer of control in the gene regulatory network in the mouse brain. *Open Biology*, *7*(9), 170166. https://doi.org/10.1098/rsob.170166

Chen, C.-X., Cho, D.-S. C., Wang, Q., Lai, F., Carter, K. C., & Nishikura, K. (2000). A third member of the RNA-specific adenosine deaminase gene family, ADAR3, contains both single- and double-stranded RNA binding domains. *RNA*, *6*(5), S1355838200000170. https://doi.org/10.1017/S1355838200000170

Chen, K., Luo, G. Z., & He, C. (2015). High-resolution mapping of N6-methyladenosine in transcriptome and genome using a photo-crosslinking-assisted strategy. *Methods in Enzymology*, 560, 161–185. https://doi.org/10.1016/bs.mie.2015.03.012

Chi, L., & Delgado-Olguín, P. (2013). Expression of NOL1/NOP2/sun domain (Nsun) RNA methyltransferase family genes in early mouse embryogenesis. *Gene Expression Patterns*, *13*(8), 319–327. https://doi.org/10.1016/j.gep.2013.06.003

Chokkalla, A. K., Mehta, S. L., Kim, T., Chelluboina, B., Kim, J., & Vemuganti, R. (2019). Transient focal ischemia significantly alters the m 6 A epitranscriptomic tagging of

RNAs in the brain. *Stroke*, *50*(10), 2912–2921. https://doi.org/10.1161/STROKEAHA.119.026433

Chou, H.-J., Donnard, E., Gustafsson, H. T., Garber, M., & Rando, O. J. (2017). Transcriptome-wide analysis of roles for tRNA modifications in translational regulation. *Molecular Cell*, *68*(5), 978–992.e4. https://doi.org/10.1016/j.molcel.2017.11.002

Cioni, J.-M., Koppers, M., & Holt, C. E. (2018). Molecular control of local translation in axon development and maintenance. *Current Opinion in Neurobiology*, 51, 86–94. https://doi.org/10.1016/j.conb.2018.02.025

Cohen, J. S., Srivastava, S., Farwell, K. D., Lu, H.-M., Zeng, W., Lu, H.,... Fatemi, A. (2015). ELP2 is a novel gene implicated in neurodevelopmental disabilities. *American Journal of Medical Genetics Part A*, *167*(6), 1391–1395. https://doi.org/10.1002/ajmg.a.36935

Cohn, W. E. (1960). Pseudouridine, a carbon-carbon linked ribonucleoside in ribonucleic acids: isolation, structure, and chemical characteristics. *The Journal of Biological Chemistry*, 235, 1488–1498. http://www.ncbi.nlm.nih.gov/pubmed/13811056

Coots, R. A., Liu, X.-M., Mao, Y., Dong, L., Zhou, J., Wan, J.,... Qian, S.-B. (2017). m6A facilitates eIF4F-independent mRNA translation. *Molecular Cell*, *68*(3), 504–514.e7. https://doi.org/10.1016/j.molcel.2017.10.002

Cozen, A. E., Quartley, E., Holmes, A. D., Hrabeta-Robinson, E., Phizicky, E. M., & Lowe, T. M. (2015). ARM-seq: AlkB-facilitated RNA methylation sequencing reveals a complex landscape of modified tRNA fragments. *Nature Methods*, *12*(9), 879–884. https://doi.org/10.1038/nmeth.3508

Dai, Q., Moshitch-Moshkovitz, S., Han, D., Kol, N., Amariglio, N., Rechavi, G.,... He, C. (2017). Nm-seq maps 2′-O-methylation sites in human mRNA with base precision. *Nature Methods*, *14*(7), 695–698. https://doi.org/10.1038/nmeth.4294

Davarniya, B., Hu, H., Kahrizi, K., Musante, L., Fattahi, Z., Hosseini, M.,... Najmabadi, H. (2015). The role of a novel TRMT1 gene mutation and rare GRM1 gene defect in intellectual disability in two Azeri families. *PLOS ONE*, *10*(8), e0129631. https://doi.org/10.1371/journal.pone.0129631

Davis, D. R. (1995). Stabilization of RNA stacking by pseudouridine. *Nucleic Acids Research*, *23*(24), 5020–5026. https://doi.org/10.1093/nar/23.24.5020

Davis, F. F., & Allen, F. W. (1957). Ribonucleic acids from yeast which contain a fifth nucleotide. *Journal of Biological Chemistry*, *227*(2), 907–915. http://www.ncbi.nlm.nih.gov/pubmed/13463012

De Pietri Tonelli, D., Pulvers, J. N., Haffner, C., Murchison, E. P., Hannon, G. J., & Huttner, W. B. (2008). miRNAs are essential for survival and differentiation of newborn neurons but not for expansion of neural progenitors during early neurogenesis in the mouse embryonic neocortex. *Development*, *135*(23), 3911–3921. https://doi.org/10.1242/dev.025080

Delatte, B., Wang, F., Ngoc, L. V., Collignon, E., Bonvin, E., Deplus, R.,... Fuks, F. (2016a). Transcriptome-wide distribution and function of RNA hydroxymethylcytosine. *Science*, *351*(6270), 282–285. https://doi.org/10.1126/science.aac5253

Delatte, B., Wang, F., Ngoc, L. V., Collignon, E., Bonvin, E., Deplus, R.,... Fuks, F. (2016b). Transcriptome-wide distribution and function of RNA hydroxymethylcytosine. *Science*, *351*(6270), 282–285. https://doi.org/10.1126/science.aac5253

Derrien, T., Johnson, R., Bussotti, G., Tanzer, A., Djebali, S., Tilgner, H.,... Guigo, R. (2012). The GENCODE v7 catalog of human long noncoding RNAs: Analysis of their gene structure, evolution, and expression. *Genome Research*, *22*(9), 1775–1789. https://doi.org/10.1101/gr.132159.111

Desrosiers, R., Friderici, K., & Rottman, F. (1974). Identification of methylated nucleosides in messenger RNA from Novikoff hepatoma cells. *Proceedings of the National Academy of Sciences USA*, *71*(10), 3971–3975. https://doi.org/10.1073/pnas.71.10.3971

Doll, A., & Grzeschik, K.-H. (2001). Characterization of two novel genes, WBSCR20 and WBSCR22, deleted in Williams-Beuren syndrome. *Cytogenetic and Genome Research*, *95*(1–2), 20–27. https://doi.org/10.1159/000057012

Dominissini, D., Moshitch-Moshkovitz, S., Schwartz, S., Salmon-Divon, M., Ungar, L., Osenberg, S., . . . Rechavi, G. (2012a). Topology of the human and mouse m6A RNA methylomes revealed by m6A-seq. *Nature*, *485*(7397), 201–206. https://doi.org/10.1038/nature11112

Dominissini, D., Moshitch-Moshkovitz, S., Schwartz, S., Salmon-Divon, M., Ungar, L., Osenberg, S., . . . Rechavi, G. (2012b). Topology of the human and mouse m6A RNA methylomes revealed by m6A-seq. *Nature*, *485*(7397), 201–206. Retrieved July 10, 2014 from https://doi.org/10.1038/nature11112

Dominissini, D., Nachtergaele, S., Moshitch-Moshkovitz, S., Peer, E., Kol, N., Ben-Haim, M. S., . . . He, C. (2016). The dynamic N1-methyladenosine methylome in eukaryotic messenger RNA. *Nature*, *530*(7591), 441–446. https://doi.org/10.1038/nature16998

Dominissini, D., & Rechavi, G. (2017). Loud and clear epitranscriptomic m1A signals: Now in single-base resolution. *Molecular Cell*. https://doi.org/10.1016/j.molcel.2017.11.029

Donega, V., Marcy, G., Lo Giudice, Q., Zweifel, S., Angonin, D., Fiorelli, R., . . . Raineteau, O. (2018). Transcriptional dysregulation in postnatal glutamatergic progenitors contributes to closure of the cortical neurogenic period. *Cell Reports*, *22*(10), 2567–2574. https://doi.org/10.1016/j.celrep.2018.02.030

Dong, Z.-W., Shao, P., Diao, L.-T., Zhou, H., Yu, C.-H., & Qu, L.-H. (2012). RTL-P: A sensitive approach for detecting sites of 2′-O-methylation in RNA molecules. *Nucleic Acids Research*, *40*(20), e157–e157. https://doi.org/10.1093/nar/gks698

Du, Y., Hou, G., Zhang, H., Dou, J., He, J., Guo, Y., . . . Yu, J. (2018). SUMOylation of the m6A-RNA methyltransferase METTL3 modulates its function. *Nucleic Acids Research*, *46*(10), 5195–5208. https://doi.org/10.1093/nar/gky156

Eberwine, J., Belt, B., Kacharmina, J. E., & Miyashiro, K. (2002). Analysis of subcellularly localized mRNAs using in situ hybridization, mRNA amplification, and expression profiling. *Neurochemical Research*, *27*(10), 1065–1077. https://doiorg/10.1023/a:1020956805307

Edelheit, S., Schwartz, S., Mumbach, M. R., Wurtzel, O., & Sorek, R. (2013). Transcriptome-wide mapping of 5-methylcytidine RNA modifications in bacteria, archaea, and yeast reveals m5C within archaeal mRNAs. *PLoS Genetics*, *9*(6), e1003602. https://doi.org/10.1371/journal.pgen.1003602

Edens, B. M., Vissers, C., Su, J., Arumugam, S., Xu, Z., Shi, H., . . . Ma, Y. C. (2019). FMRP modulates neural differentiation through m6A-dependent mRNA nuclear export. *Cell Reports*, *28*(4), 845–854.e5. https://doi.org/10.1016/j.celrep.2019.06.072

Edupuganti, R. R., Geiger, S., Lindeboom, R. G. H., Shi, H., Hsu, P. J., Lu, Z., . . . Vermeulen, M. (2017). N6-methyladenosine (m6A) recruits and repels proteins to regulate mRNA homeostasis. *Nature Structural and Molecular Biology*, *24*(10), 870–878. https://doi.org/10.1038/nsmb.3462

Edvardson, S., Prunetti, L., Arraf, A., Haas, D., Bacusmo, J. M., Hu, J. F., . . . Elpeleg, O. (2017). tRNA N6-adenosine threonylcarbamoyltransferase defect due to KAE1/TCS3 (OSGEP) mutation manifest by neurodegeneration and renal tubulopathy. *European Journal of Human Genetics*, *25*(5), 545–551. https://doi.org/10.1038/ejhg.2017.30

Engel, M., Eggert, C., Kaplick, P. M., Eder, M., Röh, S., Tietze, L., . . . Chen, A. (2018). The role of m6A/m-RNA methylation in stress response regulation. *Neuron*, *99*(2), 389–403.e9. https://doi.org/10.1016/j.neuron.2018.07.009

Fahiminiya, S., Almuriekhi, M., Nawaz, Z., Staffa, A., Lepage, P., Ali, R., . . . Ben-Omran, T. (2014). Whole exome sequencing unravels disease-causing genes in consanguineous families in Qatar. *Clinical Genetics*, *86*(2), 134–141. https://doi.org/10.1111/cge.12280

Flores, J. V., Cordero-Espinoza, L., Oeztuerk-Winder, F., Andersson-Rolf, A., Selmi, T., Blanco, S.,...Frye, M. (2017). Cytosine-5 RNA methylation regulates neural stem cell differentiation and motility. *Stem Cell Reports, 8*(1), 112–124. https://doi.org/10.1016/j.stemcr.2016.11.014

Frye, M., Harada, B. T., Behm, M., & He, C. (2018). RNA modifications modulate gene expression during development. *Science, 361*(6409), 1346–1349. https://doi.org/10.1126/science.aau1646

Fu, H.-L., Ma, Y., Lu, L.-G., Hou, P., Li, B.-J., Jin, W.-L., & Cui, D.-X. (2014). TET1 exerts its tumor suppressor function by interacting with p53-EZH2 pathway in gastric cancer. *Journal of Biomedical Nanotechnology, 10*(7), 1217–1230. https://doi.org/10.1166/jbn.2014.1861

Gallagher, R. C., Pils, B., Albalwi, M., & Francke, U. (2002). Evidence for the role of PWCR1/HBII-85 C/D box small nucleolar RNAs in Prader-Willi syndrome. *The American Journal of Human Genetics, 71*(3), 669–678. https://doi.org/10.1086/342408

Gao, X., Shin, Y.-H., Li, M., Wang, F., Tong, Q., & Zhang, P. (2010). The fat mass and obesity associated gene FTO functions in the brain to regulate postnatal growth in mice. *PLoS ONE, 5*(11), e14005. https://doi.org/10.1371/journal.pone.0014005

Garcia-Campos, M. A., Edelheit, S., Toth, U., Safra, M., Shachar, R., Viukov, S.,...Schwartz, S. (2019). Deciphering the "m6A Code" via antibody-independent quantitative profiling. *Cell, 178*(3), 731–747.e16. https://doi.org/10.1016/j.cell.2019.06.013

Gasch, A. P., Spellman, P. T., Kao, C. M., Carmel-Harel, O., Eisen, M. B., Storz, G.,...Brown, P. O. (2000). Genomic expression programs in the response of yeast cells to environmental changes. *Molecular Biology of the Cell, 11*(12), 4241–4257. https://doi.org/10.1091/mbc.11.12.4241

Gerber, A. P. (1999). An adenosine deaminase that generates inosine at the wobble position of tRNAs. *Science, 286*(5442), 1146–1149. https://doi.org/10.1126/science.286.5442.1146

Geula, S., Moshitch-Moshkovitz, S., Dominissini, D., Mansour, A. A., Kol, N., Salmon-Divon, M.,...Hanna, J. H. (2015). m 6 A mRNA methylation facilitates resolution of naïve pluripotency toward differentiation. *Science, 347*(6225), 1002–1006. https://doi.org/10.1126/science.1261417

Grozhik, A. V., Linder, B., Olarerin-George, A. O., & Jaffrey, S. R. (2017). Mapping m6A at individual-nucleotide resolution using crosslinking and immunoprecipitation (MiCLIP). In *Methods in Molecular Biology,* 1562, 55–78). https://doi.org/10.1007/978-1-4939-6807-7_5

Gumienny, R., Jedlinski, D. J., Schmidt, A., Gypas, F., Martin, G., Vina-Vilaseca, A., & Zavolan, M. (2016). High-throughput identification of C/D box snoRNA targets with CLIP and RiboMeth-seq. *Nucleic Acids Research*, gkw1321. https://doi.org/10.1093/nar/gkw1321

Guy, M. P., Podyma, B. M., Preston, M. A., Shaheen, H. H., Krivos, K. L., Limbach, P. A.,...Phizicky, E. M. (2012). Yeast Trm7 interacts with distinct proteins for critical modifications of the tRNAPhe anticodon loop. *RNA, 18*(10), 1921–1933. https://doi.org/10.1261/rna.035287.112

Ha, N., Lai, L.-T., Chelliah, R., Zhen, Y., Yi Vanessa, S. P., Lai, S.-K.,...Zhang, L.-F. (2018). Live-cell imaging and functional dissection of Xist RNA reveal mechanisms of X chromosome inactivation and reactivation. *iScience*, 8, 1–14. https://doi.org/10.1016/j.isci.2018.09.007

Harley, C. B. (1987). Hybridization of Oligo(dT) to RNA on nitrocellulose. *Gene Analysis Techniques, 4*(2), 17–22. https://doi.org/10.1016/0735-0651(87)90013-6

Heo, S., Diering, G. H., Na, C. H., Nirujogi, R. S., Bachman, J. L., Pandey, A., & Huganir, R. L. (2018). Identification of long-lived synaptic proteins by proteomic analysis of synaptosome protein turnover. *Proceedings of the National Academy of Sciences USA, 115*(16), E3827–E3836. https://doi.org/10.1073/pnas.1720956115

Hess, M. E., Hess, S., Meyer, K. D., Verhagen, L. a W., Koch, L., Brönneke, H. S.,...Brüning, J. C. (2013). The fat mass and obesity associated gene (Fto) regulates activity of the dopaminergic midbrain circuitry. *Nature Neuroscience*, *16*(8), 1042–1048. https://doi.org/10.1038/nn.3449

Higuchi, M., Maas, S., Single, F. N., Hartner, J., Rozov, A., Burnashev, N., . . . Seeburg, P. H. (2000). Point mutation in an AMPA receptor gene rescues lethality in mice deficient in the RNA-editing enzyme ADAR2. *Nature*, *406*(6791), 78–81. https://doi.org/10.1038/35017558

Holt, C. E., Martin, K. C., & Schuman, E. M. (2019). Local translation in neurons: visualization and function. *Nature Structural and Molecular Biology*, 26, 557–566. https://doi.org/10.1038/s41594-019-0263-5

Hoopengardner, B. (2003). Nervous system targets of RNA editing identified by comparative genomics. *Science*, *301*(5634), 832–836. https://doi.org/10.1126/science.1086763

Horowitz, S., Horowitz, A., Nilsen, T. W., Munns, T. W., & Rottman, F. M. (1984). Mapping of N6-methyladenosine residues in bovine prolactin mRNA. *Proceedings of the National Academy of Sciences USA*, *81*(18), 5667–5671. https://doi.org/10.1073/pnas.81.18.5667

Houwing, S., Kamminga, L. M., Berezikov, E., Cronembold, D., Girard, A., van den Elst, H., . . . Ketting, R. F. (2007). A role for Piwi and piRNAs in germ cell maintenance and transposon silencing in zebrafish. *Cell*, *129*(1), 69–82. https://doi.org/10.1016/j.cell.2007.03.026

Hsu, P. J., Shi, H., Zhu, A. C., Lu, Z., Miller, N., Edens, B. M., . . . He, C. (2019). The RNA-binding protein FMRP facilitates the nuclear export of N 6 -methyladenosine–containing mRNAs. *Journal of Biological Chemistry*, *294*(52), 19889–19895. https://doi.org/10.1074/jbc.AC119.010078

Huang, B. (2005). An early step in wobble uridine tRNA modification requires the Elongator complex. *RNA*, *11*(4), 424–436. https://doi.org/10.1261/rna.7247705

Huang, H., Weng, H., Zhou, K., Wu, T., Zhao, B. S., Sun, M., . . . Chen, J. (2019). Histone H3 trimethylation at lysine 36 guides m6A RNA modification co-transcriptionally. *Nature*, *567*(7748), 414–419. https://doi.org/10.1038/s41586-019-1016-7

Huber, C., von Watzdorf, J., & Marx, A. (2016). 5-methylcytosine-sensitive variants of Thermococcus kodakaraensis DNA polymerase. *Nucleic Acids Research*, gkw812. https://doi.org/10.1093/nar/gkw812

Hussain, S., Sajini, A. A., Blanco, S., Dietmann, S., Lombard, P., Sugimoto, Y., . . . Frye, M. (2013). NSun2-mediated cytosine-5 methylation of vault noncoding RNA determines its processing into regulatory small RNAs. *Cell Reports*, *4*(2), 255–261. https://doi.org/10.1016/j.celrep.2013.06.029

Huttenhofer, A. (2001). RNomics: an experimental approach that identifies 201 candidates for novel, small, non-messenger RNAs in mouse. *EMBO Journal*, *20*(11), 2943–2953. https://doi.org/10.1093/emboj/20.11.2943

Incarnato, D., Anselmi, F., Morandi, E., Neri, F., Maldotti, M., Rapelli, S., . . . Oliviero, S. (2017). High-throughput single-base resolution mapping of RNA 2″-O-methylated residues. *Nucleic Acids Research*, *45*(3), 1433–1441. https://doi.org/10.1093/nar/gkw810

Issler, O., van der Zee, Y. Y., Ramakrishnan, A., Wang, J., Tan, C., Loh, Y.-H. E., . . . Nestler, E. J. (2020). Sex-specific role for the long non-coding RNA LINC00473 in depression. *Neuron*, *106*(6), 912–926.e5. https://doi.org/10.1016/j.neuron.2020.03.023

Jarrous, N., & Reiner, R. (2007). Human RNase P: a tRNA-processing enzyme and transcription factor. *Nucleic Acids Research*, *35*(11), 3519–3524. https://doi.org/10.1093/nar/gkm071

Jia, G., Fu, Y., Zhao, X., Dai, Q., Zheng, G., Yang, Y., . . . He, C. (2011a). N6-methyladenosine in nuclear RNA is a major substrate of the obesity-associated FTO. *Nature Chemical Biology*, *7*(12), 885–887. https://doi.org/10.1038/nchembio.687

Jia, G., Fu, Y., Zhao, X., Dai, Q., Zheng, G., Yang, Y., . . . He, C. (2011b). N6-methyladenosine in nuclear RNA is a major substrate of the obesity-associated FTO. *Nature Chemical Biology*, *7*(12), 885–887. https://doi.org/10.1038/nchembio.687

Jin, P. (2000a). Understanding the molecular basis of fragile X syndrome. *Human Molecular Genetics*, *9*(6), 901–908. https://doi.org/10.1093/hmg/9.6.901

Jin, P. (2000b). Understanding the molecular basis of fragile X syndrome. *Human Molecular Genetics*, *9*(6), 901–908. https://doi.org/10.1093/hmg/9.6.901

Karlsborn, T., Tükenmez, H., Mahmud, A. K. M. F., Xu, F., Xu, H., & Byström, A. S. (2014). Elongator, a conserved complex required for wobble uridine modifications in Eukaryotes. *RNA Biology*, *11*(12), 1519–1528. https://doi.org/10.4161/15476286.2014.992276

Kawahara, Y., Zinshteyn, B., Sethupathy, P., Iizasa, H., Hatzigeorgiou, A. G., & Nishikura, K. (2007). Redirection of Silencing Targets by Adenosine-to-Inosine Editing of miRNAs. *Science*, *315*(5815), 1137–1140. https://doi.org/10.1126/science.1138050

Kawahara, Yukio, Zinshteyn, B., Chendrimada, T. P., Shiekhattar, R., & Nishikura, K. (2007). RNA editing of the microRNA-151 precursor blocks cleavage by the Dicer-TRBP complex. *EMBO Reports*, *8*(8), 763–769. https://doi.org/10.1038/sj.embor.7401011

Ke, S., Pandya-Jones, A., Saito, Y., Fak, J. J., Vågbø, C. B., Geula, S., . . . Darnell, R. B. (2017). m 6 A mRNA modifications are deposited in nascent pre-mRNA and are not required for splicing but do specify cytoplasmic turnover. *Genes & Development*, *31*(10), 990–1006. https://doi.org/10.1101/gad.301036.117

Khan, M. A., Rafiq, M. A., Noor, A., Hussain, S., Flores, J. V., Rupp, V., . . . Vincent, J. B. (2012). Mutation in NSUN2, which encodes an RNA methyltransferase, causes autosomal-recessive intellectual disability. *American Journal of Human Genetics*, *90*(5), 856–863. https://doi.org/10.1016/j.ajhg.2012.03.023

Khoddami, V., & Cairns, B. R. (2013). Identification of direct targets and modified bases of RNA cytosine methyltransferases. *Nature Biotechnology*, *31*(5), 458–464. https://doi.org/10.1038/nbt.2566

Khvorova, A., & Watts, J. K. (2017). The chemical evolution of oligonucleotide therapies of clinical utility. *Nature Biotechnology*, *35*(3), 238–248. https://doi.org/10.1038/nbt.3765

Kim, D. D. Y. (2004). Widespread RNA editing of embedded Alu elements in the human transcriptome. *Genome Research*, *14*(9), 1719–1725. https://doi.org/10.1101/gr.2855504

Kirino, Y., & Mourelatos, Z. (2007). Mouse Piwi-interacting RNAs are 2′-O-methylated at their 3′ termini. *Nature Structural & Molecular Biology*, *14*(4), 347–348. https://doi.org/10.1038/nsmb1218

Knight, Z. A., Tan, K., Birsoy, K., Schmidt, S., Garrison, J. L., Wysocki, R. W., . . . Friedman, J. M. (2012). Molecular profiling of activated neurons by phosphorylated ribosome capture. *Cell*. *151*(5), 1126-37. https://doi.org/10.1016/j.cell.2012.10.039

Komara, M., Al-Shamsi, A. M., Ben-Salem, S., Ali, B. R., & Al-Gazali, L. (2015). A novel single-nucleotide deletion (c.1020delA) in NSUN2 Causes intellectual disability in an Emirati child. *Journal of Molecular Neuroscience*, *57*(3), 393–399. https://doi.org/10.1007/s12031-015-0592-8

Koranda, J. L., Dore, L., Shi, H., Patel, M. J., Vaasjo, L. O., Rao, M. N., . . . Zhuang, X. (2018). Mettl14 is essential for epitranscriptomic regulation of striatal function and learning. *Neuron*, *99*(2), 283–292.e5. https://doi.org/10.1016/j.neuron.2018.06.007

Kretschmer, J., Rao, H., Hackert, P., Sloan, K. E., Höbartner, C., & Bohnsack, M. T. (2018). The m6A reader protein YTHDC2 interacts with the small ribosomal subunit and the 5′-3′ exoribonuclease XRN1. *RNA*. https://doi.org/10.1261/rna.064238.117

Kuksa, P. P., Leung, Y. Y., Vandivier, L. E., Anderson, Z., Gregory, B. D., & Wang, L.-S. (2017). In Silico Identification of RNA modifications from high-throughput sequencing data using HAMR. In Lusser A. (eds), *Methods in molecular biology*. New York, NY: Humana Press. https://doi.org/10.1007/978-1-4939-6807-7_14

Laguesse, S., Creppe, C., Nedialkova, D. D., Prévot, P.-P., Borgs, L., Huysseune, S., . . . Nguyen, L. (2015). A dynamic unfolded protein response contributes to the control of cortical neurogenesis. *Developmental Cell, 35*(5), 553–567. https://doi.org/10.1016/j.devcel.2015.11.005

Legrand, C., Tuorto, F., Hartmann, M., Liebers, R., Jacob, D., Helm, M., & Lyko, F. (2017). Statistically robust methylation calling for whole-transcriptome bisulfite sequencing reveals distinct methylation patterns for mouse RNAs. *Genome Research, 27*(9), 1589–1596. https://doi.org/10.1101/gr.210666.116

Lennox, K. A., & Behlke, M. A. (2011). Chemical modification and design of anti-miRNA oligonucleotides. *Gene Therapy, 18*(12), 1111–1120. https://doi.org/10.1038/gt.2011.100

Li, L., Zang, L., Zhang, F., Chen, J., Shen, H., Shu, L., . . . Li, X. (2017). Fat mass and obesity-associated (FTO) protein regulates adult neurogenesis. *Human Molecular Genetics, 26*(13), 2398–2411. https://doi.org/10.1093/hmg/ddx128

Li, M., Zhao, X., Wang, W., Shi, H., Pan, Q., Lu, Z., . . . Klungland, A. (2018). Ythdf2-mediated m6A mRNA clearance modulates neural development in mice. *Genome Biology, 19*(1), 69. https://doi.org/10.1186/s13059-018-1436-y

Li, X., Xiong, X., Wang, K., Wang, L., Shu, X., Ma, S., & Yi, C. (2016). Transcriptome-wide mapping reveals reversible and dynamic N1-methyladenosine methylome. *Nature Chemical Biology, 12*(5), 311–316. https://doi.org/10.1038/nchembio.2040

Li, X., Xiong, X., Zhang, M., Wang, K., Chen, Y., Zhou, J., . . . Yi, C. (2017). Base-resolution mapping reveals distinct m1A methylome in nuclear- and mitochondrial-encoded transcripts. *Molecular Cell, 68*(5), 993–1005.e9. https://doi.org/10.1016/j.molcel.2017.10.019

Li, X., Zhu, P., Ma, S., Song, J., Bai, J., Sun, F., & Yi, C. (2015). Chemical pulldown reveals dynamic pseudouridylation of the mammalian transcriptome. *Nature Chemical Biology, 11*(8), 592–597. https://doi.org/10.1038/nchembio.1836

Lin, S., Liu, Q., Lelyveld, V. S., Choe, J., Szostak, J. W., & Gregory, R. I. (2018). Mettl1/Wdr4-Mediated m7G tRNA Methylome Is Required for Normal mRNA Translation and Embryonic Stem Cell Self-Renewal and Differentiation. *Molecular Cell, 71*(2), 244–255.e5. https://doi.org/10.1016/j.molcel.2018.06.001

Lin, Y., & Kielkopf, C. L. (2008). X-ray Structures of U2 snRNA–Branchpoint Duplexes Containing Conserved Pseudouridines. *Biochemistry, 47*(20), 5503–5514. https://doi.org/10.1021/bi7022392

Linder, B., Grozhik, A. V, Olarerin-George, A. O., Meydan, C., Mason, C. E., & Jaffrey, S. R. (2015). Single-nucleotide-resolution mapping of m6A and m6Am throughout the transcriptome. *Nature Methods, 12*(8), 767–772. https://doi.org/10.1038/nmeth.3453

Liu, F., Clark, W., Luo, G., Wang, X., Fu, Y., Wei, J., . . . He, C. (2016). ALKBH1-mediated tRNA demethylation regulates translation. *Cell, 167*(3), 816–828.e16. https://doi.org/10.1016/j.cell.2016.09.038

Liu, H., Begik, O., Lucas, M. C., Ramirez, J. M., Mason, C. E., Wiener, D., . . . Novoa, E. M. (2019). Accurate detection of m6A RNA modifications in native RNA sequences. *Nature Communications, 10*(1), 4079. https://doi.org/10.1038/s41467-019-11713-9

Liu, N., Dai, Q., Zheng, G., He, C., Parisien, M., & Pan, T. (2015). N^6-methyladenosine-dependent RNA structural switches regulate RNA–protein interactions. *Nature, 518*(7540), 560–564. https://doi.org/10.1038/nature14234

Liu, N., Zhou, K. I., Parisien, M., Dai, Q., Diatchenko, L., & Pan, T. (2017). N6-methyladenosine alters RNA structure to regulate binding of a low-complexity protein. *Nucleic Acids Research, 45*(10), 6051–6063. https://doi.org/10.1093/nar/gkx141

Livneh, I., Moshitch-Moshkovitz, S., Amariglio, N., Rechavi, G., & Dominissini, D. (2020). The m6A epitranscriptome: transcriptome plasticity in brain development and function. *Nature Reviews. Neuroscience, 21*(1), 36–51. https://doi.org/10.1038/s41583-019-0244-z

Ma, C., Chang, M., Lv, H., Zhang, Z.-W., Zhang, W., He, X.,... Tong, W.-M. (2018). RNA m6A methylation participates in regulation of postnatal development of the mouse cerebellum. *Genome Biology*, *19*(1), 68. https://doi.org/10.1186/s13059-018-1435-z

Maden, B. E. H. (2001). Mapping 2′-O-methyl groups in ribosomal RNA. *Methods*, *25*(3), 374–382. https://doi.org/10.1006/meth.2001.1250

Marchand, V., Ayadi, L., Ernst, F. G. M., Hertler, J., Bourguignon-Igel, V., Galvanin, A.,... Motorin, Y. (2018). Alkaniline-seq: Profiling of m 7 G and m 3 C RNA modifications at single nucleotide resolution. *Angewandte Chemie International Edition*, *57*(51), 16785–16790. https://doi.org/10.1002/anie.201810946

Marchand, V., Blanloeil-Oillo, F., Helm, M., & Motorin, Y. (2016). Illumina-based RiboMethSeq approach for mapping of 2′-O-Me residues in RNA. *Nucleic Acids Research*, *44*(16), e135–e135. https://doi.org/10.1093/nar/gkw547

Martin, S., Zekri, L., Metz, A., Maurice, T., Chebli, K., Vignes, M., & Tazi, J. (2013). Deficiency of G3BP1, the stress granules assembly factor, results in abnormal synaptic plasticity and calcium homeostasis in neurons. *Journal of Neurochemistry*, *125*(2), 175–184. https://doi.org/10.1111/jnc.12189

Martinez, F. J., Lee, J. H., Lee, J. E., Blanco, S., Nickerson, E., Gabriel, S.,... Gleeson, J. G. (2012). Whole exome sequencing identifies a splicing mutation in NSUN2 as a cause of a Dubowitz-like syndrome. *Journal of Medical Genetics*, *49*(6), 380–385. https://doi.org/10.1136/jmedgenet-2011-100686

Massenet, S., Motorin, Y., Lafontaine, D. L. J., Hurt, E. C., Grosjean, H., & Branlant, C. (1999). Pseudouridine mapping in the saccharomyces cerevisiae spliceosomal u small nuclear RNAs (snRNAs) reveals that pseudouridine synthase PUS1p exhibits a dual substrate specificity for U2 snRNA and tRNA. *Molecular and Cellular Biology*, *19*(3), 2142–2154. https://doi.org/10.1128/MCB.19.3.2142

Mauer, J., & Jaffrey, S. R. (2018). FTO, m 6 A m, and the hypothesis of reversible epitranscriptomic mRNA modifications. *FEBS Letters*, *592*(12), 2012–2022. https://doi.org/10.1002/1873-3468.13092

Mauer, J., Luo, X., Blanjoie, A., Jiao, X., Grozhik, A. V., Patil, D. P.,... Jaffrey, S. R. (2017). Reversible methylation of m6Am in the 5′ cap controls mRNA stability. *Nature*, *541*(7637), 371–375. https://doi.org/10.1038/nature21022

Mauer, J., Sindelar, M., Despic, V., Guez, T., Hawley, B. R., Vasseur, J.-J.,... Jaffrey, S. R. (2019). FTO controls reversible m6Am RNA methylation during snRNA biogenesis. *Nature Chemical Biology*, *15*(4), 340–347. https://doi.org/10.1038/s41589-019-0231-8

Meier, F., Suter, B., Grosjean, H., Keith, G., & Kubli, E. (1985). Queuosine modification of the wobble base in tRNAHis influences "in vivo" decoding properties. *The EMBO Journal*, *4*(3), 823–827. https://doi.org/10.1002/j.1460-2075.1985.tb03704.x

Merkurjev, D., Hong, W.-T., Iida, K., Oomoto, I., Goldie, B. J., Yamaguti, H.,... Wang, D. O. (2018). Synaptic N6-methyladenosine (m6A) epitranscriptome reveals functional partitioning of localized transcripts. *Nature Neuroscience*, *21*(7), 1004–1014. https://doi.org/10.1038/s41593-018-0173-6

Meyer, K. D. (2019). DART-seq: an antibody-free method for global m6A detection. *Nature Methods*, 16(12), 1275–1280. https://doi.org/10.1038/s41592-019-0570-0

Meyer, K. D., Patil, D. P., Zhou, J., Zinoviev, A., Skabkin, M. A., Elemento, O.,... Jaffrey, S. R. (2015). 5′ UTR m6A Promotes Cap-Independent Translation. *Cell*, *163*(4), 999–1010. https://doi.org/10.1016/j.cell.2015.10.012

Meyer, K. D., Saletore, Y., Zumbo, P., Elemento, O., Mason, C. E., & Jaffrey, S. R. (2012a). Comprehensive analysis of mRNA methylation reveals enrichment in 3′ UTRs and near stop codons. *Cell, 149*(7), 1635–1646. Retrieved July 15, 2014 from https://doi.org/10.1016/j.cell.2012.05.003

Meyer, K. D., Saletore, Y., Zumbo, P., Elemento, O., Mason, C. E., & Jaffrey, S. R. (2012b). Comprehensive analysis of mRNA methylation reveals enrichment in 3′ UTRs and near stop codons. *Cell, 149*(7), 1635–1646. https://doi.org/10.1016/j.cell.2012.05.003

Meyer, K. D., Saletore, Y., Zumbo, P., Elemento, O., Mason, C. E., & Jaffrey, S. R. (2012c). Comprehensive analysis of mRNA methylation reveals enrichment in 3′ UTRs and near stop codons. *Cell, 149*(7), 1635–1646. https://doi.org/10.1016/j.cell.2012.05.003

Misra, M., Edmund, H., Ennis, D., Schlueter, M. A., Marot, J. E., Tambasco, J., . . . Gavis, E. R. (2016). A genome-wide screen for dendritically localized RNAs identifies genes required for dendrite morphogenesis. *G3: Genes|Genomes|Genetics, 6*(8), 2397–2405. https://doi.org/10.1534/g3.116.030353

Molinie, B., Wang, J., Lim, K. S., Hillebrand, R., Lu, Z., Van Wittenberghe, N., . . . Giallourakis, C. C. (2016). m6A-LAIC-seq reveals the census and complexity of the m6A epitranscriptome. *Nature Methods, 13*(8), 692–698. https://doi.org/10.1038/nmeth.3898

Motorin, Y., & Helm, M. (2019). Methods for RNA modification mapping using deep sequencing: Established and new emerging technologies. *Genes, 10*(1), 35. https://doi.org/10.3390/genes10010035

Müller, M., Hartmann, M., Schuster, I., Bender, S., Thüring, K. L., Helm, M., . . . Ehrenhofer-Murray, A. E. (2015). Dynamic modulation of Dnmt2-dependent tRNA methylation by the micronutrient queuine. *Nucleic Acids Research, 43*(22), 10952–10962. https://doi.org/10.1093/nar/gkv980

Najmabadi, H., Hu, H., Garshasbi, M., Zemojtel, T., Abedini, S. S., Chen, W., . . . Ropers, H. H. (2011). Deep sequencing reveals 50 novel genes for recessive cognitive disorders. *Nature, 478*(7367), 57–63. https://doi.org/10.1038/nature10423

Niswender, C. M., Copeland, S. C., Herrick-Davis, K., Emeson, R. B., & Sanders-Bush, E. (1999). RNA editing of the human serotonin 5-hydroxytryptamine 2c receptor silences constitutive activity. *Journal of Biological Chemistry, 274*(14), 9472–9478. https://doi.org/10.1074/jbc.274.14.9472

Patil, D. P., Chen, C.-K., Pickering, B. F., Chow, A., Jackson, C., Guttman, M., & Jaffrey, S. R. (2016). m6A RNA methylation promotes XIST-mediated transcriptional repression. *Nature, 537*(7620), 369–373. https://doi.org/10.1038/nature19342

Paul, M. S. (1998a). Inosine exists in mRNA at tissue-specific levels and is most abundant in brain mRNA. *EMBO Journal, 17*(4), 1120–1127. https://doi.org/10.1093/emboj/17.4.1120

Paul, M. S. (1998b). Inosine exists in mRNA at tissue-specific levels and is most abundant in brain mRNA. *The EMBO Journal, 17*(4), 1120–1127. https://doi.org/10.1093/emboj/17.4.1120

Pendleton, K. E., Chen, B., Liu, K., Hunter, O. V., Xie, Y., Tu, B. P., & Conrad, N. K. (2017). The U6 snRNA m 6 A methyltransferase METTL16 regulates SAM synthetase intron retention. *Cell, 169*(5), 824–835.e14. https://doi.org/10.1016/j.cell.2017.05.003

Perche-Letuvée, P., Molle, T., Forouhar, F., Mulliez, E., & Atta, M. (2014). Wybutosine biosynthesis: Structural and mechanistic overview. *RNA Biology, 11*(12), 1508–1518. https://doi.org/10.4161/15476286.2014.992271

Perry, R. P., & Kelley, D. E. (1976). Kinetics of formation of 5′ terminal caps in mRNA. *Cell, 8*(3), 433–442. https://doi.org/10.1016/0092-8674(76)90156-2

Ries, R. J., Zaccara, S., Klein, P., Olarerin-George, A., Namkoong, S., Pickering, B. F., . . . Jaffrey, S. R. (2019). m6A enhances the phase separation potential of mRNA. *Nature, 571*(7765), 424–428. https://doi.org/10.1038/s41586-019-1374-1

Rogelj, B., & Giese, K. P. (2004). Expression and function of brain specific small RNAs. *Reviews in the Neurosciences, 15*(3). https://doi.org/10.1515/REVNEURO.2004.15.3.185

Rogelj, B., Hartmann, C. E. A., Yeo, C. H., Hunt, S. P., & Giese, K. P. (2003). Contextual fear conditioning regulates the expression of brain-specific small nucleolar RNAs in hippocampus. *European Journal of Neuroscience, 18*(11), 3089–3096. https://doi.org/10.1111/j.1460-9568.2003.03026.x

Rosenberg, B. R., Hamilton, C. E., Mwangi, M. M., Dewell, S., & Papavasiliou, F. N. (2011). Transcriptome-wide sequencing reveals numerous APOBEC1 mRNA-editing targets in transcript 3′ UTRs. *Nature Structural & Molecular Biology, 18*(2), 230–236. https://doi.org/10.1038/nsmb.1975

Roy, R., Shiina, N., & Wang, D. O. (2020). More dynamic, more quantitative, unexpectedly intricate: Advanced understanding on synaptic RNA localization in learning and memory. *Neurobiology of Learning and Memory*, 168, 107149. https://doi.org/10.1016/j.nlm.2019.107149

Ryvkin, P., Leung, Y. Y., Silverman, I. M., Childress, M., Valladares, O., Dragomir, I., ... Wang, L.-S. (2013). HAMR: high-throughput annotation of modified ribonucleotides. *RNA, 19*(12), 1684–1692. https://doi.org/10.1261/rna.036806.112

Safra, M., Sas-Chen, A., Nir, R., Winkler, R., Nachshon, A., Bar-Yaacov, D., ... Schwartz, S. (2017). The m1A landscape on cytosolic and mitochondrial mRNA at single-base resolution. *Nature, 551*, 251–255. https://doi.org/10.1038/nature24456

Sahoo, P. K., Lee, S. J., Jaiswal, P. B., Alber, S., Kar, A. N., Miller-Randolph, S., ... Twiss, J. L. (2018). Axonal G3BP1 stress granule protein limits axonal mRNA translation and nerve regeneration. *Nature Communications, 9*(1), 3358. https://doi.org/10.1038/s41467-018-05647-x

Saito, K., Sakaguchi, Y., Suzuki, T., Suzuki, T., Siomi, H., & Siomi, M. C. (2007). Pimet, the Drosophila homolog of HEN1, mediates 2′-O-methylation of Piwi- interacting RNAs at their 3′ ends. *Genes & Development, 21*(13), 1603–1608. https://doi.org/10.1101/gad.1563607

Salehi Chaleshtori, A. R., Miyake, N., Ahmadvand, M., Bashti, O., Matsumoto, N., & Noruzinia, M. (2018). A novel 8-bp duplication in ADAT3 causes mild intellectual disability. *Human Genome Variation, 5*(1), 7. https://doi.org/10.1038/s41439-018-0007-9

Sanjana, N. E., Levanon, E. Y., Hueske, E. A., Ambrose, J. M., & Li, J. B. (2012). Activity-dependent A-to-I RNA editing in rat cortical neurons. *Genetics, 192*(1), 281–287. https://doi.org/10.1534/genetics.112.141200

Schaefer, M., Pollex, T., Hanna, K., & Lyko, F. (2008). RNA cytosine methylation analysis by bisulfite sequencing. *Nucleic Acids Research, 37*(2), e12–e12. https://doi.org/10.1093/nar/gkn954

Schwartz, S., Bernstein, D. A., Mumbach, M. R., Jovanovic, M., Herbst, R. H., León-Ricardo, B. X., ... Regev, A. (2014). Transcriptome-wide mapping reveals widespread dynamic-regulated pseudouridylation of ncRNA and mRNA. *Cell, 159*(1), 148–162. https://doi.org/10.1016/j.cell.2014.08.028

Sendinc, E., Valle-Garcia, D., Dhall, A., Chen, H., Henriques, T., Navarrete-Perea, J., ... Shi, Y. (2019). PCIF1 Catalyzes m6Am mRNA methylation to regulate gene expression. *Molecular Cell, 75*(3), 620–630.e9. https://doi.org/10.1016/j.molcel.2019.05.030

Shaheen, R., Abdel-Salam, G. M. H., Guy, M. P., Alomar, R., Abdel-Hamid, M. S., Afifi, H. H., ... Alkuraya, F. S. (2015). Mutation in WDR4 impairs tRNA m7G46 methylation and causes a distinct form of microcephalic primordial dwarfism. *Genome Biology, 16*(1), 210. https://doi.org/10.1186/s13059-015-0779-x

Shaheen, R., Han, L., Faqeih, E., Ewida, N., Alobeid, E., Phizicky, E. M., & Alkuraya, F. S. (2016). A homozygous truncating mutation in PUS3 expands the role of tRNA modification in normal cognition. *Human Genetics, 135*(7), 707–713. https://doi.org/10.1007/s00439-016-1665-7

Shi, H., Wang, X., Lu, Z., Zhao, B. S., Ma, H., Hsu, P. J., . . . He, C. (2017). YTHDF3 facilitates translation and decay of N6-methyladenosine-modified RNA. *Cell Research*, *27*(3), 315–328. https://doi.org/10.1038/cr.2017.15

Shi, H., Wei, J., & He, C. (2019). Where, when, and how: context-dependent functions of RNA methylation writers, readers, and erasers. *Molecular Cell*, *74*(4), 640–650. https://doi.org/10.1016/j.molcel.2019.04.025

Shi, H., Zhang, X., Weng, Y.-L., Lu, Z., Liu, Y., Lu, Z., . . . Zhou, T. (2018). m6A facilitates hippocampus-dependent learning and memory through YTHDF1. *Nature*, *563*(7730), 249–253. https://doi.org/10.1038/s41586-018-0666-1

Simpson, C. L., Lemmens, R., Miskiewicz, K., Broom, W. J., Hansen, V. K., van Vught, P. W. J., . . . Al-Chalabi, A. (2009). Variants of the elongator protein 3 (ELP3) gene are associated with motor neuron degeneration. *Human Molecular Genetics*, *18*(3), 472–481. https://doi.org/10.1093/hmg/ddn375

Sloan, K. E., Warda, A. S., Sharma, S., Entian, K.-D., Lafontaine, D. L. J., & Bohnsack, M. T. (2017). Tuning the ribosome: The influence of rRNA modification on eukaryotic ribosome biogenesis and function. *RNA Biology*, *14*(9), 1138–1152. https://doi.org/10.1080/15476286.2016.1259781

Slobodin, B., Han, R., Calderone, V., Vrielink, J. A. F. O., Loayza-Puch, F., Elkon, R., & Agami, R. (2017). Transcription Impacts the Efficiency of mRNA Translation via Co-transcriptional N6-adenosine Methylation. *Cell*, *169*(2), 326–337.e12. https://doi.org/10.1016/j.cell.2017.03.031

Sommer, B., Köhler, M., Sprengel, R., & Seeburg, P. H. (1991). RNA editing in brain controls a determinant of ion flow in glutamate-gated channels. *Cell*, *67*(1), 11–19. https://doi.org/10.1016/0092-8674(91)90568-J

Sossin, W. S., & Costa-Mattioli, M. (2019). Translational control in the brain in health and disease. *Cold Spring Harbor Perspectives in Biology*, *11*(8), a032912. https://doi.org/10.1101/cshperspect.a032912

Squires, J. E., Patel, H. R., Nousch, M., Sibbritt, T., Humphreys, D. T., Parker, B. J., . . . Preiss, T. (2012). Widespread occurrence of 5-methylcytosine in human coding and non-coding RNA. *Nucleic Acids Research*, *40*(11), 5023–5033. https://doi.org/10.1093/nar/gks144

Stoilov, P., Rafalska, I., & Stamm, S. (2002). YTH: a new domain in nuclear proteins. *Trends in Biochemical Sciences*, 27(10), 495–497. https://doi.org/10.1016/S0968-0004(02)02189-8

Stuart, K., & Panigrahi, A. K. (2002). RNA editing: complexity and complications. *Molecular Microbiology*, *45*(3), 591–596. https://doi.org/10.1046/j.1365-2958.2002.03028.x

Suzuki, T., Ueda, H., Okada, S., & Sakurai, M. (2015). Transcriptome-wide identification of adenosine-to-inosine editing using the ICE-seq method. *Nature Protocols*, *10*(5), 715–732. https://doi.org/10.1038/nprot.2015.037

Taoka, M., Nobe, Y., Hori, M., Takeuchi, A., Masaki, S., Yamauchi, Y., . . . Isobe, T. (2015). A mass spectrometry-based method for comprehensive quantitative determination of post-transcriptional RNA modifications: the complete chemical structure of Schizosaccharomyces pombe ribosomal RNAs. *Nucleic Acids Research*, *43*(18), e115–e115. https://doi.org/10.1093/nar/gkv560

Taoka, M., Nobe, Y., Yamaki, Y., Sato, K., Ishikawa, H., Izumikawa, K., . . . Isobe, T. (2018). Landscape of the complete RNA chemical modifications in the human 80S ribosome. *Nucleic Acids Research*, *46*(18), 9289–9298. https://doi.org/10.1093/nar/gky811

Tourrière, H., Chebli, K., Zekri, L., Courselaud, B., Blanchard, J. M., Bertrand, E., & Tazi, J. (2003). The RasGAP-associated endoribonuclease G3BP assembles stress granules. *Journal of Cell Biology*, *160*(6), 823–831. https://doi.org/10.1083/jcb.200212128

Trimouille, A., Lasseaux, E., Barat, P., Deiller, C., Drunat, S., Rooryck, C.,... Lacombe, D. (2018). Further delineation of the phenotype caused by biallelic variants in the WDR4 gene. *Clinical Genetics*, *93*(2), 374–377. https://doi.org/10.1111/cge.13074

Tuorto, F., Liebers, R., Musch, T., Schaefer, M., Hofmann, S., Kellner, S.,... Lyko, F. (2012a). RNA cytosine methylation by Dnmt2 and NSun2 promotes tRNA stability and protein synthesis. *Nature Structural & Molecular Biology*, *19*(9), 900–905. https://doi.org/10.1038/nsmb.2357

Tuorto, F., Liebers, R., Musch, T., Schaefer, M., Hofmann, S., Kellner, S.,... Lyko, F. (2012b). RNA cytosine methylation by Dnmt2 and NSun2 promotes tRNA stability and protein synthesis. *Nature Structural & Molecular Biology*, *19*(9), 900–905. https://doi.org/10.1038/nsmb.2357

Tushev, G., Glock, C., Heumüller, M., Biever, A., Jovanovic, M., & Schuman, E. M. (2018). Alternative 3′ UTRs modify the localization, regulatory potential, stability, and plasticity of mRNAs in neuronal compartments. *Neuron*, *98*(3), 495–511.e6. https://doi.org/10.1016/j.neuron.2018.03.030

Verheggen, C. (2001). Box C/D small nucleolar RNA trafficking involves small nucleolar RNP proteins, nucleolar factors and a novel nuclear domain. *EMBO Journal*, *20*(19), 5480–5490. https://doi.org/10.1093/emboj/20.19.5480

Vessey, J. P., Schoderboeck, L., Gingl, E., Luzi, E., Riefler, J., Di Leva, F.,... Macchi, P. (2010). Mammalian Pumilio 2 regulates dendrite morphogenesis and synaptic function. *Proceedings of the National Academy of Sciences*, *107*(7), 3222–3227. https://doi.org/10.1073/pnas.0907128107

Vilfan, I. D., Tsai, Y.-C., Clark, T. A., Wegener, J., Dai, Q., Yi, C.,... Korlach, J. (2013). Analysis of RNA base modification and structural rearrangement by single-molecule real-time detection of reverse transcription. *Journal of Nanobiotechnology*, *11*(1), 8. https://doi.org/10.1186/1477-3155-11-8

Vinayak, M., & Pathak, C. (2010). Queuosine modification of tRNA: its divergent role in cellular machinery. *Bioscience Reports*, *30*(2), 135–148. https://doi.org/10.1042/BSR20090057

Vitali, P., Basyuk, E., Le Meur, E., Bertrand, E., Muscatelli, F., Cavaillé, J., & Huttenhofer, A. (2005). ADAR2-mediated editing of RNA substrates in the nucleolus is inhibited by C/D small nucleolar RNAs. *Journal of Cell Biology*, *169*(5), 745–753. https://doi.org/10.1083/jcb.200411129

Wahlstedt, H., Daniel, C., Enstero, M., & Ohman, M. (2009). Large-scale mRNA sequencing determines global regulation of RNA editing during brain development. *Genome Research*, *19*(6), 978–986. https://doi.org/10.1101/gr.089409.108

Walters, B. J., Mercaldo, V., Gillon, C. J., Yip, M., Neve, R. L., Boyce, F. M.,... Josselyn, S. A. (2017). The role of the RNA demethylase FTO (fat mass and obesity-associated) and mRNA methylation in hippocampal memory formation. *Neuropsychopharmacology*, *42*(7), 1502–1510. https://doi.org/10.1038/npp.2017.31

Wang, C.-X., Cui, G.-S., Liu, X., Xu, K., Wang, M., Zhang, X.-X.,... Zhou, Q. (2018). METTL3-mediated m6A modification is required for cerebellar development. *PLOS Biology*, *16*(6), e2004880. https://doi.org/10.1371/journal.pbio.2004880

Wang, D. O., Kim, S. M., Zhao, Y., Hwang, H., Miura, S. K., Sossin, W. S., & Martin, K. C. (2009). Synapse- and stimulus-specific local translation during long-term neuronal plasticity. *Science*, *324*(5934), 1536–1540. https://doi.org/10.1126/science.1173205

Wang, Dan O., Ninomiya, K., Mori, C., Koyama, A., Haan, M., Kitabatake, M.,... Kataoka, N. (2017). Transport granules bound with nuclear cap binding protein and exon junction complex are associated with microtubules and spatially separated from eIF4E granules and p bodies in human neuronal processes. *Frontiers in Molecular Biosciences*, 4. https://doi.org/10.3389/fmolb.2017.00093

Wang, Dan Ohtan, Martin, K. C., & Zukin, R. S. (2010). Spatially restricting gene expression by local translation at synapses. *Trends in Neurosciences*, *33*(4), 173–182. https://doi.org/10.1016/j.tins.2010.01.005

Wang, X., Lu, Z., Gomez, A., Hon, G. C., Yue, Y., Han, D., … He, C. (2014). N6-methyladenosine-dependent regulation of messenger RNA stability. *Nature, 505*(7481), 117–120. https://doi.org/10.1038/nature12730

Wang, X., Zhao, B. S., Roundtree, I. A., Lu, Z., Han, D., Ma, H., … He, C. (2015). N6-methyladenosine modulates messenger RNA translation efficiency. *Cell, 161*(6), 1388–1399. https://doi.org/10.1016/j.cell.2015.05.014

Wang, Y., Li, Y., Yue, M., Wang, J., Kumar, S., Wechsler-Reya, R. J., … Zhao, J. C. (2018). N6-methyladenosine RNA modification regulates embryonic neural stem cell self-renewal through histone modifications. *Nature Neuroscience, 21*(2), 195–206. https://doi.org/10.1038/s41593-017-0057-1

Wang, Z., Gerstein, M., & Snyder, M. (2009). RNA-Seq: a revolutionary tool for transcriptomics. *Nature Reviews. Genetics, 10*(1), 57–63.

Warda, A. S., Freytag, B., Haag, S., Sloan, K. E., Görlich, D., & Bohnsack, M. T. (2016). Effects of the Bowen-Conradi syndrome mutation in EMG1 on its nuclear import, stability and nucleolar recruitment. *Human Molecular Genetics, 25*(24), 5353–5364. https://doi.org/10.1093/hmg/ddw351

Wei, J., Liu, F., Lu, Z., Fei, Q., Ai, Y., He, P. C., … He, C. (2018). Differential m6A, m6Am, and m1A demethylation mediated by FTO in the cell nucleus and cytoplasm. *Molecular Cell, 71*(6), 973–985.e5. https://doi.org/10.1016/j.molcel.2018.08.011

Wei, Z., Panneerdoss, S., Timilsina, S., Zhu, J., Mohammad, T. A., Lu, Z.-L., … Meng, J. (2018). Topological characterization of human and mouse m5C epitranscriptome revealed by bisulfite sequencing. *International Journal of Genomics*, 2018, 1–19. https://doi.org/10.1155/2018/1351964

Weng, Y.-L., Wang, X., An, R., Cassin, J., Vissers, C., Liu, Y., … Ming, G. (2018). Epitranscriptomic m6A regulation of axon regeneration in the adult mammalian nervous system. *Neuron, 97*(2), 313–325.e6. https://doi.org/10.1016/j.neuron.2017.12.036

Widagdo, X. J., Zhao, X. Q., Kempen, X. M., Tan, X. M. C., Ratnu, V. S., Wei, W., … Bredy, X. T. W. (2016). Experience-dependent accumulation of N6-methyladenosine in the prefrontal cortex is associated with memory processes in mice. *Journal of Neuroscience, 36*(25), 6771–6777. https://doi.org/10.1523/JNEUROSCI.4053-15.2016

Wu, R., Li, A., Sun, B., Sun, J.-G., Zhang, J., Zhang, T., … Yuan, Z. (2019). A novel m6A reader Prrc2a controls oligodendroglial specification and myelination. *Cell Research, 29*(1), 23–41. https://doi.org/10.1038/s41422-018-0113-8

Xiang, Y., Laurent, B., Hsu, C.-H., Nachtergaele, S., Lu, Z., Sheng, W., … Shi, Y. (2017). RNA m6A methylation regulates the ultraviolet-induced DNA damage response. *Nature, 543*(7646), 573–576. https://doi.org/10.1038/nature21671

Xiang, Y., Laurent, B., Hsu, C.-H., Sheng, W., Wang, S., Ling, D., … Shi, Y. (2017). RNA m^{6}A methylation regulates the ultraviolet-induced DNA damage response. *Nature Publishing Group, 543*(7646), 573–576. https://doi.org/10.1038/nature21671

Xiao, W., Adhikari, S., Dahal, U., Chen, Y.-S., Hao, Y.-J., Sun, B.-F., … Yang, Y.-G. (2016). Nuclear m(6)A reader YTHDC1 regulates mRNA splicing. *Molecular Cell, 61*(4), 507–519. https://doi.org/10.1016/j.molcel.2016.01.012

Yang, W., Chendrimada, T. P., Wang, Q., Higuchi, M., Seeburg, P. H., Shiekhattar, R., & Nishikura, K. (2006). Modulation of microRNA processing and expression through RNA editing by ADAR deaminases. *Nature Structural & Molecular Biology, 13*(1), 13–21. https://doi.org/10.1038/nsmb1041

Yang, X., Yang, Y., Sun, B.-F., Chen, Y.-S., Xu, J.-W., Lai, W.-Y., … Yang, Y.-G. (2017). 5-methylcytosine promotes mRNA export—NSUN2 as the methyltransferase and ALYREF as an m5C reader. *Cell Research, 27*(5), 606–625. https://doi.org/10.1038/cr.2017.55

Yang, Y., Fan, X., Mao, M., Song, X., Wu, P., Zhang, Y., . . . Wang, Z. (2017). Extensive translation of circular RNAs driven by N6-methyladenosine. *Cell Research*, *27*(5), 626–641. https://doi.org/10.1038/cr.2017.31

Yi, C., & Pan, T. (2011). Cellular dynamics of RNA modification. *Accounts of Chemical Research*, *44*(12), 1380–1388. https://doi.org/10.1021/ar200057m

Yoon, K.-J., Ringeling, F. R., Vissers, C., Jacob, F., Pokrass, M., Jimenez-Cyrus, D., . . . Song, H. (2017). Temporal Control of Mammalian Cortical Neurogenesis by m6A Methylation. *Cell*, *171*(4), 877–889.e17. https://doi.org/10.1016/j.cell.2017.09.003

Yu, J., Chen, M., Huang, H., Zhu, J., Song, H., Zhu, J., . . . Ji, S.-J. (2018). Dynamic m6A modification regulates local translation of mRNA in axons. *Nucleic Acids Research*, *46*(3), 1412–1423. https://doi.org/10.1093/nar/gkx1182

Zeng, Y., Wang, S., Gao, S., Soares, F., Ahmed, M., Guo, H., . . . He, H. H. (2018). Refined RIP-seq protocol for epitranscriptome analysis with low input materials. *PLOS Biology*, *16*(9), e2006092. https://doi.org/10.1371/journal.pbio.2006092

Zhang, C., & Jia, G. (2018). Reversible RNA Modification N1-methyladenosine (m1A) in mRNA and tRNA. *Genomics, Proteomics and Bioinformatics*, *16*(3), 155–161. https://doi.org/10.1016/j.gpb.2018.03.003

Zhang, X., Wei, L.-H., Wang, Y., Xiao, Y., Liu, J., Zhang, W., . . . Jia, G. (2019). Structural insights into FTO's catalytic mechanism for the demethylation of multiple RNA substrates. *Proceedings of the National Academy of Sciences USA*, *116*(8), 2919–2924. https://doi.org/10.1073/pnas.1820574116

Zhang, Zeyu, Wang, M., Xie, D., Huang, Z., Zhang, L., Yang, Y., . . . Wang, X.-J. (2018). METTL3-mediated N6-methyladenosine mRNA modification enhances long-term memory consolidation. *Cell Research*, *28*(11), 1050–1061. https://doi.org/10.1038/s41422-018-0092-9

Zhang, Zhang, Chen, L.-Q., Zhao, Y.-L., Yang, C.-G., Roundtree, I. A., Zhang, Z., . . . Luo, G.-Z. (2019). Single-base mapping of m 6 A by an antibody-independent method. *Science Advances*, *5*(7), eaax0250. https://doi.org/10.1126/sciadv.aax0250

Zhao, X., Yang, Y., Sun, B.-F., Shi, Y., Yang, X., Xiao, W., . . . Yang, Y.-G. (2014). FTO-dependent demethylation of N6-methyladenosine regulates mRNA splicing and is required for adipogenesis. *Cell Research*, *24*(12), 1403–1419. https://doi.org/10.1038/cr.2014.151

Zhao, Y., Yu, Y., Zhai, J., Ramachandran, V., Dinh, T. T., Meyers, B. C., . . . Chen, X. (2012). The arabidopsis nucleotidyl transferase HESO1 uridylates unmethylated small RNAs to trigger their degradation. *Current Biology*, *22*(8), 689–694. https://doi.org/10.1016/j.cub.2012.02.051

Zheng, G., Dahl, J. A., Niu, Y., Fedorcsak, P., Huang, C. M., Li, C. J., . . . He, C. (2013). ALKBH5 is a mammalian RNA demethylase that impacts RNA metabolism and mouse fertility. *Molecular Cell*, *49*(1), 18–29. Retrieved July 11, 2014 from https://doi.org/10.1016/j.molcel.2012.10.015

Zheng, G., Qin, Y., Clark, W. C., Dai, Q., Yi, C., He, C., . . . Pan, T. (2015). Efficient and quantitative high-throughput tRNA sequencing. *Nature Methods*, *12*(9), 835–837. https://doi.org/10.1038/nmeth.3478

Zhou, J., Wan, J., Gao, X., Zhang, X., Jaffrey, S. R., & Qian, S.-B. (2015). Dynamic m6A mRNA methylation directs translational control of heat shock response. *Nature*, *526*(7574), 591–594. https://doi.org/10.1038/nature15377

Zhu, Y., Pirnie, S. P., & Carmichael, G. G. (2017). High-throughput and site-specific identification of 2′-O-methylation sites using ribose oxidation sequencing (RibOxi-seq). *RNA*, *23*(8), 1303–1314. https://doi.org/10.1261/rna.061549.117

Zhuang, M., Li, X., Zhu, J., Zhang, J., Niu, F., Liang, F., . . . Ji, S.-J. (2019). The m6A reader YTHDF1 regulates axon guidance through translational control of Robo3.1 expression. *Nucleic Acids Research*, *47*(9), 4765–4777. https://doi.org/10.1093/nar/gkz157

Zimmerman, A. J., Hafez, A. K., Amoah, S. K., Rodriguez, B. A., Dell'Orco, M., Lozano, E., . . . Mellios, N. (2020). A psychiatric disease-related circular RNA controls synaptic gene expression and cognition. *Molecular Psychiatry*. https://doi.org/10.1038/s41380-020-0653-4

CHAPTER 9

ROLE OF CPEB-FAMILY PROTEINS IN MEMORY

KAUSIK SI

THE cellular basis of memory storage is believed to be experience-dependent changes in the synaptic efficacy, often referred to as synaptic plasticity (Kandel, Dudai, & Mayford, 2014; S. J. Martin, Grimwood, & Morris, 2000; S. J. Martin & Morris, 2002). The synapse is a communication bridge between neurons and the minimal unit of change. Since strength of an individual synapse influences the nature of the communication in a network of neurons, it was envisioned that one of the ways information could be encoded in the nervous system is by changing the function of the synapse in an experience dependent manner (Kandel et al., 2014). The discovery of long-term facilitation and long-term depression of synaptic function in response to behavioral training provided experimental credence to the synaptic plasticity hypothesis of memory (Bliss & Collingridge, 1993; Malenka & Bear, 2004; S. J. Martin et al., 2000; Takeuchi, Duszkiewicz, & Morris, 2014). The synaptic plasticity hypothesis of long-term memory also raised some fundamental questions (Takeuchi et al., 2014). How does a synapse change its function? How does the selectivity of synaptic change arise in neurons which may connect at thousands of synapses? If indeed changes in synaptic strength underlie memory, how does the synaptic change persist for long durations, particularly for months and years?

PROTEIN SYNTHESIS AT THE SYNAPSE

Synapses are made of functionally diverse proteins and these proteins are required not only for synaptic transmission and regulation of synaptic transmission but also to build and maintain a synapse (Miniaci et al., 2008). Since proteins have a finite half-life compared to the life-time of a synapse, the synapse requires continuous feedback of proteins to maintain synaptic protein content (Cohen et al., 2013). In addition to this homeostatic maintenance of synaptic content, synaptic plasticity involves alteration in both function and size of the synapse and therefore a significant change in the protein composition of the synapse (Cline, 2003; G. W. Davis & Bezprozvanny, 2001). The maintenance of overall synaptic protein homeostasis and the production of synapse-specific changes are particularly challenging for a neuron because of its organization (G. W. Davis & Bezprozvanny, 2001). The number of synapses a given neuron forms is relatively large. A typical mammalian neuron on average forms 1000 synapses (Cernuda-Cernuda, Garcia-Fernandez, Gordijn, Bovee-Geurts, & DeGrip, 2003). The relative distance of these synapses from the cell body varies significantly. Moreover, within a neuron, synapses that are just a few microns apart can be altered functionally and morphologically independently of each other (K. C. Martin et al., 1997a; Miniaci et al., 2008). These neuronal features raise several questions: how does a neuron maintain the protein composition of a synapse? How does it restrict changes in protein composition in a subset of synapses when they share the same cell body—the source of mRNA and most proteins? Most proteins are synthesized in neuronal cell bodies (Steward & Schuman, 2003). However, over time, it has become evident that in neurons, new proteins are synthesized outside the cell body in axons, dendrites and even in the synapse (Steward, Farris, Pirbhoy, Darnell, & Driesche, 2014). As the evidence of synaptic protein synthesis is discussed in other chapters, I will focus on the function of synaptic protein synthesis, particularly in the context of synaptic plasticity and memory.

Role of Local Protein Synthesis in Synapse-Specific Plasticity

The requisite features of the synapse-based model of long-term memory are selectivity, specificity, and persistence of synaptic change. The first functional role of local protein synthesis in synaptic plasticity came from the work of Erin Schuman and colleagues in the mouse hippocampus (Kang & Schuman, 1996). They noticed that local application of the growth factor BDNF (brain-derived neurotrophic factor) not only causes local protein synthesis in the dendrites, but also that this newly synthesized protein is required for the rapid enhancement of synaptic transmission. BDNF-mediated and protein-synthesis-dependent increases in synaptic efficacy were also observed in preparations where cell bodies were dissected from the neurites. Since the neurites were detached from the primary source of proteins—the cell body—this study provided the first causal link between local protein synthesis and synaptic plasticity. Around the same time, elegant experiments by Kelsey Martin, Eric Kandel, and their colleagues in *Aplysia* sensory-motor neuron co-cultures demonstrated that local protein synthesis is required for both synapse-specific long-term facilitation and stabilization of newly formed synapses (K. C. Martin et al., 1997a). In these experiments, a bifurcated *Aplysia* sensory neuron, which made synaptic contact with two separate target motor neurons, was used. Repeated application of serotonin (5-HT) to one set of synapses produced

CREB-dependent, synapse-specific, long-term facilitation, which can be captured at the opposite synapse by a single pulse of 5-HT (Casadio et al., 1999). Perfusion of protein synthesis inhibitors locally in the synaptic region did not change short-term facilitation, but selectively interfered with long-term facilitation. This system also revealed that repeated application of 5-HT only to the cell body of the sensory neuron produced a transcription-dependent, cell-wide long-term facilitation. However, this cell wide facilitation was not associated with any new synaptic growth and the synapse-specific facilitation did not persist more than 24 hours. However, this transient cell-wide facilitation can be converted to persistent synapse-specific facilitation by a single pulse of 5-HT at the synapse (Casadio et al., 1999), and this synaptic capture of cell-wide facilitation also requires local protein synthesis. Concomitant to these studies in *Aplysia*, Frey and Morris, using mouse hippocampal neurons, found that a weak tetanic stimulus that by itself would not produce long-term potentiation, if coupled with a strong stimulus in another branch of the same neuron, would now lead to long-term potentiation. Importantly, long-term potentiation in both branches required synthesis of new proteins (Alarcon, Barco, & Kandel, 2006). A more unifying picture of role of synaptic protein synthesis emerged from "synaptic capture" hypothesis proposed by Martin et.al and Morris et.al (Frey & Morris, 1997; K. C. Martin et al., 1997b). This hypothesis, also referred as "synaptic tagging," proposed two possibilities of how a synapse-agnostic global gene-expression program can produce synapse-specific alterations. First, the products of gene expression are delivered to all the synapses throughout the cell but are functionally incorporated in only those synapses that have been molecularly altered or "tagged" by previous synaptic activity. Second, the products of activated transcription and translation are selectively delivered to the synapse that has been "tagged" by synaptic activity. Subsequent work in *Aplysia* revealed that newly synthesized mRNAs are distributed to both stimulated and unstimulated branches of neurons, not only to the stimulated site (S. Kim & Martin, 2015), leading to the notion that synaptic protein synthesis serves two important functions (K. C. Martin, Barad, & Kandel, 2000). First, it synthesizes a retrograde signal that activates a nuclear transcriptional program, and activation of nuclear program is enough to alter synaptic function that persists for at least 24 hours. The second function of protein synthesis is to stabilize the synaptic changes beyond 24 hours (Frey & Morris, 1997; Reymann & Frey, 2007). I will discuss below how CPEB serves an essential function in this stabilizing component of synaptic protein synthesis.

Role of Local Protein Synthesis in Persistence of Synaptic Change

A central feature of LTM is persistence. Since proteins are short-lived compared to the duration of memory, how synaptic changes created by a transient initial stimulation persist (Lynch & Baudry, 1984; Roberson & Sweatt, 1999)? Many kinds of mechanisms of memory stability in the face of molecular instability have been proposed, and for a long time the likely solution believed to be structural changes, such as growth of new synapses (Yuste & Bonhoeffer, 2001). As a corollary it was assumed that the requirement for activity-dependent molecular changes, such as synthesis of new protein, was transient. However, studies in a number of model systems suggest that the maintenance of behavioral memory as well as structural alterations, requires ongoing macromolecular synthesis (Kandel, 2001; K. C. Martin et al., 1997a; Nader, 2015). For example, in *Aplysia*, stimulation of sensorimotor neuron culture with 5-HT results in increase of both synaptic transmission and formation of new synapses, both of which persist for days. Remarkably, application of protein synthesis inhibitors 1 day or 2 days after initial stimulation reverses not only enhanced synaptic transmission, but also the new synapses. This posed a problem: how a transient experience produces such a persistent state of macromolecular synthesis. A potential solution to this problem of molecular turnover is either to have proteins with unusually long-half lives or invoking self-sustaining biochemical reactions. In 1984, Crick first addressed the possibility of a self-sustained molecular alteration as the basis of long-term memory storage using protein phosphorylation as a candidate mechanism (Crick, 1984). Subsequently, a number of plausible candidate mechanisms has been proposed including autocatalytic Calcium-calmodulin dependent KinaseII or CamKII or a constitutively active brain-specific form of atypical protein kinase C, PKMξ, (Dudai, 2004; Lisman, 1994). Another proposed mechanism of stabilizing LTM over time has been epigenetic modifications to the genome, including both histone acetylation and DNA methylation (Chatterjee, 2017; Zovkic, Guzman-Karlsson, & Sweatt, 2013). However, the consequence of genomic modifications is global, and therefore, even if there is a genomic substrate of memory, the manifestation of this global program necessitates a persistent synapse-specific program (Li et al. 2016). As I will discuss below, accumulating evidence indicates that the ability of group of CPEB proteins to create a self-sustaining molecular state could indeed be such a synapse-specific program. CPEB is noteworthy in this context not only for its unique, stability-conferring properties, but also for its central role in synaptic translation regulation via cytoplasmic polyadenylation.

Translational Control by Cytoplasmic Polyadenylation

Even from this brief discussion, it is evident that local protein synthesis is a key molecular event in generating synapse-specific long-term changes. This raises several questions. How is local protein synthesis in the synapse regulated? What proteins are locally synthesized and how do they stabilize synaptic growth and function? Several pathways and factors have been studied in the context of translation-dependent long-term synaptic plasticity and memory (Costa-Mattioli, Sossin, Klann, & Sonenberg, 2009). For example, the translation initiation factor eIF2α is dephosphorylated in response to neuronal stimulation and its inhibition disrupts memory (Costa-Mattioli et al., 2009). Similarly, the phosphorylation of translational regulators, such as mTOR and eIF4E-binding protein 4EBP is associated with synaptic protein synthesis and various forms of memory. These regulatory processes directly impact translational machinery. The other way to control protein synthesis is to change the availability, localization or modification of the mRNA that is translated (L. Davis, Banker, & Steward, 1987; Si, Giustetto, et al., 2003; Steward & Levy, 1982; Steward & Reeves, 1988). Among these plausible mechanisms, this chapter specially focuses on regulation via changing the polyA tail length of the mRNA, because it has emerged as one of the principal mechanisms of synaptic protein synthesis regulation that influences the maintenance of long-lived memories (Si, Giustetto, et al., 2003; Udagawa et al., 2012).

All newly made mRNAs receive a polyA tail in the nucleus through transcription-coupled polyadenylation (Bilger, Fox, Wahle, & Wickens, 1994; Rigo & Martinson, 2009). Polyadenylation in the nucleus requires a cis-acting polyadenylation signal sequence, AAUAAA at 3′UTR, and a polyadenylation complex, comprised of cleavage and polyadenylation specificity factors (CPSF) and the enzyme poly (A) polymerase (PAP; Murthy & Manley, 1995; Proudfoot & Brownlee, 1976). CPSF directly binds to the hexanucleotide sequence at the 3′UTR and recruit PAP to form an active polyadenylation complex (Bienroth, Wahle, Suter-Crazzolara, & Keller, 1991; Murthy & Manley, 1995). As a result of generalized polyadenylation in the nucleus, most pre-mRNAs have a poly (A) tail of approximately ~70–90 nucleotides long in *Saccharomyces cerevisiae* (Brown & Sachs, 1998; Groner, Hynes, & Phillips, 1974) and ~250 nucleotides in mammalian cells (Brawerman, 1981). Once transported out of the nucleus to the cytoplasm, the mRNA can undergo additional modifications. Some of these modifications in the cytoplasm control mRNA translation (Barnard, Ryan, Manley, & Richter, 2004; J. H. Kim & Richter, 2006). One such cytoplasmic modification of mRNA is regulation of the polyA tail. The length of polyA tail in the cytoplasm is regulated by opposing activity of deadenylating and polyadenylating enzymes. Recruitment of these enzymes to mRNAs by sequence-specific RNA binding protein, such as CPEB, therefore can control specific mRNA.

That polyA tail length is important for translation was discovered in the *Xenopus* oocyte by Joel Richter and colleagues in an attempt to understand how translationally dormant maternal mRNAs are translated upon exposure to oocyte maturating factors (Groisman, Jung, Sarkissian, Cao, & Richter, 2002). In immature oocytes, dormant mRNAs usually have short poly (A) tail of ~40 bases, but when these mRNAs are polyadenylated to ~150 bases, they become translationally active, which results in meiotic division and oocyte maturation (Ivshina, Lasko, & Richter, 2014). Richter and colleagues discovered that cytoplasmic polyadenylation is dependent on the presence of U-rich sequence element UUUUAU at the 3′UTR (cytoplasmic polyadenylation element, or CPE) and a CPE-binding protein (CPEB; Hake, Mendez, & Richter, 1998). Pre-mRNAs with a CPE at their 3′UTR upstream of the AAUAAA poly (A) selection site are bound by CPSF and transported to the cytoplasm (Lin, Evans, Shen, Xing, & Richter, 2010). In the cytoplasm complete polyadenylation complex is assembled upon joining of symplekin, a scaffolding protein, germline development 2 poly (A) polymerase, or Gld2 and the deadenylating enzyme poly (A) ribonuclease (PARN; Barnard et al., 2004; J. H. Kim & Richter, 2006). Both Gld2 and PARN are enzymatically active, but more robust nuclease activity of PARN keeps the mRNA translationally silent by maintaining a short poly(A) tail (J. H. Kim & Richter, 2006). In addition, Maskin, which is a poly (A) binding protein, simultaneously binds CPEB and eukaryotic initiation factor 4E (eIF4E) and prevents eIF4E's interaction with 5′ cap structure to stop the assembly of the translation initiation complex (Cao & Richter, 2002; Richter & Sonenberg, 2005; Stebbins-Boaz, Cao, de Moor, Mendez, & Richter, 1999). Hormones such as progesterone cause unmasking of dormant mRNA by removing the inhibitory constraints and elongating the polyA tail end (Sarkissian, Mendez, & Richter, 2004). The first step in this process is the activation of Aurora A kinase which phosphorylates CPEB protein (Mendez, Hake, et al., 2000; Sarkissian et al., 2004). CPEB phosphorylation has multiple effects: first, it causes PARN expulsion from the RNA-protein complex, thereby stopping deadenylation; second, it enhances interaction between CPEB and the CPSF complex (J. H. Kim & Richter, 2006; Mendez, Murthy, Ryan, Manley, & Richter, 2000); finally, phosphorylated CPEB has low affinity for Maskin, resulting in dissociation of Maskin from CPEB-eIF4E complex and allowing the formation of the translation initiation complex (Cao, Kim, & Richter, 2006; J. H. Kim & Richter, 2007). A similar CPEB1-dependent mechanism also manifests in activity-dependent synaptic modification, except that in the nervous system, Maskin is likely substituted by another eIF4E- and CPE- binding protein, neuroguidin. Neuroguidin is widely expressed in the nervous system and when injected in xenopus oocytes also represses protein synthesis in a CPE-dependent manner. Like Maskin, neuroguidin likely blocks translation initiation by inhibiting the interaction between mRNA cap-binding proteins eIF4E and eIF4G.

In eukaryotes polyA tail length at the 3′ end of the mRNA is an important regulatory step in translation (Sachs, Sarnow, & Hentze, 1997). A longer polyA tail can stabilize mRNA or enhance translation of the mRNA via circularization (Munroe & Jacobson, 1990). The translation-potentiating effect is mediated by poly(A) binding protein (PABP; Bernstein & Ross, 1989). PABP, bound to the 3′ tail of the mRNA, interacts with eIF4G, a component of 5′ cap binding complex (L. Wu et al., 1998), leading to circularization of mRNA. The mRNA loop could potentiate translation by allowing translation machinery to re-initiate protein synthesis or ensuring that only complete mRNAs possessing both cap and tail attract translation machinery (Lopez-Lastra, Rivas, & Barria, 2005). PABP is implicated in memory formation; knockout of its inhibitor PAIP2A, which is normally degraded by calpains upon neuronal stimulation, increases CamKII translation and reduces the amount of training required to form a long term memory (Khoutorsky et al., 2013). Furthermore, mutation in proteins like polyA polymerase GLD2 are known to interfere specifically with long-term memory (Kwak et al., 2008). However, outside the nervous system and developing embryos, the relationship between polyA tail length and translation is less clear (Subtelny, Eichhorn, Chen, Sive, & Bartel, 2014). A crucial advantage of cytoplasmic polyadenylation as a mechanism for translational regulation is that it allows another regulatory step that further dissociates transcription and translation in time and space. Certain transcripts can be held in a "dormant" state, awaiting translational activation. In the context of neural plasticity, this means stimulation of the synapse and activation of transcripts awaiting translation as well as selective translation of globally distributed mRNAs at specific synapse.

CPEB Family across the Tree of Life

In mammals, cytoplasmic polyadenylation element binding protein (CPEB) has four family members, CPEB1–4. Although all of them are expressed in the brain, the four members are divided into two groups based on amino acid sequence similarity and function (Huang, Kan, Lin, & Richter, 2006): CPEB1 and CPEB2-4. They all contain a common RNA binding domain and a Zn-finger domain at the C-terminal end, but they differ in the N-terminal end (Richter, 2007). Intriguingly, CPEB2 and CPEB3 contain an N-terminal unstructured domain similar to prion-like domains (discussed in detail later). While it is evident that CPEB1 controls protein synthesis by regulating polyA tail length, how CPEB2 and CPEB3 control protein synthesis is mechanistically less clear. An in vitro iterative binding assay suggested that CPEB2-4 do not interact with the canonical CPE, instead binding to a U-rich stem-loop structure (Huang, Kan, Lin, & Richter, 2006). However, some studies suggest CPEB2 and CPEB3 also regulate mRNAs containing CPE elements (L. Fioriti et al., 2015; Lu, Yeh, & Huang, 2017; Pavlopoulos et al., 2011; Turimella et al., 2015). CPEB3 does not require AAUAAA polyadenylation signal, nor does it bind to CPSF polyadenylation factor (Huang et al., 2006). There are two CPEB proteins in *Aplysia* (Liu, Hu, Wu, Schwartz, & Schacher, 2006). Like mammalian CPEB, the two ApCPEB differ in the N-terminal end, one of them having a polyglutamine (Q) stretch at the N-terminus and the propensity to form an amyloid-like structure (Raveendra et al., 2013). In *Drosophila*, there are two CPEB orthologues, Orb1 and Orb2, and both of them are expressed in the nervous system (Hafer, Xu, Bhat, & Schedl, 2011; Pai et al., 2013). Orb2 also possesses a polyglutamine (Q) stretch at the N-terminus and clearly utilizes a polyA tail-dependent mechanism to control protein synthesis(Khan et al., 2015).

Mammalian CPEB Proteins and Their Role in Synaptic Plasticity and Memory

As previously stated, the plasticity underlying LTM requires dramatic changes at the synapse in terms of function and morphology. All these changes are underwritten by changes in protein expression, and CPEB family members across taxa have been shown to play crucial roles in this process. Among four CPEB family members in mammals, CPEB1 has been most intensely studied. Relatively less is known about CPEB2-4. However, due to the role of CPEB1 and apCPEB in synaptic plasticity, recent efforts have focused on understanding the function of the other family members. The goal of these mammalian studies is to determine if CPEB1-4 are critical for long-term memory. The main approach has been to generate knockout mice of each family member and assess synaptic plasticity and long-term memory formation. Detailed next is what we have learned from these studies by gene.

CPEB1

The possible role of CPEB protein in the synaptic protein was first observed by Richter and colleagues. Although earlier studies in mouse suggested CPEB1 (mCPEB, a 62 KDa protein) is expressed in testis, ovary, and kidney (Gebauer & Richter, 1996), subsequent studies found moderate level of CPEB protein in the brain (Huber, Kayser, & Bear, 2000; Kang & Schuman, 1996; K. C. Martin et al., 2000; K. C. Martin et al., 1997a; L. Wu et al., 1998). Ca^{2+}-calmodulin dependent kinase (α-CaMKII), a molecule that is important for LTP (Du & Richter, 2005; Huang, Carson, Barbarese, & Richter, 2003; Huang, Jung, Sarkissian, & Richter, 2002; Wells et al., 2001; L. Wu et al., 1998), has 2 CPE elements in its 3′UTR, both of which are bound by CPEB1. Consistent with their role in binding of CPEB, both CPE elements of CamKII are necessary for activity-dependent polyadenylation. Furthermore, visual experience causes both the polyadenylation of CamKII mRNA and its increased translation in the visual cortex. This provided the first indication of cytoplasmic polyadenylation as a mechanism of activity-dependent translation in the nervous system (Wells et al., 2001). Studies of CPEB1-regulated translation in the nervous system suggest overlap in the molecular complexes shared by oocyte and post-synaptic dendrite. CPEB1 is present in both soma and dendritic areas and bidirectionally regulates translation of dendritic mRNA, such as α-CamKII and NMDA receptor subunit NR2A, in an activity dependent manner (Huang et al., 2002; Udagawa et al., 2012; L. Wu et al., 1998). In turn, α-CamKII phosphorylates and activates CPEB1 (Atkins, Nozaki, Shigeri, & Soderling, 2004).

Despite well-understood molecular mechanism of CPEB1 in localized translation, the necessity of it in LTP and memory is less clear. Curiously, only selective forms of LTP (single train of theta-burst) are affected in CPEB1-deficient mice (Alarcon et al., 2004). In CPEB1 KO mice, growth hormone levels are dramatically decreased in the nervous system (Zearfoss, Alarcon, Trifilieff, Kandel, & Richter, 2008) and growth hormone can induce LTP in hippocampal slices (Mahmoud & Grover, 2006). Growth hormone is not a CPEB1 target; however, a transcription factor, c-jun, is a CPEB1 target and c-jun regulates growth hormone expression. CPEB1 knockout mice have intact memory consolidation and instead shows impairment in memory extinction (Berger-Sweeney, Zearfoss, & Richter, 2006). CPEB1 phenotypes are different from other mammalian and *Drosophila* CPEBs, for which perturbation tends to result in deficits in LTP and long-term memory (discussed below). These observations suggest either CPEB1 is not the critical molecule that sustains active translation over long time, or other CPEBs compensate the absence of CPEB1. However, CPEB2-4, targets do not overlap with CPEB1 (Huang et al., 2006).

Can CPEB1 anyway contribute to persistence of synapse-specific change? Since phosphorylation transforms CPEB1 from repressor to an activator, the level and duration of phosphorylated CPEB1 following synaptic stimulation is an important issue to consider. Brief stimulation (1X100hz) of hippocampal slices that produces protein-synthesis independent short-live synaptic potentiation (E-LTP) increases phopho-CPEB1 level for couple of minutes. Longer stimulation (4X100hz), which produces protein synthesis dependent L-LTP, increases phospho-CPEB1 level for approximately 30 min. The phosphorylation of CPEB1 depends on activation of CamKII and inhibition of protein phosphatase PP1 (Atkins, Davare, Oh, Derkach, & Soderling, 2005). Therefore, a balance of CamKII and PP1 activity, or a self-sustained feedback loop composed of CPEB1 and CamKII could prolong CPEB1-mediated translation beyond the stimulation period. However, available evidence shows although CPEB1 can be rapidly phosphorylated, the phosphorylated CPEB1 does not persist beyond 30 min, suggesting activated CPEB1 cannot self-maintain and is prone to be deactivated (Atkins et al., 2004). Therefore, CPEB1, despite specifically regulating translation of mRNAs that are critical for memory consolidation, may not be a candidate for a molecular memory trace.

CPEB2

CPEB2 was first discovered in mouse germ cells. The murine *Cpeb2* gene has 12 exons, predicted to encode broadly two classes of polypeptides of approximately 110KDa (CPEB2A), and ~58KDA (CPEB2B). In the adult brain, prion-like (PD) domain-containing CPEB2A is the prevalent form. (Hagele, Kuhn, Boning, & Katschinski, 2009; Theis, Si, & Kandel, 2003) and shares most homology with CPEB3 and CPEB4 (Kurihara et al., 2003). Although it was initially thought that CPEB2 does not bind CPE, later studies found CPEB2 interacts with previously identified CPE-element containing mRNAs such as β-catenin and CamKII. CPEB2 also represses translation, but through a mechanism distinct from CPEB1. CPEB2 physically interacts with translation eukaryotic elongation factor eEF2, and the CPEB2-eEF2 interaction reduces eEF2-ribsosome-triggered GTP hydrolysis. The GTP hydrolysis is important for peptide translocation and therefore binding of CPEB2 attenuates translation elongation. Dissociation of CPEB2 from mRNA releases the CPEB2-target mRNAs from this translation repression (Turimella et al., 2015). However, how CPEB2-eEF2 interaction is regulated to activate CPEB2 target mRNAs remains unclear.

What is the function of CPEB2 in synaptic potentiation and memory? Unfortunately, of the entire CPEB family, only CPEB2 KO mice die shortly after birth when CPEB2 is removed from all tissues (Lai et al., 2016), thus it was not possible to assess synaptic plasticity or behavior in these animals. However, at the prenatal stage, CPEB2 KO mice have hyper-activation of parasympathetic cholinergic neurons that results in fatal bronchoconstriction. These mice have increased levels of choline acetyltransferase (ChAT), the enzyme responsible for the synthesis of neurotransmitter acetycholine, and in Neuro-2a neuroblastoma cell line, CPEB2 binds to and represses the translation of ChAT (Lai et al., 2016). The increase in ChAT correlates with the hyper-activation of cholinergic neurons. Together these findings suggest CPEB2 acts as a translational repressor in the parasympathetic nervous system.

An alternative approach that restricted the excision of loxP-flanked CPEB2 by expressing Cre recombinase only in excitatory neurons of the cerebral cortex, amygdala, and hippocampus provided some interesting clues to the possible role CPEB2 in synaptic plasticity and memory (Lu et al., 2017). The restricted excision of CPEB2 (CPEB2 cKO) results in normal development but impaired long-term potentiation and disrupted long-term memory. The CPEB2 cKO mice also exhibit a significant increase in the number of immature spines and a decrease in surface expression of AMPARs. The decrease in AMPARs was attributed to increased expression of the GRIP-associated protein 1 (GRASP1; Lu et al., 2017). GRASP1 regulates the endosomal recycling of AMPA receptors through its association with glutamate receptor-interacting protein 1 (GRIP1) by coupling the GRASP1-GRIP1 complex to endosomes (Hoogenraad et al., 2010). This study established a plausible role of CPEB2-dependent translation in long-term plasticity and long-term memory. However, it is unclear whether a specific or all CPEB2 variants are required for memory. It is also unclear which aspect of memory—consolidation, maintenance, or recall—is dependent on CPEB2 activity and how CPEB2 activity is regulated in an activity-dependent manner.

CPEB3

CPEB3 is second most-studied CPEB family member behind CPEB1 with respect to synaptic plasticity and long-term memory. Early studies demonstrated that CPEB3 expression is upregulated in mouse hippocampal slices upon induction of LTP and in cultured hippocampal neurons upon addition of glutamate or glycine(L. Fioriti et al., 2015; Theis et al., 2003). Thus far, two approaches have been used to remove CPEB3 from the mouse genome. The first approach generated a global knockout. Surprisingly, these mice displayed normal synaptic potentiation, impaired synaptic depression and actually performed better in some memory-related tasks (Chao et al., 2013). The second, more restricted approach, selectively removed CPEB3 in the forebrain (cortex, amygdala, and hippocampus). Conditional removal of CPEB3 showed impaired long-term memory in spatial object recognition and the Morris water maze, suggesting that CPEB3 mediated processes are required for hippocampal-based spatial learning tasks. Interestingly, weeks after memory acquisition, removal of the CPEB3 gene disrupts behavioral display of the memory (L. Fioriti et al., 2015). Whether this is due to loss of memory (via impaired maintenance) or failure to recall remains unclear. At the cellular level, late phase long-term potentiation was also reduced upon CPEB3 conditional knockout. Both memory and LTP deficits were rescued by exogenous expression of CPEB3 using adenovirus. Curiously, like CPEB2, CPEB3 targets also include AMPA-receptor. However, in the case of CPEB3 it is AMPA receptors subunits such as GluA1 and GluA2 (Huang et al., 2006; Pavlopoulos et al., 2011). The protein levels of both GluA1 and GluA2 were significantly increased in the hippocampus of the CPEB3 conditional knockout. In accordance with CPEB3 regulating translation and not the transcription of its targets, the mRNA levels of GluA1 and GluA2 were not significantly different from wild-type animals. Together, the conditional KO studies suggested that CPEB3 acts as a translational regulator and is required for LTP and for maintenance of memory. However, the discrepancies in phenotype between global and conditional KO remain to be resolved. The most common explanation is molecular compensation in global KO and the fact that CPEB2 and CPEB3 both converge on the AMPA receptor provides some credence to this possibility. However, such a compensatory mechanism would also imply both proteins are expressed in the same neuronal population and can functionally substitute each other.

CPEB4

The final member of the mammalian CPEB family, CPEB4, is also expressed in neurons (Theis et al., 2003). Similar to CPEB1,CPEB4 induces polyadenylation and translation in Xenopus oocytes (Novoa, Gallego, Ferreira, & Mendez, 2010). However, no overt phenotype is evident with complete removal of the CPEB4 gene in mice (Tsai et al., 2013). Also, CPEB4 KO mice have normal synaptic plasticity and memory. These findings suggest CPEB4 is the only CPEB family member not required for plasticity and learning. However, the phenotype for conditional removal of CPEB4 is unknown.

Invertebrate CPEB Proteins and Their Role in Synaptic Plasticity and Memory

The role of CPEB in synaptic plasticity was also explored in the *Aplysia* bifurcated sensory neuron culture. The Aplysia CPEB (ApCPEB) is upregulated exclusively in the stimulated synapse after a single pulse of 5-HT. This is a transcriptionally and even cell-body independent process, as blocking transcription pharmacologically and physically severing the connection to the cell body does not impair serotonin-mediated synapse-specific increase of ApCPEB. It is, however, dependent on rapamycin-dependent, P13-regulated translation. ApCPEB is indispensable in the stimulated synapse for persistent serotonin-induced long-term facilitation, as its localized knockdown impairs the persistence, but not the formation, of local long-term facilitation. The loss of long-term facilitation upon ApCPEB knockdown at 24 hours correlated with a loss of newly formed varicosities in the stimulated branch. This is true of knockdown initiated before stimulation and at 24 and 48 hours after stimulation, but not at 72 hours after stimulation (Miniaci et al., 2008). Similarly, a general translation inhibitor applied at 72 hours after stimulation also did not impair LTF thereafter. This raises the question of whether CPEB creates a state that doesn't need to be translated further at least in the time scale assessed or another mechanism eventually takes over for the initially CPEB-regulated translation-dependent memory. Nonetheless, the following observations suggest that *Aplysia* CPEB has properties required of a synaptic tag that stabilizes synapse-specific long-term facilitation: 1) ApCPEB is upregulated only at the activated synapse in a stimulus-dependent manner (Si, Giustetto, et al., 2003), 2) It is required not for initiation but for the persistence of long-term facilitation and stabilization of newly formed synapses, 3) ApCPEB can activate translation of dormant mRNAs (Bally-Cuif, Schatz, & Ho, 1998; Schroeder, Condic, Eisenberg, & Yost, 1999; Stebbins-Boaz, Hake, & Richter, 1996; Tan, Chang, Costa, & Schedl, 2001), and finally 4) CPEB target mRNAs are involved in synaptic growth (Chang, Tan, Wolf, & Schedl, 2001; Groisman et al., 2002).

In *Drosophila*, CPEB has two family members: Orb1 and Orb2. Orb1 is known to be involved in establishing polarity axes in the developing egg as well as early embryo (Hafer et al., 2011). Beyond germline, the role of Orb1 in somatic tissues is unclear (Hafer et al., 2011). However, one recent study reported that *Drosophila* Orb1 is expressed in the adult nervous system and is required for conversion of labile olfactory memory into stable long-term memory (Pai et al., 2013). Unlike Orb1, Orb2 mRNA and protein are expressed throughout development and in somatic tissues (Hafer et al., 2011). Orb2 localization in the embryonic central nervous system (CNS) is restricted to cell bodies and mostly absent in the axonal tracts (Hafer et al., 2011). However, in the adult fly brain Orb2 protein can be detected in the cell bodies, axonal tracts, and dendritic terminals (Hafer et al., 2011; Krüttner et al., 2015; Majumdar et al., 2012). Orb2 has been implicated in the development of the embryonic nervous system and mesoderm through asymmetric division of stem cells and other precursor cells (Hafer et al., 2011). In adult flies, removal of Orb2 does not interfere with short-term memory but impairs long-term memory (Majumdar et al., 2012).

Prion-like Properties of CPEB Proteins and Persistence of Memory

The existence of biochemical switches that can undergo persistent changes in activity in response to a transient experience, and thus serving as a memory trace, has been postulated for a long time. However, experimental evidence of the existence of such biochemical switches, and, more importantly, what such switches are made of remain virtually unknown. Three types of evidence should be considered in testing the biochemical trace: (1) must produce a persistent neurobiological change of a type that can account for the behavioral manifestation of memory; (2) must be utilized during memory retrieval; and (3) must link the processes of memory encoding and retrieval. Unlike the circuit substrate of memory that has endurable cellular material, proteins have short lives; therefore, a biochemical substrate of memory must be immune to turnover of individual molecules.

A potential solution to this problem emerged from the discovery of a Q/N rich prion-like domain in the N-terminus of CPEB family members, namely ApCPEB in *Aplysia* (Si, Lindquist, & Kandel, 2003), Orb2 in *Drosophila* (Keleman, Krüttner, Alenius, & Dickson, 2007; Krüttner et al., 2012; Krüttner et al., 2015; Majumdar et al., 2012; Si, Choi, White-Grindley, Majumdar, & Kandel, 2010) and CPEB2-4(Luana Fioriti et al., 2015;

Huang et al., 2006; Stephan et al., 2015) in mice. Prions were first discovered as the causative agent for number of neurodegenerative diseases (Prusiner, 2013). However, studies primarily in yeast revealed that proteins can adopt a prion-like state and be non-pathogenic, even serving as the basis of functional phenotypic variations (Shorter & Lindquist, 2005; Wickner, Edskes, Shewmaker, & Nakayashiki, 2007). Prions are proteins that can assume at least two distinct physical states: a monomeric state and self-templating amyloidogenic aggregated state (Prusiner, 1998). Once the aggregate state is formed it induces the conformational conversion of the monomeric state and initiates a dominant autocatalytic state. This creates a protein "conformational memory" that outlives individual components, like a molecular "Ship of Theseus" that can replace individual building blocks while maintaining its formal identity. Intrinsic and extrinsic modulation of *Drosophila* Orb2 assembly has provided unique insights into how a prion-like amyloid-forming protein can act as a substrate for long-term memory. The first of these studies involved deletion of the prion domain of Orb2, leaving the RNA-binding domain intact, which resulted in long-term, but not short-term, memory impairments in the courtship suppression paradigm (Keleman et al., 2007). It was subsequently found that a single mutation (F5Y) in the prion-like domain of Orb2 impaired both its ability to assemble and the persistence of memory (Majumdar et al., 2012). Further work, employing extrinsic modulators of Orb2 further demonstrated how modification of its prion-like properties can affect memory. Exogenous expression of the anti-amyloid QBP1 peptide, which impedes Orb2 aggregation, blocked LTM, but not STM (Hervas et al., 2016). A screen in yeast identified the low-expressed non-essential Hsp40/DnaJ chaperone JJJ2 as a factor that promotes Orb2 aggregation. When expressed in the fly system, JJJ2 specifically reduced the threshold for training for LTM without reducing the absolute strength of the memory. This effect was consistent across different memory paradigms (Li et al 2016). These findings are corroborated in *Aplysia*; an antibody that specifically binds ApCPEB oligomer/aggregates inhibited 48h long-term facilitation, but not 24h long-term facilitation (Si et al., 2010), suggesting aggregation of CPEB specifically contributes to stabilizing component of translation regulation.

What is the biochemical consequence of aggregation of CPEB proteins? Interestingly, all CPEBs studied seem to have dual functions, switching between translational activation and repression in response to cell signaling and post-translational modifications. Aurora kinase A phosphorylates CPEB1 and remodels the CPEB1-associated ribonucleoprotein complex from a repressor to an activator. Aggregation of CPEB3 and homologs in *Aplysia* (ApCPEB) and *Drosophila* (Orb2) via their prion-like domains may be involved in the functional switch. Work from *Drosophila* Orb2 suggests it forms distinct protein complexes in monomeric and aggregated states: monomeric Orb2 associates with a deadenylation complex and shortens the polyA tail of target mRNA, while aggregated Orb2 associates with a polyadenylation complex to preserve the target mRNA's elongated polyA tail (Khan et al., 2015). In mammalian neurons, CPEB3's switch between oligomeric translational activation and monomeric translational repression is regulated by SUMOylation (small ubiquitin-related modifier) and ubiquitination, respectively (Drisaldi et al., 2015; Pavlopoulos et al., 2011). In its basal state CPEB3 is

SUMOylated and in its SUMOylated form CPEB3 is monomeric and acts as a repressor of translation. On the other hand, mono-ubiquitination of CPEB3 by the ubiquitin ligase, Neuralized, promotes the translation of the AMPA receptor. CPEB-3 interacts with Neuralized1 via its N-terminal, prion-like domain, and this interaction leads to the monoubiquitination and consequent activation of CPEB-3 (Pavlopoulos et al., 2011). Moreover, CPEB3 loses its ability to maintain long-term synaptic plasticity and long-term memory if its prion-like N-terminus domain is deleted (Luana Fioriti et al., 2015).

The aggregated CPEB proteins persist through different time scales in different models: in *Aplysia* and *Drosophila*, aggregated CPEB is still elevated 24h post stimulation relative to naïve conditions, while in mice CPEB3 aggregates can sustain to 24h, albeit with a reduced level compared to 1h post training. This suggests that in higher eukaryotes, a stringent regulation might take place to prevent the overgrowth of aggregation. Nevertheless, the time scale of aggregate persistence, even in the mammalian system, is beyond post-translational modification that are induced within 1h after training.

How could an amyloid-like state of CPEB serve as a memory substrate? CPEB protein has three domains: a prion-like domain, an mRNA binding domain, and a protein-protein interaction domain. The current model is that the prion-like domain allows the protein to alter its conformation and assemble into an ordered amyloid-fold. A consequence of the monomer to amyloid transition is a change in protein-protein interaction. In case of *Drosophila* Orb2, the monomer-binding protein CG13928 recruits the deadenylation complex and reduces translation (Khan et al., 2015). When the protein forms an amyloid, CG13928 does not bind and hence the deadenylation complex is not recruited. In contrast, CG4612, which has a weak affinity for monomer, binds more readily to the amyloid state. CG4612 recruits polyA polymerase complex, which increases polyA tail length and enhances translation (Khan et al., 2015). Therefore, before synaptic stimulation, Orb2 target mRNAs, which include proteins involved in both synapse formation and function, are maintained in a translationally dormant state by the Orb2 monomer. When a synapse is stimulated, Orb2 proteins attain an amyloid state. Now, in that synapse, instead of being repressed, Orb2 target mRNAs are translated more readily. This results in enhanced synaptic activity as well as stabilization of morphological changes of the synapse. Owing to the ability of the amyloid-fold to seed and self-perpetuate once engaged, any newly arriving Orb2-target mRNAs, instead of being repressed, are now activated in a synapse-specific manner. This allows the synapse to continue to maintain its altered activity state even in the absence of any further synaptic stimulation.

Unresolved Issues and Future Directions

In silico analysis of mammalian proteome using the PLAAC (proteome prion-like amino acid composition analysis), shows that in mouse and in human, CPEB2 ranks in the 99th percentile and CPEB3 ranks in the 98.8th percentile. In addition to possessing a prion-like domain, CPEB3 has already been shown to have some prion-like properties in a heterologous system and is reported to form SDS-resistant aggregates in the adult brain, which is one of the features of amyloid-like aggregates (Stephan et al., 2015). However, to what extent aggregation of mammalian CPEB2 and CPEB3 contributes to long-term synaptic potentiation and persistence of memory remains unclear.

Of most low-complexity sequence RNA binding proteins identified so far, aggregation usually functions to inhibit protein translation. Prion-like CPEB appears to be an example whose aggregation is associated with active translation. The cryo-EM structure of *Drosophila* Orb2 suggests that the amyloid-core structure resides within the assembly, while the mRNA locate in the outer layer of the assembly (Hervas et al., 2020). As memory consolidation during different time windows might call for different mRNA substrates, this organized structure can allow dynamic exchange of mRNA targets, keeping the core translation machinery the same. However, although various biochemical assays indicate mouse CPEB3 form stable aggregates reminiscent of amyloid, there is no direct structural analysis of endogenous CPEB from mouse or human brains that either supports or refutes the existence of amyloid-like fibers of CPEB proteins, or a specific requirement of the amyloidogenic variants of CPEBs in consolidation and/or recall of memory. A solution to this problem would require three types of studies. First, a detailed biophysical characterization of endogenous CPEB2 and CPEB3 aggregates from the brain, preferably at atomic resolution; second, determination of the consequence of CPEB aggregation at biochemical, cellular and behavioral levels; finally, altering the biophysical properties of CPEB aggregates and assessing the consequence in memory formation, consolidation and retrieval.

Lastly, an important question raised by these findings is how amyloidogenic aggregation could be non-toxic or how memory can be amenable to modification since amyloids are generally believed to "irreversible"? While this aspect of CPEB biology has been largely unexplored, some insight has been gained by studying the kinetics of Orb2 assembly compared to those of pathological amyloids (Hervas et al., 2016). Like pathological amyloids, Orb2 forms A11-reactive oligomers which give rise to OC-reactive fibrils. However, this process is more rapid for Orb2 than for pathological amyloid-forming proteins and switching the aggregation-promoting domains of Orb2 and disease-associated huntingtin confers toxicity on Orb2 and non-toxicity on Huntingtin. Importantly, pharmacologically arresting Orb2 in its A11-reactive state renders it toxic to cells, and this toxicity is rescued with the A11 antibody. This indicates that one differentiating factor between toxic and functional amyloids may be the kinetics of their respective assemblies. However, these studies carried outside the nervous system can only provide small clues to what happens in the nervous system.

How can memory be reversible or dynamic if amyloids are irreversible in physiological conditions (Harrison, Chan, Prusiner, & Cohen, 2001)? Or is it possible that some memories are indeed irreversible, and they only appear lost due to inability to retrieve them? First, amyloids are not completely irreversible; they exist in a dynamic equilibrium with the available monomer. Use of a photoconvertible GFP-tag revealed that

ApCPEB aggregates are self-renewing, as photoconverted assemblies changed color as they captured monomers from the surrounding cytosol (Si et al., 2010). Second, studies of Orb2 suggest that the amyloid core of functional amyloids have an amino acid composition distinct from toxic amyloids, being composed of hydrophilic, rather than hydrophobic residues (Hervas et al., 2020). Recently, new classes of protein assemblies formed by prion-like proteins have been reported, such as non-amyloid filaments of mitochondrial antiviral-signaling protein (Hou et al., 2011), or the gel-like state of Fus (H. Wu & Fuxreiter, 2016), physical states that are believed to be more amenable to regulation (Cereghetti, Saad, Dechant, & Peter, 2018; Hughes et al., 2018). Indeed, a recent study of ectopically expressed CPEB3 in non-neuronal cells suggested formation of gel-like state. Therefore, much remains to be learned with respect to function of neuronal CPEB as a regulator of synaptic protein synthesis and as a substrate of long-lived memories, from the biophysical to the behavioral.

Acknowledgments

I thank my current student Kyle Patton for his help in preparing this book chapter. I also thank my former students Liying Li and Mohammed Repon Khan, whose work contributed to the content of this book chapter. This work was supported by the Stowers Institute for Medical Research.

References

Alarcon, J. M., Barco, A., & Kandel, E. R. (2006). Capture of the late phase of long-term potentiation within and across the apical and basilar dendritic compartments of CA1 pyramidal neurons: synaptic tagging is compartment restricted. *Journal of Neuroscience, 26*(1), 256–264. https://doi.org/10.1523/JNEUROSCI.3196-05.2006

Alarcon, J. M., Hodgman, R., Theis, M., Huang, Y. S., Kandel, E. R., & Richter, J. D. (2004). Selective modulation of some forms of Schaffer collateral-CA1 synaptic plasticity in mice with a disruption of the CPEB-1 gene. *Learning and Memory, 11*(3), 318–327. https://doi.org/10.1101/lm.72704

Atkins, C. M., Davare, M. A., Oh, M. C., Derkach, V., & Soderling, T. R. (2005). Bidirectional regulation of cytoplasmic polyadenylation element-binding protein phosphorylation by Ca2+/calmodulin-dependent protein kinase II and protein phosphatase 1 during hippocampal long-term potentiation. *Journal of Neuroscience, 25*(23), 5604–5610. https://doi.org/10.1523/JNEUROSCI.5051-04.2005

Atkins, C. M., Nozaki, N., Shigeri, Y., & Soderling, T. R. (2004). Cytoplasmic polyadenylation element binding protein-dependent protein synthesis is regulated by calcium/calmodulin-dependent protein kinase II. *Journal of Neuroscience, 24*(22), 5193–5201. https://doi.org/10.1523/JNEUROSCI.0854-04.2004

Bally-Cuif, L., Schatz, W. J., & Ho, R. K. (1998). Characterization of the zebrafish Orb/CPEB-related RNA binding protein and localization of maternal components in the zebrafish oocyte. *Mechanisms of Development, 77*, 31–47.

Barnard, D. C., Ryan, K., Manley, J. L., & Richter, J. D. (2004). Symplekin and xGLD-2 are required for CPEB-mediated cytoplasmic polyadenylation. *Cell, 119*(5), 641–651.

Berger-Sweeney, J., Zearfoss, N. R., & Richter, J. D. (2006). Reduced extinction of hippocampal-dependent memories in CPEB knockout mice. *Learning and Memory, 13*(1), 4–7. https://doi.org/10.1101/lm.73706

Bernstein, P., & Ross, J. (1989). Poly(A), poly(A) binding protein and the regulation of mRNA stability. *Trends in Biochemical Sciences, 14*(9), 373–377.

Bienroth, S., Wahle, E., Suter-Crazzolara, C., & Keller, W. (1991). Purification of the cleavage and polyadenylation factor involved in the 3′-processing of messenger RNA precursors. *Journal of Biological Chemistry, 266*(29), 19768–19776.

Bilger, A., Fox, C. A., Wahle, E., & Wickens, M. (1994). Nuclear polyadenylation factors recognize cytoplasmic polyadenylation elements. *Genes & Development, 8*(9), 1106–1116.

Bliss, T. V., & Collingridge, G. L. (1993). A synaptic model of memory: long-term potentiation in the hippocampus. *Nature, 361*(6407), 31–39. https://doi.org/10.1038/361031a0

Brawerman, G. (1981). The Role of the poly(A) sequence in mammalian messenger RNA. *CRC Critical Reviews in Biochemistry, 10*(1), 1–38.

Brown, C. E., & Sachs, A. B. (1998). Poly(A) tail length control in Saccharomyces cerevisiae occurs by message-specific deadenylation. *Molecular and Cellular Biology, 18*(11), 6548–6559.

Cao, Q., Kim, J. H., & Richter, J. D. (2006). CDK1 and calcineurin regulate Maskin association with eIF4E and translational control of cell cycle progression. *Nature Structural & Molecular Biology, 13*(12), 1128–1134. https://doi.org/10.1038/nsmb1169

Cao, Q., & Richter, J. D. (2002). Dissolution of the maskin-eIF4E complex by cytoplasmic polyadenylation and poly(A)-binding protein controls cyclin B1 mRNA translation and oocyte maturation. *EMBO Journal, 21*(14), 3852–3862. https://doi.org/10.1093/emboj/cdf353

Casadio, A., Martin, K. C., Giustetto, M., Zhu, H., Chen, M., Bartsch, D., . . . Kandel, E. R. (1999). A transient, neuron-wide form of CREB-mediated long-term facilitation can be stabilized at specific synapses by local protein synthesis. *Cell, 99*, 221–237.

Cereghetti, G., Saad, S., Dechant, R., & Peter, M. (2018). Reversible, functional amyloids: towards an understanding of their regulation in yeast and humans. *Cell Cycle, 17*(13), 1545–1558. https://doi.org/10.1080/15384101.2018.1480220

Cernuda-Cernuda, R., Garcia-Fernandez, J. M., Gordijn, M. C., Bovee-Geurts, P. H., & DeGrip, W. J. (2003). The eye of the African mole-rat Cryptomys anselli: to see or not to see? *European Journal of Neuroscience, 17*(4), 709–720.

Chang, J. S., Tan, L., Wolf, M. R., & Schedl, P. (2001). Functioning of the Drosophila orb gene in gurken mRNA localization and translation. *Development, 128*, 3169–3177.

Chao, H. W., Tsai, L. Y., Lu, Y. L., Lin, P. Y., Huang, W. H., Chou, H. J., . . . Huang, Y. S. (2013). Deletion of CPEB3 enhances hippocampus-dependent memory via increasing expressions of PSD95 and NMDA receptors. *Journal of Neuroscience, 33*(43), 17008–17022. https://doi.org/10.1523/JNEUROSCI.3043-13.2013

Chatterjee, S. A., T. (2017). Transcriptional regulation of memory formation. In Sara, S. J. (ed.), *Learning and memory: a comprehensive reference* (2nd ed., 329–343). Oxford: Academic Press.

Cline, H. (2003). Synaptic plasticity: importance of proteasome-mediated protein turnover. *Current Biology, 13*(13), R514–R516.

Cohen, L. D., Zuchman, R., Sorokina, O., Muller, A., Dieterich, D. C., Armstrong, J. D., . . . Ziv, N. E. (2013). Metabolic turnover of synaptic proteins: kinetics, interdependencies and implications for synaptic maintenance. *PLoS One, 8*(5), e63191. https://doi.org/10.1371/journal.pone.0063191

Costa-Mattioli, M., Sossin, W. S., Klann, E., & Sonenberg, N. (2009). Translational control of long-lasting synaptic plasticity and memory. *Neuron, 61*(1), 10–26. https://doi.org/10.1016/j.neuron.2008.10.055

Crick, F. (1984). Memory and molecular turnover. *Nature, 312*(5990), 101.

Davis, G. W., & Bezprozvanny, I. (2001). Maintaining the stability of neural function: a homeostatic hypothesis. *Annual Review of Physiology, 63*, 847–869. https://doi.org/10.1146/annurev.physiol.63.1.847

Davis, L., Banker, G. A., & Steward, O. (1987). Selective dendritic transport of RNA in hippocampal neurons in culture. *Nature, 330*(6147), 477–479. https://doi.org/10.1038/330477a0

Drisaldi, B., Colnaghi, L., Fioriti, L., Rao, N., Myers, C., Snyder, A. M., . . . Kandel, E. R. (2015). SUMOylation is an inhibitory constraint that regulates the prion-like aggregation and activity of CPEB3. *Cell Reports, 11*(11), 1694–1702. https://doi.org/10.1016/j.celrep.2015.04.061

Du, L., & Richter, J. D. (2005). Activity-dependent polyadenylation in neurons. *RNA, 11*, 1340–1347.

Dudai, Y. (2004). The neurobiology of consolidations, or, how stable is the engram? *Annual Review of Psychology, 55*, 51–86. https://doi.org/10.1146/annurev.psych.55.090902.142050

Fioriti, L., Myers, C., Huang, Y.-Y., Li, X., Stephan, Joseph S., Trifilieff, P., . . . Kandel, E. R. (2015). The persistence of hippocampal-based memory requires protein synthesis mediated by the prion-like protein CPEB3. *Neuron, 86*(6), 1433–1448. https://doi.org/10.1016/j.neuron.2015.05.021

Frey, U., & Morris, R. G. (1997). Synaptic tagging and long-term potentiation. *Nature, 385*(6616), 533–536. https://doi.org/10.1038/385533a0

Gebauer, F., & Richter, J. D. (1996). Mouse cytoplasmic polyadenylylation element binding protein: an evolutionarily conserved protein that interacts with the cytoplasmic polyadenylylation elements of c-mos mRNA. *Proceedings of the National Academy of Sciences of the United States of America, 93*, 14602–14607.

Groisman, I., Jung, M. Y., Sarkissian, M., Cao, Q., & Richter, J. D. (2002). Translational control of the embryonic cell cycle. *Cell, 109*, 473–483.

Groner, B., Hynes, N., & Phillips, S. (1974). Length heterogeneity in the poly (adenylic acid) region of yeast messenger ribonucleic acid. *Biochemistry, 13*(26), 5378–5383.

Hafer, N., Xu, S., Bhat, K. M., & Schedl, P. (2011). The Drosophila CPEB protein Orb2 has a novel expression pattern and is important for asymmetric cell division and nervous system function. *Genetics, 189*(3), 907–921. https://doi.org/10.1534/genetics.110.123646

Hagele, S., Kuhn, U., Boning, M., & Katschinski, D. M. (2009). Cytoplasmic polyadenylation-element-binding protein (CPEB) 1 and 2 bind to the HIF-1alpha mRNA 3′-UTR and modulate HIF-1alpha protein expression. *Biochemistry Journal, 417*(1), 235–246. https://doi.org/10.1042/BJ20081353

Hake, L. E., Mendez, R., & Richter, J. D. (1998). Specificity of RNA binding by CPEB: requirement for RNA recognition motifs and a novel zinc finger. *Molecular and Cellular Biology, 18*, 685–693.

Harrison, P. M., Chan, H. S., Prusiner, S. B., & Cohen, F. E. (2001). Conformational propagation with prion-like characteristics in a simple model of protein folding. *Protein Science, 10*(4), 819–835. https://doi.org/10.1110/ps.38701

Hervas, R., Li, L., Majumdar, A., Fernandez-Ramirez, M. del C., Unruh, J. R., Slaughter, B. D., . . . Carrion-Vazquez, M. (2016). Molecular Basis of Orb2 Amyloidogenesis and

Blockade of Memory Consolidation. *PLoS Biology, 14*(1), e1002361. https://doi.org/10.1371/journal.pbio.1002361

Hervas, R., Rau, M. J., Park, Y., Zhang, W., Murzin, A. G., Fitzpatrick, J. A. J., . . . Si, K. (2020). Cryo-EM structure of a neuronal functional amyloid implicated in memory persistence in Drosophila. *Science, 367*(6483), 1230–1234. https://doi.org/10.1126/science.aba3526

Hoogenraad, C. C., Popa, I., Futai, K., Martinez-Sanchez, E., Wulf, P. S., van Vlijmen, T., . . . van der Sluijs, P. (2010). Neuron specific Rab4 effector GRASP-1 coordinates membrane specialization and maturation of recycling endosomes. *PLoS Biology, 8*(1), e1000283. https://doi.org/10.1371/journal.pbio.1000283

Hou, F., Sun, L., Zheng, H., Skaug, B., Jiang, Q. X., & Chen, Z. J. (2011). MAVS forms functional prion-like aggregates to activate and propagate antiviral innate immune response. *Cell, 146*(3), 448–461. https://doi.org/10.1016/j.cell.2011.06.041

Huang, Y. S., Carson, J. H., Barbarese, E., & Richter, J. D. (2003). Facilitation of dendritic mRNA transport by CPEB. *Genes & Development, 17*(5), 638–653.

Huang, Y. S., Jung, M. Y., Sarkissian, M., & Richter, J. D. (2002). N-methyl-D-aspartate receptor signaling results in Aurora kinase-catalyzed CPEB phosphorylation and alpha CaMKII mRNA polyadenylation at synapses. *EMBO Journal, 21*(9), 2139–2148. https://doi.org/10.1093/emboj/21.9.2139

Huang, Y. S., Kan, M. C., Lin, C. L., & Richter, J. D. (2006). CPEB3 and CPEB4 in neurons: analysis of RNA-binding specificity and translational control of AMPA receptor GluR2 mRNA. *EMBO Journal, 25*(20), 4865–4876. https://doi.org/10.1038/sj.emboj.7601322

Huber, K. M., Kayser, M. S., & Bear, M. F. (2000). Role for rapid dendritic protein synthesis in hippocampal mGluR-dependent long-term depression. *Science, 288*(5469), 1254–1257.

Hughes, M. P., Sawaya, M. R., Boyer, D. R., Goldschmidt, L., Rodriguez, J. A., Cascio, D., . . Eisenberg, D. S. (2018). Atomic structures of low-complexity protein segments reveal kinked beta sheets that assemble networks. *Science, 359*(6376), 698–701. https://doi.org/10.1126/science.aan6398

Ivshina, M., Lasko, P., & Richter, J. D. (2014). Cytoplasmic polyadenylation element binding proteins in development, health, and disease. *Annual Review of Cell and Developmental Biology, 30*, 393–415. https://doi.org/10.1146/annurev-cellbio-101011-155831

Kandel, E. R. (2001). The molecular biology of memory storage: a dialogue between genes and synapses. *Science, 294*(5544), 1030–1038. https://doi.org/10.1126/science.1067020

Kandel, E. R., Dudai, Y., & Mayford, M. R. (2014). The molecular and systems biology of memory. *Cell, 157*(1), 163–186. https://doi.org/10.1016/j.cell.2014.03.001

Kang, H., & Schuman, E. M. (1996). A requirement for local protein synthesis in neurotrophin-induced hippocampal synaptic plasticity. *Science, 273*, 1402–1406.

Keleman, K., Krüttner, S., Alenius, M., & Dickson, B. J. (2007). Function of the Drosophila CPEB protein Orb2 in long-term courtship memory. *Nature Neuroscience, 10*(12), 1587–1593. https://doi.org/10.1038/nn1996

Khan, M. R., Li, L., Perez-Sanchez, C., Saraf, A., Florens, L., Slaughter, B. D., . . . Si, K. (2015). Amyloidogenic oligomerization transforms Drosophila Orb2 from a translation repressor to an activator. *Cell, 163*(6), 1468–1483. https://doi.org/10.1016/j.cell.2015.11.020

Khoutorsky, A., Yanagiya, A., Gkogkas, C. G., Fabian, M. R., Prager-Khoutorsky, M., Cao, R., . . . Sonenberg, N. (2013). Control of synaptic plasticity and memory via suppression of poly(A)-binding protein. *Neuron, 78*(2), 298–311. https://doi.org/10.1016/j.neuron.2013.02.025

Kim, J. H., & Richter, J. D. (2006). Opposing polymerase-deadenylase activities regulate cytoplasmic polyadenylation. *Molecular Cell, 24*(2), 173–183.

Kim, J. H., & Richter, J. D. (2007). RINGO/cdk1 and CPEB mediate poly(A) tail stabilization and translational regulation by ePAB. *Genes & Development, 21*(20), 2571–2579. https://doi.org/10.1101/gad.1593007

Kim, S., & Martin, K. C. (2015). Neuron-wide RNA transport combines with netrin-mediated local translation to spatially regulate the synaptic proteome. *Elife, 4*. https://doi.org/10.7554/eLife.04158

Krüttner, S., Stepien, B., Noordermeer, Jasprina N., Mommaas, Mieke A., Mechtler, K., Dickson, Barry J., & Keleman, K. (2012). Drosophila CPEB Orb2A mediates memory independent of its RNA-binding domain. *Neuron, 76*(2), 383–395.

Krüttner, S., Traunmuller, L., Dag, U., Jandrasits, K., Stepien, B., Iyer, N., ... Keleman, K. (2015). Synaptic Orb2A bridges memory acquisition and late memory consolidation in drosophila. *Cell Reports, 11*(12), 1953–1965. https://doi.org/10.1016/j.celrep.2015.05.037

Kurihara, Y., Tokuriki, M., Myojin, R., Hori, T., Kuroiwa, A., Matsuda, Y., ... Uesugi, S. (2003). CPEB2, a novel putative translational regulator in mouse haploid germ cells. *Biology of Reproduction, 69*(1), 261–268. https://doi.org/10.1095/biolreprod.103.015677

Kwak, J. E., Drier, E., Barbee, S. A., Ramaswami, M., Yin, J. C., & Wickens, M. (2008). GLD2 poly(A) polymerase is required for long-term memory. *Proceedings of the National Academy of Sciences of the United States of America, 105*(38), 14644–14649. https://doi.org/10.1073/pnas.0803185105

Lai, Y. T., Su, C. K., Jiang, S. T., Chang, Y. J., Lai, A. C., & Huang, Y. S. (2016). Deficiency of CPEB2-confined choline acetyltransferase expression in the dorsal motor nucleus of vagus causes hyperactivated parasympathetic signaling-associated bronchoconstriction. *Journal of Neuroscience, 36*(50), 12661–12676. https://doi.org/10.1523/JNEUROSCI.0557-16.2016

Li, L., Sanchez, C. P., Slaughter, B. D., Zhao, Y., Khan, M. R., Unruh, J. R., ... Si, K. (2016). A putative biochemical engram of long-term memory. *Current Biology, 26*(23), 3143–3156.

Lin, C. L., Evans, V., Shen, S., Xing, Y., & Richter, J. D. (2010). The nuclear experience of CPEB: implications for RNA processing and translational control. *RNA, 16*(2), 338–348. https://doi.org/10.1261/rna.1779810

Lisman, J. (1994). The CaM kinase II hypothesis for the storage of synaptic memory. *Trends in Neuroscience, 17*(10), 406–412.

Liu, J., Hu, J. Y., Wu, F., Schwartz, J. H., & Schacher, S. (2006). Two mRNA-binding proteins regulate the distribution of syntaxin mRNA in Aplysia sensory neurons. *Journal of Neuroscience, 26*(19), 5204–5214. https://doi.org/10.1523/JNEUROSCI.4917-05.2006

Lopez-Lastra, M., Rivas, A., & Barria, M. I. (2005). Protein synthesis in eukaryotes: the growing biological relevance of cap-independent translation initiation. *Biological Research, 38*(2–3), 121–146. https://doi.org/10.4067/s0716-97602005000200003

Lu, W. H., Yeh, N. H., & Huang, Y. S. (2017). CPEB2 activates GRASP1 mRNA Translation and Promotes AMPA receptor surface expression, long-term potentiation, and memory. *Cell Reports, 21*(7), 1783–1794. https://doi.org/10.1016/j.celrep.2017.10.073

Lynch, G., & Baudry, M. (1984). The biochemistry of memory: a new and specific hypothesis. *Science, 224*(4653), 1057–1063. Retrieved from http://www.ncbi.nlm.nih.gov/pubmed/6144182

Mahmoud, G. S., & Grover, L. M. (2006). Growth hormone enhances excitatory synaptic transmission in area CA1 of rat hippocampus. *Journal of Neurophysiology, 95*(5), 2962–2974. https://doi.org/10.1152/jn.00947.2005

Majumdar, A., Cesario, W. C., White-Grindley, E., Jiang, H., Ren, F., Khan, M. R., ... Si, K. (2012). Critical role of amyloid-like oligomers of Drosophila Orb2 in the persistence of memory. *Cell, 148*(3), 515–529. https://doi.org/10.1016/j.cell.2012.01.004

Malenka, R. C., & Bear, M. F. (2004). LTP and LTD: an embarrassment of riches. *Neuron, 44*(1), 5–21. https://doi.org/10.1016/j.neuron.2004.09.012

Martin, K. C., Barad, M., & Kandel, E. R. (2000). Local protein synthesis and its role in synapse-specific plasticity. *Current Opinion in Neurobiology, 10*(5), 587–592.

Martin, K. C., Casadio, A., Zhu, H., Yaping, E., Rose, J. C., Chen, M.,... Kandel, E. R. (1997a). Synapse-specific, long-term facilitation of Aplysia sensory to motor synapses: a function for local protein synthesis in memory storage. *Cell, 91*, 927–938.

Martin, K. C., Casadio, A., Zhu, H., Yaping, E., Rose, J. C., Chen, M.,... Kandel, E. R. (1997b). Synapse-specific, long-term facilitation of Aplysia sensory to motor synapses: a function for local protein synthesis in memory storage. *Cell, 91*(7), 927–938.

Martin, S. J., Grimwood, P. D., & Morris, R. G. (2000). Synaptic plasticity and memory: an evaluation of the hypothesis. *Annual Review of Neuroscience, 23*, 649–711. https://doi.org/10.1146/annurev.neuro.23.1.649

Martin, S. J., & Morris, R. G. (2002). New life in an old idea: the synaptic plasticity and memory hypothesis revisited. *Hippocampus, 12*(5), 609–636. https://doi.org/10.1002/hipo.10107

Mendez, R., Hake, L. E., Andresson, T., Littlepage, L. E., Ruderman, J. V., & Richter, J. D. (2000). Phosphorylation of CPE binding factor by Eg2 regulates translation of c-mos mRNA. *Nature, 404*(6775), 302–307.

Mendez, R., Murthy, K. G., Ryan, K., Manley, J. L., & Richter, J. D. (2000). Phosphorylation of CPEB by Eg2 mediates the recruitment of CPSF into an active cytoplasmic polyadenylation complex. *Molecular Cell, 6*, 1253–1259.

Miniaci, M. C., Kim, J. H., Puthanveettil, S. V., Si, K., Zhu, H., Kandel, E. R., & Bailey, C. H. (2008). Sustained CPEB-dependent local protein synthesis is required to stabilize synaptic growth for persistence of long-term facilitation in Aplysia. *Neuron, 59*(6), 1024–1036. https://doi.org/10.1016/j.neuron.2008.07.036

Munroe, D., & Jacobson, A. (1990). mRNA poly(A) tail, a 3′ enhancer of translational initiation. *Molecular and Cellular Biology, 10*(7), 3441–3455.

Murthy, K. G., & Manley, J. L. (1995). The 160-kD subunit of human cleavage-polyadenylation specificity factor coordinates pre-mRNA 3′-end formation. *Genes & Development, 9*(21), 2672–2683.

Nader, K. (2015). Reconsolidation and the dynamic nature of memory. *Cold Spring Harbor Perspectives in Biology, 7*(10). https://doi.org/10.1101/cshperspect.a021782

Novoa, I., Gallego, J., Ferreira, P. G., & Mendez, R. (2010). Mitotic cell-cycle progression is regulated by CPEB1 and CPEB4-dependent translational control. *Nature Cell Biology, 12*(5), 447–456. https://doi.org/10.1038/ncb2046

Pai, T. P., Chen, C. C., Lin, H. H., Chin, A. L., Lai, J. S., Lee, P. T.,... Chiang, A. S. (2013). Drosophila ORB protein in two mushroom body output neurons is necessary for long-term memory formation. *Proceedings of the National Academy of Sciences of the United States of America, 110*(19), 7898–7903. https://doi.org/10.1073/pnas.1216336110

Pavlopoulos, E., Trifilieff, P., Chevaleyre, V., Fioriti, L., Zairis, S., Pagano, A.,... Kandel, E. R. (2011). Neuralized1 activates CPEB3: a function for nonproteolytic ubiquitin in synaptic plasticity and memory storage. *Cell, 147*(6), 1369–1383. https://doi.org/10.1016/j.cell.2011.09.056

Proudfoot, N. J., & Brownlee, G. G. (1976). 3′ non-coding region sequences in eukaryotic messenger RNA. *Nature, 263*(5574), 211–214.

Prusiner, S. B. (1998). Prions. *Proceedings of the National Academy of Sciences of the United States of America, 95*(23), 13363–13383.

Prusiner, S. B. (2013). Biology and genetics of prions causing neurodegeneration. *Annual Review of Genetics, 47*, 601–623. https://doi.org/10.1146/annurev-genet-110711-155524

Raveendra, B. L., Siemer, A. B., Puthanveettil, S. V., Hendrickson, W. A., Kandel, E. R., & McDermott, A. E. (2013). Characterization of prion-like conformational changes of the neuronal isoform of Aplysia CPEB. *Nature Structural and Molecular Biology, 20*(4), 495–501. https://doi.org/10.1038/nsmb.2503

Reymann, K. G., & Frey, J. U. (2007). The late maintenance of hippocampal LTP: requirements, phases, "synaptic tagging," "late-associativity" and implications. *Neuropharmacology, 52*(1), 24–40. https://doi.org/10.1016/j.neuropharm.2006.07.026

Richter, J. D. (2007). CPEB: a life in translation. *Trends in Biochemical Sciences, 32*(6), 279–285. https://doi.org/10.1016/j.tibs.2007.04.004

Richter, J. D., & Sonenberg, N. (2005). Regulation of cap-dependent translation by eIF4E inhibitory proteins. *Nature, 433*(7025), 477–480. https://doi.org/10.1038/nature03205

Rigo, F., & Martinson, H. G. (2009). Polyadenylation releases mRNA from RNA polymerase II in a process that is licensed by splicing. *RNA, 15*(5), 823–836. https://doi.org/10.1261/rna.1409209

Roberson, E. D., & Sweatt, J. D. (1999). A biochemical blueprint for long-term memory. *Learning & Memory, 6*(4), 381–388.

Sachs, A. B., Sarnow, P., & Hentze, M. W. (1997). Starting at the beginning, middle, and end: translation initiation in eukaryotes. *Cell, 89*(6), 831–838.

Sarkissian, M., Mendez, R., & Richter, J. D. (2004). Progesterone and insulin stimulation of CPEB-dependent polyadenylation is regulated by Aurora A and glycogen synthase kinase-3. *Genes & Development, 18*(1), 48–61.

Schroeder, K. E., Condic, M. L., Eisenberg, L. M., & Yost, H. J. (1999). Spatially regulated translation in embryos: asymmetric expression of maternal Wnt-11 along the dorsal-ventral axis in Xenopus. *Developmental Biology, 214*, 288–297.

Shorter, J., & Lindquist, S. (2005). Prions as adaptive conduits of memory and inheritance. *Nature Reviews Genetics, 6*(6), 435–450.

Si, K., Choi, Y. B., White-Grindley, E., Majumdar, A., & Kandel, E. R. (2010). Aplysia CPEB can form prion-like multimers in sensory neurons that contribute to long-term facilitation. *Cell, 140*(3), 421–435. https://doi.org/10.1016/j.cell.2010.01.008

Si, K., Giustetto, M., Etkin, A., Hsu, R., Janisiewicz, A. M., Miniaci, M. C., . . . Kandel, E. R. (2003). A neuronal isoform of CPEB regulates local protein synthesis and stabilizes synapse-specific long-term facilitation in Aplysia. *Cell, 115*(7), 893–904.

Si, K., Lindquist, S., & Kandel, E. R. (2003). A neuronal isoform of the Aplysia CPEB has prion-like properties. *Cell, 115*(7), 879–891.

Stebbins-Boaz, B., Cao, Q., de Moor, C. H., Mendez, R., & Richter, J. D. (1999). Maskin is a CPEB-associated factor that transiently interacts with elF-4E. *Molecular Cell, 4*, 1017–1027.

Stebbins-Boaz, B., Hake, L. E., & Richter, J. D. (1996). CPEB controls the cytoplasmic polyadenylation of cyclin, Cdk2 and c-mos mRNAs and is necessary for oocyte maturation in Xenopus. *EMBO Journal, 15*(10), 2582–2592.

Stephan, J. S., Fioriti, L., Lamba, N., Colnaghi, L., Karl, K., Derkatch, I. L., & Kandel, E. R. (2015). The CPEB3 Protein Is a Functional Prion that Interacts with the Actin Cytoskeleton. *Cell Rep, 11*(11), 1772–1785. https://doi.org/10.1016/j.celrep.2015.04.060

Steward, O., Farris, S., Pirbhoy, P. S., Darnell, J., & Driesche, S. J. (2014). Localization and local translation of Arc/Arg3.1 mRNA at synapses: some observations and paradoxes. *Frontiers in Molecular Neuroscience, 7*, 101. https://doi.org/10.3389/fnmol.2014.00101

Steward, O., & Levy, W. B. (1982). Preferential localization of polyribosomes under the base of dendritic spines in granule cells of the dentate gyrus. *Journal of Neuroscience, 2*, 284–291.

Steward, O., & Reeves, T. M. (1988). Protein-synthetic machinery beneath postsynaptic sites on CNS neurons: association between polyribosomes and other organelles at the synaptic site. *Journal of Neuroscience, 8*(1), 176–184.

Steward, O., & Schuman, E. M. (2003). Compartmentalized synthesis and degradation of proteins in neurons. *Neuron, 40*(2), 347–359.

Subtelny, A. O., Eichhorn, S. W., Chen, G. R., Sive, H., & Bartel, D. P. (2014). Poly(A)-tail profiling reveals an embryonic switch in translational control. *Nature, 508*(7494), 66–71. https://doi.org/10.1038/nature13007

Takeuchi, T., Duszkiewicz, A. J., & Morris, R. G. (2014). The synaptic plasticity and memory hypothesis: encoding, storage and persistence. *Philosophical Transactions of the Royal Society of London B: Biological Sciences, 369*(1633), 20130288. https://doi.org/10.1098/rstb.2013.0288

Tan, L., Chang, J. S., Costa, A., & Schedl, P. (2001). An autoregulatory feedback loop directs the localized expression of the Drosophila CPEB protein Orb in the developing oocyte. *Development, 128*, 1159–1169.

Theis, M., Si, K., & Kandel, E. R. (2003). Two previously undescribed members of the mouse CPEB family of genes and their inducible expression in the principal cell layers of the hippocampus. *Proceedings of the National Academy of Sciences of the United States of America, 100*(16), 9602–9607. https://doi.org/10.1073/pnas.1133424100

Tsai, L. Y., Chang, Y. W., Lin, P. Y., Chou, H. J., Liu, T. J., Lee, P. T., . . . Huang, Y. S. (2013). CPEB4 knockout mice exhibit normal hippocampus-related synaptic plasticity and memory. *PLoS One, 8*(12), e84978. https://doi.org/10.1371/journal.pone.0084978

Turimella, S. L., Bedner, P., Skubal, M., Vangoor, V. R., Kaczmarczyk, L., Karl, K., . . . Theis, M. (2015). Characterization of cytoplasmic polyadenylation element binding 2 protein expression and its RNA binding activity. *Hippocampus, 25*(5), 630–642. https://doi.org/10.1002/hipo.22399

Udagawa, T., Swanger, S. A., Takeuchi, K., Kim, J. H., Nalavadi, V., Shin, J., . . . Richter, J. D. (2012). Bidirectional control of mRNA translation and synaptic plasticity by the cytoplasmic polyadenylation complex. *Molecular Cell, 47*(2), 253–266. https://doi.org/10.1016/j.molcel.2012.05.016

Wells, D. G., Dong, X., Quinlan, E. M., Huang, Y. S., Bear, M. F., Richter, J. D., & Fallon, J. R. (2001). A role for the cytoplasmic polyadenylation element in NMDA receptor-regulated mRNA translation in neurons. *Journal of Neuroscience, 21*, 9541–9548.

Wickner, R. B., Edskes, H. K., Shewmaker, F., & Nakayashiki, T. (2007). Prions of fungi: inherited structures and biological roles. *Nature Reviews Microbiology, 5*(8), 611–618. https://doi.org/10.1038/nrmicro1708

Wu, H., & Fuxreiter, M. (2016). The structure and dynamics of higher-order assemblies: amyloids, signalosomes, and granules. *Cell, 165*(5), 1055–1066. https://doi.org/10.1016/j.cell.2016.05.004

Wu, L., Wells, D., Tay, J., Mendis, D., Abbott, M. A., Barnitt, A., . . . Richter, J. D. (1998). CPEB-mediated cytoplasmic polyadenylation and the regulation of experience-dependent translation of alpha-CaMKII mRNA at synapses. *Neuron, 21*(5), 1129–1139.

Yuste, R., & Bonhoeffer, T. (2001). Morphological changes in dendritic spines associated with long-term synaptic plasticity. *Annual Reviews Neuroscience, 24*, 1071–1089. https://doi.org/10.1146/annurev.neuro.24.1.1071

Zearfoss, N. R., Alarcon, J. M., Trifilieff, P., Kandel, E., & Richter, J. D. (2008). A molecular circuit composed of CPEB-1 and c-Jun controls growth hormone-mediated synaptic plasticity in the mouse hippocampus. *Journal of Neuroscience, 28*(34), 8502–8509. https://doi.org/10.1523/JNEUROSCI.1756-08.2008

Zovkic, I. B., Guzman-Karlsson, M. C., & Sweatt, J. D. (2013). Epigenetic regulation of memory formation and maintenance. *Learning & Memory, 20*(2), 61–74. https://doi.org/10.1101/lm.026575.112

CHAPTER 10

FMRP AND MICRORNAS IN NEURONAL PROTEIN SYNTHESIS

MONICA C. LANNOM AND STEPHANIE CEMAN

Introduction

The critical role for new protein synthesis in learning and memory was first discovered through the use of protein synthesis inhibitors like puromycin to disrupt memory formation in mice during a specific behavioral task close to 60 years ago (Flexner, Flexner, Stellar, De La Haba, & Roberts, 1962; Flexner, Flexner, & Stellar, 1963). Since this seminal discovery, inhibition of neuronal protein synthesis with a wide variety of chemicals has been performed in many organisms from humans, birds, invertebrates such as *Aplysia* and even goldfish with striking results in the areas of learning and memory (reviewed in Sweatt, 2016). A key insight into the role of local translation came when Oswald Steward and colleagues discovered that ribosomes were present in the synapse (Steward & Levy, 1982). Localized protein synthesis allows for a rapid response upon stimulation and is considered indispensable for learning and memory. For protein synthesis at the synapse to take place there needs to be a system for organization, delivery and accurate regulation of the many factors involved: from localization of mRNAs, recruitment of RNA-protein-containing granules, accessory RNA binding proteins, polysome formation and miRNA-mediated regulation. We'll begin our exploration of neuronal protein synthesis by discussing how this process is regulated.

miRNA Biogenesis and Its Consequences on mRNA Expression

Small ncRNAs represent an important subset of non-coding RNA transcripts. First discovered in 1993, they are loosely defined by their small size (typically less than 30 nucleotides), association with members of the Argonaute (AGO) family of proteins and their ability to regulate gene expression (Lee et al., 1993). This review will focus on miRNAs as a type of small non-coding RNA transcript that is now estimated to regulate over 50% of all human genes (Friedman, Burge, & Bartel, 2008).

Regulation of miRNAs can occur at the level of transcription as well as by the rate of cleavage or downstream processing (Gulyaeva & Kushlinskiy, 2016). Once fully processed, miRNAs are typically between 22 and 26 nucleotides. They form canonically when the primary transcript of a miRNA gene (pri-miRNA), a larger RNA precursor, is cleaved in the nucleus by the microprocessing complex DROSHA/DGCR8 into an approximately 70 nucleotide local hairpin structure called a precursor-miRNA (pre-miRNA) (Denli, Tops, Plasterk, Ketting, & Hannon, 2004; Gregory et al., 2004; Han, 2004; Lee, 2002, 2003; see Figure 10.1). After this cleavage, pre-miRNAs are exported into the cytoplasm and further processed by the RNAse III DICER to generate the final length miRNA, which is bound by AGO2 (Chendrimada et al., 2005; Lee, 2002; Lee et al., 2006).

miRNAs regulate gene expression through binding of the seed sequence to the complementary target sequence in the mRNA called the miRNA Recognition Element (MRE). MREs are primarily in the 3′ untranslated region (UTR) but have also been reported in 5′UTRs and in coding regions (reviewed in Brodersen & Voinnet, 2009). In *Drosophila* and many plants, miRNA base pairing to target mRNAs is perfectly complementary whereas in other eukaryotes, the presence of a shorter seed sequence still allows for mRNA regulation upon miRNA binding (Bartel, 2009; Meister & Tuschl, 2004). Furthermore, although miRNAs can bind to any portion of the mRNA sequence that possesses a degree of complementarity, mRNAs with longer 3′ UTRs have been shown to contain a greater number of miRNA binding sites, and therefore have a higher potential for post-transcriptional regulation (Brummer & Hauser, 2014; Fang & Rawjeski, 2011; Kebaara & Atkin, 2009).

The most common mode of regulation is at the level of transcript stability when complete base pairing between the miRNA and its target mRNA leads to cleavage of the transcript by AGO2. The mRNA transcript itself can be degraded at differential rates, thereby reducing the amount of template available for protein translation (Bagga et al., 2005; Hutvagner, 2002; Nottrodd, Simard, & Richter, 2006). In the case of incomplete base pairing, translational suppression, followed ultimately by transcript degradation takes place (Béthune, Artus-Revel, & Filipowicz, 2012).

miRNA Biogenesis Factors

The importance of the miRNA pathway is underscored through the characterization of the knockout mice made of the individual components. Ablation of DGCR8, part of the nuclear microprocessor complex which generates pre-miRNAs (Figure 10.1), results in embryonic lethality at E6.5 (Wang, Medvid, Melton, Jaenisch, & Blelloch, 2007), establishing the importance of miRNA function in normal development. DROSHA, the ribonuclease III that forms the other half of the microprocessing complex, also leads to embryonic lethality when ablated in mice (Fan et al., 2013). DICER deficient mice are embryonic lethal at day E8.5 (Krill, Gurdziel, Heaton, Simon, & Hammer, 2013). Functional DICER is also crucial for maintaining neuronal integrity, as absence of DICER in brain leads to neurodegeneration, most notably cell shrinkage (Hebert et al., 2010).

Although AGO2 is one of four main Argonaute proteins in cells, when individual Argonautes are knocked out in mice, only loss of AGO2 causes embryonic lethality (Liu, 2004). The other AGO proteins are dispensable for animal development once again highlighting the importance of proper miRNA regulation of target mRNAs in embryonic tissue and organ development (Liu, 2004; Morita et al., 2007). AGO2 is also the only family member with mRNA slicing activity (Liu, 2004; Meister et al., 2005) making it the main effector of miRNA-mediated regulation.

The embryonic lethal phenotypes due to loss of miRNA biogenesis and effector components are not limited to this core group of proteins. Genetic studies have demonstrated that alterations in the levels of RISC accessory proteins and in the miRNAs themselves can lead to a wide variety of aberrant phenotypes both during development

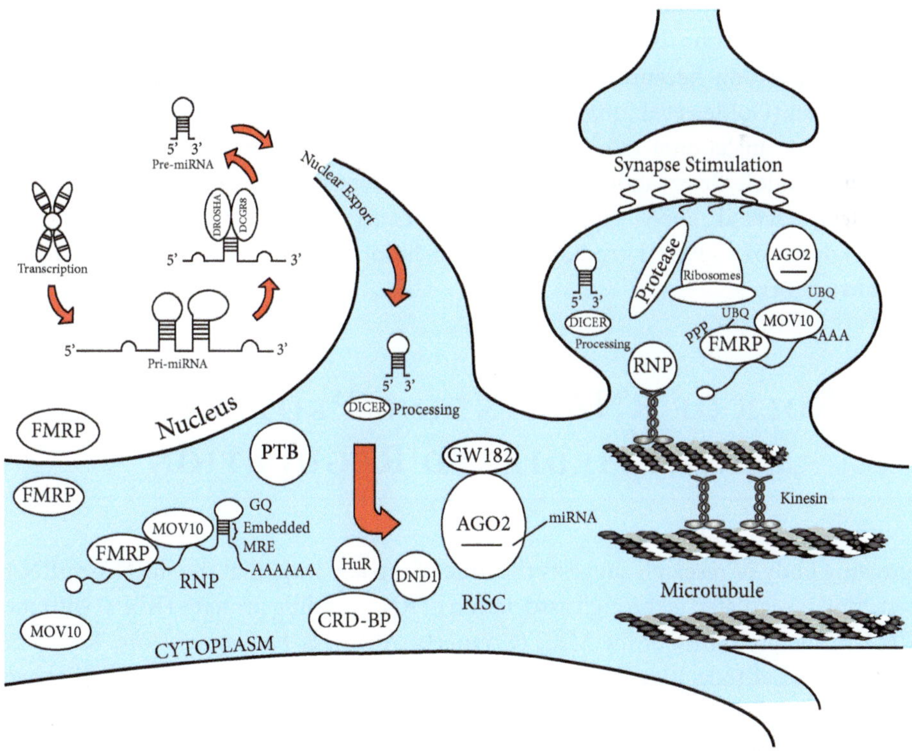

FIGURE 10.1. miRNA biogenesis and neuronal protein translation regulation of RNP-associated mRNAs. See text for explanations. Abbreviations: P, phosphorylation; UBQ, ubiquitination.

and post-embryonically (Bicker and Schratt, 2008; Edbauer et al., 2010; Schratt et al., 2006; Skariah et al., 2017). One example of this is AGO2-associated protein MOV10. Although more than one type of helicase can be found in the brain, MOV10 is particularly highly expressed embryonically and during early postnatal stages in mice, decreasing around P14, with a further drop in adulthood (Skariah et al., 2017). MOV10 ablation results in embryonic lethality before E9.5 (Skariah et al., 2017), and in *Xenopus* embryos it has been shown to be essential for completion of gastrulation (Skariah, Perry, Drnevich, Henry, & Ceman, 2018).

Regulation of AGO2 Function

Although miRNAs select their targets based on base pairing of their 2–7 nucleotide seed regions to the MRE in the target mRNA, the question remains of how a particular miRNA, which is vastly outnumbered by mRNAs in a given mammalian cell, is still able to functionally repress its targets? Recent CRISPR/Cas9 screenings have given insight by revealing novel regulators of miRNA-mediated silencing. Phosphorylation of AGO2 by CSNK1A1 triggers positive mRNA target engagement, which is then relieved upon dephosphorylation by the ANKRD52-PPP6C phosphatase. This cycle of phosphorylation/dephosphorylation is crucial for maintaining the global efficiency of miRNA-mediated repression because dephosphorylation of AGO2 allows it to engage new mRNA targets (Golden et al., 2017).

Other posttranslational modifications of AGO2 that influence miRNA-mediated silencing, include hydroxylation at residue 700 by the type I collagen prolyl-4-hydroxylase [C-P4H(I)] (Qi et al., 2008). Hydroxylation appears to enhance the stability of AGO2 as knockdown of C-P4H(I) reduces the translational suppression activity of RISC (Jee & Lai, 2014).

Role of RNA Binding Proteins in miRNA-mediated Regulation

A growing body of research suggests that one of the key ways that the mRNA-miRNA interaction is regulated is through association of RNA binding proteins (RBPs) with the mRNA in the proximity of the MRE (Connerty, Ahadi, & Hutvagner, 2015; Kenny & Ceman, 2016; Loffreda, Rigamonti, Barabino, & Lenzken, 2015). Hundreds of RBPs have been discovered to play a key role in post-transcriptional regulation (Collins & Penny, 2009; Glisovic, Bachorik, Yong, & Dreyfuss, 2008; Keene, 2007; Mata, Marguerat, & Bähler, 2005) due to their unique ability to interact with RNA while forming protein-protein interactions with key players in the cell such as AGO2. RBPs are generally conserved between yeast and humans (Beckmann et al., 2015; Gerstberger, Hafner, Ascano, &

Tuschl, 2014) yet can be functionally diverse and while many do not possess canonical RNA binding sequences, they still exert a great deal of control over their targets. We will focus now on RBPs that facilitate miRNA-mediated regulation.

FMRP

One RNA binding protein that has emerged as a key interactor with the miRNA-mediated silencing pathway is the fragile X mental retardation protein (FMRP; Ashley, Wilkinson, Reines, & Warren, 1993). FMRP is the protein product of the *FMR1* gene, whose loss leads to fragile X syndrome (FXS), a triplet-repeat expansion disease which is the leading cause of inherited cognitive impairment affecting 1:4000 males and 1:8000 females (Rajaratnam et al., 2017). Loss of FMRP leads to defects in synaptic plasticity and cognition (Bolduc, Bell, Cox, Broadie, & Tully, 2008; Kelleher & Bear, 2008; Liu-Yesucevitz et al., 2011).

FMRP is primarily a cytoplasmic protein although it does possess a nuclear localization signal (NLS; Devys, Lutz, Rouyer, Bellocq, & Mandel, 1993; Feng et al., 1997; Kim, Bellini, & Ceman, 2009; Vanderklish & Edelman, 2005; Willemsen, Oostra, Bassell, & Dictenberg, 2004). In mammals, FMRP possesses conserved functional domains, three of which are RNA binding domains (Siomi, Siomi, Nussbaum, & Dreyfuss, 1993). Two of the RNA-binding motifs are homology to hnRNP K (KH) domains and the third is an arginine-glycine-glycine (RGG) box, which is thought to bind G-rich secondary structures, such as G-quadraplexes (GQs; Darnell et al., 2001; Phan et al., 2011; Schaeffer, 2001). FMRP primarily binds within the coding sequence of its target mRNAs (Darnell et al., 2011) although it has also been shown to associate with the 3′UTR (Ascano et al., 2012; Ashley et al., 1993; Kenny et al., 2014).

Over 800 mRNA targets of FMRP have been identified, primarily in brain (Brown et al., 2001; Darnell et al., 2011; Miyashiro et al., 2003). FMRP functions in translational regulation and silencing of its mRNA targets, although recent research has shown in a small subset of mRNAs, along with other RNA binding proteins, FMRP promotes the opposite effect, leading to enhanced expression (Kenny et al., 2014).

FMRP was first implicated in miRNA-mediated regulation of transcript expression in two independent studies using its *Drosophila* homolog *dFMR1* (Caudy, 2002; Ishizuka, 2002). In the first study, dFMRP was identified as co-purifying with AGO2. Knockdown of *dFMR1* resulted in impaired RNA interference of a reporter, suggesting a functional association of dFMRP with RISC activity (Caudy, 2002). In the second study, dFMRP was also identified in an AGO2-containing messenger ribonucleoprotein particle (mRNP) that also contained ribosomal proteins L5 and L11. In contrast to the first study, though, dFMRP was not required for silencing of a reporter. One possible explanation for this difference is that FMRP may only be effective in silencing an mRNA when it also binds that mRNA, which may have occurred in the first study but not in the second study. These results were extended in mammalian cells when FMRP was shown to associate with endogenous miRNAs, DICER activity and AGO2 by co-immunoprecipitation

(Jin et al., 2004). In 2010, miRNA-mediated regulation by FMRP was explored in brain when FMRP was shown to co-immunoprecipitate with a number of miRNAs important in neuronal function (Edbauer et al., 2010). Interestingly, this study made a case for AGO1 being the primary effector of the regulatory activity of miR-125b on the NR2A mRNA, which was also bound by FMRP. The participation of AGO2 was not examined in that work. The following year, FMRP was shown again to co-immunoprecipitate with AGO2 (Lee et al., 2010) in a study of regulation of the Amyloid Precursor Protein mRNA *APP*. Both FMRP and AGO1 and AGO2 were shown to bind *APP* and both were required for FMRP-mediated translation suppression of *APP*. These investigators also proposed that an FMRP-AGO complex competed with hnRNP C for *APP* binding and subsequent translational fate regulation where the former suppressed it and the latter activated it. Cross-linking immunoprecipitation (CLIP)-seq analysis of brain FMRP in 2011 showed that FMRP bound primarily in coding sequence (Darnell et al., 2011). However, another study the following year in HEK293 cells described the FMRP CLIP sites as being distributed between coding sequence and 3′UTR (Ascano, et al., 2012). FMRP was identified as interacting with the MID domain of AGO2 (Li, Tang, Zhang, & Zhang, 2014), which is interesting because that is the same domain that interacts with another RNA binding protein, fused in sarcoma (FUS). The authors speculate that by recruiting different RBPs through its MID domain, AGO2 could potentially enhance miRNA silencing of specific targets (Zhang et al., 2018).

MOV10

The RNA helicase MOV10 was first implicated in the miRNA pathway in a mass spectrometry screen to identify proteins associated with AGO2. Importantly, knockdown of MOV10 eliminated miRNA-mediated suppression of a reporter construct, suggesting that this helicase played an important role in the miRNA pathway (Meister et al., 2005).

MOV10 possesses 5′–3′ helicase activity (Gregersen et al., 2014). The majority of MOV10 is localized to the cytoplasm and AGO2-containing cytoplasmic foci (Goodier, Cheung, & Kazazian, 2012; Messaoudi-Aubert et al., 2010), although MOV10 has been described in the nucleus in some cultured cell lines (Messaoudi-Aubert et al., 2010) and in early mouse hippocampal neurons, where it has been implicated in protection against retroviral elements such as LINE-1 (Skariah et al., 2017).

MOV10 is highly expressed in the murine brain, primarily in embryonic and early postnatal stages (Skariah et al., 2017). MOV10 is also required for the completion of gastrulation and for neural tube formation (Skariah et al., 2018). In the 3′UTR where MOV10 primarily binds its mRNA targets, it is found in G-C rich regions that include stable secondary structures such as GQs (Kenny et al., 2014). MOV10 has been implicated in the miRNA pathway as its activity-stimulated degradation leads to upregulation of bound targets (Banerjee, Neveu, & Kosik, 2009). MOV10 also colocalizes with FMRP and AGO2 *in vivo* in cultured neurons (Liu-Yesucevitz et al., 2011; Wulczyn et al., 2007). FMRP and MOV10 associate in brain and in cell lines in a partially

RNA-dependent manner, while possessing a protein-protein interaction, as well, based on experiments with purified recombinant MOV10 and FMRP (Kenny et al., 2014).

In addition to its role in facilitating AGO2-mediated silencing, MOV10 is also able to block miRNA-mediated translational suppression in some mRNAs by directly interacting with FMRP on the same site in the mRNA. This is a significant observation because it suggests that the protein complex that assembles at an MRE can control AGO2 association (Kenny et al., 2014; Kenny & Ceman, 2016). In the absence of FMRP, MOV10 had reduced association with the subset of mRNAs that are shared between the two proteins (Kenny et al., 2014), suggesting that FMRP binds the mRNAs first and then recruits MOV10 to the mRNA. However, whether MOV10 serves as an agonist or antagonist in the miRNA pathway (Kenny et al., 2014) is directly related to its binding with FMRP in the 3′UTR. If FMRP and MOV10 bind the same 3′UTR at different sites, those mRNA targets are suppressed by AGO2. We hypothesize that this miRNA-mediated translational suppression occurs when MOV10 recruited by FMRP proceeds to remodel the 3′UTR landscape by unwinding secondary structures, potentially revealing embedded or previously inaccessible MREs for AGO2 recognition and binding. However, if FMRP and MOV10 bind the same site in the 3′ UTR, the direct proximity of these proteins leads to a protein-protein interaction that results in the MRE being protected from AGO2 association—likely by blocking AGO2 recognition of the MRE (Kenny et al. 2014; Kenny & Ceman, 2016; see Figure 10.2).

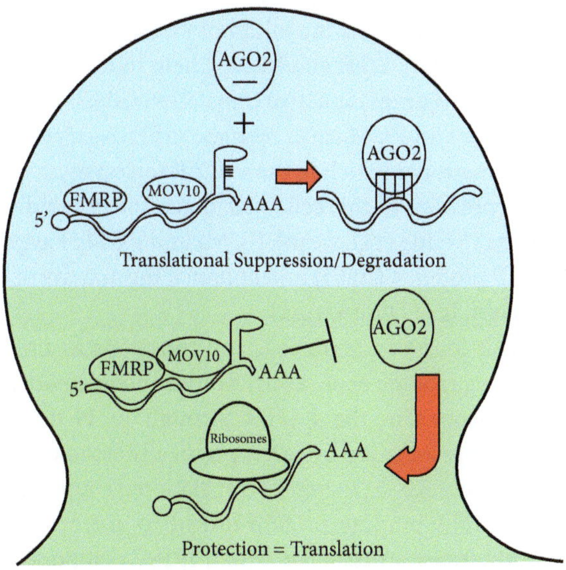

FIGURE 10.2. Translational fate of an mRNA depends on its association with RNA binding proteins FMRP and MOV10 and where they are bound relative to an MRE within the mRNA. Top. FMRP and MOV10 bind in discrete, non-overlapping regions on the mRNA, which results in AGO2 access to the MRE, likely facilitated by MOV10-mediated unwinding. Bottom. FMRP and MOV10 bind the same site on the mRNA, which blocks AGO2 association.

In brain, a 50% reduction in MOV10 such as in heterozygous mice has been shown to have a dramatic impact on proper neuronal morphology, with MOV10-deficient hippocampal neurons exhibiting reduced dendritic arborization (Skariah et al., 2017).

Other important RNA binding proteins that participate in miRNA-mediated translational regulation include but are not limited to polypyrimidine tract-binding protein (PTB), human antigen R (HuR), Pumilio (PUM1), dead end 1 (DND1), FUS, and coding region determinant-binding protein (CRD-BP).

PTB was originally described as a repressor of nervous system-specific splicing (Black, 2003; Sharma, Falick, & Black, 2005; Spellman & Smith, 2006; Wagner & Garcia-Blanco, 2001). Suppression of PTB by miR-124 led to neuronal differentiation (Boutz et al., 2007; Makeyev, Zhang, Carrasco, & Maniatis, 2007; Xue et al., 2013; Zheng et al., 2012). To understand how PTB mediates its effect, the PTB binding sites in regulated RNAs were examined to reveal that PTB binding could block miRNA association in the 3′UTR. However, in other target mRNAs, PTB binding facilitated miRNA association. By carefully examining the secondary structure of PTB-bound mRNAs, the authors found that modulation of RNA secondary structure by PTB may enhance or shield miRNA target sites in adjacent regions, thus affecting RNA stability in both directions. This was confirmed by AGO2-CLIP seq in the presence or absence of PTB (Xue et al., 2013).

HUR, a member of the ELAVL protein family possesses three RNA-binding domains (Ma, Cheng, Campbell, Wright, & Furneaux, 1996) and has been implicated in the regulation of stability and translation of over one hundred mRNAs in mammalian cells (Abdelmohsen et al., 2007; Brennan & Steitz, 2001; Cherry et al., 2006; Meisner et al., 2004). Binding of HUR may suppress the inhibitory effect of miRNAs interacting with regulatory regions such as the 3′UTR and instead help to lift repression by facilitating mRNA association with polysomes for active translation (Meisner & Filipowicz, 2010).

Pumilio proteins PUM1 and PUM2 also regulate miRNA-dependent gene silencing. PUM2 induces a conformational switch in the 3′ UTR region of *p27* mRNA, which leads to miRNA-mediated repression of this cell-cycle regulator in rapidly proliferating cells through miR-121 and 122 (Kedde et al., 2010). PUM1 and PUM2 have also been shown to bind the 3′ UTR of E2F protein 3 (E2F3) and enhance the activity of miRNAs targeting E2F3 (Miles, Tschop, Herr, Ji, & Dyson, 2012).

Dead end 1 (DND1) is found in germ cells and functions as a negative regulator of miRNA-induced silencing (Kedde et al., 2007; Kouwenhove, Kedde, & Agami, 2011). It binds to uridine-rich regions in the 3′UTR through its N-terminal RNA-binding domain and blocks AGO2 association by either physically associating with an mRNA to block AGO2 or by displacing AGO2 to relieve translational suppression.

FUS is a dual DNA/RNA binding protein linked to amyotrophic lateral sclerosis (ALS). FUS was recently shown to associate with miRISC components AGO2, miRNAs and their target transcripts (Zhang et al., 2018). FUS and AGO2 bind approximately 85% of the same mRNA targets and both are localized to approximately the same locations in the 3′ UTR (Zhang et al., 2018).

Coding region determinant-binding protein (CRD-BP) is an RRM and KH-domain-containing RNA-binding protein primarily expressed in fetal tissues and primary tumors (Ioannidis et al., 2003). CRD-BP binds to the 5′UTR of the *c-Myc* mRNA, and protects it from repression (Noubissi et al., 2006). It also binds in the coding region of *βTrCP1* mRNA and protects it from miR-183-induced degradation (Elcheva, Goswami, Noubissi, & Spiegelman, 2009; Noubissi et al., 2006). Regulation of *c-Myc* and *βTrCP1* mRNA by CDB-BP is induced in response to β-catenin signaling (Noubissi et al., 2006). Dysregulation in CDB-BP levels leads to changes in c-Myc expression, which is widely considered a factor for the development of 75% of primary tumors (Ioannidis et al., 2003).

In summary, RBPs are important for and can facilitate AGO2-mediated regulation. RBPs that block AGO2 function may do so through one of three possible ways: (1) RBP binding to the MRE directly competes with miRNA binding to that same site; (2) RBPs may bind in proximity to the MRE in such a way that alters the mRNA structure to either facilitate or decrease access to the MRE; (3) the RBP may directly interact with AGO2 to either recruit AGO2 to the MRE or to prevent AGO2 association with the MRE.

Role of miRNAs in Nervous System Development

The first miRNA found to directly regulate dendritic spine morphology was miR-134 (Schratt et al., 2006). Overexpression of miR-134 in rat hippocampal neurons reduces the size of dendritic spines by mediating the translational suppression of Lim-domain containing protein kinase 1 (LIMK1) at the synapse (Bicker & Schratt, 2008; Schratt et al., 2006). Translational suppression of LIMK1 could be relieved by addition of brain derived neurotrophic factor (BDNF). miR-134 also regulates synaptic plasticity when overexpressed in mouse hippocampus, impairing long-term potentiation (LTP) and long-term memory formation in a contextual-fear conditioning paradigm (Gao et al., 2010).

Another miRNA of high importance in neurons is miR-132, which is a positive regulator of dendritic outgrowth that has been shown both *in vitro* and *in vivo* to translationally suppress Rho GTPase activating protein p250GAP (Vo et al., 2005). miR-132 also decreases spine density upon overexpression (Cheng et al., 2007; Hotulainen & Hoogenraad, 2010; Vo et al., 2005; Wayman et al., 2008). A different miRNA, miR-125b yields a similar phenotype consisting of long and thin spines upon overexpression. Importantly, both miR-132 and miR-125b interact with FMRP—probably in a complex that includes AGO2 (Lee et al., 2010), as there is no evidence to date that FMRP directly binds miRNAs. Upon short-hairpin RNA (shRNA)–mediated knockdown of FMRP, the spine phenotype is reversed (Edbauer et al., 2010).

FMRP, through its association with molecular motors as part of an RNP, mediates axonal delivery of miR-181d (Wang et al., 2015). Specifically, this study demonstrates that FMRP is required to transport the *Map1b* mRNA into the axon from the cell body. Loss of FMRP expression results in decreased protein levels of MAP1B in the axons, but not in the cell bodies, thus supporting FMRP's role in mRNA localization and subsequent translation regulation. During the long distance axonal delivery of RNPs, FMRP is proposed to have dual roles wherein it anchors *Map1b* and *Calm1* to cytoskeletal motors and functions in miR-181d-mediated RISC assembly to repress gene translation. Upon stimulation with NGF (Nerve Growth Factor), *Map1b* and *Calm1* are released from FMRP granules and made available for local translation. The authors hypothesize that the NGF stimulation leads to reorganization of the granule to mediate this effect. There is precedent for translation repression by miRNAs without immediate degradation (Béthune et al., 2012). This would be particularly important in neurons where localized translation occurs in response to stimulation, requiring that target mRNAs are repressed translationally without major mRNA decay (Bhattacharyya, Habermacher, Martine, Closs, & Filipowicz, 2006; Muddashetty et al., 2011; Schratt et al., 2006).

Control of mRNA Localization for Proper Neuronal Protein Synthesis

Some mRNAs are specifically enriched at the synapse, dendrites or in the distal axon. A number of studies have yielded important insights into how mRNAs are targeted to the synapse for local translation regulation (Cohen, Lee, Chen, Li, & Fields, 2011; Corbin, Olsson-Carter, & Slack, 2009; Fu, Shah, & Baraban, 2016; Kiebler & Bassell, 2006; Liegro, Schiera, & Liegro, 2014; Mahmoudi & Cairns, 2016; Schratt, 2011; Wang, Martin, & Zukin, 2010; Ye, Xu, Su, & He, 2016). While the majority of mRNA translation in neurons takes place in the cell body, a subset of mRNAs upon being exported from the nucleus are coated with RNA binding proteins to form RNPs which are packaged into RNP granules. Neuronal granules typically consist of one mRNA and a large composition of RNA binding proteins (Batish, Bogaard, Kramer, & Tyagi, 2012). mRNAs in neurons are packaged into granules to transport them to sites of local protein translation (Krichevsky & Kosik, 2001). Granules are trafficked along microtubule and actin filament tracks via motor proteins to distinct localizations both in axons and along dendrites (Hirokawa, Niwa, & Tanaka, 2010). During travel, the neuronal granules contain translationally silent mRNAs. To achieve translational suppression during transport, associated RNA binding proteins contact the mRNA, essentially preventing it from associating with pro-translation factors that are located within the same RNP. In addition, RNA granules are not translationally competent because they do not include eIF4E, 4G and tRNAs (Krichevsky & Kosik, 2001).

Once an RNP has reached its target destination, translation initiation factors such as eIF4E and eIF2A promote assembly of polysomes on the mRNA (Krichevsky & Kosik, 2001; Smart, Edelman, & Vanderklish, 2003). RNA-binding proteins such as STAUFEN and FMRP have been shown to stall polysomes on a given mRNA (Darnell et al., 2011; Thomas, Tosar, Desbats, Leishman, & Boccaccio, 2009) and the RISC complex itself can also associate with polysomes and lead to ribosomal pausing (Maroney, Yu,

Fisher, & Nilsen, 2006; Nottrodd et al., 2006; Petersen, Bordeleau, Pelletier, & Sharp, 2006). The first study describing miRNAs and their regulatory role showed that they were on polysomes (Lee et al., 1993). Barman and Bhattacharyya showed that mRNAs are recruited to the ER membrane where they are translated and then AGO2/ miRNA is recruited for repression and degradation (Barman & Bhattacharyya, 2015). Although there is evidence that miRNAs block elongation (Petersen et al., 2006), including in the first description of small RNA-mediated translation regulation (Lee et al., 1993), in the majority of cases, miRNAs seem to block translation at initiation through AGO2 association with GW182, which binds eukaryotic initiation factors (Gu & Kay, 2010).

Cis-sequences within the mRNA itself may direct its transport such as with the β-*actin* mRNA which expresses a "zipcode" in its 3′ UTR that is recognized by the KH domains of the RNA binding protein zipcode binding protein 1 (ZBP1), allowing it to be translocated to specific regions within the cell (Driesche & Martin, 2018; Jambhekar & Derisi, 2007; Kindler, Wang, Richter, & Tiedge, 2005). When β-*actin* transcripts reach the periphery of the cell, Src-dependent phosphorylation of ZBP1 releases the mRNA and allows the synthesis of β-actin protein (Huttelmaier et al., 2005).

FMRP is also crucial for the proper localization of mRNA cargo to various regions along the axon and dendrites of neurons by binding to *cis*-factors within its target mRNAs. Intramolecular G-quadruplexes (GQs) have been implicated in mRNA targeting (Subramanian et al., 2011). FMRP binds to RNA *in vitro* with high affinity to kissing complex RNA secondary structure and GQs through its KH2 and RGG-binding domains, respectively (Darnell et al., 2005; Menon, Mader, & Mihailescu, 2008). GQs in mRNAs seem to be involved with the assembly of the FMRP/MOV10 complex that blocks AGO2 association (Kenny et al., 2014) and importantly all of these components such as FMRP and MOV10 are in found in neuronal granules (Fritzsche et al., 2013; Kiebler & Bassell, 2006). FMRP also has a protein-protein interaction with kinesin motors within neurons, and it serves as a linker between the motor proteins along the dendrites and the RNP carrying the mRNA cargo itself, along with other RNA regulatory factors (Davidovic et al., 2007; Dictenberg, Antar, Singer, & Bassell, 2008; Ling, Fahrner, Greenough, & Gelfand, 2004).

Activity-Dependent Translation Regulation in the Nervous System

Local regulation of miRNAs at the synapse is believed to depend on whether the synapse is undergoing chemical or electrical stimulation. DICER generates the final functional miRNA from pre-miRNAs and can be dendritically and synaptically localized (Bicker et al., 2013; Mikl, Vendra, Doyle, & Kiebler, 2010), suggesting a mechanism in place for rapid control of miRNA synthesis upon signal stimulation. An example is when miRNA biogenesis factor DICER is upregulated in dendrites upon synaptic stimulation leading to an increase in miRNA production and the eventual local regulation of CamKII (Sambandan et al., 2017).

In the unstimulated state, synaptically localized miRNA is associated with AGO2. AGO2-mediated regulation has been particularly well studied in the synaptically localized translation of PSD-95 (Miyoshi, Okada, Siomi, & Siomi, 2009; Muddashetty et al., 2011; Okamura, 2004; Pratt & Macrae, 2009) and involves the rapid phosphorylation and dephosphorylation of FMRP. Dephosphorylation of FMRP leads to the release of AGO2 upon group 1 metabotropic glutamate receptor (mGluR) stimulation, thus, leading to the release of suppression of the cobound PSD-95 mRNA. There is also evidence that stimulation of the NDMA receptor leads to ubiquitination of MOV10, which leads to degradation, releasing its bound mRNAs for translation (Banerjee et al., 2009; Fu et al., 2016).

Interestingly, the *FMR1* mRNA is itself dendritically localized and its translation regulated by group 1 mGluR signaling (Bassell & Warren, 2008). Stimulation of the group 1 mGluRs leads to a transient increase in FMRP followed by its rapid degradation by the ubiquitin proteasome pathway and release of mRNAs for translation (Figure 10.2; Hou et al., 2006).

FMRP blocks translation through ribosome stalling, which takes place during the elongation phase (Ceman et al., 2003; Darnell et al., 2011). FMRP blocks mRNA translation by binding to the translation machinery itself (Chen, Sharma, Shi, Agrawal, & Joseph, 2014). Upon receipt of a signal, ribosome stalling can be relieved—perhaps by dephosphorylation of FMRP (Ceman et al., 2003; Narayanan et al., 2007), and local protein translation can take place. Moreover, it is hypothesized that the presence of stalled ribosomes on a given mRNA during the transport process may actually decrease the likelihood of the mRNA being degraded as it is essentially being protected by the presence of the ribosomes.

Conclusion

Local protein translation is required for learning and memory by altering the synapse after stimulation. RNA binding proteins like FMRP dynamically control protein translation by facilitating mRNA localization, by being locally translated and degraded by stimulation and finally by modulating mRNA stability and translation through the miRNA pathway.

Future Directions

1. What is the identity of individual proteins that form a complex on an MRE?
2. What is the identity of *cis* sequences and/or the secondary structures in the mRNA that recruit RBPs in proximity to MREs?
3. How is assembly of the RBP complex on the MRE regulated?

4. Which post-translational modifications regulate assembly and disassembly of this complex?
5. How are the enzymes that mediate the post-translational modifications regulated?
6. How is the RNP assembled in the nucleus and cell body that is ultimately targeted to the dendritic spines?
7. How are activated spines "tagged" or recognized by transport granules?
8. How do activated synapses communicate with the cell body and nucleus?

Acknowledgments

Grants NIH/NIMH MH093661 (to SC) and Kiwanis Neuroscience Research Foundation (to SC)

References

Abdelmohsen, K., Pullman, R., Lal, A., Kim, H. H., Galban, S., Yang, X., Blethrow, J. D.,... Gorospe, M. (2007). Phosphorylation of HuR by Chk2 regulates SIRT1 expression. *Molecular Cell*, *23*(25), 543–557.

Ascano, M., Mukherjee, N., Bandaru, P., Miller, J. B., Nusbaum, J. D., Corcoran, D. L., & Tuschl, T. (2012). FMRP targets distinct mRNA sequence elements to regulate protein expression. *Nature*, *492*(7429), 382–386.

Ashley, C., Wilkinson, K., Reines, D., & Warren, S. (1993). FMR1 protein: Conserved RNP family domains and selective RNA binding. *Science*, *262*(5133), 563–566.

Bagga, S., Bracht, J., Hunter, S., Massirer, K., Holtz, J., Eachus, R., & Pasquinelli, A. E. (2005). Regulation by let-7 and lin-4 miRNAs results in target mRNA degradation. *Cell*, *122*(4), 553–563.

Banerjee, S., Neveu, P., & Kosik, K. S. (2009). A coordinated local translational control point at the synapse involving relief from silencing and MOV10 degradation. *Neuron*, *64*(6), 871–884.

Barbee, S. A., Estes, P. S., Cziko, A., Hillebrand, J., Luedeman, R. A., Coller, J. M., & Ramaswami, M. (2006). Staufen- and FMRP-containing neuronal RNPs are structurally and functionally related to somatic p bodies. *Neuron*, *52*(6), 997–1009.

Barman, B., & Bhattacharyya, S. N. (2015). MRNA targeting to endoplasmic reticulum precedes ago protein interaction and microRNA (miRNA)-mediated translation repression in mammalian cells. *Journal of Biological Chemistry*, *290*(41), 24650–24656.

Bartel, D. P. (2009). MicroRNAs: Target recognition and regulatory functions. *Cell*, *136*(2), 215–233.

Bassell, G. J., & Warren, S. T. (2008). Fragile X syndrome: Loss of local mRNA regulation alters synaptic development and function. *Neuron*, *60*(2), 201–214.

Batish, M., Bogaard, P. V., Kramer, F. R., & Tyagi, S. (2012). Neuronal mRNAs travel singly into dendrites. *Proceedings of the National Academy of Sciences*, *109*(12), 4645–4650.

Beckmann, B. M., Horos, R., Fischer, B., Castello, A., Eichelbaum, K., Alleaume, A. M., ... Hentze, M. W. (2015). The RNA-binding proteomes from yeast to man harbor conserved enigmRBPs. *Nature Communications*, *3*(6), 10127.

Béthune, J., Artus-Revel, C. G., & Filipowicz, W. (2012). Kinetic analysis reveals successive steps leading to miRNA-mediated silencing in mammalian cells. *EMBO Reports*, *13*(8), 716–723.

Bhattacharyya, S. N., Habermacher, R., Martine, U., Closs, E. I., & Filipowicz, W. (2006). Relief of microRNA-mediated translational repression in human cells subjected to stress. *Cell, 125*(6), 1111–1124.

Bicker, S., Khudayberdiev, S., Weiss, K., Zocher, K., Baumeister, S., & Schratt, G. (2013). The DEAH-box helicase DHX36 mediates dendritic localization of the neuronal precursor-microRNA-134. *Genes & Development, 27*(9), 991–996.

Bicker, S., & Schratt, G. (2008). MicroRNAs: Tiny regulators of synapse function in development and disease. *Journal of Cellular and Molecular Medicine, 12*(5a), 1466–1476.

Black, D. L. (2003). Mechanisms of Alternative Pre-Messenger RNA Splicing. *Annual Review of Biochemistry, 72*, 291–336.

Bolduc, F. V., Bell, K., Cox, H., Broadie, K. S., & Tully, T. (2008). Excess protein synthesis in Drosophila fragile X mutants impairs long-term memory. *Nature Neuroscience, 11*(10), 1143–1145.

Boutz, P. L., Stoilov, P., Li, Q., Lin, C., Chawla, G., Ostrow, K., & Black, D. L. (2007). A post-transcriptional regulatory switch in polypyrimidine tract-binding proteins reprograms alternative splicing in developing neurons. *Genes & Development, 21*(13), 1636–1652.

Bramham, C. R., Alme, M. N., Bittins, M., Kuipers, S. D., Nair, R. R., Pai, B., Panja, D.,... Wibrand, K. (2010). The Arc of synaptic memory. *Experimental Brain Research, 200*(2), 125–140.

Bramham, C. R., & Wells, D. G. (2010). Dendritic mRNA: Transport, translation and function. *Nature Reviews Neuroscience, 8*(10), 776–789.

Brennan, C. M., & Steitz, J. A. (2001). HuR and mRNA stability. *Cellular and Molecular Life Sciences, 58*(2), 266–277.

Brodersen, P., & Voinnet, O. (2009). Revisiting the principles of microRNA target recognition and mode of action. *Nature Reviews Molecular Cell Biology, 10*(2), 141–148.

Brown, V., Jin, P., Ceman, S., Darnell, J. C., Odonnell, W. T., Tenenbaum, S. A., & Warren, S. T. (2001). Microarray identification of FMRP-associated brain mRNAs and altered mRNA translational profiles in fragile X syndrome. *Cell, 107*(4), 477–487.

Brummer A., & Hausser, J. (2014). MicroRNA binding sites in the coding region of mRNAs: Extending the repertoire of post-transcriptional gene regulation. *Bioessays, 36*(6), 617–626.

Caudy, A. A. (2002). Fragile X-related protein and VIG associate with the RNA interference machinery. *Genes & Development, 16*(19), 2491–2496.

Ceman, S., Odonnell, W. T., Reed, M., Patton, S., Pohl, J., & Warren, S. T. (2003). Phosphorylation influences the translation state of FMRP-associated polyribosomes. *Human Molecular Genetics, 12*(24), 3295–3305.

Chen, E., Sharma, M., Shi, X., Agrawal, R., & Joseph, S. (2014). Fragile X mental retardation protein regulates translation by binding directly to the ribosome. *Molecular Cell, 54*(3), 407–417.

Chendrimada, T. P., Gregory, R. I., Kumaraswamy, E., Norman, J., Cooch, N., Nishikura, K., & Shiekhattar, R. (2005). TRBP recruits the Dicer complex to Ago2 for microRNA processing and gene silencing. *Nature, 436*(7051), 740–744.

Cheng, H. M., Papp, J. W., Varlamova, O., Dziema, H., Russell, B., Curfman, J. P., & Obrietan, K. (2007). MicroRNA modulation of circadian-clock period and entrainment. *Neuron, 54*(5), 813–829.

Cherry, S., Kunte, A., Wang, H., Coyne, C., Rawson, R. B., & Perrimon, N. (2006). COPI activity coupled with fatty acid biosynthesis is required for viral replication. *PLoS Pathogens, 2*(10), e102.

Cohen, J. E., Lee, P. R., Chen, S., Li, W., & Fields, R. D. (2011). MicroRNA regulation of homeostatic synaptic plasticity. *Proceedings of the National Academy of Sciences, 108*(28), 11650–11655.

Collins, L. J., & Penny, D. (2009). The RNA infrastructure: Dark matter of the eukaryotic cell? *Trends in Genetics, 25*(3), 120–128.

Connerty, P., Ahadi, A., & Hutvagner, G. (2015). RNA binding proteins in the miRNA pathway. *International Journal of Molecular Sciences, 17*(1), pii: E31.

Corbin, R., Olsson-Carter, K., & Slack, F. (2009). The role of microRNAs in synaptic development and function. *BMB Reports, 42*(3), 131–135.

Coyne, A. N., Yamada, S. B., Siddegowda, B. B., Estes, P. S., Zaepfel, B. L., Johannesmeyer, J. S., & Zarnescu, D. C. (2015). Fragile X protein mitigates TDP-43 toxicity by remodeling RNA granules and restoring translation. *Human Molecular Genetics, 24*(24), 6886–6898.

Darnell, J. C., Fraser, E. C., Mostovetsky, O., Stefani, G., Jones, T. A., Eddy, S. R., & Darnell, R. B. (2005). Kissing complex RNAs mediate interaction between the Fragile-X mental retardation protein KH2 domain and brain polyribosomes. *Genes & Development, 19*(8), 903–918.

Darnell, J. C., Jensen, K. B., Jin, P., Brown, V., Warren, S. T., & Darnell, R. B. (2001). Fragile X mental retardation protein targets g quartet mRNAs important for neuronal function. *Cell, 107*(4), 489–499.

Darnell, J., Van Driesche, S., Zhang, C., Hung, K., Mele, A., Fraser, C., & Darnell, R. (2011). FMRP stalls ribosomal translocation on mRNAs linked to synaptic function and autism. *Cell, 146*(2), 247–261.

Davidovic, L., Jaglin, X. H., Lepagnol-Bestel, A., Tremblay, S., Simonneau, M., Bardoni, B., & Khandjian, E. W. (2007). The fragile X mental retardation protein is a molecular adaptor between the neurospecific KIF3C kinesin and dendritic RNA granules. *Human Molecular Genetics, 16*(24), 3047–3058.

Denli, A. M., Tops, B. B., Plasterk, R. H., Ketting, R. F., & Hannon, G. J. (2004). Processing of primary microRNAs by the Microprocessor complex. *Nature, 432*(7014), 231–235.

Devys, D., Lutz, Y., Rouyer, N., Bellocq, J., & Mandel, J. (1993). The FMR-1 protein is cytoplasmic, most abundant in neurons and appears normal in carriers of a fragile X premutation. *Nature Genetics, 4*(4), 335–340.

Dictenberg, J., Antar, L., Singer, R., & Bassell, G. (2008). A direct role for FMRP in activity-dependent dendritic mRNA transport links filopodial-spine morphogenesis to fragile X syndrome. *International Journal of Developmental Neuroscience, 26*(8), 831.

Driesche, S. J., & Martin, K. C. (2018). New frontiers in RNA transport and local translation in neurons. *Developmental Neurobiology, 78*(3), 331–339.

Edbauer, D., Neilson, J. R., Foster, K. A., Wang, C., Seeburg, D. P., Batterton, M. N., & Sheng, M. (2010). Regulation of synaptic structure and function by FMRP-Associated MicroRNAs miR-125b and miR-132. *Neuron, 65*(3), 373–384.

Elcheva, I., Goswami, S., Noubissi, F. K., & Spiegelman, V. S. (2009). CRD-BP protects the coding region of βTrCP1 mRNA from miR-183-mediated degradation. *Molecular Cell, 35*(2), 240–246.

El-Shami, M., Pontier, D., Lahmy, S., Braun, L., Picart, C., Vega, D., & Lagrange, T. (2007). Reiterated WG/GW motifs form functionally and evolutionarily conserved ARGONAUTE-binding platforms in RNAi-related components. *Genes & Development, 21*(20), 2539–2544.

Fan, P., Chen, Z., Tian, P., Liu, W., Jiao, Y., Xue, Y., & Yue, J. (2013). MiRNA biogenesis enzyme drosha is required for vascular smooth muscle cell survival. *PLoS One, 8*(4), e60888.

Fang, Z., & Rajewsky, N. (2011). The impact of miRNA target sites in coding sequences and in 3′UTRs. *PLoS One, 6*(3), e18067.

Feng, Y., Gutekunst, C. A., Eberhart, D. E., Yi, H., Warren, S. T., & Hersch, S. M. (1997). Fragile X mental retardation protein: Nucleocytoplasmic shuttling and association with somato-dendritic ribosomes. *Journal of Neuroscience, 17*(5),1539–1547.

Filipowicz, W., Bhattacharyya, S. N., & Sonenberg, N. (2008). Mechanisms of post-transcriptional regulation by microRNAs: Are the answers in sight? *Nature Reviews Genetics, 9*(2), 102–114.

Flexner, J. B., Flexner, L. B., Stellar, E., De La Haba, G., & Roberts, R. B. 1962. Inhibition of protein synthesis in brain and learning and memory following puromycin. *Journal of Neurochemistry, 9*, 595–605.

Flexner, L. B., Flexner, J. B., & Stellar, E. 1963. Memory in mice as affected by intracerebral puromycin. *Science, 141*(3575), 57–59.

Friedman, R. C., Farh, K. K., Burge, C. B., & Bartel, D. P. (2008). Most mammalian mRNAs are conserved targets of microRNAs. *Genome Research, 19*(1), 92–105.

Fritzsche, R., Karra, D., Bennett, K., Ang, F., Heraud-Farlow, J., Tolino, M., & Kiebler, M. (2013). Interactome of two diverse RNA granules links mRNA localization to translational repression in neurons. *Cell Reports, 5*(6), 1749–1762.

Fu, X., Shah, A., & Baraban, J. M. (2016). Rapid reversal of translational silencing: Emerging role of microRNA degradation pathways in neuronal plasticity. *Neurobiology of Learning and Memory, 133*, 225–232.

Fusco, D., Accornero, N., Lavoie, B., Shenoy, S. M., Blanchard, J., Singer, R. H., & Bertrand, E. (2003). Single mRNA molecules demonstrate probabilistic movement in living mammalian cells. *Current Biology, 13*(2), 161–167.

Gao, J., Wang, W., Mao, Y., Gräff, J., Guan, J., Pan, L., & Tsai, L. (2010). A novel pathway regulates memory and plasticity via SIRT1 and miR-134. *Nature, 466*(7310), 1105–1109.

Gerstberger, S., Hafner, M., Ascano, M., & Tuschl, T. (2014). Evolutionary conservation and expression of human RNA-binding proteins and their role in human genetic disease. *Systems Biology of RNA Binding Proteins, 825*, 1–55.

Gibbings, D. J., Ciaudo, C., Erhardt, M., & Voinnet, O. (2009). Multivesicular bodies associate with components of miRNA effector complexes and modulate miRNA activity. *Nature Cell Biology, 11*(9), 1143–1149.

Glisovic, T., Bachorik, J. L., Yong, J., & Dreyfuss, G. (2008). RNA-binding proteins and post-transcriptional gene regulation. *FEBS Letters, 582*(14), 1977–1986.

Golden, R. J., Chen, B., Li, T., Braun, J., Manjunath, H., Chen, X., & Mendell, J. T. (2017). An Argonaute phosphorylation cycle promotes microRNA-mediated silencing. *Nature, 542*(7640), 197–202.

Goodier, J. L., Cheung, L. E., & Kazazian, H. H. (2012). MOV10 RNA Helicase is a potent inhibitor of retrotransposition in cells. *PLoS Genetics, 8*(10), e100294.

Gregersen, L., Schueler, M., Munschauer, M., Mastrobuoni, G., Chen, W., Kempa, S., & Landthaler, M. (2014). MOV10 Is a 5′ to 3′ RNA helicase contributing to UPF1 mRNA target degradation by translocation along 3′ UTRs. *Molecular Cell, 54*(4), 573–585.

Gregory, R. I., Yan, K., Amuthan, G., Chendrimada, T., Doratotaj, B., Cooch, N., & Shiekhattar, R. (2004). The microprocessor complex mediates the genesis of microRNAs. *Nature, 432*(7014), 235–240.

Gu, S., & Kay, M. A. (2010). How do miRNAs mediate translational repression? *Silence, 1*(1), 11.

Gulyaeva, L. F., & Kushlinskiy, N. E. (2016). Regulatory mechanisms of microRNA expression. *Journal of Translational Medicine, 14*(1), 143.

Han, J. (2004). The Drosha-DGCR8 complex in primary microRNA processing. *Genes & Development, 18*(24), 3016–3027.

Hebert, S. S., Papadopoulou, A. S., Smith, P., Galas, M. C., Planel, E., Silahtaroglu, A. N., ... De Strooper, B. (2010). Genetic ablation of Dicer in adult forebrain neurons results in abnormal tau hyperphosphorylation and neurodegeneration. *Human Molecular Genetics, 19*(20), 3959–3969.

Hirokawa, N., Niwa, S., & Tanaka, Y. (2010). Molecular motors in neurons: Transport mechanisms and roles in brain function, development, and disease. *Neuron, 68*(4), 610–638.

Hotulainen, P., & Hoogenraad, C. C. (2010). Actin in dendritic spines: Connecting dynamics to function. *The Journal of Cell Biology, 189*(4), 619–629.

Hou, L., Antion, M. D., Hu, D., Spencer, C. M., Paylor, R., & Klann, E. (2006). Dynamic translational and proteasomal regulation of fragile x mental retardation protein controls mGluR-dependent long-term depression. *Neuron, 51*(4), 441–454.

Huttelmaier, S., Zenklusen, D., Lederer, M., Dictenberg, J., Lorenz, M., Meng, X., ... Singer, R. H. (2005). Spatial regulation of beta-actin translation by Src-dependent phosphorylation of ZBP1. *Nature, 438*(7067), 512–515.

Hutvagner, G. (2002). A microRNA in a multiple-turnover RNAi enzyme complex. *Science, 297*(5589), 2056–2060.

Ioannidis, P., Mahaira, L., Papadopoulou, A., Teixeira, M. R., Heim, S., Andersen, J. A. Evangelou, E., ... Trangas, T. (2003). CDB-BP: A c-Myc mRNA stabilizing protein with an oncofetal pattern of expression. *Anticancer Research, 23*(3A), 2179–2183.

Ishizuka, A. (2002). A Drosophila fragile X protein interacts with components of RNAi and ribosomal proteins. *Genes & Development, 16*(19), 2497–2508.

Jambhekar, A., & Derisi, J. L. (2007). Cis-acting determinnants of asymmetric, cytoplasmic RNA transport. *RNA, 13*(5), 625–642.

Jee, D., & Lai, E. C. (2014). Alteration of miRNA activity via context-specific modifications of Argonaute proteins. *Trends in Cell Biology, 24*(9), 546–553.

Jin, P., Zarnescu, D. C., Ceman, S., Nakamoto, M., Mowrey, J., Jongens, T. A., & Warren, S. T. (2004). Biochemical and genetic interaction between the fragile X mental retardation protein and the microRNA pathway. *Nature Neuroscience, 7*(2), 113–117.

Kebaara, B. W., & Atkin, A. L. (2009). Long 3'-UTRs target wild-type mRNAs for nonsense-mediated mRNA decay in *Saccharomyces cerevisiae. Nucleic Acids Research, 37*(9), 2771–2778.

Kedde, M., Kouwenhove, M. V., Zwart, W., Vrielink, J. A., Elkon, R., & Agami, R. (2010). A Pumilio-induced RNA structure switch in p27-3′ UTR controls miR-221 and miR-222 accessibility. *Nature Cell Biology, 12*(10), 1014–1020.

Kedde, M., Strasser, M. J., Boldajipour, B., Vrielink, J. A., Slanchev, K., Sage, C. L., & Agami, R. (2007). RNA-binding protein Dnd1 inhibits MicroRNA access to target mRNA. *Cell, 131*(7), 1273–1286.

Keene, J. D. (2007). RNA regulons: Coordination of post-transcriptional events. *Nature Reviews Genetics, 8*(7), 533–543.

Kelleher, R. J., & Bear, M. F. (2008). The autistic neuron: Troubled translation? *Cell, 135*(3), 401–406.

Kenny, P., & Ceman, S. (2016). RNA secondary structure modulates FMRP's bi-functional role in the MicroRNA pathway. *International Journal of Molecular Sciences, 17*(12), 985.

Kenny, P., Zhou, H., Kim, M., Skariah, G., Khetani, R., Drnevich, J., & Ceman, S. (2014). MOV10 and FMRP regulate AGO2 association with microrna recognition elements. *Cell Reports*, *9*(5), 1729–1741.

Kiebler, M. A., & Bassell, G. J. (2006). Neuronal RNA granules: Movers and makers. *Neuron*, *51*(6), 685–690.

Kim, M, Bellini, M., & Ceman S. (2009). Fragile X mental retardation protein FMRP binds mRNAs in the nucleus. *Molecular and Cellular Biology*, *29*(1), 214–228.

Kindler, S., Wang, H., Richter, D., & Tiedge, H. (2005). RNA transport and local control of translation. *Annual Review of Cell and Developmental Biology*, *21*, 223–245.

Korb, E., & Finkbeiner, S. (2011). Arc in synaptic plasticity: From gene to behavior. *Trends in Neurosciences*, *34*(11), 591–598.

Kouwenhove, M. V., Kedde, M., & Agami, R. (2011). MicroRNA regulation by RNA-binding proteins and its implications for cancer. *Nature Reviews Cancer*, *11*(9), 644–656.

Krichevsky, A. M., & Kosik, K. M. (2001). Neuronal RNA granules: A link between RNA localization and stimulation-dependent translation. *Neuron* *32*(4), 693–696.

Krill, K. T., Gurdziel, K., Heaton, J. H., Simon, D. P., & Hammer, G. D. (2013). Dicer deficiency reveals microRNAs predicted to control gene expression in the developing adrenal cortex. *Molecular Endocrinology*, *27*(5), 754–768.

Lee, E. K., Kim, H. J., Kuwano, Y., Abdelmohsen, K., Srikantan, S., Subaran, S. S., . . . Gorospe, M. (2010). hnRNP C promotes APP translation by competing with FMRP for APP mRNA recruitment to P bodies. *Nature Structural & Molecular Biology*, *17*(6): 732–739.

Lee, R. C., Feinbaum, R. L., & Ambros, V. (1993). The C. elegans heterochronic gene lin-4 encodes small RNAs with antisense complementarity to lin-14. *Cell*, *75*(5), 843–854.

Lee, Y. (2002). MicroRNA maturation: Stepwise processing and subcellular localization. *The EMBO Journal*, *21*(17), 4663–4670.

Lee, Y., Ahn, C., Han, J., Choi, H., Kim, J., Yim, J., & Kim, V. N. (2003). The nuclear RNase III Drosha initiates microRNA processing. *Nature*, *425*(6956), 415–419.

Lee, Y., Hur, I., Park, S., Kim, Y., Suh, M. R., & Kim, V. N. (2006). The role of PACT in the RNA silencing pathway. *The EMBO Journal*, *25*(3), 522–532.

Li, Y., Tang, W., Zhang, L.-R., & Zhang, C.-Y. (2014). FMRP regulates miR196a-mediated repression of HOXB8 via interaction with the AGO2 MID domain. *Molecular BioSystems* *10*(7), 1757–1764.

Liegro, C. M., Schiera, G., & Liegro, I. D. (2014). Regulation of mRNA transport, localization and translation in the nervous system of mammals (Review). *International Journal of Molecular Medicine*, *33*(4), 747–762.

Ling, S., Fahrner, P. S., Greenough, W. T., & Gelfand, V. I. (2004). Transport of Drosophila fragile X mental retardation protein-containing ribonucleoprotein granules by kinesin-1 and cytoplasmic dynein. *Proceedings of the National Academy of Sciences*, *101*(50), 17428–17433.

Liu, J. (2004). Argonaute2 is the catalytic engine of mammalian RNAi. *Science*, *305*(5689), 1437–1441.

Liu-Yesucevitz, L., Bassell, G. J., Gitler, A. D., Hart, A. C., Klann, E., Richter, J. D., & Wolozin, B. (2011). Local RNA translation at the synapse and in disease. *Journal of Neuroscience*, *31*(45), 16086–16093.

Loffreda, A., Rigamonti, A., Barabino, S., & Lenzken, S. (2015). RNA-binding proteins in the regulation of miRNA activity: A focus on neuronal functions. *Biomolecules*, *5*(4), 2363–2387.

Ma, W., Cheng, S., Campbell, C., Wright, A., & Furneaux, H. (1996). Cloning and characterization of HuR, a ubiquitously expressed Elav-like protein. *Journal of Biological Chemistry, 271*(14), 8144–8151.

Mahmoudi, E., & Cairns, M. J. (2016). MiR-137: An important player in neural development and neoplastic transformation. *Molecular Psychiatry, 22*(1), 44–55.

Makeyev, E. V., Zhang, J., Carrasco, M. A., & Maniatis, T. (2007). The MicroRNA miR-124 promotes neuronal differentiation by triggering brain-specific alternative pre-mRNA splicing. *Molecular Cell, 27*(3), 435–448.

Maroney, P. A., Yu, Y., Fisher, J., & Nilsen, T. W. (2006). Evidence that microRNAs are associated with translating messenger RNAs in human cells. *Nature Structural & Molecular Biology, 13*(12), 1102–1107.

Mata, J., Marguerat, S., & Bähler, J. (2005). Post-transcriptional control of gene expression: A genome-wide perspective. *Trends in Biochemical Sciences, 30*(9), 506–514.

Meisner, N., & Filipowicz, W. (2010). Properties of the regulatory RNA-binding protein HuR and its role in controlling miRNA repression. *Advances in Experimental Medicine and Biology Regulation of MicroRNAs, 700*, 106–123. (Abstract)

Meisner, N. C., Hackermuller, J., Uhl, V., Aszodi, A., Jaritz, M., & Auer, M. (2004). mRNA openers and closers: Modulating AU-rich element-controlled mRNA stability by a molecular switch in mRNA secondary structure. *Chembiochem, 5*(10), 1432–1447.

Meister, G., Landthaler, M., Peters, L., Chen, P. Y., Urlaub, H., Luhrmann, R., & Tuschl, T. (2005). Identification of novel argonautre-associated proteins. *Current Biology, 15*(23), 2149–2155.

Meister, G., & Tuschl, T. (2004). Mechanisms of gene silencing by double-stranded RNA. *Nature, 431*(7006), 343–349.

Menon, L., Mader, S. A., & Mihailescu, M. (2008). Fragile X mental retardation protein interactions with the microtubule associated protein 1B RNA. *RNA, 14*(8), 1644–1655.

Messaoudi-Aubert, S. E., Nicholls, J., Maertens, G. N., Brookes, S., Bernstein, E., & Peters, G. (2010). Role for the MOV10 RNA helicase in Polycomb-mediated repression of the INK4a tumor suppressor. *Nature Structural & Molecular Biology, 17*(7), 862–868.

Mikl, M., Vendra, G., Doyle, M., & Kiebler, M. A. (2010). RNA localization in neurite morphogenesis and synaptic regulation: Current evidence and novel approaches. *Journal of Comparative Physiology A, 196*(5), 321–334.

Miles, W. O., Tschop, K., Herr, A., Ji, J., & Dyson, N. J. (2012). Pumilio facilitates miRNA regulation of the E2F3 oncogene. *Genes & Development, 26*(4), 356–368.

Miyashiro, K. Y., Beckel-Mitchener, A., Purk, T., Becker, K. G., Barret, T., Liu, L., & Eberwine, J. (2003). RNA cargoes associating with FMRP Reveal deficits in cellular functioning in Fmr1 null mice. *Neuron, 37*(3), 417–431.

Miyoshi, K., Okada, T. N., Siomi, H., & Siomi, M. C. (2009). Characterization of the miRNA-RISC loading complex and miRNA-RISC formed in the Drosophila miRNA pathway. *RNA, 15*(7), 1282–1291.

Morita, S., Horii, T., Kimura, M., Goto, Y., Ochiya, T., & Hatada, I. (2007). One Argonaute family member, Eif2c2 (Ago2), is essential for development and appears not to be involved in DNA methylation. *Genomics, 89*(6), 687–696.

Muddashetty, R., Nalavadi, V., Gross, C., Yao, X., Xing, L., Laur, O.,... Bassell, G. (2011). Reversible inhibition of PSD-95 mRNA translation by miR-125a, FMRP phosphorylation, and mGluR signaling. *Molecular Cell, 42*(5), 673–688.

Nalavadi, V. C., Muddashetty, R. S., Gross, C., & Bassell, G. J. (2012). Dephosphorylation-induced ubiquitination and degradation of FMRP in dendrites: A role in immediate early mGluR-stimulated translation. *Journal of Neuroscience, 32*(8), 2582–2587.

Narayanan, U., Nalavadi, V., Nakamoto, M., Pallas, D. C., Ceman, S., Bassell, G. J., & Warren, S. T. (2007). FMRP phosphorylation reveals an immediate-early signaling pathway triggered by group I mGluR and mediated by PP2A. *Journal of Neuroscience, 27*(52), 14349–14357.

Noubissi, F. K., Elcheva, I., Bhatia, N., Shakoori, A., Ougolkov, A., Liu, J., & Spiegelman, V. S. (2006). CRD-BP mediates stabilization of βTrCP1 and c-myc mRNA in response to β-catenin signalling. *Nature, 441*(7095), 898–901.

Nottrodd, S., Simard, M. J., & Richter, J. D. (2006). Human let-7a miRNA blocks protein production on actively translating polyribosomes. *Nature Structural & Molecular Biology, 13*(12), 1108–1114.

Okamura, K. (2004). Distinct roles for Argonaute proteins in small RNA-directed RNA cleavage pathways. *Genes & Development, 18*(14), 1655–1666.

Pastuzyn, E. D., Day, C. E., Kearns, R. B., Kyrke-Smith, M., Taibi, A. V., Mccormick, J., . . . Shepherd, J. D. (2018). The neuronal gene arc encodes a repurposed retrotransposon gag protein that mediates intercellular RNA transfer. *Cell, 173*(1), 275.

Petersen, C. P., Bordeleau, M., Pelletier, J., & Sharp, P. A. (2006). Short RNAs repress translation after initiation in mammalian cells. *Molecular Cell, 21*(4), 533–542.

Phan, A. T., Kuryavyi, V., Darnell, J. C., Serganov, A., Majumdar, A., Ilin, S., & Patel, D. J. (2011). Structure-function studies of FMRP RGG peptide recognition of an RNA duplex-quadruplex junction. *Nature Structural & Molecular Biology, 18*(7), 796–804.

Pratt, A. J., & Macrae, I. J. (2009). The RNA-induced silencing complex: A versatile gene-silencing machine. *Journal of Biological Chemistry, 284*(27), 17897–17901.

Qi, H. H., Ongusaha, P. P., Myllyharju, J., Cheng, D., Pakkanen, O., Shi, Y., & Shi, Y. (2008). Prolyl 4-hydroxylation regulates Argonaute 2 stability. *Nature, 455*(7211), 421–424.

Rajaratnam, A., Shergill, J., Salcedo-Arellano, M., Saldarriaga, W., Duan, X., & Hagerman, R. (2017). Fragile X syndrome and fragile X-associated disorders. *F1000Research, 6*, 2112.

Sambandan, S., Akbalik, G., Kochen, L., Rinne, J., Kahlstatt, J., Glock, C., & Schuman, E. M. (2017). Activity-dependent spatially localized miRNA maturation in neuronal dendrites. *Science, 355*(6325), 634–637.

Schaeffer, C. (2001). The fragile X mental retardation protein binds specifically to its mRNA via a purine quartet motif. *The EMBO Journal, 20*(17), 4803–4813.

Schratt, G. (2011). MicroRNAs at the synapse. *Nature Reviews Neuroscience, 10*(12), 842–849.

Schratt, G. M., Tuebing, F., Nigh, E. A., Kane, C. G., Sabatini, M. E., Kiebler, M., & Greenberg, M. E. (2006). A brain-specific microRNA regulates dendritic spine development. *Nature, 439*(7074), 283–289.

Sharma, S., Falick, A. M., & Black, D. L. (2005). Polypyrimidine tract binding protein blocks the 5′ splice site-dependent assembly of U2AF and the prespliceosomal E complex. *Molecular Cell, 19*(4), 485–496.

Siomi, H., Siomi, M. C., Nussbaum, R. L., & Dreyfuss, G. (1993). The protein product of the fragile X gene, FMR1, has characteristics of an RNA-binding protein. *Cell, 74*(2), 291–298.

Skariah, G., Perry, K. J., Drnevich, J., Henry, J. J., & Ceman, S. (2018). RNA helicase Mov10 is essential for gastrulation and central nervous system development. *Developmental Dynamics, 247*(4), 660–671.

Skariah, G., Seimetz, J., Norsworthy, M., Lannom, M. C., Kenny, P. J., Elrakhawy, M., & Ceman, S. (2017). Mov10 suppresses retroelements and regulates neuronal development and function in the developing brain. *BMC Biology, 15*(1), 54.

Smart, F. M., Edelman, G. M., & Vanderklish, P. W. (2003). BDNF induces translocation of initiation factor 4E to mRNA granules: Evidence for a role of synaptic microfilaments and integrins. *Proceedings of the National Academy of Sciences of the United States of America, 100*(24), 14403–14408.

Spellman, R., & Smith, C. W. (2006). Novel modes of splicing repression by PTB. *Trends in Biochemical Sciences, 31*(2), 73–76.

Steward, O., & Levy, W. B. (1982). Preferential localization of polyribosomes under the base of dendritic spines in granule cells of the dentate gyrus. *Journal of Neuroscience, 2*(3), 284–291.

Subramanian, M., Rage, F., Tabet, R., Flatter, E., Mandel, J., & Moine, H. (2011). G-quadruplex RNA structure as a signal for neurite mRNA targeting. *EMBO Reports, 12*(7), 697–704.

Sweatt, J. D. (2016). Neural plasticity and behavior—sixty years of conceptual advances. *Journal of Neurochemistry, 139(2)*, 179–199.

Thomas, M. G., Tosar, L. J., Desbats, M. A., Leishman, C. C., & Boccaccio, G. L. (2009). Mammalian Staufen 1 is recruited to stress granules and impairs their assembly. *Journal of Cell Science, 122* (4), 563–573.

Vanderklish, P. W., & Edelman, G. M. (2005). Differential translation and fragile X syndrome. *Genes, Brain and Behavior, 4*(6), 360–384.

Vo, N., Klein, M. E., Varlamova, O., Keller, D. M., Yamamoto, T., Goodman, R. H., & Impey, S. (2005). From the cover: A cAMP-response element binding protein-induced microRNA regulates neuronal morphogenesis. *Proceedings of the National Academy of Sciences, 102*(45), 16426–16431.

Wagner, E. J., & Garcia-Blanco, M. A. (2001). Polypyrimidine tract binding protein antagonizes exon definition. *Molecular and Cellular Biology, 21*(10), 3281–3288.

Wang, B., Pan, L., Wei, M., Wang, Q., Liu, W., Wang, N., & Bao, L. (2015). FMRP-mediated axonal delivery of miR-181d regulates axon elongation by locally targeting Map1b and calm1. *Cell Reports, 13*(12), 2794–2807.

Wang, D. O., Martin, K. C., & Zukin, R. S. (2010). Spatially restricting gene expression by local translation at synapses. *Trends in Neurosciences, 33*(4), 173–182.

Wang, K., & Chang, H. (2011). Molecular mechanisms of long noncoding RNAs. *Molecular Cell, 43*(6), 904–914.

Wang, Y., Medvid, R., Melton, C., Jaenisch, R., & Blelloch, R. (2007). DGCR8 is essential for microRNA biogenesis and silencing of embryonic stem cell self-renewal. *Nature Genetics, 39*(3), 380–385.

Wayman, G. A., Davare, M., Ando, H., Fortin, D., Varlamova, O., Cheng, H. M., & Impey, S. (2008). An activity-regulated microRNA controls dendritic plasticity by down-regulating p250GAP. *Proceedings of the National Academy of Sciences, 105*(26), 9093–9098.

Willemsen, R., Oostra, B. A., Bassell, G. J., & Dictenberg, J. (2004). The fragile X syndrome: From molecular genetics to neurobiology. *Mental Retardation and Developmental Disabilities Research Reviews, 10*(1), 60–67.

Wulczyn, F. G., Smirnova, L., Rybak, A., Brandt, C., Kwidzinski, E., Ninnemann, O., & Nitsch, R. (2007). Post-transcriptional regulation of the let-7 microRNA during neural cell specification. *The FASEB Journal, 21*(2), 415–426.

Xu, J., Zhang, R., Shen, Y., Liu, G., Lu, X., & Wu, C. (2013). The evolution of evolvability in microRNA target sites in vertebrates. *Genome Research*, *23*(11), 1810–1816.

Xue, Y., Ouyang, K., Huang, J., Zhou, Y., Ouyang, H., Li, H., & Fu, X. (2013). Direct conversion of fibroblasts to neurons by reprogramming PTB-regulated MicroRNA circuits. *Cell*, *152*(1–2), 82–96.

Ye, Y., Xu, H., Su, X., & He, X. (2016). Role of MicroRNA in governing synaptic plasticity. *Neural Plasticity*, *2016*, 1–13.

Zekri, L., Huntzinger, E., Heimstadt, S., & Izaurralde, E. (2009). The silencing domain of GW182 Interacts with PABPC1 to promote translational repression and degradation of MicroRNA targets and is required for target release. *Molecular and Cellular Biology*, *29*(23), 6220–6231.

Zhang, T., Wu, Y.-C., Mullane, P., Ji, Y. J., Liu, H., He, L., Arora, A., . . . Wang, J. (2018). FUS regulates activity of MicroRNA-mediated gene silencing. *Molecular Cell*, *69*(5), 787–801, e788.

Zhang, Y. Q., Bailey, A. M., Matthies, H. J., Renden, R. B., Smith, M. A., Speese, S. D., & Broadie, K. (2001). Drosophila fragile x-related gene regulates the MAP1B homolog Futsch to control synaptic structure and function. *Cell*, *107*(5), 591–603.

Zheng, S., Gray, E. E., Chawla, G., Porse, B. T., Odell, T. J., & Black, D. L. (2012). PSD-95 is post-transcriptionally repressed during early neural development by PTBP1 and PTBP2. *Nature Neuroscience*, *15*(3), 381–388.

Zhou, F., Zhou, J., Dedhar, S., Wu, Y., & Snider, W. D. (2004). NGF-induced axon growth is mediated by localized inactivation of GSK-3β and functions of the microtubule plus end binding protein APC. *Neuron*, *42*(6), 897–912.

CHAPTER 11

FOCUSING ON mRNA GRANULES AND STALLED POLYSOMES AMIDST DIVERSE MECHANISMS UNDERLYING mRNA TRANSPORT, mRNA STORAGE, AND LOCAL TRANSLATION

MINA N. ANADOLU AND WAYNE S. SOSSIN

Introduction

The control of local mRNA translation in distal dendrites and axons is an important feature of neuronal signaling. Local translation plays multiple roles in neuronal function, including (i) determining the proteome of distal sites (Glock et al., 2017; Miller et al., 2002), (ii) allowing local homeostasis in the face of local hyper- or hypo-excitability (Mori et al., 2019; Sutton & Schuman, 2006), (iii) responding to guidance cues important for growth-cone dynamics during neuronal development (Cioni et al., 2018), and (iv) providing the plasticity-related proteins required for translation-dependent forms of synaptic plasticity (Costa-Mattioli et al., 2009; Sossin & Costa-Mattioli, 2019). These distinct roles for local translation are likely to be implemented by different mechanisms of translational control. Of particular interest, since multiple forms of plasticity require local translation, it is likely that distinct control mechanisms exist for each distinct type of synaptic plasticity.

Local translation is a common feature in many cellular settings and general principles of local translation have been largely elucidated (Huang & Richter, 2004; Palacios & St Johnston, 2001; Wilhelm & Vale, 1993). These include (i) a mechanism for repression of mRNA translation during transport, (ii) a mechanism for active transport of mRNAs to local sites using actin or microtubule tracks, and (iii) a mechanism for de-repression of the mRNA either when reaching the appropriate target or when appropriate signals are received. In particular, detailed mechanisms for all these features have been elucidated for particularly well-studied examples, such as Ash1 in yeast daughter cells (Paquin & Chartrand, 2008), localized translation of morphology-determining factors during Drosophila development (Becalska & Gavis, 2009; Palacios & St Johnston, 2001), and the local translation of beta-actin mRNA in neurons (Eliscovich et al., 2013; Huttelmaier et al., 2005; Wong et al., 2017).

In these examples, transport, repression, and activation are understood at the level of the individual mRNAs that are regulated. However, mRNAs are also transported to distal sites of neurons in bulk within specialized RNA granules (Kiebler & Bassell, 2006). This chapter will focus on these granules, their components, their mechanisms of repression and their mechanisms of reactivation, and how these granules are critical for both synaptic plasticity and neurodevelopment. There has been growing interest in the macro-organization of RNAs in liquid-liquid-phase separated structures in cells, such as P bodies and stress granules (Anderson & Kedersha, 2006; Protter & Parker, 2016). These structures, particularly stress granules, have been implicated in multiple neurological disorders (Wolozin & Ivanov, 2019). This chapter will also address the relationship of the RNA granules important for transport and other types of RNA structures present in neurons.

Evidence for mRNA Transport and Local Translation at Distal Sites in the Nervous System

The first evidence for localized translation in the central nervous system came from a study by Colman and colleagues in 1982, in which they studied the incorporation of myelin basic protein (MBP) into newly myelinated membranes in the developing rat brain. They observed that MBP was synthesized in ribosomes highly enriched in the processes of myelinating oligodendrocytes where the MBP was being incorporated into the myelin sheath (Colman et al., 1982). Next, the observation of clusters of ribosomes localized within dendritic spines of granule cells, identified by electron microscopy (EM) in the rat dentate gyrus, suggested that dendritic spines could be hot spots for localized translation of mRNAs (Steward and Levy, 1982). The first mRNA to be localized to dendrites was microtubule-associated protein 2 (MAP-2), while other mRNAs were localized exclusively to the soma (Garner et al., 1988). Since this discovery, thousands of mRNAs have been shown to be present in dendrites (Cajigas et al., 2012), although some mRNAs are much more abundant than others. The most quantitative analysis of mRNA enrichment using nanostring technology showed that the top 10 most abundant mRNAs, including calcium-calmodulin kinase II (CAMKII), dendrin, and beta-actin make up over 40 percent of the total mRNA in neuronal processes (Cajigas et al., 2012). Many of these mRNAs encode proteins of particular abundance in dendrites, suggesting that a major role of their transport is to keep the levels of these proteins high in distal processes (Glock et al., 2017). In addition, distinct subsets of mRNAs have also been found in axons (Ostroff et al., 2019; Shigeoka et al., 2016; Taylor et al., 2009; Willis et al., 2005).

The large number of mRNAs that can be found in neuronal processes raises the question of whether all these mRNAs have dendritic sorting signals or whether the sorting mechanism is not precise and some of the mRNAs found in dendrites represent this low fidelity sorting process. Indeed, the absence of some mRNAs in dendrites may be due to active retention mechanisms that prevent some mRNAs from being transported (Martinez et al., 2019; Mori et al., 2000; Vicario et al., 2015), as opposed to lacking a distal targeting signal. Moreover, many studies using artificial constructs, designed to produce mRNAs with no known targeting sequences, find that these mRNAs are found in dendrites at similar levels to targeted mRNAs, although cryptic signals in the vectors cannot be ruled out (Bauer et al., 2019; Langille et al., 2019; Lebeau et al., 2011). While the localization signal for some mRNAs (e.g., the beta-actin zipcode; heterogeneous nuclear ribonucleoprotein A2 (hnRNP-A2) binding site), have been elucidated (Kislauskis et al., 1997; Shan et al., 2003) there appears to be many distinct signals that allow for sorting of mRNAs to neuronal processes, often with redundant and dispersed signals present in the mRNA (Mori et al., 2000). The mechanism by which most mRNAs are selected for targeting, and what determines the percentage of a transcript that is to be transported to local sites in neurons is still largely unknown.

It is widely believed that mRNA localization is driven by recognition of specific sequences in the mRNA by RNA binding proteins (RBPs). Protein-protein interactions between these RBPs allows for combinatorial interactions and the formation of complexes that then interact with active transport motors (Jansen & Niessing, 2012), translational activators and translational repressors, creating a framework for a hierarchical organization for mRNA localization and abundance (Mayya & Duchaine, 2019). Multiple RBPs can bind to the same or similar sequences on the mRNA, often competing for binding, and this competition can determine the fate of the mRNA. For example, the embryonic lethal abnormal visual system (ELAV) family of RBPs bind to AU rich elements (often termed AREs) and stabilize mRNAs by competing with other RBPs that bind to the same ARE and lead to degradation (Simone & Keene, 2013). Similar competitions can determine transport versus retention (Gardiner et al., 2015). Thus, the competition between RBPs is important for post-transcriptional RNA control. Each type of neuron expresses a distinct cohort of RBPs suggesting that the mRNAs transported in each neuron could be distinct. Indeed, mRNAs enriched in the dendrites of Purkinje neurons are quite different from those found in hippocampal neurons (Bian et al., 1996).

Even with the multiplicity of RBPs and their interactions, there remain a large number of other factors that determine the fate of mRNAs. At the level of the transcript, post-transcriptional modifications of mRNAs, such as alternative 3′ends (Fontes et al., 2017) and RNA modifications (methylation, etc.; Zhang et al., 2018) can alter the

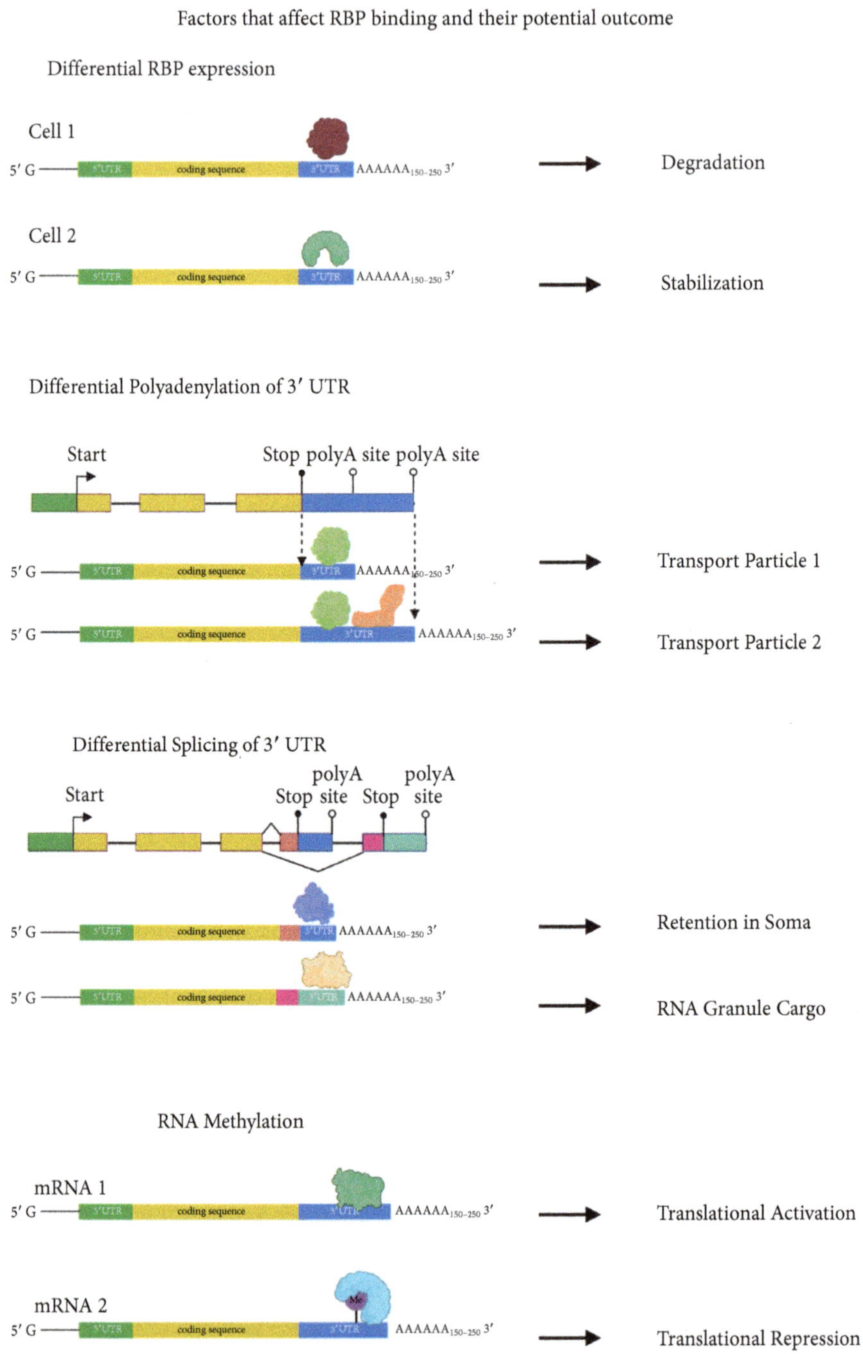

FIGURE 11.1. Factors that affect RBP binding to mRNA and their potential outcome. RBP binding to mRNAs plays an important role in the localization and fate of mRNAs. It should be noted that we do not mean to imply that the specific differences in RBPs or the mRNA on the left necessarily lead to the different regulation on the right. These are just examples; and any of the different regulations could be linked to the differences in RBP binding. (A) Differential RBP

complement of RBPs that bind to mRNAs. All these moving parts change during development and in response to recent neuronal activity. Thus, there is great room for complexity in post-transcriptional regulation of mRNA stability, localization and translational activation (Figure 11.1).

Evidence for Stimulus Dependent Translation of Localized mRNAs and Synaptic Plasticity

Following the evidence supporting the transport and translation of specific mRNAs in neuronal processes and sub-compartments, the first direct evidence linking stimulus-dependent local translation of dendritic mRNAs to synaptic plasticity came from Kang and Schuman (Kang & Schuman, 1996) who showed that increases in synaptic strength mediated by brain derived neurotrophic factor (BDNF) requires local translation. Another pioneering study in *Aplysia californica* sensory neurons showed that serotonin could stimulate local translation in neurites and that this played multiple roles in the increase in synaptic strength induced by serotonin, including activation of transcription back in the cell soma (Martin et al., 1997). This was also one of the first systems in which specific mechanisms of translational control in neurons was examined, as the increase in translation induced by serotonin was blocked by rapamycin, implicating the mechanistic target of rapamycin complex 1 (mTORC1) pathway in this process (Casadio et al., 1999; Yanow et al., 1998). Another influential finding was that long-term depression (LTD) induced by activation of metabotropic glutamate receptors (mGLUR-LTD) in rat hippocampal neurons, required local protein synthesis and could be achieved even after

---

FIGURE 11.1. Continued
expression in different cells can determine the outcome for the same mRNA, for example, the binding of one RBP in Cell 1 may lead to degradation while the binding of another RBP in Cell 2 may lead to stabilization of that mRNA. (B) Differential polyadenylation of an mRNA based on different polyA sites in the 3′ UTR can result in longer 3′ UTRs, as well as one or more RBP binding sites to be retained in the final mRNA. Based on this, one or more RBPs can bind to the transcript to determine which transport particle the mRNA will be recruited into. It is worth noting that the interaction of two RBPs bound to the same transcript may be important for the outcome. (C) Differential splicing of the 3′ UTR of transcripts may also affect the binding of different RBPs. One splice variant may lead to interaction with an RBP leading to the retention of the mRNA in the soma, while another splice variant may allow the binding of an RBP that will recruit the mRNA to RNA Granules as cargo to be shipped to distal sites. (D) Post-transcriptional modifications such as the Methylation of a transcript may facilitate the interaction of different RBPs with the Methylated versus non-methylated mRNA. The non-methylated mRNA may be translationally active, whereas the methylated mRNA may be translationally repressed by the RBP that is bound to it.

dendrites were separated from the cell soma (Huber et al., 2000). Soon afterward, examples of local translation of specific mRNAs during synaptic plasticity were found, including CAMKII (Ouyang et al., 1999; Scheetz et al., 2000), S6 (Khan et al., 2001) and activity-related cytoskeleton-associated protein (ARC; Zalfa et al., 2003). The use of fluorescent protein synthesis reporters firmly established that synaptic plasticity-inducing stimuli were associated with increases in local translation (Aakalu et al., 2001; Job & Eberwine, 2001; Wang et al., 2009).

The role of local translation in homeostatic plasticity has also been strongly established by a number of studies (Mori et al., 2019 Chapter 13 in this volume). Pioneering work by the Holt group showed the importance of local translation in the growth cone for responding to guidance cues (Cioni et al., 2018; Yoon et al., 2009). A number of recent studies have firmly established that local axonal translation is present at the basal state (Hafner et al., 2019) and is important for response to injury (Hanz et al., 2003) and presynaptic forms of plasticity (Younts et al., 2016). Thus, the requirement of local translation for many forms of synaptic plasticity in both axons and dendrites is strongly supported by recent and past findings.

Neuronal Ribosome-Containing RNA Granules

Similar to localization of mRNAs, the first evidence for the granular transport of mRNAs in the nervous system came from oligodendrocytes and studies of MBP mRNA, where MBP mRNA and translational machinery (elongation factor 1 a (EF1a) and rRNA) were found to be colocalized to large supramolecular structures in the oligodendrocyte periphery (Barbarese et al., 1995). This was soon followed by similar findings in neurons where Kosik and colleagues (Knowles et al., 1996), who first defined the term neuronal RNA granule for these granules containing both mRNA and ribosomes, demonstrated their microtubule dependent transport in dendrites. Large ribosome-containing granules could be purified by sucrose gradient sedimentation (Krichevsky & Kosik, 2001) as they were denser than both monosomes and polysomes, thus passing through the gradients to form a pellet. This highlights a potential technical issue when examining neuronal translation, since many studies use sucrose sedimentation to determine which mRNAs are on polysomes, but mostly ignore the pellet fraction, despite numerous studies showing the presence of RNA granules containing collections of polysomes in the dense pellet (Aschrafi et al., 2005; El Fatimy et al., 2016; Elvira et al., 2006; Krichevsky & Kosik, 2001). Importantly, despite the abundance of ribosomes in the pellet, this fraction was translationally repressed (Krichevsky & Kosik, 2001). However, neuronal activity was able to dissociate these granules, and lead to an increase in ribosomes now being found in the polysome fraction (Aschrafi et al., 2005;

Krichevsky & Kosik, 2001). Another indication that these RNA granule associated polysomes were repressed is that the ribosomes in these granules were aggregated and compacted compared to normal polysomes (El Fatimy et al., 2016; Elvira et al., 2006). This has been observed even when the granules were formed in an in-vitro translational assay from neuronal extracts (Darnell et al., 2011). These granules were also shown to be resistant to nuclease treatment suggesting that their compact structure can protect the contents from degradation (Darnell et al., 2011). One important unresolved issue is that the large, aggregated compacted ribosomes are not easily seen in electron microscopy (EM) of dendrites that have been used to characterize polysomes in dendrites. It may be that it is difficult to distinguish a circular translating polysome rosette from a compacted RNA granule in EM. Interestingly, polysomes found in dendrites are somewhat insensitive to puromycin (Dynes & Steward, 2012) consistent with these polysomes either representing ribosomes not involved in translation, or representing stalled polysomes (discussed later in the section Evidence for Stalled Polysomes in RNA Granules). However, the large collections of polysomes as seen in some studies (Krichevsky & Kosik, 2001) are not seen in EMs. These aggregates could form after homogenization perhaps because the granules have a higher propensity for aggregation than normal polysomes. Another possibility is that the aggregates are masked in neurons and are not identifiable in EM. The best evidence for masking comes from the Singer lab with the observation that many transported mRNAs were resistant to probes before protease treatment (Buxbaum et al., 2014). Interestingly, neuronal stimulation opened up these granules to probes, consistent with the idea that RNA granules dissociate to free mRNAs for translation (Buxbaum et al., 2014).

Components of Neuronal RNA Granules

A number of proteomics studies have identified a large number of proteins found in these large aggregates of ribosomes isolated from neurons. More specifically, two proteomic studies were done based on purification by sedimentation (El Fatimy et al., 2016; Elvira et al., 2006) and one based on a pull-down by tail of kinesin5a (KIF5a; Kanai et al., 2004). While the KIF5a study also purified a densely sedimenting particle (1000S), it should be noted that unlike the other studies granules identified by KIF5a pulldown only identified one ribosomal protein. This is likely due to technical aspects of the proteomics (bands were isolated from 2-D gels where the extremely positively charged ribosomal proteins had likely run off, and only one of the most negatively charged ribosomal proteins, L3, was found at the very edge of the gel; Kanai et al., 2004). Other than the ribosomal proteins, the shared proteomics results of this study as well as the high density of the purified particle suggests that the same structure was identified. The major protein components (other than ribosomal proteins) detected in these proteomic studies are RBPs and some RBPs were shared between the three studies (fragile X mental retardation protein [FMRP], pur-alpha and beta, staufen, hnRNP-U, synaptotagmin binding cytoplasmic RNA interacting protein [SYNCRIP], also known as hnRNP-Q, despite the different developmental stages of the tissues used in the three studies (E18 vs. P5 vs. adult). Many more proteins were found in at least two studies, including cytoplasmic activation/proliferation-associated protein 1 (CAPRIN-1, also known as RNA granule protein 105), RAS-GTPase-activating protein-binding proteins (G3BPs), ELAVs, interleukin-enhancer binding protein, insulin-like growth factor mRNA binding proteins (IMPs; also known as zipcode binding protein [ZBP]) and other hnRNPs. There are some additional proteins, not necessarily RBPs, that have been implicated in RNA biology, such as dead box RNA helicases (DDX)1 and DDX 3, RNA transcription, translation, and transport factor (RTRAF, also known as CGI-99)—a protein in a complex with DDX1-, up frameshift mutation 1 (UPF1), plasminogen activator inhibitor 1 RNA binding protein (PAI1 RBP, also known as serpine1 mRNA binding protein (SERBP1), that were found in at least two of the studies. While some of these proteins can also be found in generic proteomic studies of polysomes (Reschke et al., 2013) their relative enrichment in these preparations suggests specific roles in these granules, either in selecting mRNA cargo, forming the compact ribosome bundles, connecting to motors, or in the actual stalling process itself. Another striking finding in all these studies is the absence of most translation factors implicated in translation initiation, including any component of the initiating complex, eIF4G, consistent with the status of these structures as translationally repressed.

RNA Granules and Liquid-Liquid Phase Separation

Many RNA structures (nucleolus, P bodies, stress granules) are known to retain their structure through liquid-liquid phase separation (Langdon & Gladfelter, 2018; Ryan & Fawzi, 2019). This phase separation of RNA structures is achieved in part by RBPs with low complexity disordered domains that are found abundantly in these structures (Harrison & Shorter, 2017; Mittag & Parker, 2018). Indeed, many of the proteins with disordered domains important for stress granules (Harrison & Shorter, 2017) are also present in RNA granules based on the proteomic studies outlined above (Figure 11.2). Live imaging of RNA granules demonstrated that they were visible in differential interference contrast microscopy (DIC), indicating that a phase separation had occurred (Miller et al., 2009). It is not clear if the phase separation seen in these

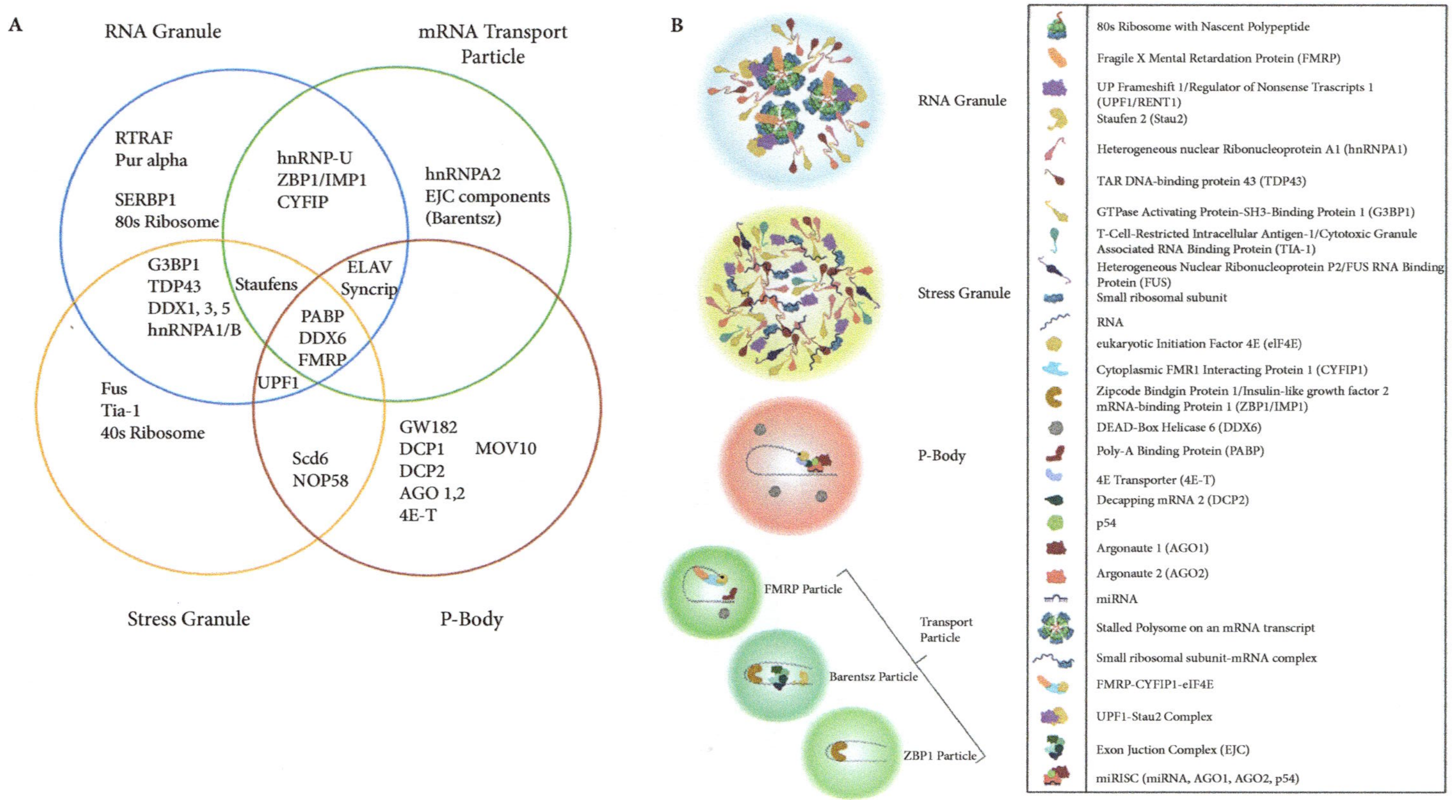

FIGURE 11.2. Structures containing mRNAs. (A) Venn diagram showing overlap of selected proteins from RNA transport particles, RNA granules, Stress granules and P bodies. Although each structure has a distinct purpose, there is an overlap of some proteins that are shared among these structures. (B) Structure and relative organization of each mRNA containing structure. Proteins with disordered domains are shown in a cloud surrounding the cargo indicating their role in generating a liquid-liquid phase separated structure for RNA granules and Stress granules. The disordered proteins form P bodies have not been described. On the right is a key for the icons used in the diagrams on the left. This also includes abbreviations for many of the proteins shown in the VENN diagram in A.

structures is also important for the repression of translation seen in RNA granules, perhaps if several factors important for translation are segregated out of these structures. Also, how liquid-liquid phase separated structures can be tethered to and transported on microtubules is an issue that has not yet been addressed. Interestingly, relationships between stress granules and the endoplasmic reticulum (ER) have generated a model for how membraneless organelles can interact with other structures through protein-protein interactions (Lee et al., 2020). It is likely that a true understanding of the transport of these granules and their repression, and removal of repression, may require a better understanding of how liquid-liquid phase interactions are regulated.

RNA Transport Particles

There have been a number of proteomic studies examining the mRNA transport particles that contain mRNAs but lack ribosomes (Sossin & DesGroseillers, 2006) often purified through a specific RBP that marks this particle. Two distinct particles and associated proteins were isolated using affinity purification with antibodies against either staufen 2 (Stau2) or barentsz (Fritzsche et al., 2013). It was observed that less than 50 percent of the 150 proteins identified in this study were present in both particles, supporting the idea of heterogeneity in these transport particles based on the RBPs that are present. The mRNAs contained within the two isolated particles were also distinct. The barentsz particle contained components of the exon junction complex (EJC), suggesting that the mRNA contained within was not previously translated since translation is known to displace the EJC from the message. It is also possible that this particle may be specific to mRNAs that contain EJC in their 3′ untranslated region, such as the mRNA for ARC that would be the target of nonsense-mediated decay if fully translated (Giorgi et al., 2007). The Stau2 particle shared a number of features with stalled polysomes (that also contain Stau2), including UPF1, FMRP, pur-alpha (Figure 11.2). One complication of this study is the exclusion of many proteins that were immunoprecipitated as they were also present in control immunoprecipitations; this included ribosomal proteins and many other proteins that were previously identified in the stalled polysome purifications (Fritzsche et al., 2013). Since the RNA granules are very dense, it is possible that they would pellet in the control precipitations, making it difficult to distinguish the relationship between the Stau2 particle isolated here and the RNA granules described previously containing Stau2 (Elvira et al., 2006). HnRNPA2 is another RBP that appears to be critical for mRNA transport to distal sites and many transported mRNAs have a binding site for hnRNPA2 (Muslimov et al., 2011; Shan et al., 2003) however, the composition of these transport particles have not been well characterized. FMRP binds to cytoplasmic FMR1 interacting protein family (CYFIP) that represses specific mRNAs by repressing translation initiation through eIF4E and represses translation of the specific mRNAs (Napoli et al., 2008; Figure 11.2). Another RBP implicated in transport in both axons and dendrites is ZBP1/IMP1. It has been most closely associated with the transport and translational regulation of beta-actin mRNA in dendrites (Huttelmaier et al., 2005) and axons (Welshhans & Bassell, 2011). Purification of ZBP1/IMP1 granules from cell lines using antibodies to this RBP revealed a structure containing small ribosomal subunits, EJC, and a number of RBPs (Jonson et al., 2007). Although these granules resembled stress granules (see below) in containing small, but not large ribosomal subunits, they did not colocalize with stress granules in these cells as marked by G3BP1 and FMRP (Jonson et al., 2007). Their relationship to the barentsz particle that also contains EJC is not clear, perhaps due to the different cell types (neurons vs HEK293 cells) from which the particles were isolated. In summary, there are a wide variety of RNA transport particles that have been identified, mainly based on immunoprecipitation with a particular RBP. It is not clear, however, whether specific RBPs can define a particular transport particle and how distinct these identifications are, as well as how many distinct transport particles exist. It is conceivable that every transported mRNA, based on its binding to an individual set of RBPs, defines its own transport particle.

Comparison to Other mRNA Containing Granules

There are other mRNA containing granules, such as stress granules, P-bodies and RNA-induced silencing complexes (RISC) particles specialized for RNA decay and storage. These granules can also be defined by the RBPs they contain, such as those with low complexity disordered domains that allows the liquid-liquid phase separation of the granules from the surrounding cytoplasm (also found in RNA granules), as well as other RBPs that make them distinct from RNA granules.

Stress Granules

Stress granules are formed as part of the integrated stress response which can be induced by multiple types of stress (proteostatic stress, oxidative stress, starvation, etc.) that converge on the phosphorylation of initiation factor eIF2 alpha, and store mRNAs stalled at initiation until the stress is resolved. As a consequence of eIF2 alpha phosphorylation, initiation is blocked before subunit joining, and thus stress granules contain mRNAs bound to the small ribosomal subunit but lacking the large ribosomal subunit (Anderson & Kedersha, 2006). The lack of the large ribosomal subunit is the clearest distinction between neuronal RNA granules and stress granules. Stress granules are organized by specific RBPs, such as G3BP and TIA1, with low complexity disordered domains that are important for the liquid-liquid phase separation (Protter & Parker, 2016).

Notably, some of these proteins such as G3BPs are shared with neuronal RNA granules (Figure 11.2), perhaps due to both structures requiring the liquid-liquid phase separation. It thus brings up a concern that some characterization of stress granules using co-localization with shared proteins like G3BP in neurons (Martin & Tazi, 2014; Sahoo et al., 2018) may confuse stress granules and neuronal RNA granules. It would be preferable that co-localization with a large ribosomal subunit protein should be used to distinguish the two. Since transported mRNAs can also be blocked at initiation after association with the small ribosomal subunit, some transport particles may also contain small but not large ribosomal subunits. The composition of these transport granules may be quite similar to stress granules, although in one case an RNA transport particle containing the small subunit did not contain G3BP1 (Jonson et al., 2007). In contrast, if mRNA transport particles contain mRNAs for which initiation was blocked before 43 S joining (e.g., if eIF4E is prevented from binding to eIF4G), these would not be expected to be associated with small subunits during transport. Thus, not all RNA transport particles would be expected to contain small ribosomal subunits.

P Bodies and RISC Particles

Two other RNA-related structures in neurons are P bodies and RISC particles. P bodies are sites of RNA degradation, and indeed are defined by the presence of decapping enzymes (Franks & Lykke-Andersen, 2008) and ribonucleases important for RNA degradation (Siwaszek et al., 2014). However, they may act as storage granules as well (Luo et al., 2018; Parker & Sheth, 2007). RISC particles represent complexes in which miRNAs repress translation. While miRNA binding usually leads to degradation of the mRNA (Fabian & Sonenberg, 2012), there are a number of examples in neurons for the reversibility of miRNA-mediated repression, particularly at distal sites in neurons, suggesting local regulation by miRNAs (Banerjee et al., 2009; Kenny et al., 2014; Muddashetty et al., 2011). This could be due to either transport of RISC particles, or local production of miRNAs. Interestingly, markers of the RISC complex, including argonaute (AGO) and glycine-tryptophan repeat (GW) protein of 182 kD (GW182), are also found in P bodies (Liu et al., 2005), making it difficult to differentiate putative P bodies and RISC complexes. A granule with similarity to P bodies has been implicated in mRNA transport and translational control at distal sites in Drosophila neurons (Barbee et al., 2006; Hillebrand et al., 2010). While most of the P body/RISC proteins are not found in neuronal RNA granules, they do share some components, such as DDX6 and FMRP (Figure 11.2). One study examined the co-localization of RNA granule markers and P body markers in neuronal processes, which also showed some overlap (Miller et al., 2009). There are a number of possibilities concerning the relationship of these structures. It is possible that mRNA granules containing mRNAs and ribosomes also contain mRNAs with RISC complexes on their target mRNAs that would regulate the mRNAs once they are released from the granule. It is also possible that there is a distinct mRNA transport particle in neurons where mRNAs are repressed by miRNA-mediated mechanisms in a P body-like transport particle that shares components with neuronal RNA granules. Further studies are required to differentiate these possibilities.

Evidence for Stalled Polysomes in RNA Granules

Large RNA granules in neurons are made up of a mixture of transported mRNAs and ribosomes (Krichevsky & Kosik, 2001). While it is possible that this is due to co-transport of mRNAs stalled before elongation and ribosomal subunits not engaged in translation, multiple lines of evidence below suggest that these granules are made up of stalled polysomes, where translation has been stalled during elongation. Thus, some local translation can be mediated by reactivation of stalled polysomes as opposed to activating translation initiation pathways (Graber et al., 2013). What is the purpose for using stalled ribosomes for local translation in neurons? Neurons are highly dynamic cells that exhibit fast synaptic responses to incoming stimuli in order to process information at a physiologically relevant rate. This process often requires the production of new proteins at local synaptic sites in a fast and efficient manner, which then alters synaptic strength. One mechanism for translating proteins quickly is to bypass initiation, the rate limiting step for translation, by assembling multiple ribosomes on an mRNA and then stalling at elongation or termination, allowing for fast production of proteins after this stall is removed.

Because the critical distinction of RNA granule-mediated translation is the lack of requirement for initiation, specific initiation inhibitors have been key to providing evidence for these structures. These compounds only block the initial formation of polysomes, not their elongation. One class of inhibitors, including pateamine A and hippuristanol, are dominant negative inhibitors of eIF4A that trap the mRNA scanning complex, preventing initiation before subunit joining (Shen & Pelletier, 2019). Another class, exemplified by homoharringtonine and bruceantin, blocks only the very first step of elongation, allowing ribosomes that have already passed the first step of translation elongation to continue elongating (Robert et al., 2009). These compounds allow for identification of stalled polysomes in-situ as they allow run-off of translating polysomes but prevent formation of new polysomes. Thus, polysomes that remain after prolonged treatment with initiation inhibitors are, by definition, stalled. Tools are then required to distinguish stalled polysomes from ribosomes not in the process of translation. A major distinguishing feature is that stalled polysomes contain nascent polypeptide chains, but ribosomes not involved in translation do not. There are two methods to identify nascent chains: nascent chain ribopuromycylation (RPM) and SunTagging, which are described below.

Puromycin covalently attaches to the nascent polypeptide chain on translating ribosomes and eventually leads to the dissociation of the polypeptide and ribosome.

However, if puromycin is added in conjunction with an elongation inhibitor such as emetine, the nascent chain becomes trapped on the ribosome, thus one can label ribosomes with nascent chains using this ribopuromycylation (RPM) technique (David et al., 2013). By inducing ribosomal run-off with initiation inhibitors prior to the addition of puromycin, RPM will only label stalled polysomes. Indeed, in hippocampal neurites, large puncta of polysomes resistant to run-off were observed. The number of these puncta decreased after activation of mGLURs, consistent with the reactivation of stalled polysomes and their subsequent disappearance (Graber et al., 2013). Interestingly, emetine is not necessary to prevent the dissociation of the nascent chain and stalled ribosomes in the initiation-inhibitor resistant RPM puncta, presumably since these ribosomes are already stalled and puromycin dissociation is prevented (Langille et al., 2019).

Another way to image nascent peptides is the SunTag method, in which multiple copies of an antigen for a fluorescently-tagged nanobody are incorporated into a message along with a degradation signal, allowing for the imaging of the translated nascent chains, only before translation is complete (Yan et al., 2016). The co-localization of Sun-Tag and RPM puncta after run-off of translating polysomes firmly established the presence of stalled polysomes in neurites (Langille et al., 2019).

One can also distinguish translation that comes from stalled polysomes by examining translation in the presence of initiation inhibitors in conjunction with the nascent protein synthesis reporter Click-IT™ AHA (L-Azidohomoalanine). Using this technique, it was shown that in processes of cultured hippocampal neurons, activation of mGLURs leads to a strong initiation-independent increase in new protein synthesis consistent with reactivation of stalled polysomes (Graber et al., 2013). If stalled polysomes are important for production of proteins and plasticity, then these proteins and the types of synaptic plasticity they are associated with should be distinguished by their dependence of elongation inhibitors (such as cycloheximide or anisomycin), and their independence of initiation inhibitors. Indeed, production of microtubule-associated protein (MAP)-1b (Graber et al., 2013), and Arc (Na et al., 2016), as well as mGLUR-LTD (Graber et al., 2013) have all been shown to depend on elongation, but not initiation. A form of synaptic plasticity called intermediate term facilitation (ITF) in the invertebrate *Aplysia californica* is also blocked by elongation, but not by initiation inhibitors (McCamphill et al., 2015), thus suggesting that this method of fast and local translation is evolutionarily conserved.

One mRNA versus Multiple mRNAs in Transport

Imaging of mRNAs in dendrites using sensitive in-situ hybridization has strongly indicated that most mRNAs are present individually and do not co-localize with other mRNAs or even with separate mRNAs encoding the same protein (Batish et al., 2012;

Mikl et al., 2011). This is in contrast to the evidence that large RNA granules appear to consist of multiple polysomes. Imaging of stalled polysomes appears to show aggregates of large structures (Miller et al., 2009). When imaging tagged FMRP, there appears to be coalescence and dissociation of smaller particles from these larger structures, indicating units or individual arrays of stalled polysomes, coming on or off (El Fatimy et al., 2016). RPM puncta of stalled polysomes also appears more consistent with the presence of multiple stalled polysomes in one structure, particularly when using super resolution STORM microscopy the number of puromycin associated with an individual granule appears larger than what would be expected from a single polysome (Graber et al., 2017). If one compares the density of in-situ hybridization derived puncta with the density of RPM puncta, it is clear that there are many more individual molecules of mRNA than RPM puncta in dendrites (at least 10-fold higher numbers of a single mRNA than total number of RPM puncta in Graber et al, 2017). Thus, RNA granules containing stalled polysomes are sparse compared to transport particles containing individual mRNAs. This is consistent with some puzzling findings. The first is that initiation inhibitors did not significantly decrease the number of initiation inhibitor resistant RPM puncta seen in dendrites, despite the ability of these inhibitors to strongly decrease basal translation in dendrites (Graber et al., 2013). This could be due to translation in processes coming mainly from mRNAs with one or only a few ribosomes attached (Biever et al., 2020) that are under the detection limit for RPM puncta. Second, when expressing an mRNA containing SunTag epitopes, there was a high percentage of RPM puncta that overlapped with the nascent peptides (Langille et al., 2019). If polysomes were transported individually, then one would expect only the polysomes containing the mRNA encoding the SunTag epitope to show colocalization of RPM and SunTag signal, while all other polysomes containing other messages to label only for RPM. In contrast, virtually all RPM puncta were also positive for Sun-Tag, thus suggesting that either under these conditions this mRNA competed out all others for generation of granules or that there were many polysomes in each granule (Langille et al., 2019). Thus, these data suggest that in dendrites while most mRNAs for a particular message are either in transport particles or individual translating ribosomes, a small subset of mRNAs may be found in larger aggregates of stalled polysomes, available for quick release upon appropriate stimuli.

How Are Stalled Polysomes Stalled?

If stalled polysomes represent a source of mRNAs underlying local translation, it is important to understand how they are stalled and unstalled. There are several signals that appear to be important for this process (Figure 11.3) and these are detailed below. However, despite some clues from proteins that appear to be required for stalling and unstalling, the fundamental mechanisms involved in these processes are still unknown.

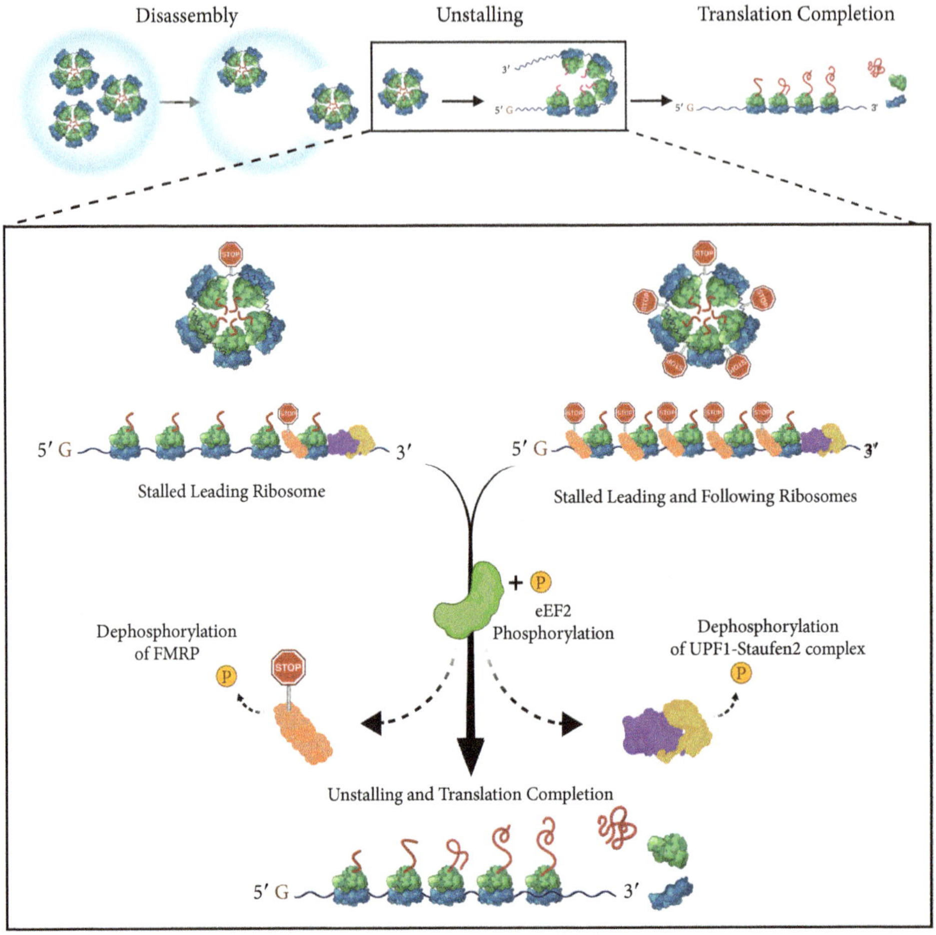

FIGURE 11.3. Release of RNA-Granule Associated Stalled Polysomes. RNA Granule associated stalled polysomes are a specialized mechanism for neurons to regulate the expression of mRNAs important for synaptic plasticity during development. Once RNA Granules arrive at their designated local sites, appropriate cues induce the disassembly of one or more stalled polysomes to dissociate from the liquid-liquid phase separated granule structure and unravel to expose the contents to translation factors. Unstalling of polysomes on the mRNA occurs when the repression induced by RBPs such as FMRP and UPF1-Staufen2 complex, is removed and this is facilitated by signals such as the activation of mGluRs. It has been established that the phosphorylation of eEF2 is required for the release of stalled polysomes, however the exact mechanism is unclear. In terms of stalling, there are two possibilities: (A) Only the leading ribosome is stalled, causing a pile up of the trailing ribosomes on the mRNA, and (B) Both leading and trailing ribosomes are stalled. The activation of mGluRs leads to the phosphorylation of eEF2 and the dephosphorylation of FMRP and the UPF1-Staufen2 complex, resulting in unstalling and completion of translation.

Fragile-X Mental Retardation Protein

FMRP is enriched in neuronal RNA granules (El Fatimy et al., 2016; Elvira et al., 2006; Kanai et al., 2004). This protein is encoded for by an X-linked gene subject to amplification of a CGG repeat in the 5′ untranslated region of the fragile-X gene that leads to (i) hypermethylation, (ii) a lack of production of the mRNA encoding FMRP, and (iii) the absence of functional FMRP protein (Garber et al., 2008). Loss of FMRP leads to the Fragile-X syndrome, including severe intellectual disability, epilepsy, and often symptoms of autism spectrum disorder (Garber et al., 2008). One common finding in animal models of Fragile-X is an increase in overall translation, suggesting a normal role for FMRP in repressing translation (Richter et al., 2015). The FMRP protein was found to segregate with polysomes in sucrose gradients, suggesting a role in elongation (Khandjian et al., 1996). Run-off of translating polysomes left a fraction of polysomes still associated with FMRP, suggesting that FMRP was specifically associated with stalled polysomes (Ceman et al., 2003). This form of FMRP was phosphorylated at a specific residue (serine 499; Ceman et al., 2003), initially thought to be due to S6 kinase, but now shown to be partly due to CK2 and other kinases (Bartley et al., 2016; Bartley et al., 2014). Dephosphorylation of this residue is thought to be important for mGLUR-LTD (Niere et al., 2012) and this is consistent with the phosphorylation playing a role in reactivation of stalled polysomes and the production of proteins that are also required for mGLUR-LTD (Graber et al., 2013; Figure 11.3). Perhaps, the most direct association between FMRP and stalled polysomes are studies where FMRP was required for stalling of transcripts shown separately to be associated with FMRP (Darnell et al., 2011). Unlike most RBPs that associate with the 3′ untranslated regions of their mRNA targets, the majority of sites for FMRP association were located in the coding region (Darnell et al., 2011), consistent with FMRP's association with the ribosome as opposed to with specific sites on its target mRNAs (Chen et al., 2014). Loss of FMRP was also associated with increased elongation rates for a number of messages, consistent with a role for FMRP in the regulation of elongation (Udagawa et al., 2013).

UPF1–Staufen 2 Interactions

UPF1 is an RNA helicase that plays a major role in nonsense-mediated decay (Kim & Maquat, 2019). It is recruited when the mRNA reaches the stop codon through an interaction with eukaryotic release factors 1 and 3 and stalls termination. If an EJC containing UPF2/3 exists downstream of the stop codon, indicating that the stop codon is premature, then interactions between UPF1 and UPF2/3 leads to the phosphorylation of UPF1 and subsequent decay of the message (Kim & Maquat, 2019). Staufen binding to UPF1 can also lead to its phosphorylation and can result in staufen-mediated decay for some messages that harbor staufen-binding sites in their 3′ untranslated region (Park & Maquat, 2013). Since both UPF1 and Stau2 have been found in proteomics of RNA granules (El Fatimy et al., 2016; Elvira et al., 2006), it suggests that the UPF1/Stau2 interaction may be repurposed to generate stalled polysomes. Indeed, removal of UPF1 or Stau leads to a decrease in the number of stalled polysomes, initiation-independent translation and mGLUR-LTD (Graber et al., 2017). Moreover, both initiation-independent translation and mGLUR-LTD can be rescued with re-expression of Stau2, but not with a form of Stau2 that does not bind UPF1, demonstrating the importance of their interaction for the regulation of stalled polysomes (Graber et al., 2017). UPF1 dephosphorylation was also induced by mGLUR-LTD, similar to FMRP (Graber et al., 2017; Figure. 11.3). In this study, the role of UPF1 and Stau2 were also examined for mRNA transport and production of a specific protein, MAP1b, which is required for mGLUR-LTD (Davidkova & Carroll, 2007). Both UPF1 and Stau2 were required for the transport of the mRNA encoding Map1b, as well as the induced translation of the protein following unstalling (Graber et al., 2017). Importantly, MAP1b is also a transcript highly regulated by FMRP (Lu et al., 2004), linking the two distinct regulators of stalled polysomes.

eEF2 Phosphorylation

There is a correlation between synaptic plasticity events that appear to require reactivation of stalled polysomes (i.e., elongation-dependent, initiation-independent) and plasticity events that require phosphorylation of the elongation factor eEF2, reviewed in (Sossin & Costa-Mattioli, 2019; Figure 11.3). eEF2 is the elongation factor important for ratcheting the mRNA forward through the ribosome and phosphorylation inactivates eEF2 through removing its ability to bind the ribosome (Proud, 2015). What underlies this correlation is not clear, but may suggest some relationship between the structure of the stalled polysome and the role of eEF2.

Prevention of Nonsense-Mediated and No-go Decay

No-go decay is a dedicated pathway in cells for the recovery of stalled polysomes and degradation of their partially translated nascent chains (Buskirk & Green, 2017). The stability of stalled polysomes in neuronal processes would require a mechanism to prevent this form of decay. Indeed, none of the proteins specific to no-go decay process are present in the proteomics of stalled polysomes. Similarly, UPF1 phosphorylation normally leads to nonsense-mediated decay, as does the association of staufen with an mRNA (Schoenberg & Maquat, 2012). Thus, stalled polysomes in neurons containing phosphorylated UPF1 would need a mechanism to prevent this form of decay as well. Similar to no-go Decay, none of the proteins (other than UPF1) normally associated with nonsense-mediated decay are present in the proteomics of RNA granules.

Stalled Polysomes and Neurodevelopmental Disorders

The loss of FMRP is the leading genetic cause of intellectual deficiency and autism (Garber et al., 2008). Recently, a number of other proteins implicated in stalled polysomes have been found to cause intellectual deficiency when mutated. This includes the RBPs pur-alpha (Reijnders et al., 2018), SYNCRIP (Lelieveld et al., 2016), hnRNP-U (Bramswig et al., 2017), and the most abundant RNA helicase in the polysomes DDX3 (Snijders Blok et al., 2015; Wang et al., 2018). While most of these RBPs have also been implicated in other granules or other translational regulatory steps, it is still striking that so many of the proteins central to the proteome of the stalled polysome, when altered, result in neurodevelopmental disorders. mRNAs associated with FMRP are highly enriched for those encoding proteins whose mutation causes intellectual deficiency and Autism (Darnell et al., 2011). If this represents the mRNAs mostly regulated through stalled polysomes, it would be consistent with the dysregulation of many of the proteins implicated in these granules causing neurodevelopmental disorders. Moreover, stalled polysomes have mostly been characterized in developmental settings, using neuronal cultures or purified from tissues of embryonic or developing brains. While a quantitative developmental profile has not been accomplished, it is possible that this form of mRNA transport is much more important in developmental stages than in the mature brain.

Conclusions

There are many structures implicated in the transport of mRNAs to distal sites in neurons and their local translational regulation. One particular neuronal RNA granule consisting of stalled polysomes appears to play a unique role in the control of mRNAs produced for certain forms of plasticity such as mGLUR-LTD, and loss of this regulation is associated with neurodevelopmental disorders. Since many of the proteins are shared between this granule and other structures such as mRNA transport particles, stress granules and P bodies, it is still difficult to ascertain particular roles for any one RNA-containing structure. Nevertheless, it is clear that a better understanding of these different modes of transport will be required for a deeper understanding of local mRNA translation and its role in the healthy and diseased brain. In the case of fragile-X syndrome, the restoration of homeostasis through the regulation of stalled polysome dependent translation even in the absence of FMRP, holds great promise.

Acknowledgments

This work was funded by a CIHR grant to WSS and WSS is a James McGill Professor. We thank Arkady Khotoursky and Keith Murai for their helpful comments. The figures were created with biorender.com

References

Aakalu, G., Smith, W. B., Nguyen, N., Jiang, C., & Schuman, E. M. (2001). Dynamic visualization of local protein synthesis in hippocampal neurons. *Neuron, 30*(2), 489–502. https://doi.org/10.1016/s0896-6273(01)00295-1

Anderson, P., & Kedersha, N. (2006). RNA granules. *Journal of Cell Biology, 172*(6), 803–808. https://doi.org/10.1083/jcb.200512082

Aschrafi, A., Cunningham, B. A., Edelman, G. M., & Vanderklish, P. W. (2005). The fragile X mental retardation protein and group I metabotropic glutamate receptors regulate levels of mRNA granules in brain. *Proceedings of the National Academy of Science of the United States of America, 102*(6), 2180–2185. https://doi.org/10.1073/pnas.0409803102

Banerjee, S., Neveu, P., & Kosik, K. S. (2009). A coordinated local translational control point at the synapse involving relief from silencing and MOV10 degradation. *Neuron, 64*(6), 871–884. https://doi.org/10.1016/j.neuron.2009.11.023

Barbarese, E., Koppel, D. E., Deutscher, M. P., Smith, C. L., Ainger, K., Morgan, F., & Carson, J. H. (1995). Protein translation components are colocalized in granules in oligodendrocytes. *Journal of Cell Science, 108*(8), 2781–2790.

Barbee, S. A., Estes, P. S., Cziko, A. M., Hillebrand, J., Luedeman, R. A., Coller, J. M.,... Ramaswami, M. (2006). Staufen- and FMRP-containing neuronal RNPs are structurally and functionally related to somatic P bodies. *Neuron, 52*(6), 997–1009. https://doi.org/10.1016/j.neuron.2006.10.028

Bartley, C. M., O'Keefe, R. A., Blice-Baum, A., Mihailescu, M. R., Gong, X., Miyares, L.,... Bordey, A. (2016). Mammalian FMRP S499 Is Phosphorylated by CK2 and Promotes Secondary Phosphorylation of FMRP. *eNeuro, 3*(6). https://doi.org/10.1523/ENEURO.0092-16.2016

Bartley, C. M., O'Keefe, R. A., & Bordey, A. (2014). FMRP S499 is phosphorylated independent of mTORC1-S6K1 activity. *PLoS One, 9*(5), e96956. https://doi.org/10.1371/journal.pone.0096956

Batish, M., van den Bogaard, P., Kramer, F. R., & Tyagi, S. (2012). Neuronal mRNAs travel singly into dendrites. *Proceedings of the National Academy of Science of the United States of America, 109*(12), 4645–4650. https://doi.org/10.1073/pnas.1111226109

Bauer, K. E., Segura, I., Gaspar, I., Scheuss, V., Illig, C., Ammer, G.,... Kiebler, M. A. (2019). Live cell imaging reveals 3′-UTR dependent mRNA sorting to synapses. *Nature Communications, 10*(1), 3178. https://doi.org/10.1038/s41467-019-11123-x

Becalska, A. N., & Gavis, E. R. (2009). Lighting up mRNA localization in Drosophila oogenesis. *Development, 136*(15), 2493–2503. https://doi.org/10.1242/dev.032391

Bian, F., Chu, T., Schilling, K., & Oberdick, J. (1996). Differential mRNA transport and the regulation of protein synthesis: selective sensitivity of Purkinje cell dendritic mRNAs to translational inhibition. *Molecular and Cellular Neuroscience, 7*(2), 116–133. https://doi.org/10.1006/mcne.1996.0009

Biever, A., Glock, C., Tushev, G., Ciirdaeva, E., Dalmay, T., Langer, J. D., & Schuman, E. M. (2020). Monosomes actively translate synaptic mRNAs in neuronal processes. *Science, 367*(6477). https://doi.org/10.1126/science.aay4991

Bramswig, N. C., Ludecke, H. J., Hamdan, F. F., Altmuller, J., Beleggia, F., Elcioglu, N. H., ... Wieczorek, D. (2017). Heterozygous HNRNPU variants cause early onset epilepsy and severe intellectual disability. *Human Genetics, 136*(7), 821–834. https://doi.org/10.1007/s00439-017-1795-6

Buskirk, A. R., & Green, R. (2017). Ribosome pausing, arrest and rescue in bacteria and eukaryotes. *Philosophical Transactions of the Royal Society of London Series B: Biological Sciences, 372*(1716). https://doi.org/10.1098/rstb.2016.0183

Buxbaum, A. R., Wu, B., & Singer, R. H. (2014). Single beta-actin mRNA detection in neurons reveals a mechanism for regulating its translatability. *Science, 343*(6169), 419–422. https://doi.org/10.1126/science.1242939

Cajigas, I. J., Tushev, G., Will, T. J., tom Dieck, S., Fuerst, N., & Schuman, E. M. (2012). The local transcriptome in the synaptic neuropil revealed by deep sequencing and high-resolution imaging. *Neuron, 74*(3), 453–466. https://doi.org/10.1016/j.neuron.2012.02.036

Casadio, A., Martin, K. C., Giustetto, M., Zhu, H., Chen, M., Bartsch, D., ... Kandel, E. R. (1999). A transient, neuron-wide form of CREB-mediated long-term facilitation can be stabilized at specific synapses by local protein synthesis. *Cell, 99*(2), 221–237. https://doi.org/10.1016/s0092-8674(00)81653-0

Ceman, S., O'Donnell, W. T., Reed, M., Patton, S., Pohl, J., & Warren, S. T. (2003). Phosphorylation influences the translation state of FMRP-associated polyribosomes. *Human Molecular Genetics, 12*(24), 3295–3305. https://doi.org/10.1093/hmg/ddg350

Chen, E., Sharma, M. R., Shi, X., Agrawal, R. K., & Joseph, S. (2014). Fragile X mental retardation protein regulates translation by binding directly to the ribosome. *Molecular Cell, 54*(3), 407–417. https://doi.org/10.1016/j.molcel.2014.03.023

Cioni, J. M., Koppers, M., & Holt, C. E. (2018). Molecular control of local translation in axon development and maintenance. *Current Opinion in Neurobiology, 51*, 86–94. https://doi.org/10.1016/j.conb.2018.02.025

Colman, D. R., Kreibich, G., Frey, A. B., & Sabatinie, D. D. (1982) Synthesis and incorporation of myelin polypeptides into CNS myelin. *Journal of Cellular Biology, 95*(2), 598–608. https://doi.org/10.1083/jcb.95.2.598

Costa-Mattioli, M., Sossin, W. S., Klann, E., & Sonenberg, N. (2009). Translational control of long-lasting synaptic plasticity and memory. *Neuron, 61*(1), 10–26. https://doi.org/10.1016/j.neuron.2008.10.055

Darnell, J. C., Van Driesche, S. J., Zhang, C., Hung, K. Y., Mele, A., Fraser, C. E., ... Darnell, R. B. (2011). FMRP stalls ribosomal translocation on mRNAs linked to synaptic function and autism. *Cell, 146*(2), 247–261. https://doi.org/10.1016/j.cell.2011.06.013

David, A., Bennink, J. R., & Yewdell, J. W. (2013). Emetine optimally facilitates nascent chain puromycylation and potentiates the ribopuromycylation method (RPM) applied to inert cells. *Histochemistry and Cell Biology, 139*(3), 501–504. https://doi.org/10.1007/s00418-012-1063-8

Davidkova, G., & Carroll, R. C. (2007). Characterization of the role of microtubule-associated protein 1B in metabotropic glutamate receptor-mediated endocytosis of AMPA receptors in hippocampus. *Journal of Neuroscience, 27*(48), 13273–13278. https://doi.org/10.1523/JNEUROSCI.3334-07.2007

Dynes, J. L., & Steward, O. (2012). Arc mRNA docks precisely at the base of individual dendritic spines indicating the existence of a specialized microdomain for synapse-specific mRNA translation. *Journal of Comparative Neurology, 520*(14), 3105–3119. https://doi.org/10.1002/cne.23073

El Fatimy, R., Davidovic, L., Tremblay, S., Jaglin, X., Dury, A., Robert, C., . . . Khandjian, E. W. (2016). Tracking the fragile X mental retardation protein in a highly ordered neuronal ribonucleoparticles population: A link between stalled polyribosomes and RNA granules. *PLoS Genetics, 12*(7), e1006192. https://doi.org/10.1371/journal.pgen.1006192

Eliscovich, C., Buxbaum, A. R., Katz, Z. B., & Singer, R. H. (2013). mRNA on the move: The road to its biological destiny. *Journal of Biological Chemistry, 288*(28), 20361–20368. https://doi.org/10.1074/jbc.R113.452094

Elvira, G., Wasiak, S., Blandford, V., Tong, X. K., Serrano, A., Fan, X., . . . Sossin, W. S. (2006). Characterization of an RNA granule from developing brain. *Molecular and Cellular Proteomics, 5*(4), 635–651. https://doi.org/10.1074/mcp.M500255-MCP200

Fabian, M. R., & Sonenberg, N. (2012). The mechanics of miRNA-mediated gene silencing: a look under the hood of miRISC. *Nature Structural & Molecular Biology, 19*(6), 586–593. https://doi.org/10.1038/nsmb.2296

Fontes, M. M., Guvenek, A., Kawaguchi, R., Zheng, D., Huang, A., Ho, V. M., . . . Martin, K. C. (2017). Activity-Dependent Regulation of Alternative Cleavage and Polyadenylation During Hippocampal Long-Term Potentiation. *Science Reports, 7*(1), 17377. https://doi.org/10.1038/s41598-017-17407-w

Franks, T. M., & Lykke-Andersen, J. (2008). The control of mRNA decapping and P-body formation. *Molecular Cell, 32*(5), 605–615. https://doi.org/10.1016/j.molcel.2008.11.001

Fritzsche, R., Karra, D., Bennett, K. L., Ang, F. Y., Heraud-Farlow, J. E., Tolino, M., . . . Kiebler, M. A. (2013). Interactome of two diverse RNA granules links mRNA localization to translational repression in neurons. *Cell Reports, 5*(6), 1749–1762. https://doi.org/10.1016/j.celrep.2013.11.023

Garber, K. B., Visootsak, J., & Warren, S. T. (2008). Fragile X syndrome. *European Journal of Human Genetics, 16*(6), 666–672. https://doi.org/10.1038/ejhg.2008.61

Gardiner, A. S., Twiss, J. L., & Perrone-Bizzozero, N. I. (2015). Competing interactions of RNA-binding proteins, microRNAs, and their targets control neuronal development and function. *Biomolecules, 5*(4), 2903–2918. https://doi.org/10.3390/biom5042903

Garner, C. C., Tucker, R. P., & Matus, A. (1988). Selective localization of messenger RNA for cytoskeletal protein MAP2 in dendrites. *Nature 336*(6200), 674–677. https://doi.org/10.1038/336674a0

Giorgi, C., Yeo, G. W., Stone, M. E., Katz, D. B., Burge, C., Turrigiano, G., & Moore, M. J. (2007). The EJC factor eIF4AIII modulates synaptic strength and neuronal protein expression. *Cell, 130*(1), 179–191. https://doi.org/10.1016/j.cell.2007.05.028

Glock, C., Heumüller, M., & Schuman, E. M. (2017). mRNA transport & local translation in neurons. *Current Opinions in Neurobiology, 45*, 169–177. https://doi.org/10.1016/j.conb.2017.05.005

Graber, T. E., Freemantle, E., Anadolu, M. N., Hebert-Seropian, S., MacAdam, R. L., Shin, U., . . . Sossin, W. S. (2017). UPF1 governs synaptic plasticity through association with a STAU2 RNA granule. *Journal of Neuroscience, 37*(38), 9116–9131. https://doi.org/10.1523/jneurosci.0088-17.2017

Graber, T. E., Hebert-Seropian, S., Khoutorsky, A., David, A., Yewdell, J. W., Lacaille, J. C., & Sossin, W. S. (2013). Reactivation of stalled polyribosomes in synaptic plasticity. *Proceedings of the National Academy of Science of the United States of America, 110*(40), 16205–16210. https://doi.org/10.1073/pnas.1307747110

Hafner, A. S., Donlin-Asp, P. G., Leitch, B., Herzog, E., & Schuman, E. M. (2019). Local protein synthesis is a ubiquitous feature of neuronal pre- and postsynaptic compartments. *Science, 364*(6441). https://doi.org/10.1126/science.aau3644

Hanz, S., Perlson, E., Willis, D., Zheng, J. Q., Massarwa, R., Huerta, J. J.,... Fainzilber, M. (2003). Axoplasmic importins enable retrograde injury signaling in lesioned nerve. *Neuron, 40*(6), 1095–1104. https://doi.org/10.1016/s0896-6273(03)00770-0

Harrison, A. F., & Shorter, J. (2017). RNA-binding proteins with prion-like domains in health and disease. *Biochemistry Journal, 474*(8), 1417–1438. https://doi.org/10.1042/BCJ20160499

Hillebrand, J., Pan, K., Kokaram, A., Barbee, S., Parker, R., & Ramaswami, M. (2010). The Me31B DEAD-box helicase localizes to postsynaptic foci and regulates expression of a CaMKII reporter mRNA in dendrites of Drosophila olfactory projection neurons. *Frontiers in Neural Circuits, 4*, 121. https://doi.org/10.3389/fncir.2010.00121

Huang, Y. S., & Richter, J. D. (2004). Regulation of local mRNA translation. *Current Opinions in Cell Biology, 16*(3), 308–313. https://doi.org/10.1016/j.ceb.2004.03.002

Huber, K. M., Kayser, M. S., & Bear, M. F. (2000). Role for rapid dendritic protein synthesis in hippocampal mGluR-dependent long-term depression. *Science, 288*(5469), 1254–1257.

Huttelmaier, S., Zenklusen, D., Lederer, M., Dictenberg, J., Lorenz, M., Meng, X.,... Singer, R. H. (2005). Spatial regulation of beta-actin translation by Src-dependent phosphorylation of ZBP1. *Nature, 438*(7067), 512–515. https://doi.org/10.1038/nature04115

Jansen, R. P., & Niessing, D. (2012). Assembly of mRNA-protein complexes for directional mRNA transport in eukaryotes—an overview. *Current Protein & Peptide Science, 13*(4), 284–293. https://doi.org/10.2174/138920312801619493

Job, C., & Eberwine, J. (2001). Identification of sites for exponential translation in living dendrites. *Proceedings of the National Academy of Science of the United States of America, 98*(23), 13037–13042. https://doi.org/10.1073/pnas.231485698

Jonson, L., Vikesaa, J., Krogh, A., Nielsen, L. K., Hansen, T., Borup, R.,... Nielsen, F. C. (2007). Molecular composition of IMP1 ribonucleoprotein granules. *Molecular & Cellular Proteomics, 6*(5), 798–811. https://doi.org/10.1074/mcp.M600346-MCP200

Kanai, Y., Dohmae, N., & Hirokawa, N. (2004). Kinesin transports RNA: isolation and characterization of an RNA-transporting granule. *Neuron, 43*(4), 513–525. https://doi.org/10.1016/j.neuron.2004.07.022

Kang, H., & Schuman, E. M. (1996). A requirement for local protein synthesis in neurotrophin-induced hippocampal synaptic plasticity. *Science, 273*(5280), 1402–1406. https://doi.org/10.1126/science.273.5280.1402

Kenny, P. J., Zhou, H., Kim, M., Skariah, G., Khetani, R. S., Drnevich, J.,... Ceman, S. (2014). MOV10 and FMRP regulate AGO2 association with microRNA recognition elements. *Cell Reports, 9*(5), 1729–1741. https://doi.org/10.1016/j.celrep.2014.10.054

Khan, A., Pepio, A. M., & Sossin, W. S. (2001). Serotonin activates S6 kinase in a rapamycin-sensitive manner in Aplysia synaptosomes. *Journal of Neuroscience, 21*(2), 382–391.

Khandjian, E. W., Corbin, F., Woerly, S., & Rousseau, F. (1996). The fragile X mental retardation protein is associated with ribosomes. *Nature Genetics, 12*(1), 91–93. https://doi.org/10.1038/ng0196-91

Kiebler, M. A., & Bassell, G. J. (2006). Neuronal RNA granules: movers and makers. *Neuron, 51*(6), 685–690. https://doi.org/10.1016/j.neuron.2006.08.021

Kim, Y. K., & Maquat, L. E. (2019). UPFront and center in RNA decay: UPF1 in nonsense-mediated mRNA decay and beyond. *RNA, 25*(4), 407–422. https://doi.org/10.1261/rna.070136.118

Kislauskis, E. H., Zhu, X., & Singer, R. H. (1997). beta-Actin messenger RNA localization and protein synthesis augment cell motility. *Journal of Cell Biology, 136*(6), 1263–1270. https://doi.org/10.1083/jcb.136.6.1263

Knowles, R. B., Sabry, J. H., Martone, M. E., Deerinck, T. J., Ellisman, M. H., Bassell, G. J., & Kosik, K. S. (1996). Translocation of RNA granules in living neurons. *Journal of Neuroscience, 16*(24), 7812.

Krichevsky, A. M., & Kosik, K. S. (2001). Neuronal RNA granules: a link between RNA localization and stimulation-dependent translation. *Neuron, 32*(4), 683–696.

Langdon, E. M., & Gladfelter, A. S. (2018). A new lens for RNA localization: liquid-liquid phase separation. *Annual Review of Microbiology, 72*, 255–271. https://doi.org/10.1146/annurev-micro-090817-062814

Langille, J. J., Ginzberg, K., & Sossin, W. S. (2019). Polysomes identified by live imaging of nascent peptides are stalled in hippocampal and cortical neurites. *Learning & Memory, 26*(9), 351–362. https://doi.org/10.1101/lm.049965.119

Lebeau, G., Miller, L. C., Tartas, M., McAdam, R., Laplante, I., Badeaux, F., . . . Lacaille, J. C. (2011). Staufen 2 regulates mGluR long-term depression and Map1b mRNA distribution in hippocampal neurons. *Learning & Memory, 18*(5), 314–326. https://doi.org/10.1101/lm.2100611

Lee, J. E., Cathey, P. I., Wu, H., Parker, R., & Voeltz, G. K. (2020). Endoplasmic reticulum contact sites regulate the dynamics of membraneless organelles. *Science, 367*(6477). https://doi.org/10.1126/science.aay7108

Lelieveld, S. H., Reijnders, M. R., Pfundt, R., Yntema, H. G., Kamsteeg, E. J., de Vries, P., . . . Gilissen, C. (2016). Meta-analysis of 2,104 trios provides support for 10 new genes for intellectual disability. *Nature Neuroscience, 19*(9), 1194–1196. https://doi.org/10.1038/nn.4352

Liu, J., Rivas, F. V., Wohlschlegel, J., Yates, J. R., 3rd, Parker, R., & Hannon, G. J. (2005). A role for the P-body component GW182 in microRNA function. *Nature Cell Biology, 7* (12), 1261–1266. https://doi.org/10.1038/ncb1333

Lu, R., Wang, H., Liang, Z., Ku, L., O'Donnell W. T., Li, W., . . . Feng, Y. (2004). The fragile X protein controls microtubule-associated protein 1B translation and microtubule stability in brain neuron development. *Proceedings of the National Academy of Science of the United States of America, 101*(42), 15201–15206. https://doi.org/10.1073/pnas.0404995101

Luo, Y., Na, Z., & Slavoff, S. A. (2018). P-Bodies: composition, properties, and functions. *Biochemistry, 57*(17), 2424–2431. https://doi.org/10.1021/acs.biochem.7b01162

Martin, K. C., Casadio, A., Zhu, H., Yaping, E., Rose, J. C., Chen, M., . . . Kandel, E. R. (1997). Synapse-specific, long-term facilitation of aplysia sensory to motor synapses: a function for local protein synthesis in memory storage. *Cell, 91*(7), 927–938. https://doi.org/10.1016/s0092-8674(00)80484-5

Martin, S., & Tazi, J. (2014). Visualization of G3BP stress granules dynamics in live primary cells. *Journal of Visualized Experiments*, (87). https://doi.org/10.3791/51197

Martinez, J. C., Randolph, L. K., Iascone, D. M., Pernice, H. F., Polleux, F., & Hengst, U. (2019). Pum2 shapes the transcriptome in developing axons through retention of target mRNAs in the cell body. *Neuron, 104*(5), 931–946 e935. https://doi.org/10.1016/j.neuron.2019.08.035

Mayya, V. K., & Duchaine, T. F. (2019). Ciphers and executioners: how 3′-untranslated regions determine the fate of messenger RNAs. *Frontiers in Genetics, 10*, 6. https://doi.org/10.3389/fgene.2019.00006

McCamphill, P. K., Farah, C. A., Anadolu, M. N., Hoque, S., & Sossin, W. S. (2015). Bidirectional regulation of eEF2 phosphorylation controls synaptic plasticity by decoding neuronal activity patterns. *Journal of Neuroscience, 35*(10), 4403–4417. https://doi.org/10.1523/JNEUROSCI.2376-14.2015

Mikl, M., Vendra, G., & Kiebler, M. A. (2011). Independent localization of MAP2, CaMKIIalpha and beta-actin RNAs in low copy numbers. *EMBO Reports, 12*(10), 1077–1084. https://doi.org/10.1038/embor.2011.149

Miller, L. C., Blandford, V., McAdam, R., Sanchez-Carbente, M. R., Badeaux, F., DesGroseillers, L., & Sossin, W. S. (2009). Combinations of DEAD box proteins distinguish distinct types of RNA: protein complexes in neurons. *Molecular and Cellular Neuroscience, 40*(4), 485–495.

Miller, S., Yasuda, M., Coats, J. K., Jones, Y., Martone, M. E., & Mayford, M. (2002). Disruption of dendritic translation of CaMKIIalpha impairs stabilization of synaptic plasticity and memory consolidation. *Neuron, 36*(3), 507–519. https://doi.org/10.1016/s0896-6273(02)00978-9

Mittag, T., & Parker, R. (2018). Multiple modes of protein-protein interactions promote RNP granule assembly. *Journal of Molecular Biology, 430*(23), 4636–4649. https://doi.org/10.1016/j.jmb.2018.08.005

Mori, Y., Imaizumi, K., Katayama, T., Yoneda, T., & Tohyama, M. (2000). Two cis-acting elements in the 3′ untranslated region of alpha-CaMKII regulate its dendritic targeting. *Nature Neuroscience, 3*(11), 1079–1084. https://doi.org/10.1038/80591

Muddashetty, R. S., Nalavadi, V. C., Gross, C., Yao, X., Xing, L., Laur, O., . . . Bassell, G. J. (2011). Reversible inhibition of PSD-95 mRNA translation by miR-125a, FMRP phosphorylation, and mGluR signaling. *Molecular Cell, 42*(5), 673–688. https://doi.org/10.1016/j.molcel.2011.05.006

Muslimov, I. A., Patel, M. V., Rose, A., & Tiedge, H. (2011). Spatial code recognition in neuronal RNA targeting: role of RNA-hnRNP A2 interactions. *Journal of Cell Biology, 194*(3), 441–457. https://doi.org/10.1083/jcb.201010027

Na, Y., Park, S., Lee, C., Kim, D. K., Park, J. M., Sockanathan, S., . . . Worley, P. F. (2016). Real-time imaging reveals properties of glutamate-induced Arc/Arg 3.1 translation in neuronal dendrites. *Neuron, 91*(3), 561–573. https://doi.org/10.1016/j.neuron.2016.06.017

Napoli, I., Mercaldo, V., Boyl, P. P., Eleuteri, B., Zalfa, F., De Rubeis, S., . . . Bagni, C. (2008). The fragile X syndrome protein represses activity-dependent translation through CYFIP1, a new 4E-BP. *Cell, 134*(6), 1042–1054. https://doi.org/10.1016/j.cell.2008.07.031

Niere, F., Wilkerson, J. R., & Huber, K. M. (2012). Evidence for a fragile X mental retardation protein-mediated translational switch in metabotropic glutamate receptor-triggered Arc translation and long-term depression. *Journal of Neuroscience, 32*(17), 5924–5936. https://doi.org/10.1523/JNEUROSCI.4650-11.2012

Ostroff, L. E., Santini, E., Sears, R., Deane, Z., Kanadia, R. N., LeDoux, J. E., . . . Klann, E. (2019). Axon TRAP reveals learning-associated alterations in cortical axonal mRNAs in the lateral amgydala. *Elife, 8*. https://doi.org/10.7554/eLife.51607

Ouyang, Y., Rosenstein, A., Kreiman, G., Schuman, E. M., & Kennedy, M. B. (1999). Tetanic stimulation leads to increased accumulation of Ca(2+)/calmodulin-dependent protein kinase II via dendritic protein synthesis in hippocampal neurons. *Journal of Neuroscience, 19*(18), 7823–7833.

Palacios, I. M., & St Johnston, D. (2001). Getting the message across: the intracellular localization of mRNAs in higher eukaryotes. *Annual Review of Cell and Developmental Biology, 17*, 569–614. https://doi.org/10.1146/annurev.cellbio.17.1.569

Paquin, N., & Chartrand, P. (2008). Local regulation of mRNA translation: new insights from the bud. *Trends in Cell Biology, 18*(3), 105–111. https://doi.org/10.1016/j.tcb.2007.12.004

Park, E., & Maquat, L. E. (2013). Staufen-mediated mRNA decay. *Wiley Interdisciplinary Reviews RNA, 4*(4), 423–435. https://doi.org/10.1002/wrna.1168

Parker, R., & Sheth, U. (2007). P bodies and the control of mRNA translation and degradation. *Molecular Cell, 25*(5), 635–646. https://doi.org/10.1016/j.molcel.2007.02.011

Protter, D. S. W., & Parker, R. (2016). Principles and Properties of Stress Granules. *Trends in Cell Biology, 26*(9), 668–679. https://doi.org/10.1016/j.tcb.2016.05.004

Proud, C. G. (2015). Regulation and roles of elongation factor 2 kinase. *Biochemical Society Transactions, 43*(3), 328–332. https://doi.org/10.1042/BST20140323

Reijnders, M. R. F., Janowski, R., Alvi, M., Self, J. E., van Essen, T. J., Vreeburg, M.,... Baralle, D. (2018). PURA syndrome: clinical delineation and genotype-phenotype study in 32 individuals with review of published literature. *Journal of Medical Genetics, 55*(2), 104–113. https://doi.org/10.1136/jmedgenet-2017-104946

Reschke, M., Clohessy, J. G., Seitzer, N., Goldstein, D. P., Breitkopf, S. B., Schmolze, D. B.,... Pandolfi, P. P. (2013). Characterization and analysis of the composition and dynamics of the mammalian riboproteome. *Cell Rep, 4*(6), 1276–1287. https://doi.org/10.1016/j.celrep.2013.08.014

Richter, J. D., Bassell, G. J., & Klann, E. (2015). Dysregulation and restoration of translational homeostasis in fragile X syndrome. *Nature Review Neuroscience, 16*(10), 595–605. https://doi.org/10.1038/nrn4001

Robert, F., Carrier, M., Rawe, S., Chen, S., Lowe, S., & Pelletier, J. (2009). Altering chemosensitivity by modulating translation elongation. *PLoS One, 4*(5), e5428. https://doi.org/10.1371/journal.pone.0005428

Ryan, V. H., & Fawzi, N. L. (2019). Physiological, pathological, and targetable membraneless organelles in neurons. *Trends in Neuroscience, 42*(10), 693–708. https://doi.org/10.1016/j.tins.2019.08.005

Sahoo, P. K., Smith, D. S., Perrone-Bizzozero, N., & Twiss, J. L. (2018). Axonal mRNA transport and translation at a glance. *Journal of Cell Science, 131*(8). https://doi.org/10.1242/jcs.196808

Scheetz, A. J., Nairn, A. C., & Constantine-Paton, M. (2000). NMDA receptor-mediated control of protein synthesis at developing synapses. *Nature Neuroscience, 3*(3), 211–216. https://doi.org/10.1038/72915

Schoenberg, D. R., & Maquat, L. E. (2012). Regulation of cytoplasmic mRNA decay. *Nature Review of Genetics, 13*(4), 246–259. https://doi.org/10.1038/nrg3160

Shan, J., Munro, T. P., Barbarese, E., Carson, J. H., & Smith, R. (2003). A molecular mechanism for mRNA trafficking in neuronal dendrites. *Journal of Neuroscience, 23*(26), 8859–8866.

Shen, L., & Pelletier, J. (2019). Selective targeting of the DEAD-box RNA helicase eukaryotic initiation factor (eIF) 4A by natural products. *Natural Products Report,* Issue 5 (2020). https://doi.org/10.1039/c9np00052f

Shigeoka, T., Jung, H., Jung, J., Turner-Bridger, B., Ohk, J., Lin, J. Q.,... Holt, C. E. (2016). Dynamic axonal translation in developing and mature visual circuits. *Cell, 166*(1), 181–192. https://doi.org/10.1016/j.cell.2016.05.029

Simone, L. E., & Keene, J. D. (2013). Mechanisms coordinating ELAV/Hu mRNA regulons. *Current Opinion in Genetics & Development, 23*(1), 35–43. https://doi.org/10.1016/j.gde.2012.12.006

Siwaszek, A., Ukleja, M., & Dziembowski, A. (2014). Proteins involved in the degradation of cytoplasmic mRNA in the major eukaryotic model systems. *RNA Biology, 11*(9), 1122–1136. https://doi.org/10.4161/rna.34406

Snijders Blok, L., Madsen, E., Juusola, J., Gilissen, C., Baralle, D., Reijnders, M. R.,... Kleefstra, T. (2015). Mutations in DDX3X are a common cause of unexplained intellectual disability with gender-specific effects on Wnt signaling. *American Journal of Human Genetics, 97*(2), 343–352. https://doi.org/10.1016/j.ajhg.2015.07.004

Sossin, W. S., & Costa-Mattioli, M. (2019). Translational control in the brain in health and disease. *Cold Spring Harbor Perspectives in Biology, 11*(8). https://doi.org/10.1101/cshperspect.a032912

Sossin, W. S., & DesGroseillers, L. (2006). Intracellular trafficking of RNA in neurons. *Traffic, 7*(12), 1581–1589. https://doi.org/10.1111/j.1600–0854.2006.00500.x

Steward, O., & Levy, W. B. (1982). Preferential localization of polyribosomes under dendritic spines in granule cells of the dentate gyrus. *Journal of Neuroscience, 2*(3) 284–291, doi10.1523/JNEUROSCI.02-03-00284

Sutton, M. A., & Schuman, E. M. (2006). Dendritic protein synthesis, synaptic plasticity, and memory. *Cell, 127*(1), 49–58. https://doi.org/10.1016/j.cell.2006.09.014

Taylor, A. M., Berchtold, N. C., Perreau, V. M., Tu, C. H., Li Jeon, N., & Cotman, C. W. (2009). Axonal mRNA in uninjured and regenerating cortical mammalian axons. *Journal of Neuroscience, 29*(15), 4697–4707. https://doi.org/10.1523/JNEUROSCI.6130-08.2009

Udagawa, T., Farny, N. G., Jakovcevski, M., Kaphzan, H., Alarcon, J. M., Anilkumar, S., . . . Richter, J. D. (2013). Genetic and acute CPEB1 depletion ameliorate fragile X pathophysiology. *Nature Medicine, 19*(11), 1473–1477. https://doi.org/10.1038/nm.3353

Vicario, A., Colliva, A., Ratti, A., Davidovic, L., Baj, G., Gricman, L., . . . Tongiorgi, E. (2015). Dendritic targeting of short and long 3′ UTR BDNF mRNA is regulated by BDNF or NT-3 and distinct sets of RNA-binding proteins. *Frontiers in Molecular Neuroscience, 8*, 62. https://doi.org/10.3389/fnmol.2015.00062

Wang, D. O., Kim, S. M., Zhao, Y., Hwang, H., Miura, S. K., Sossin, W. S., & Martin, K. C. (2009). Synapse- and stimulus-specific local translation during long-term neuronal plasticity. *Science, 324*(5934), 1536–1540. https://doi.org/10.1126/science.1173205

Wang, X., Posey, J. E., Rosenfeld, J. A., Bacino, C. A., Scaglia, F., Immken, L., . . . Moretti, P. (2018). Phenotypic expansion in DDX3X—a common cause of intellectual disability in females. *Annals of Clinical and Translational Neurology, 5*(10), 1277–1285. https://doi.org/10.1002/acn3.622

Welshhans, K., & Bassell, G. J. (2011). Netrin-1-induced local beta-actin synthesis and growth cone guidance requires zipcode binding protein 1. *Journal of Neuroscience, 31*(27), 9800–9813. https://doi.org/10.1523/JNEUROSCI.0166-11.2011

Wilhelm, J. E., & Vale, R. D. (1993). RNA on the move: the mRNA localization pathway. *Journal of Cell Biology, 123*(2), 269–274. https://doi.org/10.1083/jcb.123.2.269

Willis, D., Li, K. W., Zheng, J. Q., Chang, J. H., Smit, A. B., Kelly, T., . . . Twiss, J. L. (2005). Differential transport and local translation of cytoskeletal, injury-response, and neurodegeneration protein mRNAs in axons. *Journal of Neuroscience, 25*(4), 778–791. https://doi.org/10.1523/JNEUROSCI.4235-04.2005

Wolozin, B., & Ivanov, P. (2019). Stress granules and neurodegeneration. *Nature Reviews Neuroscience, 20*(11), 649–666. https://doi.org/10.1038/s41583-019-0222-5

Wong, H. H., Lin, J. Q., Strohl, F., Roque, C. G., Cioni, J. M., Cagnetta, R., . . . Holt, C. E. (2017). RNA docking and local translation regulate site-specific axon remodeling in vivo. *Neuron, 95*(4), 852–868 e858. https://doi.org/10.1016/j.neuron.2017.07.016

Yan, X., Hoek, T. A., Vale, R. D., & Tanenbaum, M. E. (2016). Dynamics of translation of single mRNA molecules in vivo. *Cell, 165*(4), 976–989. https://doi.org/10.1016/j.cell.2016.04.034

Yanow, S. K., Manseau, F., Hislop, J., Castellucci, V. F., & Sossin, W. S. (1998). Biochemical pathways by which serotonin regulates translation in the nervous system of Aplysia. *Journal of Neurochemistry, 70*(2), 572–583. https://doi.org/10.1046/j.1471-4159.1998.70020572.x

Yoon, B. C., Zivraj, K. H., & Holt, C. E. (2009). Local translation and mRNA trafficking in axon pathfinding. *Results and Problems in Cell Differentiation, 48*, 269–288. https://doi.org/10.1007/400_2009_5

Younts, T. J., Monday, H. R., Dudok, B., Klein, M. E., Jordan, B. A., Katona, I., & Castillo, P. E. (2016). Presynaptic protein synthesis is required for long-term plasticity of GABA release. *Neuron, 92*(2), 479–492. https://doi.org/10.1016/j.neuron.2016.09.040

Zalfa, F., Giorgi, M., Primerano, B., Moro, A., Di Penta, A., Reis, S., . . . Bagni, C. (2003). The fragile X syndrome protein FMRP associates with BC1 RNA and regulates the translation of specific mRNAs at synapses. *Cell, 112*(3), 317–327. https://doi.org/10.1016/s0092-8674(03)00079-5

Zhang, F., Kang, Y., Wang, M., Li, Y., Xu, T., Yang, W., . . . Jin, P. (2018). Fragile X mental retardation protein modulates the stability of its m6A-marked messenger RNA targets. *Human Molecular Genetics, 27*(22), 3936–3950. https://doi.org/10.1093/hmg/ddy292

PART III

PHYSIOLOGICAL ROLES OF TRANSLATION

CHAPTER 12

PROTEIN SYNTHESIS AND SYNAPSE SPECIFICITY IN FUNCTIONAL PLASTICITY

RADHA RAGHURAMAN, AMRITA BENOY, AND SREEDHARAN SAJIKUMAR

Introduction

Synaptic plasticity, wherein synapses are weakened or strengthened in response to different input stimulation intensities and frequencies, is considered the cellular correlate of learning and memory (Bear, 1996). Long-term potentiation (LTP) is a form of synaptic plasticity whereby the synaptic strength is increased in response to strong input stimulation. A persistent potentiation of synaptic strength, termed late-LTP (L-LTP), lasting for at least eight hours in vitro (E. P. Huang, 1998; Ostroff, Fiala, Allwardt, & Harris, 2002), is protein-synthesis dependent. Plasticity-related products (PRPs) synthesized in response to the signaling cascades triggered by the L-LTP inducing stimuli, lead to the persistence of potentiation, which translates as the long-term preservation of memories. However, several theories/models exist for the protein-synthesis dependence of L-LTP. The synaptic tagging and capture model, as well as the synaptic cross-tagging model, discusses the role of tagging of active synapses in being able to capture PRPs and thereby enable the persistence of synaptic potentiation at specific synapses (Frey & Morris, 1998b; Sajikumar, Navakkode, Sacktor, & Frey, 2005). In addition, local dendritic protein synthesis, and a possible cooperation between dendritically and somatically synthesized proteins are suggested to be involved in the maintenance of LTP (Sherff & Carew, 1999; Sutton & Schuman, 2006). However, protein synthesis in the absence of a balance by proteasomal degradation of PRPs does not result in L-LTP. The balance between synthesis and degradation of PRPs is thus critical in LTP maintenance (Fonseca, Vabulas, Hartl, Bonhoeffer, & Nagerl, 2006). Epigenetic mechanisms activated during memory consolidation may poise genes for reactivation at a later point in time.

Protein synthesis in memory: Role in consolidation of LTP

LTP is the persistent strengthening of synapses that explains the cellular and molecular processes that underlie learning and memory formation (Lynch, 2004; Whitlock, Heynen, Shuler, & Bear, 2006). It was imperative to understand the role of protein synthesis in hippocampal LTP, given the hippocampal role in various memory processes (Milner, 1972; O'Keefe & Nadel, 1978; Penfield & Milner, 1958; Teyler & DiScenna, 1986; Thompson, 1986). Studies have demonstrated that the induction and maintenance of LTP is governed by pre-synaptic and post-synaptic mechanisms (Buzsáki, 1985; Frey, Krug, Reymann, & Matthies, 1988; Desmond & Levy, 1986; Dolphin, Errington, & Bliss, 1982; Krug, Brodemann, & Ott, 1982). Studies previously have shown how presynaptic mechanisms leading to an increase in transmitter release are involved at least in early-phase of LTP (E-LTP) in the dentate area while the post-synaptic changes of the membrane structure and second messenger systems engender the long-lasting maintenance of LTP (Akers & Routtenberg, 1985; T. V. Bliss, Errington, Lynch, & Williams, 1990; Dolphin et al., 1982; Reymann, Frey, Jork, & Matthies, 1988). Nevertheless, several recent studies suggest that expression of LTP involves the reinforcement of both pre and post synaptic function and is not confined to a single neuron (Lisman & Raghavachari, 2006). The requirement of protein synthesis in LTP maintenance was established by the observation that anisomycin, a protein synthesis inhibitor, completely abolished the late phase of LTP (Krug, Lossner, & Ott, 1984). The dependence of late phase LTP on intact protein synthesis has been corroborated by findings from Frey et al. (Frey et al., 1988). Further evidence for protein synthesis dependency of long-lasting plasticity was shown in *Aplysia* where protein synthesis blockers blocked long-term heterosynaptic facilitation which correlates to learning, but not short-lasting forms of facilitation (Montarolo et al., 1986).The requirement of protein synthesis for both LTP and memory is consistent with a role for LTP in the formation of memories.

The continued expression of LTP depends on activation of gene transcription and protein synthesis (Roberson, English, & Sweatt, 1996). Post-synaptic calcium influx in response to strong synaptic stimulation activates intracellular signaling cascades leading to an increase in cAMP concentration and activation of protein kinase A (PKA; Lynch, 2004). PKA catalytic subunit translocates to the nucleus and phosphorylates and activates the transcription factor cAMP response element-binding protein (CREB; Kandel, 2012). CREB can also be activated by mitogen-activated protein kinases/extracellular signal regulated kinases (MAPK/ERKs) and Ca^{2+}/calmodulin kinases (CaMKs; Kandel, 2012). Phosphorylated CREB aids in transcription of plasticity-related genes and is thereby a central component of memory storage (Bourtchuladze et al., 1994; Kida, 2012). Studies have shown that a constitutively active form of CREB, VP16-CREB, lowers the threshold for L-LTP in hippocampal CA1 neurons at Schaffer collateral synapses (Barco, Alarcon, & Kandel, 2002). cAMP response element (CRE)-driven gene products result in a cell-wide priming for LTP and synaptic capture of CRE-driven gene products enable LTP consolidation at tagged active synapses that are only weakly stimulated (Barco et al., 2002). Extracellular signal regulated kinase ERK1/2 has an important role in regulating CREB-mediated gene transcription and long-term memory (Giese & Mizuno, 2013). CaMKIV, a Ca^{2+}/calmodulin kinase found exclusively in the nucleus, is another crucial component in the regulation of CRE-dependent transcription and the sustenance of hippocampal L-LTP and memory consolidation (Kang et al., 2001). Interestingly, CREB has been proposed to be involved in neuronal memory allocation, wherein neurons with higher levels of CREB are more excitable and therefore has a higher probability to fire in response to sensory input and be represented to a greater extent in the memory trace (Zhou et al., 2009).

Studies have investigated how LTP can be reversed. One of the studies from (Hardt, Nader, & Wang, 2014) showed that the E-LTP decay is caused by homeostatic downscaling while the decay of late phase LTP is mediated by the regulators of metaplasticity phenomenon such as the NMDA receptors with their subunit composition, wherein both E-LTP and L-LTP processes mandate the activity-dependent removal of post-synaptic GluA2-containing α-amino-3-hydroxy-5-methyl-4-isoxazolepropionic acid receptors (AMPARs). The number and composition of AMPARs dictate the stability and formation of LTP where *N*-methyl-D-aspartate receptor (NMDAR) is a core regulating element. In addition, it is well known that NMDAR activation leads to an intracellular increase in calcium which in turn triggers downstream signaling processes involving Ca^{2+}/calmodulin-dependent protein kinase II (CaMKII), protein kinase C (PKC), transcription factors such as CREB/C/EBP, translation initiation factors (e.g., eIF4E) and growth factors such as brain-derived neurotrophic factor (BDNF; Rampon et al., 2000; Tsien, Huerta, & Tonegawa, 1996). Research over the years has established that the atypical PKC isoform M-zeta (PKMζ) is involved in the late phase of LTP and in the consolidation processes (Sacktor, 2011, 2012; Shema et al., 2011). The PKMζ protein is known to promote its own synthesis leading to PKMζ mediated late-LTP (Shema et al., 2011; Whitlock et al., 2006).

AMPAR with GluA2 subunits are pivotal for long-term memory formation (Collingridge, Peineau, Howland, & Wang, 2010; C. H. Kim, Chung, Lee, & Huganir, 2001; Scholz et al., 2010; Y. T. Wang & Linden, 2000). There have been findings highlighting the involvement of PKMζ with the mechanisms that regulate AMPAR/GluA2 expression. PKMζ infusions have been shown to increase the AMPAR currents and the post-synaptic insertion of AMPAR/GluA2s as well as preventing the activity-dependent endocytosis of GluA2/AMPARs (Hardt et al., 2014). These studies, thus engendered the idea that synaptic efficacy and the long-lasting changes concomitant with it, require incessant maintenance for the memory to not be perished. It has been suggested that LTP decay is a direct consequence of the effects of protein turnover (Genoux et al., 2002; C. C. Huang, Liang, & Hsu, 2001; E. P. Huang, 1998). Alternatively, some studies have brought out the significance of NMDAR activation in LTP decay and this LTP decay may have an overlap in the signaling pathways of LTD, showing decreased AMPAR levels at the post-synaptic sites on depotentiation and LTD (Cazakoff & Howland, 2011; Dalton, Wang, Floresco, & Phillips, 2008; J. Kim et al., 2007; H. K. Lee & Kirkwood, 2011; Park, Lee, Kim, & Choi, 2012; Villarreal, Do, Haddad, & Derrick, 2002). LTP decay may be mediated by both depotentiation and LTD that results in the decrease of synaptic potentiation. Recent studies have demonstrated a role for AMPAR surface diffusion in facilitating hippocampal LTP and contextual learning and that, this diffusion is governed by protein-protein interactions which are anticipated to be a fundamental trafficking mechanism for LTP (Penn et al., 2017).

In view of the critical requirement for protein synthesis in maintaining synaptic plasticity, two models for activity-induced protein synthesis in postsynaptic neurons will be discussed in detail:

1. **Somatic protein synthesis and synaptic tagging and capture:** Activated synapses are marked by synaptic tags which capture the somatically synthesized plasticity-related products synthesized from translation of plasticity-related mRNAs in the cell body in response to strong synaptic activation.
2. **Local dendritic protein synthesis:** Plasticity-related mRNAs are translated locally in neuronal dendrites and the proteins thus synthesized are used at the activated synapses to sustain LTP.

Synaptic tagging and capture

Synaptic tagging and capture is a cellular correlate of associative plasticity/memory where plasticity/memories triggered by weakly stimulating events are converted to long-term memories by means of associating the short-lasting memories with strong memories. A strong synaptic stimulation that results in protein-synthesis dependent late-LTP in synapse S1 helps transform a short-lasting E-LTP at another input S2 to the same neuron to long lasting late-LTP by virtue of this capture of PRPs synthesized in response to strong synaptic stimulation at S1. This is possible by means of the synaptic tags set at the weakly stimulated synapse S2 that capture PRPs.

The process of tagging and capture should occur within the time frame of decay of PRPs synthesized as a result of the strong stimulation at S1. Therefore, the time window between the two synaptic stimulations is crucial. PRPs are either synthesized in the soma or in the dendrites and thereafter distributed diffusely throughout the dendritic tree. The synaptic tag helps sequester these proteins and prevents the requirement for extensive protein trafficking (Frey & Morris, 1998a) and also helps facilitate input specificity. The requirements for the synthesis and trafficking of PRPs in this process demonstrates the need for protein synthesis in associative plasticity. Figure 12.1 demonstrates the protein synthesis dependence of the synaptic tagging hypothesis. Strong tetanization (STET) at synapse S1 resulting in L-LTP followed by STET at S2 in presence

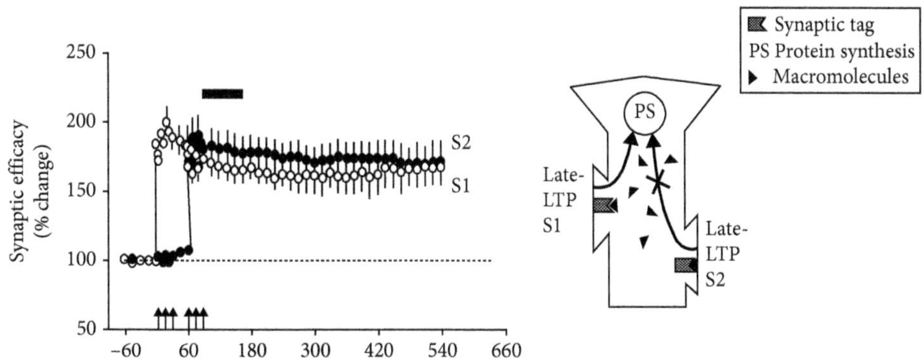

FIGURE 12.1. L-LTP induced in S1 (open circles) without protein synthesis inhibitor anisomycin. 35 minutes after tetanization in S1, anisomycin (black bar) was bath applied, and 1 hour after LTP induction in S1, L-LTP was induced in S2 (filled circles) by repeated tetanization. L-LTP was still observed in S2 despite protein synthesis inhibition, supporting the synaptic tag hypothesis. The cartoon on the right demonstrates that synaptic tag is set at both synaptic inputs S1 and S2 but protein synthesis is only activated by S1, since anisomycin prevents protein synthesis at S2.

From Frey & Morris (1998a). Reprinted from *Trends in Neurosciences*, with permission of Elsevier.

of anisomycin still results in L-LTP at S2, indicating that PRPs resulting from L-LTP at S1 were captured by the tag set at S2.

PKMζ is a critical LTP-specific PRP required for synaptic tagging. This has been proven from studies where the persistent potentiation at the tagged synapses by virtue of L-LTP at neighboring synapses, was blocked when inhibitors of PKMζ were applied after tetanization (see Figure 12.2; Sajikumar et al., 2005). Although the specificity of the inhibitor ZIP has been questioned (Glanzman, 2013) and even the role of PKMζ itself in the maintenance of memories (A. M. Lee et al., 2013; Volk, Bachman, Johnson, Yu, & Huganir, 2013), subsequent studies have demonstrated compensatory mechanisms via PKCι/λ that enable the maintenance of LTP and long-term memories in PKMζ-null mice (Tsokas et al., 2016). Thus, while PKMζ is essential for LTP in wild-type mice, compensatory mechanisms enable LTP maintenance and long-term memories in PKMζ-null mice (Tsokas et al., 2016).

The lifetime of the tagged state is considered to be approximately 90 minutes in brain slice experiments (Frey & Morris, 1998b), and PRPs arriving after tag decay cannot be added to the PSD of dendritic spines, and thus is unable to transform E-LTP to L-LTP beyond the 90 minutes tagging lifetime window. Not only is the associative maintenance of LTP dependent on the half-life of the synaptic tags, but also on the half-life of PRPs. Therefore, the interaction between the tag and the PRPs must take place within the time window dictated by their individual half-lives (Sajikumar, Navakkode, & Frey, 2007). The dependence on half-life of PRPs for the associative plasticity mechanism can be demonstrated by employing a strong-before-weak tetanization paradigm in *in vitro* slice electrophysiology experiments. The E-LTP to L-LTP transformation by virtue of capture

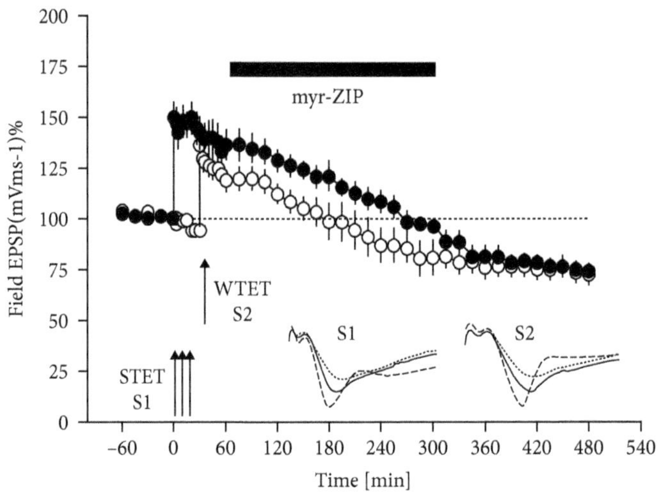

FIGURE 12.2. E-LTP to L-LTP transformation at tagged synapses is prevented by bath application of myristoylated ζ-pseudosubstrate inhibitory peptide (myr-ZIP) that inhibits PKMζ. L-LTP is induced at one set of synapses (S1 filled circles) by means of strong tetanization (STET). This is followed by induction of E-LTP in S2 (open circles) by weak tetanization (WTET) and bath application of myr-ZIP 30 minutes after WTET to S2. This results in fall of LTP at both S1 and S2 synapses.

of PRPs synthesized in response to the STET depends on how temporally spaced the WTET is from the STET. The more spaced the two tetanizations, the PRPs are subject to decay and hence cannot facilitate the associative transformation of E-LTP to L-LTP at the weakly tetanized synapses.

The STC hypothesis has been revised over the years. The current view of the nature of the synaptic tag is that it is not a single molecule, but a state of the synapse, wherein tagging should be viewed as a temporary structural state of the synapse involving interaction between several proteins rendering an "unlocking" of the synapse for stabilization of synaptic plasticity (Redondo & Morris, 2011). Tagging possibly involves an alteration of the dendritic spine architecture rendering remodeling of the postsynaptic density (PSD). A tagged synapse has the potential to transform its potentiated state to persistent long-lasting plasticity if it receives PRPs that will stabilize the functional and structural alterations of the synapse, within the time window of decay of the tagged state. For example, autophosphorylated CaMKII that moves into the PSD of activated spines in LTP (Shen & Meyer, 1999), and although less accessible to inactivation by phosphatases (Bayer et al., 2006; Bayer, De Koninck, Leonard, Hell, & Schulman, 2001), can still be inactivated over time and released from the PSD (Redondo & Morris, 2011; Strack, Choi, Lovinger, & Colbran, 1997) unless there is a PRP capture process that stabilizes the tagged state. The structural scaffold at the tagged synapse will revert to a "locked" state and over time the kinases responsible for the tagged state of the synapse become inactivated and the synapse returns to an untagged state (Redondo & Morris, 2011). The capture of PRPs culminates in spine remodeling, addition of PSD slots or scaffolding molecules available for AMPARs (Bosch et al., 2014; Luscher, Nicoll, Malenka, & Muller, 2000) and presynaptic modifications that include changes in synaptic vesicle release sites (Lisman & Raghavachari, 2006). AMPAR trafficking into and out of the PSD slots continues in a dynamic fashion and the final state of the tagged and PRP-captured synapse leads to the persistence of LTP (Redondo & Morris, 2011).

Availability of Plasticity Proteins Decides Synaptic Cooperation and Competition

Synaptic tagging and capture offers a blueprint to understand the interaction between different groups of synapses and the possible mechanisms that account for memory maintenance and establishment. Associative learning is an unceasing process that can impact how information is processed amongst the activated synapses resulting in the modulation of the ability to induce and maintain LTP and LTD (Etkin et al., 2006; Govindarajan, Israely, Huang, & Tonegawa, 2011). These dynamic interactions among synapses can result in either synaptic cooperation or synaptic competition. The synaptic tagging and capture model (STC) was successful in establishing the cooperation phenomenon by demonstrating that synapses share the PRPs (Frey & Morris, 1998a). It was shown by studies from (Redondo et al., 2010; Sajikumar et al., 2005, 2007) that "tag setting" and LTP maintenance are processes that occur separately in time and in an independent manner. Synaptic cooperation among the synapses is space-restricted. This space constraint brings about a bias for correlated neurons even at the developmental phase for establishing connections (Turney & Lichtman, 2012). Close-knit dendritic branches have a higher probability of summation and thus the induction of LTP and formation of synaptic tags. By corollary, this also increases the chances of the same being involved in synaptic competition where PRPs may be inadequately available. The occurrence of limited PRPs in the presence of multiple inputs results in these PRPs being distributed competitively among all activated synapses. During such a situation, parameters such as the tag strength, distance between activated synapse and translational initiation site and the time between the two events have a bearing, on which synapse is to be stabilized. Studies from (Sajikumar & Frey, 2004; Sajikumar, Morris, & Korte, 2014) have elucidated the significance of synaptic competition in the formation of long lasting memory. These studies showed that the potentiation of a third pathway around the same time that the synaptic potentiation is enabled on a specific pathway, given the availability of PRPs from another earlier or later event, may result in the blockade of potentiation at all the pathways. Shetty et al. (Shivarama Shetty, Gopinadhan, & Sajikumar, 2016) showed that D1/D5 receptor mediated potentiation in a dose dependent manner aids in the fine tuning of associativity processes for long-term memory formation. This study also elucidated ERK1/2 as the molecular pathway involved in the synaptic associativity process in the differential regulation of synaptic cooperation and competition (Figure 12.3 a, b).

Metaplasticity prolongs associative plasticity and rescues synaptic competition

The term *metaplasticity* coined by Abraham and Bear in 1996 (Abraham & Bear, 1996) dictates the threshold levels at which the tags can operate. In addition, the time window during which these tags are functional is also altered. The normal time window of a synaptic tag and capture phenomenon is confined to an E-LTP duration of 60 minutes (Frey & Morris, 1998a, 1998b; Redondo & Morris, 2011) which is mediated by CaMKII (Sajikumar et al., 2007). As a result, the information from various synapses is stipulated to be integrated within this restricted time period (Q. Li et al., 2014). There are evidences from studies showing that RyR (Ryanodine receptors) activation before the induction of LTP can prolong the duration of tags from normal 1 hour to more than 5 hours offering more leeway for associative interaction for an extended period of time (Q. Li et al., 2014). By virtue of switching synaptic tags from the short-lived CaMKII tag to a longer-lived tag, this process is mediated by PKMζ, wherein, a metaplasticity-enabled synaptic tagging and capture process (primed-STC) enables the primed synapses to capture PRPs across a prolonged duration of 4-5 hours (Q. Li et al., 2014). This is mediated through an extended half-life of PKMζ and is a longer-lasting tag setting process which facilitates the extended time window for both the sets of synapses to associate. It is noteworthy that this study brought out the facets of PKMζ as a molecule that can mediate synaptic tags aside being able to function as a PRP (Figure 12.4).

The metaplasticity-enabled recalibration of threshold levels is believed to facilitate learning and establish homeostasis. Studies from Li et al. (Q. Li et al., 2017) demonstrated that L-LTP and STC and cross tagging and capture are impaired in the APP/PS1 mouse model of Alzheimer's disease and could be rescued by priming synapses with activation of RyRs through metaplastic mechanisms. Some cognitive impairment in neurodegenerative and neurological disorders stems from aberrant metaplasticity-mediated mechanisms. Hence exploring these mechanisms could offer insights on understanding behavioral outcomes and harnessing clinical benefits. Activation of metabotropic glutamate receptors mGluRs is one of the widely studied forms of homosynaptic metaplasticity (Tim V. P. Bliss, Collingridge, Morris, & Reymann, 2018). mGluRs trigger the cell signaling pathways inclusive of the activation of PKC (group 1) as well as inhibition of cAMP (groups II and III). This offered the incentive to investigate the effects of mGluRs in the induction of LTP by using mGluR antagonist (Behnisch, Fjodorow, & Reymann, 1991). Collingridge, Watkins, and Jane then confirmed similar findings by developing a selective mGluR antagonist (+)-α-methyl-4-carboxyphenylglycine

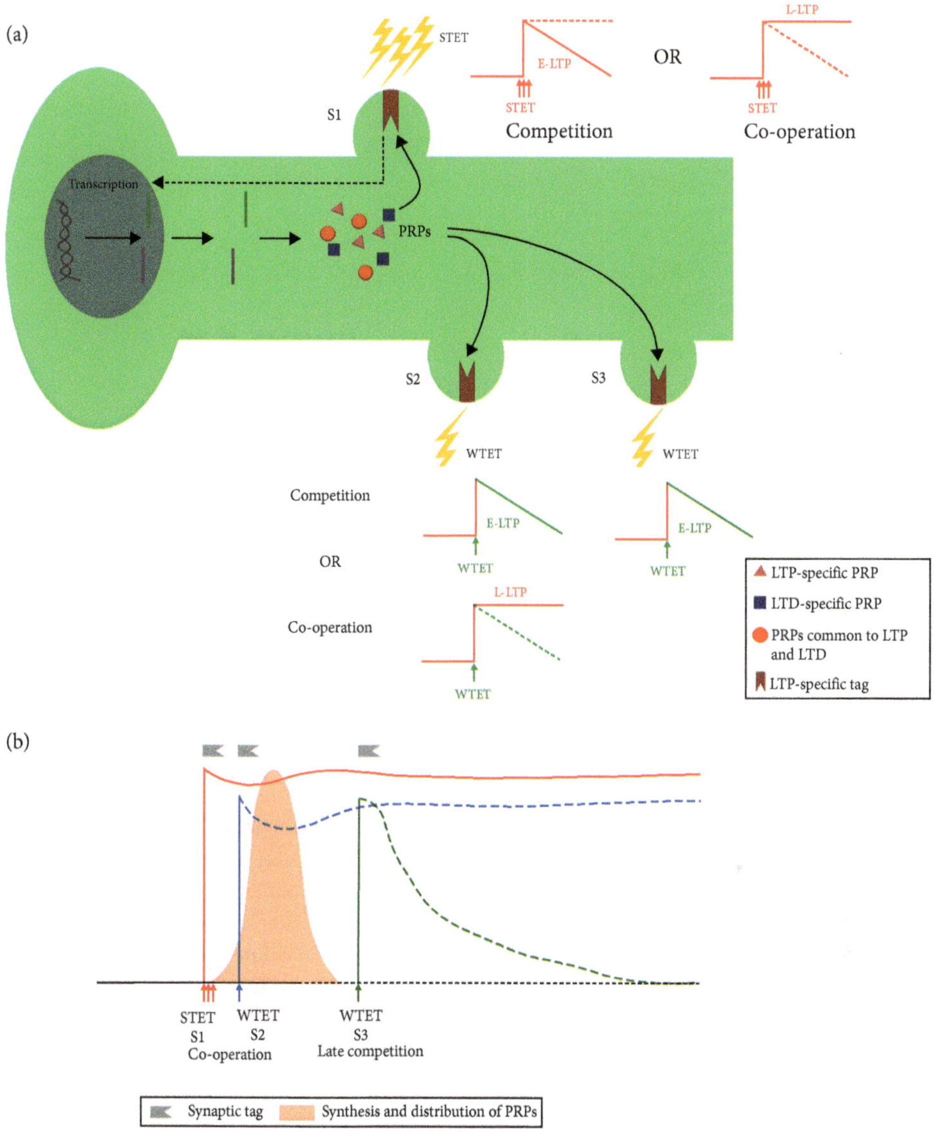

FIGURE 12.3. Synaptic tagging, competition, and cooperation model. (a) Synaptic input S1 subjected to strong tetanization (STET; triple yellow flash symbol) leads to transcription (purple and green bars representing mRNA) and thereafter translation of process-dependent and process-independent plasticity-related products (PRPs). Neighboring synaptic inputs are subjected to weak tetanization (WTET; single yellow flash symbol). Synaptic tags set at all the stimulated synapses (weak and strong) either compete or cooperate for capture of process-specific PRPs. Limited PRPs in the face of multiple synaptic inputs (S1, S2, S3) puts a constraint on sufficient synaptic strengthening at all the activated synapses. This leads to short-lasting E-LTP to be expressed at all three synapses due to the increased need but inadequate availability of PRPs (synaptic competition). Competition is fierce when multiple synaptic inputs are activated at the same time and compete for a limited pool of PRPs. If there were no third synaptic stimulation at S3 and sufficient PRPs are synthesized and available within the time window of decay of the synaptic tags, S1 and S2 would cooperate and enable L-LTP at both S1 and S2. (b) LTP maintenance by virtue of STC is possible when the products of PRP synthesis overlap in time with the availability of tag setting. If S3 is stimulated much later than S1 and S2, this leads to a late competition that would allow maintenance of LTP at S1 and S2 (cooperation) as PRP synthesis overlaps with tag setting but short-lasting E-LTP at S3 as it does not overlap with PRP synthesis.

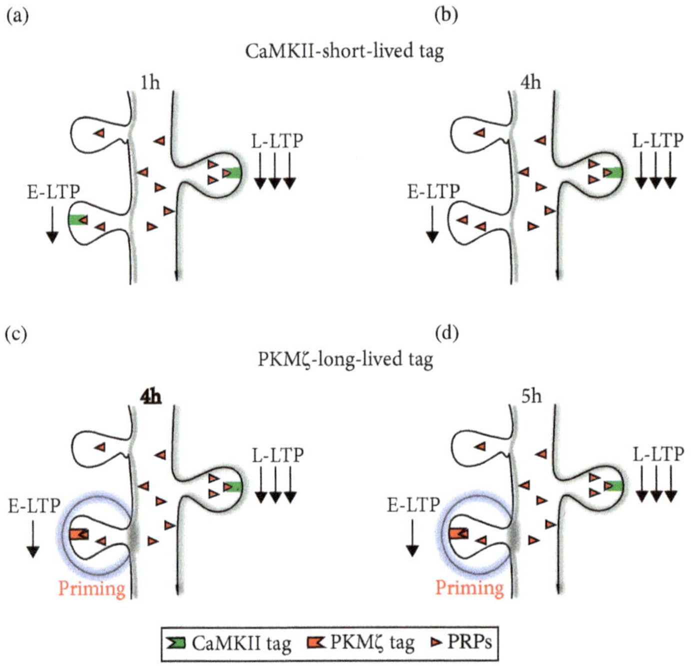

FIGURE 12.4. Metaplasticity enables LTP associativity by switching synaptic tags. (a) In non-primed STC, synapses are marked with the short-lasting CaMKII tag that lasts up to 1 hour, the time window during which the PRPs such as PKMζ synthesized in response to L-LTP inducing STET at neighboring synapses, are captured. (b) Prolonging the interval between the E-LTP-inducing WTET and L-LTP-inducing STET to 4 hours does not enable STC, as the CaMKII tag is short-lived and does not persist for a long period. (c) & (d) In primed-STC, the E-LTP tag is switched from CaMKII to PKMζ, which is a long-lived tag and can thus enable STC by capture of PRPs across 4–5 hours.

Adapted and modified from Q. Li, et al. (2014) by permission of Oxford University Press.

(MCPG) (Bortolotto et al., 1995). Studies from (Sajikumar et al., 2014) have shown how priming of mGluRs acts to prevent synaptic competition. This was shown in a three input model, where L-LTP was originally induced in one of the synaptic inputs followed by E-LTP (strong before weak paradigm) at 30 and 45 minutes in other synaptic inputs S2 and S3, respectively. Synaptic competition was observed in these synaptic clusters, in which all the synapses compete for scarce plasticity related proteins, thus preventing all forms of plasticity. Competition was prevented with the metaplastic activation of mGluR prior to the induction of L-LTP in the three input model. Synaptic competition can be overcome by increasing PRP availability over time by promoting transcriptional activation and increased PRP synthesis by metaplastic stimulation (Sajikumar et al., 2014). Metaplasticity-mediated rescue of synaptic competition is depicted in Figure 12.5. Thus, a combination of synaptic tag switching and increased PRP synthesis possibly facilitates the strengthening of synapses, prolonged associativity and rescue of synaptic competition, thereby contributing to robust forms of memory.

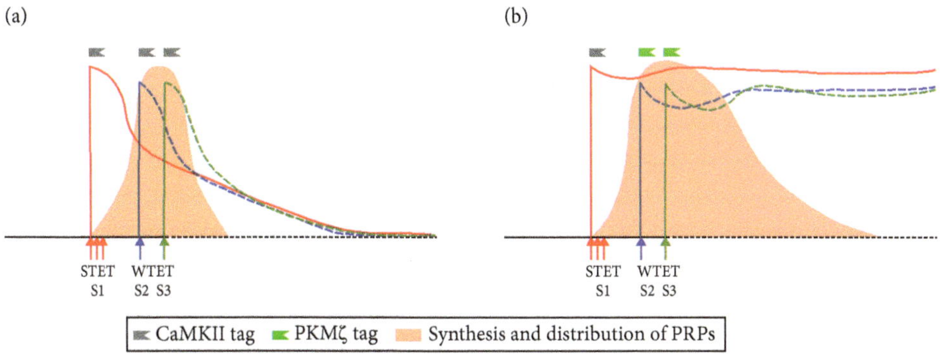

FIGURE 12.5. Model of synaptic competition and metaplasticity-mediated rescue of competition. (a) In the strong-before-weak stimulation paradigm when STET delivered first at synaptic input S1 and E-LTP-inducing WTET at S2 30 minutes and at S3 45 minutes after the STET, there is no STC-mediated transformation of E-LTP to L-LTP at S2 and S3, and all inputs, including S1 where STET was delivered, return to baseline levels of synaptic potentials. This is due to fierce competition for PRPs in the face of insufficient availability of PRPs. (b) Metaplastic stimulation rescues this synaptic competition by increasing PRP synthesis, thereby facilitating late maintenance of LTP at S1, and by prolonging associativity it enables E-LTP to L-LTP transformation at S2 and S3.

Behavioral tagging

The behavioral tagging (BT) hypothesis postulates that a learning experience triggers the setting of a tag along with the induction of PRPs. Thus the behavioral tagging requires the integration of these two processes at common neural substrates within a critical time window. This was proven by studies from (Ballarini, Moncada, Martinez, Alen, & Viola, 2009) wherein a taste recognition task that induces activation of insular cortex and a spatial learning task that activates hippocampus failed to integrate the formation of CTA-LTM (conditioned taste aversion–long-term memory) using the exposure to a novel environment; and reciprocally, the taste recognition task failed to integrate consolidation of SOR-LTM (spatial object recognition–long-term memory). These results corroborate the importance of spatial coexistence of both tags and PRPs to facilitate a long lasting memory. BT processes also exhibit input specificity wherein PRPs are captured only by the tagged sites reinforcing only the tagged sites and not every single input of a network. It has been established that novelty induces LTP reinforcement in the BT (behavioral tagging) processes and that this phenomenon depends on D1/D5 dopaminergic receptors functionality (S. Li, Cullen, Anwyl, & Rowan, 2003). Further to this evidence, there are studies supporting that these receptors may have a crucial role in triggering the synthesis of PRPs. Direct administration of SKF-38393 and adrenergic (dobutamine) agonists could replace novel experience to promote IA-LTM consolidation. Inhibitory avoidance task is a widely used behavioral task which allows investigating and manipulating the accuracy of memory in the study of fear learning and memory mechanisms in rodents. In the behavioral tagging hypothesis, wIA task which helps set the tags utilizes the PRPs induced as a result of novel open field resulting in the IA-LTM consolidation. The injection of these drugs failed to promote IA memory 180 minutes prior to weak IA training, unlike the case wherein drugs injected 70 minutes prior to the wIA training showed IA-LTM consolidation. This proves that the strong event is effective when occurring within a critical time window around the weak one and thereby the time scale of the agonists proves consistency with novelty being used as memory promoter. Activation of D1/D5 dopaminergic and beta-adrenergic receptors in the hippocampus during strong IA training is specifically involved in mediating the synthesis of PRPs imperative for the memory consolidation (Moncada, Ballarini, Martinez, Frey, & Viola, 2011).

Cross-tagging and capture: Early LTP to late LTP conversion by virtue of late-LTD at neighboring synapses

Associative plasticity is not solely limited to early and late forms LTD or LTP acting in isolation. It also involves interactions between LTP and LTD at different synaptic inputs to the same neuron, in a process called "cross-tagging." Similar to synaptic tagging and capture, synaptic "cross-tagging" is also a conceptual cellular correlate of associative memory, wherein L-LTD/L-LTP at one synaptic input S1 to a neuron helps transform an early form of the opposite plasticity event viz. E-LTP/E-LTD at another synaptic input S2 to the same neuron to its long-lasting late form (Sajikumar & Frey, 2004). In terms of E-LTP to L-LTP transformation at a synaptic input S2 by virtue of L-LTD at a separate synaptic input S1 to the same neuron, "cross-tagging" involves capture of PRPs synthesized as a result of the late plasticity event (here L-LTD) at synaptic input S1. The L-LTD at S1 leads to synthesis of a pool of PRPs- both LTD and LTP-specific. E-LTP inducing weak stimulation at S2 sets synaptic tags at S2 that are able to capture process-specific PRPs such as PKMζ (Sajikumar et al., 2005) and thereby help transform E-LTP at S2 to L-LTP. It has been shown that PKMζ inhibitors were able to block this E-LTP to L-LTP transformation of the weakly tetanized cross-tagged synapses (S2) although the L-LTD maintenance at S1 was unaffected by inhibiting PKMζ, showing that PKMζ is an LTP-specific PRP that is necessary for the persistent potentiation demonstrated at the cross-tagged synapses (Figure 12.6).

CaMKIIβ and Arc/Arg3.1 interaction in inverse synaptic tagging

"Inverse synaptic tagging" is an inactivity-dependent redistribution of synaptic weights (Okuno, Minatohara, & Bito, 2018). Similar to proteins that tag active synapses in the STC model enabling capture of PRPs and in turn maintenance of synaptic potentiation,

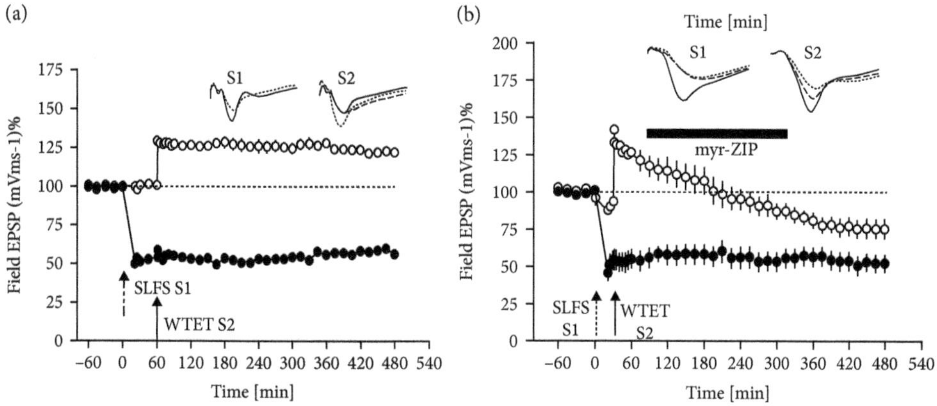

FIGURE 12.6. (a) Cross-tagging between LTD and LTP enables E-LTP synapses (S2 open circles) to be transformed to L-LTP. Strong low-frequency stimulation (SLFS) at one set of synapses (S1 filled circles) leads to L-LTD in S1 and synthesis of a common pool of PRPs. When weak tetanization(WTET) is given 1 hour later in S2, an E-LTP develops in S2. Capture of LTP-specific PRPs from the common pool of PRPs enables E-LTP in S2 to be transformed to L-LTP. (b) Inhibition of PKMζ prevents E-LTP to L-LTP transformation in S2 (open circles). Bath application of myristoylated ζ-pseudosubstrate inhibitory peptide (myr-ZIP) prevents the ability of L-LTD synapses (S1 filled circles) to prolong the LTP at S2, where field EPSP values soon fall below baseline values.

From Sajikumar et al. (2005). Republished with permission of Society for Neuroscience, conveyed through Copyright Clearance Center, Inc.

an "inverse synaptic tag" marks inactive synapses for synaptic weakening. Such differential tagging allows the activity-dependent maintenance of the contrast between strong and weak synapses. CaMKIIβ has been suggested to be an "inverse synaptic tag" that marks inactive synapses (Okuno et al., 2012). A dynamic interaction between the immediate early gene product Arc/Arg3.1 and CaMKIIβ has been shown to facilitate the inverse tagging process (Okuno et al., 2012). The inactive, calmodulin-unbound form of CaMKIIβ has a much higher affinity for Arc/Arg3.1 than the active form of CaMKIIβ (Okuno et al., 2012). It has been proposed that synaptic tagging and inverse synaptic tagging coexist and thereby enable redistribution of synaptic weights and enhance the contrast between potentiated and non-potentiated synapses (Okuno et al., 2018), thereby possibly playing a homeostatic role. A similar mechanism has been shown to be brought about by recent studies on G9a/GLP complex through its properties as a bidirectional switch that regulates mGluR-dependent plasticity. Studies from (Sharma & Sajikumar, 2018) has demonstrated the involvement of this epigenetic mechanism in facilitating the expression of mGluR-LTD when turned on and in promoting the expression of long-term potentiation when turned off and as a result, indicating a homeostatic role of the G9a/GLP complex in bringing about a delicate balance between the strong and the weak synapses during the course of long term synaptic plasticity.

The inverse tagging model involving Arc/Arg3.1 and CaMKIIβ interaction is shown in Figure 12.7.

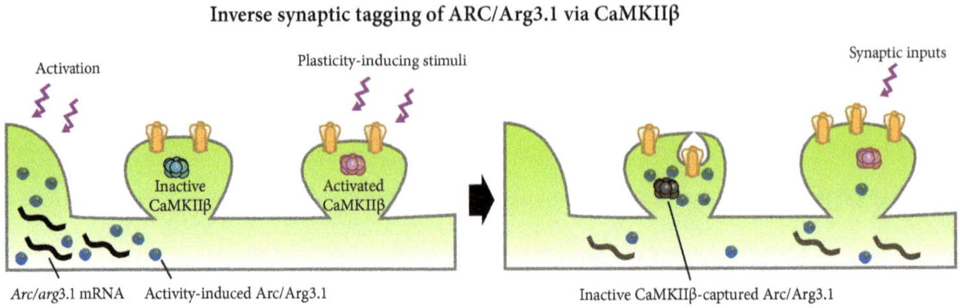

FIGURE 12.7. Plasticity-inducing stimuli such as a high frequency stimulation leads to Arc/Arg3.1 induction in the soma which is then transported to dendritic spines of both potentiated and non-potentiated synapses. However, over time Arc/Arg3.1 is lost from active synapses and accumulates in non-potentiated synapses, and it is this accumulation that has been shown to be dependent on the interaction between Arc/Arg3.1 and inactive CaMKIIβ. This also allows the AMPAR removal from the post-synaptic membrane and thereby facilitate synaptic weakening.

Adapted from Okuno et al. (2018). Reprinted with permission of Elsevier.

Protein synthesis for LTP maintenance and associative plasticity in hippocampal area CA2: Role of PKMζ and CaMKIV

The hippocampal subfield CA2, a small area sandwiched between the more well studied CA1 and CA3 subfield, has gained importance with findings indicating its role in social memory (Hitti & Siegelbaum, 2014). CA2 functional plasticity is unique in comparison to the other CA subfields in that the Schaffer collateral-CA2 (SC-CA2) synapses are resistant to activity-dependent LTP whereas the entorhinal cortical synapses onto CA2 (EC-CA2) express robust LTP (Chevaleyre & Siegelbaum, 2010; Dudek, Alexander, & Farris, 2016). However, neuromodulators have shown to induce LTP at the otherwise plasticity-resistant SC-CA2 synapses (Dudek et al., 2016). Substance P (SP) released from the supramammillary nucleus terminals synapsing at CA2 induces a slow-onset LTP at SC-CA2 and EC-CA2 synapses which is protein-synthesis dependent (Dasgupta et al., 2017). Moreover, the SP-induced lasting potentiation at SC-CA2 is capable of mediating associative plasticity in that it enables transformation of an E-LTP to L-LTP at EC-CA2 in a synaptic tagging and capture-dependent manner and hence dependent on protein synthesis (Dasgupta et al., 2017). PKMζ and CaMKIV, which have been shown to play an important role in maintenance of plasticity and STC in CA1 (Redondo et al., 2010; Sajikumar et al., 2005, 2007) were also identified as key PRPs for SP-induced LTP and associative plasticity at synapses in CA2 (Dasgupta et al., 2017).

LOCAL DENDRITIC PROTEIN SYNTHESIS IN MEMORY MAINTENANCE

Protein synthesis has been shown to be critical for the formation of long term memories. Recent studies have demonstrated the translation of proteins in neuronal dendrites, meaning local protein synthesis (Aakalu, Smith, Nguyen, Jiang, & Schuman, 2001) and studies suggest a role for local dendritic protein synthesis in memory (Sutton & Schuman, 2006). It has been reported from studies in "isolated" hippocampal slices where dendritic and cell body regions were isolated by a microsurgical cut, that high frequency stimulation-induced NMDA receptor dependent L-LTP, lasting up to 5 hours, is expressed in the "isolated" slices devoid of cell body (Vickers, Dickson, & Wyllie, 2005). Vickers et al. report that the magnitude of the fEPSPs in these isolated dendritic preparations was similar to the fEPSPs recorded in intact slices and that the late-LTP in isolated slices was blocked upon incubating the slices with mRNA translation inhibitor. However, incubation of the isolated slices with mRNA transcription inhibitor did not affect the late-LTP in the isolated slices. The results suggested that the maintenance of L-LTP in the isolated dendritic preparations was facilitated by protein synthesis from dendritically localized pre-existing mRNAs and point to a postsynaptic dendritic locus for the protein synthesis requirement of L-LTP expression. The idea of dendritic protein synthesis from translation of dendritically localized mRNAs is further supported from other studies that have identified numerous dendritically localized mRNAs and translation machinery (Jiang & Schuman, 2002; Steward & Schuman, 2003; Vickers et al., 2005). Moreover, translocation of ribosomes to active synaptic sites upon giving LTP-inducing have been reported by (Ostroff et al., 2002) further supporting local dendritic protein synthesis.

Protein synthesis inhibitors applied to local dendritic regions have been shown to decrease tetanically induced L-LTP (Bradshaw, Emptage, & Bliss, 2003). Moreover, this study showed that the protein synthesis inhibitor emetine when applied locally to the apical dendritic field of hippocampal CA1 pyramidal neurons, impaired L-LTP at only apical but not basal dendrites. Similarly, emetine when applied to basal dendrites, impaired L-LTP at only basal but not apical dendrites. Thus, the complete expression of late-LTP requires local dendritic protein synthesis (Bradshaw et al., 2003) but this does not preclude the role for somatic protein synthesis in L-LTP and perhaps requires cooperation of both somatically and dendritically synthesized proteins (Casadio et al., 1999; Sherff & Carew, 1999).

The signaling cascades that occur in conjunction with synaptic activation and local dendritic protein synthesis remain notoriously elusive. Studies have investigated the role of glutamate receptors and mTOR signaling pathways in the regulation of local dendritic protein synthesis in live neurons. It highlighted the importance of NMDAR and mTOR signaling for synaptic activity induced dendritic protein synthesis in hippocampal neurons by showing the lack of αCaMKII and MAP2 proteins through high frequency stimulations in the hippocampal slices when mTOR kinase was inhibited (Gong, Park, Abbassi, & Tang, 2006).

The idea of activity-dependent local dendritic protein synthesis supports synapse specificity that is characteristic of NMDA-receptor dependent plasticity. It could be that a cooperation of both synaptic tagging and capture of somatically synthesized proteins and activity-dependent local dendritic protein synthesis processes are required (H. Wang & Tiedge, 2004). Dendritic protein synthesis, by virtue of regulation of actin cytoskeletal dynamics, controls stabilization and long-term maintenance of both structural and functional synaptic plasticity (Bramham, 2008). The actin network can itself be considered as part of the LTP tag machinery that facilitates interaction of tagging complexes at activated synapses with PRPs, and thereby contribute to L-LTP (Ramachandran & Frey, 2009).

Balance between Proteasomal Degradation and Synthesis of Plasticity-Related Proteins

Studies have shown that protein synthesis is pivotal for late phase LTP. This stems from the fact that different experimental settings have demonstrated how the inhibitors have affected late phase LTP yet retaining the early phase of LTP. A balance between degradation and synthesis of proteins is essential for the sustenance of late-phase LTP (Fonseca et al., 2006). When protein synthesis alone is blocked, meaning translation of both positive and negative proteins is inhibited, LTP is diminished as the degradation of pre-existing positive proteins overwhelms the degradation of pre-existing negative proteins. The blockade of proteasome-dependent degradation of proteins also brings about a decrease in LTP as the pool of negative proteins that persist, counteracts the effects of positive proteins. When both degradation and synthesis is disrupted, L-LTP is sustained by the abundance of the pre-existing positive proteins over the pre-existing negative proteins. This leads to the surprising conclusion that de novo protein synthesis is not an absolute requirement for the late maintenance of LTP. Protein synthesis is required for the sustenance of LTP when protein degradation is functional. Thus, a dynamic balance between synthesis and degradation of positive plasticity proteins and negative proteins is critical for the maintenance of LTP (Fonseca et al., 2006). The possibility that tagging of activated synapses enables selective stabilization of the synapse by delaying degradation of PRPs and thereby enabling maintenance of LTP could be a mechanism for partial explanation of these results.

Role of epigenetics in the maintenance of LTP

Long-term contextual memory formation demands for transcription within the hippocampus wherein histone acetylation is known to be involved along with other mechanisms that initiate and maintain this transcription process. HDAC inhibitors, which increase histone acetylation have shown to increase long term memory when injected into the hippocampus during memory consolidation. Investigating the genes that regulate this process was believed to offer insights into the necessary parameters of the memory consolidation. Phosphorylated CREB and histone acetyltransferase CREB-binding protein (CBP) interaction is mandatory for long term memory.

CREB as an important transcription factor in hippocampus for memory formation and storage

Several studies have focused on the molecular mechanisms behind the contextual fear memory storage in hippocampus and amygdala. There happens to be two time windows in these brain regions post learning, with an increase in the phosphorylation of CREB, which is the cAMP-response element binding protein (Stanciu, Radulovic, & Spiess, 2001). There is a striking coincidence with the two time windows, 0–30 minutes and 3–6 hours post training, with that of the time windows during which inhibition of transcription or translation which impairs memory storage happens (Bourtchouladze et al., 1998; Igaz, Vianna, Medina, & Izquierdo, 2002). Despite CREB being one of the pivotal transcription factors for the formation of long-term memory, CREB phosphorylation alone is insufficient, stipulating the requirement of additional other co-activators of CREB for the expression of target genes.

The significance of histone acetyltransferases in long-term memory

Many proteins such as HATs, including CBP, p300 and p300/CBP associated factor (PCAF) interact with phosphorylated CREB. Portelli (Portelli, 1975) first proposed the concept that DNA-histone complexes regulate memory formation. These HATs have been believed to be important regulators of the transcription necessary for long-term memory wherein each HAT serves a role in a specific type of long term memory. A truncated form of p300 or conditional p300 deletion causes only a selective long-term memory deficits in both contextual fear conditioning and object recognition memory

(Oliveira, Hawk, Abel, & Havekes, 2010) suggesting that memory deficits in p300 mutant mice is effected due to a possible transcriptional effects in brain regions outside hippocampus. Histone methylation is the transfer of one, two or three methyl groups from S-adenosyl-L—methionine to lysine or arginine residues of histone proteins by histone methyltransferases (HMTs) and they regulate DNA methylation through chromatin dependent transcriptional repression or activation. G9a/GLP is a histone lysine methyltransferase complex that has been shown to be critical for brain development and goal directed learning. Some of the findings from (Sharma, Dierkes, & Sajikumar, 2017) have highlighted the key role for G9a/GLP in the maintenance and the homeostasis of neuronal transcription. The study demonstrated that the inhibition of this G9a/GLP complex reverses the amyloid-b oligomer induced deficits in late-LTP and synaptic tagging and capture and is achieved by the capture of BDNF by the weakly activated synapses. This evidence for plasticity and associativity in conditions such as Alzheimer's disease has highlighted its significance as a possible target to prevent amyloid-b oligomer-induced plasticity deficits in hippocampal neurons.

Acetylation in memory consolidation

Acetylation of particular lysine residues have been believed to control the transcription for long-term memory. Studies showed that acetylation of lysine 14 on H3 increased in bulk histone extracts post one hour of contextual learning (Levenson et al., 2004). An efficient way of fishing out acetylation marks owing to memory formation is to investigate proteins required for memory consolidation such as CBP which gets affected by histone acetyltransferases. In vitro and more in specific, some of the in-vivo works have suggested HATs role in the regulation of acetylation of specific lysine residues (Barrett et al., 2011; Jin et al., 2011). Exploring the genes that are regulated by acetylation during the memory consolidation helps unfold the targets that are important for long-term memory formation and thereby novel therapeutics that may help facilitate the formation of memory.

Histone deacetylase inhibitors increase long-term memory

Increased levels of histone acetylation during memory consolidation has suggested that unnaturally increasing histone acetylation could augment long-term memory. A delicate balance of HATs such as CBP that add acetyl groups to specific lysine residues on histone Tails and HDACs that remove acetyl groups from these lysines governs the histone acetylation as shown in Figure 12.8.

Either enhancing HAT activity or abating HDAC activity results in an increase in the histone acetylation. HDAC inhibitors enhance long term memory when administered during memory consolidation. Work from Vecsey et al. (Vecsey et al., 2007) has shown the enhancement of memory by HDAC inhibitor trichostatin A (TSA) which requires

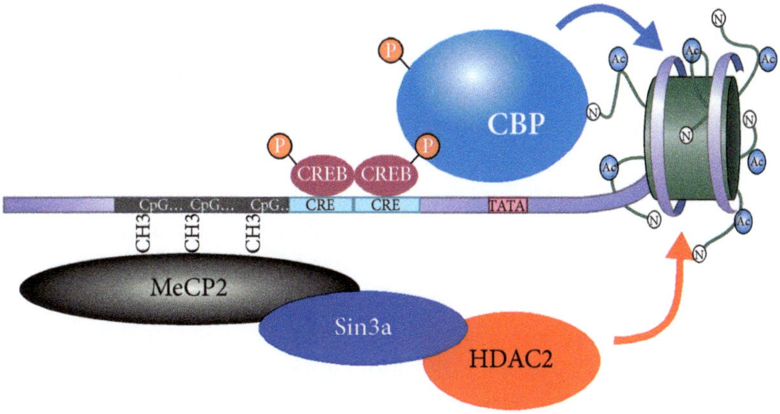

FIGURE 12.8. Role of histone acetylation in long-term memory storage. After learning, CREB is activated by phosphorylation, binds to CREB response elements (CREs) in the genome, and recruits the co-activator CBP to the region. Acetyl groups are added to lysine residues on histone tails by the histone acetyltransferase (HAT) function of CBP. Acetylation is removed by class I histone deacetylase (HDAC) proteins.

From Poplawski & Abel (2012). Reprinted by permission from Springer Nature, Springer.

CREB-CBP interaction, suggesting the stipulation of CREB target genes for memory enhancement. Studies have shown that chronic treatment of broad spectrum HDAC inhibitors such as TSA causes synaptic dysfunction (Nelson, Kavalali, & Monteggia, 2006). Therefore, targeting only specific HDAC proteins that dampen memory formation may enable the application of selective therapeutics with a reduced side effect profiles in comparison to the broad spectrum HDAC inhibitors.

DNA methylation in long-term memory formation

For memory to persist, there must be a continual replacement of that proteins that were originally responsible for its formation. It was a speculation even decades ago by Francis Crick and Robin Holliday that DNA methylation might be a self-perpetuating mechanism involved in the storage of long-term memories. The chromatin-regulating mechanism, DNA methylation, is a dynamic process that controls long-lasting changes in synaptic function and behavior. This mechanism was proposed since DNA methyltransferase inhibitors (DNMTi) applied acutely into the adult CNS alters the methylation of genes like *BDNF, PP1* and *reelin* and blocks long-term potentiation (LTP) induction in the hippocampal slice preparation through the following studies (Lubin, Roth, & Sweatt, 2008; Miller & Sweatt, 2007). DNA methylation is capable of self-regeneration and self-perpetuation which are salient features for a stable molecular mark and these processes are accomplished partly by DNA methyltransferases (DNMTs). DNMTs can acknowledge the presence of hemi-methylated C-G dinucleotide and can convert the complementary C-G on the opposite strand into a methylated C-G. As a result, DNA can be methylated perpetually in a manner that is similar to the self-perpetuating autophosphorylation of activated CaMKII that had been identified originally as a candidate for the molecular mechanism for memory storage (Roberson & Sweatt, 1999). DNA methylation has been associated both with transcriptional silencing and with activating roles (Chahrour et al., 2008). There have been results demonstrating that application of different types of DNMT inhibitors results in deficits in hippocampal LTP and deficits in memory consolidation (Levenson et al., 2006). Several such studies through biochemical evidence have also come to a similar conclusion that long-term memory consolidation is associated with altered DNA methylation in the hippocampus (Lubin et al., 2008; Miller & Sweatt, 2007). Having established that changes in transcription such as DNA methylation and histone modifications are important for learning, it is now an active question whether these transcriptional marks are required for the memory to be maintained (Miller et al., 2010). Demonstrations from the Glanzman's lab has shown that even after the loss of long-term synaptic changes, transcriptional marks allow for memories to be relearned more easily(Chen et al., 2014). Some have argued that this shows that memories are not stored at synapses (Poo et al., 2016; Trettenbrein, 2016). Recent arguments from Sossin (2018) state that long-term transcriptional marks encode changes that may aid in mechanisms for the storage of specific weights locally at synapses. It also highlights that the transcription marks will need to know the specificity of the synaptic connections that require strengthening in order for these marks to elicit these changes.

Conclusion and Future Directions

Remarkable progress has been made in understanding the regulation of proteins in bringing about long lasting molecular changes in a neuron which will partake in future plasticity. The finding that simultaneous inhibition of both protein synthesis as well as degradation did not interfere with either the induction or the maintenance of the LTP suggests the possibility of protein synthesis-independent synaptic plasticity or metaplasticity that could last for long duration of time the preponderance of evidence suggests an important role for protein synthesis in the production of PRPs. Synaptic plasticity is a pivotal learning mechanism in the brain wherein many aspects of tagging needs to be understood yet. The three models of activity induced protein synthesis in the post-synaptic neurons is discussed elaborately viz., synaptic targeting, synaptic tagging and capture and local dendritic protein synthesis. It has been discussed how dendritic protein synthesis has a role to play in the maintenance of long term memories of both structural and functional synaptic plasticity. This has helped in the conjecture that actin network could be a part of the tagging machinery that helps set tags in both early and late forms of LTP. One of the important emerging avenues in understanding the molecular mechanisms of long-term memory storage is tapping into the epigenetic control of brain function. Several studies already showing pharmacological manipulations of epigenetic markers reinstating learning and restoring remote memory after neurodegeneration provides hope in terms of clinical implications for the reversal of memory deficits.

Acknowledgments

S.S. is supported by National Medical Research Council Collaborative Research Grant (NMRC/CBRG/0099/2015 and NMRC-OFIRG-0037-2017) and NUS-Strategic and Aspiration Research Funds. R.R and A.B is supported by President Graduate Fellowship, National University of Singapore.

References

Aakalu, G., Smith, W. B., Nguyen, N., Jiang, C., & Schuman, E. M. (2001). Dynamic visualization of local protein synthesis in hippocampal neurons. *Neuron, 30*(2), 489–502.

Abraham, W. C., & Bear, M. F. (1996). Metaplasticity: The plasticity of synaptic plasticity. *Trends in Neuroscience, 19*(4), 126–130.

Akers, R. F., & Routtenberg, A. (1985). Protein kinase C phosphorylates a 47 Mr protein (F1) directly related to synaptic plasticity. *Brain Research, 334*(1), 147–151.

Ballarini, F., Moncada, D., Martinez, M. C., Alen, N., & Viola, H. (2009). Behavioral tagging is a general mechanism of long-term memory formation. *Proceedings of the National Academy of Sciences, 106*(34), 14599–14604.

Barco, A., Alarcon, J. M., & Kandel, E. R. (2002). Expression of constitutively active CREB protein facilitates the late phase of long-term potentiation by enhancing synaptic capture. *Cell, 108*(5), 689–703.

Barrett, R. M., Malvaez, M., Kramar, E., Matheos, D. P., Arrizon, A., Cabrera, S. M., . . . Wood, M. A. (2011). Hippocampal focal knockout of CBP affects specific histone modifications, long-term potentiation, and long-term memory. *Neuropsychopharmacology, 36*(8), 1545–1556. https://doi.org/10.1038/npp.2011.61

Bayer, K. U., De Koninck, P., Leonard, A. S., Hell, J. W., & Schulman, H. (2001). Interaction with the NMDA receptor locks CaMKII in an active conformation. *Nature, 411*(6839), 801–805. https://doi.org/10.1038/35081080

Bayer, K. U., LeBel, E., McDonald, G. L., O'Leary, H., Schulman, H., & De Koninck, P. (2006). Transition from reversible to persistent binding of CaMKII to postsynaptic sites and NR2B. *Journal of Neuroscience, 26*(4), 1164–1174. https://doi.org/10.1523/JNEUROSCI.3116-05.2006

Bear, M. F. (1996). A synaptic basis for memory storage in the cerebral cortex. *Proceedings of the National Academy of Sciences of the United States of America, 93*(24), 13453–13459.

Behnisch, T., Fjodorow, K., & Reymann, K. G. (1991). L-2-amino-3-phosphonopropionate blocks late synaptic long-term potentiation. *Neuroreport, 2*(7), 386–388.

Bliss, T. V., Collingridge, G. L., Morris, R. G., & Reymann, K. G. (2018). Long-term potentiation in the hippocampus: discovery, mechanisms and function. *Neuroforum, 24*(3), A103–A120.

Bliss, T. V., Errington, M. L., Lynch, M. A., & Williams, J. H. (1990). Presynaptic mechanisms in hippocampal long-term potentiation. *Cold Spring Harbor Symposium on Quantitative Biology, 55*, 119–129.

Bortolotto, Z. A., Bashir, Z. I., Davies, C. H., Taira, T., Kaila, K., & Collingridge, G. L. (1995). Studies on the role of metabotropic glutamate receptors in long-term potentiation: some methodological considerations. *Journal of Neuroscience Methods, 59*(1), 19–24.

Bosch, M., Castro, J., Saneyoshi, T., Matsuno, H., Sur, M., & Hayashi, Y. (2014). Structural and molecular remodeling of dendritic spine substructures during long-term potentiation. *Neuron, 82*(2), 444–459. https://doi.org/10.1016/j.neuron.2014.03.021

Bourtchouladze, R., Abel, T., Berman, N., Gordon, R., Lapidus, K., & Kandel, E. R. (1998). Different training procedures recruit either one or two critical periods for contextual memory consolidation, each of which requires protein synthesis and PKA. *Learning and Memory, 5*(4–5), 365–374.

Bourtchuladze, R., Frenguelli, B., Blendy, J., Cioffi, D., Schutz, G., & Silva, A. J. (1994). Deficient long-term memory in mice with a targeted mutation of the cAMP-responsive element-binding protein. *Cell, 79*(1), 59–68.

Bradshaw, K. D., Emptage, N. J., & Bliss, T. V. (2003). A role for dendritic protein synthesis in hippocampal late LTP. *European Journal of Neuroscience, 18*(11), 3150–3152.

Bramham, C. R. (2008). Local protein synthesis, actin dynamics, and LTP consolidation. *Current Opinion in Neurobiology, 18*(5), 524–531. https://doi.org/10.1016/j.conb.2008.09.013

Buzsáki, G. (1985). *Electrical activity of the archicortex.* Budapest: Akad. Kiadó.

Casadio, A., Martin, K. C., Giustetto, M., Zhu, H., Chen, M., Bartsch, D., ... Kandel, E. R. (1999). A transient, neuron-wide form of CREB-mediated long-term facilitation can be stabilized at specific synapses by local protein synthesis. *Cell, 99*(2), 221–237.

Cazakoff, B. N., & Howland, J. G. (2011). AMPA receptor endocytosis in rat perirhinal cortex underlies retrieval of object memory. *Learning and Memory, 18*(11), 688–692. https://doi.org/10.1101/lm.2312711

Chahrour, M., Jung, S. Y., Shaw, C., Zhou, X., Wong, S. T., Qin, J., & Zoghbi, H. Y. (2008). MeCP2, a key contributor to neurological disease, activates and represses transcription. *Science, 320*(5880), 1224–1229. https://doi.org/10.1126/science.1153252

Chen, S., Cai, D., Pearce, K., Sun, P. Y., Roberts, A. C., & Glanzman, D. L. (2014). Reinstatement of long-term memory following erasure of its behavioral and synaptic expression in Aplysia. *Elife, 3*, e03896.

Chevaleyre, V., & Siegelbaum, S. A. (2010). Strong CA2 pyramidal neuron synapses define a powerful disynaptic cortico-hippocampal loop. *Neuron, 66*(4), 560–572. https://doi.org/10.1016/j.neuron.2010.04.013

Collingridge, G. L., Peineau, S., Howland, J. G., & Wang, Y. T. (2010). Long-term depression in the CNS. *Nature Reviews Neuroscience, 11*(7), 459–473. https://doi.org/10.1038/nrn2867

Dalton, G. L., Wang, Y. T., Floresco, S. B., & Phillips, A. G. (2008). Disruption of AMPA receptor endocytosis impairs the extinction, but not acquisition of learned fear. *Neuropsychopharmacology, 33*(10), 2416–2426. https://doi.org/10.1038/sj.npp.1301642

Dasgupta, A., Baby, N., Krishna, K., Hakim, M., Wong, Y. P., Behnisch, T., ... Sajikumar, S. (2017). Substance P induces plasticity and synaptic tagging/capture in rat hippocampal area CA2. *Proceedings of the National Academy of Sciences of the United States of America, 114*(41), E8741–E8749. https://doi.org/10.1073/pnas.1711267114

Desmond, N. L., & Levy, W. B. (1986). Changes in the postsynaptic density with long-term potentiation in the dentate gyrus. *Journal of Comparative Neurology, 253*(4), 476–482. https://doi.org/10.1002/cne.902530405

Dolphin, A. C., Errington, M. L., & Bliss, T. V. (1982). Long-term potentiation of the perforant path in vivo is associated with increased glutamate release. *Nature, 297*(5866), 496–498.

Dudek, S. M., Alexander, G. M., & Farris, S. (2016). Rediscovering area CA2: unique properties and functions. *Nature Reviews Neuroscience, 17*(2), 89–102. https://doi.org/10.1038/nrn.2015.22

Etkin, A., Alarcon, J. M., Weisberg, S. P., Touzani, K., Huang, Y. Y., Nordheim, A., & Kandel, E. R. (2006). A role in learning for SRF: Deletion in the adult forebrain disrupts LTD and the formation of an immediate memory of a novel context. *Neuron, 50*(1), 127–143. https://doi.org/10.1016/j.neuron.2006.03.013

Fonseca, R., Vabulas, R. M., Hartl, F. U., Bonhoeffer, T., & Nagerl, U. V. (2006). A balance of protein synthesis and proteasome-dependent degradation determines the maintenance of LTP. *Neuron, 52*(2), 239–245. https://doi.org/10.1016/j.neuron.2006.08.015

Frey, U., Krug, M., Reymann, K. G., & Matthies, H. (1988). Anisomycin, an inhibitor of protein synthesis, blocks late phases of LTP phenomena in the hippocampal CA1 region in vitro. *Brain Research, 452*(1–2), 57–65.

Frey, U., & Morris, R. G. (1998a). Synaptic tagging: implications for late maintenance of hippocampal long-term potentiation. *Trends in Neuroscience, 21*(5), 181–188.

Frey, U., & Morris, R. G. (1998b). Weak before strong: Dissociating synaptic tagging and plasticity-factor accounts of late-LTP. *Neuropharmacology, 37*(4–5), 545–552.

Genoux, D., Haditsch, U., Knobloch, M., Michalon, A., Storm, D., & Mansuy, I. M. (2002). Protein phosphatase 1 is a molecular constraint on learning and memory. *Nature, 418*(6901), 970–975. https://doi.org/10.1038/nature00928

Giese, K. P., & Mizuno, K. (2013). The roles of protein kinases in learning and memory. *Learning and Memory, 20*(10), 540–552. https://doi.org/10.1101/lm.028449.112

Glanzman, D. L. (2013). PKM and the maintenance of memory. *F1000 Biology Reports, 5*, 4. https://doi.org/10.3410/B5-4

Gong, R., Park, C. S., Abbassi, N. R., & Tang, S. J. (2006). Roles of glutamate receptors and the mammalian target of rapamycin (mTOR) signaling pathway in activity-dependent dendritic protein synthesis in hippocampal neurons. *Journal of Biological Chemistry, 281*(27), 18802–18815. https://doi.org/10.1074/jbc.M512524200

Govindarajan, A., Israely, I., Huang, S. Y., & Tonegawa, S. (2011). The dendritic branch is the preferred integrative unit for protein synthesis-dependent LTP. *Neuron, 69*(1), 132–146. https://doi.org/10.1016/j.neuron.2010.12.008

Hardt, O., Nader, K., & Wang, Y. T. (2014). GluA2-dependent AMPA receptor endocytosis and the decay of early and late long-term potentiation: Possible mechanisms for forgetting of short- and long-term memories. *Philosophical Transactions of the Royal Society of London B: Biological Sciences, 369*(1633), 20130141. https://doi.org/10.1098/rstb.2013.0141

Hitti, F. L., & Siegelbaum, S. A. (2014). The hippocampal CA2 region is essential for social memory. *Nature, 508*(7494), 88–92. https://doi.org/10.1038/nature13028

Huang, C. C., Liang, Y. C., & Hsu, K. S. (2001). Characterization of the mechanism underlying the reversal of long term potentiation by low frequency stimulation at hippocampal CA1 synapses. *Journal of Biological Chemistry, 276*(51), 48108–48117. https://doi.org/10.1074/jbc.M106388200

Huang, E. P. (1998). Synaptic plasticity: Going through phases with LTP. *Current Biology, 8*(10), R350–352.

Igaz, L. M., Vianna, M. R., Medina, J. H., & Izquierdo, I. (2002). Two time periods of hippocampal mRNA synthesis are required for memory consolidation of fear-motivated learning. *Journal of Neuroscience, 22*(15), 6781–6789. https://doi.org/10.1523/JNEUROSCI.22-15-06781.2002

Jiang, C., & Schuman, E. M. (2002). Regulation and function of local protein synthesis in neuronal dendrites. *Trends in Biochemical Science, 27*(10), 506–513.

Jin, Q., Yu, L. R., Wang, L., Zhang, Z., Kasper, L. H., Lee, J. E., . . . Ge, K. (2011). Distinct roles of GCN5/PCAF-mediated H3K9ac and CBP/p300-mediated H3K18/27ac in nuclear receptor transactivation. *EMBO Journal, 30*(2), 249–262. https://doi.org/10.1038/emboj.2010.318

Kandel, E. R. (2012). The molecular biology of memory: cAMP, PKA, CRE, CREB-1, CREB-2, and CPEB. *Molecular Brain, 5*, 14. https://doi.org/10.1186/1756-6606-5-14

Kang, H., Sun, L. D., Atkins, C. M., Soderling, T. R., Wilson, M. A., & Tonegawa, S. (2001). An important role of neural activity-dependent CaMKIV signaling in the consolidation of long-term memory. *Cell, 106*(6), 771–783.

Kida, S. (2012). A functional role for CREB as a positive regulator of memory formation and LTP. *Experimental Neurobiology, 21*(4), 136–140. https://doi.org/10.5607/en.2012.21.4.136

Kim, C. H., Chung, H. J., Lee, H. K., & Huganir, R. L. (2001). Interaction of the AMPA receptor subunit GluR2/3 with PDZ domains regulates hippocampal long-term depression. *Proceedings of the National Academy of Sciences of the United States of America, 98*(20), 11725–11730. https://doi.org/10.1073/pnas.211132798

Kim, J., Lee, S., Park, K., Hong, I., Song, B., Son, G., . . . Choi, S. (2007). Amygdala depotentiation and fear extinction. *Proceedings of the National Academy of Sciences of the United States of America, 104*(52), 20955–20960. https://doi.org/10.1073/pnas.0710548105

Krug, M., Brodemann, R., & Ott, T. (1982). Blockade of long-term potentiation in the dentate gyrus of freely moving rats by the glutamic acid antagonist GDEE. *Brain Research, 249*(1), 57–62.

Krug, M., Lossner, B., & Ott, T. (1984). Anisomycin blocks the late phase of long-term potentiation in the dentate gyrus of freely moving rats. *Brain Research Bulletin, 13*(1), 39–42.

Lee, A. M., Kanter, B. R., Wang, D., Lim, J. P., Zou, M. E., Qiu, C., . . . Messing, R. O. (2013). Prkcz null mice show normal learning and memory. *Nature, 493*(7432), 416–419. https://doi.org/10.1038/nature11803

Lee, H. K., & Kirkwood, A. (2011). AMPA receptor regulation during synaptic plasticity in hippocampus and neocortex. *Seminars in Cell and Developmental Biology, 22*(5), 514–520. https://doi.org/10.1016/j.semcdb.2011.06.007

Levenson, J. M., O'Riordan, K. J., Brown, K. D., Trinh, M. A., Molfese, D. L., & Sweatt, J. D. (2004). Regulation of histone acetylation during memory formation in the hippocampus. *Journal of Biological Chemistry, 279*(39), 40545–40559. https://doi.org/10.1074/jbc.M402229200

Levenson, J. M., Roth, T. L., Lubin, F. D., Miller, C. A., Huang, I. C., Desai, P., . . . Sweatt, J. D. (2006). Evidence that DNA (cytosine-5) methyltransferase regulates synaptic plasticity in the hippocampus. *Journal of Biological Chemistry, 281*(23), 15763–15773. https://doi.org/10.1074/jbc.M511767200

Li, Q., Navakkode, S., Rothkegel, M., Soong, T. W., Sajikumar, S., & Korte, M. (2017). Metaplasticity mechanisms restore plasticity and associativity in an animal model of Alzheimer's disease. *Proceedings of the National Academy of Sciences of the United States of America, 114*(21), 5527–5532. https://doi.org/10.1073/pnas.1613700114

Li, Q., Rothkegel, M., Xiao, Z. C., Abraham, W. C., Korte, M., & Sajikumar, S. (2014). Making synapses strong: metaplasticity prolongs associativity of long-term memory by switching synaptic tag mechanisms. *Cerebral Cortex, 24*(2), 353–363. https://doi.org/10.1093/cercor/bhs315

Li, S., Cullen, W. K., Anwyl, R., & Rowan, M. J. (2003). Dopamine-dependent facilitation of LTP induction in hippocampal CA1 by exposure to spatial novelty. *Nature Neuroscience, 6*(5), 526–531. https://doi.org/10.1038/nn1049

Lisman, J., & Raghavachari, S. (2006). A unified model of the presynaptic and postsynaptic changes during LTP at CA1 synapses. *Science's STKE, 2006*(356), re11. https://doi.org/10.1126/stke.3562006re11

Lubin, F. D., Roth, T. L., & Sweatt, J. D. (2008). Epigenetic regulation of BDNF gene transcription in the consolidation of fear memory. *Journal of Neuroscience, 28*(42), 10576–10586. https://doi.org/10.1523/JNEUROSCI.1786-08.2008

Luscher, C., Nicoll, R. A., Malenka, R. C., & Muller, D. (2000). Synaptic plasticity and dynamic modulation of the postsynaptic membrane. *Nature Neuroscience, 3*(6), 545–550. https://doi.org/10.1038/75714

Lynch, M. A. (2004). Long-term potentiation and memory. *Physiology Review, 84*(1), 87–136. https://doi.org/10.1152/physrev.00014.2003

Miller, C. A., Gavin, C. F., White, J. A., Parrish, R. R., Honasoge, A., Yancey, C. R., . . . Sweatt, J. D. (2010). Cortical DNA methylation maintains remote memory. *Nature Neuroscience, 13*(6), 664–666. https://doi.org/10.1038/nn.2560

Miller, C. A., & Sweatt, J. D. (2007). Covalent modification of DNA regulates memory formation. *Neuron, 53*(6), 857–869. https://doi.org/10.1016/j.neuron.2007.02.022

Milner, B. (1972). Disorders of learning and memory after temporal lobe lesions in man. *Clinical Neurosurgery, 19*, 421–446.

Moncada, D., Ballarini, F., Martinez, M. C., Frey, J. U., & Viola, H. (2011). Identification of transmitter systems and learning tag molecules involved in behavioral tagging during memory formation. *Proceedings of the National Academy of Sciences of the United States of America, 108*(31), 12931–12936. https://doi.org/10.1073/pnas.1104495108

Montarolo, P. G., Goelet, P., Castellucci, V. F., Morgan, J., Kandel, E. R., & Schacher, S. (1986). A critical period for macromolecular synthesis in long-term heterosynaptic facilitation in Aplysia. *Science, 234*(4781), 1249–1254.

Nelson, E. D., Kavalali, E. T., & Monteggia, L. M. (2006). MeCP2-dependent transcriptional repression regulates excitatory neurotransmission. *Current Biology, 16*(7), 710–716. https://doi.org/10.1016/j.cub.2006.02.062

O'Keefe, J., & Nadel, L. (1978). *The hippocampus as a cognitive map*. Oxford, England; New York, NY: Clarendon Press; Oxford University Press.

Okuno, H., Akashi, K., Ishii, Y., Yagishita-Kyo, N., Suzuki, K., Nonaka, M., . . . Bito, H. (2012). Inverse synaptic tagging of inactive synapses via dynamic interaction of Arc/Arg3.1 with CaMKIIbeta. *Cell, 149*(4), 886–898. https://doi.org/10.1016/j.cell.2012.02.062

Okuno, H., Minatohara, K., & Bito, H. (2018). Inverse synaptic tagging: An inactive synapse-specific mechanism to capture activity-induced Arc/arg3.1 and to locally regulate spatial distribution of synaptic weights. *Seminars in Cell and Developmental Biology, 77*, 43–50. https://doi.org/10.1016/j.semcdb.2017.09.025

Oliveira, A. M., Hawk, J. D., Abel, T., & Havekes, R. (2010). Post-training reversible inactivation of the hippocampus enhances novel object recognition memory. *Learning and Memory, 17*(3), 155–160. https://doi.org/10.1101/lm.1625310

Ostroff, L. E., Fiala, J. C., Allwardt, B., & Harris, K. M. (2002). Polyribosomes redistribute from dendritic shafts into spines with enlarged synapses during LTP in developing rat hippocampal slices. *Neuron, 35*(3), 535–545.

Park, S., Lee, S., Kim, J., & Choi, S. (2012). Ex vivo depotentiation of conditioning-induced potentiation at thalamic input synapses onto the lateral amygdala requires GluN2B-containing NMDA receptors. *Neuroscience Letters, 530*(2), 121–126. https://doi.org/10.1016/j.neulet.2012.10.011

Penfield, W., & Milner, B. (1958). Memory deficit produced by bilateral lesions in the hippocampal zone. *AMA Archives of Neurology and Psychiatry, 79*(5), 475–497.

Penn, A. C., Zhang, C. L., Georges, F., Royer, L., Breillat, C., Hosy, E., . . . Choquet, D. (2017). Hippocampal LTP and contextual learning require surface diffusion of AMPA receptors. *Nature, 549*(7672), 384–388. https://doi.org/10.1038/nature23658

Poo, M. M., Pignatelli, M., Ryan, T. J., Tonegawa, S., Bonhoeffer, T., Martin, K. C., . . . Stevens, C. (2016). What is memory? The present state of the engram. *BMC Biology, 14*, 40. https://doi.org/10.1186/s12915-016-0261-6

Poplawski, S.G., & Abel, T. (2012). The role of histone acetylation in long-term memory storage. In P. Sassone-Corsi & Y. Christen (Eds.), *Epigenetics, brain, and behavior* (pp. 71–80). Berlin/Heidelberg, Germany: Springer.

Portelli, C. (1975). A model of the mechanisms of memory. *Physiologie, 12*(4), 313–316.

Ramachandran, B., & Frey, J. U. (2009). Interfering with the actin network and its effect on long-term potentiation and synaptic tagging in hippocampal CA1 neurons in slices in vitro. *Journal of Neuroscience, 29*(39), 12167–12173. https://doi.org/10.1523/JNEUROSCI.2045-09.2009

Rampon, C., Tang, Y. P., Goodhouse, J., Shimizu, E., Kyin, M., & Tsien, J. Z. (2000). Enrichment induces structural changes and recovery from nonspatial memory deficits in CA1 NMDAR1-knockout mice. *Nature Neuroscience, 3*(3), 238–244. https://doi.org/10.1038/72945

Redondo, R. L., & Morris, R. G. (2011). Making memories last: The synaptic tagging and capture hypothesis. *Nature Reviews Neuroscience, 12*(1), 17–30. https://doi.org/10.1038/nrn2963

Redondo, R. L., Okuno, H., Spooner, P. A., Frenguelli, B. G., Bito, H., & Morris, R. G. (2010). Synaptic tagging and capture: differential role of distinct calcium/calmodulin kinases in protein synthesis-dependent long-term potentiation. *Journal of Neuroscience, 30*(14), 4981–4989. https://doi.org/10.1523/JNEUROSCI.3140-09.2010

Reymann, K. G., Frey, U., Jork, R., & Matthies, H. (1988). Polymyxin B, an inhibitor of protein kinase C, prevents the maintenance of synaptic long-term potentiation in hippocampal CA1 neurons. *Brain Research, 440*(2), 305–314.

Roberson, E. D., English, J. D., & Sweatt, J. D. (1996). A biochemist's view of long-term potentiation. *Learning and Memory, 3*(1), 1–24.

Roberson, E. D., & Sweatt, J. D. (1999). A biochemical blueprint for long-term memory. *Learning and Memory, 6*(4), 381–388.

Sacktor, T. C. (2011). How does PKMzeta maintain long-term memory? *Nature Reviews Neuroscience, 12*(1), 9–15. https://doi.org/10.1038/nrn2949

Sacktor, T. C. (2012). Memory maintenance by PKMzeta: An evolutionary perspective. *Molecular Brain, 5*, 31. https://doi.org/10.1186/1756-6606-5-31

Sajikumar, S., & Frey, J. U. (2004). Late-associativity, synaptic tagging, and the role of dopamine during LTP and LTD. *Neurobiology of Learning and Memory, 82*(1), 12–25. https://doi.org/10.1016/j.nlm.2004.03.003

Sajikumar, S., Morris, R. G., & Korte, M. (2014). Competition between recently potentiated synaptic inputs reveals a winner-take-all phase of synaptic tagging and capture. *Proceedings of the National Academy of Sciences of the United States of America, 111*(33), 12217–12221. https://doi.org/10.1073/pnas.1403643111

Sajikumar, S., Navakkode, S., & Frey, J. U. (2007). Identification of compartment- and process-specific molecules required for "synaptic tagging" during long-term potentiation and long-term depression in hippocampal CA1. *Journal of Neuroscience, 27*(19), 5068–5080. https://doi.org/10.1523/JNEUROSCI.4940-06.2007

Sajikumar, S., Navakkode, S., Sacktor, T. C., & Frey, J. U. (2005). Synaptic tagging and cross-tagging: The role of protein kinase Mzeta in maintaining long-term potentiation but not long-term depression. *Journal of Neuroscience, 25*(24), 5750–5756. https://doi.org/10.1523/JNEUROSCI.1104-05.2005

Scholz, R., Berberich, S., Rathgeber, L., Kolleker, A., Kohr, G., & Kornau, H. C. (2010). AMPA receptor signaling through BRAG2 and Arf6 critical for long-term synaptic depression. *Neuron, 66*(5), 768–780. https://doi.org/10.1016/j.neuron.2010.05.003

Sharma, M., Dierkes, T., & Sajikumar, S. (2017). Epigenetic regulation by G9a/GLP complex ameliorates amyloid-beta 1–42 induced deficits in long-term plasticity and synaptic tagging/capture in hippocampal pyramidal neurons. *Aging Cell, 16*(5), 1062–1072. https://doi.org/10.1111/acel.12634

Sharma, M., & Sajikumar, S. (2018). G9a/GLP Complex acts as a bidirectional switch to regulate metabotropic glutamate receptor-dependent plasticity in hippocampal CA1 pyramidal neurons. *Cerebral Cortex*. https://doi.org/10.1093/cercor/bhy161

Shema, R., Haramati, S., Ron, S., Hazvi, S., Chen, A., Sacktor, T. C., & Dudai, Y. (2011). Enhancement of consolidated long-term memory by overexpression of protein kinase Mzeta in the neocortex. *Science, 331*(6021), 1207–1210. https://doi.org/10.1126/science.1200215

Shen, K., & Meyer, T. (1999). Dynamic control of CaMKII translocation and localization in hippocampal neurons by NMDA receptor stimulation. *Science, 284*(5411), 162–166.

Sherff, C. M., & Carew, T. J. (1999). Coincident induction of long-term facilitation in Aplysia: Cooperativity between cell bodies and remote synapses. *Science, 285*(5435), 1911–1914.

Shivarama Shetty, M., Gopinadhan, S., & Sajikumar, S. (2016). Dopamine D1/D5 receptor signaling regulates synaptic cooperation and competition in hippocampal CA1 pyramidal neurons via sustained ERK1/2 activation. *Hippocampus, 26*(2), 137–150. https://doi.org/10.1002/hipo.22497

Sossin, W. S. (2018). Memory synapses are defined by distinct molecular complexes: A proposal. *Frontiers in Synaptic Neuroscience, 10*, 5.

Stanciu, M., Radulovic, J., & Spiess, J. (2001). Phosphorylated cAMP response element binding protein in the mouse brain after fear conditioning: relationship to Fos production. *Brain Research Molecular Brain Research, 94*(1–2), 15–24.

Steward, O., & Schuman, E. M. (2003). Compartmentalized synthesis and degradation of proteins in neurons. *Neuron, 40*(2), 347–359.

Strack, S., Choi, S., Lovinger, D. M., & Colbran, R. J. (1997). Translocation of autophosphorylated calcium/calmodulin-dependent protein kinase II to the postsynaptic density. *Journal of Biological Chemistry, 272*(21), 13467–13470.

Sutton, M. A., & Schuman, E. M. (2006). Dendritic protein synthesis, synaptic plasticity, and memory. *Cell, 127*(1), 49–58. https://doi.org/10.1016/j.cell.2006.09.014

Teyler, T. J., & DiScenna, P. (1986). The hippocampal memory indexing theory. *Behavioral Neuroscience, 100*(2), 147–154.

Thompson, R. F. (1986). The neurobiology of learning and memory. *Science, 233*(4767), 941–947.

Trettenbrein, P. C. (2016). The demise of the synapse as the locus of memory: A looming paradigm shift? *Frontiers in Systems Neuroscience, 10*, 88. https://doi.org/10.3389/fnsys.2016.00088

Tsien, J. Z., Huerta, P. T., & Tonegawa, S. (1996). The essential role of hippocampal CA1 NMDA receptor-dependent synaptic plasticity in spatial memory. *Cell, 87*(7), 1327–1338.

Tsokas, P., Hsieh, C., Yao, Y., Lesburgueres, E., Wallace, E. J. C., Tcherepanov, A., . . . Sacktor, T. C. (2016). Compensation for PKMzeta in long-term potentiation and spatial long-term memory in mutant mice. *Elife, 5*. https://doi.org/10.7554/eLife.14846

Turney, S. G., & Lichtman, J. W. (2012). Reversing the outcome of synapse elimination at developing neuromuscular junctions in vivo: evidence for synaptic competition and its mechanism. *PLoS Biology, 10*(6), e1001352. https://doi.org/10.1371/journal.pbio.1001352

Vecsey, C. G., Hawk, J. D., Lattal, K. M., Stein, J. M., Fabian, S. A., Attner, M. A., . . . Wood, M. A. (2007). Histone deacetylase inhibitors enhance memory and synaptic plasticity via CREB:CBP-dependent transcriptional activation. *Journal of Neuroscience, 27*(23), 6128–6140. https://doi.org/10.1523/JNEUROSCI.0296-07.2007

Vickers, C. A., Dickson, K. S., & Wyllie, D. J. (2005). Induction and maintenance of late-phase long-term potentiation in isolated dendrites of rat hippocampal CA1 pyramidal neurones. *Journal of Physiology, 568*(Pt 3), 803–813. https://doi.org/10.1113/jphysiol.2005.092924

Villarreal, D. M., Do, V., Haddad, E., & Derrick, B. E. (2002). NMDA receptor antagonists sustain LTP and spatial memory: Active processes mediate LTP decay. *Nature Neuroscience, 5*(1), 48–52. https://doi.org/10.1038/nn776

Volk, L. J., Bachman, J. L., Johnson, R., Yu, Y., & Huganir, R. L. (2013). PKM-zeta is not required for hippocampal synaptic plasticity, learning and memory. *Nature, 493*(7432), 420–423. https://doi.org/10.1038/nature11802

Wang, H., & Tiedge, H. (2004). Translational control at the synapse. *Neuroscientist, 10*(5), 456–466. https://doi.org/10.1177/1073858404265866

Wang, Y. T., & Linden, D. J. (2000). Expression of cerebellar long-term depression requires postsynaptic clathrin-mediated endocytosis. *Neuron, 25*(3), 635–647.

Whitlock, J. R., Heynen, A. J., Shuler, M. G., & Bear, M. F. (2006). Learning induces long-term potentiation in the hippocampus. *Science, 313*(5790), 1093–1097. https://doi.org/10.1126/science.1128134

Zhou, Y., Won, J., Karlsson, M. G., Zhou, M., Rogerson, T., Balaji, J., . . . Silva, A. J. (2009). CREB regulates excitability and the allocation of memory to subsets of neurons in the amygdala. *Nature Neuroscience, 12*(11), 1438–1443. https://doi.org/10.1038/nn.2405

CHAPTER 13

REGULATION OF SYNAPTIC HOMEOSTASIS BY TRANSLATIONAL MECHANISMS

MEGUMI MORI, JAY PENNEY, AND PEJMUN HAGHIGHI

Introduction

A fundamental feature of synapses is their ability to adapt and modify their structural and functional properties in response to environmental cues and experience, a phenomenon known as synaptic plasticity. A well-studied form of synaptic plasticity manifests as changes in synaptic efficacy as a result of coordinated pre- and post-synaptic activity; this form of plasticity, first articulated by Donald Hebb in 1949, is referred to as Hebbian plasticity (Hebb, 1949). Coordinated, high-frequency presynaptic firing induces postsynaptic firing, strengthens synaptic connections, and increases the amplitude of postsynaptic responses, a process known as long-term potentiation (LTP). Conversely, other distinct patterns of presynaptic activity can weaken synaptic connections and decrease the amplitude of postsynaptic responses, a process known as long-term depression (LTD). These changes in synaptic efficacy can be long lasting and are believed to underlie learning and memory formation (Lisman, Grace, & Duzel, 2011; Nabavi et al., 2014). The strengthening of a particular synapse or neuronal circuit, during the formation of a memory, for instance, is a feed-forward process and would in effect create a positive feedback loop. As such, LTP would lead to runaway neuronal excitability and epileptic circuit activity if left unchecked. Similar problems would develop following synaptic depression, leading to little or no circuit activity. (G. G. Turrigiano & Nelson, 2000). The feed-forward nature of LTP and LTD, therefore, necessitates the presence of stabilizing forces that would curb these processes (Abbott & Nelson, 2000). These compensatory and homeostatic mechanisms are generally known as homeostatic plasticity, mechanisms that allow synapses to maintain their strength within a tightly regulated range, while functioning in concert with Hebbian forms of synaptic plasticity (G. G. Turrigiano, 2017; Vitureira & Goda, 2013; Zenke & Gerstner, 2017). Homeostatic plasticity can be detected at many central and peripheral synapses across species, and it is thought to play a major role in maintaining and stabilizing activity in neural circuits, including during sensory deprivation and disease (Braegelmann, Streeter, Fields, & Baker, 2017; Davis, 2013; Davis & Bezprozvanny, 2001; Davis & Muller, 2015; A. Goel et al., 2006; A. Goel & Lee, 2007; Maffei, Nelson, & Turrigiano, 2004; Maffei & Turrigiano, 2008; Perry, Han, Das, & Dickman, 2017; Plomp, 2017; Takamori, 2017; G. Turrigiano, 2012; G. G. Turrigiano & Nelson, 2004).

Many forms of synaptic plasticity can lead to lasting alterations in synaptic function, and a large body of experimental evidence indicates that these alterations depend critically on de novo protein synthesis and local translational regulation of synaptic components and signaling molecules (Costa-Mattioli, Sossin, Klann, & Sonenberg, 2009; Henry et al., 2018; Jakawich et al., 2010; Kauwe et al., 2016; Kelleher, Govindarajan, & Tonegawa, 2004; Penney et al., 2012; Penney et al., 2016; Schanzenbacher, Sambandan, Langer, & Schuman, 2016; Sutton et al., 2006; Sutton & Schuman, 2006). This article will review the experimental evidence for the importance of translational mechanisms in the regulation of synaptic plasticity with an emphasis on homeostatic synaptic plasticity.

Regulation of Translation

Translation in eukaryotes can be divided into four stages: initiation, elongation, termination, and ribosome recycling, of which translation initiation is thought to be the rate-limiting step. During initiation, the 40S and 60S ribosome subunits are separately recruited to the 5′ untranslated region (5′UTR) of the mRNA. First, the 40S ribosomal subunit binds to the initiator methionyl tRNA (Met-tRNA$_i^{Met}$) to form the 43S preinitiation complex. Following recruitment of the 43S complex to the mRNA and base-pairing of the Adenine Uracil Guanine (AUG) initiation codon with the Met-tRNA$_i^{Met}$ anticodon, the 60S ribosome subunit is recruited to form the 80S ribosome, thus ending the initiation stage and starting the elongation step. Two critical steps during the initiation phase are the formation of the ternary complex and binding of the cap-binding protein complex (Sonenberg & Hinnebusch, 2009).

The Ternary Complex

The ternary complex is a multi-protein complex whose function is to deliver Met-tRNA$_i^{Met}$ to the 40S ribosome subunit (Asano, Clayton, Shalev, & Hinnebusch, 2000).

The attachment of the ternary complex to the 40S subunit, along with other components, forms the 43S preinitiation complex. The activity of the ternary complex is regulated primarily by the phosphorylation state of one of its constituents, the eukaryotic initiation factor 2α (eIF2α), which regulates the GTP-dependent recycling of Met-$tRNA_i^{Met}$ (Pavitt, Ramaiah, Kimball, & Hinnebusch, 1998).

Phosphorylation of eIF2α occurs via one of four upstream kinases, and this inhibits Met-$tRNA_i^{Met}$ recycling, slowing translation initiation (Thomas E. Dever, Dar, & Sicheri, 2007). Each eIF2α kinase is activated in response to different cellular stress; GCN2 (general control nonspecific 2) upon amino acid deprivation, HRI (heme regulated inhibitor) upon heme depletion, PKR (protein kinase RNA) upon DNA damage and PERK (PRK-like endoplasmic reticulum kinase) upon endoplasmic reticulum (ER) stress (J. J. Chen et al., 1991; Clemens, 1994; T. E. Dever et al., 1992; Kaufman, Davies, Pathak, & Hershey, 1989). As Met-$tRNA_i^{Met}$ is required for all mRNA translation, phosphorylation of eIF2α can slow the rate of mRNA translation globally; counter-intuitively, however, specific mRNAs show increased translation under these conditions.

Cap-Dependent Translation

The 5′ termini of all mRNAs transcribed in the nucleus contain a 7-methylguanosine, m7GpppN (where N is any nucleotide), which is called a cap structure. A number of proteins, collectively known as initiation factors, come together to form the cap-binding protein complex, which interacts with the cap structure of mRNAs. Two of these important components are the eukaryotic initiation factor 4E (eIF4E) and 4A (eIF4A): eIF4E directly binds to the cap structure and is rate-limiting for the initiation of translation, while eIF4A is a helicase that is required for the unwinding of the secondary structure of the 5′UTR of the mRNA prior to the commencement of translation (Sonenberg & Hinnebusch, 2009). Only a small number of eukaryotic mRNAs are translated in a cap-independent manner, where the ribosome binds to the mRNA through a specialized secondary structure of the mRNA, known as an internal ribosome entry site (IRES; Pelletier & Sonenberg, 1988).

Regulation of Cap-Dependent Translation

The target of rapamycin (TOR), or mechanistic target of rapamycin (mTOR) in mammals, acts as a major promoter of cap-dependent translation. This evolutionarily conserved kinase plays a central role in linking many cellular and environmental cues to cell metabolism, growth, and proliferation in all eukaryotes, and its abnormal function is associated with a number of diseases (Buckmaster, Ingram, & Wen, 2009; Ehninger et al., 2008; Hoeffer & Klann, 2004; X. M. Ma & Blenis, 2009; Sharma et al., 2010; Swiech, Perycz, Malik, & Jaworski, 2008; Tang et al., 2002). Some of these diseases are characterized or accompanied by aberrant synaptic activity, including abnormal features in synaptic plasticity. TOR promotes cap-dependent translation primarily through phosphorylation of eIF4E binding protein (4E-BP) and S6 ribosomal protein kinase (S6K). TOR-induced phosphorylation of 4E-BP suppresses its ability to bind and inhibit eIF4E, thus enhancing the interaction of the cap-binding complex with the mRNA 5′ cap and promoting translation (Sonenberg & Hinnebusch, 2009). In parallel, phosphorylation by TOR activates S6K, in turn enhancing its ability to phosphorylate downstream targets that promote translation. S6K is best known for phosphorylating the ribosomal protein S6 (Hay & Sonenberg, 2004; X. M. Ma & Blenis, 2009). In addition, S6K activity leads to increased helicase activity of eIF4A, thereby promoting mRNA scanning and base-pairing of AUG start codon with the Met-$tRNA_i^{Met}$ anticodon (Dorrello et al., 2006; Shahbazian et al., 2010). Primarily through 4E-BP and S6K, TOR can regulate various aspects of protein translation.

Hebbian Plasticity and Protein Translation

The development and refinement of protocols to induce LTP and LTD allowed for the dissection of the mechanisms underlying these processes. Experimentally, LTP is generally induced by short bursts of high-frequency neuronal stimulation, which results in persistent enhancement in the amplitude of synaptic responses lasting hours (Andersen, Krauth, & Nabavi, 2017; Bliss & Collingridge, 1993; Lynch, 2004). LTD, on the other hand, is usually induced by low-frequency stimulation, leading to a lasting decrease in the amplitude of synaptic responses (Collingridge, Peineau, Howland, & Wang, 2010; Dudek & Bear, 1992; Luscher & Malenka, 2012; Massey & Bashir, 2007). We now know that LTP and LTD are not counterparts of a singular phenomenon, but rather they are independent classes of synaptic plasticity with distinctive molecular mechanisms (Malenka & Bear, 2004; Massey & Bashir, 2007). The considerable evidence that cap-dependent protein synthesis and translation regulation are critical for both LTP and LTD (Costa-Mattioli et al., 2009; Kelleher, Govindarajan, & Tonegawa, 2004; Sutton & Schuman, 2006) will be briefly summarized here, followed by a more thorough examination of the role of protein translation in homeostatic synaptic plasticity.

LTP and Translation

Evidence for a role of de novo protein synthesis in memory formation dates back to more than 50 years ago when it was first shown that application of protein synthesis inhibitors disrupted motor learning and memory (Flexner, Flexner, & Stellar, 1963; Flexner, Flexner, Stellar, De La Haba, & Roberts, 1962). Observations that followed, however, indicated that the dependence of memory formation on protein synthesis was more complex: short-term memory formation was shown not to require de novo protein synthesis, while long-term memory formation was found to be critically dependent (Grecksch & Matthies, 1980; Montarolo et al., 1986; Squire & Davis, 1981). This led to the use of protein synthesis-dependency to distinguish short- from long-term memory formation. We now understand that at the synaptic level, LTP also has two distinct phases, early LTP and late LTP, which are mechanistically distinct. Late LTP is now considered a neo Hebbian form of plasticity, as it does not conform to classical criteria of Hebbian plasticity (Lisman et al., 2011). At some synapses, early LTP can be induced by a single burst of high-frequency stimulation or a short train of theta-burst stimulation, resulting in increased synaptic efficacy that lasts for 3 hours or less, without a requirement for de novo protein synthesis. In contrast, late LTP requires multiple sets of high-frequency stimuli or a longer theta-burst stimulation, culminating in greater postsynaptic efficacy that can last for at least 8 hours (Kelleher, Govindarajan, Jung, Kang, & Tonegawa, 2004). Late LTP can also be induced chemically, by application of factors such as cyclic adenosine monophosphate (cAMP) and bone derived neurotrophic factor (BDNF; Frey, Huang, & Kandel, 1993; Kang & Schuman, 1996). Considering that late LTP has longer-lasting effects on synaptic potentiation and is associated with increased growth of postsynaptic densities (Bosch et al., 2014), it is perhaps not surprising that late LTP is dependent on protein translation (Frey et al., 1993; Kang & Schuman, 1996; Nguyen & Kandel, 1997).

Since the requirement of protein synthesis is a defining factor for early versus late LTP, molecular regulators of translation are expected to act as a switch between the two phases of LTP. Indeed, removing the breaks on cap-dependent protein translation *via* knockout of 4E-BP, expressing unphosphorylatable forms of eIF2α, or knockout of the eIF2α kinase GCN2 all lower the threshold for stimulation required to elicit late LTP in hippocampal synapses (Banko et al., 2005; Costa-Mattioli et al., 2005; Costa-Mattioli et al., 2007). This effectively induces late LTP under stimulation protocols that normally would only elicit early LTP (Banko et al., 2005; Costa-Mattioli et al., 2005; Costa-Mattioli et al., 2007). Stimulation of protein translation via TOR is also sufficient to induce dendritic growth following subthreshold stimulation (Henry, Hockeimer, Chen, Mysore, & Sutton, 2017). Conversely, dampening active players in protein translation initiation disrupts the manifestation of late LTP (Costa-Mattioli et al., 2007) and impairs performance in behavioral tests of memory (Costa-Mattioli et al., 2007; Jian et al., 2014). These findings suggest that regulators of cap-dependent translation can act as switches that modulate the persistence of LTP and thereby influence memory formation.

LTD and Protein Translation

While LTP and LTD are mechanistically distinct processes, sustained expression of LTD also requires de novo protein synthesis. The precise mechanism of LTD induction varies at different synapses, but in general LTD occurs through low-frequency stimulation of N-Methyl-D-aspartic acid (NMDA) receptors or activation of metabotropic glutamate receptors (mGluRs) (Collingridge et al., 2010). Like LTP, LTD at hippocampal CA1 synapses can be divided into early and late phases, of which the late phase demonstrates longer-lasting depression (Kauderer & Kandel, 2000; Linden, 1996). Both NMDA- and mGluR-dependent forms of late LTD are dependent on protein translation (Huber, Kayser, & Bear, 2000; Huber, Roder, & Bear, 2001; Kauderer & Kandel, 2000; Linden, 1996). Disruption of the cap-binding complex by introducing an artificial cap structure or pharmacological inhibition of TOR or other initiation factors eliminates LTD maintenance (Banko, Hou, Poulin, Sonenberg, & Klann, 2006; Hou & Klann, 2004; Karachot, Shirai, Vigot, Yamamori, & Ito, 2001).

Compared to LTP, the requirement for de novo protein translation in LTD might appear less intuitive, since LTD manifests as reduced synaptic strength and is associated with synapse elimination. One mechanism by which protein translation promotes sustained depression of synaptic efficacy is by changing the subunit stoichiometry of α-amino-3-hydroxy-5-methyl-4-isoxazolepropionic acid (AMPA) receptors. AMPA receptors containing GluA2 subunits are less permeable to calcium and show reduced single channel conductance compared to non-GluA2 AMPA receptors (S. Cull-Candy, Kelly, & Farrant, 2006; Isaac, Ashby, & McBain, 2007). LTD induction promotes rapid synthesis and incorporation of GluA2 subunits, resulting in reduced synaptic function and dampened postsynaptic calcium influx (Mameli, Balland, Lujan, & Luscher, 2007). LTD also results in marked endocytosis of AMPA receptors (Y. T. Wang & Linden, 2000; Xiao, Zhou, & Nicoll, 2001), and protein translation supports this process by synthesizing components of the endocytic pathway (H. Wang et al., 2016; Waung & Huber, 2009; Waung, Pfeiffer, Nosyreva, Ronesi, & Huber, 2008). Examples include Arc, which enhances endophilin and dynamin mediated endocytosis of AMPA receptors (Chowdhury et al., 2006; Waung et al., 2008); striatal-enriched protein tyrosine phosphatase (STEP), which phosphorylates AMPA receptors to signal endocytosis (Y. Zhang et al., 2008); and microtubule-associated protein 1B (MAP1B), which sequesters factors that stabilize AMPA localization in the membrane (Davidkova & Carroll, 2007). Altogether, it appears that translation pathways are required at multiple levels to allow for persistent LTD.

Homeostatic Plasticity and Protein Translation in the Central Nervous System

Homeostatic mechanisms exist to keep synaptic activity within a tight physiological range, and depending on input, they can tune synaptic strength to maintain a desirable set-point. This is crucial both to keep the feedforward nature of LTP and LTD in check, as well as to maintain the ability of synapses to respond to appropriate inputs in a Hebbian manner (Braegelmann et al., 2017; Davis, 2013; Davis & Bezprozvanny, 2001;

Davis & Muller, 2015; Perry et al., 2017; Plomp, 2017; Takamori, 2017; G. Turrigiano, 2012; G. G. Turrigiano, 2017; G. G. Turrigiano & Nelson, 2000, 2004; Vitureira & Goda, 2013; Zenke & Gerstner, 2017). Homeostatic plasticity is widely observed in both the central nervous system and the peripheral nervous system and is highly conserved from invertebrate models to human synapses. (For reviews on homeostatic plasticity, see [Davis, 2013; Davis & Bezprozvanny, 2001; Davis & Muller, 2015; G. G. Turrigiano, 2008]). In general, synaptic homeostasis can be achieved by changes in postsynaptic receptor function, presynaptic neurotransmitter release or neuronal membrane excitability (Garcia-Bereguiain, Gonzalez-Islas, Lindsly, & Wenner, 2016; A. Goel et al., 2006; Hengen, Lambo, Van Hooser, Katz, & Turrigiano, 2013; Keck et al., 2013; Kirov, Goddard, & Harris, 2004; Lissin et al., 1998; Maffei et al., 2004; Maffei & Turrigiano, 2008; O'Brien et al., 1998; Orr, Fetter, & Davis, 2017; Plomp, Morsch, Phillips, & Verschuuren, 2015; Plomp et al., 1995; Pozo & Goda, 2010; Teichert, Liebmann, Hubner, & Bolz, 2017; G. G. Turrigiano, Leslie, Desai, Rutherford, & Nelson, 1998; Tyler, Petzold, Pal, & Murthy, 2007; Vale & Sanes, 2002; W. Zhang & Linden, 2003). Experimental evidence indicates that synapses depend on protein synthesis to engage and maintain homeostatic adjustments of synaptic function. Translational regulation can be adapted to achieve highly diverse effects by fine-tuning the targets of mRNA translation; it is therefore not surprising that both presynaptic and postsynaptic adjustments to synaptic activity can be achieved by increased protein translation activities.

Homeostatic Plasticity in the Vertebrate Central Nervous System: Postsynaptic Compensation

One of the well-characterized examples of homeostatic plasticity, also known as synaptic scaling, occurs at excitatory glutamatergic synapses on hippocampal and cortical neurons. Inhibition of evoked synaptic activity by tetrodotoxin (TTX) treatment in dissociated primary cortical neurons results in a compensatory up-regulation of mEPSC (miniature excitatory postsynaptic current) amplitudes (G. G. Turrigiano et al., 1998). The increase in mEPSCs is a result of increased postsynaptic sensitivity to glutamate, rather than changes in presynaptic neurotransmitter release (G. G. Turrigiano et al., 1998). In fact, TTX application increases GluA1 and GluA2 AMPA receptor insertion into the postsynaptic dendrites (Wierenga, Ibata, & Turrigiano, 2005), a process that can be disrupted with inhibitors of protein synthesis (Schanzenbacher et al., 2016). While this experimental paradigm clearly demonstrates the importance of protein synthesis for postsynaptic homeostatic compensation in response to global activity blockade, it requires a prolonged application of blockers before homeostatic plasticity is observed (>12 hours). An alternative protocol, which includes co-application of the AMPA receptor antagonist CNQX (6-cyano-7-nitroquinoxaline-2,3-dione) and the NMDA receptor antagonist APV ((2R)-amino-5-phosphonovaleric acid), induces homeostatic compensation in the dendrite at a much shorter time scale. Under this inhibitory regime, neurons show homeostatic compensation after 2–3 hours of treatment (Sutton et al., 2006). Inhibition of postsynaptic receptors and hence miniature synaptic activity was shown to induce protein synthesis within the first hour of treatment, increasing de novo synthesis and incorporation of GluA1 containing AMPA receptors (Sutton et al., 2006). As such, inhibition of protein translation eliminates GluA1 synthesis and incorporation, thereby abrogating the homeostatic postsynaptic compensation (Ju et al., 2004; Sutton et al., 2006; Sutton, Taylor, Ito, Pham, & Schuman, 2007; Sutton, Wall, Aakalu, & Schuman, 2004). These findings demonstrate that differential responses of translational mechanisms to different synaptic perturbations can determine the degree and speed of homeostatic compensation in neurons.

A potential mechanism whereby mEPSCs could regulate local protein translation is through modulation of receptor calcium influx. NMDA receptors have been proposed to inhibit local protein translation elongation via phosphorylation of eEF2 (eukaryotic elongation factor 2) (Scheetz, Nairn, & Constantine-Paton, 2000), whose phosphorylation state is regulated by calcium dependent eEF2 kinase (Nairn & Palfrey, 1987). Blocking NMDA receptors is sufficient to dephosphorylate eEF2 and increase protein translation at the elongation level (Sutton et al., 2007). NMDA receptor function has also been tied to TOR pathway activity via SynGAP (Synaptic Ras-GTPase activating protein; C. C. Wang, Held, & Hall, 2013), which colocalizes with NMDA receptors (H. J. Chen, Rojas-Soto, Oguni, & Kennedy, 1998; Kim, Liao, Lau, & Huganir, 1998) and is activated by CaMKII (Ca^{2+}-Calmodulin protein Kinase II) in the presence of calcium (Oh, Manzerra, & Kennedy, 2004). SynGAP functions to limit protein translation via suppression of mTOR under normal conditions, and knockdown of SynGAP or the activation of the TOR pathway is sufficient to increase mEPSC amplitudes in the absence of synaptic activity blockade (C. C. Wang et al., 2013). In addition to synaptic calcium, alteration in calcium release from internal sources can also influence synaptic scaling. Blocking calcium-induced calcium release from internal storage sources triggers synaptic scaling by promoting translation initiation through an eIF2α dependent mechanism (Reese & Kavalali, 2015).

Interestingly, calcium can also specifically control GluA1 synthesis via retinoic acid (RA) signaling. Under basal conditions, calcium inhibits the production of RA, but upon synaptic perturbation, RA synthesis is disinhibited (H. L. Wang, Zhang, Hintze, & Chen, 2011). RA then activates its receptor RARα, an atypical receptor that associates with dendritic RNA granules and promotes the synthesis of GluA1 (Aoto, Nam, Poon, Ting, & Chen, 2008; N. Chen, Onisko, & Napoli, 2008; Maghsoodi et al., 2008). Consistent with a role in synaptic scaling, RA production can be induced by blockade of action potentials along with AMPA receptor blockade (Aoto et al., 2008). Conversely, disruptions in RA production or RARα expression blocks synaptic upscaling in response to activity blockade (Aoto et al., 2008). RA signaling via neuronal calcium acts as another avenue of protein synthesis control that can respond to changes in neuronal activity to induce synaptic homeostasis.

Homeostatic Plasticity in the Vertebrate Central Nervous System: Presynaptic Compensation

In addition to postsynaptic compensation, homeostatic plasticity can be achieved through presynaptic alterations of neurotransmitter release sites or the amount of neurotransmitter release at individual sites. Several lines of evidence have revealed presynaptic changes in neurotransmitter release as part of the homeostatic compensation. In particular, hippocampal neurons exhibit homeostatic plasticity in response to activity blockade not only through enhancement on mEPSC amplitudes, but also through enhancement in mEPSC frequencies (Henry et al., 2012; Henry et al., 2018; Jakawich et al., 2010; Wierenga, Walsh, & Turrigiano, 2006), indicating presynaptic modifications. The changes in mEPSC frequencies are corroborated by morphological alterations in the presynaptic terminal, including increased number of docked vesicles, size of readily releasable pool, and size of vesicles (Murthy, Schikorski, Stevens, & Zhu, 2001). Since presynaptic changes occur in response to postsynaptic activity perturbations, this suggests the presence of a retrograde signaling mechanism that relays the state of postsynaptic activity to the presynaptic neuron. Indeed, experimental evidence points to a critical role for TOR activity in postsynaptic dendrites for the ability of presynaptic terminals to undergo homeostatic compensation: block of TOR activity blocks the compensation, while activation of TOR is sufficient to upregulate presynaptic release (Henry et al., 2012; Henry et al., 2018; Jakawich et al., 2010). It appears that retrograde synaptic compensation in cultured hippocampal neurons requires BDNF, which is released in response to synaptic perturbations. Further, extrinsic application of BDNF is sufficient to induce presynaptic functional increase (Jakawich et al., 2010). In fact, the production of BDNF is eliminated as a result of pharmacological inhibition of TOR, which also eliminates presynaptic homeostatic compensation (Henry et al., 2012; Henry et al., 2018). All in all, protein translation regulation therefore mediates homeostatic plasticity in the presynaptic terminal by promoting the synthesis of retrograde signaling factor(s) during synaptic perturbation.

Homeostatic Plasticity at the Neuromuscular Junction

Like many central synapses, synapses at the neuromuscular junction (NMJ) show robust homeostatic synaptic plasticity: reduced neurotransmitter receptor activity in postsynaptic muscles can induce compensatory increases in presynaptic release, a response that is well conserved from invertebrates to humans (S. G. Cull-Candy, Miledi, & Trautmann, 1979; Frank, Kennedy, Goold, Marek, & Davis, 2006; Katz & Miledi, 1978; Miledi, Molenaar, & Polak, 1978; Petersen, Fetter, Noordermeer, Goodman, & DiAntonio, 1997; Plomp, van Kempen, & Molenaar, 1992; Tian, Prior, Dempster, &

Marshall, 1994; X. Wang, Pinter, & Rich, 2016). The ability of the NMJ to homeostatically regulate presynaptic release may be especially critical for patients with myasthenia gravis, an autoimmune disorder that targets acetylcholine receptors or related proteins at the NMJ (Gilhus & Verschuuren, 2015). The affected NMJs often have reduced levels of acetylcholine receptor expression and reduced sensitivity to acetylcholine (Albuquerque, Rash, Mayer, & Satterfield, 1976; Fambrough, Drachman, & Satyamurti, 1973; Tsujihata, Hazama, Ishii, Ide, & Takamori, 1980); however, the motor neurons exhibit a compensatory increase in presynaptic neurotransmitter release, which may be of clinical benefit during early stages of the disease (Albuquerque et al., 1976; Plomp et al., 1995).

Homeostatic enhancement in presynaptic release has been particularly well studied at the *Drosophila* NMJ, where the genetic removal of GluRIIA, one of the glutamate receptor subunits, causes a strong decrease in single channel mean open time leading to a reduction in mEPSC amplitudes (DiAntonio, Petersen, Heckmann, & Goodman, 1999). This reduction ultimately triggers a retrograde signal that leads to presynaptic enhancement of neurotransmitter release to compensate for the reduced response of neurotransmitter receptors (Petersen et al., 1997). A similar compensatory homeostatic response can also be triggered acutely through pharmacological block of GluRIIA-containing receptors (Frank et al., 2006). As one might expect, this robust synaptic compensation requires the presynaptic machinery to respond to the increased demand at the synapse. As such, many presynaptic components have been identified to be essential elements for the compensatory synaptic enhancement at the NMJ, including calcium channels, active zone components, and signaling molecules (Davis & Muller, 2015; Frank et al., 2006; Frank, Pielage, & Davis, 2009; Goold & Davis, 2007; Marie, Pym, Bergquist, & Davis, 2010; Muller & Davis, 2012; Muller, Liu, Sigrist, & Davis, 2012; Muller, Pym, Tong, & Davis, 2011; Pilgram, Potikanond, van der Plas, Fradkin, & Noordermeer, 2011; Tsurudome et al., 2010; Weyhersmuller, Hallermann, Wagner, & Eilers, 2011; Younger, Muller, Tong, Pym, & Davis, 2013).

For the presynaptic neuron to initiate a compensatory response, postsynaptic components must be engaged to trigger and/or maintain a retrograde signaling cascade downstream from glutamate receptor manipulations in the muscle (Frank et al., 2006; Haghighi et al., 2003; Petersen et al., 1997). Over the past several years, it has become clear that translational mechanisms play a central role in the regulation of this retrograde signaling (Kauwe et al., 2016; Penney et al., 2012; Penney et al., 2016). Here, we will describe these select studies in detail.

Postsynaptic TOR Regulates Presynaptic Homeostatic Plasticity

Propelled by the power of genetics, *Drosophila* larval NMJs became the first in vivo model for testing the role of translational mechanisms in the retrograde regulation of synaptic homeostasis. Based on the critical role of translation initiation in protein synthesis, limiting initiation either by removing one gene copy of eIF4E or eIF2α was hypothesized to be sufficient to influence the ability of the NMJ to undergo synaptic homeostasis. Normally, GluRIIA mutant larvae show reduced miniature amplitudes but can match the wild-type evoked response. Heterozygosity for eIF4E curtailed this ability, while heterozygosity for eIF2α failed to do so (Penney et al., 2012). This finding pointed to a potential role for TOR as a regulator of eIF4E availability in the regulation of synaptic homeostasis. Indeed, genetic experiments supported a role for TOR: TOR heterozygosity was sufficient to block synaptic homeostasis in GluRIIA mutant larvae. Interestingly, certain hypomorphic homozygous mutant combinations of TOR can live to larval stages, which allowed for a genetic examination of the role of TOR at the synapse in a tissue-specific manner. TOR mutant larvae showed no structural defects at the NMJ but, as the previous experiments would have predicted, failed to exhibit synaptic homeostasis. Rescue experiments were clear but surprising: TOR was needed in postsynaptic muscles rather than in motoneurons to restore synaptic homeostasis. If TOR were required for synaptic compensation, then one would expect pharmacological inhibition of TOR to block synaptic homeostasis at the NMJ. Living and free moving larvae were fed rapamycin in their food prior to dissection and electrophysiological analysis. Interestingly, at least 6–12 hours of feeding on rapamycin-containing food was required before any significant block could be detected in larvae, thus suggesting that the early synaptic homeostatic response was not dependent on de novo protein synthesis, but longer-term maintenance of compensation became critically dependent on de novo translation (Penney et al., 2012). This idea is supported by the fact that acute pharmacological induction of synaptic homeostasis at the larval NMJ can operate within minutes independently of de novo protein synthesis (Cheng, Locke, & Davis, 2011; Frank et al., 2006; P. Goel, Li, & Dickman, 2017). The distinction between acute and chronic forms of synaptic homeostasis is reminiscent of the early versus late forms of LTP and LTD; the synapse is well equipped to adjust synaptic activity in an acute manner, but persistent changes require protein synthesis. Together with evidence from mammalian culture systems (Henry et al., 2012; Henry et al., 2018; Sutton et al., 2006; Sutton et al., 2007; Sutton et al., 2004; C. C. Wang et al., 2013), these findings clearly demonstrate a role for postsynaptic TOR-dependent translation in maintaining synaptic homeostasis and highlight the importance of retrograde signaling in this process.

Postsynaptic 4E-BP Links Nutrient Availability to Presynaptic Homeostatic Plasticity

TOR acts to integrate a number of intracellular and extracellular signals, but perhaps its most prominent role is to link nutrient signaling to protein synthesis and cell growth (Barbet et al., 1996; Hara et al., 1998; for reviews, see Loewith & Hall, 2011; Wullschleger, Loewith, & Hall, 2006). Thus, it would be expected that nutrient scarcity could suppress TOR activity and thereby suppress retrograde synaptic homeostasis. Interestingly, however, although these predictions were supported by experimental data, the picture was more complicated than expected. Wild-type larvae moved to vials with or without food for 3–6 hours showed no apparent difference in synaptic structure or the amount of neurotransmitter release at the NMJ, but GluRIIA mutant larvae showed great sensitivity to food: acute starvation completely blocked their ability to show synaptic compensation (Kauwe et al., 2016). Indeed, the level of S6K phosphorylation (an index of TOR activity) was reduced in these larvae, so it appeared that the culprit had been identified. However, a surprising experiment suggested otherwise. Transgenic expression of a constitutively active form of S6K, which can function independently of TOR, failed to rescue synaptic homeostasis during starvation. This finding suggested that TOR independent pathways might be responsible for failure of GluRIIA mutants to exhibit synaptic homeostasis during acute starvation. In addition to inhibiting TOR activity, starvation has been shown to influence the transcription of 4E-BP through the action of the Forkhead-box-O transcription factor (Foxo); Foxo is normally under the negative control of insulin signaling under nutrient rich conditions, but starvation removes this inhibition and activates Foxo, which then can activate 4E-BP transcription as one of its target genes (Demontis & Perrimon, 2010; Junger et al., 2003; Puig, Marr, Ruhf, & Tjian, 2003; Teleman, Chen, & Cohen, 2005). Confirming this scenario, heterozygosity for 4E-BP was sufficient to maintain the ability of the NMJ to undergo synaptic homeostasis even under starvation conditions. Similarly, genetic knockdown of Foxo in muscle blocked the increase in 4E-BP transcription in postsynaptic muscles in response to acute starvation and restored synaptic homeostasis (Kauwe et al., 2016). These findings not only add to the complexity of the translational regulation of synaptic homeostasis, but they also show that the state of nutrient availability can have a significant influence on synaptic homeostasis.

Translation, Synaptic Homeostasis, and Implications for Disease

The requirement for TOR activity in postsynaptic muscles for the retrograde compensatory enhancement of presynaptic release in GluRIIA mutants (Penney et al., 2012) raised the question of whether increased postsynaptic TOR activity in an otherwise wild-type animal could hijack this retrograde pathway and abnormally enhance presynaptic release. Indeed, experimental data indicated that increased postsynaptic TOR activity was sufficient to induce a retrograde enhancement in presynaptic release (Penney et al., 2012); in other words, increased TOR activity appeared to have pitched the set point of synaptic strength higher. This abnormal increase in synaptic activity could potentially explain some aspects of disease in tuberous sclerosis complex (TSC), a multi-system disorder characterized by non-malignant tumors, and often accompanied by major neurological effects (Tsai & Sahin, 2011). About 90% of TSC patients exhibit seizures, about half meet the criteria for mental retardation, and about one-third exhibit autistic behaviors (Kwiatkowski & Manning, 2005). TSC is caused by mutation of either the TSC1 or TSC2 genes, whose products form a protein complex that negatively regulates TOR activity. The TSC protein complex is a GAP (GTPase activating protein) for the small GTPase Rheb (Ras homolog enriched in the brain), whose activity is required for activation of TOR. Mutation of either TSC1or TSC2 leads to disruption of the complex, abnormal activation of TOR and deregulation of the pathway (Hoeffer & Klann, 2010). It is conceivable that the enhanced TOR activity in the absence of TSC can lead to an abnormal increase in neurotransmission, which can in turn destabilize circuit function and lead to epileptic seizures. Indeed, rodent TSC models exhibit numerous learning and memory defects, as well as alterations in both short-term and long-term plasticity, many of which can be rescued by short-term rapamycin treatment during adulthood (Ehninger et al., 2008; Goorden, van Woerden, van der Weerd, Cheadle, & Elgersma, 2007; von der Brelie, Waltereit, Zhang, Beck, & Kirschstein, 2006). These findings further highlight the critical importance of translational mechanisms in the regulation of higher brain functions and reveal the instrumental role these mechanisms play in nervous system diseases.

Leucine Rich Repeat Kinase 2 Is a Modulator of Cap-Dependent Translation and a Regulator of Synaptic Homeostasis

Another disease related gene that has been suggested to influence cap-dependent translation is leucine-rich repeat kinase 2 (LRRK2). Mutations in the gene encoding LRRK2 were originally identified in linkage studies of familial Parkinson's disease (Funayama et al., 2002; Paisan-Ruiz et al., 2004; Zimprich et al., 2004). Today, LRRK2 mutations are considered the single most common cause of inherited Parkinson's disease (Greggio & Cookson, 2009). LRRK2 is a large protein with multiple functional domains including a kinase domain (Martin, Dawson, & Dawson, 2011; Mata, Wedemeyer, Farrer, Taylor, & Gallo, 2006). Interestingly, the most common familial mutations in LRRK2 appear to lead to an increase in autophosphorylation and kinase activity (Greggio & Cookson, 2009; Martin et al., 2011); therefore, it is plausible to assume that some aspects of the disease may be related to dysregulated and/or elevated levels of kinase activity. There has been a widespread interest in identifying LRRK2's phosphorylation substrates, and several potential candidates have been identified, including Rab GTPases (Alessi & Sammler, 2018; Imai et al., 2008; Jaleel et al., 2007). The *Drosophila* genome contains a single homologous gene to LRRK2, which we will call dLRRK to distinguish it from the human protein hLRRK2. Several studies have used transgenic hLRRK2 in *Drosophila* to study its gain-of-function in vivo (Liu et al., 2008; Ng et al., 2009; Venderova et al., 2009). A loss of function allele of dLRRK has also been generated, which has enabled insight into the endogenous function of LRRK2 in the nervous system (Imai et al., 2008;

S. B. Lee, Kim, Lee, & Chung, 2007). Moreover, the *Drosophila* model has allowed for genetic interaction experiments that have linked LRRK2 to other Parkinson's related genes, including Parkin, DJ-1, and PINK-1 (Ng et al., 2009; Tain et al., 2009; Venderova et al., 2009). Although the precise mechanism through which LRRK2 influences translational mechanisms remains unclear, there is strong evidence that LRRK2 can enhance cap-dependent translation both in *Drosophila* and in mammalian cells (Imai et al., 2008; S. Lee, Liu, Lin, Guo, & Lu, 2010; Penney et al., 2016; Tain et al., 2009). Genetic knockdown of LRRK2 in muscle blocks synaptic homeostasis at the NMJ, while its overexpression in muscle leads to a retrograde enhancement in presynaptic neurotransmitter release (Penney et al., 2016). Further, all gain-of-function effects of LRRK2 at the NMJ could be countered by feeding larvae rapamycin or cycloheximide or by genetic manipulation of translation initiation, indicating that this synaptic function of LRRK2 depends on its ability to enhance translation (Penney et al., 2016). The ability of pathogenic LRRK2 mutations to disrupt synaptic homeostasis and enhance synaptic release, reminiscent of TOR gain-of-function, is potentially relevant to the pathogenesis of Parkinson's disease and suggest that perhaps synaptic dysfunction precedes the pathological manifestations of the disease. These results open new avenues of research into the biological function of LRRK2, and its dysfunction during disease may point to alternative approaches toward tackling Parkinson's disease.

Future Directions

The experimental evidence accumulated over the last two decades has built a strong and conclusive case for the essential role of translational mechanisms in the regulation of synaptic plasticity in general and synaptic homeostasis in particular. The inevitable future challenge will be to identify synapse-specific translational targets and understand the details of how specificity is achieved. With growing evidence implicating translational mechanisms in age-dependent neurodegenerative diseases (Beckelman et al., 2016; Cestra, Rossi, Di Salvio, & Cozzolino, 2017; T. Ma et al., 2013; Moon, Sonenberg, & Parker, 2018; Shih & Hsueh, 2018; Taymans, Nkiliza, & Chartier-Harlin, 2015; Zheng et al., 2016), identifying synapse-specific translational targets will open new avenues for designing novel therapeutics aimed at tackling these devastating diseases.

References

Abbott, L. F., & Nelson, S. B. (2000). Synaptic plasticity: Taming the beast. *Nature Neuroscience, 3 Suppl*, 1178–1183. https://doi.org/10.1038/81453

Albuquerque, E. X., Rash, J. E., Mayer, R. F., & Satterfield, J. R. (1976). An electrophysiological and morphological study of the neuromuscular junction in patients with myasthenia gravis. *Experimental Neurology, 51*(3), 536–563.

Alessi, D. R., & Sammler, E. (2018). LRRK2 kinase in Parkinson's disease. *Science, 360*(6384), 36–37. https://doi.org/10.1126/science.aar5683

Andersen, N., Krauth, N., & Nabavi, S. (2017). Hebbian plasticity in vivo: Relevance and induction. *Current Opinion in Neurobiology, 45*, 188–192. https://doi.org/10.1016/j.conb.2017.06.001

Aoto, J., Nam, C. I., Poon, M. M., Ting, P., & Chen, L. (2008). Synaptic signaling by all-trans retinoic acid in homeostatic synaptic plasticity. *Neuron, 60*(2), 308–320. https://doi.org/10.1016/j.neuron.2008.08.012

Asano, K., Clayton, J., Shalev, A., & Hinnebusch, A. G. (2000). A multifactor complex of eukaryotic initiation factors, eIF1, eIF2, eIF3, eIF5, and initiator tRNA(Met) is an important translation initiation intermediate in vivo. *Genes & Development, 14*(19), 2534–2546.

Banko, J. L., Hou, L., Poulin, F., Sonenberg, N., & Klann, E. (2006). Regulation of eukaryotic initiation factor 4E by converging signaling pathways during metabotropic glutamate receptor-dependent long-term depression. *Journal of Neuroscience, 26*(8), 2167–2173. https://doi.org/10.1523/JNEUROSCI.5196-05.2006

Banko, J. L., Poulin, F., Hou, L., DeMaria, C. T., Sonenberg, N., & Klann, E. (2005). The translation repressor 4E-BP2 is critical for eIF4F complex formation, synaptic plasticity, and memory in the hippocampus. *Journal of Neuroscience, 25*(42), 9581-9590. https://doi.org/10.1523/JNEUROSCI.2423-05.2005

Barbet, N. C., Schneider, U., Helliwell, S. B., Stansfield, I., Tuite, M. F., & Hall, M. N. (1996). TOR controls translation initiation and early G1 progression in yeast. *Molecular Biology of the Cell, 7*(1), 25–42.

Beckelman, B. C., Day, S., Zhou, X., Donohue, M., Gouras, G. K., Klann, E.,...Ma, T. (2016). Dysregulation of Elongation Factor 1A Expression is Correlated with Synaptic Plasticity Impairments in Alzheimer's Disease. *Journal of Alzheimers Disease, 54*(2), 669–678. https://doi.org/10.3233/JAD-160036

Bliss, T. V., & Collingridge, G. L. (1993). A synaptic model of memory: Long-term potentiation in the hippocampus. *Nature, 361*(6407), 31–39. https://doi.org/10.1038/361031a0

Bosch, M., Castro, J., Saneyoshi, T., Matsuno, H., Sur, M., & Hayashi, Y. (2014). Structural and molecular remodeling of dendritic spine substructures during long-term potentiation. *Neuron, 82*(2), 444–459. https://doi.org/10.1016/j.neuron.2014.03.021

Braegelmann, K. M., Streeter, K. A., Fields, D. P., & Baker, T. L. (2017). Plasticity in respiratory motor neurons in response to reduced synaptic inputs: A form of homeostatic plasticity in respiratory control? *Experimental Neurology, 287*(Pt 2), 225–234. https://doi.org/10.1016/j.expneurol.2016.07.012

Buckmaster, P. S., Ingram, E. A., & Wen, X. (2009). Inhibition of the mammalian target of rapamycin signaling pathway suppresses dentate granule cell axon sprouting in a rodent model of temporal lobe epilepsy. *Journal Neuroscience, 29*(25), 8259–8269. https://doi.org/10.1523/JNEUROSCI.4179-08.2009

Cestra, G., Rossi, S., Di Salvio, M., & Cozzolino, M. (2017). Control of mRNA translation in ALS proteinopathy. *Frontiers in Molecular Neuroscience, 10*, 85. https://doi.org/10.3389/fnmol.2017.00085

Chen, H. J., Rojas-Soto, M., Oguni, A., & Kennedy, M. B. (1998). A synaptic Ras-GTPase activating protein (p135 SynGAP) inhibited by CaM kinase II. *Neuron, 20*(5), 895–904.

Chen, J. J., Pal, J. K., Petryshyn, R., Kuo, I., Yang, J. M., Throop, M. S.,...London, I. M. (1991). Amino acid microsequencing of internal tryptic peptides of heme-regulated eukaryotic initiation factor 2 alpha subunit kinase: homology to protein kinases. *Proceedings of the National Academy of Sciences of the United States of America, 88*(2), 315–319.

Chen, N., Onisko, B., & Napoli, J. L. (2008). The nuclear transcription factor RARalpha associates with neuronal RNA granules and suppresses translation. *Journal of Biological Chemistry, 283*(30), 20841–20847. https://doi.org/10.1074/jbc.M802314200

Cheng, L., Locke, C., & Davis, G. W. (2011). S6 kinase localizes to the presynaptic active zone and functions with PDK1 to control synapse development. *Journal of Cell Biology, 194*(6), 921–935. https://doi.org/10.1083/jcb.201101042

Chowdhury, S., Shepherd, J. D., Okuno, H., Lyford, G., Petralia, R. S., Plath, N., . . . Worley, P. F. (2006). Arc/Arg3.1 interacts with the endocytic machinery to regulate AMPA receptor trafficking. *Neuron, 52*(3), 445–459. https://doi.org/10.1016/j.neuron.2006.08.033

Clemens, M. J. (1994). Regulation of eukaryotic protein synthesis by protein kinases that phosphorylate initiation factor eIF-2. *Molecular Biology Reports, 19*(3), 201–210.

Collingridge, G. L., Peineau, S., Howland, J. G., & Wang, Y. T. (2010). Long-term depression in the CNS. *Nature Reviews Neuroscience, 11*(7), 459–473. https://doi.org/10.1038/nrn2867

Costa-Mattioli, M., Gobert, D., Harding, H., Herdy, B., Azzi, M., Bruno, M., . . . Sonenberg, N. (2005). Translational control of hippocampal synaptic plasticity and memory by the eIF2alpha kinase GCN2. *Nature, 436*(7054), 1166–1173. https://doi.org/10.1038/nature03897

Costa-Mattioli, M., Gobert, D., Stern, E., Gamache, K., Colina, R., Cuello, C., . . . Sonenberg, N. (2007). eIF2alpha phosphorylation bidirectionally regulates the switch from short- to long-term synaptic plasticity and memory. *Cell, 129*(1), 195–206. https://doi.org/10.1016/j.cell.2007.01.050

Costa-Mattioli, M., Sossin, W. S., Klann, E., & Sonenberg, N. (2009). Translational control of long-lasting synaptic plasticity and memory. *Neuron, 61*(1), 10–26. https://doi.org/10.1016/j.neuron.2008.10.055

Cull-Candy, S., Kelly, L., & Farrant, M. (2006). Regulation of Ca2+-permeable AMPA receptors: Synaptic plasticity and beyond. *Current Opinion in Neurobiology, 16*(3), 288–297. https://doi.org/10.1016/j.conb.2006.05.012

Cull-Candy, S. G., Miledi, R., & Trautmann, A. (1979). End-plate currents and acetylcholine noise at normal and myasthenic human end-plates. *Journal of Physiology, 287*, 247–265.

Davidkova, G., & Carroll, R. C. (2007). Characterization of the role of microtubule-associated protein 1B in metabotropic glutamate receptor-mediated endocytosis of AMPA receptors in hippocampus. *Journal of Neuroscience, 27*(48), 13273–13278. https://doi.org/10.1523/JNEUROSCI.3334-07.2007

Davis, G. W. (2013). Homeostatic signaling and the stabilization of neural function. *Neuron, 80*(3), 718–728. https://doi.org/10.1016/j.neuron.2013.09.044

Davis, G. W., & Bezprozvanny, I. (2001). Maintaining the stability of neural function: A homeostatic hypothesis. *Annual Review of Physiology, 63*, 847–869. https://doi.org/10.1146/annurev.physiol.63.1.847

Davis, G. W., & Muller, M. (2015). Homeostatic control of presynaptic neurotransmitter release. *Annual Review of Physiology, 77*, 251–270. https://doi.org/10.1146/annurev-physiol-021014-071740

Demontis, F., & Perrimon, N. (2010). FOXO/4E-BP signaling in Drosophila muscles regulates organism-wide proteostasis during aging. *Cell, 143*(5), 813–825. https://doi.org/10.1016/j.cell.2010.10.007

Dever, T. E., Dar, A. C., & Sicheri, F. (2007). The eIF2alpha kinases. *Cold Spring Harbor Monograph Series, 48*, 319.

Dever, T. E., Feng, L., Wek, R. C., Cigan, A. M., Donahue, T. F., & Hinnebusch, A. G. (1992). Phosphorylation of initiation factor 2 alpha by protein kinase GCN2 mediates gene-specific translational control of GCN4 in yeast. *Cell, 68*(3), 585–596.

DiAntonio, A., Petersen, S. A., Heckmann, M., & Goodman, C. S. (1999). Glutamate receptor expression regulates quantal size and quantal content at the Drosophila neuromuscular junction. *Journal of Neuroscience, 19*(8), 3023–3032.

Dorrello, N. V., Peschiaroli, A., Guardavaccaro, D., Colburn, N. H., Sherman, N. E., & Pagano, M. (2006). S6K1- and betaTRCP-mediated degradation of PDCD4 promotes protein translation and cell growth. *Science, 314*(5798), 467–471. https://doi.org/10.1126/science.1130276

Dudek, S. M., & Bear, M. F. (1992). Homosynaptic long-term depression in area CA1 of hippocampus and effects of N-methyl-D-aspartate receptor blockade. *Proceedings of the National Academy of Sciences of the United States of America, 89*(10), 4363–4367.

Ehninger, D., Han, S., Shilyansky, C., Zhou, Y., Li, W., Kwiatkowski, D. J., . . . Silva, A. J. (2008). Reversal of learning deficits in a Tsc2+/- mouse model of tuberous sclerosis. *Nature Medicine, 14*(8), 843–848. https://doi.org/10.1038/nm1788

Fambrough, D. M., Drachman, D. B., & Satyamurti, S. (1973). Neuromuscular junction in myasthenia gravis: decreased acetylcholine receptors. *Science, 182*(4109), 293–295.

Flexner, J. B., Flexner, L. B., & Stellar, E. (1963). Memory in mice as affected by intracerebral puromycin. *Science, 141*(3575), 57–59.

Flexner, J. B., Flexner, L. B., Stellar, E., De La Haba, G., & Roberts, R. B. (1962). Inhibition of protein synthesis in brain and learning and memory following puromycin. *Journal of Neurochemistry, 9*, 595–605.

Frank, C. A., Kennedy, M. J., Goold, C. P., Marek, K. W., & Davis, G. W. (2006). Mechanisms underlying the rapid induction and sustained expression of synaptic homeostasis. *Neuron, 52*(4), 663–677. https://doi.org/10.1016/j.neuron.2006.09.029

Frank, C. A., Pielage, J., & Davis, G. W. (2009). A presynaptic homeostatic signaling system composed of the Eph receptor, ephexin, Cdc42, and CaV2.1 calcium channels. *Neuron, 61*(4), 556–569. https://doi.org/10.1016/j.neuron.2008.12.028

Frey, U., Huang, Y. Y., & Kandel, E. R. (1993). Effects of cAMP simulate a late stage of LTP in hippocampal CA1 neurons. *Science, 260*(5114), 1661–1664.

Funayama, M., Hasegawa, K., Kowa, H., Saito, M., Tsuji, S., & Obata, F. (2002). A new locus for Parkinson's disease (PARK8) maps to chromosome 12p11.2-q13.1. *Annals of Neurology, 51*(3), 296–301. https://doi.org/10.1002/ana.10113

Garcia-Bereguiain, M. A., Gonzalez-Islas, C., Lindsly, C., & Wenner, P. (2016). Spontaneous Release Regulates Synaptic Scaling in the Embryonic Spinal Network In Vivo. *Journal of Neuroscience, 36*(27), 7268–7282. https://doi.org/10.1523/JNEUROSCI.4066-15.2016

Gilhus, N. E., & Verschuuren, J. J. (2015). Myasthenia gravis: Subgroup classification and therapeutic strategies. *Lancet Neurology, 14*(10), 1023–1036. https://doi.org/10.1016/S1474-4422(15)00145-3

Goel, A., Jiang, B., Xu, L. W., Song, L., Kirkwood, A., & Lee, H. K. (2006). Cross-modal regulation of synaptic AMPA receptors in primary sensory cortices by visual experience. *Nature Neuroscience, 9*(8), 1001–1003. https://doi.org/10.1038/nn1725

Goel, A., & Lee, H. K. (2007). Persistence of experience-induced homeostatic synaptic plasticity through adulthood in superficial layers of mouse visual cortex. *Journal of Neuroscience, 27*(25), 6692–6700. https://doi.org/10.1523/JNEUROSCI.5038-06.2007

Goel, P., Li, X., & Dickman, D. (2017). Disparate postsynaptic induction mechanisms ultimately converge to drive the retrograde enhancement of presynaptic efficacy. *Cell Reports, 21*(9), 2339–2347. https://doi.org/10.1016/j.celrep.2017.10.116

Goold, C. P., & Davis, G. W. (2007). The BMP ligand Gbb gates the expression of synaptic homeostasis independent of synaptic growth control. *Neuron, 56*(1), 109–123. https://doi.org/10.1016/j.neuron.2007.08.006

Goorden, S. M., van Woerden, G. M., van der Weerd, L., Cheadle, J. P., & Elgersma, Y. (2007). Cognitive deficits in Tsc1+/− mice in the absence of cerebral lesions and seizures. *Annals of Neurology, 62*(6), 648–655. https://doi.org/10.1002/ana.21317

Grecksch, G., & Matthies, H. (1980). Two sensitive periods for the amnesic effect of anisomycin. *Pharmacology, Biochemistry, and Behavior, 12*(5), 663–665.

Greggio, E., & Cookson, M. R. (2009). Leucine-rich repeat kinase 2 mutations and Parkinson's disease: Three questions. *ASN Neuro, 1*(1). https://doi.org/10.1042/AN20090007

Haghighi, A. P., McCabe, B. D., Fetter, R. D., Palmer, J. E., Hom, S., & Goodman, C. S. (2003). Retrograde control of synaptic transmission by postsynaptic CaMKII at the Drosophila neuromuscular junction. *Neuron, 39*(2), 255–267.

Hara, K., Yonezawa, K., Weng, Q. P., Kozlowski, M. T., Belham, C., & Avruch, J. (1998). Amino acid sufficiency and mTOR regulate p70 S6 kinase and eIF-4E BP1 through a common effector mechanism. *Journal of Biological Chemistry, 273*(23), 14484–14494.

Hay, N., & Sonenberg, N. (2004). Upstream and downstream of mTOR. *Genes and Development, 18*(16), 1926–1945. https://doi.org/10.1101/gad.1212704

Hebb, D. O. (1949). *The organization of behavior: a neuropsychological theory*. New York, NY: Wiley.

Hengen, K. B., Lambo, M. E., Van Hooser, S. D., Katz, D. B., & Turrigiano, G. G. (2013). Firing rate homeostasis in visual cortex of freely behaving rodents. *Neuron, 80*(2), 335–342. https://doi.org/10.1016/j.neuron.2013.08.038

Henry, F. E., Hockeimer, W., Chen, A., Mysore, S. P., & Sutton, M. A. (2017). Mechanistic target of rapamycin is necessary for changes in dendritic spine morphology associated with long-term potentiation. *Molecular Brain, 10*(1), 50. https://doi.org/10.1186/s13041-017-0330-y

Henry, F. E., McCartney, A. J., Neely, R., Perez, A. S., Carruthers, C. J., Stuenkel, E. L., . . . Sutton, M. A. (2012). Retrograde changes in presynaptic function driven by dendritic mTORC1. *Journal ofNeuroscience,32*(48),17128–17142.https://doi.org/10.1523/JNEUROSCI.2149-12.2012

Henry, F. E., Wang, X., Serrano, D., Perez, A. S., Carruthers, C. J. L., Stuenkel, E. L., & Sutton, M. A. (2018). A unique homeostatic signaling pathway links synaptic inactivity to postsynaptic mTORC1. *Journal of Neuroscience, 38*(9), 2207–2225. https://doi.org/10.1523/JNEUROSCI.1843-17.2017

Hoeffer, C. A., & Klann, E. (2004). mTOR signaling: At the crossroads of plasticity, memory and disease. *Trends in Neuroscience, 33*(2), 67–75. https://doi.org/10.1016/j.tins.2009.11.003

Hoeffer, C. A., & Klann, E. (2010). mTOR signaling: At the crossroads of plasticity, memory and disease. *Trends in Neuroscience, 33*(2), 67–75. https://doi.org/10.1016/j.tins.2009.11.003

Hou, L., & Klann, E. (2004). Activation of the phosphoinositide 3-kinase-Akt-mammalian target of rapamycin signaling pathway is required for metabotropic glutamate receptor-dependent long-term depression. *Journal of Neuroscience, 24*(28), 6352–6361. https://doi.org/10.1523/JNEUROSCI.0995-04.2004

Huber, K. M., Kayser, M. S., & Bear, M. F. (2000). Role for rapid dendritic protein synthesis in hippocampal mGluR-dependent long-term depression. *Science, 288*(5469), 1254–1257.

Huber, K. M., Roder, J. C., & Bear, M. F. (2001). Chemical induction of mGluR5- and protein synthesis—dependent long-term depression in hippocampal area CA1. *Journal of Neurophysiology, 86*(1), 321–325. https://doi.org/10.1152/jn.2001.86.1.321

Imai, Y., Gehrke, S., Wang, H. Q., Takahashi, R., Hasegawa, K., Oota, E., & Lu, B. (2008). Phosphorylation of 4E-BP by LRRK2 affects the maintenance of dopaminergic neurons in Drosophila. *EMBO Journal, 27*(18), 2432–2443. https://doi.org/10.1038/emboj.2008.163

Isaac, J. T., Ashby, M. C., & McBain, C. J. (2007). The role of the GluR2 subunit in AMPA receptor function and synaptic plasticity. *Neuron, 54*(6), 859–871. https://doi.org/10.1016/j.neuron.2007.06.001

Jakawich, S. K., Nasser, H. B., Strong, M. J., McCartney, A. J., Perez, A. S., Rakesh, N., ... Sutton, M. A. (2010). Local presynaptic activity gates homeostatic changes in presynaptic function driven by dendritic BDNF synthesis. *Neuron, 68*(6), 1143–1158. https://doi.org/10.1016/j.neuron.2010.11.034

Jaleel, M., Nichols, R. J., Deak, M., Campbell, D. G., Gillardon, F., Knebel, A., & Alessi, D. R. (2007). LRRK2 phosphorylates moesin at threonine-558: characterization of how Parkinson's disease mutants affect kinase activity. *Biochemistry Journal, 405*(2), 307–317. https://doi.org/10.1042/BJ20070209

Jian, M., Luo, Y. X., Xue, Y. X., Han, Y., Shi, H. S., Liu, J. F., ... Lu, L. (2014). eIF2alpha dephosphorylation in basolateral amygdala mediates reconsolidation of drug memory. *Journal of Neuroscience, 34*(30), 10010–10021. https://doi.org/10.1523/JNEUROSCI.0934-14.2014

Ju, W., Morishita, W., Tsui, J., Gaietta, G., Deerinck, T. J., Adams, S. R., ... Malenka, R. C. (2004). Activity-dependent regulation of dendritic synthesis and trafficking of AMPA receptors. *Nature Neuroscience, 7*(3), 244–253. https://doi.org/10.1038/nn1189

Junger, M. A., Rintelen, F., Stocker, H., Wasserman, J. D., Vegh, M., Radimerski, T., ... Hafen, E. (2003). The Drosophila forkhead transcription factor FOXO mediates the reduction in cell number associated with reduced insulin signaling. *Journal of Biology, 2*(3), 20. https://doi.org/10.1186/1475-4924-2-20

Kang, H., & Schuman, E. M. (1996). A requirement for local protein synthesis in neurotrophin-induced hippocampal synaptic plasticity. *Science, 273*(5280), 1402–1406.

Karachot, L., Shirai, Y., Vigot, R., Yamamori, T., & Ito, M. (2001). Induction of long-term depression in cerebellar Purkinje cells requires a rapidly turned over protein. *Journal of Neurophysiology, 86*(1), 280–289. https://doi.org/10.1152/jn.2001.86.1.280

Katz, B., & Miledi, R. (1978). A re-examination of curare action at the motor endplate. *Proceedings of the Royal Society of London B: Biological Sciences, 203*(1151), 119–133.

Kauderer, B. S., & Kandel, E. R. (2000). Capture of a protein synthesis-dependent component of long-term depression. *Proceedings of the National Academy of Sciences of the United States of America, 97*(24), 13342–13347. https://doi.org/10.1073/pnas.97.24.13342

Kaufman, R. J., Davies, M. V., Pathak, V. K., & Hershey, J. W. (1989). The phosphorylation state of eucaryotic initiation factor 2 alters translational efficiency of specific mRNAs. *Molecular and Cellular Biology, 9*(3), 946–958.

Kauwe, G., Tsurudome, K., Penney, J., Mori, M., Gray, L., Calderon, M. R., ... Haghighi, A. P. (2016). Acute Fasting Regulates Retrograde Synaptic Enhancement through a 4E-BP-Dependent Mechanism. *Neuron, 92*(6), 1204–1212. https://doi.org/10.1016/j.neuron.2016.10.063

Keck, T., Keller, G. B., Jacobsen, R. I., Eysel, U. T., Bonhoeffer, T., & Hubener, M. (2013). Synaptic scaling and homeostatic plasticity in the mouse visual cortex in vivo. *Neuron, 80*(2), 327–334. https://doi.org/10.1016/j.neuron.2013.08.018

Kelleher, R. J., 3rd, Govindarajan, A., Jung, H. Y., Kang, H., & Tonegawa, S. (2004). Translational control by MAPK signaling in long-term synaptic plasticity and memory. *Cell, 116*(3), 467–479.

Kelleher, R. J., 3rd, Govindarajan, A., & Tonegawa, S. (2004). Translational regulatory mechanisms in persistent forms of synaptic plasticity. *Neuron, 44*(1), 59–73. https://doi.org/10.1016/j.neuron.2004.09.013

Kim, J. H., Liao, D., Lau, L. F., & Huganir, R. L. (1998). SynGAP: a synaptic RasGAP that associates with the PSD-95/SAP90 protein family. *Neuron, 20*(4), 683–691.

Kirov, S. A., Goddard, C. A., & Harris, K. M. (2004). Age-dependence in the homeostatic upregulation of hippocampal dendritic spine number during blocked synaptic transmission. *Neuropharmacology, 47*(5), 640–648. https://doi.org/10.1016/j.neuropharm.2004.07.039

Kwiatkowski, D. J., & Manning, B. D. (2005). Tuberous sclerosis: a GAP at the crossroads of multiple signaling pathways. *Human Molecular Genetics, 14 Spec. No. 2*, R251–258. https://doi.org/10.1093/hmg/ddi260

Lee, S. B., Kim, W., Lee, S., & Chung, J. (2007). Loss of LRRK2/PARK8 induces degeneration of dopaminergic neurons in Drosophila. *Biochemical and Biophysical Research Communications, 358*(2), 534–539. https://doi.org/10.1016/j.bbrc.2007.04.156

Lee, S., Liu, H. P., Lin, W. Y., Guo, H., & Lu, B. (2010). LRRK2 kinase regulates synaptic morphology through distinct substrates at the presynaptic and postsynaptic compartments of the Drosophila neuromuscular junction. *Journal of Neuroscience, 30*(50), 16959–16969. https://doi.org/10.1523/JNEUROSCI.1807-10.2010

Linden, D. J. (1996). A protein synthesis-dependent late phase of cerebellar long-term depression. *Neuron, 17*(3), 483–490.

Lisman, J., Grace, A. A., & Duzel, E. (2011). A neoHebbian framework for episodic memory; role of dopamine-dependent late LTP. *Trends in Neuroscience, 34*(10), 536–547. https://doi.org/10.1016/j.tins.2011.07.006

Lissin, D. V., Gomperts, S. N., Carroll, R. C., Christine, C. W., Kalman, D., Kitamura, M., . . . von Zastrow, M. (1998). Activity differentially regulates the surface expression of synaptic AMPA and NMDA glutamate receptors. *Proceedings of the National Academy of Sciences of the United States of America, 95*(12), 7097–7102.

Liu, Z., Wang, X., Yu, Y., Li, X., Wang, T., Jiang, H., . . . Smith, W. W. (2008). A Drosophila model for LRRK2-linked parkinsonism. *Proceedings of the National Academy of Sciences of the United States of America, 105*(7), 2693–2698. https://doi.org/10.1073/pnas.0708452105

Loewith, R., & Hall, M. N. (2011). Target of rapamycin (TOR) in nutrient signaling and growth control. *Genetics, 189*(4), 1177–1201. https://doi.org/10.1534/genetics.111.133363

Luscher, C., & Malenka, R. C. (2012). NMDA receptor-dependent long-term potentiation and long-term depression (LTP/LTD). *Cold Spring Harbor Perspectives in Biology, 4*(6). https://doi.org/10.1101/cshperspect.a005710

Lynch, M. A. (2004). Long-term potentiation and memory. *Physiology Review, 84*(1), 87–136. https://doi.org/10.1152/physrev.00014.2003

Ma, T., Trinh, M. A., Wexler, A. J., Bourbon, C., Gatti, E., Pierre, P., . . . Klann, E. (2013). Suppression of eIF2alpha kinases alleviates Alzheimer's disease-related plasticity and memory deficits. *Nature Neuroscience, 16*(9), 1299–1305. https://doi.org/10.1038/nn.3486

Ma, X. M., & Blenis, J. (2009). Molecular mechanisms of mTOR-mediated translational control. *Nature Reviews: Molecular Cell Biology, 10*(5), 307–318. https://doi.org/10.1038/nrm2672

Maffei, A., Nelson, S. B., & Turrigiano, G. G. (2004). Selective reconfiguration of layer 4 visual cortical circuitry by visual deprivation. *Nature Neuroscience, 7*(12), 1353–1359. https://doi.org/10.1038/nn1351

Maffei, A., & Turrigiano, G. G. (2008). Multiple modes of network homeostasis in visual cortical layer 2/3. *Journal of Neuroscience, 28*(17), 4377–4384. https://doi.org/10.1523/JNEUROSCI.5298-07.2008

Maghsoodi, B., Poon, M. M., Nam, C. I., Aoto, J., Ting, P., & Chen, L. (2008). Retinoic acid regulates RARalpha-mediated control of translation in dendritic RNA granules during homeostatic synaptic plasticity. *Proceedings of the National Academy of Sciences of the United States of America, 105*(41), 16015–16020. https://doi.org/10.1073/pnas.0804801105

Malenka, R. C., & Bear, M. F. (2004). LTP and LTD: An embarrassment of riches. *Neuron, 44*(1), 5–21. https://doi.org/10.1016/j.neuron.2004.09.012

Mameli, M., Balland, B., Lujan, R., & Luscher, C. (2007). Rapid synthesis and synaptic insertion of GluR2 for mGluR-LTD in the ventral tegmental area. *Science, 317*(5837), 530–533. https://doi.org/10.1126/science.1142365

Marie, B., Pym, E., Bergquist, S., & Davis, G. W. (2010). Synaptic homeostasis is consolidated by the cell fate gene gooseberry, a Drosophila pax3/7 homolog. *Journal of Neuroscience, 30*(24), 8071–8082. https://doi.org/10.1523/JNEUROSCI.5467-09.2010

Martin, I., Dawson, V. L., & Dawson, T. M. (2011). Recent advances in the genetics of Parkinson's disease. *Annual Review of Genomics and Human Genetics, 12*, 301–325. https://doi.org/10.1146/annurev-genom-082410-101440

Massey, P. V., & Bashir, Z. I. (2007). Long-term depression: Multiple forms and implications for brain function. *Trends in Neuroscience, 30*(4), 176–184. https://doi.org/10.1016/j.tins.2007.02.005

Mata, I. F., Wedemeyer, W. J., Farrer, M. J., Taylor, J. P., & Gallo, K. A. (2006). LRRK2 in Parkinson's disease: protein domains and functional insights. *Trends in Neuroscience, 29*(5), 286–293. https://doi.org/10.1016/j.tins.2006.03.006

Miledi, R., Molenaar, P. C., & Polak, R. L. (1978). Alpha-Bungarotoxin enhances transmitter "released" at the neuromuscular junction. *Nature, 272*(5654), 641–643.

Montarolo, P. G., Goelet, P., Castellucci, V. F., Morgan, J., Kandel, E. R., & Schacher, S. (1986). A critical period for macromolecular synthesis in long-term heterosynaptic facilitation in Aplysia. *Science, 234*(4781), 1249–1254.

Moon, S. L., Sonenberg, N., & Parker, R. (2018). Neuronal regulation of eIF2alpha function in health and neurological disorders. *Trends in Molecular Medicine*. https://doi.org/10.1016/j.molmed.2018.04.001

Muller, M., & Davis, G. W. (2012). Transsynaptic control of presynaptic Ca(2)(+) influx achieves homeostatic potentiation of neurotransmitter release. *Current Biology, 22*(12), 1102–1108. https://doi.org/10.1016/j.cub.2012.04.018

Muller, M., Liu, K. S., Sigrist, S. J., & Davis, G. W. (2012). RIM controls homeostatic plasticity through modulation of the readily-releasable vesicle pool. *Journal of Neuroscience, 32*(47), 16574–16585. https://doi.org/10.1523/JNEUROSCI.0981-12.2012

Muller, M., Pym, E. C., Tong, A., & Davis, G. W. (2011). Rab3-GAP controls the progression of synaptic homeostasis at a late stage of vesicle release. *Neuron, 69*(4), 749–762. https://doi.org/10.1016/j.neuron.2011.01.025

Murthy, V. N., Schikorski, T., Stevens, C. F., & Zhu, Y. (2001). Inactivity produces increases in neurotransmitter release and synapse size. *Neuron, 32*(4), 673–682.

Nabavi, S., Fox, R., Proulx, C. D., Lin, J. Y., Tsien, R. Y., & Malinow, R. (2014). Engineering a memory with LTD and LTP. *Nature, 511*(7509), 348–352. https://doi.org/10.1038/nature13294

Nairn, A. C., & Palfrey, H. C. (1987). Identification of the major Mr 100,000 substrate for calmodulin-dependent protein kinase III in mammalian cells as elongation factor-2. *Journal of Biological Chemistry, 262*(36), 17299–17303.

Ng, C. H., Mok, S. Z., Koh, C., Ouyang, X., Fivaz, M. L., Tan, E. K., . . . Lim, K. L. (2009). Parkin protects against LRRK2 G2019S mutant-induced dopaminergic neurodegeneration in Drosophila. *Journal of Neuroscience, 29*(36), 11257–11262. https://doi.org/10.1523/JNEUROSCI.2375-09.2009

Nguyen, P. V., & Kandel, E. R. (1997). Brief theta-burst stimulation induces a transcription-dependent late phase of LTP requiring cAMP in area CA1 of the mouse hippocampus. *Learning and Memory, 4*(2), 230–243.

O'Brien, R. J., Kamboj, S., Ehlers, M. D., Rosen, K. R., Fischbach, G. D., & Huganir, R. L. (1998). Activity-dependent modulation of synaptic AMPA receptor accumulation. *Neuron, 21*(5), 1067–1078.

Oh, J. S., Manzerra, P., & Kennedy, M. B. (2004). Regulation of the neuron-specific Ras GTPase-activating protein, synGAP, by Ca2+/calmodulin-dependent protein kinase II. *Journal of Biological Chemistry, 279*(17), 17980–17988. https://doi.org/10.1074/jbc.M314109200

Orr, B. O., Fetter, R. D., & Davis, G. W. (2017). Retrograde semaphorin-plexin signalling drives homeostatic synaptic plasticity. *Nature, 550*(7674), 109–113. https://doi.org/10.1038/nature24017

Paisan-Ruiz, C., Jain, S., Evans, E. W., Gilks, W. P., Simon, J., van der Brug, M., ... Singleton, A. B. (2004). Cloning of the gene containing mutations that cause PARK8-linked Parkinson's disease. *Neuron, 44*(4), 595–600. https://doi.org/10.1016/j.neuron.2004.10.023

Pavitt, G. D., Ramaiah, K. V., Kimball, S. R., & Hinnebusch, A. G. (1998). eIF2 independently binds two distinct eIF2B subcomplexes that catalyze and regulate guanine-nucleotide exchange. *Genes and Development, 12*(4), 514–526.

Pelletier, J., & Sonenberg, N. (1988). Internal initiation of translation of eukaryotic mRNA directed by a sequence derived from poliovirus RNA. *Nature, 334*(6180), 320–325. https://doi.org/10.1038/334320a0

Penney, J., Tsurudome, K., Liao, E. H., Elazzouzi, F., Livingstone, M., Gonzalez, M., ... Haghighi, A. P. (2012). TOR is required for the retrograde regulation of synaptic homeostasis at the Drosophila neuromuscular junction. *Neuron, 74*(1), 166–178. https://doi.org/10.1016/j.neuron.2012.01.030

Penney, J., Tsurudome, K., Liao, E. H., Kauwe, G., Gray, L., Yanagiya, A., ... Haghighi, A. P. (2016). LRRK2 regulates retrograde synaptic compensation at the Drosophila neuromuscular junction. *Nature Communications, 7*, 12188. https://doi.org/10.1038/ncomms12188

Perry, S., Han, Y., Das, A., & Dickman, D. (2017). Homeostatic plasticity can be induced and expressed to restore synaptic strength at neuromuscular junctions undergoing ALS-related degeneration. *Human Molecular Genetics, 26*(21), 4153–4167. https://doi.org/10.1093/hmg/ddx304

Petersen, S. A., Fetter, R. D., Noordermeer, J. N., Goodman, C. S., & DiAntonio, A. (1997). Genetic analysis of glutamate receptors in Drosophila reveals a retrograde signal regulating presynaptic transmitter release. *Neuron, 19*(6), 1237–1248. https://doi.org/10.1016/S0896-6273(00)80415-8

Pilgram, G. S., Potikanond, S., van der Plas, M. C., Fradkin, L. G., & Noordermeer, J. N. (2011). The RhoGAP crossveinless-c interacts with Dystrophin and is required for synaptic homeostasis at the Drosophila neuromuscular junction. *Journal of Neuroscience, 31*(2), 492–500. https://doi.org/10.1523/JNEUROSCI.4732-10.2011

Plomp, J. J. (2017). Trans-synaptic homeostasis at the myasthenic neuromuscular junction. *Frontiers in Bioscience, Landmark, 22*, 1033–1051.

Plomp, J. J., Morsch, M., Phillips, W. D., & Verschuuren, J. J. (2015). Electrophysiological analysis of neuromuscular synaptic function in myasthenia gravis patients and animal models. *Experimental Neurology, 270*, 41–54. https://doi.org/10.1016/j.expneurol.2015.01.007

Plomp, J. J., Van Kempen, G. T., De Baets, M. B., Graus, Y. M., Kuks, J. B., & Molenaar, P. C. (1995). Acetylcholine release in myasthenia gravis: Regulation at single end-plate level. *Annals of Neurology, 37*(5), 627–636. https://doi.org/10.1002/ana.410370513

Plomp, J. J., van Kempen, G. T., & Molenaar, P. C. (1992). Adaptation of quantal content to decreased postsynaptic sensitivity at single endplates in alpha-bungarotoxin-treated rats. *Journal of Physiology, 458*, 487–499.

Pozo, K., & Goda, Y. (2010). Unraveling mechanisms of homeostatic synaptic plasticity. *Neuron, 66*(3), 337–351. https://doi.org/10.1016/j.neuron.2010.04.028

Puig, O., Marr, M. T., Ruhf, M. L., & Tjian, R. (2003). Control of cell number by Drosophila FOXO: Downstream and feedback regulation of the insulin receptor pathway. *Genes and Development, 17*(16), 2006–2020. https://doi.org/10.1101/gad.1098703

Reese, A. L., & Kavalali, E. T. (2015). Spontaneous neurotransmission signals through store-driven $Ca^{(2+)}$ transients to maintain synaptic homeostasis. *Elife, 4*. https://doi.org/10.7554/eLife.09262

Schanzenbacher, C. T., Sambandan, S., Langer, J. D., & Schuman, E. M. (2016). Nascent proteome remodeling following homeostatic scaling at hippocampal synapses. *Neuron, 92*(2), 358–371. https://doi.org/10.1016/j.neuron.2016.09.058

Scheetz, A. J., Nairn, A. C., & Constantine-Paton, M. (2000). NMDA receptor-mediated control of protein synthesis at developing synapses. *Nature Neuroscience, 3*(3), 211–216. https://doi.org/10.1038/72915

Shahbazian, D., Parsyan, A., Petroulakis, E., Topisirovic, I., Martineau, Y., Gibbs, B. F., . . . Sonenberg, N. (2010). Control of cell survival and proliferation by mammalian eukaryotic initiation factor 4B. *Molecular and Cellular Biology, 30*(6), 1478–1485. https://doi.org/10.1128/MCB.01218-09

Sharma, A., Hoeffer, C. A., Takayasu, Y., Miyawaki, T., McBride, S. M., Klann, E., & Zukin, R. S. (2010). Dysregulation of mTOR signaling in fragile X syndrome. *Journal of Neuroscience, 30*(2), 694–702. https://doi.org/10.1523/JNEUROSCI.3696-09.2010

Shih, Y. T., & Hsueh, Y. P. (2018). The involvement of endoplasmic reticulum formation and protein synthesis efficiency in VCP- and ATL1-related neurological disorders. *Journal of Biomedical Science, 25*(1), 2. https://doi.org/10.1186/s12929-017-0403-3

Sonenberg, N., & Hinnebusch, A. G. (2009). Regulation of translation initiation in eukaryotes: Mechanisms and biological targets. *Cell, 136*(4), 731–745. https://doi.org/10.1016/j.cell.2009.01.042

Squire, L. R., & Davis, H. P. (1981). The pharmacology of memory: A neurobiological perspective. *Annual Review of Pharmacology and Toxicology, 21*, 323–356. https://doi.org/10.1146/annurev.pa.21.040181.001543

Sutton, M. A., Ito, H. T., Cressy, P., Kempf, C., Woo, J. C., & Schuman, E. M. (2006). Miniature neurotransmission stabilizes synaptic function via tonic suppression of local dendritic protein synthesis. *Cell, 125*(4), 785–799. https://doi.org/10.1016/j.cell.2006.03.040

Sutton, M. A., & Schuman, E. M. (2006). Dendritic protein synthesis, synaptic plasticity, and memory. *Cell, 127*(1), 49–58. https://doi.org/10.1016/j.cell.2006.09.014

Sutton, M. A., Taylor, A. M., Ito, H. T., Pham, A., & Schuman, E. M. (2007). Postsynaptic decoding of neural activity: eEF2 as a biochemical sensor coupling miniature synaptic transmission to local protein synthesis. *Neuron, 55*(4), 648–661. https://doi.org/10.1016/j.neuron.2007.07.030

Sutton, M. A., Wall, N. R., Aakalu, G. N., & Schuman, E. M. (2004). Regulation of dendritic protein synthesis by miniature synaptic events. *Science, 304*(5679), 1979–1983. https://doi.org/10.1126/science.1096202

Swiech, L., Perycz, M., Malik, A., & Jaworski, J. (2008). Role of mTOR in physiology and pathology of the nervous system. *Biochimica et Biophysica Acta, 1784*(1), 116–132. https://doi.org/10.1016/j.bbapap.2007.08.015

Tain, L. S., Mortiboys, H., Tao, R. N., Ziviani, E., Bandmann, O., & Whitworth, A. J. (2009). Rapamycin activation of 4E-BP prevents parkinsonian dopaminergic neuron loss. *Nature Neuroscience, 12*(9), 1129–1135. https://doi.org/10.1038/nn.2372

Takamori, M. (2017). Synaptic homeostasis and its immunological disturbance in neuromuscular junction disorders. *International Journal of Molecular Science, 18*(4). https://doi.org/10.3390/ijms18040896

Tang, S. J., Reis, G., Kang, H., Gingras, A. C., Sonenberg, N., & Schuman, E. M. (2002). A rapamycin-sensitive signaling pathway contributes to long-term synaptic plasticity in the hippocampus. *Proceedings of the National Academy of Sciences of the United States of America, 99*(1), 467–472. https://doi.org/10.1073/pnas.012605299

Taymans, J. M., Nkiliza, A., & Chartier-Harlin, M. C. (2015). Deregulation of protein translation control, a potential game-changing hypothesis for Parkinson's disease pathogenesis. *Trends in Molecular Medicine, 21*(8), 466–472. https://doi.org/10.1016/j.molmed.2015.05.004

Teichert, M., Liebmann, L., Hubner, C. A., & Bolz, J. (2017). Homeostatic plasticity and synaptic scaling in the adult mouse auditory cortex. *Science Reports, 7*(1), 17423. https://doi.org/10.1038/s41598-017-17711-5

Teleman, A. A., Chen, Y. W., & Cohen, S. M. (2005). 4E-BP functions as a metabolic brake used under stress conditions but not during normal growth. *Genes and Development, 19*(16), 1844–1848. https://doi.org/10.1101/gad.341505

Tian, L., Prior, C., Dempster, J., & Marshall, I. G. (1994). Nicotinic antagonist-produced frequency-dependent changes in acetylcholine release from rat motor nerve terminals. *Journal of Physiology, 476*(3), 517–529.

Tsai, P., & Sahin, M. (2011). Mechanisms of neurocognitive dysfunction and therapeutic considerations in tuberous sclerosis complex. *Current Opinion in Neurology, 24*(2), 106–113. https://doi.org/10.1097/WCO.0b013e32834451c4

Tsujihata, M., Hazama, R., Ishii, N., Ide, Y., & Takamori, M. (1980). Ultrastructural localization of acetylcholine receptor at the motor endplate: myasthenia gravis and other neuromuscular diseases. *Neurology, 30*(11), 1203–1211.

Tsurudome, K., Tsang, K., Liao, E. H., Ball, R., Penney, J., Yang, J. S., . . . Haghighi, A. P. (2010). The Drosophila miR-310 cluster negatively regulates synaptic strength at the neuromuscular junction. *Neuron, 68*(5), 879–893. https://doi.org/10.1016/j.neuron.2010.11.016

Turrigiano, G. (2012). Homeostatic synaptic plasticity: local and global mechanisms for stabilizing neuronal function. *Cold Spring Harbor Perspectives in Biology, 4*(1), a005736. https://doi.org/10.1101/cshperspect.a005736

Turrigiano, G. G. (2008). The self-tuning neuron: Synaptic scaling of excitatory synapses. *Cell, 135*(3), 422–435. https://doi.org/10.1016/j.cell.2008.10.008

Turrigiano, G. G. (2017). The dialectic of Hebb and homeostasis. *Proceedings of the Royal Society of London B: Biological Sciences, 372*(1715). https://doi.org/10.1098/rstb.2016.0258

Turrigiano, G. G., Leslie, K. R., Desai, N. S., Rutherford, L. C., & Nelson, S. B. (1998). Activity-dependent scaling of quantal amplitude in neocortical neurons. *Nature, 391*(6670), 892–896. https://doi.org/10.1038/36103

Turrigiano, G. G., & Nelson, S. B. (2000). Hebb and homeostasis in neuronal plasticity. *Current Biochim Biophys Acta Opinion in Neurobiology, 10*(3), 358–364.

Turrigiano, G. G., & Nelson, S. B. (2004). Homeostatic plasticity in the developing nervous system. *Nature Reviews. Neuroscience, 5*(2), 97–107. https://doi.org/10.1038/nrn1327

Tyler, W. J., Petzold, G. C., Pal, S. K., & Murthy, V. N. (2007). Experience-dependent modification of primary sensory synapses in the mammalian olfactory bulb. *Journal of Neuroscience, 27*(35), 9427–9438. https://doi.org/10.1523/JNEUROSCI.0664-07.2007

Vale, C., & Sanes, D. H. (2002). The effect of bilateral deafness on excitatory and inhibitory synaptic strength in the inferior colliculus. *European Journal of Neuroscience, 16*(12), 2394–2404.

Venderova, K., Kabbach, G., Abdel-Messih, E., Zhang, Y., Parks, R. J., Imai, Y.,... Park, D. S. (2009). Leucine-Rich Repeat Kinase 2 interacts with Parkin, DJ-1 and PINK-1 in a Drosophila melanogaster model of Parkinson's disease. *Human Molecular Genetics, 18*(22), 4390–4404. https://doi.org/10.1093/hmg/ddp394

Vitureira, N., & Goda, Y. (2013). Cell biology in neuroscience: The interplay between Hebbian and homeostatic synaptic plasticity. *Journal of Cellular Biology, 203*(2), 175–186. https://doi.org/10.1083/jcb.201306030

von der Brelie, C., Waltereit, R., Zhang, L., Beck, H., & Kirschstein, T. (2006). Impaired synaptic plasticity in a rat model of tuberous sclerosis. *European Journal Neuroscience, 23*(3), 686–692. https://doi.org/10.1111/j.1460-9568.2006.04594.x

Wang, C. C., Held, R. G., & Hall, B. J. (2013). SynGAP regulates protein synthesis and homeostatic synaptic plasticity in developing cortical networks. *PLoS One, 8*(12), e83941. https://doi.org/10.1371/journal.pone.0083941

Wang, H., Ardiles, A. O., Yang, S., Tran, T., Posada-Duque, R., Valdivia, G.,... Kirkwood, A. (2016). Metabotropic glutamate receptors induce a form of LTP controlled by translation and arc signaling in the hippocampus. *Journal of Neuroscience, 36*(5), 1723–1729. https://doi.org/10.1523/JNEUROSCI.0878-15.2016

Wang, H. L., Zhang, Z., Hintze, M., & Chen, L. (2011). Decrease in calcium concentration triggers neuronal retinoic acid synthesis during homeostatic synaptic plasticity. *Journal of Neuroscience, 31*(49), 17764–17771. https://doi.org/10.1523/JNEUROSCI.3964-11.2011

Wang, X., Pinter, M. J., & Rich, M. M. (2016). Reversible recruitment of a homeostatic reserve pool of synaptic vesicles underlies rapid homeostatic plasticity of quantal content. *Journal of Neuroscience, 36*(3), 828–836. https://doi.org/10.1523/JNEUROSCI.3786-15.2016

Wang, Y. T., & Linden, D. J. (2000). Expression of cerebellar long-term depression requires postsynaptic clathrin-mediated endocytosis. *Neuron, 25*(3), 635–647.

Waung, M. W., & Huber, K. M. (2009). Protein translation in synaptic plasticity: mGluR-LTD, Fragile X. *Current Opinion in Neurobiology, 19*(3), 319–326. https://doi.org/10.1016/j.conb.2009.03.011

Waung, M. W., Pfeiffer, B. E., Nosyreva, E. D., Ronesi, J. A., & Huber, K. M. (2008). Rapid translation of Arc/Arg3.1 selectively mediates mGluR-dependent LTD through persistent increases in AMPAR endocytosis rate. *Neuron, 59*(1), 84–97. https://doi.org/10.1016/j.neuron.2008.05.014

Weyhersmuller, A., Hallermann, S., Wagner, N., & Eilers, J. (2011). Rapid active zone remodeling during synaptic plasticity. *Journal of Neuroscience, 31*(16), 6041–6052. https://doi.org/10.1523/JNEUROSCI.6698-10.2011

Wierenga, C. J., Ibata, K., & Turrigiano, G. G. (2005). Postsynaptic expression of homeostatic plasticity at neocortical synapses. *Journal of Neuroscience, 25*(11), 2895–2905. https://doi.org/10.1523/JNEUROSCI.5217-04.2005

Wierenga, C. J., Walsh, M. F., & Turrigiano, G. G. (2006). Temporal regulation of the expression locus of homeostatic plasticity. *Journal of Neurophysiology, 96*(4), 2127–2133. https://doi.org/10.1152/jn.00107.2006

Wullschleger, S., Loewith, R., & Hall, M. N. (2006). TOR signaling in growth and metabolism. *Cell, 124*(3), 471–484. https://doi.org/10.1016/j.cell.2006.01.016

Xiao, M. Y., Zhou, Q., & Nicoll, R. A. (2001). Metabotropic glutamate receptor activation causes a rapid redistribution of AMPA receptors. *Neuropharmacology, 41*(6), 664–671.

Younger, M. A., Muller, M., Tong, A., Pym, E. C., & Davis, G. W. (2013). A presynaptic ENaC channel drives homeostatic plasticity. *Neuron, 79*(6), 1183–1196. https://doi.org/10.1016/j.neuron.2013.06.048

Zenke, F., & Gerstner, W. (2017). Hebbian plasticity requires compensatory processes on multiple timescales. *Proceedings of the Royal Society of London B: Biological Sciences, 372*(1715). https://doi.org/10.1098/rstb.2016.0259

Zhang, W., & Linden, D. J. (2003). The other side of the engram: Experience-driven changes in neuronal intrinsic excitability. *Nature Reviews. Neuroscience, 4*(11), 885–900. https://doi.org/10.1038/nrn1248

Zhang, Y., Venkitaramani, D. V., Gladding, C. M., Zhang, Y., Kurup, P., Molnar, E., . . . Lombroso, P. J. (2008). The tyrosine phosphatase STEP mediates AMPA receptor endocytosis after metabotropic glutamate receptor stimulation. *Journal of Neuroscience, 28*(42), 10561–10566. https://doi.org/10.1523/JNEUROSCI.2666-08.2008

Zheng, X., Boyer, L., Jin, M., Kim, Y., Fan, W., Bardy, C., . . . Hunter, T. (2016). Alleviation of neuronal energy deficiency by mTOR inhibition as a treatment for mitochondria-related neurodegeneration. *Elife, 5*. https://doi.org/10.7554/eLife.13378

Zimprich, A., Muller-Myhsok, B., Farrer, M., Leitner, P., Sharma, M., Hulihan, M., . . . Gasser, T. (2004). The PARK8 locus in autosomal dominant parkinsonism: confirmation of linkage and further delineation of the disease-containing interval. *American Journal of Human Genetics, 74*(1), 11–19. https://doi.org/10.1086/380647

CHAPTER 14

MULTIPLE ROLES OF RNA REGULATORY FACTORS IN NEURONAL DEVELOPMENT AND FUNCTION IN *C. ELEGANS*

MATTHEW G. ANDRUSIAK
AND YISHI JIN

Introduction

The elaborate morphology of neurons requires unique regulatory mechanisms to ensure the proper localization and quantity of proteins (Malgaroli, Vallar, & Zimarino, 2006). Studies over the past several decades have uncovered broad roles for mRNA-dependent mechanisms throughout development of the nervous system (Jaffrey & Wilkinson, 2018; Jung & Holt, 2011; Vuong, Black, & Zheng, 2016). The formation and function of neuronal connections, synapses, require rapid, dynamic changes in cellular components. These dynamic changes are evident in the fluxes in transcription associated with neuronal development (Maslon et al., 2019) and the rapid induction of transcription following neuronal stimulation (Morgan, Cohen, Hempstead, & Curran, 1987). Recent evidence has highlighted further complexity of the transcriptome in nervous system formation and function, with transcript localization, modification, and stability all contributing significantly to proper functional output (Nussbacher, Tabet, Yeo, & Lagier-Tourenne, 2019). De-regulation of RNA-associated processes is linked with multiple nervous system disorders. In particular, ALS, a devastating degenerative disorder, exhibits de-regulation of RNA-binding proteins associated with RNA stability including FUS, TIA1, and TDP-43 (Kabashi et al., 2008; Kwiatkowski et al., 2009; Mackenzie et al., 2017; Sreedharan et al., 2008; Van Deerlin et al., 2008; Vance et al., 2009). Developmental neurological conditions such as autism spectrum disorders are also linked to mutations in RNA-binding proteins (Martin et al., 2007; Sebat et al., 2007; Verkerk et al., 1991).

C. elegans is an established model organism, with its nervous system fully reconstructed at the ultrastructural level (Cook et al., 2019; White, Southgate, Thomson, & Brenner, 1986). Its powerful genetics have led to groundbreaking discoveries of fundamental mechanisms underlying formation and function of the nervous system. In his pioneering phenotype-based genetic screen, Sydney Brenner isolated a large number of mutants by their uncoordinated (Unc) locomotion, under the assumption that the genes defined by these mutations are likely to regulate neuronal function (Brenner, 1974). Subsequent decades studying these mutants revealed diverse gene cohorts regulating cellular processes such as neuronal differentiation (Finney, Ruvkun, & Horvitz, 1988), axon guidance (Run et al., 1996), synapse vesicle formation (Nonet et al., 1999; Schuske et al., 2003), neurotransmitter specificity and packaging (Alfonso, Grundahl, Duerr, Han, & Rand, 1993), and synaptic transmission (Hosono et al., 1992; Maruyama & Brenner, 1991). Further screening using pharmacological agents or fluorescently labelled synaptic proteins uncovered additional pre- and post-synaptic regulators (Lewis, Wu, Berg, & Levine, 1980; Nguyen, Alfonso, Johnson, & Rand, 1995; Zhen & Jin, 1999; Pinan-Lucarre et al., 2014). Furthermore, elegant secondary or modifier genetic screens for enhancers or suppressors of these identified genes enable signaling pathway dissection (Jorgensen & Mango, 2002). With advances in microscopy and molecular biology tools, the identification of genes required for neuronal function has evolved from identification of behavioral variants to resolving transcriptional changes at single-cell resolution (Norris et al., 2014). Importantly, the *C. elegans* genome contains genes that have clear vertebrate orthologs (W. Kim, Underwood, Greenwald, & Shaye, 2018), allowing findings to be translated into higher organisms, and ultimately, an understanding of human neurobiology.

This chapter will focus on neuronal regulatory mechanisms involving RNA-binding proteins (RBP) and the critical roles they play in splicing and alternative polyadenylation during nervous system development and function (Table 14.1). In addition, non-coding RNAs and injury-relevant roles of RNA granules will be discussed.

Alternative Splicing by RBPs Gives Rise to Functional Specificity in the Nervous System

Following transcriptional initiation, pre-mRNA undergoes essential modifications such as splicing to remove intronic sequences, and polyadenylation to generate stable mRNAs (Figure 14.1). Alternative splicing can generate diversity from a single genetic locus by producing multiple isoforms from a single transcribed pre-mRNA. Alternative splicing generally involves four basic modes of actions: intron retention, exon skipping, alternative 5′ acceptor sites and alternative 3′ acceptor sites. A wealth of literature has

Table 14.1. A Select List of RNA Binding Genes in *C. elegans* and Their Human Orthologs

Function in the Nervous System		*C. elegans* Gene	Human Ortholog	mRNA Targets*	References
		mec-8	RBPMS	*unc-52, mec-2, sad-1*	Thompson, Bixby et al., 2019; Calixto, Ma et al., 2010; Tan & Fraser, 2017
		exc-7	ELAV	*unc-16, twk-39, dyn-1*	Tan and Fraser, 2017; Fujita, Hawkinson et al. 2003
		asd-1	RBFOX	*unc-32*	Kuroyanagi, Watanabe et al., 2013
	mRNA	*mbl-1*	MBNL1	*sad-1*	Thompson, Bixby et al., 2019
	Splicing	*unc-75*	CELF	*unc-75, unc-64, unc-16, twk-39, dyn-1, unc-32, tom-1, nrx-1, etc*	Kuroyanagi, Watanabe et al., 2013; Norris, Gao et al., 2014; Chen, Liu et al., 2016; Chen, Wang et al., 2011
Synapse Development and Function		*syd-9*	N/A	SV endocytosis genes (inferred)	Wang, Gracheva et al., 2006
		smn-1	SMN	*gar-2, daf-2, unc-26, unc-57, bas-1, tdc-1*	Gao, Xu et al., 2019; Dimitriadi, Derdowski et al., 2016; O'Hern, do Carmo et al., 2017;, Wu, Gagnon et al., 2018
		grld-1	RBPM15	*glr-1*	Wang, Kang et al., 2010
		ess-2	DGCR14	*dlk-1, dpy-10*	Noma, Goncharov et al., 2014
		sydn-1	N/A	*unc-44, dlk-1*	Chen, Zhou et al., 2015
	APA**	*kin-20*	CK1δ	*unc-44*	LaBella, Hujber et al., 2020
		ssup-72	SSU72	*unc-44, dlk-1*	Chen, Zhou et al., 2015; LaBella, Hujber et al., 2020
		linc-73	ND	*unc-104*	Wei, Chen et al., 2019
	lncRNA	*lep-5*	ND	*srj-54, odr-10, mab-3, daf-7, pkd-2*	Lawson, Vuong et al., 2019
		mir-2	mir-2	*gar-2*	O'Hern, do Carmo et al., 2017

(*Continued*)

Table 14.1. (Continued)

Function in the Nervous System		*C. elegans* Gene	Human Ortholog	mRNA Targets*	References
		mir-1	mir-1	*unc-29, unc-63, mef-2, nrx-1*	Simon, Madison et al., 2008; Hu, Hom et al., 2012
		mir-84	let-7	*hbl-1*	Thompson-Peer, Bai et al., 2012
		mir-51	mir-51	*glo-4*	Zhang, Fan et al., 2018
	miRNA	*lin-4*	mir-125	N/A	Armakola & Ruvkun, 2019
		lin-28	LIN28A	*srj-54, odr-10, mab-3, daf-7, pkd-2*	Lawson, Vuong et al., 2019
		mir-5550	ND	*linc-60*	Wei, Chen et al., 2019
		let-7	let-7	*srj-54, odr-10, mab-3, daf-7, pkd-2, lin-41, ced-7*	Lawson, Vuong et al., 2019; Zou, Chiu et al., 2013; Basu, Dey et al., 2017
		rtcb-1	RTCB	N/A	Kosmaczewski, Han et al., 2015
Axon	**RNA**	*tiar-2*	TIA	N/A	Andrusiak, Sharifnia et al., 2019
Regeneration	**Granules**	*car-1*	LSM14	*micu-1*	Tang, Kim et al., 2020
		cgh-1	DDX6	N/A	Tang, Kim et al., 2020
	piRNA	*prde-1*	N/A	N/A	Kim, Tang et al., 2018
		prg-1	PIWIL1	N/A	Kim, Tang et al., 2018

* Denotes only selected mRNA targets are listed.
** APA, alternative polyadenylation.

been produced using *C. elegans* as a model that show RBPs identified by genetic screens play important roles in alternative splicing events that regulate synapse formation and function. We will highlight cumulated findings of how specific RNA-binding proteins regulate splicing of target mRNA in the nervous system.

MEC-8/RBPMS Regulates Mechanosensation and Neuronal Subtype Specification

The initial identification of genetic determinants for phenotypic variants in *C. elegans* came using screens examining changes in overt animal morphology or behavior (Brenner, 1974). One specific behavior that was used for screening genetic mutants was

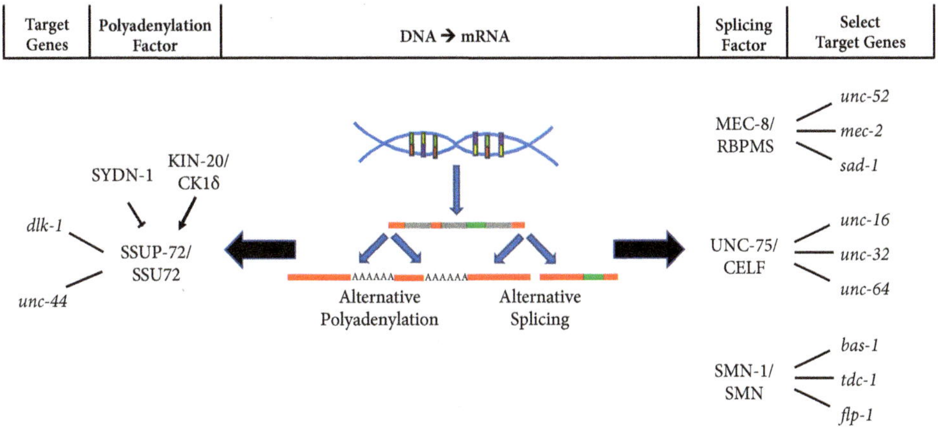

FIGURE 14.1. Examples of nuclear RNA processing in nervous system development and function in C. elegans. *Center*: Schematic of nuclear pre-mRNA processing events that generate transcript diversity. Processing events include the addition of polyA+ tails at different locations, alternative polyadenylation (APA), and the inclusion/exclusion of pre-mRNA sequence, alternative splicing, respectively. *Left*: APA has been shown to be regulated by SYDN-1 and KIN-20, which repress and activate SSUP-72, respectively, resulting in the production of different mRNA isoforms of *dlk-1* and *unc-44*. *Right*: Alternative splicing by the RNA-binding factors MEC-8, UNC-75, and SMN-1 generates different isoforms of select target genes that function in neuronal development and function.

the response to gentle touch (Chalfie & Au, 1989; Chalfie & Sulston, 1981). Upon light touch to the head or tail, the animal initiates a reverse or forward moment, respectively, to avoid further contact (Chalfie et al., 1985). A large set of mutant animals were identified based upon these mechanosensory abnormalities (Mec). Studies of these mutants uncovered many genes that fall into a broad range of functional and molecular distinct classes, including *mec-8*, encoding an ortholog of RNA-binding proteins with multiple splicing (RBPMS) (Lundquist et al., 1996). RBPMS proteins are characterized by the presence of two RNA recognition motifs (RRM) and a disordered N-terminal region. MEC-8 and its vertebrate (RBPMS) and Drosophila (Couch potato) orthologs preferentially bind a tandem GCAC motif by isothermal titration calorimetry (Soufari & Mackereth, 2017), and this motif is also enriched among mRNA targets bound by RBPMS in human cells (Farazi et al., 2014).

Searching for mRNA targets processed by MEC-8 was aided by the observation that compound mutants of *mec-8* and *unc-52* produce a synthetic lethal phenotype (Lundquist & Herman, 1994). *unc-52* encodes an extracellular matrix associated protein Perlecan, mutations of which were first identified by Sydney Brenner based on its Unc phenotype, and were later shown to be required for the formation of presynaptic boutons (Qin, Liang, & Ding, 2014). Alternative splicing leads to the production of several UNC-52 isoforms (Rogalski, Gilchrist, Mullen, & Moerman, 1995). The synthetic lethal phenotype of *mec-8; unc-52* suggests that MEC-8 may regulate *unc-52* mRNA isoform diversity. RT-PCR analyses of *unc-52* exon 15-19 mRNA showed altered patterns in *mec-8* mutants, compared to control, corresponding to the production of altered UNC-52 isoforms (Lundquist et al., 1996). A subsequent study, using tissue-specific splicing reporter constructs, identified *mec-8* mediated alternative splicing of *unc-52* in the hypodermis (Spike, Davies, Shaw, & Herman, 2002). This provides evidence that *mec-8*-dependent *unc-52* mRNA alternative splicing can regulate facets of pre-synaptic development (Figure 14.1).

mec-8 is ubiquitously expressed during early embryogenesis, and becomes more restricted in select tissues, including the six light-touch sensing mechanosensory neurons, later in development (Spike et al., 2002). Investigation of mRNAs encoding proteins mediating mechanosensation uncovered *mec-8* dependent alternative splicing of *mec-2* (Calixto, Ma, & Chalfie, 2010; Figure 14.1). MEC-2, a Stomatin ortholog, links mechanosensory channels to the cytoskeleton to mediate mechanosensation (M. Huang, Gu, Ferguson, & Chalfie, 1995). *mec-2* produces three mRNA isoforms and the longest (designated as isoform A), consisting of 13 exons, is absent in *mec-8* mutants, suggesting that the Mec phenotype associated with loss of *mec-8* may be partly due to perturbed expression of *mec-2*. It was further shown that the presence of the 9th intron of *mec-2* isoform A was sufficient to produce *mec-8* dependent splicing. The splicing of the 9th intron by *mec-8* occurs in a temperature dependent manner, with splicing occurring at the permissive temperature of 15°C not the restrictive 25°C. The temperature dependence of 9th intron splicing was then coupled with mutations in the RNAi regulatory gene *rde-1* to generate conditional RNAi sensitivity, providing a technology-based application for this temperature dependence for *mec-8* splicing of *mec-2* (Calixto et al., 2010).

Transcriptome analysis using deep sequencing of mRNAs from wild type and multiple splice factor gene mutants provided a regulatory landscape of alternative splicing dependent on MEC-8 and its interacting factors (Tan & Fraser, 2017). *mec-8* dependent splicing targets shared significant overlap with other splicing factors, including the ELAV homolog *exc-7* and the RbFox homolog *asd-1*. Loss of both *mec-8* and *exc-7* resulted in an additive effect on their overlapping targets, such as *unc-52*, suggesting combinatorial regulation by these splicing factors. Supporting this idea, the sequence features surrounding cassette exons, which can be differentially included or excluded, showed an enrichment for MEC-8 and EXC-7 consensus sites. Moreover, gene ontology (GO) analysis of MEC-8 and EXC-7 co-regulated proteins revealed an enrichment in synapse-related categories, such as calcium ion transmembrane transporter activity. This is suggestive that *mec-8* can regulate a variety of synaptic genes and that this can occur in conjunction with other RNA-binding proteins.

Another neuronal mRNA target of *mec-8* is the *sad-1* gene, which encodes a conserved kinase identified based on its roles in sensory and motor neuron synaptogenesis (Crump, Zhen, Jin, & Bargmann, 2001) and whose vertebrate ortholog can phosphorylate the presynaptic active zone protein RIM (Inoue et al., 2006). SAD-1 isoforms differ in their capacity to interact with the F-actin binding protein NAB-1 (Hung, Hwang, Po, & Zhen, 2007). During asymmetric differentiation of the ALM and BDU neurons, *sad-1* is differentially spliced such that an alternative exon is included in the *sad-1* mRNA expressed in ALM, but not in BDU, providing an important paradigm to dissect developmentally regulated splicing (Thompson et al., 2019). Examination of sequence features in the alternatively spliced exon cassette of *sad-1* revealed consensus binding sites for MEC-8 and MBL-1, the latter an RNA-binding protein of the Muscleblind family (Spilker, Wang, Tugizova, & Shen, 2012). Loss of either *mec-8* or *mbl-1* impacted *sad-1* mRNA splicing, and mutating the consensus sequences for either *mec-8* or *mbl-1* within the intron prevented splicing (Thompson et al., 2019). Double mutants of *mec-8* and *mbl-1* showed a significant additive effect impairing *sad-1* mRNA splicing. Although the functional impact of perturbed *sad-1* splicing in *mec-8; mbl-1* double mutants was not assessed, it is conceivable that synaptic related phenotypes associated with *sad-1* would be exacerbated in these animals. Furthermore, a forward genetic screen using a two-color *sad-1* splicing reporter in the ALM neuron unexpectedly uncovered mutations in three transcription factors (*unc-86*, *mec-3* and *alr-1*). Both *mbl-1* and *mec-8* expression was regulated by *mec-3*, providing a mechanistic link to a transcription factor identified by unbiased screening (Thompson et al., 2019). Similarly, expression of *mec-8* or *mbl-1* in *alr-1* (an ALX/ARX family transcription factor) mutants was able to restore expected splicing of *sad-1*. This study provides both transcriptional and splicing dependent mechanisms regulating SAD-1 isoforms during neuronal development (Figure 14.1). It is unclear at this point how SAD-1 isoforms are differentially utilized within each neuron type to regulate synapse formation and function.

UNC-75/CELF Regulates Neuronal Development, Synaptic Transmission, and Axon Regeneration

Sydney Brenner's pioneering genetic screen provided the foundation for the use of *C. elegans* as an experimental organism (Brenner, 1974). Characterization of mutants with Unc movement identified genes spanning a broad spectrum of cellular functions. One such gene, *unc-75*, encodes a member of the CELF (CUG-binding and ELAV factor) family (Loria, Duke, Rand, & Hobert, 2003). CELF family proteins contain 3 RNA-recognition motifs and regulate a diverse range of RNA-centric processes, including mRNA decay, deadenylation, C to U editing, and splicing (Dasgupta & Ladd, 2012). Mammals express six CELF genes, whereas *C. elegans* has two orthologs, *unc-75* and *etr-1*. *unc-75* expression is enriched in the nervous system and localizes to nuclear speckles, consistent with a role in RNA splicing (Loria et al., 2003). In addition to its Unc movement, loss of *unc-75* decreases sensitivity to the cholinesterase aldicarb and some neurons display ectopic neurite spouting (Loria et al., 2003). The broad effects of loss of function mutations in *unc-75* suggest that it likely regulates multiple mRNA dependent processes beyond splicing.

A transcriptome analysis using mRNA deep-sequencing supported *unc-75*'s role in splicing (Kuroyanagi, Watanabe, Suzuki, & Hagiwara, 2013; Figure 14.1). To avoid loss of errant transcripts in *unc-75* mutant animals by nonsense-mediated decay (NMD), *unc-75* mutation was combined with a mutation in *smg-2*, a gene essential for NMD for its ability to detect truncated open reading frames (Page, Carr, Anders, Grimson, & Anderson, 1999). 24 putative UNC-75 splicing targets were experimentally validated with bi-chromatic fluorescent reporter constructs. These transgenes were constructed such that inclusion or exclusion of alternative exons leads to alternative reading frames, resulting in changes in the ratio of green to red fluorescent protein (Kuroyanagi, Watanabe, & Hagiwara, 2013; Norris et al., 2014; Orengo, Bundman, & Cooper, 2006). Transcripts showing changes in *unc-75* mutants included multiple genes functioning in synapses such as *tom-1, nrx-1, unc-16, unc-32,* and *unc-75* (Kuroyanagi, Watanabe, Suzuki et al., 2013). UNC-75 splicing targets were enriched for a (G/U)UGUUGUG motif at both upstream and downstream introns. Mutations in this sequence significantly inhibited splicing of mRNAs expressed from fluorescent reporter transgenes and reduced UNC-75 binding in vitro by electrophoretic mobility shift assay (EMSA). Examination of splicing of *unc-32* identified regulation of a single exon (Exon 7a) by two separate RNA binding proteins, ASD-1, an Rbfox1 orthologue, and UNC-75 (Kuroyanagi, Watanabe, & Hagiwara, 2013). EMSA analysis using truncated variants of UNC-75 revealed that RRM3 is sufficient to bind the specified intronic sequence of *unc-32*. This study provided a proof of principle on the use of dual color fluorescent reporters as a read-out for splicing in the nervous system of *C. elegans*. Additional analysis also revealed that another target *unc-16* displayed tissue-specific splicing, with the hypodermis and nervous system retaining exon 16, which is perturbed in *unc-75* mutants (Kuroyanagi, Watanabe, Suzuki et al., 2013; Figure 14.1).

Development of a fluorescence reporter for *unc-16* splicing in the nervous system (Norris et al., 2014) has provided a platform to assess, in an unbiased manner, alternative splicing in a live animal through forward genetic screening. Mutant animals showing altered splicing of *unc-16* were isolated and shown to harbor mutations in *unc-75* and *exc-7*, the latter encoding an RNA-binding protein of the ELAV family (Fujita et al., 2003). Both UNC-75 and EXC-7 bind consensus sequences in the intron downstream of exon 16 and regulate splicing of *unc-16* in GABAergic neurons. This data lends itself to the notion that they may regulate splicing in additional genes at similar locations. Analysis of RNA-sequencing data revealed that a portion of targets overlap between these RNA-binding proteins, however, they do act independently and antagonistically of each other. The shared targets between *unc-75* and *exc-7* include the synapse relevant genes *twk-39* and *dyn-1*. Electrophysiological studies of neuromuscular junctions showed decreased EPSC peak current and current density in *unc-75* mutant animals, which is further reduced in combination with *exc-7* mutations. Additional synergy on neuromuscular transmission was also observed in double mutants of *unc-75* and *syd-9*, a nematode specific Zn-finger protein that localizes to nuclear speckles and likely functions in mRNA splicing (Wang et al., 2006). These data suggest that combinatorial interaction of UNC-75 with other RNA binding proteins impacts neuromuscular signaling.

Another functionally relevant mRNA target of UNC-75 is the Syntaxin ortholog *unc-64*, a SNARE complex protein essential for exocytosis. *unc-64* mutants have locomotion defects and are resistant to aldicarb (Hosono & Kamiya, 1991). *unc-64* produces two

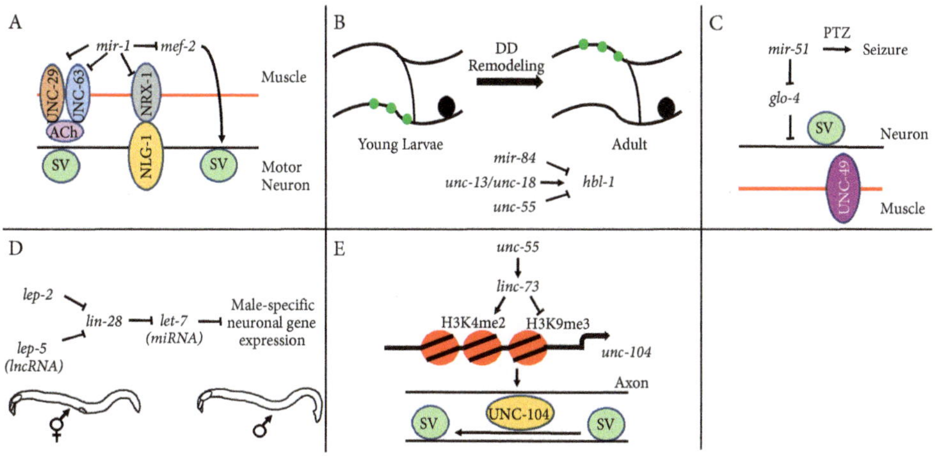

FIGURE 14.2. Multifaced roles of RNA-associated factors in adult axon regeneration following injury. Schematic summarizing the role of RNA regulatory factors during axon regeneration. The piRNA pathway (PRDE-1/PRG-1) and the RNA ligase *rtcb-1* both inhibit axon regeneration cell autonomously; and the direct targets remain unknown. The CELF RNA splicing protein UNC-75 promotes regeneration by alternative splicing of *unc-64*, encoding syntaxin. The mRNA decay factor CAR-1/LSM14 regulates axon regeneration by directly binding the mRNA of the mitochondrial calcium uniporter auxiliary factor *micu-1*, thereby controlling the levels of mitochondrial calcium. The PRLD-containing protein TIAR-2 forms liquid-liquid phase separated granules following axon injury which inhibit axon regeneration.

mRNA isoforms as a result of alternative splicing of the last exon. CLIP-sequencing identified an UNC-75 binding site important for the alternative splicing of *unc-64* (Chen et al., 2016; Figure 14.1). Both UNC-64 isoforms are sufficient to rescue aldicarb sensitivity, however, only the B isoform restores normal locomotion. *unc-75* dependent *unc-64* splicing in mechanosensory neurons is required for axon regeneration following injury, however, expression of UNC-64A, not B, was sufficient to alleviate the regeneration deficits of *unc-75* mutants (Chen et al., 2011; Figure 14.2). Thus, *unc-75* mediated *unc-64* splicing plays an important role during normal neuromuscular function, as well as following axon injury, providing important insight into non-overlapping, isoform specific functions of *unc-64*. These analyses of *unc-75* reveal its broad roles in regulation of RNA homeostasis, however; it is currently unclear to which extent splicing perturbations in the absence of *unc-75* lead to the manifestation of its uncoordinated locomotion.

SMN-1/SMN Regulates Motor Neuron Function

Spinal muscular atrophy is caused by mutations in the survival motor neuron gene SMN1, which is a ubiquitously expressed cytoplasmic and nuclear RNA-binding protein (Nussbacher et al., 2019). In the nucleus, SMN proteins regulate the formation of spliceosomal U snRNPs (Liu, Fischer, Wang, & Dreyfuss, 1997; Sharpe, Williams,

Norton, & Latchman, 1989), and in the cytoplasm, have diverse roles including local translation and RNA granule formation (Hua & Zhou, 2004; Kye et al., 2014). The *C. elegans* genome encodes a single SMN ortholog, *smn-1,* which has been shown to be both structurally and functionally conserved, as it also interacts with the spliceosome component U2AF (Gao et al., 2014). Functional studies of *smn-1* have provided novel insights into the biological role of SMN proteins in the nervous system.

Genetic null mutants of *smn-1* exhibit several neuronal-associated behavioral dysfunctions (Briese et al., 2009), including decreased food intake related pharyngeal pumping, and a progressive decline in mobility. Although no overt defects in neuronal number, morphology or synapse number were identified, later studies revealed, using more quantitative approaches, that cholinergic synapses are decreased in both width and fluorescence intensity relative to control in this *smn-1* deletion (O'Hern et al., 2017). Moreover, neuronal expression of *smn-1* was sufficient to partially rescue the movement defects, suggesting that SMN-1 has neuronal roles in addition to its function in muscle. Highlighting the importance of SMN-1 mediated RNA splicing in synaptic homeostasis, these observed phenotypes of *smn-1* mutants are dependent upon its interaction with U2AF (Gao et al., 2014); and a loss of function of *mel-46*, the ortholog of Gemin3, a binding partner of SMN, shared similar phenotypes (O'Hern et al., 2017). Analysis of RNA-sequencing data from *smn-1* null mutants identified multiple splicing defects, including transcripts involved in monoaminergic transmission such as *bas-1* and *tdc-1* (Gao et al., 2019). Examination of specific splicing defects revealed increased intron retention in *abt-5* mRNA and splicing at an alternative 3′ splice site in *flp-1* mRNA (Figure 14.1).

smn-1 is essential for overall development, as null mutants show larval growth arrest which present confounds in studying its role in the mature nervous system. Animals expressing a SMN-1(D44V) mutation, which is present in a patient with type IIIb SMA (Sleigh et al., 2011; Y. Sun et al., 2005), display similar, albeit much milder, neuronal phenotypes and do not exhibit developmental arrest or delay. This D44V mutation provides the opportunity to examine neuronal function without gross developmental defects. *smn-1*(D44V) mutants were resistant to an acetylcholinesterase inhibitor, suggesting defects in neuromuscular transmission. Using automated movement analysis, a screen was performed on chemical compounds that improve locomotion of *smn-1*(D44V) animals. Although it remains to be understood which mRNA targets are directly regulated by SMN-1 dependent splicing, this screen identified multiple compounds which could later be trialed in higher organisms for potential therapeutic use. Additional studies of genetic interactions revealed that the pumping-dependent defects associated with *smn-1* mutation are due to impaired synaptic vesicle recycling (Dimitriadi et al., 2016). Double mutants of *smn-1* with *unc-26* and *unc-57,* two genes that function in synaptic vesicle recycling (Harris, Hartwieg, Horvitz, & Jorgensen, 2000; Schuske et al., 2003), behaved similarly to each single mutant, with no additive effect, suggesting they are in the same pathway. Endocytic defects were also seen in SMN depleted cells in culture. Moreover, the localization of several presynaptic proteins was also perturbed in *smn-1* mutant animals. Analysis of motor neuron pre-synaptic compartments showed a decrease in synaptic vesicles. This important study provides a mechanistic link between synaptic vesicle recycling defects and locomotor impairment in *smn-1* mutants. Future studies will need to assess the specific regulatory mechanism linking *smn-1* and vesicle recycling.

Another study linked miRNA-mediated regulation to *smn-1*'s function in the nervous system (O'Hern et al., 2017; C. Y. Wu et al., 2018). *mir-2* is a miRNA enriched in neurons and *mir-2* loss of function animals are resistant to aldicarb, similar to *smn-1* mutants (O'Hern et al., 2017). A computationally predicted target of *mir-2* is the muscarinic acetylcholine receptor GAR-2. In *smn-1* loss of function animals, *mir-2* no longer represses the expression of *gar-2*. Mechanistically, *smn-1* has been shown to interact with *mel-46*, a putative RISC complex regulatory protein (Hock et al., 2007; Murashov et al., 2007), which suggests that an interaction between SMN-1 and MEL-46 could shape miRNA activity. Moreover, loss of *gar-2* eliminates all neuromuscular phenotypes associated with loss of *smn-1*, suggesting that de-regulation of *gar-2* may likely account for the neuronal phenotypes associated with loss of *smn-1*. These data present a model in which increased GAR-2 levels in *smn-1* mutants lead to its motor dysfunction. Similar epistasis experiments were performed using *daf-2*, the *C. elegans* insulin-like growth factor receptor (Wu et al., 2018), loss of function of which extends lifespan and improves locomotor behavior of *smn-1*. Analysis of pre- and post-synaptic markers in *daf-2; smn-1* double mutants revealed that the restoration of locomotion is coincident with preservation of these synaptic compartments. This provides an important link between the longevity associated insulin-signaling axis and *smn-1* in aging and neuronal homeostasis.

Recent clinical advances in the treatment of spinal muscle atrophy (Ramos et al., 2019; Wood, Talbot, & Bowerman, 2017), a genetic disorder resulting from mutations in the *SMN1* gene, highlight the importance of understanding the biological functions regulated by RNA-binding proteins. To date, a wealth of knowledge has been obtained regarding the genetic pathways that interact with *smn-1*, however, little has been done to reconcile these pathways with SMN-1 activity. In the future, more refined studies examining the role of SMN-1 in splicing and other RNA regulatory processes needs to be performed to better understand its true mechanistic roles.

GRLD-1/RBM15B Regulates AMPA Receptor Levels with Neuronal-Type Specific Outcome

The *C. elegans grld-1* gene encodes the ortholog of RNA-binding motif protein 15B (RBM15B) which is characterized by two RNA recognition motifs (RRM) and a Spen paralog and ortholog C-terminal domain (SPOC); (Table 14.1). Loss of function mutations in *grld-1* were found in a genetic screen looking at decreased expression of the AMPA receptor GLR-1 in the interneuron AVE, visualized using a fluorescently labelled GLR-1 transgene that contains endogenous introns and untranslated regions (UTR); (Wang et al., 2010). *grld-1* mutants are nose-touch defective, as the AVE neuron drives backward locomotion, and have a decrease in glutamate-gated current. GRLD-1 is nuclear localized and acts cell-autonomously within AVE. Expression of the two RRM domains of GRLD-1 is sufficient for its ability to restore GLR-1 levels. Moreover, when GLR-1 is expressed using cDNA, with heterologous introns and UTRs, it bypasses regulation from *grld-1*. The use of alternative UTR's did not affect GRLD-1 regulation of GLR-1, whereas the removal of endogenous introns of *glr-1* did, supporting the specific role of GRLD-1 on regulating GLR-1 by its intronic sequences. Mammalian RBM15B has been shown to regulate the methylation of RNA through recruitment of the m(6)A complex (Patil et al., 2016). It would be of interest to test if GRLD-1 may regulate *glr-1* processing by a similar mechanism. This study highlights the utility of forward genetic screens in the identification of novel targets for synaptic proteins such as AMPA type receptors.

ESS-2/DGCR14 Regulates Cryptic Splicing of DLK-1 During Synapse Formation

The ESS-2/DGCR14 proteins are evolutionarily conserved from yeast to human (Hegele et al., 2012; Table 14.1). Deletion of a chromosomal region containing DGCR14, the human ESS-2 ortholog, is associated with DiGeorge syndrome (Rizzu et al., 1996). A genetic modifier screen for synapse defects in *C. elegans* provided initial functional evidence for this protein family in mRNA splicing. Two genes, *rpm-1* and *syd-2* were first identified in a visual screen for defective presynaptic morphology (Hallam, Goncharov, McEwen, Baran, & Jin, 2002; Zhen, Huang, Bamber, & Jin, 2000; Zhen & Jin, 1999). While loss of function in each gene affects different aspects of synapse assembly and causes subtle behavior defects, double mutants of *rpm-1; syd-2* results in paralysis with severely reduced synapse size and number. This synthetic mutant phenotype was used to identify genes acting in the signaling pathways of RPM-1 and SYD-2 (Liao, Hung, Abrams, & Zhen, 2004; Nakata et al., 2005; Trujillo, Nakata, Yan, Maruyama, & Jin, 2010). Studies of a large number of genetic suppressors of synapse phenotypes of *rpm-1* mutants led to the discovery of the DLK-1 MAPK cascade that is negatively regulated by RPM-1 during synapse formation (Nakata et al., 2005). During the analyses of a splice acceptor site mutation of *dlk-1* (CAG to CAA), it was observed that this mutation caused strong loss of function only in the absence of the *ess-2* gene (Noma, Goncharov, & Jin, 2014). ESS-2 localizes to the nucleus, similar to its human ortholog. RT-PCR analysis of splice variants further showed a specific role for *ess-2* in *dlk-1* spice acceptor mutants, and such effects were also observed for a splice acceptor mutation in an unrelated gene. This provides a physiological consequence of the results of cryptic splicing, a process whereby a potential spliceosomal recognition site is unmasked by the presence of a mutation (Kapustin et al., 2011). This study highlights a specific role of ESS-2 in mRNA splicing for proper establishment of synaptic connectivity mediated by the DLK signaling axis.

SYDN-1 and KIN-20/CK2delta Regulate Alternative Polyadenylation to Generate the Neuronal Ankyrin Isoform

An alternative means of generating transcript diversity is a mechanism referred to as alternative polyadenylation (APA). APA results in distinct 3′ ends for RNA polymerase II transcripts which requires distinct components compared to splicing. APA can result in changes to the 3′ UTR of transcripts, which can alter stability, translational regulation and localization (Gruber & Zavolan, 2019; Figure 14.1). Additionally, APA upstream of the 3′ UTR can be coupled with splicing to produce distinct protein isoforms.

Previous screens for regulators of synaptogenesis yielded multiple genes, including the E3 ligase *rpm-1* and the presynaptic active zone organizer *syd-2* (Zhen et al., 2000; Zhen & Jin, 1999). Loss of function in each gene causes subtle movement defects, while double mutants exhibit severe deficit in synapses resulting in paralyzed locomotion, providing an easily identifiable phenotype for genetic modifier screens. A genetic screen was carried out to search for mutations enhancing behavioral defects of *rpm-1,* and yielded a gene called *sydn-1* (SYnapse-Defective eNhancer-1). Loss of function in *sydn-1* alone results in mild synapse and movement defects, but in combination with *rpm-1* or *syd-2* mutations, produced severe movement deficits, due to excessive neurite branching and dramatic loss of synapses (Chen et al., 2015; Van Epps, Dai, Qi, Goncharov, & Jin, 2010). *sydn-1* has no clear orthologs in other species, and localizes to nuclear speckles, suggesting a likely role in pre-mRNA processing. Indeed, a subsequent suppressor screen of *sydn-1* mutants yielded numerous loss of function mutations in a set of genes encoding components of pre-mRNA 3′ end processing machinery and provided an important clue that SYDN-1 negatively regulates polyadenylation of pre-mRNA.

Pre-mRNA 3′ end processing is coupled with transcriptional termination by RNA polymerase II. The phosphorylation state of the C-terminal domain (CTD) of RNA Pol II is tightly regulated by coordinated action of kinases and phosphatases (Harlen & Churchman, 2017). Loss of function in *ssup-72,* which encodes the ortholog of the CTD Ser5 phosphatase Ssu72 (Krishnamurthy, He, Reyes-Reyes, Moore, & Hampsey, 2004), completely suppressed neuronal phenotypes of *sydn-1* mutants (F. Chen et al., 2015). The phosphatase activity of SSUP-72 was shown to be essential for its interaction with *sydn-1*-mediated processes. To identify the specific targets regulated by the SYDN-1 and SSUP-72 pathway, a genome-wide analysis of RNA pol II occupancy was performed in control and *sydn-1* mutant animals, which revealed a rather selective number of genes displaying changes. The most significant change was reduced RNA Pol II occupancy in a 10kb exon of the *unc-44* gene, which encodes several isoforms of *C. elegans* ankyrin, with the 10 kb exon being expressed exclusively in neurons (Otsuka et al., 1995). The production of this neuronal isoform of UNC-44, which is composed of numerous repeat peptides unique to *C. elegans,* depends on the regulation of an internal polyadenylation site (PAS) preceding the 10 kb exon. Further analysis of the 3′ end of *unc-44* transcripts revealed that SYDN-1 selectively regulates 3′ end processing at this internal PAS. Removing this PAS site by CRISPR/Cas9 editing increased production of neuronal

UNC-44 and completely suppressed the ectopic branching of *sydn-1* mutants (Chen et al., 2015). Besides acting on the *unc-44* PAS, SYDN-1 also selectively regulates *dlk-1* alternative polyadenylation to promote neurite stability. This comprehensive study provided genetic and biochemical insight into the role of SYDN-1/SSUP-72 dependent regulation of alternative polyadenylation during neurite outgrowth and synaptogenesis (Figure 14.1).

Another study revealed that animals with loss of function mutations in the gene *kin-20,* a casein kinase, exhibited ectopic growth cones and progressive paralysis (LaBella et al., 2020). The effects of *kin-20* were confined to the nervous system, as its expression in neurons was sufficient to rescue the ectopic growth cone and Unc phenotype. Moreover, *kin-20* function was required for axon maintenance, as developmental axon guidance and synapse formation were intact during early larval development. A large-scale suppressor screen of *kin-20* mutants identified mutations in multiple genes involved in alternative polyadenylation, as well as *unc-44* mutations specifically affecting APA to produce its neuronal isoform. These data support a model whereby *kin-20* interaction with the APA machinery modulates the production of the neuronal isoform of UNC-44. The mammalian *kin-20* ortholog CK1δ was shown to phosphorylate *C. elegans* SSUP-72 in vitro, and a phosphomimic allele of *ssup-72* (S39E) was sufficient to repress the effects of *kin-20* mutation. The neuronal isoform of UNC-44 (~517kDa) is repressed in all non-neuronal tissues (Otsuka et al., 1995). *kin-20* mutants expressed significantly less of the neuronal *unc-44* isoform than control animals by RT-PCR, whereas the small and medium isoforms were similar to control. In addition, overexpression of the neuronal UNC-44 isoform in *kin-20* mutants significantly reduced the ectopic branches and growth cones. This study indicates that KIN-20 mediated phosphorylation of SSUP-72 negatively regulates APA production of the neuron-specific isoform of UNC-44 (Figure 14.1). This neuronal ankyrin isoform contains an extended C-terminal domain with unknown function, although it has been shown to modulate formation of gap junctions in *C. elegans* (Meng, Chen, & Yan, 2016) and may have analogous function as the vertebrate Ankyrin G in the formation of the axon initial segment (Rasband, 2010).

Emerging Roles of Non-coding RNAs in Synapse Formation and Function

The discovery of miRNA in *C. elegans* has opened a new field to examine non-coding RNAs across eukaryotic organisms and biological processes (Gebert & MacRae, 2019; Lee, Feinbaum, & Ambros, 1993). Subsequent studies have led to the identification and characterization of other classes of non-coding RNA including piRNA, and lncRNAs. Though each of these non-coding RNA species have different properties and functions, a common theme is their ability to regulate the levels of mRNA. miRNA recruitment of the Dicer and RISC complexes results in binding their mRNA targets, ultimately resulting in mRNA degradation or translational repression (Hutvagner & Zamore, 2002;

Lee et al., 2004; Pham, Pellino, Lee, Carthew, & Sontheimer, 2004). Together with computational approaches, a wealth of synapse-related mRNA have been identified as potential miRNA target genes (Sun, Zhao, & Wang, 2006) and have been further validated in experiments, discussed next.

mir-1 Regulates Synaptic Homeostasis

The *C. elegans mir-1* is perfectly conserved in sequence with mammalian *mir-1*. *mir-1* expression is confined to the body wall muscle of *C. elegans* (Ibanez-Ventoso et al., 2006). Contraction of body muscles in *C. elegans* is primarily induced by acetylcholine; upon exposure to exogeneous levamisole, an agonist of nematode-specific acetylcholine receptors, animals become paralyzed and eventually die (Thienpont et al., 1966). Loss of *mir-1* diminishes muscle responses to levamisole (Simon et al., 2008). mRNAs for two muscle nicotinic acetylcholine receptor (nAChR) subunits, *unc-29* and *unc-63*, contain predicted *mir-1* binding sequences (Figure 14.3A). Loss of *mir-1* results in their increased expression, and overexpression of UNC-29 and UNC-63 decreases sensitivity to levamisole, supporting the functional relevance of *mir-1*-dependent regulation on nAChR mRNAs. Electrophysiological analysis of *mir-1* mutants also revealed a decrease in pre-synaptic acetylcholine release. Overexpression of UNC-29 and UNC-63 had no impact on the presynaptic phenotypes, suggesting an alternative regulatory mechanism. In mammals, the transcription factor MEF-2 is a critical regulator of synapse formation (Flavell et al., 2006; Shalizi et al., 2006). *C. elegans mef-2* contains a predicted *mir-1* binding sequence, and *mef-2* acts downstream of *mir-1* in acetylcholine release (Figure 14.3A). *mir-1* mutants showed increases in the pre-synaptic vesicle protein RAB-3, which was restored to normal levels when combined with loss of *mef-2*. Expression of MEF-2 in muscle of *mir-1; mef-2* animals was sufficient to restore the *mir-1* phenotype, confirming the post-synaptic regulation of pre-synaptic vesicles. Acute treatment with levamisole also induced MEF-2 transcriptional changes. Thus, the effects on MEF-2 expression via *mir-1* ultimately result in retrograde changes in presynaptic RAB-3. This study implies a cell non-autonomous role for the miRNA, *mir-1*, in the regulation of pre-synaptic termini by its regulation of the transcription factor MEF-2.

A direct mediator of retrograde synaptic homeostasis involves the interaction of neuroligin (*nlg-1*) and neurexin (*nrx-1*; Figure 14.3A), which are trans-synaptic adhesion molecules required for establishment of proper pre- and post-synaptic interactions (Ichtchenko et al., 1995; Yamagata, Sanes, & Weiner, 2003). Elimination of either *nlg-1* or *nrx-1* had no gross effects on morphology or number of pre-synapses (Hu et al., 2012). In combination with *mir-1* mutation, however, loss of either *nlg-1* or *nrx-1* suppresses the diminished post-synaptic currents seen in *mir-1* mutant animals. NRX-1 levels were significantly increased in *mir-1* mutant animals, which was eliminated in *mir-1; mef-2* double mutant animals. As neuroligin and neurexin mutations are linked to human neurological conditions (Feng et al., 2006; Jamain et al., 2003), this study's finding that a *mir-1, mef-2* retrograde signal regulates their function in neurotransmitter release, provides important insight into the mechanisms underlying these conditions.

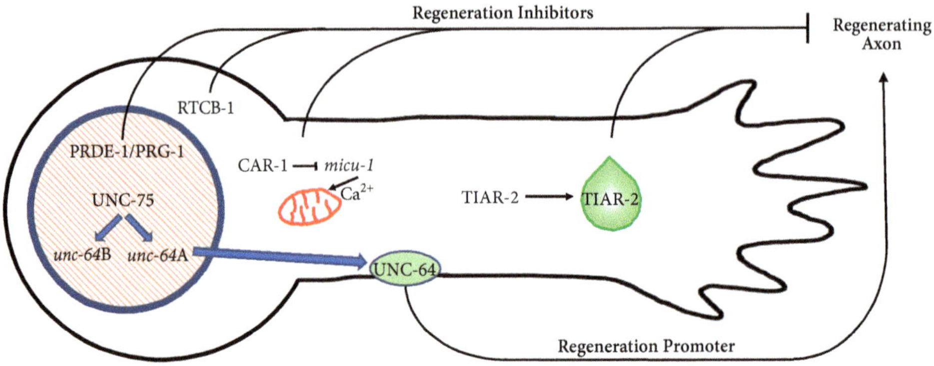

FIGURE 14.3. Non-coding RNAs play diverse roles in the nervous system of *C. elegans*. Schematic summaries of select functions of non-coding RNAs in neuronal development and function. (A) In muscle, *mir-1* directly binds and downregulates *unc-29*, *unc-63*, *nrx-1*, and *mef-2*. The transcription factor, MEF-2, acts downstream of *mir-1* to regulate acetylcholine release. SV for synaptic vesicle, ACh for acetylcholine. (B) Regulation of *hbl-1*, which encodes a Hunchback transcription factor, by miRNA and other factors promotes DD neuron synapse remodeling during development. The miRNA *mir-84* and transcription factor *unc-55* both negatively regulate *hbl-1* levels. Genes functioning in synaptic transmission, e.g. *unc-13* and *unc-18*, positively regulate *hbl-1* levels. Green dots represent synapses; black dots represent neuronal cell body; lines illustrate DD neuron axon morphology. (C) *mir-51* negatively regulates GEF *glo-4*, which in turn influences the levels of synaptic proteins, such as the presynaptic vesicle (SV) protein SNB-1 and postsynaptic muscle receptor UNC-49. PTZ for pentylenetetrazole. (D) The specification of a male-specific neuronal gene expression profile is regulated by the lncRNA *lep-5* which acts upstream of the heterochronic genes *lin-28/let-7*. (E) UNC-55 positively regulates the transcription of *linc-73*, which in turn regulates the expression of kinesin *unc-104* by influencing the modification of histones in its promoter region. SV for synaptic vesicles.

mir-84 Modulates Initiation of Developmental Synapse Remodeling of Motor Neurons

mir-84 is a member of the *let-7* miRNA family, and has been shown to regulate expression of the hunchback transcription factor, *hbl-1*, in non-neuronal tissues (Abrahante et al., 2003; Lin et al., 2003). In the locomotor circuit of *C. elegans*, several embryonically born motor neurons known as DD (for Dorsal type D) undergo a complete connectivity remodeling during early larval development, a process known as DD remodeling (White, Albertson, & Anness, 1978; Figure 14.3B). Concurrent with DD remodeling, post-embryonic neuroblast division generates a set of neurons known as VD (ventral type D), which adopt the synaptic connectivity of embryonic DD neurons and do not undergo remodeling. The COUP-family nuclear hormone receptor *unc-55* is expressed in VD neurons and critical for VD neurons to form synapses to ventral muscles (Walthall & Plunkett, 1995). In *unc-55* mutants, VD neurons do not form synapses to ventral body muscles, and instead make innervation to dorsal synapses. Consistently, electrophysiology recording detected an absence of inhibitory post-synaptic currents

(IPSC) in ventral muscles, and an increased IPSCs in dorsal muscles (Thompson-Peer, Bai, Hu, & Kaplan, 2012). The promoter of *hbl-1* contains several predicted UNC-55 binding sites (Figure 14.3B) and loss of *hbl-1* prevents ectopic dorsal synapses of VD neurons while restoring ventral synapses caused by loss of *unc-55*. Moreover, loss of *hbl-1* alone caused a delay in remodeling of DD synapses, suggesting it plays a role in normal DD remodeling. The 3′ UTR of *hbl-1* contains several predicted miRNA binding sites, with one corresponding to *mir-84*. In the absence of *mir-84*, *hbl-1* level in larval DD neurons was increased, leading to an earlier onset of DD remodeling. *hbl-1* expression in DD neurons also depends on neuronal activity (Figure 14.3B), as it is decreased in mutants of *unc-13* and *unc-18*, both of which impair synaptic vesicle release (Richmond, Davis, & Jorgensen, 1999; Weimer et al., 2003). These studies provide evidence that circuit activity and miRNA regulation precisely regulate the levels of *hbl-1* to facilitate neuronal remodeling.

mir-51 Regulates the Formation of GABAergic Synapses

The conserved family of *mir-51* miRNAs includes several miRNAs that are redundantly required for the development of multiple tissues in *C. elegans* (Shaw, Armisen, Lehrbach, & Miska, 2010). A study of GABAergic synaptic function investigated six specific miRNAs as they were predicted to bind a significant number of synaptic related genes (Zhang et al., 2018). These miRNA mutants were treated with the GABA receptor antagonist pentylenetetrazole (PTZ) or the acetylcholinesterase inhibitor aldicarb. Treatment with PTZ in *mir-51* mutants induced seizures (Zhang et al., 2018), a phenotype associated with motor circuit imbalance (Jospin et al., 2009). *mir-51* mutants had decreased GABA synapses, measured by the pre-synaptic marker SNB-1 and post-synaptic UNC-49 $GABA_A$ receptor (Figure 14.3C). Eight potential targets of *mir-51* were selected based on several prediction algorithms and expression data analyses. Their 3′ UTRs were fused to a luciferase reporter, and upon overexpression of *mir-51*, the reporter with the 3′ UTR of *glo-4* showed significant downregulation, which was partially rescued by deletion of the *mir-51* binding site. GLO-4 is a Rab GEF that regulates lysosomal sorting through GLO-1 and APM-3 (AP3 components; Hermann et al., 2005) and implicated in axon extension and synapse formation (Grill et al., 2007). Analysis of *mir-51; glo-4* mutants revealed a restoration of GABA receptor and synapses to control levels, also through the GLO/AP-3 complex.

The Heterochronic miRNA Pathway Regulates Cholinergic Synapses

The cholinergic DA9 neuron in *C. elegans* shows stereotyped synaptic connectivity that is regulated by a number of signaling pathways, including netrin (Poon, Klassen, &

Shen, 2008). An RNAi screen targeting genes that alter the aging phenotypes of the insulin receptor *daf-2* mutants was performed by examining the presynaptic marker GFP-RAB-3 in DA9 neurons (Armakola & Ruvkun, 2019). Downregulation of the developmental timing regulator, *blmp-1*, which encodes an ortholog of PRDM1 and is widely known for its role in the heterochronic gene network (Horn et al., 2014), produced an increase in dendritic RAB-3 puncta (Armakola & Ruvkun, 2019). The miRNA heterochronic pathway components LIN-28, an RNA-binding protein and *let-7*, a conserved miRNA, were identified as crucial regulators of late developmental timing in hermaphrodite animals (Ambros & Horvitz, 1984; Reinhart et al., 2000). Further analysis of *let-7* mutant animals and its regulator *lin-28*, revealed similar increases of RAB-3 puncta in the DA9 dendrite, while loss of the miRNA *lin-4* caused an overall decreased GFP-RAB-3 fluorescent signal in DA9. This study provides insight into the role of the heterochronic pathway in the regulation of synapse formation in the cholinergic DA9 neuron. Further investigation is required to determine what specific transcripts are regulated by *lin-4* and *let-7* during synaptogenesis. Moreover, the *C. elegans* genome contains more than 5,000 predicted miRNAs, and individual deletion of most does not cause overt, visibly detectable phenotypes. Nonetheless, the current progress towards understanding the *in vivo* function of miRNA's indicates that many play an important role in shaping the nervous system by finely tuning the expression of specific developmental and synaptic components. Whereas loss of specific genes involved in aspects of synaptic function can eliminate their presence or activity, modulating miRNAs results in dysfunction due to more modest changes in gene expression.

lncRNA in Regulating Sexual Dimorphic Neuronal Gene Expression

Long non-coding RNAs (lncRNA) comprise a class of non-coding RNA that is autonomously transcribed and is greater than 200 nucleotides (Batista & Chang, 2013). Unlike miRNAs, which act to limit mRNA levels through defined RNA-protein complexes (Gebert & MacRae, 2019), lncRNAs have a diverse mechanistic impact on gene expression (Batista & Chang, 2013). lncRNAs have been shown to regulate chromatin structure (Brockdorff et al., 1992), mRNA stability (Gong & Maquat, 2011), translational regulation (Eom et al., 2014), and act as protein and RNA decoys (Lee et al., 2016; Wu, Wang, Wang, & Wang, 2013). Emerging studies offer a glimpse at the phenotypic consequences of perturbations in specific lncRNAs in the *C. elegans* nervous system.

C. elegans exist as a predominantly hermaphrodite population capable of self-fertilization and propagation. Male animals are produced at a frequency of 0.2 percent in laboratory strains due to defects in X chromosomal segregation (Herman, Kari, & Hartman, 1982; Hodgkin, Horvitz, & Brenner, 1979). Male animals have many fundamental differences in cellular composition, locomotion and mating behavior (Barr & Garcia, 2006; Emmons, 2005; Zarkower, 2006). Importantly, males have significant differences in both neuronal subtypes and connectivity, many of which occur at the juvenile to adult transition, wherein the animals become sexually mature.

To examine the regulatory mechanisms involved in sexual maturation, perturbations in the expression of five different male-specific genes that regulate aspects of neuronal function in five neuronal subtypes were examined (Lawson et al., 2019). miRNAs have been shown to regulate male-specific gene expression programs (Pereira et al., 2019). Mutants of *lin-28* or *let-7* displayed perturbed expression of all five genes male-specific genes examined. Previous studies identified *lep-2* and *lep-5* as key regulators of male tail morphogenesis, acting upstream of *lin-28* (Herrera, Kiontke, & Fitch, 2016; Kiontke et al., 2019; Figure 14.3D). *lep-5* encodes a lncRNA, and *lep-2* encodes an ortholog of MKRN2 that has predicted metal ion binding activity. *lep-2* and *lep-5* deficient animals displayed alterations in male-specific nervous system genes, with animals retaining an early larval expression pattern into adulthood. Epistatic analysis revealed that *lin-28* acts downstream of *lep-2*, with GFP-LIN-28 showing precocious expression in *lep-5* mutant L3 animals, whereas its expression was absent in control animals at this stage. Similarly, *let-7*, an established miRNA target of *lin-28*, acts downstream of *lin-28* and *lep-2/lep-5* in the nervous system. This study provides evidence that lncRNA and miRNA can act in concert to regulate sexual dimorphic changes in the nervous system.

lincRNA Interacts with miRNA and Histone Modification to Regulate Locomotion and Vesicle Trafficking

Long intergenic coding RNA (lincRNA) is a specific class of lncRNA, as lincRNA exist independently in the genome, whereas IncRNA are generally present in the coding region of a protein coding gene (Ransohoff, Wei, & Khavari, 2018). A recent study reported the analysis of genetic mutants of 155 lincRNA, generated by CRISPR/Cas9 mediated gene editing (Wei et al., 2019). These lincRNA were chosen due to their prediction as a lincRNA and analysis of their histone-modifications by chromatin-immunoprecipitation followed by sequencing to indicate their expression. 23 lincRNA mutants showed changes in overt phenotypes, of which 6 exhibited uncoordinated movement and 5 with pharyngeal pumping defects, two phenotypes with strong neuronal components (Wei et al., 2019). Many lincRNAs have been shown to act by blocking the binding of miRNA to their target sequence within an mRNA (Ransohoff et al., 2018). Loss of *linc-60* results in an Unc phenotype and *linc-60* contains a predicted seed sequence for *mir-5550*. Overexpressing *mir-5550* diminished levels of *linc-60*, compared to control. Loss of *linc-73* also causes a specific uncoordinated phenotype reminiscent of mutants in the *unc-55* COUP/TF nuclear hormone receptor (Petersen et al., 2011; Walthall & Plunkett, 1995), and expression of *linc-73* was decreased in *unc-55* mutants (Figure 14.3E). In *linc-73* mutants, expression of the *C. elegans* kinesin gene, *unc-104*, was significantly reduced. Consistent with the role of *unc-104* in vesicle trafficking, *linc-73* mutants displayed decreases in motor neuron synaptic vesicles. *linc-73* mutants also displayed a decrease in the repressive histone modification H3K4me2 and an increase in the activation mark H3K9me3 at the *unc-104* genomic locus (Figure 14.3E). Mutating binding sites for UNC-55 in the *linc-73* promoter resulted in increased expression of *unc-104*. This study establishes a model whereby *linc-73* modulates the epigenetic status of *unc-104*, downstream of its transcriptional regulation by UNC-55.

RNA-Mediated Signaling in Neuronal Injury and Axon Regeneration

The development of neuronal processes can occur under two different contexts: during normal organismal growth or following damage to an established neuron. Though these two processes share several common regulatory mechanisms, they are largely governed by disparate factors (Hilton & Bradke, 2017). Axon injury therefore represents a novel stress paradigm to dissect the role of proteins in a unique form of neuronal growth and function.

The RNA Ligase RTCB-1 Inhibits Axon Regeneration

An ancient RNA repair and splicing pathway consists of RtCA (an RNA 3′ terminal phosphate cyclase), RtcB (an RNA ligase) and Archease (a chaperone; Table 14.1). In eukaryotes, this pathway is shown to mediate mRNA splicing of the cellular stress sensor XBP1 (Calfon et al., 2002). The processing of *Xbp1* mRNA then activates the unfolded protein response (Englert, Sheppard, Aslanian, Yates, & Soll, 2011; Jurkin et al., 2014; Kosmaczewski et al., 2014; Popow et al., 2011). The RtCB protein is the only identified metazoan RNA ligase and regulates the maturation of intron containing tRNAs. In *C. elegans*, following axon injury in GABAergic motor neurons, animals with a loss of function mutation in *rtcb-1* display enhanced regenerative capacity (Kosmaczewski et al., 2015; Figure 14.2). However, RTCB-1's function in axon regeneration appears to be independent of its role in tRNA ligation, as several other genes essential for this function did not display a regeneration phenotype. Moreover, RTCB-1 function also occurs independently of its role in regulating *xbp-1* processing, as neither loss nor constitutive activation of XBP-1 modulated its regeneration function, and loss of *rtcb-1* did not impact the induction of the unfolded protein response following tunicamycin treatment. Nonetheless, we note that independent studies from Drosophila uncovered roles for RtCA in inhibiting axon regeneration, however; this was shown to occur through both ligase and *Xbp1* processing (Song et al., 2015). This study highlighted the importance of non-canonical RTCB-1 activity in axon regeneration and further studies will likely identify new mechanisms in neuronal stress response.

RNA Granule-Associated Genes Have Distinct Impacts on Axon Regeneration

mRNAs are dynamic molecules, with their regulation occurring at the level of transcription, stability, and localization. RNA granules are mRNA/protein complexes which regulate multiple facets of mRNA dynamics under both normal and stress conditions (Guzikowski, Chen, & Zid, 2019). The most well characterized class of RNA granules in *C. elegans*, the P-granule, exist in the germline and are thought to partition mRNA during cell division and help aid in establishing a germline specific transcriptional program (Seydoux, 2018). Recent evidence has suggested a role for RNA binding proteins associated with two other classes of RNA granule, processing bodies and stress granules, as important for axon growth following injury.

TIAR-2 is one of two *C. elegans* orthologs of the mammalian TIA family and functions as an inhibitor of axon regrowth (Andrusiak et al., 2019; Table 14.1). TIA family proteins have been shown to be core components of stress granules (Kedersha, Gupta, Li, Miller, & Anderson, 1999). Loss of *tiar-2* results in a significant increase in regeneration capacity following laser-induced axon injury (Andrusiak et al., 2019). Axons of *tiar-2* mutant animals develop normally, suggesting this role is specific to regrowth. Following axon injury, TIAR-2 rapidly forms structures reminiscent of RNA granules. TIAR-2 function in regeneration is dependent upon a C-terminal prion-like domain (PrLD), a class of intrinsically disordered domain characterized by its aggregation propensity. The RNA-binding regions of TIAR-2 are not required for its function in regeneration, suggesting that its role is either negatively regulated by RNA, or may interact with RNA through other RNA-binding proteins within the granule. Finally, TIAR-2 undergoes liquid-liquid phase separation, a process utilized in the formation of non-membrane bound cellular compartments (Figure 14.2). This study provides comprehensive insight into the functional role of the stress granule related protein TIAR-2 and liquid-liquid phase separation in the context of axon regeneration.

The mRNA decay factor CAR-1/LSM14 also acts as an inhibitor of axon regrowth (Tang et al., 2020). LSM14 and CAR-1 are components of RNA granules known as processing bodies, which are multi-protein complexes which include mRNA decapping enzymes (DCP) and the RNA helicase DDX6 (Boag, Nakamura, & Blackwell, 2005; Brandmann et al., 2018). Examination of genetic null mutants for this complex revealed differential effects on axon regeneration such that loss of *cgh-1*, the orthologue of DDX6, results in an increase in regrowth, whereas loss of *dcap/DCP* genes reduced axon regrowth. Analysis of mRNA targets of CAR-1 by enhanced crosslinking and immunoprecipitation (eCLIP) revealed binding of the mitochondrial transporter subunit MICU-1 (Figure 14.2). Calcium homeostasis plays an important role following axon injury (Ding & Hammarlund, 2018; Ghosh-Roy, Wu, Goncharov, Jin, & Chisholm, 2010; Kulbatski, Cook, & Tator, 2004; L. Sun et al., 2014; Tedeschi et al., 2016; D. Yan & Jin, 2012), however, its role in the mitochondria has not been defined. Using mitochondrial directed calcium reporters revealed a significant increase in Ca^{2+} in *car-1* mutant animals, which is restored in the absence of *micu-1*. Similarly, loss of *micu-1* in the background of *car-1* mutants is able to restore control level regrowth. This study suggests a model that CAR-1 regulates injury induced regrowth by regulating the balance of mitochondrial calcium through MICU-1.

miRNA-Dependent Regulation of Axon Regeneration

A functional role for miRNAs in axon regeneration was suggested by the observation that loss of the Argonaut gene *alg-1* reduced axon regeneration (L. Chen et al., 2011). Subsequent studies have elucidated different signaling pathways regulated by *let-7* in neuron-type and stage-dependent manner. Regeneration of the AVM axon following laser injury declines as animals age from young larvae to adults (Zou et al., 2013). *let-7* was shown to inhibit AVM axon regeneration in older adults, by down-regulating *lin-41*, an ortholog of the translational regulator Trim71 (Table 14.1). In PLM axon regeneration following injury, some axons undergo fusion and restore function (Abay et al., 2017; Basu et al., 2017; Ghosh-Roy et al., 2010). This injury-induced axon fusion requires the fusogen protein EFF-1 and phosphatidylserine-dependent signaling (Neumann et al., 2015). Loss of *let-7* results in a higher frequency of axon fusion in both larvae and adults (Basu et al., 2017). This effect of *let-7* is mediated through regulation of CED-7, an ortholog of human ABCA3. Importantly, evidence has emerged that the *let-7* family of miRNAs also regulate mammalian peripheral nerve regeneration via regulation of nerve growth factors and axon guidance molecules (Li et al., 2015; X. Wang et al., 2019).

piRNA-Mediated Regulation of Axon Regeneration

piRNA are a class of non-coding RNA involved in the regulation of gene expression (Ozata, Gainetdinov, Zoch, O'Carroll, & Zamore, 2019). The expression and impact of piRNA was originally thought to be confined to the germline, however, roles in somatic tissues are now being described (S. Huang et al., 2019; Jones et al., 2016; Perera et al., 2019; Z. Yan et al., 2011). Loss of critical regulators of piRNA activity, including *prde-1* and *prg-1*, result in an increase in regeneration (K. W. Kim et al., 2018; Table 14.1; Figure 14.2). These phenotypes could be rescued by neuronal specific expression of PRDE-1, and PRDE-1 mRNA was detected in neurons via single molecule fluorescent in situ hybridization. In addition, the role following injury was independent of the germline, as its ablation in early development had no impact on its adult axon regeneration capacity. This study identified an important role for piRNA in a somatic tissue following the acute stress of axon injury. Additional evidence from other species points to a significant role of piRNA in both normal and stress-related neuronal processes (Phay, Kim, & Yoo, 2018; Rajasethupathy et al., 2012; Sohn, Jo, & Park, 2019; Zhao et al., 2015). Further dissection of specific transcripts regulated by piRNA during axon regeneration will provide great insight into the role of piRNA in transcriptional regulation in a somatic cell type.

Conclusion

The regulation of nervous system development and function is dependent on balancing the levels and locations of mRNA. To this end, RNA-binding proteins such as MEC-8 and UNC-75 have been extensively studied in *C. elegans*, with multiple aspects of nervous system activity dependent on their ability to alternatively splice transcripts such as *mec-2* and *unc-64* (Calixto et al., 2010; L. Chen et al., 2016; Norris et al., 2014). Additionally, isoform diversity generated by alternative polyadenylation specifically impacts UNC-44, with significant consequences on nervous system development and maintenance (F. Chen et al., 2015; LaBella et al., 2020). The regulation of transcript levels by non-coding RNA plays a pervasive role throughout the nervous system ranging from developmental regulation of synapse plasticity via *mir-84* (Thompson-Peer et al., 2012) or male-specific nervous system establishment by the lncRNA *lep-5* (Lawson, et al. 2019). The use of axon regeneration as a novel stress paradigm has identified roles for RNA granule proteins such as CAR-1 and TIAR-2 (Andrusiak et al., 2019; Tang et al., 2020), and a somatic role for piRNAs (K. W. Kim et al., 2018). Taken together, the shaping of the nervous system requires a complex orchestration of gene expression, with a significant component occurring post-transcriptionally at the level of splicing and abundance.

Future Perspectives

The use of *C. elegans* as a model organism has been firmly rooted in its versatility in performing unbiased genetic screens. The saturation of existing phenotypes has necessitated the exploration of new means to identify gene function. Importantly, the evolving role of mRNA and improved in vivo visualization techniques will provide additional platforms to identify mRNA-centric genes and processes important in synaptic function and beyond. The identification and characterization of novel RNA-associated genes in neuronal development and function in *C. elegans* further enhances the illustrious portfolio created using this powerful invertebrate model.

Acknowledgments

Work in the Jin lab is supported by NIH grants (NS R37 035536 and NS R01 093588). MGA was an CIHR fellow; and Y. J. is a holder of Junior Seau Foundation Endowed Chair in Traumatic Brain Injury. We thank E. Jen Jin, Steve Blazie for helpful comments, as well as editorial input on the manuscript.

References

Abay, Z. C., Wong, M. Y., Teoh, J. S., Vijayaraghavan, T., Hilliard, M. A., & Neumann, B. (2017). Phosphatidylserine save-me signals drive functional recovery of severed axons in

Caenorhabditis elegans. *Proceedings of the National Academy of Sciences of the United States of America, 114*(47), E10196-E10205. https://doi.org/10.1073/pnas.1703807114

Abrahante, J. E., Daul, A. L., Li, M., Volk, M. L., Tennessen, J. M., Miller, E. A., & Rougvie, A. E. (2003). The Caenorhabditis elegans hunchback-like gene lin-57/hbl-1 controls developmental time and is regulated by microRNAs. *Developmental Cell, 4*(5), 625–637. https://doi.org/10.1016/s1534-5807(03)00127-8 [pii]

Alfonso, A., Grundahl, K., Duerr, J. S., Han, H. P., & Rand, J. B. (1993). The Caenorhabditis elegans unc-17 gene: a putative vesicular acetylcholine transporter. *Science, 261*(5121), 617–619. https://doi.org/10.1126/science.8342028

Ambros, V., & Horvitz, H. R. (1984). Heterochronic mutants of the nematode Caenorhabditis elegans. *Science, 226*(4673), 409–416. https://doi.org/10.1126/science.6494891

Andrusiak, M. G., Sharifnia, P., Lyu, X., Wang, Z., Dickey, A. M., Wu, Z., . . . Jin, Y. (2019). Inhibition of axon regeneration by liquid-like TIAR-2 granules. *Neuron, 104*(2), 290–304 e298. https://doi.org/10.1016/j.neuron.2019.07.004 [pii]

Armakola, M., & Ruvkun, G. (2019). Regulation of Caenorhabditis elegans neuronal polarity by heterochronic genes. *Proceedings of the National Academy of Sciences of the United States of America, 116*(25), 12327–12336. https://doi.org/10.1073/pnas.1820928116

Barr, M. M., & Garcia, L. R. (2006). Male mating behavior. *WormBook*, 1–11. https://doi.org/10.1895/wormbook.1.78.1

Basu, A., Dey, S., Puri, D., Das Saha, N., Sabharwal, V., Thyagarajan, P., . . . Ghosh-Roy, A. (2017). let-7 miRNA controls CED-7 homotypic adhesion and EFF-1-mediated axonal self-fusion to restore touch sensation following injury. *Proceedings of the National Academy of Sciences of the United States of America, 114*(47), E10206–E10215. https://doi.org/10.1073/pnas.1704372114

Batista, P. J., & Chang, H. Y. (2013). Long noncoding RNAs: cellular address codes in development and disease. *Cell, 152*(6), 1298–1307. https://doi.org/10.1016/j.cell.2013.02.012

Boag, P. R., Nakamura, A., & Blackwell, T. K. (2005). A conserved RNA-protein complex component involved in physiological germline apoptosis regulation in C. elegans. *Development, 132*(22), 4975–4986. https://doi.org/10.1242/dev.02060 [pii]

Brandmann, T., Fakim, H., Padamsi, Z., Youn, J. Y., Gingras, A. C., Fabian, M. R., & Jinek, M. (2018). Molecular architecture of LSM14 interactions involved in the assembly of mRNA silencing complexes. *EMBO Journal, 37*(7). https://doi.org/10.15252/embj.201797869 [pii]

Brenner, S. (1974). The genetics of Caenorhabditis elegans. *Genetics, 77*(1), 71–94.

Briese, M., Esmaeili, B., Fraboulet, S., Burt, E. C., Christodoulou, S., Towers, P. R., . . . Sattelle, D. B. (2009). Deletion of smn-1, the Caenorhabditis elegans ortholog of the spinal muscular atrophy gene, results in locomotor dysfunction and reduced lifespan. *Human Molecular Genetics, 18*(1), 97–104. https://doi.org/10.1093/hmg/ddn320

Brockdorff, N., Ashworth, A., Kay, G. F., McCabe, V. M., Norris, D. P., Cooper, P. J., . . . Rastan, S. (1992). The product of the mouse Xist gene is a 15 kb inactive X-specific transcript containing no conserved ORF and located in the nucleus. *Cell, 71*(3), 515–526. https://doi.org/10.1016/0092-8674(92)90519-i [pii]

Calfon, M., Zeng, H., Urano, F., Till, J. H., Hubbard, S. R., Harding, H. P., . . . Ron, D. (2002). IRE1 couples endoplasmic reticulum load to secretory capacity by processing the XBP-1 mRNA. *Nature, 415*(6867), 92–96. https://doi.org/10.1038/415092a

Calixto, A., Ma, C., & Chalfie, M. (2010). Conditional gene expression and RNAi using MEC-8-dependent splicing in C. elegans. *Nature Methods, 7*(5), 407–411. https://doi.org/10.1038/nmeth.1445

Chalfie, M., & Au, M. (1989). Genetic control of differentiation of the Caenorhabditis elegans touch receptor neurons. *Science, 243*(4894 Pt 1), 1027–1033. https://doi.org/10.1126/science.2646709

Chalfie, M., & Sulston, J. (1981). Developmental genetics of the mechanosensory neurons of Caenorhabditis elegans. *Developmental Biology, 82*(2), 358–370. https://doi.org/10.1016/0012-1606(81)90459-0 [pii]

Chalfie, M., Sulston, J. E., White, J. G., Southgate, E., Thomson, J. N., & Brenner, S. (1985). The neural circuit for touch sensitivity in Caenorhabditis elegans. *Journal of Neuroscience, 5*(4), 956–964.

Chen, F., Zhou, Y., Qi, Y. B., Khivansara, V., Li, H., Chun, S. Y., . . . Jin, Y. (2015). Context-dependent modulation of Pol II CTD phosphatase SSUP-72 regulates alternative polyadenylation in neuronal development. *Genes & Development, 29*(22), 2377–2390. https://doi.org/10.1101/gad.266650.115

Chen, L., Liu, Z., Zhou, B., Wei, C., Zhou, Y., Rosenfeld, M. G., . . . Jin, Y. (2016). CELF RNA binding proteins promote axon regeneration in C. elegans and mammals through alternative splicing of Syntaxins. *Elife, 5*. https://doi.org/10.7554/eLife.16072

Chen, L., Wang, Z., Ghosh-Roy, A., Hubert, T., Yan, D., O'Rourke, S., . . . Chisholm, A. D. (2011). Axon regeneration pathways identified by systematic genetic screening in C. elegans. *Neuron, 71*(6), 1043–1057. https://doi.org/10.1016/j.neuron.2011.07.009

Cook, S. J., Jarrell, T. A., Brittin, C. A., Wang, Y., Bloniarz, A. E., Yakovlev, M. A., . . . Emmons, S. W. (2019). Whole-animal connectomes of both Caenorhabditis elegans sexes. *Nature, 571*(7763), 63–71. https://doi.org/10.1038/s41586-019-1352-7

Crump, J. G., Zhen, M., Jin, Y., & Bargmann, C. I. (2001). The SAD-1 kinase regulates presynaptic vesicle clustering and axon termination. *Neuron, 29*(1), 115–129. https://doi.org/10.1016/s0896-6273(01)00184-2 [pii]

Dasgupta, T., & Ladd, A. N. (2012). The importance of CELF control: molecular and biological roles of the CUG-BP, Elav-like family of RNA-binding proteins. *Wiley Interdisciplinary Reviews RNA, 3*(1), 104–121. https://doi.org/10.1002/wrna.107

Dimitriadi, M., Derdowski, A., Kalloo, G., Maginnis, M. S., O'Hern, P., Bliska, B., . . . Hart, A. C. (2016). Decreased function of survival motor neuron protein impairs endocytic pathways. *Proceedings of the National Academy of Sciences of the United States of America, 113*(30), E4377-4386. https://doi.org/10.1073/pnas.1600015113

Ding, C., & Hammarlund, M. (2018). Aberrant information transfer interferes with functional axon regeneration. *eLife, 7*. https://doi.org/10.7554/eLife.38829

Emmons, S. W. (2005). Male development. *WormBook*, 1–22. https://doi.org/10.1895/wormbook.1.33.1

Englert, M., Sheppard, K., Aslanian, A., Yates, J. R., 3rd, & Soll, D. (2011). Archaeal 3′-phosphate RNA splicing ligase characterization identifies the missing component in tRNA maturation. *Proceedings of the National Academy of Sciences of the United States of America, 108*(4), 1290–1295. https://doi.org/10.1073/pnas.1018307108

Eom, T., Muslimov, I. A., Tsokas, P., Berardi, V., Zhong, J., Sacktor, T. C., & Tiedge, H. (2014). Neuronal BC RNAs cooperate with eIF4B to mediate activity-dependent translational control. *Journal of Cell Biology, 207*(2), 237–252. https://doi.org/10.1083/jcb.201401005

Farazi, T. A., Leonhardt, C. S., Mukherjee, N., Mihailovic, A., Li, S., Max, K. E., . . . Tuschl, T. (2014). Identification of the RNA recognition element of the RBPMS family of RNA-binding proteins and their transcriptome-wide mRNA targets. *RNA, 20*(7), 1090–1102. https://doi.org/10.1261/rna.045005.114

Feng, J., Schroer, R., Yan, J., Song, W., Yang, C., Bockholt, A., ... Sommer, S. S. (2006). High frequency of neurexin 1beta signal peptide structural variants in patients with autism. *Neuroscience Letters, 409*(1), 10–13. https://doi.org/10.1016/j.neulet.2006.08.017 [pii]

Finney, M., Ruvkun, G., & Horvitz, H. R. (1988). The C. elegans cell lineage and differentiation gene unc-86 encodes a protein with a homeodomain and extended similarity to transcription factors. *Cell, 55*(5), 757–769. https://doi.org/10.1016/0092-8674(88)90132-8 [pii]

Flavell, S. W., Cowan, C. W., Kim, T. K., Greer, P. L., Lin, Y., Paradis, S., ... Greenberg, M. E. (2006). Activity-dependent regulation of MEF2 transcription factors suppresses excitatory synapse number. *Science, 311*(5763), 1008–1012. https://doi.org/10.1126/science.1122511 [pii]

Fujita, M., Hawkinson, D., King, K. V., Hall, D. H., Sakamoto, H., & Buechner, M. (2003). The role of the ELAV homologue EXC-7 in the development of the Caenorhabditis elegans excretory canals. *Developmental Biology, 256*(2), 290–301. https://doi.org/10.1016/s0012-1606(03)00040-x [pii]

Gao, X., Teng, Y., Luo, J., Huang, L., Li, M., Zhang, Z., ... Ma, L. (2014). The survival motor neuron gene smn-1 interacts with the U2AF large subunit gene uaf-1 to regulate Caenorhabditis elegans lifespan and motor functions. *RNA Biology, 11*(9), 1148–1160. https://doi.org/10.4161/rna.36100

Gao, X., Xu, J., Chen, H., Xue, D., Pan, W., Zhou, C., ... Ma, L. (2019). Defective expression of mitochondrial, vacuolar H(+)-ATPase and histone genes in a C. elegans model of SMA. *Frontiers in Genetics, 10*, 410. https://doi.org/10.3389/fgene.2019.00410

Gebert, L. F. R., & MacRae, I. J. (2019). Regulation of microRNA function in animals. *Nature Reviews Molecular Cell Biology, 20*(1), 21–37. https://doi.org/10.1038/s41580-018-0045-7

Ghosh-Roy, A., Wu, Z., Goncharov, A., Jin, Y., & Chisholm, A. D. (2010). Calcium and cyclic AMP promote axonal regeneration in Caenorhabditis elegans and require DLK-1 kinase. *Journal of Neuroscience, 30*(9), 3175–3183. https://doi.org/10.1523/JNEUROSCI.5464-09.2010

Gong, C., & Maquat, L. E. (2011). lncRNAs transactivate STAU1-mediated mRNA decay by duplexing with 3′ UTRs via Alu elements. *Nature, 470*(7333), 284–288. https://doi.org/10.1038/nature09701

Grill, B., Bienvenut, W. V., Brown, H. M., Ackley, B. D., Quadroni, M., & Jin, Y. (2007). C. elegans RPM-1 regulates axon termination and synaptogenesis through the Rab GEF GLO-4 and the Rab GTPase GLO-1. *Neuron, 55*(4), 587–601. https://doi.org/10.1016/j.neuron.2007.07.009 [pii]

Gruber, A. J., & Zavolan, M. (2019). Alternative cleavage and polyadenylation in health and disease. *Nature Reviews Genetics, 20*(10), 599–614. https://doi.org/10.1038/s41576-019-0145-z

Guzikowski, A. R., Chen, Y. S., & Zid, B. M. (2019). Stress-induced mRNP granules: form and function of processing bodies and stress granules. *Wiley Interdisciplinary Reviews RNA, 10*(3), e1524. https://doi.org/10.1002/wrna.1524

Hallam, S. J., Goncharov, A., McEwen, J., Baran, R., & Jin, Y. (2002). SYD-1, a presynaptic protein with PDZ, C2 and rhoGAP-like domains, specifies axon identity in C. elegans. *Nature Neuroscience, 5*(11), 1137–1146. https://doi.org/10.1038/nn959

Hallam, S. J., & Jin, Y. (1998). lin-14 regulates the timing of synaptic remodelling in Caenorhabditis elegans. *Nature, 395*(6697), 78–82. https://doi.org/10.1038/25757

Harlen, K. M., & Churchman, L. S. (2017). The code and beyond: transcription regulation by the RNA polymerase II carboxy-terminal domain. *Nature Reviews Molecular Cell Biology, 18*(4), 263–273. https://doi.org/10.1038/nrm.2017.10

Harris, T. W., Hartwieg, E., Horvitz, H. R., & Jorgensen, E. M. (2000). Mutations in synaptojanin disrupt synaptic vesicle recycling. *Journal of Cell Biology, 150*(3), 589–600. https://doi.org/10.1083/jcb.150.3.589

Hegele, A., Kamburov, A., Grossmann, A., Sourlis, C., Wowro, S., Weimann, M.,... Stelzl, U. (2012). Dynamic protein-protein interaction wiring of the human spliceosome. *Molecular Cell, 45*(4), 567–580. https://doi.org/10.1016/j.molcel.2011.12.034

Herman, R. K., Kari, C. K., & Hartman, P. S. (1982). Dominant X-chromosome nondisjunction mutants of Caenorhabditis elegans. *Genetics, 102*(3), 379–400.

Hermann, G. J., Schroeder, L. K., Hieb, C. A., Kershner, A. M., Rabbitts, B. M., Fonarev, P... Priess, J. R. (2005). Genetic analysis of lysosomal trafficking in Caenorhabditis elegans. *Molecular Biology of the Cell, 16*(7), 3273–3288. https://doi.org/10.1091/mbc.e05-01-0060 [pii]

Herrera, R. A., Kiontke, K., & Fitch, D. H. (2016). Makorin ortholog LEP-2 regulates LIN-28 stability to promote the juvenile-to-adult transition in Caenorhabditis elegans. *Development, 143*(5), 799–809. https://doi.org/10.1242/dev.132738

Hilton, B. J., & Bradke, F. (2017). Can injured adult CNS axons regenerate by recapitulating development? *Development, 144*(19), 3417–3429. https://doi.org/10.1242/dev.148312

Hock, J., Weinmann, L., Ender, C., Rudel, S., Kremmer, E., Raabe, M.,... Meister, G. (2007). Proteomic and functional analysis of Argonaute-containing mRNA-protein complexes in human cells. *EMBO Reports, 8*(11), 1052–1060. https://doi.org/10.1038/sj.embor.7401088 [pii]

Hodgkin, J., Horvitz, H. R., & Brenner, S. (1979). Nondisjunction mutants of the nematode Caenorhabditis elegans. *Genetics, 91*(1), 67–94.

Horn, M., Geisen, C., Cermak, L., Becker, B., Nakamura, S., Klein, C.,... Antebi, A. (2014). DRE-1/FBXO11-dependent degradation of BLMP-1/BLIMP-1 governs C. elegans developmental timing and maturation. *Developmental Cell, 28*(6), 697–710. https://doi.org/10.1016/j.devcel.2014.01.028

Hosono, R., Hekimi, S., Kamiya, Y., Sassa, T., Murakami, S., Nishiwaki, K.,... Kodaira, K. I. (1992). The unc-18 gene encodes a novel protein affecting the kinetics of acetylcholine metabolism in the nematode Caenorhabditis elegans. *Journal of Neurochemistry, 58*(4), 1517–1525. https://doi.org/10.1111/j.1471-4159.1992.tb11373.x

Hosono, R., & Kamiya, Y. (1991). Additional genes which result in an elevation of acetylcholine levels by mutations in Caenorhabditis elegans. *Neuroscience Letters, 128*(2), 243–244. https://doi.org/10.1016/0304-3940(91)90270-4 [pii]

Hu, Z., Hom, S., Kudze, T., Tong, X. J., Choi, S., Aramuni, G.,... Kaplan, J. M. (2012). Neurexin and neuroligin mediate retrograde synaptic inhibition in C. elegans. *Science, 337*(6097), 980–984. https://doi.org/10.1126/science.1224896

Hua, Y., & Zhou, J. (2004). Survival motor neuron protein facilitates assembly of stress granules. *FEBS Letters, 572*(1-3), 69–74. https://doi.org/10.1016/j.febslet.2004.07.010

Huang, M., Gu, G., Ferguson, E. L., & Chalfie, M. (1995). A stomatin-like protein necessary for mechanosensation in C. elegans. *Nature, 378*(6554), 292–295. https://doi.org/10.1038/378292a0

Huang, S., Ichikawa, Y., Igarashi, Y., Yoshitake, K., Kinoshita, S., Omori, F.,... Asakawa, S. (2019). Piwi-interacting RNA (piRNA) expression patterns in pearl oyster (Pinctada fucata) somatic tissues. *Science Reports, 9*(1), 247. https://doi.org/10.1038/s41598-018-36726-0

Hung, W., Hwang, C., Po, M. D., & Zhen, M. (2007). Neuronal polarity is regulated by a direct interaction between a scaffolding protein, Neurabin, and a presynaptic SAD-1 kinase in Caenorhabditis elegans. *Development, 134*(2), 237–249. https://doi.org/10.1242/dev.02725 [pii]

Hutvagner, G., & Zamore, P. D. (2002). A microRNA in a multiple-turnover RNAi enzyme complex. *Science, 297*(5589), 2056–2060. https://doi.org/10.1126/science.1073827

Ibanez-Ventoso, C., Yang, M., Guo, S., Robins, H., Padgett, R. W., & Driscoll, M. (2006). Modulated microRNA expression during adult lifespan in Caenorhabditis elegans. *Aging Cell, 5*(3), 235–246. https://doi.org/10.1111/j.1474-9726.2006.00210.x [pii]

Ichtchenko, K., Hata, Y., Nguyen, T., Ullrich, B., Missler, M., Moomaw, C., & Sudhof, T. C. (1995). Neuroligin 1: a splice site-specific ligand for beta-neurexins. *Cell, 81*(3), 435–443. https://doi.org/10.1016/0092-8674(95)90396-8 [pii]

Inoue, E., Mochida, S., Takagi, H., Higa, S., Deguchi-Tawarada, M., Takao-Rikitsu, E., . . . Takai, Y. (2006). SAD: a presynaptic kinase associated with synaptic vesicles and the active zone cytomatrix that regulates neurotransmitter release. *Neuron, 50*(2), 261–275. https://doi.org/10.1016/j.neuron.2006.03.018 [pii]

Jaffrey, S. R., & Wilkinson, M. F. (2018). Nonsense-mediated RNA decay in the brain: emerging modulator of neural development and disease. *Nature Reviews Neuroscience, 19*(12), 715–728. https://doi.org/10.1038/s41583-018-0079-z

Jamain, S., Quach, H., Betancur, C., Rastam, M., Colineaux, C., Gillberg, I. C., . . . Bourgeron, T. (2003). Mutations of the X-linked genes encoding neuroligins NLGN3 and NLGN4 are associated with autism. *Nature Genetics, 34*(1), 27–29. https://doi.org/10.1038/ng1136

Jones, B. C., Wood, J. G., Chang, C., Tam, A. D., Franklin, M. J., Siegel, E. R., & Helfand, S. L. (2016). A somatic piRNA pathway in the Drosophila fat body ensures metabolic homeostasis and normal lifespan. *Nature Communications, 7*, 13856. https://doi.org/10.1038/ncomms13856

Jorgensen, E. M., & Mango, S. E. (2002). The art and design of genetic screens: Caenorhabditis elegans. *Nature Reviews Genetics, 3*(5), 356–369. https://doi.org/10.1038/nrg794

Jospin, M., Qi, Y. B., Stawicki, T. M., Boulin, T., Schuske, K. R., Horvitz, H. R., . . . Jin, Y. (2009). A neuronal acetylcholine receptor regulates the balance of muscle excitation and inhibition in Caenorhabditis elegans. *PLoS Biol, 7*(12), e1000265. https://doi.org/10.1371/journal.pbio.1000265

Jung, H., & Holt, C. E. (2011). Local translation of mRNAs in neural development. *Wiley Interdisciplinary Reviews RNA, 2*(1), 153–165. https://doi.org/10.1002/wrna.53

Jurkin, J., Henkel, T., Nielsen, A. F., Minnich, M., Popow, J., Kaufmann, T., . . . Martinez, J. (2014). The mammalian tRNA ligase complex mediates splicing of XBP1 mRNA and controls antibody secretion in plasma cells. *EMBO Journal, 33*(24), 2922–2936. https://doi.org/10.15252/embj.201490332

Kabashi, E., Valdmanis, P. N., Dion, P., Spiegelman, D., McConkey, B. J., Vande Velde, C., . . . Rouleau, G. A. (2008). TARDBP mutations in individuals with sporadic and familial amyotrophic lateral sclerosis. *Nature Genetics, 40*(5), 572–574. https://doi.org/10.1038/ng.132

Kapustin, Y., Chan, E., Sarkar, R., Wong, F., Vorechovsky, I., Winston, R. M., . . . Dibb, N. J. (2011). Cryptic splice sites and split genes. *Nucleic Acids Research, 39*(14), 5837–5844. https://doi.org/10.1093/nar/gkr203

Kedersha, N. L., Gupta, M., Li, W., Miller, I., & Anderson, P. (1999). RNA-binding proteins TIA-1 and TIAR link the phosphorylation of eIF-2 alpha to the assembly of mammalian stress granules. *Journal of Cell Biology, 147*(7), 1431–1442. https://doi.org/10.1083/jcb.147.7.1431

Kim, K. W., Tang, N. H., Andrusiak, M. G., Wu, Z., Chisholm, A. D., & Jin, Y. (2018). A neuronal piRNA pathway inhibits axon regeneration in C. elegans. *Neuron, 97*(3), 511–519 e516. https://doi.org/10.1016/j.neuron.2018.01.014 [pii]

Kim, W., Underwood, R. S., Greenwald, I., & Shaye, D. D. (2018). OrthoList 2: A new comparative genomic analysis of human and Caenorhabditis elegans genes. *Genetics, 210*(2), 445–461. https://doi.org/10.1534/genetics.118.301307

Kiontke, K. C., Herrera, R. A., Vuong, E., Luo, J., Schwarz, E. M., Fitch, D. H. A., & Portman, D. S. (2019). The long non-coding RNA lep-5 promotes the juvenile-to-adult transition by destabilizing LIN-28. *Developmental Cell, 49*(4), 542–555 e549. https://doi.org/10.1016/j.devcel.2019.03.003 [pii]

Kosmaczewski, S. G., Edwards, T. J., Han, S. M., Eckwahl, M. J., Meyer, B. I., Peach, S., . . . Hammarlund, M. (2014). The RtcB RNA ligase is an essential component of the metazoan unfolded protein response. *EMBO Reports, 15*(12), 1278–1285. https://doi.org/10.15252/embr.201439531

Kosmaczewski, S. G., Han, S. M., Han, B., Irving Meyer, B., Baig, H. S., Athar, W., . . . Hammarlund, M. (2015). RNA ligation in neurons by RtcB inhibits axon regeneration. *Proceedings of the National Academy of Sciences of the United States of America, 112*(27), 8451–8456. https://doi.org/10.1073/pnas.1502948112

Krishnamurthy, S., He, X., Reyes-Reyes, M., Moore, C., & Hampsey, M. (2004). Ssu72 Is an RNA polymerase II CTD phosphatase. *Molecular Cell, 14*(3), 387–394. https://doi.org/10.1016/s1097-2765(04)00235-7 [pii]

Kulbatski, I., Cook, D. J., & Tator, C. H. (2004). Calcium entry through L-type calcium channels is essential for neurite regeneration in cultured sympathetic neurons. *Journal of Neurotrauma, 21*(3), 357–374. https://doi.org/10.1089/089771504322972130

Kuroyanagi, H., Watanabe, Y., & Hagiwara, M. (2013). CELF family RNA-binding protein UNC-75 regulates two sets of mutually exclusive exons of the unc-32 gene in neuron-specific manners in Caenorhabditis elegans. *PLoS Genetics, 9*(2), e1003337. https://doi.org/10.1371/journal.pgen.1003337

Kuroyanagi, H., Watanabe, Y., Suzuki, Y., & Hagiwara, M. (2013). Position-dependent and neuron-specific splicing regulation by the CELF family RNA-binding protein UNC-75 in Caenorhabditis elegans. *Nucleic Acids Research, 41*(7), 4015–4025. https://doi.org/10.1093/nar/gkt097

Kwiatkowski, T. J., Jr., Bosco, D. A., Leclerc, A. L., Tamrazian, E., Vanderburg, C. R., Russ, C., . . . Brown, R. H., Jr. (2009). Mutations in the FUS/TLS gene on chromosome 16 cause familial amyotrophic lateral sclerosis. *Science, 323*(5918), 1205–1208. https://doi.org/10.1126/science.1166066

Kye, M. J., Niederst, E. D., Wertz, M. H., Goncalves Ido, C., Akten, B., Dover, K. Z., . . . Sahin, M. (2014). SMN regulates axonal local translation via miR-183/mTOR pathway. *Human Molecular Genetics, 23*(23), 6318–6331. https://doi.org/10.1093/hmg/ddu350

LaBella, M. L., Hujber, E. J., Moore, K. A., Rawson, R. L., Merrill, S. A., Allaire, P. D., . . . Jorgensen, E. M. (2020). Casein kinase 1delta stabilizes mature axons by inhibiting transcription termination of ankyrin. *Developmental Cell, 52*(1), 88–103 e106. https://doi.org/10.1016/j.devcel.2019.12.005 [pii]

Lawson, H., Vuong, E., Miller, R. M., Kiontke, K., Fitch, D. H., & Portman, D. S. (2019). The Makorin lep-2 and the lncRNA lep-5 regulate lin-28 to schedule sexual maturation of the C. elegans nervous system. *eLife, 8*. https://doi.org/10.7554/eLife.43660

Lee, R. C., Feinbaum, R. L., & Ambros, V. (1993). The C. elegans heterochronic gene lin-4 encodes small RNAs with antisense complementarity to lin-14. *Cell, 75*(5), 843–854. https://doi.org/10.1016/0092-8674(93)90529-y [pii]

Lee, S., Kopp, F., Chang, T. C., Sataluri, A., Chen, B., Sivakumar, S., . . . Mendell, J. T. (2016). Noncoding RNA NORAD regulates genomic stability by sequestering PUMILIO proteins. *Cell, 164*(1-2), 69–80. https://doi.org/10.1016/j.cell.2015.12.017 [pii]

Lee, Y. S., Nakahara, K., Pham, J. W., Kim, K., He, Z., Sontheimer, E. J., & Carthew, R. W. (2004). Distinct roles for Drosophila Dicer-1 and Dicer-2 in the siRNA/miRNA silencing pathways. *Cell, 117*(1), 69–81. https://doi.org/10.1016/s0092-8674(04)00261-2 [pii]

Lewis, J. A., Wu, C. H., Berg, H., & Levine, J. H. (1980). The genetics of levamisole resistance in the nematode Caenorhabditis elegans. *Genetics, 95*(4), 905–928.

Li, S., Wang, X., Gu, Y., Chen, C., Wang, Y., Liu, J., . . . Gu, X. (2015). Let-7 microRNAs regenerate peripheral nerve regeneration by targeting nerve growth factor. *Molecular Therapy, 23*(3), 423–433. https://doi.org/10.1038/mt.2014.220

Liao, E. H., Hung, W., Abrams, B., & Zhen, M. (2004). An SCF-like ubiquitin ligase complex that controls presynaptic differentiation. *Nature, 430*(6997), 345–350. https://doi.org/10.1038/nature02647

Lin, S. Y., Johnson, S. M., Abraham, M., Vella, M. C., Pasquinelli, A., Gamberi, C., . . . Slack, F. J. (2003). The C. elegans hunchback homolog, hbl-1, controls temporal patterning and is a probable microRNA target. *Developmental Cell, 4*(5), 639–650. https://doi.org/10.1016/s1534-5807(03)00124-2 [pii]

Liu, Q., Fischer, U., Wang, F., & Dreyfuss, G. (1997). The spinal muscular atrophy disease gene product, SMN, and its associated protein SIP1 are in a complex with spliceosomal snRNP proteins. *Cell, 90*(6), 1013–1021. https://doi.org/10.1016/s0092-8674(00)80367-0 [pii]

Loria, P. M., Duke, A., Rand, J. B., & Hobert, O. (2003). Two neuronal, nuclear-localized RNA binding proteins involved in synaptic transmission. *Current Biology, 13*(15), 1317–1323. https://doi.org/10.1016/s0960-9822(03)00532-3 [pii]

Lundquist, E. A., & Herman, R. K. (1994). The mec-8 gene of Caenorhabditis elegans affects muscle and sensory neuron function and interacts with three other genes: unc-52, smu-1 and smu-2. *Genetics, 138*(1), 83–101.

Lundquist, E. A., Herman, R. K., Rogalski, T. M., Mullen, G. P., Moerman, D. G., & Shaw, J. E. (1996). The mec-8 gene of C. elegans encodes a protein with two RNA recognition motifs and regulates alternative splicing of unc-52 transcripts. *Development, 122*(5), 1601–1610.

Mackenzie, I. R., Nicholson, A. M., Sarkar, M., Messing, J., Purice, M. D., Pottier, C., . . . Rademakers, R. (2017). TIA1 mutations in amyotrophic lateral sclerosis and frontotemporal dementia promote phase separation and alter stress granule dynamics. *Neuron, 95*(4), 808–816 e809. https://doi.org/10.1016/j.neuron.2017.07.025 [pii]

Malgaroli, A., Vallar, L., & Zimarino, V. (2006). Protein homeostasis in neurons and its pathological alterations. *Current Opinion in Neurobiology, 16*(3), 270–274. https://doi.org/10.1016/j.conb.2006.05.009 [pii]

Martin, C. L., Duvall, J. A., Ilkin, Y., Simon, J. S., Arreaza, M. G., Wilkes, K., . . . Geschwind, D. H. (2007). Cytogenetic and molecular characterization of A2BP1/FOX1 as a candidate gene for autism. *American Journal of Medical Genetics B Neuropsychiatric Genetics, 144B*(7), 869–876. https://doi.org/10.1002/ajmg.b.30530

Maruyama, I. N., & Brenner, S. (1991). A phorbol ester/diacylglycerol-binding protein encoded by the unc-13 gene of Caenorhabditis elegans. *Proceedings of the National Academy of Sciences of the United States of America, 88*(13), 5729–5733. https://doi.org/10.1073/pnas.88.13.5729

Maslon, M. M., Braunschweig, U., Aitken, S., Mann, A. R., Kilanowski, F., Hunter, C. J., . . . Caceres, J. F. (2019). A slow transcription rate causes embryonic lethality and perturbs kinetic coupling of neuronal genes. *EMBO Journal, 38*(9). https://doi.org/10.15252/embj.2018101244 [pii]

Meng, L., Chen, C. H., & Yan, D. (2016). Regulation of Gap Junction Dynamics by UNC-44/ankyrin and UNC-33/CRMP through VAB-8 in C. elegans Neurons. *PLoS Genetics, 12*(3), e1005948. https://doi.org/10.1371/journal.pgen.1005948

Morgan, J. I., Cohen, D. R., Hempstead, J. L., & Curran, T. (1987). Mapping patterns of c-fos expression in the central nervous system after seizure. *Science, 237*(4811), 192–197. https://doi.org/10.1126/science.3037702

Murashov, A. K., Chintalgattu, V., Islamov, R. R., Lever, T. E., Pak, E. S., Sierpinski, P. L., . . . Van Scott, M. R. (2007). RNAi pathway is functional in peripheral nerve axons. *FASEB J, 21*(3), 656–670. https://doi.org/10.1096/fj.06-6155com [pii]

Nakata, K., Abrams, B., Grill, B., Goncharov, A., Huang, X., Chisholm, A. D., & Jin, Y. (2005). Regulation of a DLK-1 and p38 MAP kinase pathway by the ubiquitin ligase RPM-1 is required for presynaptic development. *Cell, 120*(3), 407–420. https://doi.org/10.1016/j.cell.2004.12.017 [pii]

Neumann, B., Coakley, S., Giordano-Santini, R., Linton, C., Lee, E. S., Nakagawa, A., . . . Hilliard, M. A. (2015). EFF-1-mediated regenerative axonal fusion requires components of the apoptotic pathway. *Nature, 517*(7533), 219–222. https://doi.org/10.1038/nature14102

Nguyen, M., Alfonso, A., Johnson, C. D., & Rand, J. B. (1995). Caenorhabditis elegans mutants resistant to inhibitors of acetylcholinesterase. *Genetics, 140*(2), 527–535.

Noma, K., Goncharov, A., & Jin, Y. (2014). Systematic analyses of rpm-1 suppressors reveal roles for ESS-2 in mRNA splicing in Caenorhabditis elegans. *Genetics, 198*(3), 1101–1115. https://doi.org/10.1534/genetics.114.167841

Nonet, M. L., Holgado, A. M., Brewer, F., Serpe, C. J., Norbeck, B. A., Holleran, J., . . . Alfonso, A. (1999). UNC-11, a Caenorhabditis elegans AP180 homologue, regulates the size and protein composition of synaptic vesicles. *Molecular Biology of the Cell, 10*(7), 2343–2360. https://doi.org/10.1091/mbc.10.7.2343

Norris, A. D., Gao, S., Norris, M. L., Ray, D., Ramani, A. K., Fraser, A. G., . . . Calarco, J. A. (2014). A pair of RNA-binding proteins controls networks of splicing events contributing to specialization of neural cell types. *Molecular Cell, 54*(6), 946–959. https://doi.org/10.1016/j.molcel.2014.05.004

Nussbacher, J. K., Tabet, R., Yeo, G. W., & Lagier-Tourenne, C. (2019). Disruption of RNA Metabolism in Neurological Diseases and Emerging Therapeutic Interventions. *Neuron, 102*(2), 294–320. https://doi.org/10.1016/j.neuron.2019.03.014 [pii]

O'Hern, P. J., do Carmo, G. Goncalves I., Brecht, J., Lopez Soto, E. J., Simon, J., Chapkis, N., . . . Hart, A. C. (2017). Decreased microRNA levels lead to deleterious increases in neuronal M2 muscarinic receptors in Spinal Muscular Atrophy models. *eLife, 6*. https://doi.org/10.7554/eLife.20752

Orengo, J. P., Bundman, D., & Cooper, T. A. (2006). A bichromatic fluorescent reporter for cell-based screens of alternative splicing. *Nucleic Acids Research, 34*(22), e148. https://doi.org/10.1093/nar/gkl967 [pii]

Otsuka, A. J., Franco, R., Yang, B., Shim, K. H., Tang, L. Z., Zhang, Y. Y., . . . et al. (1995). An ankyrin-related gene (unc-44) is necessary for proper axonal guidance in Caenorhabditis elegans. *Journal of Cell Biology, 129*(4), 1081–1092. https://doi.org/10.1083/jcb.129.4.1081

Ozata, D. M., Gainetdinov, I., Zoch, A., O'Carroll, D., & Zamore, P. D. (2019). PIWI-interacting RNAs: small RNAs with big functions. *Nature Reviews Genetics, 20*(2), 89–108. https://doi.org/10.1038/s41576-018-0073-3

Page, M. F., Carr, B., Anders, K. R., Grimson, A., & Anderson, P. (1999). SMG-2 is a phosphorylated protein required for mRNA surveillance in Caenorhabditis elegans and related to Upf1p of yeast. *Molecular Cell Biology, 19*(9), 5943–5951. https://doi.org/10.1128/mcb.19.9.5943

Patil, D. P., Chen, C. K., Pickering, B. F., Chow, A., Jackson, C., Guttman, M., & Jaffrey, S. R. (2016). m(6)A RNA methylation promotes XIST-mediated transcriptional repression. *Nature, 537*(7620), 369–373. https://doi.org/10.1038/nature19342

Pereira, L., Aeschimann, F., Wang, C., Lawson, H., Serrano-Saiz, E., Portman, D. S., . . . Hobert, O. (2019). Timing mechanism of sexually dimorphic nervous system differentiation. *eLife, 8*. https://doi.org/10.7554/eLife.42078

Perera, B. P. U., Tsai, Z. T., Colwell, M. L., Jones, T. R., Goodrich, J. M., Wang, K.,...Dolinoy, D. C. (2019). Somatic expression of piRNA and associated machinery in the mouse identifies short, tissue-specific piRNA. *Epigenetics, 14*(5), 504–521. https://doi.org/10.1080/15592294.2019.1600389

Petersen, S. C., Watson, J. D., Richmond, J. E., Sarov, M., Walthall, W. W., & Miller, D. M., 3rd. (2011). A transcriptional program promotes remodeling of GABAergic synapses in Caenorhabditis elegans. *Journal of Neuroscience, 31*(43), 15362–15375. https://doi.org/10.1523/JNEUROSCI.3181-11.2011

Pham, J. W., Pellino, J. L., Lee, Y. S., Carthew, R. W., & Sontheimer, E. J. (2004). A Dicer-2-dependent 80s complex cleaves targeted mRNAs during RNAi in Drosophila. *Cell, 117*(1), 83–94. https://doi.org/10.1016/s0092-8674(04)00258-2 [pii]

Phay, M., Kim, H. H., & Yoo, S. (2018). Analysis of piRNA-Like Small Non-coding RNAs Present in Axons of Adult Sensory Neurons. *Molecular Neurobiology, 55*(1), 483–494. https://doi.org/10.1007/s12035-016-0340-2

Pinan-Lucarre, B., Tu, H., Pierron, M., Cruceyra, P. I., Zhan, H., Stigloher, C.,...Bessereau, J. L. (2014). C. elegans Punctin specifies cholinergic versus GABAergic identity of postsynaptic domains. *Nature, 511*(7510), 466–470. https://doi.org/10.1038/nature13313

Poon, V. Y., Klassen, M. P., & Shen, K. (2008). UNC-6/netrin and its receptor UNC-5 locally exclude presynaptic components from dendrites. *Nature, 455*(7213), 669–673. https://doi.org/10.1038/nature07291

Popow, J., Englert, M., Weitzer, S., Schleiffer, A., Mierzwa, B., Mechtler, K.,...Martinez, J. (2011). HSPC117 is the essential subunit of a human tRNA splicing ligase complex. *Science, 331*(6018), 760–764. https://doi.org/10.1126/science.1197847

Qin, J., Liang, J., & Ding, M. (2014). Perlecan antagonizes collagen IV and ADAMTS9/GON-1 in restricting the growth of presynaptic boutons. *Journal of Neuroscience, 34*(31), 10311–10324. https://doi.org/10.1523/JNEUROSCI.5128–13.2014

Rajasethupathy, P., Antonov, I., Sheridan, R., Frey, S., Sander, C., Tuschl, T., & Kandel, E. R. (2012). A role for neuronal piRNAs in the epigenetic control of memory-related synaptic plasticity. *Cell, 149*(3), 693–707. https://doi.org/10.1016/j.cell.2012.02.057

Ramos, D. M., d'Ydewalle, C., Gabbeta, V., Dakka, A., Klein, S. K., Norris, D. A.,...Sumner, C. J. (2019). Age-dependent SMN expression in disease-relevant tissue and implications for SMA treatment. *Journal of Clinical Investigations, 129*(11), 4817–4831. https://doi.org/10.1172/JCI124120

Ransohoff, J. D., Wei, Y., & Khavari, P. A. (2018). The functions and unique features of long intergenic non-coding RNA. *Nature Reviews Molecular and Cellular Biology, 19*(3), 143–157. https://doi.org/10.1038/nrm.2017.104

Rasband, M. N. (2010). The axon initial segment and the maintenance of neuronal polarity. *Nature Reviews Neuroscience, 11*(8), 552–562. https://doi.org/10.1038/nrn2852

Reinhart, B. J., Slack, F. J., Basson, M., Pasquinelli, A. E., Bettinger, J. C., Rougvie, A. E.,...Ruvkun, G. (2000). The 21-nucleotide let-7 RNA regulates developmental timing in Caenorhabditis elegans. *Nature, 403*(6772), 901–906. https://doi.org/10.1038/35002607

Richmond, J. E., Davis, W. S., & Jorgensen, E. M. (1999). UNC-13 is required for synaptic vesicle fusion in C. elegans. *Nature Neuroscience, 2*(11), 959–964. https://doi.org/10.1038/14755

Rizzu, P., Lindsay, E. A., Taylor, C., O'Donnell, H., Levy, A., Scambler, P., & Baldini, A. (1996). Cloning and comparative mapping of a gene from the commonly deleted region of DiGeorge and Velocardiofacial syndromes conserved in C. elegans. *Mammalian Genome, 7*(9), 639–643. https://doi.org/10.1007/s003359900197

Rogalski, T. M., Gilchrist, E. J., Mullen, G. P., & Moerman, D. G. (1995). Mutations in the unc-52 gene responsible for body wall muscle defects in adult Caenorhabditis elegans are located in alternatively spliced exons. *Genetics, 139*(1), 159–169.

Run, J. Q., Steven, R., Hung, M. S., van Weeghel, R., Culotti, J. G., & Way, J. C. (1996). Suppressors of the unc-73 gene of Caenorhabditis elegans. *Genetics, 143*(1), 225–236.

Schaefer, A. M., Hadwiger, G. D., & Nonet, M. L. (2000). rpm-1, a conserved neuronal gene that regulates targeting and synaptogenesis in C. elegans. *Neuron, 26*(2), 345–356. https://doi.org/10.1016/s0896-6273(00)81168-x [pii]

Schuske, K. R., Richmond, J. E., Matthies, D. S., Davis, W. S., Runz, S., Rube, D. A., . . . Jorgensen, E. M. (2003). Endophilin is required for synaptic vesicle endocytosis by localizing synaptojanin. *Neuron, 40*(4), 749–762. https://doi.org/10.1016/s0896-6273(03)00667-6 [pii]

Sebat, J., Lakshmi, B., Malhotra, D., Troge, J., Lese-Martin, C., Walsh, T., . . . Wigler, M. (2007). Strong association of de novo copy number mutations with autism. *Science, 316*(5823), 445–449. https://doi.org/10.1126/science.1138659 [pii]

Seydoux, G. (2018). The P granules of C. elegans: a genetic model for the study of RNA-protein condensates. *Journal of Molecular Biology, 430*(23), 4702–4710. https://doi.org/10.1016/j.jmb.2018.08.007 [pii]

Shalizi, A., Gaudilliere, B., Yuan, Z., Stegmuller, J., Shirogane, T., Ge, Q., . . . Bonni, A. (2006). A calcium-regulated MEF2 sumoylation switch controls postsynaptic differentiation. *Science, 311*(5763), 1012–1017. https://doi.org/10.1126/science.1122513 [pii]

Sharpe, N. G., Williams, D. G., Norton, P., & Latchman, D. S. (1989). Expression of the SmB' splicing protein in rodent cells capable of following an alternative RNA splicing pathway. *FEBS Lett, 243*(2), 132–136. https://doi.org/10.1016/0014-5793(89)80114-0 [pii]

Shaw, W. R., Armisen, J., Lehrbach, N. J., & Miska, E. A. (2010). The conserved miR-51 microRNA family is redundantly required for embryonic development and pharynx attachment in Caenorhabditis elegans. *Genetics, 185*(3), 897–905. https://doi.org/10.1534/genetics.110.117515

Simon, D. J., Madison, J. M., Conery, A. L., Thompson-Peer, K. L., Soskis, M., Ruvkun, G. B., . . . Kim, J. K. (2008). The microRNA miR-1 regulates a MEF-2-dependent retrograde signal at neuromuscular junctions. *Cell, 133*(5), 903–915. https://doi.org/10.1016/j.cell.2008.04.035

Sleigh, J. N., Buckingham, S. D., Esmaeili, B., Viswanathan, M., Cuppen, E., Westlund, B. M., & Sattelle, D. B. (2011). A novel Caenorhabditis elegans allele, smn-1(cb131), mimicking a mild form of spinal muscular atrophy, provides a convenient drug screening platform highlighting new and pre-approved compounds. *Human Molecular Genetics, 20*(2), 245–260. https://doi.org/10.1093/hmg/ddq459

Sohn, E. J., Jo, Y. R., & Park, H. T. (2019). Downregulation MIWI-piRNA regulates the migration of Schwann cells in peripheral nerve injury. *Biochemical and Biophysical Research Communications, 519*(3), 605–612. https://doi.org/10.1016/j.bbrc.2019.09.008 [pii]

Song, Y., Sretavan, D., Salegio, E. A., Berg, J., Huang, X., Cheng, T., . . . Jan, Y. N. (2015). Regulation of axon regeneration by the RNA repair and splicing pathway. *Nature Neuroscience, 18*(6), 817–825. https://doi.org/10.1038/nn.4019

Soufari, H., & Mackereth, C. D. (2017). Conserved binding of GCAC motifs by MEC-8, couch potato, and the RBPMS protein family. *RNA, 23*(3), 308–316. https://doi.org/10.1261/rna.059733.116

Spike, C. A., Davies, A. G., Shaw, J. E., & Herman, R. K. (2002). MEC-8 regulates alternative splicing of unc-52 transcripts in C. elegans hypodermal cells. *Development, 129*(21), 4999–5008.

Spilker, K. A., Wang, G. J., Tugizova, M. S., & Shen, K. (2012). Caenorhabditis elegans Muscleblind homolog mbl-1 functions in neurons to regulate synapse formation. *Neural Development, 7*, 7. https://doi.org/10.1186/1749-8104-7-7

Sreedharan, J., Blair, I. P., Tripathi, V. B., Hu, X., Vance, C., Rogelj, B., . . . Shaw, C. E. (2008). TDP-43 mutations in familial and sporadic amyotrophic lateral sclerosis. *Science, 319*(5870), 1668–1672. https://doi.org/10.1126/science.1154584

Sun, L., Shay, J., McLoed, M., Roodhouse, K., Chung, S. H., Clark, C. M., . . . Gabel, C. V. (2014). Neuronal regeneration in C. elegans requires subcellular calcium release by ryanodine receptor channels and can be enhanced by optogenetic stimulation. *Journal of Neuroscience, 34*(48), 15947–15956. https://doi.org/10.1523/JNEUROSCI.4238-13.2014

Sun, Y., Grimmler, M., Schwarzer, V., Schoenen, F., Fischer, U., & Wirth, B. (2005). Molecular and functional analysis of intragenic SMN1 mutations in patients with spinal muscular atrophy. *Human Mutation, 25*(1), 64–71. https://doi.org/10.1002/humu.20111

Sun, Y., Zhao, Y. N., & Wang, D. Y. (2006). Computational analysis of genetic loci required for synapse structure and function and their corresponding microRNAs in C. elegans. *Neuroscience Bulletin, 22*(6), 339–349.

Tan, J. H., & Fraser, A. G. (2017). The combinatorial control of alternative splicing in C. elegans. *PLoS Genet, 13*(11), e1007033. https://doi.org/10.1371/journal.pgen.1007033

Tang, N. H., Kim, K.W., Xu, S., Blazie, S.M., Yee, B.A., Yeo, G. W., . . . Chisholm, A. D. (2020). The mRNA decay factor CAR-1/LSM14 regulates axon regeneration via mitochondrial calcium dynamics. *Current Biology,* 30(5):865–876. https://doi.org/10.1016/j.cub.2019.12.061

Tedeschi, A., Dupraz, S., Laskowski, C. J., Xue, J., Ulas, T., Beyer, M., . . . Bradke, F. (2016). The calcium channel subunit alpha2delta2 suppresses axon regeneration in the adult CNS. *Neuron, 92*(2), 419–434. https://doi.org/10.1016/j.neuron.2016.09.026 [pii]

Thienpont, D., Vanparijs, O. F., Raeymaekers, A. H., Vandenberk, J., Demoen, J. A., Allewijn, F. T., . . . Janssen, P. A. (1966). Tetramisole (R 8299), a new, potent broad spectrum anthelmintic. *Nature, 209*(5028), 1084–1086. https://doi.org/10.1038/2091084a0

Thompson, M., Bixby, R., Dalton, R., Vandenburg, A., Calarco, J. A., & Norris, A. D. (2019). Splicing in a single neuron is coordinately controlled by RNA binding proteins and transcription factors. *eLife, 8*. https://doi.org/10.7554/eLife.46726

Thompson-Peer, K. L., Bai, J., Hu, Z., & Kaplan, J. M. (2012). HBL-1 patterns synaptic remodeling in C. elegans. *Neuron, 73*(3), 453–465. https://doi.org/10.1016/j.neuron.2011.11.025

Trujillo, G., Nakata, K., Yan, D., Maruyama, I. N., & Jin, Y. (2010). A ubiquitin E2 variant protein acts in axon termination and synaptogenesis in Caenorhabditis elegans. *Genetics, 186*(1), 135–145. https://doi.org/10.1534/genetics.110.117341

Van Deerlin, V. M., Leverenz, J. B., Bekris, L. M., Bird, T. D., Yuan, W., Elman, L. B., . . . Yu, C. E. (2008). TARDBP mutations in amyotrophic lateral sclerosis with TDP-43 neuropathology: a genetic and histopathological analysis. *Lancet Neurology, 7*(5), 409–416. https://doi.org/10.1016/S1474-4422(08)70071-1

Van Epps, H., Dai, Y., Qi, Y., Goncharov, A., & Jin, Y. (2010). Nuclear pre-mRNA 3′-end processing regulates synapse and axon development in C. elegans. *Development, 137*(13), 2237–2250. https://doi.org/10.1242/dev.049692

Vance, C., Rogelj, B., Hortobagyi, T., De Vos, K. J., Nishimura, A. L., Sreedharan, J., . . . Shaw, C. E. (2009). Mutations in FUS, an RNA processing protein, cause familial amyotrophic lateral sclerosis type 6. *Science, 323*(5918), 1208–1211. https://doi.org/10.1126/science.1165942

Verkerk, A. J., Pieretti, M., Sutcliffe, J. S., Fu, Y. H., Kuhl, D. P., Pizzuti, A., . . . et al. (1991). Identification of a gene (FMR-1) containing a CGG repeat coincident with a breakpoint cluster region exhibiting length variation in fragile X syndrome. *Cell, 65*(5), 905–914. https://doi.org/10.1016/0092-8674(91)90397-h [pii]

Vuong, C. K., Black, D. L., & Zheng, S. (2016). The neurogenetics of alternative splicing. *Nature Reviews Neuroscience, 17*(5), 265–281. https://doi.org/10.1038/nrn.2016.27

Walthall, W. W., & Plunkett, J. A. (1995). Genetic transformation of the synaptic pattern of a motoneuron class in Caenorhabditis elegans. *Journal of Neuroscience, 15*(2), 1035–1043.

Wang, G. J., Kang, L., Kim, J. E., Maro, G. S., Xu, X. Z., & Shen, K. (2010). GRLD-1 regulates cell-wide abundance of glutamate receptor through post-transcriptional regulation. *Nature Neuroscience, 13*(12), 1489–1495. https://doi.org/10.1038/nn.2667

Wang, X., Chen, Q., Yi, S., Liu, Q., Zhang, R., Wang, P., ... Li, S. (2019). The microRNAs let-7 and miR-9 down-regulate the axon-guidance genes Ntn1 and Dcc during peripheral nerve regeneration. *Journal of Biological Chemistry, 294*(10), 3489–3500. https://doi.org/10.1074/jbc.RA119.007389

Wang, Y., Gracheva, E. O., Richmond, J., Kawano, T., Couto, J. M., Calarco, J. A., ... Zhen, M. (2006). The C2H2 zinc-finger protein SYD-9 is a putative posttranscriptional regulator for synaptic transmission. *Proceedings of the National Academy of Sciences of the United States of America, 103*(27), 10450–10455. https://doi.org/10.1073/pnas.0602073103

Wei, S., Chen, H., Dzakah, E. E., Yu, B., Wang, X., Fu, T., ... Shan, G. (2019). Systematic evaluation of C. elegans lincRNAs with CRISPR knockout mutants. *Genome Biology, 20*(1), 7. https://doi.org/10.1186/s13059-018-1619-6

Weimer, R. M., Richmond, J. E., Davis, W. S., Hadwiger, G., Nonet, M. L., & Jorgensen, E. M. (2003). Defects in synaptic vesicle docking in unc-18 mutants. *Nature Neuroscience, 6*(10), 1023–1030. https://doi.org/10.1038/nn1118

White, J. G., Albertson, D. G., & Anness, M. A. (1978). Connectivity changes in a class of motoneurone during the development of a nematode. *Nature, 271*(5647), 764–766. https://doi.org/10.1038/271764a0

White, J. G., Southgate, E., Thomson, J. N., & Brenner, S. (1986). The structure of the nervous system of the nematode Caenorhabditis elegans. *Philosophical Transactions of the Royal Society of London, B: Biological Science, 314*(1165), 1–340. https://doi.org/10.1098/rstb.1986.0056

Wood, M. J. A., Talbot, K., & Bowerman, M. (2017). Spinal muscular atrophy: antisense oligonucleotide therapy opens the door to an integrated therapeutic landscape. *Human Molecular Genetics, 26*(R2), R151–R159. https://doi.org/10.1093/hmg/ddx215

Wu, C. Y., Gagnon, D. A., Sardin, J. S., Barot, U., Telenson, A., Arratia, P. E., & Kalb, R. G. (2018). Enhancing GABAergic transmission improves locomotion in a Caenorhabditis elegans model of spinal muscular atrophy. *eNeuro, 5*(6). https://doi.org/10.1523/eneuro.0289-18.2018 [pii]

Wu, H. J., Wang, Z. M., Wang, M., & Wang, X. J. (2013). Widespread long noncoding RNAs as endogenous target mimics for microRNAs in plants. *Plant Physiology, 161*(4), 1875–1884. https://doi.org/10.1104/pp.113.215962

Yamagata, M., Sanes, J. R., & Weiner, J. A. (2003). Synaptic adhesion molecules. *Current Opinion in Cell Biology, 15*(5), 621–632. https://doi.org/10.1016/s0955-0674(03)00107-8 [pii]

Yan, D., & Jin, Y. (2012). Regulation of DLK-1 kinase activity by calcium-mediated dissociation from an inhibitory isoform. *Neuron, 76*(3), 534–548. https://doi.org/10.1016/j.neuron.2012.08.043

Yan, Z., Hu, H. Y., Jiang, X., Maierhofer, V., Neb, E., He, L., ... Khaitovich, P. (2011). Widespread expression of piRNA-like molecules in somatic tissues. *Nucleic Acids Research, 39*(15), 6596–6607. https://doi.org/10.1093/nar/gkr298

Zarkower, D. (2006). Somatic sex determination. *WormBook* (Feb. 10), 1–12. https://doi.org/10.1895/wormbook.1.84.1

Zhang, S., Fan, Z., Qiao, P., Zhao, Y., Wang, Y., Jiang, D.,... Li, X. (2018). miR-51 regulates GABAergic synapses by targeting Rab GEF GLO-4 and lysosomal trafficking-related GLO/AP-3 pathway in Caenorhabditis elegans. *Developmental Biology, 436*(1), 66–74. https://doi.org/10.1016/j.ydbio.2018.02.009 [pii]

Zhao, P. P., Yao, M. J., Chang, S. Y., Gou, L. T., Liu, M. F., Qiu, Z. L., & Yuan, X. B. (2015). Novel function of PIWIL1 in neuronal polarization and migration via regulation of microtubule-associated proteins. *Molecular Brain, 8*, 39. https://doi.org/10.1186/s13041-015-0131-0

Zhen, M., Huang, X., Bamber, B., & Jin, Y. (2000). Regulation of presynaptic terminal organization by C. elegans RPM-1, a putative guanine nucleotide exchanger with a RING-H2 finger domain. *Neuron, 26*(2), 331–343. https://doi.org/10.1016/s0896-6273(00)81167-8 [pii]

Zhen, M., & Jin, Y. (1999). The liprin protein SYD-2 regulates the differentiation of presynaptic termini in C. elegans. *Nature, 401*(6751), 371–375. https://doi.org/10.1038/43886

Zou, Y., Chiu, H., Zinovyeva, A., Ambros, V., Chuang, C. F., & Chang, C. (2013). Developmental decline in neuronal regeneration by the progressive change of two intrinsic timers. *Science, 340*(6130), 372–376. https://doi.org/10.1126/science.1231321

CHAPTER 15

REGULATION OF mRNA TRANSLATION IN AXONS

PRIYANKA PATEL, PABITRA K. SAHOO,
AMAR N. KAR, AND JEFFERY L. TWISS

INTRODUCTION

NEURONS are highly polarized asymmetric cells that extend cytoplasmic processes, axons and dendrites, long distances from their cell body or soma. These electrically excitable cells communicate to other neurons and to target tissues through synapses where propagated action potentials stimulate release of neurotransmitters. Dendrites are post-synaptic and extend a few millimeters in length. Axons are pre-synaptic and extend much longer distances, several centimeters length in rodents to a meter or more length in larger organisms. The dendrites and axons constitute highly organized circuits for trans-cellular communication that underlie brain and spinal cord function. However, the geographic separation of the distal reaches of dendrites and axons from the neuron's soma necessitates that neurons evolved means to support the growth and unique function of their axonal and dendritic processes. Proteins and organelles are transported along axons and dendrites by motor proteins. This provides axons and dendrites with the building blocks needed for growth and function, but transport is not immediate and requires that the soma anticipate needs of these subcellular domains. Axons and dendrites maintain a high level of autonomy that is needed during growth, for plasticity, and for responses to injury. Localized synthesis of proteins was initially suggested to occur in dendrites based on ultrastructural studies showing polyribosomes at the base of dendritic spines in adult hippocampus (Steward & Levy, 1982). Dendritically synthesized proteins are now known to play key roles in synaptic plasticity (Kang & Schuman, 1996; Klann & Dever, 2004; Sutton & Schuman, 2006). In contrast, axons were initially suggested to be incapable of synthesizing proteins based on in vivo absence of ribosomes by electron microscopy (EM) studies of adult rodent central nervous system (CNS) axons (Steward & Levy, 1982).

With axons extending much further from the soma than dendrites, it certainly seemed like a similar "on demand" means to change the local proteome would be an appealing cellular mechanism for the distal axon to rapidly respond to different stimuli. There had been reports of ribosomes in some axons in the peripheral nervous system (PNS; Pannese & Ledda, 1991; Zelena, 1972a, 1972b), and axons isolated from invertebrate neurons were shown to synthesize proteins (Casadio et al., 1999; Giuditta, Hunt, & Santella, 1986; Giuditta et al., 1991; Martin et al., 1997; Schacher, Wu, Panyko, Sun, & Wang, 1999; Van Minnen et al., 1997). Further, it appeared that axons of vertebrate olfactory and hypothalamic neurons contained some mRNAs but ribosomes were not observed in these axons (Denis-Donini, Branduardi, Campiglio, & Carnevali, 1998; Mohr, Fehr, & Richter, 1991; Mohr & Richter, 1992). Without clear evidence for ribosomes, localization of mRNAs into axons was of uncertain biological significance. However, subsequent work in cultured neurons from a variety of different experimental organisms has shown that axonal mRNAs are indeed used to make new proteins in the PNS and central nervous system (CNS). There now is compelling evidence for axonal mRNA localization in vivo, both in the PNS and CNS (Brittis, Lu, & Flanagan, 2002; C. J. Donnelly et al., 2011; Hafner, Donlin-Asp, Leitch, Herzog, & Schuman, 2019b; Perry et al., 2012; Shigeoka et al., 2016; Terenzio et al., 2018; Walker, Hengst, et al., 2012; D. E. Willis et al., 2011), and evidence for ribosomes in hypothalamic and olfactory axons was uncovered (Dubacq, Jamet, & Trembleau, 2009). Much of the work on axonal protein synthesis has focused on growing axons, both in development and after injury, proteins synthesized in mature axons in vivo help to maintain axon survival, are used to respond to neuropathic pain-inducing stimuli, are needed for retrograde transport of some viral species, and likely contribute to synaptic plasticity (Biever, Donlin-Asp, & Schuman, 2019; Sahoo, Smith, Perrone-Bizzozero, & Twiss, 2018). Proteins generated in axons can also contribute to neurodegeneration (Baleriola & Hengst, 2015). RNA profiling studies, including in vivo work for ribosome-bound mRNAs, has advanced knowledge of the axonal transcriptome from hundreds to thousands of mRNAs (Briese et al., 2016; Gumy et al., 2011; Hafner et al., 2019b; Kar, Lee, & Twiss, 2018; Minis et al., 2014; Rotem et al., 2017; Shigeoka et al., 2016; Zivraj et al., 2010). Given the diverse functions that proteins encoded by these mRNAs serve, it is not surprising that translation of these mRNAs is specifically and tightly regulated. Here, we will outline the current state of knowledge for the mechanisms underlying this translational regulation in axons, as well as point out areas that are ripe for future study.

Pathways to Regulate Localized Protein Synthesis in Axons

mRNA localization provides an efficient mechanism for regulating localized translation in subcellular regions, including dendrites and axons (Doyle & Kiebler, 2011). A single mRNA can typically be subjected to multiple rounds of translation. A number of factors

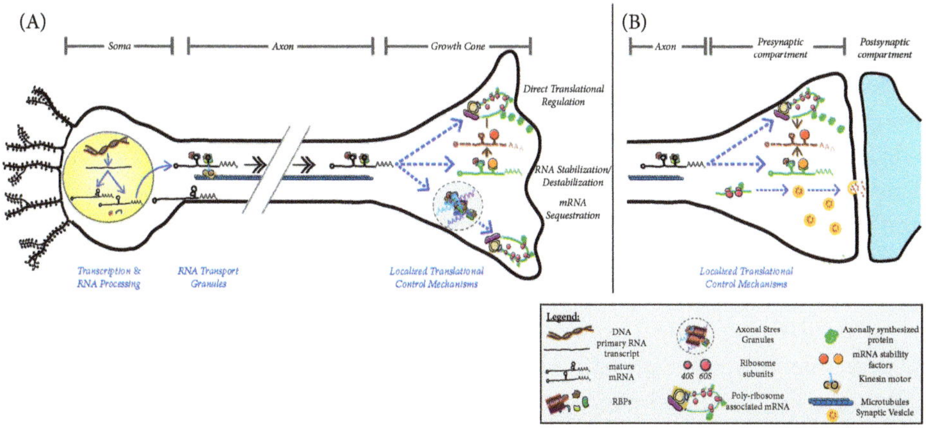

FIGURE 15.1. Mechanisms converging on axonal mRNA translation. Schematic of a neuron with growing axon (**A**) is shown to illustrate subcellular sites and general mechanisms of translational control for axonal mRNAs. It is clear that protein synthesis occurs in mature axons, probably both at synaptic terminals and along the axon shaft (**B**), and similar mechanisms will undoubtedly regulate protein synthesis in these axonal regions. Branch points are also sites of active protein synthesis in axons (not shown). Axonal mRNA translation is driven by the amount of the mRNA transported into axons, stability of that mRNA within axons, and availability of the mRNA for interaction with translational machinery. Once delivered to the axon, an mRNA can be translated immediately or stored. Storage mechanisms include stress granule-like structures and sequestration within RNP associated with receptor proteins. mRNA interactions with RBPs in axons can also facilitate or block the mRNA's translation. Finally, the stability of the mRNA determines how long it could be available for translation into new proteins.

impinge on how much protein can be generated from a given axonal mRNA. Translation of an axonal mRNA is determined by the amount of that mRNA transported into the axonal compartment, the steady state levels of the localized mRNA, whether the mRNA has access to axonal translation factors and ribosomes, and activity of those translational machineries (Figure 15.1). Thus, in addition to direct translational regulation of axonal mRNAs, their transport into axons, stability within axons and storage or sequestration away from translational machinery provide means for regulating how much, which and where different proteins are generated in axons. Further, activity of the translation machinery can help to determine which mRNAs are translated when (i.e., "translational specificity"). We present each of these mechanisms in the sections that follow and close with what is known about the signaling pathways that modulate activity of translational machinery in axons.

Mechanisms Regulating mRNA Transport into Axons

A central theme for spatial control of protein synthesis in polarized cells is that the mRNA has to be actively transported to the subcellular domain (Figure 15.1). RNA localization allows an mRNA to be utilized at or near the destination where its protein product is needed (Jung, Yoon, & Holt, 2012). Neurons actively select which mRNAs are localized into axons, dendrites, or both, and this selection can change based on developmental phase and physiological conditions. We now know that a few thousand different mRNAs are transported into axons (Kar et al., 2018), and specific and precise mechanism are involved in targeting these mRNAs into axonal and dendritic compartments. For example, although β-actin and γ-actin are remarkably similar proteins, β-*actin* (*Actb*) mRNA localizes into axons and dendrites while γ-*actin* (*Actg*) mRNA is restricted to the soma of sensory and cortical neurons (Bassell et al., 1998; Tiruchinapalli et al., 2003). The mRNA encoding Tau protein (*Mapt*) selectively localizes into axonal processes of differentiated P19 cells, while that encoding microtubule-associated protein 2 (*Map2*) selectively localizes into dendritic processes (Garner, 1988; Kleiman, Banker, & Steward, 1990; Litman, Barg, Rindzoonski, & Ginzburg, 1993). As we will outline, this specificity for mRNA localization is inherent to the individual mRNA. However, it should be noted that instances of differential mRNA localization between different neuron types (and potentially different physiological conditions) is likely to increase as our knowledge of axonal transcriptomes grows (Kar et al., 2018). For example, although *Actg* mRNA apparently does not localize into axons of cortical or sensory neurons (Bassell et al., 1998; D. Willis et al., 2005), recent work from the Sendtner lab surprisingly showed localization of *Actg* mRNA in motor axons (Moradi et al., 2017).

Targeting mRNAs for transport into subcellular compartments involves recognition of sequence motifs within the mRNAs (*cis*-acting elements) by RNA-binding proteins (RBPs; *trans*-acting factors). The localizing regions within the mRNA have been variably referred to as "zip-codes," but this designation is also used to selectively refer to a 54 nucleotide (nt) motif in *Actb* mRNA that is recognized by zip-code binding protein 1 (ZBP1; also called insulin-like growth factor 2 mRNA binding protein 1 [IGF2BP1]; Kislauskis, Zhu, & Singer, 1994; Zheng et al., 2001). *Actg* mRNA, which seems to be restricted to the soma in neurons and perinuclear region in fibroblasts, does not contain the 54 nucleotide (nt) zip-code motif that drives axonal localization of *Actb* mRNA (Kislauskis et al., 1994; Zheng et al., 2001). Similarly, nucleolin protein (Ncl) was shown to bind to the axonal localizing motif in Importin β1 mRNA (*Kpnb1*), a 34 nt stem-loop structure in the 3′UTR of *Kpnb1* referred to as the MAIL motif (motif for axonal importin β1 localization; Perry et al., 2016). Thus, we will use the term "axonal localization motif" here unless specifically referring to the *Actb* motif. Typically, axonal localization motifs reside in untranslated regions (UTRs) of an mRNA, particularly the 3′UTR (Gomes, Merianda, Lee, Yoo, & Twiss, 2014). Axonal localizing motifs have been detected in 5′ UTRs (Merianda, Gomes, Yoo, Vuppalanchi, & Twiss, 2013), and, at least in yeast, the coding sequences of an mRNA has been shown to drive subcellular localization (Kilchert & Spang, 2011). Identification of axonal localization motifs has relied on demonstrating that a region of the axonal mRNA is both necessary and sufficient for transport into axons, typically, by deletion analysis of the UTR sequence placed into a reporter mRNA or testing potential motifs in driving axonal transport of a non-localizing UTR (Ben-Yaakov et al., 2012; Vuppalanchi et al., 2010; Wu et al., 2005; Yoo et al., 2013;

Yudin et al., 2008). Unfortunately, there have been no unifying themes in the primary sequences of these *cis*-elements that allow us to predict potential localizing motifs in other mRNAs. Sequence conservation of motifs between different species has provided some guidance for pointing to regions of UTRs to focus on for motif discovery work (Lee et al., 2018; Vuppalanchi et al., 2010). These observations point to the importance of secondary and higher order RNA structures for RNA-protein interactions, and also raises the possibility that an mRNA's interaction with specific RBPs is conserved between those species with conserved UTRs.

Since RBP interaction with an mRNA is thought to be required for the mRNA's localization into axons, RBPs play a central role in regulating the levels of different axonal mRNAs and hence the levels of proteins that can potentially be generated from those mRNAs. RBP(s) binding to a localizing mRNA generates a ribonucleotide protein (RNP) complex that can interact with motor proteins to travel along the axonal cytoskeleton. A moving RNP is frequently referred to as a "transport particle" or "transport granule." Recent affinity mass spectrometry studies using axonal localizing motifs as bait to isolate interacting RBPs from PNS axons uncovered a surprising number of axonal RBPs, but also showed that multiple RBPs can interact with the same RNA motif (Lee et al., 2018). This included many hnRNPs that were thought to strictly function in the nucleus. The combinations of RBP interactions define distinct RNP populations in axons and dendrites. This is seen for Fragile X Mental Retardation protein (FMRP) interacting Fragile X related proteins 1 and/or 2 (FXR1 and FXR2, respectively) for CNS axons as well as DEAD box protein combinations for localizing to dendrites of CNS neurons (Akins et al., 2017; S. B. Christie, Akins, Schwob, & Fallon, 2009; Chyung, LeBlanc, Fallon, & Akins, 2018; Elvira et al., 2006; Miller et al., 2009). Further, RBPs in addition to ZBP have been shown to participate in the transport and translational regulation of axonal *Actb* mRNA, including HuD and hnRNP R (also called Syncrip; C. J. Donnelly et al., 2011; Glinka et al., 2010; Yoo et al., 2013). Thus, it is appealing to hypothesize that RNP complex-defining RBP combinations regulate transport and/or translation of different pools of axonal mRNAs. mRNA co-immunoprecipitations (RIP) and crosslinking immunoprecipitations (CLIP) type approaches are uncovering the RNA cohorts that individual RBPs bind to, but these do not distinguish between different RNPs that these RBPs may be complexed with. Also, it should be noted that the RNA affinity approach used by Lee et al. (2018) did not identify HuD or ZBP1 as binding to *Actb* and *Gap43* mRNAs' axonal localizing motifs, so there is clearly more to be done for unraveling RNA-protein interactions (Lee et al., 2018). Moreover, it is highly likely that axonal RNP makeup is dynamic, with RBP combinations changing over time and space. Such complexity would provide a means for both spatial and temporal regulation of axonal protein production.

Transcriptional and co-transcriptional events can obviously impact the axonal transcriptome, and hence indirectly influence translational regulation in axons. The genes encoding Importin-β1, Ran binding protein 1 (*Ranbp1*), and Signal transducer and activator of transcription 3 (*Stat3α*) can generate mRNAs with 3′UTRs of varying lengths, where differential polyadenylation site usage generates long mRNA populations that can localize into axons (Ben-Yaakov et al., 2012; Hanz et al., 2003; Yudin et al., 2008). For *Kpbn1*, this long 3′UTR containing mRNA includes the 34 nt stem loop structure that is recognized by Nucleolin protein (Ncl), where the *Kpnb1*-Ncl RNP interacts with kinesin proteins for axonal localization (Perry et al., 2016). Similarly differential poly-A site usage was reported for localization of *brain-derived neurotrophic factor* (*BDNF*) mRNA into dendrites (An et al., 2008). Differential splicing has also been shown to generate localizing vs. non-localizing mRNA variants from the same gene. Large scale RNA-sequencing studies showed that a shift in alternative splicing that favors mRNA isoforms with localizing 3′UTRs occurs during neuronal differentiation of catecholaminergic neuronal tumor (CAD) cells (Taliaferro et al., 2016). Further, an unbiased assessment of neuropil-enriched mRNAs from adult rodent hippocampus showed the presence of significantly longer 3′UTRs as compared to soma-enriched mRNAs (Tushev et al., 2018), and, with recent evidence for presynaptic mRNAs in this same experimental system (Hafner et al., 2019b), it will be intriguing to determine which of those mRNAs came from axons. The transcriptional events that determine how these neuronal gene products are alternatively spliced to generate localizing vs. soma-restricted mRNAs are not known, and it will be of great interest to determine if these splicing events are coordinated by the axon's developmental or physiological states.

Changing gene transcription rates can obviously impact how much of an mRNA could be available for axonal localization. This is seen for *Gap43* mRNA, where axotomy-induced transcription of the *GAP43* gene results in a commensurate increase in axonal levels of *Gap43* mRNA (Yoo et al., 2013). Axonal localization of *Gap43* mRNA requires both HuD (also called Elavl4) and ZBP1 proteins, and the injury-induced upregulation of *Gap43* mRNA levels competes with *Actb* mRNA for ZBP1 interaction (Donnelly et al., 2011). This idea of a single RBP (or groups of RBPs) interacting with an mRNA cohort has been termed an "RNA regulon" (Keene, 2007), and, though initially described for stabilization of mRNAs by Hu proteins, localization and translational regulation of axonal mRNA cohorts by shared RBPs likely defines new RNA regulons. For *Gap43* and *Actb* mRNA, the levels of proteins generated from these mRNAs in axons helps to determine the growth morphology of the regenerating axon (C. J. Donnelly et al., 2013). So which mRNA wins the competition for ZBP1 interaction helps shape the axon's morphology when ZBP1 is in limited supply. Work from our lab has shown that both the abundance and affinity of competing mRNAs can impact which target mRNA an RBP engages (Gomes et al., 2017), so affinity of RBP-mRNA interaction must also be considered in addition to overall levels of individual mRNAs. The affinity of RNA-RBP interactions has not received much attention from the field. As outlined in the next section, HuD binds to many other mRNAs through AU-rich elements (AREs), including axonal mRNAs like Neuritin1 (*Nrn1*, also called *Cpg15*; Merianda et al., 2013), and this interaction stabilizes target mRNAs (Gardiner, Twiss, & Perrone-Bizzozero, 2015). Thus, RBP interaction can effectively impact delivery of the mRNA into axons as well as determine the fate of the localized mRNA. As we outline in the translational regulation section, RBP interactions can also directly influence translation of target mRNAs in axons (Figure 15.2).

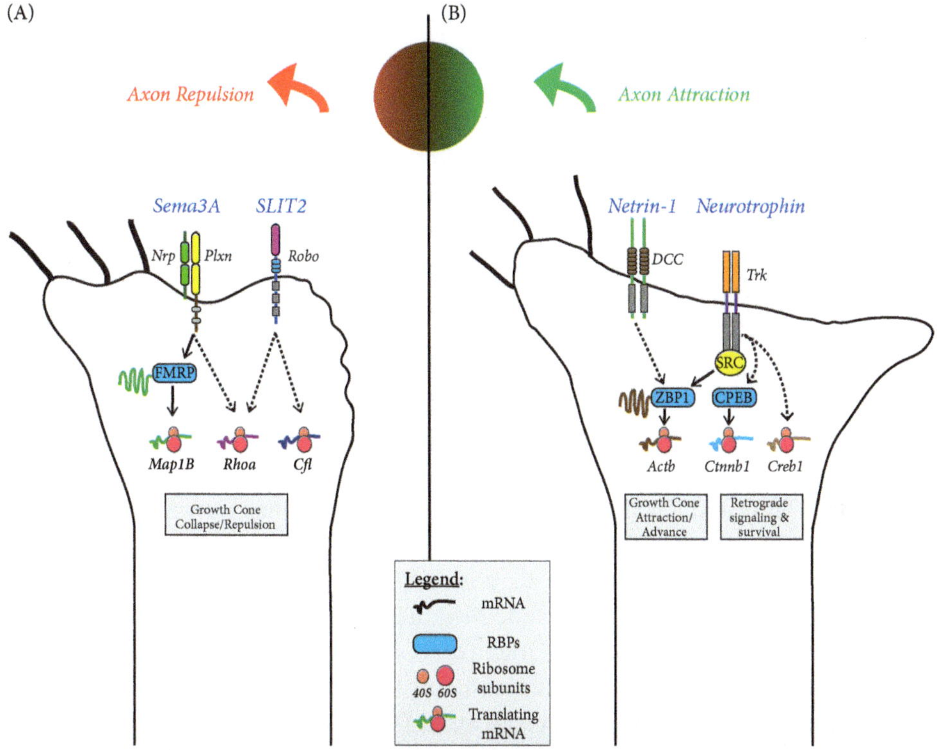

FIGURE 15.2. Axon guidance cues regulate axonal translation of specific mRNAs. Examples of chemo-repulsive (**A**) and chemo-attractant/neurotropic stimuli (**B**) that activate translation of mRNAs in distal axons. As outlined in the text, translation of different mRNAs can be matched to the stimulus thereby bringing specificity to axonal protein synthesis at the mRNA level. Where known, axonal RBPs binding to those axonal mRNAs are indicated. Notably neurotrophin signaling in axons has been reported to trigger retrograde transport of newly translated CREB1 protein that supports neuronal survival. Neurotrophins and other stimuli have also been reported to support axon survival/maintenance through localized protein synthesis.

Mechanisms Regulating mRNA Stability in Axons

Since RNA is susceptible to degradation by ribonucleases (RNase) that can be tenaciously difficult to inactivate, mRNAs are notoriously unstable. But half-lives of mRNAs vary substantially within cells, with some being very short-lived and others being relatively stable (Mitchell & Tollervey, 2000). Unfortunately, we do not have good systematic assessments for stability of mRNA populations in subcellular compartments. However, there is evidence that stability of some localized mRNAs can be impacted by interactions with RBPs and micro-RNAs (miRNAs). We have recently shown that axonal mRNAs can be stored in stress granule-like compartments (to be discussed further), and these stored mRNAs are presumably also sequestered from the RNA degradation machinery in axons (Sahoo et al., 2018). Furthermore, mRNAs are likely to be protected from degradation when they reside in transport granules. The ability to alternatively stabilize or destabilize different axonal mRNAs brings an appealing mechanism for stimulus-dependent alteration of mRNA levels available for spatially controlling protein expression. This also brings the potential to indirectly regulate transport or translation of other mRNAs by altering axonal RBP availability.

HuD is well characterized for stabilizing ARE containing mRNAs, including several that localize into axons. Yoo et al. (2013) showed that the axotomy-induced increase in axonal levels of the ARE-containing *Gap43* mRNA is accompanied by an increase in axonal levels of HuD protein (Yoo et al., 2013). KH-type splicing regulatory protein (KHSRP; also called Far Upstream Element-Binding Protein 2 [FUBP2]) can also bind to the ARE of *Gap43* mRNA, and this interaction destabilizes *Gap43* mRNA. This differential interaction of HuD and KHSRP extends into axons since expression of an axonally localizing *Gap43* mRNA lacking the ARE rescues axon growth deficits attributed to KHSRP (Bird et al., 2013). It is not clear how many axonal mRNAs constitute the RNA regulons for HuD and KHSRP, but axonal RNA datasets contain several mRNAs that were published as part of a brain HuD-RNA interactome (Bolognani, Contente-Cuomo, & Perrone-Bizzozero, 2010; Gumy et al., 2011; Minis et al., 2014; Saal, Briese, Kneitz, Glinka, & Sendtner, 2014). These include *Nrn1* that we have shown is stabilized by HuD interaction in vitro similar to *Gap43* (Akten et al., 2011; Gomes et al., 2017). Interestingly, *Nrn1* and *Gap43* mRNAs can compete for binding to HuD, and such a competition could effectively impact RNA survival locally in axons (Gomes et al., 2017), particularly since Nrn1 can localize through its 5′ UTR using motifs not recognized by HuD (Lee et al., 2018; Merianda et al., 2013). Extension of this competition to other mRNAs as well as to HuD and KHSRP competing with one another for RNA interactions could profoundly impact metabolism of axonal mRNAs. However, it should be noted that KHSRP has 4 separate KH domains for RNA binding and it has known nuclear roles in RNA splicing and cytoplasmic roles in micro-RNA (miRNA) maturation that are separate from this mRNA destabilization function (Briata et al., 2015; Garcia-Mayoral et al., 2007; Gherzi et al., 2004; Gherzi et al., 2006). RNA destabilization by KHSRP interaction is thought to target mRNAs to the cytoplasmic exosome complex for degradation (Briata, Chen, Ramos, & Gherzi, 2013).

Nonsense mediated decay (NMD) of mRNAs was originally discovered as an RNA surveillance mechanism for degrading mRNAs containing premature stop codons generated due to genetic mutations (Kurosaki, Popp, & Maquat, 2019). Several lines of evidence indicate that NMD extends to cellular mRNAs with stop codons in close upstream proximity to an exon-exon junction. This necessitates that the encoding gene's 3′ UTR region is composed of multiple exons, but it could also derive from alternative splice site usage in some mRNAs. A number of NMD effector proteins including Upstream frameshift protein1 (UPF1, also called Rent1), UPF3b (also known as UFP3X), Serine/threonine-protein kinase (SMG1) and eukaryotic initiation factor 4A3 (EIF4A3) localize to both axons and dendrites (Alrahbeni et al., 2015; Colak, Ji, Porse, & Jaffrey, 2013; Giorgi et al., 2007; Jolly, Homan, Jacob, Barry, & Gecz, 2013; J. Y. Kim, Deglincerti, & Jaffrey, 2017). In support of NMD mechanisms impacting axonal translation, axonal

Robo 3.2 mRNA was shown to be a target for NMD in developing spinal cord neurons after their axons cross the mid-plate; in this case, NMD of *Robo3.2* mRNA ultimately alters axon responsiveness to pathfinding cues (Colak et al., 2013). NMD of *Activity Regulated Cytoskeleton Associated Protein* mRNA (*Arc*, also known as *Arg3.1*) serves to spatially restrict Arc protein expression in dendrites distal to activated synaptic laminae (Farris, Lewandowski, Cox, & Steward, 2014). Of interest, mRNAs with short reading frames upstream of the main open reading frames or mRNAs with long 3′ untranslated regions (UTRs) can also be subjected to NMD (Jaffrey & Wilkinson, 2018), and recent mRNA next generation sequencing work has shown that the mRNAs containing longer 3′ UTR are often enriched in axons and dendrites (Hafner, Donlin-Asp, Leitch, Herzog, & Schuman, 2019a; Tushev et al., 2018). Thus, NMD may be a much more prominent feature for axonal mRNA regulation than we have recognized thus far.

In addition to NMD, UPF1 protein has also been shown to function in Staufen-mediated decay, a process in which mRNAs with either Staufen 1 (STAU1) or Staufen 2 (STAU2) binding sites in their 3′ UTR are targeted for decay (Kim, Furic, Desgroseillers, & Maquat, 2005;. Kim et al., 2007; Kim & Maquat, 2019; E. Park & Maquat, 2013). The Jaffrey lab reported axonal localization of STAU1- and UPF1-containing RNA granules in embryonic DRG neurons and depletion of STAU1 increased levels of a subset of axonal mRNAs, including *Actb* and *Rac1* (Kim et al., 2017). Importantly, the increase in *Actb* mRNA was axon-specific (Kim et al., 2017), suggesting a role for Staufen-mediated decay in locally regulating axonal mRNA levels. However, it should be noted that Staufen and UPF1 protein can have functions beyond RNA degradation. The Sossin lab showed that STAU2 and UPF1 contribute to dendritic localization of mRNAs bound by stalled ribosomes (i.e., stalled polysomal mRNAs), with UPF1 also contributing to metabotropic glutamate receptor-dependent translational activation of dendritic *Map1b* mRNA (Graber et al., 2017). UPF1 was also shown to suppress translation of *Arc* mRNA in Neuro2A cells and neurites of hippocampal neurons (Ryu et al., 2019). With these proteins localizing to axons as well, it is likely that Staufen and UPF proteins may alter mRNA transport into and/or translation within axons.

In contrast to the RNA decay pathways just described, miRNAs may provide a more direct route for stimulus dependent RNA degradation. miRNAs are a class of non-coding RNAs (ncRNAs), 21-23 nt in length, that post-transcriptionally regulate gene expression by binding to target sites in 3′UTRs to either induce degradation or silence translation of the target mRNA (Djuranovic, Nahvi, & Green, 2011). miRNAs bind to members of the Argonaute (Ago) protein family to form an effector complex known as the "RNA-induced silencing complex" (RISC), and the activated miRNA-containing RISC is then recruited to target mRNA (Krol, Loedige, & Filipowicz, 2010). RISC also contains RBPs including protein kinase RNA activator (PACT), transactivation response RNA binding protein (TRBP), Dicer, and fragile-X mental retardation protein (FMRP; Lee, Zhou, Smith, Noland, & Doudna, 2013; Tang, 2005), many of which have been shown to localize into axons in cultured neurons and in vivo (Hengst, Cox, Macosko, & Jaffrey, 2006; Murashov et al., 2007). KHSRP is also reported to have roles in miRNA processing (Trabucchi et al., 2009), and this RBP is also present in axons

(Bird et al., 2013). Application of synthetic small-interfering RNAs (siRNAs) to axons of cultured neurons and PNS in vivo show that the axonal RISC machinery is indeed functional (Hengst et al., 2006; Murashov et al., 2007; Toth et al., 2009), and several lines of evidence indicate that endogenous miRNAs function in axons. Initial profiling studies of miRNAs in primary sympathetic neurons showed that there is considerable complexity in the axonal miRNA population, with several miRNA enriched in the axons compared to the cell bodies (Natera-Naranjo, Aschrafi, Gioio, & Kaplan, 2010). These findings raise the appealing possibility that subsets of miRNAs are selectively transported into axons; however, it is not clear what mechanisms underlie the specificity for axonal localization specific miRNAs or their precursors (see below). At least some of these miRNAs have been shown to regulate the fate of axonal mRNAs. For example, axonal levels of the mRNA encoding Cytochrome oxidase IV (COXIV) are affected by axonal miR-338 localization in sympathetic neurons, and elevation of axonal miR-338 decreases mitochondrial ATP production in axons (Aschrafi et al., 2008). Interestingly, axonal translation of the mRNAs encoding components of eukaryotic translation initiation pathway, eIF2B2 and eIF4G2, are regulated by axonal miR-16 (Kar, Macgibeny, Gervasi, Gioio, & Kaplan, 2013). miRNA-regulated decrease in axonal *eIF2B2* and *eIF4G2* mRNAs effectively alters translation of other axonal mRNAs and thereby modulates axon growth (Kar et al., 2013). Similarly, miR-19a modulates local levels of phosphatase and tensin homologue deleted on chromosome ten (PTEN) protein in distal axons of primary embryonic cortical neurons (Zhang et al., 2013), and decreased PTEN activity supports a generalized increase in cap-dependent protein synthesis through increased activation of mTOR (see below). Thus, miRNAs could effectively turn on or turn off translation of axonal mRNAs beyond the direct miRNA targets.

The presence of Dicer in axons raises the possibility for on demand activation of miRNA function in axons, as Dicer is needed for maturation of miRNAs from pre-miRNAs (Hengst et al., 2006; Kim, Han, & Siomi, 2009). Work from the Yoo lab has shown that several pre-miRNAs localize into axons of peripheral nerves (Kim, Kim, Phay, & Yoo, 2015). Interestingly, processing of these pre-miRNAs increased (let 7c-a and pre-miRs-16, -23a, -25, -125b-1, -433, and -541) or decreased (pre-miRs-138-2, -185, and -221) following nerve injury (Kim et al., 2015; Phay, Kim, & Yoo, 2015). Target scan analyses for these miRNAs points to many potential targets among the known axonal mRNAs (Phay et al., 2015). These findings raise the exciting possibility for stimulus-dependent miRNA maturation as a means to regulate protein synthesis in axons, as was recently demonstrated for activity dependent maturation of miR-181a in dendrites (Sambandan et al., 2017).

It should be noted that other ncRNAs localize into axons and some have been shown to have functions in axons. Phay et al. (2018) showed that PIWI-interacting RNAs (piRNAs), 25–34 nt long ncRNAs that were thought to be restricted to germline cells, and other piRNA-like ncRNAs have been detected in axons (Phay, Kim, & Yoo, 2018). These may have similar functions to miRNAs, and the Jin lab recently showed that *C. elegans* piRNAs inhibit axon regeneration (Kim et al., 2018). Interestingly, work from the Sendtner lab indicates that the non-coding 7SK RNA is highly represented in hnRNP R CLIP analyses, with hnRNP R and 7SK RNA in close association in axons of motor neurons (Briese et al., 2018). Disruption of the hnRNP R-7SK RNA association alters the axonal transcriptome and disrupts axon growth. Finally, Crerar et al. (2019) very recently showed that an axonal mRNA, *Tp53inp2*, modulates NGF-dependent axon growth in the absence of the mRNA's translation in axons (Crerar et al., 2019). Though the unifying mechanism(s) driving *Tp53inp2* mRNA's function in axons are not clear, it is intriguing to speculate that non-coding functions could be awaiting discovery for some of the few thousand other axonal mRNAs that are presumed to function as protein coding templates in axons.

Based on our previous studies that show that repertoire of axonally localized mRNAs is in part determined by the interplay of differential affinity of mRNAs to a cognate RBP and the limited availability of RBPs in axon (Donnelly et al., 2013; Donnelly et al., 2011; Gomes et al., 2017), it would be of interest to investigate how NMD, NMD-associated protein interactions, and ncRNA interactions may indirectly impact translation of other non-target mRNAs by freeing up shared RPBs for interacting with those mRNAs. It is also not clear whether NMD, RBP-mediated mRNA destabilization, or ncRNA mechanisms outlined earlier respond to physiological stimuli, but it is very appealing to speculate that these could provide on demand means to modulate the axonal transcriptome.

Mechanisms Directly Impacting mRNA Translation in Axons

Axons are capable of both cap-dependent and cap-independent translation through internal ribosome entry sites (IRES). Just as in other cellular regions, the well-characterized mechanisms of translation factor and ribosome subunit interaction with mRNAs for translation occur in the axonal compartment. As we outline below, activity of translation factors can be modified in axons through post-translational modifications for stimulus dependent changes in the axonal proteome. Interactions with RBPs, including those RBPs used for transport of mRNAs into axons discussed previously, can both attenuate and activate translation of specific mRNAs. Further, sequestration of some axonal mRNAs from translational machinery, either through association with stress granule-like structures or receptor interactions, provides a means for stimulus-dependent release of mRNAs for translation. We discuss each of these mechanisms in the following paragraphs emphasizing biological relevance of these intra-axonal translational control mechanisms.

Axonal mRNA Translational Regulation through Interactions with RNA Binding Proteins

In addition to prominent roles for RBPs in mRNA transport and stability, a number of axonal RBPs have been shown to directly impact translation of axonal mRNAs (Figure 15.2). mRNAs in transport granules are thought to be translationally silent, such

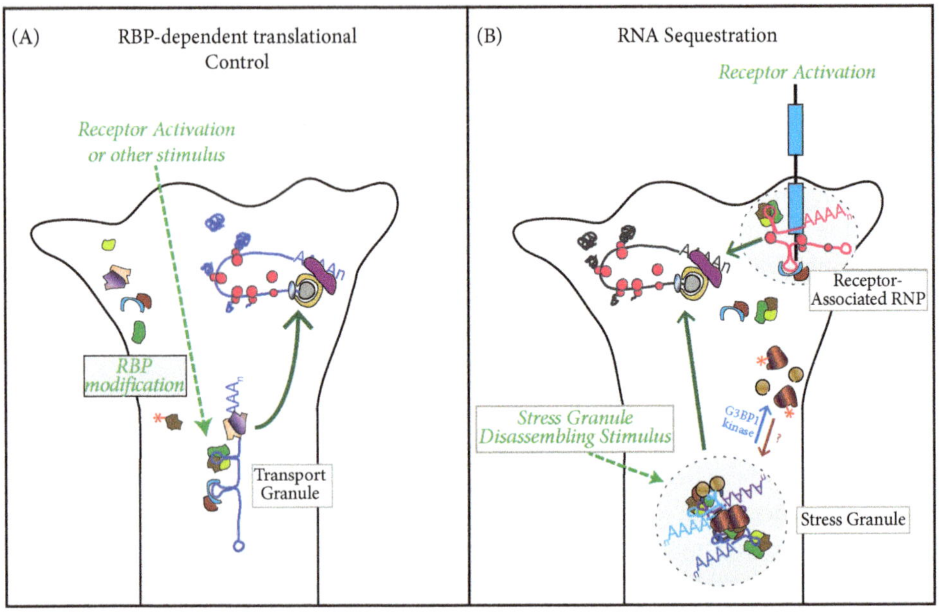

FIGURE 15.3. Translational regulation in axons through axon-RBP interactions and RNA sequestration. **(A)** mRNAs are transported into axons as RNPs termed "transport particles" or "transport granules." These RNPs are established by RBPs binding to specific mRNAs. Translation of mRNAs is likely suppressed during their transport such that the RNPs provide a means for translational regulation. Post-translational modifications of RBPs in axons can release this translational suppression allowing the mRNA to be translated upon arrival into distal axons or in response to specific extracellular cues. Similarly, stimulus-activated signals can modify RBPs for interactions with axonal mRNAs and activate their translation. **(B)** Axonal mRNAs can also be stored in axons until needed. In this case a physiological condition or extracellular cue releases the mRNAs from these storage depots so they are available for translation. Axonal "stress granule-like" structures and RNPs assembled along the cytoplasmic tail of cell surface receptors are two known mechanisms. The stress granule-like structures are present in naïve PNS axons and can be disassembled on demand where phosphorylation of G3BP1 protein is associated with decrease in axonal stress granule protein aggregation. For receptor-mediated storage, DCC and neuropilin-1 proteins have been shown to serve as a site of RNP and ribosome subunit aggregation.

that translation of the mRNA has to be activated once reaching its destination (Figure 15.3A; Wells, 2006). Evidence of such roles of axonal RBP is seen in the interaction of ZBP1 with *Actb* mRNA, which blocks the access to *Actb* by the translational machinery. Phosphorylation of ZBP1 tyrosine 396 by an Src-family protein kinase decreases the protein's affinity for RNA binding, thereby releasing *Actb* mRNA for translation (Huttelmaier et al., 2005). Decreased axonal translation of *Actb* mRNA in neurons transfected with tyrosine 496 to phenylalanine ZBP1, a non-phosphorylatable ZBP1 mutant, emphasizes that this phosphorylation event is biologically relevant for translation in CNS and PNS axon growth (Donnelly et al., 2011; Welshhans & Bassell, 2011). However, ZBP1 serine 181 phosphorylation was recently shown to alter dendritic mRNA transport (Urbanska et al., 2017), so other post-translational modifications may affect ZBP1 target mRNAs in different ways, and the potential contributions of post-translational modifications to other RBPs in axonal mRNA transport and translation need further study. For example, FMRP is well characterized as a translational suppressor in dendrites in addition to its links with miRNA-mediated mRNA regulation outlined earlier (Banerjee, Ifrim, Valdez, Raj, & Bassell, 2018), and FMRP localizes to axons where it has been shown to regulate translation of *Map1b* in growth cones and *Munc19-1* in synapses (Li, Bassell, & Sasaki, 2009; Parvin et al., 2018). Liquid phase separations of RBPs, which occurs through low complexity domains (LCD; also referred to as "intrinsically disordered domains") can promote protein aggregation by inclusion of mRNAs (Kim et al., 2013; Kwiatkowski et al., 2009; Mackenzie et al., 2017; Neumann et al., 2006). This effectively sequesters mRNAs from translation, including translationally stalled mRNAs. Phosphorylation and arginine methylation of FMRP was very recently shown to conversely regulate its phase separation in vitro, hence implying ability to inhibit mRNA translation (Tsang et al., 2019). It is not clear if this mechanism occurs in axons, but as outlined below such aggregation of other RBPs in axons has been shown to regulate translation of axonal mRNA cohorts during axon regeneration.

Localized modulation of the poly-adenylate (A) tail of mRNAs, termed "cytoplasmic polyadenylation," brings another RBP-mediated mechanism to modulate axonal mRNA translation. Cytoplasmic element binding protein (CPEB) was initially shown to bind to the cytoplasmic polyadenylation element (CPE) in 3′ UTRs of maternal mRNAs in oocytes, where this is used to activate translation of those mRNAs through extension of their poly-A tails and CPEB's interaction with translation factors (Figure 15.2; Ivshina, Lasko, & Richter, 2014). There are 4 known vertebrate CPEB proteins, CPEBs 1–4, with varying numbers of orthologs in invertebrate model organisms (Ito et al., 2000). For vertebrate neurons, CPEB1 was initially shown to regulate translation of CPE-containing dendritic mRNAs, where translation of those mRNAs can be activated by phosphorylation of CPEB and subsequent recruitment of cap-binding initiation factor complex (see below) to those mRNAs (Huang, Kan, Lin, & Richter, 2006). CPEB1 does localize into axons where its activity has been shown to modulate mRNA translation. Translation of CPEB-regulated mRNAs is required for *Xenopus* retinal axon growth cone collapse in response to Sema3A, suggesting that regulation of poly(A) tail length may regulate guidance-cue-induced local translation of specific mRNAs (Lin et al., 2009). Local translation of β-catenin mRNA in growth cones of hippocampal neurons is enhanced by neurotrophin (NT3)-dependent phosphorylation of CPEB1 (Kundel, Jones, Shin, & Wells, 2009). Translation of the mRNA encoding the Down syndrome cell adhesion molecule similarly shows CPEB-dependent translation in axonal growth cones in response to stimulation with the guidance cue Netrin 1 (Jain & Welshhans, 2016). Systematic analyses for CPEB-regulated axonal mRNAs has yet to be published, but there undoubtedly will be more axonal mRNAs that are regulated by CPEBs. For example, many translationally regulated mRNAs seen after murine spinal cord injury contain CPEs, but it is not clear if their translational regulation is CPEB-dependent (Lou et al., 2017). It is also not known if the CPEB 2, 3, or 4 contribute to axonal mRNA translation.

Interestingly, Netrin 1 and Sema3A illicit opposite responses from growth cones (attraction vs. repulsion; Figure 15.2), and uncovering how CPEB1 can be driving these morphologically distinct outcomes will be of high interest.

Storage of Axonal mRNAs for "on Demand" Protein Synthesis

Several lines of evidence indicate that some axonal mRNAs can be recruited into translation on demand. Evidence of this is seen after injury of PNS axons, where translation of mRNAs encoding injury response proteins like Importin β1, RanBP1, Stat3α, and Vimentin are rapidly translated following axotomy (Ben-Yaakov et al., 2012; Hanz et al., 2003; Perlson et al., 2005; Yudin et al., 2008). Chemorepulsive and growth inhibitory stimuli can selectively activate translation of *RhoA* and *Cofilin* mRNAs in growing axons (Piper et al., 2006; Wu et al., 2005). These two proteins attenuate polymerization of actin filaments that are needed for axon extension, so their translation is presumably low in axons until the axon encounters a growth-attenuating stimulus. Extrapolating from the examples of activity-dependent translation in dendrites, it is likely that many axonal mRNAs are subjected to stimulus-dependent translational control in axons and these could be either translationally silent or inefficiently translated until a stimulus specifically activates their translation. Consistent with this idea, Younts et al. (2016) reported that presynaptic local protein synthesis is required for long term plasticity of GABA release from type 1 cannabinoid receptor 1 (CB_1)-expressing axons. This long term depression of inhibitory transmission increased protein synthesis in an mTOR-dependent manner (Younts et al., 2016). Further, recent work from the Schuman lab shows distinct patterns of pre-synaptic mRNA translation in response to metabotropic glutamate receptor agonist, CB1 agonists or BDNF (Hafner et al., 2019b). This provides a very effective temporal mechanism for modulating the levels of proteins locally in distal axons, and it is clear that axonal mRNAs can be recruited into translation by specific stimuli. Sequestration of mRNAs in stress granule like structures and in RNP complexes near cytoplasmic domains of growth cone receptors are mechanisms that have clearly been shown for storing mRNAs axons until needed (Figure 15.3B).

Similar to the translational suppression of axonal mRNAs by RBPs in transport granules, axons have been shown to contain proteins that associate with stress granules in other cellular systems (Gilks et al., 2004; Nonhoff et al., 2007). Stress granules are used to store translationally stalled mRNAs during periods of cellular stress, such as metabolic or oxidative stress (Anderson & Kedersha, 2008). This allows the cell to translate mRNAs whose protein products are needed to respond to the stress and not translate those mRNAs whose protein products serve other functions (Khong et al., 2017). The phase separation of FMRP and other RBPs outlined earlier represents one mechanism for RBP aggregation in neurons mediated by LCDs (Tsang et al., 2019), and aggregating stress granule proteins similarly contain LCDs. We recently showed that the RasGAP SH3 domain binding protein 1 (G3BP1) and T-cell-restricted intracellular antigen-1 protein (TIA1), two LCD-containing containing stress granule proteins, aggregate in PNS axons under basal conditions (Sahoo et al., 2018). These axonal stress granule-like structures also contain FMRP and FXR1 as well as colocalizing mRNAs. The colocalizing mRNAs are translationally suppressed by overexpression of G3BP1, and their translation is increased by triggering disassembly of the granules by expressing the acidic domain of G3BP1 or applying a cell permeable peptide for rodent G3BP1 amino acids 190–208. These granule-disassembling agents selectively increase axonal mRNA translation in PNS neurons and accelerate axon growth in both PNS and CNS neurons (Sahoo et al., 2018). The Jin lab recently showed that LCD-mediated aggregation of *C. elegans* TIAR-2 protein, the orthologue of TIA1, in injured axons attenuates axon regeneration (Andrusiak et al., 2019).

Phosphorylation of G3BP1 on serine 149 has been shown to trigger disassembly of G3BP1 aggregates (Tourriere et al., 2003), and kinetic analyses of G3BP1 and TIA1 aggregation in axons show decreased aggregation of these axonal proteins during nerve regeneration (Figure 15.3B; Sahoo et al., 2018). This suggests that axons have an intrinsic mechanism for stress granule disassembly. Both Akt pathway and Casein kinase 2 activity have been implicated in phosphorylation of G3BP1 (Kwok et al., 2017; Reineke et al., 2017), but it is not clear if axons use these signaling cascades for modulating translation through stress granule disassembly. Interestingly, the *Kpnb1* mRNA is translationally suppressed by association with axonal G3BP1, but it requires additional signals for translational activation after release from G3BP1 aggregates (Sahoo et al., 2018). Specificity for translation of axonal stress granule-associated mRNAs is likely coordinated through signaling pathways that converge on the translational machinery, with translation of a released mRNA dependent on the physiological state of the axon. Thus, further studies to determine which axonal mRNAs are sequestered in the stress granule like aggregates and what other mechanisms may contribute to their translational regulation once released from these aggregates is clearly needed. The phosphorylations of translation factors outlined below are prime candidates for translational regulation of released mRNAs. Notably, a recent paper from the Anderson lab refuted the idea that serine 149 phosphorylation of G3BP1 contributes to its aggregation status based on additional N-terminal mutations in phosphomimetic and non-phosphorylatable G3BP1 constructs used by many groups (Panas et al., 2019; Tourriere & Tazi, 2019). Future studies will need to determine whether this is the case in neurons, where there seems to be high aggregation of G3BP1 in apparent absence of stress stimuli (Sahoo et al., 2018).

The adenomatous polyposis coli (APC) protein, which is well characterized for its microtubule plus-end binding functions and also regulates β-catenin activity downstream of Wnt (Etienne-Manneville, 2009), has been implicated as an RBP that concentrates mRNAs in pseudopodia of fibroblasts for translation near the migratory front of these cells (Mili, Moissoglu, & Macara, 2008). In axons, APC binds to mRNAs near microtubule plus-ends, thus providing a means to spatially concentrate mRNAs in growth cones (Preitner et al., 2014). With regards to storage of axonal mRNAs, Tcherkezian et al. (2010) showed that cytoplasmic tail of deleted in colon cancer protein (DCC), which serves as a signaling receptor for the axon guidance cue Netrin 1, provides a scaffold to store RNP complexes containing APC, translation initiation factors, ribosome subunits, and monosome-bound mRNAs (Tcherkezian, Brittis, Thomas, Roux, & Flanagan, 2010). Binding of Netrin 1 to DCC dissociates the RNP complex and promotes translational activation of the DCC-interacting mRNAs. Koppers et al. (2019) very recently showed that this mechanism extends to Neuropilin 1 but not to Robo2 receptors (Koppers et al., 2019). This suggests receptor-mediated RNP storage may be broadly used to match translational responses to individual stimuli by selectively matching mRNA cohorts to different receptor complexes. Along this line of logic, the Hengst lab showed that the mRNA encoding Lis1 protein (*Pafah1b1*) can interact with APC in growing axons, and whether axonal *Pafah1b1* is APC-bound or not determines whether the mRNA is translated in response to nerve growth factor (NGF) stimulation vs. withdrawal (Villarin, McCurdy, Martinez, & Hengst, 2016). NGF stimulation promotes axon growth and survival while NGF withdrawal has the opposite effects. Since Lis1 protein alters activity of the microtubule-dependent motor protein dynein (Baumbach et al., 2017), introduction of Lis1 locally in axons could provide a means to alter retrograde axon-to-soma signaling and specificity of this signal is likely determined by whether or not the mRNA is locally sequestered by APC.

Direct Regulation of Axonal Translational Machinery

mRNA translation is generally comprised of three phases, translational initiation, elongation, and termination. Obviously, these translation phases must also occur in axons. Translational initiation in eukaryotes is a rate limiting step for protein synthesis, and hence is a target for intense regulation (Sonenberg & Hinnebusch, 2009). This is also highly relevant for translational regulation in axons. Cap-dependent translation requires that the "cap-binding complex" of translation factors termed the eukaryotic initiation factor (eIF) 4F complex, which consists of eIF4E, eIF4A and the scaffold protein eIF4G. This eIF4F complex binds to the mRNA's 5′ 7-methyl guanosine (m^7G) cap that then allows recruitment of the pre-initiation complex consisting of the 40S ribosome subunit and other translation factors to the mRNA (Hershey, Sonenberg, & Mathews, 2012; Hinnebusch & Lorsch, 2012). Sequestration of eIF4E protein by the 4E binding proteins 1 and 2 (4EBP1/2) provides a means to limit cap-dependent initiation of protein synthesis and post-translational modification of 4EBP1/2 is used to activate cap-dependent protein synthesis (Gingras et al., 1999). Likewise, stress-induced post-translational modification of eIF2α, which is a component of the eIF2 complex needed for bringing the methionine-charged methionyl tRNA to the preinitiation complex, has been shown to generally block protein synthesis (Knutsen et al., 2015; Pakos-Zebrucka et al., 2016). Together, phosphorylation of 4EBP1/2 and eIF2α provide opposing outcomes, increasing or decreasing overall protein synthesis, respectively, but also bring opportunities for regulating translation of some mRNAs over others. That is, translational specificity can be derived by individual mRNAs having varying efficiency of translation in the face of eIF2α phosphorylation. Translation of most cellular mRNAs is repressed by phospho-eIF2α, but translation of some mRNAs is paradoxically upregulated under this condition and this is oftentimes for mRNAs with upstream open reading frames (Hinnebusch, Ivanov, & Sonenberg, 2016). Translation of mRNAs encoding the stress-induced transcription factors ATF4, ATF6, XBP-1 and CHOP is increased by phospho-eIF2α, with their protein products subsequently inducing transcription genes for ER chaperones and enzymes needed to cope with the load of unfolded proteins in the ER or trigger apoptosis of damaged cells (Calfon et al., 2002; Hetz, Chevet, & Oakes, 2015; Palam, Baird, & Wek, 2011; Teske et al., 2011). Translation of ER chaperone protein mRNAs in axons, *Calr* and *Grp78*, is similarly increased by phospho-eIF2α (Vuppalanchi et al., 2012), and translation of axonal *Calr* mRNA helps to initiate axon regeneration after an in vitro axotomy (Pacheco, Merianda, Twiss, & Gallo, 2020).

Neurotrophic factors, axon guidance cues, and axonal injury have been shown to induce 4EBP1/2 phosphorylation locally in distal axons through activation of the mammalian target of rapamycin protein (mTOR Campbell & Holt, 2001; Gouveia Roque & Holt, 2018; Nie et al., 2010; Terenzio et al., 2018). Phosphorylation of 4EBP1/2 disrupts its interaction with eIF4E, freeing eIF4E to interact with 5′ m^7G cap of mRNAs (Gingras et al., 1999). mTOR activation also leads to S6 kinase activation also can promote protein synthesis. mTOR activity is a major effector of cell growth and proliferation in non-neuronal cells, and neuronal mTOR activity has been shown to support growth of neurons, where it has gained particular interest for its role in axon regeneration after injury (Park, Liu, Hu, Kanter, & He, 2010). mTOR is a serine/threonine kinase that exists in two different complexes: the rapamycin-sensitive mTOR complex 1 (mTORC1) that includes the RAPTOR protein and the rapamycin-insensitive mTOR complex 2 (mTORC2) that includes RICTOR protein (Guertin & Sabatini, 2007; Ma & Blenis, 2009). Work in optic nerve injury indicates that mTORC1 and not mTORC2 increases axon regeneration after injury (Miao et al., 2016). On the other hand, work in PNS neurons points to role for mTORC2 and not mTORC1 for the enhanced axon regeneration seen following a conditioning crush nerve lesion (Chen et al., 2016); still other works have pointed to roles for mTORC1 in axon growth (Abe, Borson, Gambello, Wang, & Cavalli, 2010; Nie et al., 2010). Though these differential effects of the two mTOR complexes will need to be sorted out, disruptions of upstream regulators of mTOR activity have been linked to autism spectrum disorders, epilepsy, and tuberous sclerosis through altered activity of upstream mediators of mTOR activity (Crino, Nathanson, & Henske, 2006; Jeste, Sahin, Bolton, Ploubidis, & Humphrey, 2008; Scheffer et al., 2014; Zhou & Parada, 2012). These disease connections indicate that disruption of mTOR signaling has the potential to alter neuronal connectivity, but it is not clear the extent to which, or if, axonal translation is involved in these phenotypes.

Upstream regulators of mTOR protein activity, including Akt, tuberous sclerosis complex 1 and 2 (TSC1/TSC2), and Ras homolog enriched brain (Rheb) proteins, enable cells to sense levels of growth factors, ATP, nutrients, and stress to inhibit or activate mTOR (Figure 15.4; Inoki, Li, Xu, & Guan, 2003). Phosphate-inositol 3 Kinase (PI3K) is a major trigger for activating the Akt à mTOR pathway by cell surface receptors for axon guidance cues and growth factors in axons (Brazil & Hemmings, 2001; Brunet et al., 1999; Kandel & Hay, 1999). PI3K signaling is effectively turned off by activity of the

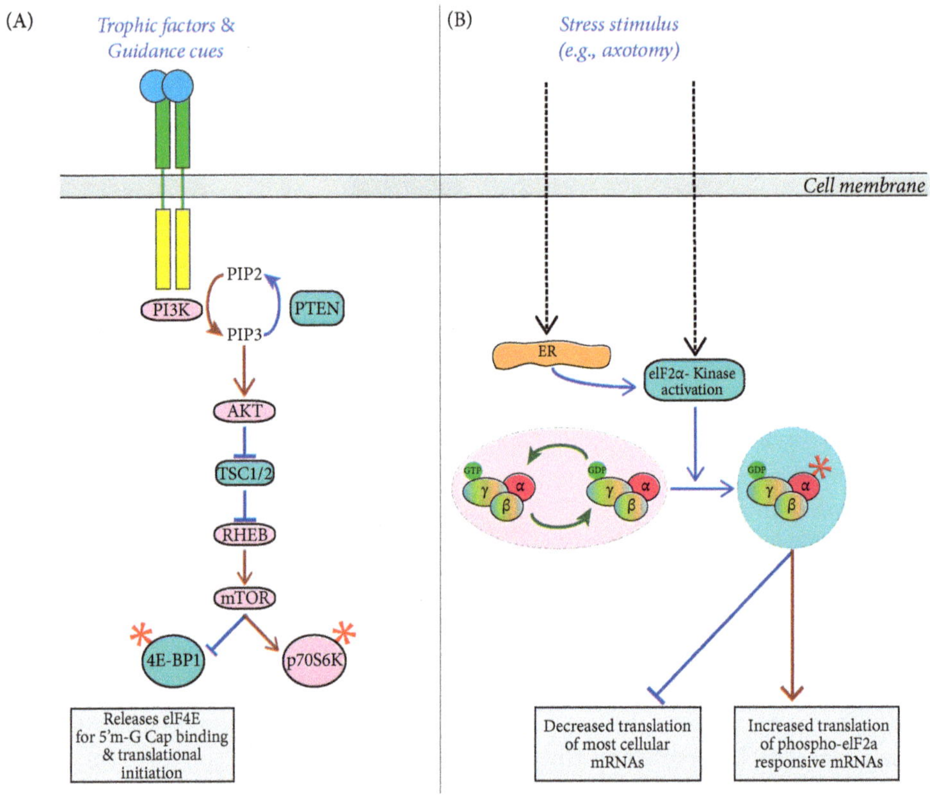

FIGURE 15.4. Regulation of axonal mRNA translation through translation factor modifications. Activation of the mTOR and eIf2α kinase pathways has opposing effects on axonal protein synthesis. On the one hand, Akt à mTOR pathway activation in axons occurs in response to trophic factors (e.g., neurotrophins) and axon guidance cue exposure (e.g., Netrin 1 and Semaphorin). This results in phosphorylation of 4EBP1 that sequesters eIF4E from mRNA cap-binding, a rate limiting step in cap-dependent protein synthesis (**A**). Additionally, mTOR activates S6 kinase to phosphorylate ribosomal protein (RP) S6, which may further increase ribosome function. On the other hand, stress stimuli including axotomy can converge on the family of eIF2α protein kinases to inactivate eIF2α through phosphorylation on serine 51 (**B**). This shuts off general protein synthesis but results in a paradoxical increase in translation of chaperone protein, stress pathway transcription factor and (potentially) other axonal mRNAs. Unfolded protein response with activation of PERK to phosphorylate eIF2α has been shown to be activated in axons with axonal injury (Pacheco et al., 2020; Vuppalanchi et al., 2012), and recent work indicates that Semaphorins can induce phosphorylation of axonal eIF2α (Cagnetta et al., 2019). Thus, both the mTOR and eIF2α kinase pathways could be activated simultaneously in response to an axonal stimulus.

tumor suppressor PTEN (Georgescu, 2010; Stambolic et al., 1998). Genetic deletion of *PTEN* genes or global inhibition of PTEN function has been shown to increase axon regeneration in both the CNS and PNS (K. J. Christie, Webber, Martinez, Singh, & Zochodne, 2010'; Lewandowski & Steward, 2014; Liu et al., 2010; Ning et al., 2010; K. K. Park et al., 2008; Sun et al., 2011; Zukor et al., 2013). Reports showing that *mTor* mRNA can be translated locally in both developing and mature axons bring a potential feed forward mechanism to upregulate translation of many other mRNAs in axons (Figure 15.3; Kye et al., 2014; Terenzio et al., 2018). This further emphasizes that the PTEN pathway linked to axon regeneration functions locally in axons. For PNS axons, translation of *mTor* mRNA is increased by axonal injury in an mTOR-dependent fashion, suggesting that axonal mTOR levels are not sufficient to support the protein synthesis needs of the regenerating axon. Consistent with this, decreased axonal localization of murine *mTor* mRNA through deletion of its 3′ UTR decreased intra-axonal mRNA translation and regeneration after PNS axotomy (Terenzio et al., 2018). It is not clear how translational specificity is imparted by this generalized increase in mTOR activity in axons. As noted previously, both Sema3A and Netrin 1 increase axonal 4EBP1 phosphorylation (Campbell & Holt, 2001, 2003), yet they invoke distinct morphological responses from the growth cone.

Evidence from different systems suggests that specificity for translating mRNAs can be driven by phosphorylation of eIF2α. eIF2α is a key regulator of generalized translation through its inhibition by serine 51 phosphorylation (Baird & Wek, 2012; Sonenberg & Hinnebusch, 2009). Phosho-eIF2α increases under cell stress conditions, such as ER stress and amino acid starvation, and this reduces general translation initiation but also facilitates preferential translation of selected mRNAs that promote adaptation to or recovery from specific stressors. This paradoxical increase in translation of some mRNAs that are needed for the stress response is how eIF2α phosphorylation brings translational specificity for different mRNAs (Figure 15.4). For example, eIF2α phosphorylation leads to increased translation of mRNAs encoding activating transcription factors 4 and 6 (ATF4 and ATF6, respectively), X-binding protein 1 (XBP-1), and CCAAT/enhancer-binding protein homologous protein (CHOP), transcription factors that increase transcription of genes encoding chaperone proteins and enzymes that are needed to cope with the stressor (Calfon et al., 2002; Hetz et al., 2015; Palam et al., 2011; Teske et al., 2011). Translation of existing chaperone protein mRNAs has long known to be upregulated by stress (Brostrom & Brostrom, 1990), which implies that these chaperone protein mRNAs can also be translationally-activated when eIF2α is phosphorylated. Consistent with this, there is clear evidence that ER chaperone protein mRNA translation can be activated in isolated axons by phospho-eIF2α (Vuppalanchi et al., 2012). Four different protein kinases have been shown to phosphorylate eIF2α: general control nonderepressible 2 (GCN2), heme-regulated eIF2 kinase (HRI), protein kinase R (PKR), and PKR-like endoplasmic reticulum kinase (PERK; Donnelly, Gorman, Gupta, & Samali, 2013). Each of these is regulated by different stressors, with GCN2 being activated during metabolic stress, HRI by oxidative stress, PKR by double stranded RNAs, and PERK by increased levels of unfolded proteins typical of ER stress in the unfolded protein response (UPR; Bravo et al., 2013; Lemaire, Anderson, Lary, & Cole, 2008; Peng et al., 2012; Suragani et al., 2012). It is appealing to speculate that these four eIF2α kinases provide stressor-specific means for distal axons to modulate their translational activity, but potential contribution of these kinases to regulating axonal protein synthesis has not been well studied. Recent work from Pacheco et al. (2020) has shown that PERK activity is needed for translation of axonal *Calr* mRNA acutely after axotomy. Activity of GCN2 has been linked to neurite growth (Roffe, Hajj, Azevedo, Alves, & Castilho, 2013), and activity of HRI has been shown to impact synaptogenesis (Ill-Raga et al., 2015), but more work is clearly needed to determine if these activities impact the axonal translatome and if these kinases consistently localize into axons.

Elevations in axoplasmic Ca^{2+} have been shown to regulate several axonal mRNAs following peripheral nerve injury (e.g., *Kpnb1, Ranbp1, Stat3α, Calr*, and *Grp78*; Ben-Yaakov et al., 2012; Hanz et al., 2003; Vuppalanchi et al., 2012; Yudin et al., 2008). These elevations in axoplasmic Ca^{2+} initially occur through membrane injury (Detrait et al., 2000; Gitler & Spira, 1998), but this may also involve local release of Ca^{2+} stores from the ER and other axonal organelles. This injury-induced increase in axoplasmic Ca^{2+} contributes to growth cone formation, but also has to be buffered to allow the axon to start regenerating (Bradke, Fawcett, & Spira, 2012). Translation of *Calr* and *Grp78* mRNAs can be increased by blocking protein folding in isolated sensory axons, with a commensurate increase in axonal phospho-eIF2α that was also mimicked by release of ER Ca^{2+} in axons (Vuppalanchi et al., 2012). Translation assays using axonally targeted fluorescent reporter proteins further proved the functional role of eIF2α phosphorylation in this process and showed specificity for different mRNAs since increased phospho-eIF2α decreased axonal translation of *Actb* mRNA (Vuppalanchi et al., 2012). A sustained elevation of axoplasmic Ca^{2+} by release of mitochondrial and ER stores has been associated with axonal degeneration (Barrientos et al., 2011; Villegas et al., 2014). Interestingly, axon survival has been linked to intra-axonal translation of *Bcl1l2* mRNA (Cosker, Fenstermacher, Pazyra-Murphy, Elliott, & Segal, 2016), and the encoded Bclw protein prevents axon degeneration by inhibiting inositol 1,4,5-trisphosphate receptor (IP3R) that has been linked to ER-Ca^{2+} release during axonal degeneration (Pease-Raissi et al., 2017). Consequently, intra-axonal translation of *Bclw* mRNA may help to regulate translation of other axonal mRNAs by impacting axonal Ca^{2+} levels.

Axotomy has been shown to activate UPR in both PNS and CNS neurons. Sciatic nerve crush induces activation of IRE1a leading to increase in the spliced isoform of *Xbp1*, which can be used to generate active XBP1 protein, and nuclear accumulation of ATF4 (Onate et al., 2016). Overexpression of XBP1 was shown to increase motor recovery following PNS injury suggests that activation of UPR after injury of PNS axons in vivo is pro-regenerative (Onate et al., 2016). Sensory axons contain the mRNA encoding the ER transmembrane protein Luman (also called cAMP response element binding protein 3 [CREB3]); *Luman* mRNA is translated locally through activation of UPR after axotomy, and its protein product is then retrogradely transported to the nucleus (Ying, Misra, & Verge, 2014). Axonally derived Luman supports subsequent nerve regeneration by altering expression of genes needed for cholesterol synthesis (Ying et al., 2015). For CNS neurons, the Hetz group showed that a rapid UPR activation within a few hours after spinal cord injury that seems to be sustained for several days (Hetz, 2012; Hetz et al., 2015;

Hetz & Glimcher, 2009) and preventing UPR activation significantly increased numbers of damaged axons (Valenzuela et al., 2012). Axotomy in cultured neurons also requires activation of PERK to initiate axon growth (Pacheco et al., 2020). However, sustained UPR could also be detrimental to axon growth since work in optic nerve injury points to UPR-dependent activation of *CHOP* gene expression in promoting death of retinal ganglion cell neurons (Hu et al., 2012) and *Atf4* mRNA translation in axons has been linked to neurodegeneration in Alzheimer's disease (Baleriola et al., 2014). Together these observations point to signals converging on eIF2α as a key regulator of axonal mRNA translation but also emphasize the need for a balance in eIF2α's activation and inactivation.

Importantly, recent work from the Holt lab shows that the axon guidance cue, Sema3A, can regulate axonal mRNA translation by phosphorylation of axonal eIF2α protein (Cagnetta et al., 2019). Alterations in axonal Ca^{2+} levels are known to be regulated by both attractant and repellant axon guidance cues (Henley, Huang, Wang, & Poo, 2004; Hong, Nishiyama, Henley, Tessier-Lavigne, & Poo, 2000; Zheng, 2000), and stimuli that block axon growth in the injured spinal cord that are known to modulate axonal mRNA translation have been linked to increases in axonal Ca^{2+} (Kalinski et al., 2019; Snow, Atkinson, Hassinger, Letourneau, & Kater, 1994; Walker, Ji, & Jaffrey, 2012). Thus, it is intriguing to speculate that eIF2α phosphorylation regulates axonal mRNA translation for these cues as well. Interestingly, Sema3A also increases phosphorylation of axonal 4EBP1/2 to activate cap-dependent protein synthesis (Campbell & Holt, 2001). This could effectively bring greater level of translational specificity by targeting a subset of phospho-eIF2α-responsive mRNAs that are initiated through cap-dependent rather than cap-independent mechanisms. Of note, both axonal *Calr* and *Grp78* mRNAs were shown to be phospho-eIF2α-responsive, but *Grp78* initiates through an IRES while *Calr*'s 5′ UTR showed no IRES activity (Pacheco & Twiss, 2012).

Conclusion and Future Directions

The concept of mRNA translation occurring in axons, both during development and in mature neurons, has gained more acceptance by the neuroscience community. Subcellularly localized protein synthesis is likely to occur in many polarized cell types, and the distances separating axon terminals from the neuronal soma bring an obvious advantage for locally generating proteins in axons. This gives autonomy to the distal axon for responding to its environment well before new proteins could be delivered from the soma by anterograde transport. With increasing acceptance of axonal protein synthesis, the field has moved from trying to prove existence of mRNAs and translational machinery in axons to understanding which proteins are generated in axons, what these locally translated proteins do, and how transport and translation of their mRNAs are regulated. Advances on these fronts have been fueled by the tenacity of investigators plus technical advances in RNA detection, cell biology tools and detection methods, and emerging animal models for visualizing protein synthetic events and profiling locally translated mRNAs. When and how much of specific proteins are generated in axons can be regulated by the transport of the mRNAs into axons, sequestration of the mRNA, stability of the mRNA and translational regulation. Just like transport of proteins and organelles, transport of mRNAs down axons takes time and several lines of evidence indicate that mRNAs are stored in axons until their proteins are needed. RBPs, RNPs, miRNAs and activity of the translational machinery provide convergent mechanisms that together influence which, when, and where individual proteins are generated in axons. Each brings some specificity with regard to individual mRNA cohorts that can be translated, and combinatorial use of these mechanisms likely brings a high level of translational specificity allowing the axon to choose which proteins it generates when. Such would imply collaboration between the different translation controlling mechanisms outlined herein. Interestingly, recent work points to interactions between translation mechanisms, mitochondria and lysosomes in axons and dendrites (Cioni et al., 2019; Rangaraju, Lauterbach, & Schuman, 2019), such translation controlling mechanisms may extend to other cellular organelles beyond the pathways outlined here. Crosstalk between these mechanisms and how these might contribute to translational specificity for different mRNAs needs further study from the community. Also, the possibility that different regions of an axon can regulate translation of different mRNAs needs to be considered.

Acknowledgments

Research related to the topic of this chapter has been supported by Wings for Life—Spinal Cord Research Foundation (WFL-US-09/18 to PP), South Carolina Spinal Cord Injury Research Fund (#2018 PD-01 and 2019 PD-02 to PKS), ASPIRE award from the Office of the Vice President for Research at the University of South Carolina (#13010-18-47809 to ANK), National Institutes of Health (R01-NS041596, R01-NS089633, P01- NS055976 to JLT), National Science Foundation (MCB-1020970 to JLT), Department of Defense (W81XWH-13-1-0308 to JLT), US-Israel Binational Science Foundation (2011329 to JLT), and the Dr. Miriam and Sheldon G. Adelson Medical Research Foundation (to JLT). JLT is the incumbent South Carolina SmartState Chair in Childhood Neurotherapeutics at the University of South Carolina.

References

Abe, N., Borson, S. H., Gambello, M. J., Wang, F., & Cavalli, V. (2010). mTOR activation increases axonal growth capacity of injured peripheral nerves. *Journal of Biological Chemistry, 285*, 28034–28043. https://doi.org/10.1074/jbc.M110.125336

Akins, M. R., Berk-Rauch, H. E., Kwan, K. Y., Mitchell, M. E., Shepard, K. A., Korsak, L. I., . . . Fallon, J. R. (2017). Axonal ribosomes and mRNAs associate with fragile X granules in adult rodent and human brains. *Human Molecular Genetics, 26*(1), 192–209. https://doi.org/10.1093/hmg/ddw381

Akten, B., Kye, M. J., Hao le, T., Wertz, M. H., Singh, S., Nie, D., . . . Sahin, M. (2011). Interaction of survival of motor neuron (SMN) and HuD proteins with mRNA cpg15 rescues motor neuron axonal deficits. *Proceedings of the National Academy of Sciences of the United States of America, 108*(25), 10337–10342. https://doi.org/10.1073/pnas.1104928108

Alrahbeni, T., Sartor, F., Anderson, J., Miedzybrodzka, Z., McCaig, C., & Muller, B. (2015). Full UPF3B function is critical for neuronal differentiation of neural stem cells. *Molecular Brain, 8*, 33. https://doi.org/10.1186/s13041-015-0122-1 [pii]

An, J. J., Gharami, K., Liao, G. Y., Woo, N. H., Lau, A. G., Vanevski, F., ... Xu, B. (2008). Distinct role of long 3′ UTR BDNF mRNA in spine morphology and synaptic plasticity in hippocampal neurons. *Cell, 134*(1), 175–187. https://doi.org/10.1016/j.cell.2008.05.045

Anderson, P., & Kedersha, N. (2008). Stress granules: the Tao of RNA triage. *Trends in Biochemical Science, 33*(3), 141–150. https://doi.org/10.1016/j.tibs.2007.12.003

Andrusiak, M. G., Sharifnia, P., Lyu, X., Wang, Z., Dickey, A. M., Wu, Z., ... Jin, Y. (2019). Inhibition of axon regeneration by liquid-like TIAR-2 granules. *Neuron, 104*(2), 290–304. https://doi.org/10.1016/j.neuron.2019.07.004

Aschrafi, A., Schwechter, A. D., Mameza, M. G., Natera-Naranjo, O., Gioio, A. E., & Kaplan, B. B. (2008). MicroRNA-338 regulates local cytochrome c oxidase IV mRNA levels and oxidative phosphorylation in the axons of sympathetic neurons. *Journal of Neuroscience, 28*(47), 12581–12590. https://doi.org/10.1523/JNEUROSCI.3338-08.2008

Baird, T. D., & Wek, R. C. (2012). Eukaryotic initiation factor 2 phosphorylation and translational control in metabolism. *Advances in Nutrition, 3*(3), 307–321. https://doi.org/10.3945/an.112.002113

Baleriola, J., & Hengst, U. (2015). Targeting axonal protein synthesis in neuroregeneration and degeneration. *Neurotherapeutics, 12*(1), 57–65. https://doi.org/10.1007/s13311-014-0308-8

Baleriola, J., Walker, C. A., Jean, Y. Y., Crary, J. F., Troy, C. M., Nagy, P. L., & Hengst, U. (2014). Axonally synthesized ATF4 transmits a neurodegenerative signal across brain regions. *Cell, 158*(5), 1159–1172. https://doi.org/10.1016/j.cell.2014.07.001

Banerjee, A., Ifrim, M. F., Valdez, A. N., Raj, N., & Bassell, G. J. (2018). Aberrant RNA translation in fragile X syndrome: From FMRP mechanisms to emerging therapeutic strategies. *Brain Research, 1693*(Pt A), 24–36. https://doi.org/10.1016/j.brainres.2018.04.008

Barrientos, S. A., Martinez, N. W., Yoo, S., Jara, J. S., Zamorano, S., Hetz, C., ... Court, F. A. (2011). Axonal degeneration is mediated by the mitochondrial permeability transition pore. *Journal of Neuroscience, 31*(3), 966–978. https://doi.org/10.1523/JNEUROSCI.4065-10.2011

Bassell, G. J., Zhang, H., Byrd, A. L., Femino, A. M., Singer, R. H., Taneja, K. L., ... Kosik, K. S. (1998). Sorting of beta-actin mRNA and protein to neurites and growth cones in culture. *Journal of Neuroscience, 18*(1), 251–265. https://doi.org/10.1523/JNEUROSCI.18-01-00251.1998

Baumbach, J., Murthy, A., McClintock, M. A., Dix, C. I., Zalyte, R., Hoang, H. T., & Bullock, S. L. (2017). Lissencephaly-1 is a context-dependent regulator of the human dynein complex. *Elife, 6*. https://doi.org/10.7554/eLife.21768

Ben-Yaakov, K., Dagan, S. Y., Segal-Ruder, Y., Shalem, O., Vuppalanchi, D., Willis, D. E., ... Fainzilber, M. (2012). Axonal transcription factors signal retrogradely in lesioned peripheral nerve. *EMBO Journal, 31*(6), 1350–1363. https://doi.org/10.1038/emboj.2011.494

Biever, A., Donlin-Asp, P. G., & Schuman, E. M. (2019). Local translation in neuronal processes. *Current Opinion in Neurobiology, 57*, 141–148. https://doi.org/10.1016/j.conb.2019.02.008

Bird, C. W., Gardiner, A. S., Bolognani, F., Tanner, D. C., Chen, C. Y., Lin, W. J., ... Perrone-Bizzozero, N. (2013). KSRP modulation of GAP-43 mRNA stability restricts axonal outgrowth in embryonic hippocampal neurons. *PLoS One, 8*(11), e79255. https://doi.org/10.1371/journal.pone.0079255

Bolognani, F., Contente-Cuomo, T., & Perrone-Bizzozero, N. I. (2010). Novel recognition motifs and biological functions of the RNA-binding protein HuD revealed by genome-wide identification of its targets. *Nucleic Acids Research, 38*(1), 117–130. https://doi.org/10.1093/nar/gkp863

Bradke, F., Fawcett, J. W., & Spira, M. E. (2012). Assembly of a new growth cone after axotomy: the precursor to axon regeneration. *Nature Reviews Neuroscience, 13*(3), 183–193. https://doi.org/10.1038/nrn3176

Bravo, R., Parra, V., Gatica, D., Rodriguez, A. E., Torrealba, N., Paredes, F., . . . Lavandero, S. (2013). Endoplasmic reticulum and the unfolded protein response: dynamics and metabolic integration. *International Review of Cell and Molecular Biology, 301*, 215–290. https://doi.org/10.1016/B978-0-12-407704-1.00005-1

Brazil, D. P., & Hemmings, B. A. (2001). Ten years of protein kinase B signalling: a hard Akt to follow. *Trends in Biochemical Sciences, 26*(11), 657–664. https://doi.org/S0968-0004(01)01958-2 [pii]

Briata, P., Bordo, D., Puppo, M., Gorlero, F., Rossi, M., Perrone-Bizzozero, N., & Gherzi, R. (2015). Diverse roles of the nucleic acid-binding protein KHSRP in cell differentiation and disease. *Wiley Interdisciplinary Reviews RNA, 7*(2), 227–240. https://doi.org/10.1002/wrna.1327

Briata, P., Chen, C. Y., Ramos, A., & Gherzi, R. (2013). Functional and molecular insights into KSRP function in mRNA decay. *Biochimica Biophysica Acta, 1829*(6-7), 689–694. https://doi.org/10.1016/j.bbagrm.2012.11.003

Briese, M., Saal, L., Appenzeller, S., Moradi, M., Baluapuri, A., & Sendtner, M. (2016). Whole transcriptome profiling reveals the RNA content of motor axons. *Nucleic Acids Research, 44*(4), e33. https://doi.org/10.1093/nar/gkv1027

Briese, M., Saal-Bauernschubert, L., Ji, C., Moradi, M., Ghanawi, H., Uhl, M., . . . Sendtner, M. (2018). hnRNP R and its main interactor, the noncoding RNA 7SK, coregulate the axonal transcriptome of motoneurons. *Proceedings of the National Academy of Sciences of the United States of America, 115*(12), E2859–E2868. https://doi.org/10.1073/pnas.1721670115

Brittis, P. A., Lu, Q., & Flanagan, J. G. (2002). Axonal protein synthesis provides a mechanism for localized regulation at an intermediate target. *Cell, 110*(2), 223–235. https://doi.org/10.1016/s0092-8674(02)00813-9

Brostrom, C. O., & Brostrom, M. A. (1990). Calcium-dependent regulation of protein synthesis in intact mammalian cells. *Annual Review of Physiology, 52*, 577–590. https://doi.org/10.1146/annurev.ph.52.030190.003045

Brunet, A., Bonni, A., Zigmond, M. J., Lin, M. Z., Juo, P., Hu, L. S., Greenberg, M. E. (1999). Akt promotes cell survival by phosphorylating and inhibiting a Forkhead transcription factor. *Cell, 96*(6), 857–868. https://doi.org/S0092-8674(00)80595-4 [pii]

Cagnetta, R., Wong, H. H., Frese, C. K., Mallucci, G. R., Krijgsveld, J., & Holt, C. E. (2019). Noncanonical modulation of the eIF2 pathway controls an increase in local translation during neural wiring. *Molecular Cell, 73*(3), 474–489 e475. https://doi.org/10.1016/j.molcel.2018.11.013

Calfon, M., Zeng, H., Urano, F., Till, J. H., Hubbard, S. R., Harding, H. P., . . . Ron, D. (2002). IRE1 couples endoplasmic reticulum load to secretory capacity by processing the XBP-1 mRNA. *Nature, 415*(6867), 92–96. https://doi.org/10.1038/415092a

Campbell, D. S., & Holt, C. E. (2001). Chemotropic responses of retinal growth cones mediated by rapid local protein synthesis and degradation. *Neuron, 32*(6), 1013–1026. https://doi.org/10.1016/s0896-6273(01)00551-7

Campbell, D. S., & Holt, C. E. (2003). Apoptotic pathway and MAPKs differentially regulate chemotropic responses of retinal growth cones. *Neuron, 37*(6), 939–952. https://doi.org/10.1016/s0896-6273(03)00158-2

Casadio, A., Martin, K. C., Giustetto, M., Zhu, H., Chen, M., Bartsch, D.,... Kandel, E. R. (1999). A transient, neuron-wide form of CREB-mediated long-term facilitation can be stabilized at specific synapses by local protein synthesis. *Cell, 99*(2), 221–237. https://doi.org/10.1016/s0092-8674(00)81653-0

Chen, W., Lu, N., Ding, Y., Wang, Y., Chan, L. T., Wang, X.,... Liu, K. (2016). Rapamycin-resistant mTOR activity is required for sensory axon regeneration induced by a conditioning lesion. *eNeuro, 3*(6), ENEURO.0358-0316.2016. https://doi.org/10.1523/ENEURO.0358-16.2016

Christie, K. J., Webber, C. A., Martinez, J. A., Singh, B., & Zochodne, D. W. (2010). PTEN inhibition to facilitate intrinsic regenerative outgrowth of adult peripheral axons. *Journal of Neuroscience, 30*(27), 9306–9315. https://doi.org/10.1523/JNEUROSCI.6271-09.2010

Christie, S. B., Akins, M. R., Schwob, J. E., & Fallon, J. R. (2009). The FXG: a presynaptic fragile X granule expressed in a subset of developing brain circuits. *Journal of Neuroscience, 29*(5), 1514–1524. https://doi.org/10.1523/JNEUROSCI.3937-08.2009

Chyung, E., LeBlanc, H. F., Fallon, J. R., & Akins, M. R. (2018). Fragile X granules are a family of axonal ribonucleoprotein particles with circuit-dependent protein composition and mRNA cargos. *Journal of Comparative Neurology, 526*(1), 96–108. https://doi.org/10.1002/cne.24321

Cioni, J. M., Lin, J. Q., Holtermann, A. V., Koppers, M., Jakobs, M. A. H., Azizi, A.,... Holt, C. E. (2019). Late endosomes act as mRNA translation platforms and sustain mitochondria in axons. *Cell, 176*(1-2), 56–72 e15. https://doi.org/10.1016/j.cell.2018.11.030

Colak, D., Ji, S. J., Porse, B. T., & Jaffrey, S. R. (2013). Regulation of axon guidance by compartmentalized nonsense-mediated mRNA decay. *Cell, 153*(6), 1252–1265. https://doi.org/10.1016/j.cell.2013.04.056

Cosker, K. E., Fenstermacher, S. J., Pazyra-Murphy, M. F., Elliott, H. L., & Segal, R. A. (2016). The RNA-binding protein SFPQ orchestrates an RNA regulon to promote axon viability. *Nature Neuroscience, 19*(5), 690–696. https://doi.org/10.1038/nn.4280

Crerar, H., Scott-Solomon, E., Bodkin-Clarke, C., Andreassi, C., Hazbon, M., Logie, E.,... Riccio, A. (2019). Regulation of NGF signaling by an axonal untranslated mRNA. *Neuron, 102*(3), 553–563. https://doi.org/10.1016/j.neuron.2019.02.011

Crino, P. B., Nathanson, K. L., & Henske, E. P. (2006). The tuberous sclerosis complex. *New England Journal of Medicine, 355*(13), 1345–1356. https://doi.org/10.1056/NEJMra055323

Denis-Donini, S., Branduardi, P., Campiglio, S., & Carnevali, M. D. (1998). Localization of calcitonin gene-related peptide mRNA in developing olfactory axons. *Cell Tissue Research, 294*(1), 81–91. https://doi.org/10.1007/s004410051158

Detrait, E., Yoo, S., Eddleman, C., Fukada, M., Bittner, G., & Fishman, H. (2000). Plasmalemmal repair of severed neurites of PC12 cells requires Ca^{2+} and synaptotagmin. *Journal of Neuroscience Research, 62*, 566–573.doi 10.1002/1097-4547(20001115)62:4<566::AID-JNR11>3.0.CO;2-4

Djuranovic, S., Nahvi, A., & Green, R. (2011). A parsimonious model for gene regulation by miRNAs. *Science, 331*(6017), 550–553. https://doi.org/10.1126/science.1191138

Donnelly, C. J., Park, M., Spillane, M., Yoo, S., Pacheco, A., Gomes, C., Twiss, J. L. (2013). Axonally synthesized beta-actin and GAP-43 proteins support distinct modes of axonal growth. *Journal of Neuroscience, 33*(8), 3311–3322. https://doi.org/10.1523/JNEUROSCI.1722-12.2013

Donnelly, C. J., Willis, D. E., Xu, M., Tep, C., Jiang, C., Yoo, S., . . . Twiss, J. L. (2011). Limited availability of ZBP1 restricts axonal mRNA localization and nerve regeneration capacity. *EMBO Journal, 30*(22), 4665–4677. https://doi.org/10.1038/emboj.2011.347

Donnelly, N., Gorman, A. M., Gupta, S., & Samali, A. (2013). The eIF2alpha kinases: their structures and functions. *Cellular and Molecular Life Sciences, 70*(19), 3493–3511. https://doi.org/10.1007/s00018-012-1252-6

Doyle, M., & Kiebler, M. A. (2011). Mechanisms of dendritic mRNA transport and its role in synaptic tagging. *EMBO Journal, 30*(17), 3540–3552. https://doi.org/10.1038/emboj.2011.278

Dubacq, C., Jamet, S., & Trembleau, A. (2009). Evidence for developmentally regulated local translation of odorant receptor mRNAs in the axons of olfactory sensory neurons. *Journal of Neuroscience, 29*(33), 10184–10190. https://doi.org/10.1523/JNEUROSCI.2443-09.2009

Elvira, G., Wasiak, S., Blandford, V., Tong, X. K., Serrano, A., Fan, X., Sossin, W. S. (2006). Characterization of an RNA granule from developing brain. *Molecular and Cellular Proteomics, 5*(4), 635–651. https://doi.org/10.1074/mcp.M500255-MCP200

Etienne-Manneville, S. (2009). APC in cell migration. *Advances in Experimental Medicine and Biology, 656*, 30–40 https://doi.org/10.1007/978-1-4419-1145-2_3

Farris, S., Lewandowski, G., Cox, C. D., & Steward, O. (2014). Selective localization of arc mRNA in dendrites involves activity- and translation-dependent mRNA degradation. *Journal of Neuroscience, 34*(13), 4481–4493. https://doi.org/10.1523/JNEUROSCI.4944-13.2014

Garcia-Mayoral, M. F., Hollingworth, D., Masino, L., Diaz-Moreno, I., Kelly, G., Gherzi, R., . . . Ramos, A. (2007). The structure of the C-terminal KH domains of KSRP reveals a noncanonical motif important for mRNA degradation. *Structure, 15*(4), 485–498. https://doi.org/10.1016/j.str.2007.03.006

Gardiner, A. S., Twiss, J. L., & Perrone-Bizzozero, N. I. (2015). Competing interactions of RNA-binding proteins, microRNAs, and their targets control neuronal development and function. *Biomolecules, 5*(4), 2903–2918. https://doi.org/10.3390/biom5042903

Garner, J. A. (1988). Differential turnover of tubulin and neurofilament proteins in central nervous system neuron terminals. *Brain Res, 458*(2), 309–318. https://doi.org/10.1016/0006-8993(88)90473-8

Georgescu, M. M. (2010). PTEN tumor suppressor network in PI3K-Akt pathway control. *Genes Cancer, 1*(12), 1170–1177. https://doi.org/10.1177_1947601911407325 [pii]

Gherzi, R., Lee, K. Y., Briata, P., Wegmuller, D., Moroni, C., Karin, M., & Chen, C. Y. (2004). A KH domain RNA binding protein, KSRP, promotes ARE-directed mRNA turnover by recruiting the degradation machinery. *Molecular Cell, 14*(5), 571–583. https://doi.org/10.1016/j.molcel.2004.05.002

Gherzi, R., Trabucchi, M., Ponassi, M., Ruggiero, T., Corte, G., Moroni, C., . . . Briata, P. (2006). The RNA-binding protein KSRP promotes decay of beta-catenin mRNA and is inactivated by PI3K-AKT signaling. *PLoS Biology, 5*(1), e5. https://doi.org/10.1371/journal.pbio.0050005

Gilks, N., Kedersha, N., Ayodele, M., Shen, L., Stoecklin, G., Dember, L. M., & Anderson, P. (2004). Stress granule assembly is mediated by prion-like aggregation of TIA-1. *Molecular Biology of the Cell, 15*(12), 5383–5398. https://doi.org/10.1091/mbc.e04-08-0715E04-08-0715 [pii]

Gingras, A. C., Gygi, S. P., Raught, B., Polakiewicz, R. D., Abraham, R. T., Hoekstra, M. F., . . . Sonenberg, N. (1999). Regulation of 4E-BP1 phosphorylation: a novel two-step mechanism. *Genes & Development, 13*(11), 1422–1437. https://doi.org/10.1101/gad.13.11.1422

Giorgi, C., Yeo, G. W., Stone, M. E., Katz, D. B., Burge, C., Turrigiano, G., & Moore, M. J. (2007). The EJC factor eIF4AIII modulates synaptic strength and neuronal protein expression. *Cell, 130*(1), 179–191. https://doi.org/10.1016/j.cell.2007.05.028

Gitler, D., & Spira, M. E. (1998). Real time imaging of calcium induced localized proteolytic activity after axotomy and its relation to growth cone formation. *Neuron, 20*, 1123–1135 https://doi.org/10.1016/s0896-6273(00)80494-8

Giuditta, A., Hunt, T., & Santella, L. (1986). Rapid important paper Messenger RNA in squid axoplasm. *Neurochemistry International, 8*(3), 435–442. https://doi.org/10.1016/0197-0186(86)90019-7

Giuditta, A., Menichini, E., Perrone Capano, C., Langella, M., Martin, R., Castigli, E., & Kaplan, B. B. (1991). Active polysomes in the axoplasm of the squid giant axon. *Journal of Neuroscience Research, 28*(1), 18–28. https://doi.org/10.1002/jnr.490280103

Glinka, M., Herrmann, T., Funk, N., Havlicek, S., Rossoll, W., Winkler, C., & Sendtner, M. (2010). The heterogeneous nuclear ribonucleoprotein-R is necessary for axonal beta-actin mRNA translocation in spinal motor neurons. *Human Molecular Genetics, 19*(10), 1951–1966. https://doi.org/10.1093/hmg/ddq073

Gomes, C., Lee, S. J., Gardiner, A. S., Smith, T., Sahoo, P. K., Patel, P., Twiss, J. L. (2017). Axonal localization of neuritin/CPG15 mRNA is limited by competition for HuD binding. *Journal of Cell Science, 130*(21), 3650–3662. https://doi.org/10.1242/jcs.201244

Gomes, C., Merianda, T. T., Lee, S. J., Yoo, S., & Twiss, J. L. (2014). Molecular determinants of the axonal mRNA transcriptome. *Developmental Neurobiology, 74*(3), 218–232. https://doi.org/10.1002/dneu.22123

Gouveia Roque, C., & Holt, C. E. (2018). Growth cone Tctp is dynamically regulated by guidance cues. *Frontiers in Molecular Neuroscience, 11*, 399. https://doi.org/10.3389/fnmol.2018.00399

Graber, T. E., Freemantle, E., Anadolu, M. N., Hebert-Seropian, S., MacAdam, R. L., Shin, U., . . . Sossin, W. S. (2017). UPF1 governs synaptic plasticity through association with a STAU2 RNA granule. *Journal of Neuroscience, 37*(38), 9116–9131. https://doi.org/10.1523/JNEUROSCI.0088-17.2017

Guertin, D. A., & Sabatini, D. M. (2007). Defining the role of mTOR in cancer. *Cancer Cell, 12*(1), 9–22. https://doi.org/10.1016/j.ccr.2007.05.008

Gumy, L. F., Yeo, G. S., Tung, Y. C., Zivraj, K. H., Willis, D., Coppola, G., . . . Fawcett, J. W. (2011). Transcriptome analysis of embryonic and adult sensory axons reveals changes in mRNA repertoire localization. *RNA, 17*(1), 85–98. https://doi.org/10.1261/rna.2386111

Hafner, A. S., Donlin-Asp, P. G., Leitch, B., Herzog, E., & Schuman, E. M. (2019a). Local protein synthesis is a ubiquitous feature of neuronal pre- and postsynaptic compartments. *Science, 364*(6441). https://doi.org/10.1126/science.aau3644

Hafner, A. S., Donlin-Asp, P. G., Leitch, B., Herzog, E., & Schuman, E. M. (2019b). Local protein synthesis is a ubiquitous feature of neuronal pre- and postsynaptic compartments. *Science, 364*(6441), eaau3644. https://doi.org/10.1126/science.aau3644

Hanz, S., Perlson, E., Willis, D., Zheng, J. Q., Massarwa, R., Huerta, J. J., Fainzilber, M. (2003). Axoplasmic importins enable retrograde injury signaling in lesioned nerve. *Neuron, 40*(6), 1095–1104. https://doi.org/10.1016/s0896-6273(03)00770-0

Hengst, U., Cox, L. J., Macosko, E. Z., & Jaffrey, S. R. (2006). Functional and selective RNA interference in developing axons and growth cones. *Journal of Neuroscience, 26*(21), 5727–5732. https://doi.org/10.1523/JNEUROSCI.5229-05.2006

Henley, J. R., Huang, K. H., Wang, D., & Poo, M. M. (2004). Calcium mediates bidirectional growth cone turning induced by myelin-associated glycoprotein. *Neuron, 44*(6), 909–916. https://doi.org/10.1016/j.neuron.2004.11.030

Hershey, J. W., Sonenberg, N., & Mathews, M. B. (2012). Principles of translational control: an overview. *Cold Spring Harbor Perspectives in Biology, 4*(12), pii: a011528. https://doi.org/10.1101/cshperspect.a011528

Hetz, C. (2012). The unfolded protein response: controlling cell fate decisions under ER stress and beyond. *Nature Reviews Molecular Cell Biology, 13*(2), 89–102. https://doi.org/10.1038/nrm3270

Hetz, C., Chevet, E., & Oakes, S. A. (2015). Proteostasis control by the unfolded protein response. *Nature Cell Biology, 17*(7), 829–838. https://doi.org/10.1038/ncb3184

Hetz, C., & Glimcher, L. H. (2009). Fine-tuning of the unfolded protein response: assembling the IRE1alpha interactome. *Molecular Cell, 35*(5), 551–561. https://doi.org/10.1016/j.molcel.2009.08.021

Hinnebusch, A. G., Ivanov, I. P., & Sonenberg, N. (2016). Translational control by 5′-untranslated regions of eukaryotic mRNAs. *Science, 352*(6292), 1413–1416. https://doi.org/10.1126/science.aad9868

Hinnebusch, A. G., & Lorsch, J. R. (2012). The mechanism of eukaryotic translation initiation: new insights and challenges. *Cold Spring Harbor Perspectives in Biology, 4*(10), pii: a011544. https://doi.org/10.1101/cshperspect.a011544

Hong, K., Nishiyama, M., Henley, J., Tessier-Lavigne, M., & Poo, M. (2000). Calcium signalling in the guidance of nerve growth by netrin-1. *Nature, 403*(6765), 93–98. https://doi.org/10.1038/47507

Hu, Y., Park, K. K., Yang, L., Wei, X., Yang, Q., Cho, K. S., . . . He, Z. (2012). Differential effects of unfolded protein response pathways on axon injury-induced death of retinal ganglion cells. *Neuron, 73*(3), 445–452. https://doi.org/10.1016/j.neuron.2011.11.026

Huang, Y. S., Kan, M. C., Lin, C. L., & Richter, J. D. (2006). CPEB3 and CPEB4 in neurons: analysis of RNA-binding specificity and translational control of AMPA receptor GluR2 mRNA. *EMBO Journal, 25*(20), 4865–4876. https://doi.org/10.1038/sj.emboj.7601322

Huttelmaier, S., Zenklusen, D., Lederer, M., Dictenberg, J., Lorenz, M., Meng, X., Singer, R. H. (2005). Spatial regulation of beta-actin translation by Src-dependent phosphorylation of ZBP1. *Nature, 438*(7067), 512–515. https://doi.org/10.1038/nature04115

Ill-Raga, G., Tajes, M., Busquets-Garcia, A., Ramos-Fernandez, E., Vargas, L. M., Bosch-Morato, M., . . . Munoz, F. J. (2015). Physiological control of nitric oxide in neuronal BACE1 translation by heme-regulated eIF2alpha kinase HRI induces synaptogenesis. *Antioxid Redox Signal, 22*(15), 1295–1307. https://doi.org/10.1089/ars.2014.6080

Inoki, K., Li, Y., Xu, T., & Guan, K. L. (2003). Rheb GTPase is a direct target of TSC2 GAP activity and regulates mTOR signaling. *Genes & Development, 17*(15), 1829–1834. https://doi.org/10.1101/gad.1110003

Ito, T., Kagoshima, M., Sasaki, Y., Li, C., Udaka, N., Kitsukawa, T., . . . Goshima, Y. (2000). Repulsive axon guidance molecule Sema3A inhibits branching morphogenesis of fetal mouse lung. *Mechanisms of Development, 97*(1-2), 35–45. https://doi.org/10.1016/s0925-4773(00)00401-9

Ivshina, M., Lasko, P., & Richter, J. D. (2014). Cytoplasmic polyadenylation element binding proteins in development, health, and disease. *Annual Review of Cell and Developmental Biology, 30*, 393–415. https://doi.org/10.1146/annurev-cellbio-101011-155831

Jaffrey, S. R., & Wilkinson, M. F. (2018). Nonsense-mediated RNA decay in the brain: emerging modulator of neural development and disease. *Nature Reviews Neuroscience, 19*(12), 715–728. https://doi.org/10.1038/s41583-018-0079-z

Jain, S., & Welshhans, K. (2016). Netrin-1 induces local translation of down syndrome cell adhesion molecule in axonal growth cones. *Developmental Neurobiology, 76*(7), 799–816. https://doi.org/10.1002/dneu.22360

Jeste, S. S., Sahin, M., Bolton, P., Ploubidis, G. B., & Humphrey, A. (2008). Characterization of autism in young children with tuberous sclerosis complex. *Journal of Child Neurology, 23*(5), 520–525. https://doi.org/10.1177/0883073807309788

Jolly, L. A., Homan, C. C., Jacob, R., Barry, S., & Gecz, J. (2013). The UPF3B gene, implicated in intellectual disability, autism, ADHD and childhood onset schizophrenia regulates neural progenitor cell behaviour and neuronal outgrowth. *Human Molecular Genetics, 22*(23), 4673–4687. https://doi.org/10.1093/hmg/ddt315

Jung, H., Yoon, B. C., & Holt, C. E. (2012). Axonal mRNA localization and local protein synthesis in nervous system assembly, maintenance and repair. *Nature Reviews Neuroscience, 13*(5), 308–324. https://doi.org/10.1038/nrn3210

Kalinski, A. L., Kar, A. N., Craver, J., Tosolini, A. P., Sleigh, J. N., Lee, S. J., . . . Twiss, J. L. (2019). Deacetylation of Miro1 by HDAC6 blocks mitochondrial transport and mediates axon growth inhibition. *Journal of Cell Biology, 218*, 1871–1890. https://doi.org/10.1083/jcb.201702187

Kandel, E. S., & Hay, N. (1999). The regulation and activities of the multifunctional serine/threonine kinase Akt/PKB. *Experimental Cell Research, 253*(1), 210–229. https://doi.org/10.1006/excr.1999.4690S0014-4827(99)94690-6 [pii]

Kang, H., & Schuman, E. M. (1996). A requirement for local protein synthesis in neurotrophin-induced hippocampal synaptic plasticity. *Science, 273*(5280), 1402–1406. https://doi.org/10.1126/science.273.5280.1402

Kar, A., Lee, S., & Twiss, J. (2018). Expanding axonal transcriptome brings new functions for axonally synthesized proteins in health and disease. *The Neuroscientist, 24*(2), 111–129. https://doi.org/10.1177/1073858417712668

Kar, A. N., Macgibeny, M. A., Gervasi, N. M., Gioio, A. E., & Kaplan, B. B. (2013). Intra-axonal Synthesis of Eukaryotic Translation Initiation Factors Regulates Local Protein Synthesis and Axon Growth in Rat Sympathetic Neurons. *Journal of Neuroscience, 33*(17), 7165–7174. https://doi.org/10.1523/JNEUROSCI.2040-12.2013

Keene, J. D. (2007). RNA regulons: coordination of post-transcriptional events. *Nature Reviews Genetics, 8*(7), 533–543. https://doi.org/10.1038/nrg2111

Khong, A., Matheny, T., Jain, S., Mitchell, S. F., Wheeler, J. R., & Parker, R. (2017). The stress granule transcriptome reveals principles of mRNA accumulation in stress granules. *Molecular Cell, 68*(4), 808–820 e805. https://doi.org/10.1016/j.molcel.2017.10.015

Kilchert, C., & Spang, A. (2011). Cotranslational transport of ABP140 mRNA to the distal pole of S. cerevisiae. *EMBO Journal, 30*(17), 3567–3580. https://doi.org/10.1038/emboj.2011.247

Kim, H. H., Kim, P., Phay, M., & Yoo, S. (2015). Identification of precursor microRNAs within distal axons of sensory neuron. *Journal Neurochemistry, 134*(2), 193–199. https://doi.org/10.1111/jnc.13140

Kim, H. J., Kim, N. C., Wang, Y. D., Scarborough, E. A., Moore, J., Diaz, Z., . . . Taylor, J. P. (2013). Mutations in prion-like domains in hnRNPA2B1 and hnRNPA1 cause multisystem proteinopathy and ALS. *Nature, 495*(7442), 467–473. https://doi.org/10.1038/nature11922

Kim, J. Y., Deglincerti, A., & Jaffrey, S. R. (2017). A Staufen1-mediated decay pathway influences the local transcriptome in axons. *Translation (Austin), 5*(2), e1414016. https://doi.org/10.1080/21690731.2017.1414016

Kim, K. W., Tang, N. H., Andrusiak, M. G., Wu, Z., Chisholm, A. D., & Jin, Y. (2018). A neuronal piRNA pathway inhibits axon regeneration in C. elegans. *Neuron, 97*(3), 511–519 e516. https://doi.org/10.1016/j.neuron.2018.01.014

Kim, V. N., Han, J., & Siomi, M. C. (2009). Biogenesis of small RNAs in animals. *Nature Reviews Molecular Cell Biology, 10*(2), 126–139. https://doi.org/10.1038/nrm2632

Kim, Y. K., Furic, L., Desgroseillers, L., & Maquat, L. E. (2005). Mammalian Staufen1 recruits Upf1 to specific mRNA 3′ UTRs so as to elicit mRNA decay. *Cell, 120*(2), 195–208. https://doi.org/10.1016/j.cell.2004.11.050

Kim, Y. K., Furic, L., Parisien, M., Major, F., DesGroseillers, L., & Maquat, L. E. (2007). Staufen1 regulates diverse classes of mammalian transcripts. *EMBO Journal, 26*(11), 2670–2681. https://doi.org/10.1038/sj.emboj.7601712

Kim, Y. K., & Maquat, L. E. (2019). UPFront and center in RNA decay: UPF1 in nonsense-mediated mRNA decay and beyond. *RNA, 25*(4), 407–422. https://doi.org/10.1261/rna.070136.118

Kislauskis, E. H., Zhu, X., & Singer, R. H. (1994). Sequences responsible for intracellular localization of beta-actin messenger RNA also affect cell phenotype. *Journal of Cell Biology, 127*(2), 441–451. https://doi.org/10.1083/jcb.127.2.441

Klann, E., & Dever, T. E. (2004). Biochemical mechanisms for translational regulation in synaptic plasticity. *Nature Review Neurosciences, 5*(12), 931–942. https://doi.org/10.1038/nrn1557

Kleiman, R., Banker, G., & Steward, O. (1990). Differential subcellular localization of particular mRNAs in hippocampal neurons in culture. *Neuron, 5*(6), 821–830. https://doi.org/10.1016/0896-6273(90)90341-C

Knutsen, J. H., Rodland, G. E., Boe, C. A., Haland, T. W., Sunnerhagen, P., Grallert, B., & Boye, E. (2015). Stress-induced inhibition of translation independently of eIF2alpha phosphorylation. *Journal of Cell Science, 128*(23), 4420–4427. https://doi.org/10.1242/jcs.176545

Koppers, M., Cagnetta, R., Shigeoka, T., Wunderlich, L. C., Vallejo-Ramirez, P., Qiaojin Lin, J., . . . Holt, C. E. (2019). Receptor-specific interactome as a hub for rapid cue-induced selective translation in axons. *Elife, 8*. https://doi.org/10.7554/eLife.48718

Krol, J., Loedige, I., & Filipowicz, W. (2010). The widespread regulation of microRNA biogenesis, function and decay. *Nature Reviews Genetics, 11*(9), 597–610. https://doi.org/10.1038/nrg2843

Kundel, M., Jones, K. J., Shin, C. Y., & Wells, D. G. (2009). Cytoplasmic polyadenylation element-binding protein regulates neurotrophin-3-dependent beta-catenin mRNA translation in developing hippocampal neurons. *Journal of Neuroscience, 29*(43), 13630–13639. https://doi.org/10.1523/JNEUROSCI.2910-08.2009

Kurosaki, T., Popp, M. W., & Maquat, L. E. (2019). Quality and quantity control of gene expression by nonsense-mediated mRNA decay. *Nature Reviews Molecular Cell Biology, 20*(7), 406–420. https://doi.org/10.1038/s41580-019-0126-2

Kwiatkowski, T. J., Jr., Bosco, D. A., Leclerc, A. L., Tamrazian, E., Vanderburg, C. R., Russ, C., . . . Brown, R. H., Jr. (2009). Mutations in the FUS/TLS gene on chromosome 16 cause familial amyotrophic lateral sclerosis. *Science, 323*(5918), 1205–1208. https://doi.org/10.1126/science.1166066

Kwok, H. H., Poon, P. Y., Mak, K. H., Zhang, L. Y., Liu, P., Zhang, H., . . . Wong, R. N. (2017). Role of G3BP1 in glucocorticoid receptor-mediated microRNA-15b and microRNA-23a biogenesis in endothelial cells. *Cellular and Molecular Life Sciences, 74*(19), 3613–3630. https://doi.org/10.1007/s00018-017-2540-y

Kye, M. J., Niederst, E. D., Wertz, M. H., Goncalves Ido, C., Akten, B., Dover, K. Z., . . . Sahin, M. (2014). SMN regulates axonal local translation via miR-183/mTOR pathway. *Human Molecular Genetics, 23*(23), 6318–6331. https://doi.org/10.1093/hmg/ddu350

Lee, H. Y., Zhou, K., Smith, A. M., Noland, C. L., & Doudna, J. A. (2013). Differential roles of human Dicer-binding proteins TRBP and PACT in small RNA processing. *Nucleic Acids Res, 41*(13), 6568–6576. https://doi.org/10.1093/nar/gkt361

Lee, S., Sahoo, P., Oses-Prieto, J., Kawaguchi, R., Kar, A., Oliver, D., ... Twiss, J. (2018). hnRNPs binding to the axonal localization motifs of Nrn1 and HMGB1 mRNAs define growth-associated RNA regulons. *Molecular and Cellular Proteomics, 17*, 2091–2106. https://doi.org/10.1074/mcp.RA118.000603

Lee, S. J., Oses-Prieto, J. A., Kawaguchi, R., Sahoo, P. K., Kar, A. N., Rozenbaum, M., ... Twiss, J. L. (2018). hnRNPs interacting with mRNA Localization motifs define axonal RNA regulons. *Molecular and Cellular Proteomics omics, 17*(11), 2091–2106. https://doi.org/10.1074/mcp.RA118.000603

Lemaire, P. A., Anderson, E., Lary, J., & Cole, J. L. (2008). Mechanism of PKR activation by dsRNA. *Journal of Molecular Biology, 381*(2), 351–360. https://doi.org/10.1016/j.jmb.2008.05.056

Lewandowski, G., & Steward, O. (2014). AAVshRNA-mediated suppression of PTEN in adult rats in combination with salmon fibrin administration enables regenerative growth of corticospinal axons and enhances recovery of voluntary motor function after cervical spinal cord injury. *Journal of Neuroscience, 34*(30), 9951–9962. https://doi.org/10.1523/JNEUROSCI.1996-14.2014

Li, C., Bassell, G. J., & Sasaki, Y. (2009). Fragile X mental retardation protein is involved in protein synthesis-dependent collapse of growth cones induced by Semaphorin-3A. *Frontiers in Neural Circuits, 3*, 11. https://doi.org/10.3389/neuro.04.011.2009

Lin, A. C., Tan, C. L., Lin, C. L., Strochlic, L., Huang, Y. S., Richter, J. D., & Holt, C. E. (2009). Cytoplasmic polyadenylation and cytoplasmic polyadenylation element-dependent mRNA regulation are involved in Xenopus retinal axon development. *Neural Development, 4*, 8. https://doi.org/10.1186/1749-8104-4-8

Litman, P., Barg, J., Rindzoonski, L., & Ginzburg, I. (1993). Subcellular localization of tau mRNA in differentiating neuronal cell culture: implications for neuronal polarity. *Neuron, 10*(4), 627–638. https://doi.org/0896-6273(93)90165-N [pii]

Liu, K., Lu, Y., Lee, J. K., Samara, R., Willenberg, R., Sears-Kraxberger, I., ... He, Z. (2010). PTEN deletion enhances the regenerative ability of adult corticospinal neurons. *Nature Neuroscience, 13*(9), 1075–1081. https://doi.org/10.1038/nn.2603

Lou, W. P., Mateos, A., Koch, M., Klussman, S., Yang, C., Lu, N., ... Martin-Villalba, A. (2017). Regulation of adult CNS axonal regeneration by the post-transcriptional regulator Cpeb1. *Frontiers in Molecular Neuroscience, 10*, 445. https://doi.org/10.3389/fnmol.2017.00445

Ma, X. M., & Blenis, J. (2009). Molecular mechanisms of mTOR-mediated translational control. *Nature Reviews Molecular Cell Biology, 10*(5), 307–318. https://doi.org/10.1038/nrm2672

Mackenzie, I. R., Nicholson, A. M., Sarkar, M., Messing, J., Purice, M. D., Pottier, C., ... Rademakers, R. (2017). TIA1 mutations in amyotrophic lateral sclerosis and frontotemporal dementia promote phase separation and alter stress granule dynamics. *Neuron, 95*(4), 808–816 e809. https://doi.org/10.1016/j.neuron.2017.07.025

Martin, K. C., Casadio, A., Zhu, H., Yaping, E., Rose, J. C., Chen, M., ... Kandel, E. R. (1997). Synapse-specific, long-term facilitation of Aplysia sensory to motor synapses: a function for local protein synthesis in memory storage. *Cell, 91*(7), 927–938. https://doi.org/S0092-8674(00)80484-5 [pii]

Merianda, T. T., Gomes, C., Yoo, S., Vuppalanchi, D., & Twiss, J. L. (2013). Axonal localization of neuritin/CPG15 mRNA in neuronal populations through distinct 5′ and 3′ UTR elements. *Journal of Neuroscience, 33*(34), 13735–13742. https://doi.org/10.1523/JNEUROSCI.0962-13.2013

Miao, L., Yang, L., Huang, H., Liang, F., Ling, C., & Hu, Y. (2016). mTORC1 is necessary but mTORC2 and GSK3beta are inhibitory for AKT3-induced axon regeneration in the central nervous system. *Elife, 5*, e14908. https://doi.org/10.7554/eLife.14908

Mili, S., Moissoglu, K., & Macara, I. G. (2008). Genome-wide screen reveals APC-associated RNAs enriched in cell protrusions. *Nature, 453*(7191), 115–119. https://doi.org/10.1038/nature06888

Miller, L. C., Blandford, V., McAdam, R., Sanchez-Carbente, M. R., Badeaux, F., DesGroseillers, L., & Sossin, W. S. (2009). Combinations of DEAD box proteins distinguish distinct types of RNA: protein complexes in neurons. *Molecular and Cellular Neuroscience, 40*(4), 485–495. https://doi.org/10.1016/j.mcn.2009.01.007

Minis, A., Dahary, D., Manor, O., Leshkowitz, D., Pilpel, Y., & Yaron, A. (2014). Subcellular transcriptomics-dissection of the mRNA composition in the axonal compartment of sensory neurons. *Developmental Neurobiology, 74*(3), 365–381. https://doi.org/10.1002/dneu.22140

Mitchell, P., & Tollervey, D. (2000). mRNA stability in eukaryotes. *Current Opinion in Genetics and Development, 10*(2), 193–198. https://doi.org/10.1016/s0959-437x(00)00063-0

Mohr, E., Fehr, S., & Richter, D. (1991). Axonal transport of neuropeptide encoding mRNAs within the hypothalamo-hypophyseal tract of rats. *EMBO Journal, 10*(9), 2419–2424. https://doi.org/10.1016/s0959-437x(00)00063-0

Mohr, E., & Richter, D. (1992). Diversity of mRNAs in the axonal compartment of peptidergic neurons in the rat. *European Journal of Neuroscience, 4*(9), 870–876. https://doi.org/10.1111/j.1460-9568.1992.tb00197.x

Moradi, M., Sivadasan, R., Saal, L., Luningschror, P., Dombert, B., Rathod, R. J., . . . Sendtner, M. (2017). Differential roles of alpha-, beta-, and gamma-actin in axon growth and collateral branch formation in motoneurons. *Journal of Cell Biology, 216*(3), 793–814. https://doi.org/10.1083/jcb.201604117

Murashov, A. K., Chintalgattu, V., Islamov, R. R., Lever, T. E., Pak, E. S., Sierpinski, P. L., . . . Van Scott, M. R. (2007). RNAi pathway is functional in peripheral nerve axons. *FASEB Journal, 21*(3), 656–670. https://doi.org/10.1096/fj.06-6155com

Natera-Naranjo, O., Aschrafi, A., Gioio, A. E., & Kaplan, B. B. (2010). Identification and quantitative analyses of microRNAs located in the distal axons of sympathetic neurons. *RNA, 16*(8), 1516–1529. https://doi.org/10.1261/rna.1833310

Neumann, M., Sampathu, D. M., Kwong, L. K., Truax, A. C., Micsenyi, M. C., Chou, T. T., . . . Lee, V. M. (2006). Ubiquitinated TDP-43 in frontotemporal lobar degeneration and amyotrophic lateral sclerosis. *Science, 314*(5796), 130–133. https://doi.org/10.1126/science.1134108

Nie, D., Di Nardo, A., Han, J. M., Baharanyi, H., Kramvis, I., Huynh, T., . . . Sahin, M. (2010). Tsc2-Rheb signaling regulates EphA-mediated axon guidance. *Nature Neuroscience, 13*(2), 163–172. https://doi.org/10.1038/nn.2477

Ning, K., Drepper, C., Valori, C. F., Ahsan, M., Wyles, M., Higginbottom, A., . . . Sendtner, M. (2010). PTEN depletion rescues axonal growth defect and improves survival in SMN-deficient motor neurons. *Human Molecular Genetics, 19*(16), 3159–3168. https://doi.org/10.1093/hmg/ddq226

Nonhoff, U., Ralser, M., Welzel, F., Piccini, I., Balzereit, D., Yaspo, M. L., . . . Krobitsch, S. (2007). Ataxin-2 interacts with the DEAD/H-box RNA helicase DDX6 and interferes with P-bodies and stress granules. *Molecular Biology of the Cell, 18*(4), 1385–1396. https://doi.org/10.1091/mbc.e06-12-1120

Onate, M., Catenaccio, A., Martinez, G., Armentano, D., Parsons, G., Kerr, B., . . . Court, F. A. (2016). Activation of the unfolded protein response promotes axonal regeneration after peripheral nerve injury. *Science Reports, 6*, 21709. https://doi.org/10.1038/srep21709

Pacheco, A., Merianda, T. T., Twiss, J. L., & Gallo, G. (2020). Mechanism and role of the intra-axonal Calreticulin translation in response to axonal injury. *Experimental Neurology, 323*, 113072. https://doi.org/10.1016/j.expneurol.2019.113072

Pacheco, A., & Twiss, J. L. (2012). Localized IRES-dependent translation of ER chaperone protein mRNA in sensory axons. *PLoS One, 7*(7), e40788. https://doi.org/10.1371/journal.pone.0040788

Pakos-Zebrucka, K., Koryga, I., Mnich, K., Ljujic, M., Samali, A., & Gorman, A. M. (2016). The integrated stress response. *EMBO Reports, 17*(10), 1374–1395. https://doi.org/10.15252/embr.201642195

Palam, L. R., Baird, T. D., & Wek, R. C. (2011). Phosphorylation of eIF2 facilitates ribosomal bypass of an inhibitory upstream ORF to enhance CHOP translation. *Journal of Biological Chemistry, 286*(13), 10939–10949. https://doi.org/10.1074/jbc.M110.216093

Panas, M. D., Kedersha, N., Schulte, T., Branca, R. M., Ivanov, P., & Anderson, P. (2019). Phosphorylation of G3BP1-S149 does not influence stress granule assembly. *Journal of Cell Biology, 218*, 2425–2432. https://doi.org/10.1083/jcb.201801214

Pannese, E., & Ledda, M. (1991). Ribosomes in myelinated axons of the rabbit spinal ganglion neurons. *Journal of Submicroscopic Cytology and Pathology, 23*(1), 33–38.

Park, E., & Maquat, L. E. (2013). Staufen-mediated mRNA decay. *Wiley Interdisciplinary Reviews RNA, 4*(4), 423–435. https://doi.org/10.1002/wrna.1168

Park, K. K., Liu, K., Hu, Y., Kanter, J. L., & He, Z. (2010). PTEN/mTOR and axon regeneration. *Experimental Neurology, 223*(1), 45–50. https://doi.org/10.1016/j.expneurol.2009.12.032

Park, K. K., Liu, K., Hu, Y., Smith, P. D., Wang, C., Cai, B., . . . He, Z. (2008). Promoting axon regeneration in the adult CNS by modulation of the PTEN/mTOR pathway. *Science, 322*(5903), 963–966. https://doi.org/10.1126/science.1161566

Parvin, S., Takeda, R., Sugiura, Y., Neyazaki, M., Nogi, T., & Sasaki, Y. (2018). Fragile X mental retardation protein regulates accumulation of the active zone protein Munc18-1 in presynapses via local translation in axons during synaptogenesis. *Neuroscience Research, 146*, 36–47. https://doi.org/10.1016/j.neures.2018.09.013

Pease-Raissi, S. E., Pazyra-Murphy, M. F., Li, Y., Wachter, F., Fukuda, Y., Fenstermacher, S. J., . . . Segal, R. A. (2017). Paclitaxel reduces axonal Bclw to initiate IP3R1-dependent axon degeneration. *Neuron, 96*(2), 373–386, e376. https://doi.org/10.1016/j.neuron.2017.09.034

Peng, W., Robertson, L., Gallinetti, J., Mejia, P., Vose, S., Charlip, A., . . . Mitchell, J. R. (2012). Surgical stress resistance induced by single amino acid deprivation requires Gcn2 in mice. *Science Translational Medicine, 4*(118), 118ra111. https://doi.org/10.1126/scitranslmed.3002629

Perlson, E., Hanz, S., Ben-Yaakov, K., Segal-Ruder, Y., Seger, R., & Fainzilber, M. (2005). Vimentin-dependent spatial translocation of an activated MAP kinase in injured nerve. *Neuron, 45*(5), 715–726. https://doi.org/10.1016/j.neuron.2005.01.023

Perry, R. B., Doron-Mandel, E., Iavnilovitch, E., Rishal, I., Dagan, S. Y., Tsoory, M., . . . Fainzilber, M. (2012). Subcellular knockout of importin beta1 perturbs axonal retrograde signaling. *Neuron, 75*(2), 294–305. https://doi.org/10.1016/j.neuron.2012.05.033

Perry, R. B., Rishal, I., Doron-Mandel, E., Kalinski, A. L., Medzihradszky, K. F., Terenzio, M., . . . Fainzilber, M. (2016). Nucleolin-mediated RNA localization regulates neuron growth and cycling cell size. *Cell Reports, 16*(6), 1664–1676. https://doi.org/10.1016/j.celrep.2016.07.005

Phay, M., Kim, H. H., & Yoo, S. (2015). Dynamic change and target prediction of axon-specific microRNAs in regenerating sciatic nerve. *PLoS One, 10*(9), e0137461. https://doi.org/10.1371/journal.pone.0137461

Phay, M., Kim, H. H., & Yoo, S. (2018). Analysis of piRNA-Like small non-coding RNAs present in axons of adult sensory neurons. *Molecular Neurobiology, 55*(1), 483–494. https://doi.org/10.1007/s12035-016-0340-2

Piper, M., Anderson, R., Dwivedy, A., Weinl, C., van Horck, F., Leung, K. M., & Holt, C. (2006). Signaling mechanisms underlying Slit2-induced collapse of Xenopus retinal growth cones. *Neuron, 49*(2), 215–228. https://doi.org/10.1016/j.neuron.2005.12.008

Preitner, N., Quan, J., Nowakowski, D. W., Hancock, M. L., Shi, J., Tcherkezian, J., . . . Flanagan, J. G. (2014). APC is an RNA-binding protein, and its interactome provides a link to neural development and microtubule assembly. *Cell, 158*(2), 368–382. https://doi.org/10.1016/j.cell.2014.05.042

Rangaraju, V., Lauterbach, M., & Schuman, E. M. (2019). Spatially stable mitochondrial compartments fuel local translation during plasticity. *Cell, 176*(1-2), 73–84 e15. https://doi.org/10.1016/j.cell.2018.12.013

Reineke, L. C., Tsai, W. C., Jain, A., Kaelber, J. T., Jung, S. Y., & Lloyd, R. E. (2017). Casein Kinase 2 Is Linked to Stress Granule Dynamics through Phosphorylation of the Stress Granule Nucleating Protein G3BP1. *Molecular and Cellular Biology, 37*(4), e00596–00516. https://doi.org/10.1128/MCB.00596-16

Roffe, M., Hajj, G. N., Azevedo, H. F., Alves, V. S., & Castilho, B. A. (2013). IMPACT is a developmentally regulated protein in neurons that opposes the eukaryotic initiation factor 2alpha kinase GCN2 in the modulation of neurite outgrowth. *Journal of Biological Chemistry, 288*(15), 10860–10869. https://doi.org/10.1074/jbc.M113.461970

Rotem, N., Magen, I., Ionescu, A., Gershoni-Emek, N., Altman, T., Costa, C. J., . . . Perlson, E. (2017). ALS along the axons: expression of coding and noncoding RNA differs in axons of ALS models. *Science Reports, 7*, 44500. https://doi.org/10.1038/srep44500

Ryu, H. G., Seo, J. Y., Jung, Y., Kim, S. W., Kim, E., Jang, S. K., & Kim, K. T. (2019). Upf1 regulates neurite outgrowth and branching by transcriptional and post-transcriptional modulation of Arc. *Journal of Cell Science, 132*(2). https://doi.org/10.1242/jcs.224055

Saal, L., Briese, M., Kneitz, S., Glinka, M., & Sendtner, M. (2014). Subcellular transcriptome alterations in a cell culture model of spinal muscular atrophy point to widespread defects in axonal growth and presynaptic differentiation. *RNA, 20*(11), 1789–1802. https://doi.org/10.1261/rna.047373.114

Sahoo, P. K., Lee, S. J., Jaiswal, P. B., Alber, S., Kar, A. N., Miller-Randolph, S., Twiss, J. L. (2018). Axonal G3BP1 stress granule protein limits axonal mRNA translation and nerve regeneration. *Nature Communications, 9*, 3358. https://doi.org/10.1038/s41467-018-05647-x

Sahoo, P. K., Smith, D. S., Perrone-Bizzozero, N., & Twiss, J. L. (2018). Axonal mRNA transport and translation at a glance. *Journal of Cell Science, 131 pii*(8), jcs196808. https://doi.org/10.1242/jcs.196808

Sambandan, S., Akbalik, G., Kochen, L., Rinne, J., Kahlstatt, J., Glock, C., . . . Schuman, E. M. (2017). Activity-dependent spatially localized miRNA maturation in neuronal dendrites. *Science, 355*(6325), 634–637. https://doi.org/10.1126/science.aaf8995

Schacher, S., Wu, F., Panyko, J. D., Sun, Z. Y., & Wang, D. (1999). Expression and branch-specific export of mRNA are regulated by synapse formation and interaction with specific postsynaptic targets. *Journal of Neuroscience, 19*(15), 6338–6347. https://doi.org/10.1523/JNEUROSCI.19-15-06338.1999

Scheffer, I. E., Heron, S. E., Regan, B. M., Mandelstam, S., Crompton, D. E., Hodgson, B. L., . . . Dibbens, L. M. (2014). Mutations in mammalian target of rapamycin regulator DEPDC5 cause focal epilepsy with brain malformations. *Annals of Neurology, 75*(5), 782–787. https://doi.org/10.1002/ana.24126

Shigeoka, T., Jung, H., Jung, J., Turner-Bridger, B., Ohk, J., Lin, J. Q., ... Holt, C. E. (2016). Dynamic axonal translation in developing and mature visual circuits. *Cell, 166*(1), 181–192. https://doi.org/10.1016/j.cell.2016.05.029

Snow, D. M., Atkinson, P. B., Hassinger, T. D., Letourneau, P. C., & Kater, S. B. (1994). Chondroitin sulfate proteoglycan elevates cytoplasmic calcium in DRG neurons. *Developmental Biology, 166*(1), 87–100. https://doi.org/10.1006/dbio.1994.1298

Sonenberg, N., & Hinnebusch, A. G. (2009). Regulation of translation initiation in eukaryotes: mechanisms and biological targets. *Cell, 136*(4), 731–745. https://doi.org/10.1016/j.cell.2009.01.042

Stambolic, V., Suzuki, A., de la Pompa, J. L., Brothers, G. M., Mirtsos, C., Sasaki, T., ... Mak, T. W. (1998). Negative regulation of PKB/Akt-dependent cell survival by the tumor suppressor PTEN. *Cell, 95*(1), 29–39. https://doi.org/S0092-8674(00)81780-8 [pii]

Steward, O., & Levy, W. B. (1982). Preferential localization of polyribosomes under the base of dendritic spines in granule cells of the dentate gyrus. *Journal of Neuroscience, 2*(3), 284–291. https://doi.org/10.1523/JNEUROSCI.02-03-00284.1982

Sun, F., Park, K. K., Belin, S., Wang, D., Lu, T., Chen, G., ... He, Z. (2011). Sustained axon regeneration induced by co-deletion of PTEN and SOCS3. *Nature, 480*(7377), 372–375. https://doi.org/10.1038/nature10594

Suragani, R. N., Zachariah, R. S., Velazquez, J. G., Liu, S., Sun, C. W., Townes, T. M., & Chen, J. J. (2012). Heme-regulated eIF2alpha kinase activated Atf4 signaling pathway in oxidative stress and erythropoiesis. *Blood, 119*(22), 5276–5284. https://doi.org/10.1182/blood-2011-10-388132

Sutton, M. A., & Schuman, E. M. (2006). Dendritic protein synthesis, synaptic plasticity, and memory. *Cell, 127*(1), 49–58. https://doi.org/10.1016/j.cell.2006.09.014

Taliaferro, J. M., Vidaki, M., Oliveira, R., Olson, S., Zhan, L., Saxena, T., ... Burge, C. B. (2016). Distal Alternative Last Exons Localize mRNAs to Neural Projections. *Molecular Cell, 61*(6), 821–833. https://doi.org/10.1016/j.molcel.2016.01.020

Tang, G. (2005). siRNA and miRNA: an insight into RISCs. *Trends in Biochemical Sciences, 30*(2), 106–114. https://doi.org/10.1016/j.tibs.2004.12.007

Tcherkezian, J., Brittis, P. A., Thomas, F., Roux, P. P., & Flanagan, J. G. (2010). Transmembrane receptor DCC associates with protein synthesis machinery and regulates translation. *Cell, 141*(4), 632–644. https://doi.org/10.1016/j.cell.2010.04.008

Terenzio, M., Koley, S., Samra, N., Rishal, I., Zhao, Q., Sahoo, P. K., ... Fainzilber, M. (2018). Locally translated mTOR controls axonal local translation in nerve injury. *Science, 359*(6382), 1416–1421. https://doi.org/10.1126/science.aan1053

Teske, B. F., Wek, S. A., Bunpo, P., Cundiff, J. K., McClintick, J. N., Anthony, T. G., & Wek, R. C. (2011). The eIF2 kinase PERK and the integrated stress response facilitate activation of ATF6 during endoplasmic reticulum stress. *Molecular Biology of the Cell, 22*(22), 4390–4405. https://doi.org/10.1091/mbc.E11-06-0510

Tiruchinapalli, D. M., Oleynikov, Y., Kelic, S., Shenoy, S. M., Hartley, A., Stanton, P. K., ... Bassell, G. J. (2003). Activity-dependent trafficking and dynamic localization of zipcode binding protein 1 and beta-actin mRNA in dendrites and spines of hippocampal neurons. *Journal of Neuroscience, 23*(8), 3251–3261. https://doi.org/10.1523/JNEUROSCI.23-08-03251.2003

Toth, C. C., Willis, D., Twiss, J. L., Walsh, S., Martinez, J. A., Liu, W. Q., ... Zochodne, D. W. (2009). Locally synthesized calcitonin gene-related peptide has a critical role in peripheral nerve regeneration. *Journal of Neuropathology & Experimental Neurology, 68*(3), 326–337. https://doi.org/10.1097/NEN.0b013e31819ac71b

Tourriere, H., Chebli, K., Zekri, L., Courselaud, B., Blanchard, J. M., Bertrand, E., & Tazi, J. (2003). The RasGAP-associated endoribonuclease G3BP assembles stress granules. *Journal of Cell Biology, 160*(6), 823–831. https://doi.org/10.1083/jcb.200212128

Tourriere, H., & Tazi, J. (2019). Reply to "Phosphorylation of G3BP1-S149 does not influence stress granule assembly." *Journal of Cell Biology, 218*, 2433–2434. https://doi.org/10.1083/jcb.201905105

Trabucchi, M., Briata, P., Garcia-Mayoral, M., Haase, A. D., Filipowicz, W., Ramos, A., .Â… Rosenfeld, M. G. (2009). The RNA-binding protein KSRP promotes the biogenesis of a subset of microRNAs. *Nature, 459*(7249), 1010–1014. https://doi.org/10.1038/nature08025

Tsang, B., Arsenault, J., Vernon, R. M., Lin, H., Sonenberg, N., Wang, L. Y.,…Forman-Kay, J. D. (2019). Phosphoregulated FMRP phase separation models activity-dependent translation through bidirectional control of mRNA granule formation. *Proceedings of the National Academy of Sciences of the United States of America, 116*(10), 4218–4227. https://doi.org/10.1073/pnas.1814385116

Tushev, G., Glock, C., Heumuller, M., Biever, A., Jovanovic, M., & Schuman, E. M. (2018). Alternative 3′ UTRs Modify the Localization, Regulatory Potential, Stability, and Plasticity of mRNAs in Neuronal Compartments. *Neuron, 98*(3), 495–511, e496. https://doi.org/10.1016/j.neuron.2018.03.030

Urbanska, A. S., Janusz-Kaminska, A., Switon, K., Hawthorne, A. L., Perycz, M., Urbanska, M.,… Jaworski, J. (2017). ZBP1 phosphorylation at serine 181 regulates its dendritic transport and the development of dendritic trees of hippocampal neurons. *Science Reports, 7*(1), 1876. https://doi.org/10.1038/s41598-017-01963-2

Valenzuela, V., Collyer, E., Armentano, D., Parsons, G. B., Court, F. A., & Hetz, C. (2012). Activation of the unfolded protein response enhances motor recovery after spinal cord injury. *Cell Death & Disease, 3*, e272. https://doi.org/10.1038/cddis.2012.8

Van Minnen, J., Bergman, J. J., Van Kesteren, E. R., Smit, A. B., Geraerts, W. P., Lukowiak, K., Syed, N. I. (1997). De novo protein synthesis in isolated axons of identified neurons. *Neuroscience, 80*(1), 1–7. https://doi.org/10.1016/s0306-4522(97)00137-1

Villarin, J. M., McCurdy, E. P., Martinez, J. C., & Hengst, U. (2016). Local synthesis of dynein cofactors matches retrograde transport to acutely changing demands. *Nature Communications, 7*, 13865. https://doi.org/10.1038/ncomms13865

Villegas, R., Martinez, N. W., Lillo, J., Pihan, P., Hernandez, D., Twiss, J. L., & Court, F. A. (2014). Calcium release from intra-axonal endoplasmic reticulum leads to axon degeneration through mitochondrial dysfunction. *Journal of Neuroscience, 34*(21), 7179–7189. https://doi.org/10.1523/JNEUROSCI.4784-13.2014https://doi.org/10.1523/JNEUROSCI.4784-13.2014

Vuppalanchi, D., Coleman, J., Yoo, S., Merianda, T. T., Yadhati, A. G., Hossain, J., & Twiss, J. L. (2010). Conserved 3′-untranslated region sequences direct subcellular localization of chaperone protein mRNAs in neurons. *Journal of Biological Chemistry, 285*(23), 18025–18038. https://doi.org/10.1074/jbc.M109.061333

Vuppalanchi, D., Merianda, T. T., Donnelly, C., Pacheco, A., Williams, G., Yoo, S., & Twiss, J. L. (2012). Lysophosphatidic acid differentially regulates axonal mRNA translation through 5′ UTR elements. *Molecular and Cellular Neuroscience, 50*(2), 136–146. https://doi.org/10.1016/j.mcn.2012.04.001

Walker, B. A., Hengst, U., Kim, H. J., Jeon, N. L., Schmidt, E. F., Heintz, N.,…Jaffrey, S. R. (2012). Reprogramming axonal behavior by axon-specific viral transduction. *Gene Therapy, 19*(9), 947–955. https://doi.org/10.1038/gt.2011.217

Walker, B. A., Ji, S. J., & Jaffrey, S. R. (2012). Intra-axonal translation of RhoA promotes axon growth inhibition by CSPG. *Journal Neuroscience, 32*(41), 14442–14447. https://doi.org/10.1523/JNEUROSCI.0176-12.2012

Wells, D. G. (2006). RNA-binding proteins: a lesson in repression. *Journal of Neuroscience, 26*(27), 7135–7138. https://doi.org/10.1523/JNEUROSCI.1795-06.2006

Welshhans, K., & Bassell, G. J. (2011). Netrin-1-induced local beta-actin synthesis and growth cone guidance requires zipcode binding protein 1. *Journal of Neuroscience, 31*(27), 9800–9813. https://doi.org/10.1523/JNEUROSCI.0166-11.2011

Willis, D., Li, K. W., Zheng, J. Q., Chang, J. H., Smit, A. B., Kelly, T., . . . Twiss, J. L. (2005). Differential transport and local translation of cytoskeletal, injury-response, and neurodegeneration protein mRNAs in axons. *Journal Neuroscience, 25*(4), 778–791. https://doi.org/10.1523/JNEUROSCI.4235-04.2005

Willis, D. E., Xu, M., Donnelly, C. J., Tep, C., Kendall, M., Erenstheyn, M., . . . Twiss, J. L. (2011). Axonal Localization of transgene mRNA in mature PNS and CNS neurons. *Journal of Neuroscience, 31*(41), 14481–14487. https://doi.org/10.1523/JNEUROSCI.2950-11.2011

Wu, K. Y., Hengst, U., Cox, L. J., Macosko, E. Z., Jeromin, A., Urquhart, E. R., & Jaffrey, S. R. (2005). Local translation of RhoA regulates growth cone collapse. *Nature, 436*(7053), 1020–1024. https://doi.org/10.1038/nature03885

Ying, Z., Misra, V., & Verge, V. M. (2014). Sensing nerve injury at the axonal ER: activated Luman/CREB3 serves as a novel axonally synthesized retrograde regeneration signal. *Proceedings of the National Academy of Sciences of the United States of America, 111*(45), 16142–16147. https://doi.org/10.1073/pnas.1407462111

Ying, Z., Zhai, R., McLean, N. A., Johnston, J. M., Misra, V., & Verge, V. M. (2015). The unfolded protein response and cholesterol biosynthesis link luman/CREB3 to regenerative axon growth in sensory neurons. *Journal of Neuroscience, 35*(43), 14557–14570. https://doi.org/10.1523/JNEUROSCI.0012-15.2015

Yoo, S., Kim, H. H., Kim, P., Donnelly, C. J., Kalinski, A. L., Vuppalanchi, D., . . . Twiss, J. L. (2013). A HuD-ZBP1 ribonucleoprotein complex localizes GAP-43 mRNA into axons through its 3′ untranslated region AU-rich regulatory element. *Journal of Neurochemistry, 126*(6), 792–804. https://doi.org/10.1111/jnc.12266

Younts, T. J., Monday, H. R., Dudok, B., Klein, M. E., Jordan, B. A., Katona, I., & Castillo, P. E. (2016). Presynaptic protein synthesis is required for long-term plasticity of GABA release. *Neuron, 92*(2), 479–492. https://doi.org/10.1016/j.neuron.2016.09.040

Yudin, D., Hanz, S., Yoo, S., Iavnilovitch, E., Willis, D., Gradus, T., . . . Fainzilber, M. (2008). Localized regulation of axonal RanGTPase controls retrograde injury signaling in peripheral nerve. *Neuron, 59*(2), 241–252. https://doi.org/10.1016/j.neuron.2008.05.029

Zelena, J. (1972a). Ribosomes in myelinated axons of dorsal root ganglia. *Zeitschrift für Zellforschung und mikroskopische Anatomie, 124*(2), 217–229. https://doi.org/10.1007/bf00335680

Zelena, J. (1972b). Ribosomes in the axoplasm of myelinated nerve fibres. *Folia Morphologica (Praha), 20*(1), 91–93. 1972;20(1):91–93.

Zhang, Y., Ueno, Y., Liu, X. S., Buller, B., Wang, X., Chopp, M., & Zhang, Z. G. (2013). The MicroRNA-17-92 cluster enhances axonal outgrowth in embryonic cortical neurons. *JournalofNeuroscience,33*(16),6885–6894.https://doi.org/10.1523/JNEUROSCI.5180-12.2013

Zheng, J. Q. (2000). Turning of nerve growth cones induced by localized increases in intracellular calcium ions. *Nature, 403*(6765), 89–93. https://doi.org/10.1038/47501

Zheng, J. Q., Kelly, T. K., Chang, B., Ryazantsev, S., Rajasekaran, A. K., Martin, K. C., & Twiss, J. L. (2001). A functional role for intra-axonal protein synthesis during axonal regeneration from adult sensory neurons. *Journal of Neuroscience, 21*(23), 9291–9303 https://doi.org/10.1523/JNEUROSCI.21-23-09291.2001

Zhou, J., & Parada, L. F. (2012). PTEN signaling in autism spectrum disorders. *Current Opinion in Neurobiology, 22*(5), 873–879. https://doi.org/10.1016/j.conb.2012.05.004

Zivraj, K. H., Tung, Y. C., Piper, M., Gumy, L., Fawcett, J. W., Yeo, G. S., & Holt, C. E. (2010). Subcellular profiling reveals distinct and developmentally regulated repertoire of growth cone mRNAs. *Journal of Neuroscience, 30*(46), 15464–15478. https://doi.org/10.1523/JNEUROSCI.1800-10.2010

Zukor, K., Belin, S., Wang, C., Keelan, N., Wang, X., & He, Z. (2013). Short hairpin RNA against PTEN enhances regenerative growth of corticospinal tract axons after spinal cord injury. *Journal of Neuroscience, 33*(39), 15350–15361. https://doi.org/10.1523/JNEUROSCI.2510-13.2013

CHAPTER 16

PROTEIN SYNTHESIS AND TRANSLATIONAL CONTROL IN NEURAL STEM CELL DEVELOPMENT AND NEUROGENESIS

LAMEES MOHAMMAD, JOSCELYN WISEMAN, SARAH ERICKSON, AND GUANG YANG

Introduction

The mammalian central nervous system originates from embryonic neural stem/progenitor cells (NSCs) that are capable of self-renewal or differentiation into all major neural lineages (neurons and glia; Taverna, Gotz, & Huttner, 2014). The earliest population of NSCs consists of neuroepithelial cells (NECs) surrounding the ventricle of the neural tube, which elongate radially to the pial surface. These highly proliferative cells divide symmetrically to make identical daughter stem cells, thereby expanding the pool of NSCs. At the onset of neurogenesis, NECs transform into radial glial cells (RGCs). This transition is accompanied by the switch of NSC division mode from symmetric, proliferative to asymmetric, self-renewing (Delaunay et al., 2017). In this latter mode of division, one RGC gives rise to a daughter RGC, maintaining the stem cell pool, and producing a more differentiated cell, such as a neuron (direct neurogenesis) or a fate-restricted intermediate progenitor cell (IPC) that undergo transient proliferation to make multiple neurons (indirect neurogenesis). After neurogenesis has ceased, NSCs continue to make other cell types (e.g., astrocytes and oligodendrocytes; Taverna,

Gotz, & Huttner, 2014). A few NSCs persist into adulthood, becoming adult NSCs capable of generating neurons throughout life (Bond, Ming, & Song, 2015).

Neurogenic differentiation and lineage transition of NSCs is accompanied by dynamic, rapid and coordinated changes in gene expression (Albert & Huttner, 2018, Lennox, Mao, & Silver, 2018). Most research has focused on transcriptional control of fate determinants in NSCs (Albert & Huttner, 2018, Lennox, Mao, & Silver, 2018). However, it has become increasingly apparent that the transcriptome profile of many cell types does not always correlate with the proteome, and gene-specific correlations vary significantly among different cell types at different times (Liu, Beyer, & Aebersold, 2016; Ghazalpour et al., 2011; Edfors et al., 2016; Haider & Pal, 2013). The discordance between mRNA and protein levels is particularly evident in cells and tissues that are engaged in growth and differentiation processes, such as embryonic stem cells (Corsini et al., 2018; Bulut-Karslioglu et al., 2018) as well as the developing mouse limb bud (Fujii et al., 2017). Similarly, NSCs in the developing mouse cerebral cortex have been found to robustly transcribe many proneural genes that promote differentiation, despite the absence of corresponding proteins (Yang et al., 2014). While proteostasis is controlled by both the production and degradation of proteins (Tuoc & Stoykova, 2010), the rate of protein synthesis is found to play a predominant role in determining cell differentiation (Kristensen, Gsponer, & Foster, 2013).

Protein synthesis is most commonly regulated at the step of mRNA translation (Larsson, Tian, & Sonenberg, 2013). At the global level, translation determines the dynamics of the proteome (Sonenberg & Hinnebusch, 2009). At the individual gene level, it fine-tunes protein production using target-specific mechanisms, which frequently involve *cis*-regulatory elements present in mRNA and *trans*-acting factors such as RNA-binding proteins (RBPs; Shi & Barna, 2015, Gebauer, Preiss, & Hentze, 2012). In this review, we will discuss the dynamics of protein synthesis in NSCs during neurogenesis and the regulatory mechanisms at both global and gene-specific levels. Examples of RBPs (Pilaz & Silver, 2015: Popovitchenko & Rasin, 2017), microRNAs (Lennox, Mao, & Silver, 2018) and long non-coding RNAs (Quan, Zheng, & Qing, 2017; Aprea & Calegari, 2015) involved in brain development and NSC activation (Baser, Skabkin, & Martin-Villalba, 2017) have been discussed in excellent recent reviews. Here, we will discuss the latest findings on protein synthesis in NSCs and its translational regulation, highlighting examples to illustrate the close relationship between protein synthesis and NSC development and neurogenesis.

Global Protein Synthesis in NSCs

The synthesis of proteins from mRNA is a fundamental process in all living cells. Rather than just a housekeeping function as is often assumed, studies from multiple stem cell systems indicate that changes in overall protein synthesis rates play a specific role in stem cell fate decision (Buszczak, Signer, & Morrison, 2014). For example, in the hematopoietic system, slow-dividing stem cells synthesized fewer proteins than their more differentiated but fast-dividing progenitors (Signer et al., 2014). Although cells that undergo active cell division generally demand more proteins and thus synthesize proteins more rapidly, it was found that when slow-dividing hematopoietic stem cells were induced to divide, they still maintained a lower rate of protein synthesis (Signer et al., 2014). This suggests that low protein synthesis rates in hematopoietic stem cells are not a nonspecific consequence that reflects merely the cell cycle and proliferation status, but rather a stem cell fate determinant. Indeed, a modest decrease or increase of global protein synthesis in hematopoietic stem cells perturbed their fate decision between self-renewal and differentiation (Signer et al., 2014). Similar observations have been made in quiescent skin and muscle stem cells where they displayed low protein synthesis rates independent of the proliferation status, which was required for stem cell maintenance (Blanco et al., 2016; Zismanov et al., 2016). Therefore, low protein synthesis appears to represent a hallmark trait of adult somatic stem cells and is essential for their overall homeostasis.

Low Protein Synthesis Rates as a Hallmark of Adult NSCs

Somatic NSCs in the adult brain are also largely quiescent (Bond, Ming, & Song, 2015). During neurogenesis, quiescent adult NSCs undergo an activation step and re-enter the cell cycle to generate fast-dividing IPCs, which in turn give rise to neuroblasts and neurons (Bond, Ming, & Song, 2015). Similar to hematopoietic, skin and muscle stem cells, quiescent NSCs were found to have markedly lower rates of protein synthesis compared to activated NSCs as well as more differentiated fast-dividing IPCs (Figure 16.1; Llorens-Bobadilla et al., 2015; Baser et al., 2019; Hwang et al., 2018). Abnormal increase of protein synthesis in quiescent NSCs by deleting the tumor suppressor gene *Pten* led to aberrant activation of NSCs and ultimately depleted the adult NSC pool due to premature differentiation (Bonaguidi et al., 2011). These findings are consistent with the idea that overall protein synthesis rates may act in a specific way to determine NSC quiescence and homeostasis.

Changes in Protein Synthesis during NSC Lineage Progression

Quiescent adult NSCs are derived from embryonic NSCs (Bond, Ming, & Song, 2015; Taverna, Gotz, & Huttner, 2014). Given that low protein synthesis is an intrinsic property of quiescent adult NSCs, an interesting question arises: Is lowering protein synthesis rates in a subset of embryonic NSCs required and sufficient for their entry into quiescence to become adult NSCs? While a direct comparison between embryonic and adult NSCs is lacking, Chau et al. (2018) showed a drop in protein synthesis rates along the lineage progression of embryonic NSCs. It was found that as the earliest NSCs, NECs had much higher proteins synthesis rates compared to RGCs, which are derived from

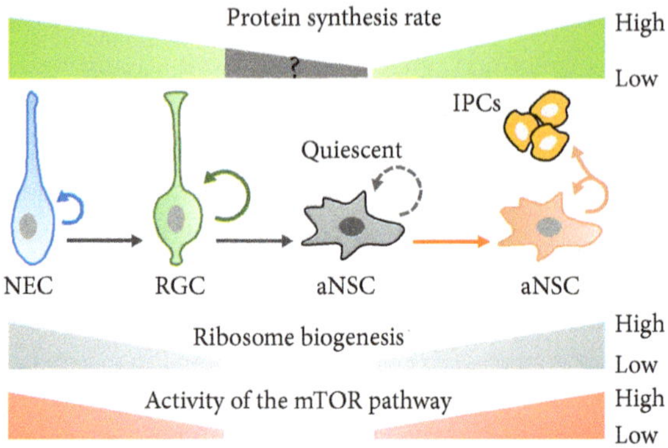

FIGURE 16.1. Control of global protein synthesis during NSC lineage progression. The transition from neuroepithelial cells (NECs) to radial glial cells (RGCs) in the embryonic brain is associated with extended cell cycle and accompanied by a decrease of global protein synthesis rates, as well as a reduction of mTOR signaling and ribosome biogenesis. Adult NSCs (aNSCs) originated from embryonic RGCs are quiescent with low protein synthesis rates. Activation of aNSCs and their differentiation into intermediate progenitors (IPCs) promote ribosome biogenesis, activate mTOR signaling, and enhance protein synthesis.

NECs (Chau et al., 2018). This reduction in protein synthesis in RGCs was unlikely due to their extended cell cycle length and reduced proliferation, since increasing or decreasing the expression of Myc, a master regulator of cell proliferation, did not change global protein synthesis rates in RGCs (Chau et al., 2018). Nonetheless, an aberrant increase of protein synthesis in RGCs through the deletion of *Pten* led to increased proliferation (Groszer et al., 2001). This suggests that global protein synthesis is not a simple adaptation to the cell cycle status of NSCs but may play a specific role in controlling NSC lineage progression (Figure 16.1). It will be interesting to determine whether protein synthesis rates further decrease in some or all RGCs over the developmental time and whether this reduction contributes to the establishment of adult NSCs.

Translational Control of Global Protein Synthesis in Neurogenesis

Dynamic changes in protein synthesis during NSC activation are accompanied by changes in the major components of the translational machinery, such as translation initiation factors and ribosomal proteins. For example, upon the differentiation of NSCs to neurons, enhanced protein synthesis was correlated with an increase in the transcriptional upregulation of ribosomal subunits and in the phosphorylation of the eukaryotic translation initiation factor 4E (eIF4E; Shin et al., 2015; Chau et al., 2018; Hwang et al., 2018), which synergistically promote global translation. This suggests that the regulation of protein synthesis at the step of translation may play a significant role in NSC maintenance and differentiation.

Cap-Dependent Translation, Initiation Factors and mTORC1

Most mammalian mRNA undergoes cap-dependent translation (Sonenberg & Hinnebusch, 2009). As the rate-limiting factor, eIF4E controls translational initiation by binding to the 5′cap of mRNA and interacts with eIF4G to recruit the translational machinery (Sonenberg & Hinnebusch, 2009; Jackson, Hellen, & Pestova, 2010). eIF4E can be sequestered from eIF4G by eIF4E-binding proteins (4E-BPs), thus preventing translation (Jackson, Hellen, & Pestova, 2010). In adult NSCs, Hartman et al. (2013) linked cap-dependent translation to NSC homeostasis by showing that the loss of 4E-BP enhanced NSC differentiation into neurons, and conversely, expressing a mutant form of 4E-BP that constantly bound to eIF4E to block cap-dependent translation promoted NSC self-renewal at the expense of differentiation. A critical upstream regulator of 4E-BPs and the cap-dependent translation is mechanistic target of rapamycin complex 1 (mTORC1), a central signaling hub that coordinates protein synthesis and cell growth (Meng, Frank, & Jewell, 2018). Activated mTORC1 phosphorylates 4E-BPs, releasing them from eIF4E and thus enhances global translation (Meng, Frank, & Jewell 2018). In quiescent NSCs, mTORC1 activity was found to be maintained at low levels, corresponding to their low protein synthesis rates (Paliouras et al., 2012). In contrast, more differentiated transient-amplifying IPCs possessed higher mTORC1 activity, consistent with the observation that NSC differentiation was accompanied by an upregulation of protein synthesis (Llorens-Bobadilla et al., 2015; Baser et al., 2019; Blair et al., 2017). Forced activation of mTORC1 depleted NSCs in the adult mouse brain due to aberrant differentiation (Hartman et al., 2013; Bonaguidi et al., 2011), whereas mTORC1 inactivation caused a lengthening of the NSC cell cycle and induced a quiescence-like phenotype, leading to impaired neurogenesis and microcephaly (Han et al., 2008; Paliouras et al., 2012; Cloëtta et al., 2013). Thus, the regulation of cap-dependent translation through the mTORC1-eIF4E pathway is essential for the balance of NSC maintenance and differentiation.

Another well-known regulator of global protein synthesis and stem cell homeostasis is eIF2, an initiation factor that brings methionyl-tRNA$_i$ (Met-tRNA$_i$) to ribosomes to initiate translation (Jackson, Hellen, & Pestova, 2010). Phosphorylation of the eIF2 α-subunit prevents reloading Met-tRNA$_i$ and attenuates translational initiation (Jackson, Hellen, & Pestova, 2010). In muscle stem cells, hyperphosphorylation of eIF2α maintained global protein synthesis at low levels to preserve stem cell quiescence (Zismanov et al., 2016). Interestingly, eIF2α was also found to be hyperphosphorylated in NSCs, correlating with low protein synthesis rates (Hwang et al., 2018). Upon induction of neurogenic differentiation, a decrease of eIF2α phosphorylation was observed, accompanying an increase in protein synthesis (Hwang et al., 2018). Nonetheless, it has yet to be determined whether eIF2α mediates the translational changes in NSCs and contributes to neurogenesis.

Transfer RNA and RNA Modifications

Like other RNA species, tRNA molecules undergo various post-transcriptional modifications that contribute to the regulation of global protein synthesis (Roundtree et al., 2017; Pan, 2018). Loss of appropriate modifications can cause tRNA degradation, resulting in reduced translational accuracy and protein synthesis (Pan, 2018; Roundtree et al., 2017). For example, some cytoplasmic tRNAs bare a 5-methoxycarbonylmethyl (mcm^5) or 5-carbamoylmethyl (ncm^5) modification at the wobble uridine site (U_{34}), loss of which weakens the efficiency and fidelity of translation (Tuorto & Lyko, 2016). The mcm^5 and ncm^5 modifications are catalyzed by the Elongator (Elp) complex (Tuorto & Lyko, 2016). Laguesse et al. (2015) showed that deletion of the catalytic subunit (*Elp3*) of the Elp complex in mice elicited codon-specific translational pausing, which impaired NSC differentiation and ultimately led to microcephaly. Codon-specific translation defects can trigger protein aggregation (Nedialkova & Leidel, 2015). In *Elp3* deleted NSCs, misfolded proteins were found to accumulate in the ER and induced unfolded protein response (UPR; Laguesse et al., 2015). It was thus proposed that upregulated UPR may account for neurogenesis defects after *Elp3* deletion (Laguesse et al., 2015). While the specific mechanisms that mediate the effect of UPR on NSCs are still not fully understood, it likely involves eIF2α. It was found that *Elp3* deletion led to increased phosphorylation of eIF2α (Laguesse et al., 2015), consistent with the fact that a main transducer of UPR is protein kinase RNA (PKR)-like ER kinase (PERK) that phosphorylates eIF2α to attenuate global protein synthesis and relieve the ER stress (Pavitt & Ron, 2012).

Cytosine-5 methylation (m^5C) is also essential for tRNA maturation; loss of m^5C modification causes increased cleavage and fragmentation of tRNA (Pan, 2018; Tuorto et al., 2012). The vast majority of tRNAs are methylated by NOP2/Sun RNA methyltransferase family member 2 (*NSUN2*), of which deletion was shown to downregulate global protein synthesis in the developing mouse brain (Tuorto et al., 2012, Tuorto & Lyko, 2016; Blanco et al., 2014). The loss of *NSUN2* perturbed neurogenesis and led to microcephaly (Blanco et al., 2014, Flores et al., 2017). Despite a similar reduction in protein synthesis to that seen in the mcm^5 and ncm^5 deficiency, perturbations in tRNA m^5C modification resulted in an accumulation of IPCs that failed to differentiate into neurons, rather than a depletion of IPCs (Flores et al., 2017; Laguesse et al., 2015). The opposing effects of tRNA processing defects induced by *Elp3* and *NSUN2* deletions may be due to additional targets of *Elp3* and/or *NSUN2* since RNA modification enzymes frequently display substrate promiscuity (Roundtree et al., 2017). Indeed, NSUN2 is known to methylate both mRNA and long non-coding RNA (Roundtree et al., 2017, Hussain et al., 2013, Khoddami & Cairns, 2013, Yang et al. 2017). In this regard, methylation of mRNA at the N6 position of adenosine (m^6A) was shown to regulate NSC self-renewal and neurogenic differentiation (Wang et al., 2018; Yoon et al. 2017).

Ribosome Biogenesis

Ribosomes are the center of the whole protein synthesis machinery and key for fine-tuning the proteome. While the biogenesis of ribosomes has been linked to stem cell homeostasis in other systems (Sanchez et al., 2016; Khajuria et al., 2018), it remains inconclusive as to whether and to what extent ribosome biogenesis affects NSC development in the mammalian brain. For example, a reduced transcription of ribosomal RNA and ribosomal protein genes (e.g., *Rpl11* and *Rps12*, two factors crucial for ribosomal assembly) was associated with NSC lineage progression in the embryonic cortex (transition from NECs to RGCs), although cytoplasmic ribosome density in NSCs was found to be unchanged (Chau et al., 2018). Similarly, the activation of quiescent adult NSCs was correlated with an initial upregulation of ribosomal genes before cell cycle entry (Dulken et al., 2017; Llorens-Bobadilla et al., 2015; Shin et al., 2015). However, when NSCs were forced to proliferate, protein synthesis rates were not affected despite an enhancement in the transcription of ribosomal protein genes (Chau et al., 2018), suggesting additional levels of regulation. Indeed, it was recently found that in proliferative NSCs, the expression of many ribosomal protein genes was themselves coordinated at the translational level according to NSC developmental states (Blair et al., 2017). Therefore, the role of ribosome biogenesis in NSC development and neurogenesis is likely to be more complex and involve multiple regulatory mechanisms.

One such mechanism includes the changes in the composition of ribosomes. Often assumed to be an invariant, homogeneous set of molecular machinery, recent evidence shows that ribosomes are heterogeneous regarding their molecular composition (Shi & Barna, 2015; Guo, 2018). They are made up of different ribosomal proteins within and across different tissues, allowing the ribosomes to meet the varying needs of protein synthesis in the cell (Slavov et al., 2015, Shi et al., 2017). In the mammalian brain, the protein composition of ribosomes was found to be cell-type specific (e.g., *Rpl7l1* in NSCs), showing dynamic changes throughout development and in response to extracellular signals (Kraushar et al., 2015; Kraushar et al., 2016). Further studies will need to determine to what extent and how the spatiotemporal expression of distinct ribosomal components differentially affects protein synthesis, NSC lineage progression and neurogenesis.

Target-Specific Translational Control of Neurogenesis

The translation of mRNA can be regulated at a global level (as discussed in the previous sections), changing the whole proteome, or a transcript-specific level, affecting the synthesis of a single or a subset of proteins to achieve specific cellular functions. An interesting observation in NSCs was that they not only transcribed genes essential for self-renewal (e.g., *Hes1*, *Sox2*) but also constantly made mRNA from many genes that induce neuronal differentiation (e.g., *NeuroD1*, *Neurogenin1*; Yang et al., 2014; Zahr et al., 2018). Simultaneous transcription of genes that encode proteins with conflicting functions requires gene-specific control at translational and/or post-translational levels to ensure proper neurogenesis. Target-specific translational control provides several advantages in this regard. First, and perhaps the most apparent, it allows rapid production of specific proteins from pre-existing mRNA to elicit immediate actions. This bypasses the need for the transcription, splicing and export of new mRNA. Rapid synthesis of fate determinant proteins may facilitate timely neuronal fate commitment. Second, target-specific translational control can fine-tune gene expression regarding the onset, termination and level of expression of select genes that may have related or conflicting functions. Third, it allows for local production of select proteins to deliver spatial-specific functions by transporting translationally silenced mRNAs to different compartments of NSCs. This is particularly relevant to NSC fate decision, as translationally controlled cell fate determinants can be asymmetrically segregated during NSC self-renewing division (Delaunay et al., 2017). RBPs are essential players that elicit target-specific regulation of mRNAs. Several RBPs involved in NSC homeostasis have been recently reviewed elsewhere (Pilaz & Silver, 2015; Popovitchenko & Rasin, 2017; Lennox, Mao, & Silver, 2018; Baser, Skabkin, & Martin-Villalba, 2017). In the following sections, we will focus on the latest findings and a few classic examples to discuss how NSCs employ spatiotemporal translational regulation to instruct neurogenesis.

RBP-Mediated Translational Activation and Repression in NSCs

Modulating the translation of specific mRNA by RBPs before or after they are needed is a rapid mechanism to control differentiation. The embryonic lethal abnormal vision-like 4 (Elavl4/HuD) is one such RBP that activates translation of selective transcripts to promote neurogenesis. HuD was found to be expressed in NSCs throughout development and into adulthood (Bronicki & Jasmin, 2013). Loss of HuD in mice enhanced NSC self-renewal at the expense of neurogenic differentiation, whereas ectopically expressing HuD promoted neurogenesis and neurite development (DeBoer et al., 2014). Using CRAC (cross-linking and analysis of cDNAs), Tebaldi et al. (2018) recently found that HuD predominantly bound to the 3′ untranslated regions (UTRs) of a select set of mRNAs that encode regulators of protein synthesis, including mTORC1-responsive ribosomal proteins (e.g., *Rps20*, *Rpl32*) and translation factors (e.g., *Eif4a1*, *Eef1a1*). Several proneural genes (e.g., *Ascl4*, *Msi1*) were also HuD targets. Upon HuD overexpression, the translation of these mRNAs was upregulated to enhance neural differentiation (Tebaldi et al., 2018). HuD relied on both the poly-A tail and cap structure of mRNA to enhance translation, suggesting that HuD may work with the translational initiation machinery to promote mRNA translation (Fukao et al., 2009).

In addition to initiating translation, components of the translation initiation machinery can also be recruited for repressing the translation in a target-specific manner. For example, the initiation factor eIF4E can interact with a translational repressor eIF4E-transporter (4E-T) in self-renewing NSCs to selectively target proneural mRNAs (e.g., *Neurogenin1* and *NeuroD1*) to the processing bodies (P-bodies), cytoplasmic foci where mRNA repression and decay occur (Standart & Weil, 2018; Luo, Na, & Slavoff, 2018; Yang et al., 2014). Disruption of the eIF4E/4E-T repressive complex or the P-bodies resulted in abnormal activation of proneural mRNAs, leading to compromised NSC self-renewal and premature neurogenesis (Yang et al., 2014). This suggests that translational repression plays a critical role in maintaining the stem cell state of NSCs. While the 4E-T complex regulates specific mRNAs, 4E-T itself cannot directly recognize mRNA (Kamenska et al., 2014; Ozgur et al., 2015), but is likely to corporate with target-specific RBPs for this function (Figure 16.2). In NSCs, one such RBP is Smaug2, which was shown to interact with 4E-T and repress the translation of *Nanos1* mRNA, a proneural factor (Amadei et al., 2015). Knocking down *Smaug2* in NSCs caused an aberrant increase of *Nanos1* translation, leading to premature neurogenesis and phenocopied *4E-T* loss of function (Amadei et al., 2015; Yang et al., 2014). This suggests that 4E-T forms a complex with Smaug2 and likely other RBPs to regulate neurogenesis. A recent interactome study has brought new insights into this by identifying several RBPs that interact with 4E-T in human kidney embryonic cells (Youn et al., 2018). It is thus likely that 4E-T forms various sub-complexes with discrete RBPs to orchestrate distinct groups of functionally related mRNAs (i.e., mRNA regulons) to control different aspects of neurogenesis. Assessing the functional similarities and differences of 4E-T sub-complexes in NSCs would be valuable for understanding how RBP-mediated regulation of mRNA coordinates the diverse aspects of neurogenesis. In this regard, some progress has been recently made revealing the mechanisms by which RBPs regulate the temporal genesis of neurons with distinct identities.

Translational Control for the Temporal Genesis of Neuronal Subtypes

The mammalian neocortex is perhaps one of the best and most-studied parts of the brain with respect to the genesis of different subtypes of neurons, owing to its highly organized structure and the stereotypic pattern of temporal neurogenesis (Taverna, Gotz, & Huttner 2014). Typically, the neocortex is organized into six layers (I–VI), characterized by discrete subtypes of projection neurons that express unique repertoires of genes (Custo Greig et al., 2013). For example, upper-layer (II–IV) neurons express POU class III homeobox 3 (*Brn1*) and cut-like homeobox 1 (*Cux1*), whereas deep-layer (V/VI) neurons express Fez family zinc finger protein 2 (*Fezf2*) or transducing-like enhancer of split 4 (*Tle4*; Custo Greig et al., 2013). Notably, distinct projection neuron subtypes arise from a common pool of NSCs, but at different developmental time points (Toma & Hanashima, 2015). Lineage tracing and transplantation experiments suggest that at least some NSCs can change their potential to produce multiple neuronal subtypes for both deep and upper cortical layers, a potential that becomes progressively restricted over time (Gao et al., 2014). It is generally thought that NSCs assign the initial identity to neurons by expressing specific subtype genes, such as *Tle4* and *Brn1* and that a switch in the expression of these genes over the course of development directs a temporal transition from the production of deep-layer to upper-layer neurons (Toma & Hanashima, 2015). Interestingly, Zahr et al. (2018) showed that incompatible subtype identity genes were frequently transcribed together. For example, in earlier-stage NSCs, when deep-layer neurons were being generated, identity genes for later-stage upper-layer neurons, such as *Brn1* and *Cux1*, were transcribed before they were needed. Conversely, at a later developmental stage, when upper-layer neurons were being generated, several identity genes for deep-layer neurons, such as *Tle4*, were still actively transcribed. Although the transcription of incompatible genes seems paradoxical, a subsequent level of translational control ensures appropriate spatial and temporal protein expression. The authors showed that this translational regulation was partly mediated by an RBP Pumilio2 (*Pum2*) and its binding partner *4E-T* (Zahr et al., 2018). Knockdown of *4E-T* or *Pum2* in NSCs resulted in misspecification of neuronal subtype identities, a phenotype that was not observed when *Smaug2* was perturbed (Zahr et al., 2018, Amadei et al., 2015). This suggests that Pum2 organizes a distinct 4E-T sub-complex in NSCs that prevents promiscuous translation of incompatible identity genes to ensure genesis of neurons with the correct subtype identities (Figure 16.2).

Another RBP that interacts with the translational machinery to regulate neuronal subtype identity is HuR (Kraushar et al., 2014). Kraushar et al. (2014) showed that HuR deletion in the mouse brain perturbed the assembly of polyribosomes (polysomes), where multiple ribosomes are recruited and active mRNA translation occurs, and compromised ribosome specificity for selective transcripts. This caused aberrant translation of genes specific to layer II, III, and V, leading to an abnormal laminar distribution of deep-layer neurons. Interestingly, the impact of HuR deletion varied in a temporal-dependent manner (Kraushar et al., 2014), suggesting that mRNA regulatory programs coordinated by HuR, and possibly other RBPs, are under dynamic temporal control. It is likely that at different developmental stages, RBPs are organized into discrete complexes with different protein-binding partners and target mRNAs to precisely control NSC lineage progression and neuronal differentiation (Buchan, 2014). The mechanisms that underlie the dynamic assembly of distinct RBP complexes in NSCs are still not well understood.

uORFs and Translational Regulation

Upstream open reading frames (uORFs) are small ORFs that reside upstream of main ORFs, frequently acting as *cis*-acting repressor elements (Barbosa, Peixeiro, &

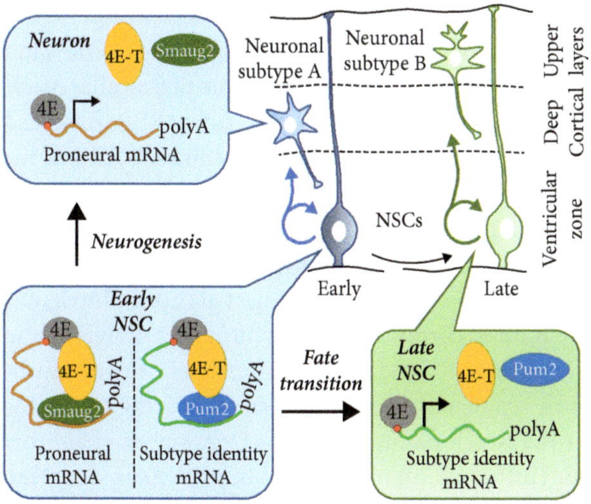

FIGURE 16.2. Translational repression control of neurogenesis and NSC fate transition. In NSCs, the expression of genes encoding proneural factors and subtype identity genes are suppressed at the translational level by repression complexes. A group of such repression complexes comprise of eukaryotic initiation factor 4E (4E), 4E-Transporter (4E-T), and target-specific RBPs Smaug2 and Pumilio2 (Pum2) to recognize and translationally suppress distinct mRNAs. Dissociation of the 4E-T/Smaug2 complex releases proneural mRNA for translation and induces neurogenesis. Over the developmental time, subtype identity mRNA suppressed by the 4E-T/Pum2 complex is released in late-stage NSCs, allowing the translation of specific subtype identity determinant to direct NSC fate transition for making different subtypes of neurons.

Romão, 2013; McGeachy & Ingolia, 2016; Johnstone, Bazzini, & Giraldez, 2016). When uORFs are recognized and translated, fewer ribosomes are available for translation of the main ORFs downstream, resulting in translational repression. In other cases, the stop codon of a uORF can be recognized as premature and trigger nonsense-mediated mRNA decay, thus serving as another mean to downregulate translation (Somers, Pöyry, & Willis, 2013). uORF-mediated translational control is a widespread regulatory mechanism for gene expression in mammals (Chew, Pauli, & Schier, 2016; Barbosa, Peixeiro, & Romão, 2013; Johnstone, Bazzini, & Giraldez, 2016). In mouse limb buds and the neural tube, Fujii et al. (2017) showed that genes encoding critical components of core signaling pathways (e.g., Shh, Wnt) were selectively repressed at the level of translation, which was mediated by uORFs. Intriguingly, a recent study by Blair et al. (2017) further showed that uORF might also play a role in controlling translational efficiency during embryonic stem cell differentiation into NSCs. For instance, the translation of an essential NSC gene *Sox2* was upregulated upon neural differentiation and was correlated with less ribosome occupancy at its uORF versus the main ORF downstream. Similarly, the distribution of ribosomes in uORFs versus main ORFs changed for a different set of genes upon differentiation of NSCs into neurons (Blair et al., 2017).

The mechanisms that underlie gene-specific uORF regulation in neural stem cell maintenance and neurogenesis remain mostly unknown. Findings from muscle stem cells suggest that phosphorylation of eIF2α may be a critical node of uORF regulation for stem cell maintenance by inducing ribosome bypass of uORFs in the transcripts of stemness genes (Zismanov et al., 2016). It is plausible that similar mechanisms may contribute to uORF regulation in NSC homeostasis. A few RBPs have also been implicated in orchestrating translation of the uORF versus the main ORF. In a recent study, Zhang et al. (2016) showed that the heterogeneous nuclear ribonucleoproteins hnRNPA2B1 and hnRNPA0, as well as Elavl1/HuR bound to and enhanced the translation of the uORFs of target genes to repress the translation of main ORFs. Downregulation of these RBPs increased target protein synthesis (Zhang et al., 2016). Intriguingly, HuR deletion in the developing mouse cortex reduced neurogenic differentiation of NSCs and decreased cortical thickness (García-Domínguez et al., 2011; Kraushar et al., 2014). It was found that HuR regulated neurogenesis by stabilizing the mRNA of Delta-like 1 (*Dll1*), a ligand for the Notch pathway (García-Domínguez et al., 2011; Kraushar et al., 2014). Nonetheless, HuR can bind to hundreds of mRNAs in a context-specific manner (García-Domínguez et al., 2011; Kraushar et al., 2014; Calaluce et al., 2010). In the future, it would be interesting to assess whether uORF-mediated translational repression of mRNA targets may be a mechanism used by HuR to regulate neurogenesis.

Asymmetric Division and Spatial Control of Translation in NSCs

Another driving force underlying NSC fate decision is the asymmetric segregation of cell fate determinants between two daughter cells (Figure 16.3). As fate determinants, RBPs and mRNAs have been best-understood in model organisms. For example, the asymmetric localization of translationally repressed *Bicoid* mRNA and its RBP Staufen in the fly oocytes sets the morphogen gradients essential for the proper spatial patterning of the developing embryo (Ferrandon et al., 1994). Two recent studies showed that similar mechanisms governed asymmetric fate decision of NSCs in the developing mouse brain. It was found that in dividing NSCs, mRNAs encoding proneural factors, such as *Prox1*, *Trim32* and *Bbs2*, were translationally repressed and preferentially distributed to one daughter cell during asymmetric division (Kusek et al., 2012; Vessey et al., 2012). These proneural mRNAs were shown to be bound and repressed by Staufen2 (Stau2), a human homolog of fly Staufen, which was similarly segregated asymmetrically during cell division (Kusek et al., 2012; Vessey et al., 2012). Knockdown of *Stau2* consistently led to symmetric distribution and aberrant translation of *Prox1* in daughter cells, resulting in compromised NSC self-renewal and premature neurogenesis (Kusek et al., 2012; Vessey et al., 2012).

The segregation of fate determinants can occur not only in the cell body of NSCs but also other cellular structures, such as the basal process that is extended radially and contacts the basal lamina (Pilaz & Silver, 2017; Kosodo & Huttner, 2009). At the symmetric proliferation stage, the basal process of NSCs is split into two with each inherited by one

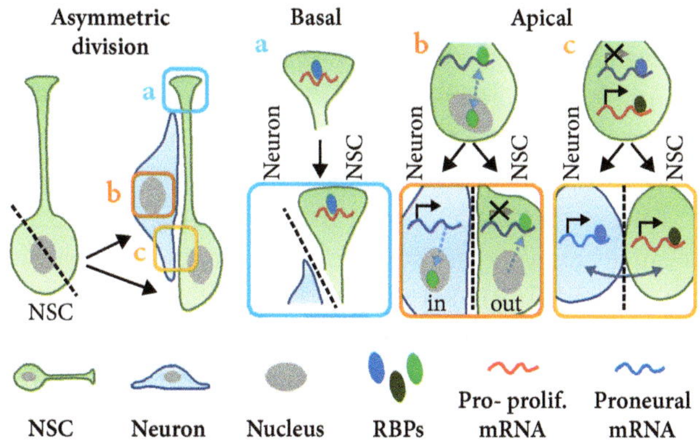

FIGURE 16.3. Translational regulatory mechanisms of asymmetric fate decision. NSCs undergo asymmetric division to give rise to one daughter NSC and one neuron. At the basal side of NSCs, specific mRNA (e.g., *Ccnd2*) as cell fate determinants can be transported by RBPs (e.g., FMRP) to the endfoot of the basal process. Asymmetric inheritance of the basal process and thus cell fate determinants instructs the maintenance of the NSC fate in one daughter cell. At the apical side, nucleocytoplasmic shuttling of specific RBPs may provide another strategy to control the translation of mRNA encoding proneural factors. mRNAs encoding pro-proliferation or proneural factors can also be segregated in the cell body between daughter cells to direct fate specification, which is controlled by specific RBPs (e.g., Stau2).

daughter cell (Kosodo et al., 2008). In contrast, the basal process of self-renewing NSCs at the neurogenic stage is retained by only one daughter cell during mitosis (Miyata et al., 2001; Noctor et al., 2001). It is thus proposed that cell fate determinants stored in the basal process control the fate decision of the daughter cell that retains the process after cell division. Several groups have identified specific mRNAs showing polarized distribution biased to the basal process (Figure 16.3). For example, the mRNA of *Transitin* was found to be present along the basal process to the endfeet in chicken NSCs, although the functional significance of this distribution is not well understood (Lee & Cole, 2000). In the mouse cortex, Tsunekawa et al. (2012) showed that the mRNA of Cyclin D2 (*Ccnd2*), a cell cycle regulator, was mainly localized near the basal lamina throughout development. The inheritance of the basal process was frequently associated with stronger *Ccnd2* expression in daughter cells (Tsunekawa et al., 2012; Pilaz et al., 2016). Using loss- and gain-of-function approaches, the authors showed that ectopic expression of *Ccnd2* in daughter cells inhibited neuronal differentiation, whereas ablation of *Ccnd2* promoted it. This is consistent with the idea that the asymmetrical inheritance of the basal process and the cell fate regulators such as *Ccnd2* instructs the maintenance of NSC fate in one daughter cell (Tsunekawa et al., 2012). Nonetheless, it is still not clear whether the fate decision is mediated by the transport of basally synthesized Ccnd2 proteins back to the cell body or by other factors that induce local upregulation of *Ccnd2* expression.

Given that the localized translation of *Ccnd2* mRNA is a critical control point, it first must be repressed and localized to the cellular region that is asymmetrically distributed. A *cis*-regulatory element in the 3′ UTR of *Ccnd2* and a *trans*-acting RBP, fragile X mental retardation protein (FMRP), were found to mediate the localization of *Ccnd2* mRNA to the endfeet (Pilaz et al., 2016). In the absence of FMRP, the basal localization of *Ccnd2* mRNA in embryonic NSCs was impaired (Pilaz et al., 2016). However, deletion of *Fmr1*, the gene that encodes FMRP, caused NSC depletion and premature differentiation in the embryonic cortex, a phenotype opposite to that seen in *Ccnd2* overexpression (Tsunekawa et al., 2012, Saffary & Xie, 2011). This suggests that distinct *cis*- or *trans*-acting factors may differentially regulate the transport and translation of specific mRNA. In this regard, another RBP, IGF2BP1 (IMP1) can also bind to *Ccnd2* mRNA (as well as *Ccnd1* mRNA) and regulate their expression (Nishino et al., 2013). While *Ccnd2* displayed a biased distribution to the basal process, *Ccnd1* mRNA and IMP1 were predominantly located in the cell body of NSCs (Nishino et al., 2013; Tsunekawa et al., 2012). Similarly, *IMP1* loss-of-function resulted in compromised NSC self-renewal and aberrant neurogenesis (Nishino et al., 2013). Elucidating the mechanisms that underlie the complex interaction between *cis*-regulatory elements on mRNAs and discrete RBPs will be important in understanding how translational programs coordinate NSC homeostasis and neurogenesis (Figure 16.3).

Protein Synthesis and Neurogenesis in Human Neurodevelopmental Disorders

Growing evidence suggests that the dysregulation of protein synthesis represents a shared feature of several complex neurodevelopmental disorders, such as autism spectrum disorder (ASD) and schizophrenia (Louros & Osterweil, 2016; Tahmasebi et al., 2018). This frequently involves mutations that directly or indirectly affect mRNA translation in a global or target-specific manner (Tahmasebi et al., 2018). Given that NSCs are tightly regulated at the translational level, it is not surprising that NSC self-renewal and neurogenesis are often perturbed in these disorders. This perturbation in NSC self-renewal and neurogenesis can cause macroscopic abnormalities of the brain structure, such as focal patches of disorganized neurons that are seen in some ASD patients (Geschwind & Levitt, 2007; Willsey et al., 2013; Stoner et al., 2014; Bailey et al., 1998; Hutsler, Love, & Zhang, 2007).

Dysregulation of Global Protein Synthesis in NSCs

A downregulation of the mTOR pathway and a reduction in the expression of genes involved in protein synthesis have been found in postmortem brain tissues from idiopathic ASD cases (Ginsberg et al., 2012, Nicolini et al., 2015). In contrast, schizophrenia cases showed upregulation of genes involved in mRNA translation (Darby, Yolken, & Sabunciyan, 2016). Similar perturbations in the mTOR signaling pathway were also observed in syndromic ASD and other neurodevelopmental disorders (Troca-Marín,

Alves-Sampaio, & Montesinos 2012), such as the fragile X syndrome (FXS; Jacquemont et al., 2018) and tuberous sclerosis complex (TSC; Tsai et al., 2014). While insightful, postmortem analysis at the postnatal stage allows only a static measurement at the endpoint of the disease and cannot provide information on the developmental trajectory that leads to such outcomes. Many ASD susceptibility genes identified in human genetic studies are expressed in the early phases of brain development when NSCs undergo active neurogenesis (Iossifov et al., 2014). This suggests that perturbations in protein synthesis may already show adverse impacts on NSCs and neurogenesis during early brain development, which likely contributes to the pathogenesis of these diseases (Kaushik & Zarbalis, 2016; Packer, 2016).

The use of patient-derived induced pluripotent stem cells (iPSCs) offers a unique opportunity to understand the underlying mechanisms and the developmental trajectory perturbed in neurodevelopmental disorders (Ardhanareeswaran et al., 2017). In a recent study, Boland et al. (2017) generated iPSC-derived NSCs and neurons from several individuals with FXS. They found that patient-derived NSCs showed reduced protein synthesis and delayed early neurogenesis (Boland et al., 2017), a phenotype opposite to that seen in *Fmr1* mutant mice (Castrén et al., 2005, Saffary & Xie, 2011). This discrepancy may reflect differences in culturing and experimental conditions. In another study, Topol et al. (2015) generated iPSCs from schizophrenia patients and showed that protein synthesis was upregulated in iPSC-derived NSCs, leading to an increase in total protein levels. This global enhancement of protein synthesis appeared to be a direct result of upregulation in the translational machinery, as both ribosomal proteins and initiation and elongation factors were markedly increased in patient-derived NSCs (Topol et al., 2015). The alterations in protein synthesis were only seen in NSCs; neither iPSCs nor iPSC-derived neurons showed different protein synthesis rates and total protein levels compared to healthy controls (Topol et al., 2015). Intriguingly, cells derived from a nasal biopsy of the olfactory mucosa in schizophrenia patients, in contrast, showed a significant reduction in protein synthesis rates, accompanied by a decrease in ribosomal biogenesis, downregulated mTOR signaling and reduced levels of initiation factors (e.g., eIF2; English et al., 2015). These context-dependent differences in protein synthesis were also observed in a cell model for Rett syndrome, where the causal gene methyl CpG binding protein 2 (*MECP2*) was deleted in human ESCs (Li et al., 2013). A global reduction of nascent protein synthesis and mTOR signaling was found in ESC-derived neurons but not in ESC-derived NSCs (Li et al., 2013). These findings indicate that disease-related alterations in protein synthesis occur in a manner specific to cell-type and their developmental stage. Given the alterations in global protein synthesis, it will be interesting to assess further if the balance of self-renewal and neurogenesis is affected in patient-derived NSCs.

Perturbation in the Translational Machinery

Human genetic studies of neurodevelopmental disorders have identified mutations in many genes that encode essential components of the translational machinery (Scheper, van der Knaap, & Proud, 2007). In some individuals with microcephaly, intellectual disability and ASD, *de novo* missense mutations in the ribosomal protein gene *RPS23* (*uS12*) were found to impair polysome formation and the accuracy of mRNA translation (Paolini et al., 2017). Mutations in genes encoding other ribosomal proteins, such as *RPL10* (*uL16*, or *AUTSX5*) have also been linked to intellectual disability and ASD (Klauck et al., 2006, Brooks et al., 2014). Furthermore, it was found that the copy number of active ribosomal genes was significantly lower in ASD cases, whereas patients with schizophrenia had more ribosomal genes (Porokhovnik et al., 2015, Malinovskaya et al., 2018). Nonetheless, the pathological impact of these changes on NSCs and neurogenesis remains to be determined.

In addition to ribosomal defects, some mutations affect translational initiation. For example, mutations that lead to eIF4E overproduction were found in some ASD patients (Neves-Pereira et al., 2009). Modeling of this eIF4E perturbation in adult mice showed an increase in the translation of several synaptic proteins in neurons, including the ASD-risk factor *Neuroligin* (Gkogkas et al., 2013; Santini et al., 2013). In line with the aberrant increase of translation, inhibiting cap-dependent initiation reversed ASD-related behavior seen in these mice (Santini et al., 2013; Gkogkas et al., 2013). Intriguingly, in the developing cortex, ectopic expression of eIF4E was found to increase NSC proliferation at the expense of neurogenesis, likely by inducing target-specific translational repression of proneural genes (Yang et al., 2014). This suggests that perturbed translation caused by genetic alterations in eIF4E may have an adverse impact on brain development at multiple stages. Another example of perturbation in translational initiation comes from TSC, a rare genetic disorder characterized by benign tumors (Tsai et al., 2014). TSC is caused by mutations in *TSC1* and *TSC2* genes, which are upstream negative regulators of mTORC1. Deletion of *Tsc1* or *Tsc2* in mice led to aberrant activation of mTORC1 and a global increase in protein synthesis in adult NSCs, compromising the quiescent state of NSCs and causing their premature differentiation (Magri et al., 2013; Mahoney et al., 2016).

RBPs in Neurodevelopmental Disorders

Several RBPs have been linked to neurodevelopmental disorders (Sartor et al., 2015). These could represent effects through specific mRNA rather than global changes. These RBPs are either found to be genetically perturbed in human patients such as *FMRP* (mutated in FXS/ASD) or differentially expressed under pathological conditions to induce a broader impact on mRNA that encode disease risk factors. In a recent study, Parras et al. (2018) showed that cytoplasmic polyadenylation element binding protein 4 (CPEB4) orchestrated mRNAs that encode many high-confidence ASD risk factors (e.g., *Dyrk1a*, *Ptchd1*). In a group of idiopathic ASD patients, the expression of *CPEB4* was aberrantly regulated; a *CPEB4* isoform without exon four was increased with a concomitant decrease of total CPEB4 protein levels (Parras et al., 2018). The altered expression of *CPEB4* resulted in a reduction of translational efficiency for its bound mRNAs that are encoded by ASD risk genes, leading to ASD-like changes at the neuroanatomical, electrophysiological and behavioral levels in mouse models (Parras et al., 2018). Given that *CPEBs* are expressed throughout development (Ivshina, Lasko, & Richter, 2014), it remains unknown whether altered *CPEB4* expression affects brain development at earlier stages when neurogenesis occurs. Nonetheless, deletion of *Orb2*, the fly homolog of human *CPEB2-4* isoforms, caused premature neural differentiation of NSCs (Hafer et al., 2011), suggesting a potential role of *CPEBs* in developing neurogenesis in mammals.

Alteration in cis-Regulatory Elements

In some cases, pathological features arise from mutations in *cis*-regulatory elements of specific genes that affect their translation. Suhl et al. (2015) identified a variant located in *FMR1* 3′UTR. This mutation reduced translation of *FMR1* mRNA due to its compromised interaction with the *trans*-acting factor HuR (Suhl et al., 2015). Another example is *MECP2*, the causal gene for Rett syndrome. Mutations in *MECP2* 3′UTR caused a reduction in *MECP2* translation and may contribute to the clinical expression of pathological phenotypes (McGowan & Pang, 2015). Interestingly, *MECP2* mRNA underwent 3′UTR-based stabilization and translational activation during neural differentiation of human ESCs (Rodrigues et al., 2016). This process was controlled by the combinatory actions of *trans*-acting factors including the RBPs Tia1, HuC, Pum1, and pluripotent-specific miRNAs (Rodrigues et al., 2016). It will be interesting to investigate if and how these ASD-related *trans*-acting factors, together with CPEB4, act in post-transcriptional networks to coordinate NSC lineage progression and neurogenesis.

Concluding Remarks

Over the past few years, growing evidence indicates that protein synthesis orchestrated by the translational machinery, and RBPs play a critical role in NSC maintenance and differentiation in the mammalian brain. Despite significant recent advances, our understanding of how coordinated translational events regulate NSCs and neurogenesis is far from complete. Several outstanding questions remain unresolved. For example, how do translational programs instruct NSC fate transition to make different neuronal subtypes and glial cells throughout development? Do translational programs contribute to the developmental establishment of adult NSCs? How are discrete mRNA-protein complexes organized and regulated in NSCs to coordinate the translation of functionally related mRNAs to instruct neurogenesis? During asymmetric division in NSCs, what upstream mechanisms trigger the differential activation or repression of fate determinant mRNAs to help daughter cells acquire different cellular fates, and at what cell cycle stage is the fate decision made? How do translational programs integrate niche signals to instruct NSC fate decision? The use and development of new molecular and biochemical techniques will bring novel insights into these questions. For example, genome-wide identification and comparison of 4E-T-interacting RBPs in NSCs and other cell types will help elucidate functional similarities and differences of discrete translational repression 4E-T sub-complexes and their roles in neurogenesis. Furthermore, temporal translatome analyses of developing NSCs will provide valuable information to understand the role of translational control in NSC lineage progression. Addressing these and other questions related to translational mechanisms will ultimately help reveal a comprehensive picture of how gene expression instructs brain development and function under normal and disease conditions.

Acknowledgments

This work was supported by the Natural Sciences and Engineering Research Council (RGPIN-2018-04246). L.M. is supported by the Alberta Children's Hospital Research Institute Graduate Scholarship. We thank Drs. Savraj Grewal, Tim Shutt, and Katherine Gratton for helpful discussions and reading the manuscript. We apologize to authors whose work could not be cited due to space limitation.

References

Albert, M., & Huttner, W. B. (2018). Epigenetic and transcriptional pre-patterning: An emerging theme in cortical neurogenesis. *Frontiers in Neuroscience, 12*, 359. https://doi.org/10.3389/fnins.2018.00359.

Amadei, G., Zander, M. A., Yang, G., Dumelie, J. G., Vessey, J. P., Lipshitz, H. D., . . . Miller, F. D. (2015). A Smaug2-based translational repression complex determines the balance between precursor maintenance versus differentiation during mammalian neurogenesis. *Journal of Neuroscience, 35*(47), 15666–15681. https://doi.org/10.1523/JNEUROSCI.2172-15.2015.

Aprea, J., & Calegari, F. (2015). Long non-coding RNAs in corticogenesis: deciphering the non-coding code of the brain. *EMBO Journal, 34*(23), 2865–2884. https://doi.org/10.15252/embj.201592655.

Ardhanareeswaran, K., Mariani, J., Coppola, G., Abyzov, A., & Vaccarino, F. M. (2017). Human induced pluripotent stem cells for modelling neurodevelopmental disorders" *Nature Reviews Neurology, 13*(5), 265–278. https://doi.org/10.1038/nrneurol.2017.45.

Bailey, A., Luthert, P., Dean, A., Harding, B., Janota, I., Montgomery, M., . . . Lantos, P. (1998). A clinicopathological study of autism. *Brain, 121*(Pt 5), 889–905.

Barbosa, C., Peixeiro, I., & Romão, L. (2013). Gene expression regulation by upstream open reading frames and human disease. *PLoS Genetics, 9*(8), e1003529. https://doi.org/10.1371/journal.pgen.1003529.

Baser, A., Skabkin, M., Kleber, S., Dang, Y., Gülce Gülcüler Balta, G., Kalamakis, G., . . . Martin-Villalba, A. (2019). Onset of differentiation is post-transcriptionally controlled in adult neural stem cells. *Nature*, 566, 100–104.

Baser, A., Skabkin, M., & Martin-Villalba, A. (2017). Neural stem cell activation and the role of protein synthesis. *Brain Plasticity, 3*(1), 27–41. https://doi.org/10.3233/BPL-160038.

Blair, J. D., Hockemeyer, D., Doudna, J. A., Bateup, H. S., & Floor, S. N. (2017). Widespread translational remodeling during human neuronal differentiation. *Cell Reports, 21*(7), 2005–2016. https://doi.org/10.1016/j.celrep.2017.10.095.

Blanco, S., Bandiera, R., Popis, M., Hussain, S., Lombard, P., Aleksic, J.,...Frye, M. (2016). Stem cell function and stress response are controlled by protein synthesis. *Nature*, *534*(7607), 335–340. https://doi.org/10.1038/nature18282.

Blanco, S., Dietmann, S., Flores, J. V., Hussain, S., Kutter, C., Humphreys, P....Frye, M. (2014). Aberrant methylation of tRNAs links cellular stress to neuro-developmental disorders. *EMBO Journal* *33*(18), 2020–2039. https://doi.org/10.15252/embj.201489282.

Boland, M. J., Nazor, K. L. Tran, H. T., Szücs, A., Lynch, C. L., Paredes, R.,...Loring, J. F. (2017). Molecular analyses of neurogenic defects in a human pluripotent stem cell model of fragile X syndrome. *Brain* *140*(3), 582–598. https://doi.org/10.1093/brain/aww357.

Bonaguidi, M. A., Wheeler, M. A., Shapiro, J. S., Stadel, R. P., Sun, G. J., Ming, G. L., Song, H. (2011). In vivo clonal analysis reveals self-renewing and multipotent adult neural stem cell characteristics. *Cell* *145*(7), 1142–1155. https://doi.org/10.1016/j.cell.2011.05.024.

Bond, A. M., Ming, G. L., & Song, H. (2015). Adult mammalian neural stem cells and neurogenesis: Five decades later. *Cell Stem Cell* *17*(4), 385–395. https://doi.org/10.1016/j.stem.2015.09.003.

Bronicki, L. M., & Jasmin, B. J. 2013. Emerging complexity of the HuD/ELAVl4 gene: Implications for neuronal development, function, and dysfunction. *RNA*, *19*(8):1019–1037. https://doi.org/10.1261/rna.039164.113.

Brooks, S. S., Wall, A. L., Golzio, C., Reid, D. W., Kondyles, A., Willer, J. R.,...Davis, E. E. (2014). A novel ribosomopathy caused by dysfunction of RPL10 disrupts neurodevelopment and causes X-linked microcephaly in humans. *Genetics*, *198*(2), 723–733. https://doi.org/10.1534/genetics.114.168211.

Buchan, J. R. (2014). mRNP granules. Assembly, function, and connections with disease. *RNA Biology*, *11*(8), 1019–1030. https://doi.org/10.4161/15476286.2014.972208.

Bulut-Karslioglu, A., Macrae, T. A., Oses-Prieto, J. A., Covarrubias, S., Percharde, M., Ku, G.,...Ramalho-Santos, M. (2018). The transcriptionally permissive chromatin state of embryonic stem cells is acutely tuned to translational output. *Cell Stem Cell*, *22*(3), 369–383.e8. https://doi.org/10.1016/j.stem.2018.02.004.

Buszczak, M., Signer, R. A., & Morrison, S. J. (2014). Cellular differences in protein synthesis regulate tissue homeostasis. *Cell*, *159*(2), 242–251. https://doi.org/10.1016/j.cell.2014.09.016.

Calaluce, R., Gubin, M. M., Davis, J. W., Magee, J. D., Chen, J., Kuwano, Y., &...Atasoy, U. (2010). The RNA binding protein HuR differentially regulates unique subsets of mRNAs in estrogen receptor negative and estrogen receptor positive breast cancer. *BMC Cancer*, *10*, 126. https://doi.org/10.1186/1471-2407-10-126.

Castrén, M., Tervonen, T., Kärkkäinen, V., Heinonen, S., Castrén, E., Larsson, K.,...Akerman, K. (2005). Altered differentiation of neural stem cells in fragile X syndrome. *Proceedings of the National Academy of Sciences of the United States of America*, *102*(49), 17834–17839. https://doi.org/10.1073/pnas.0508995102.

Chau, K. F., Shannon, M. L., Fame, R. M., Fonseca, E., Mullan, H., Johnson, M. B.,...Lehtinen, M. K. (2018). Downregulation of ribosome biogenesis during early forebrain development. *Elife*, *7*. https://doi.org/10.7554/eLife.36998.

Chew, G. L., Pauli, A., & Schier, A. F. (2016). Conservation of uORF repressiveness and sequence features in mouse, human and zebrafish." *Nature Communications*, *7*, 11663. https://doi.org/10.1038/ncomms11663.

Cloëtta, D., Thomanetz, V., Baranek, C., Lustenberger, R. M., Lin, S., Oliveri, F.,...Rüegg. M. A. (2013). Inactivation of mTORC1 in the developing brain causes microcephaly and affects gliogenesis. *Journal of Neuroscience*, *33*(18), 7799–7810. https://doi.org/10.1523/JNEUROSCI.3294-12.2013.

Corsini, N. S., Peer, A. M., Moeseneder, P., Roiuk, M., Burkard, T. R., Theussl, H. C., Moll, I., & Knoblich, J. A. (2018). Coordinated control of mRNA and rRNA processing controls embryonic stem cell pluripotency and differentiation. *Cell Stem Cell*, *22*(4), 543–558. e12. https://doi.org/10.1016/j.stem.2018.03.002.

Custo Greig, L. F., Woodworth, M. B., Galazo, M. J. H. Padmanabhan, H., & Macklis, J. D. (2013). Molecular logic of neocortical projection neuron specification, development and diversity. *Nature Reviews Neuroscience*, *14*(11), 755–769. https://doi.org/10.1038/nrn3586.

Darby, M. M., Yolken, R. H., & Sabunciyan, S. (2016). Consistently altered expression of gene sets in postmortem brains of individuals with major psychiatric disorders. *Translational Psychiatry*, *6*(9), e890. https://doi.org/10.1038/tp.2016.173.

DeBoer, E. M., Azevedo, R., Vega, T. A., Brodkin, J., Akamatsu, W., Okano, H.,... Rasin. M. R. (2014). Prenatal deletion of the RNA-binding protein HuD disrupts postnatal cortical circuit maturation and behavior. *Journal of Neuroscience*, *34*(10), 3674–3686. https://doi.org/10.1523/JNEUROSCI.3703-13.2014.

Delaunay, D., Kawaguchi, A., Dehay, C., & Matsuzaki, F. (2017). Division modes and physical asymmetry in cerebral cortex progenitors. *Current Opinion in Neurobiology*, *42*, 75–83. https://doi.org/10.1016/j.conb.2016.11.009.

Dulken, B. W., Leeman, D. S., Boutet, S. C., Hebestreit, K., & Brunet, A. (2017). Single-cell transcriptomic analysis defines heterogeneity and transcriptional dynamics in the adult neural stem cell lineage. *Cell Reports*, *18*(3), 777–790. https://doi.org/10.1016/j.celrep.2016.12.060.

Edfors, F., Danielsson, F., Hallström, B. M., Käll, L., Lundberg, E., Pontén, F.,... Uhlén, M. (2016). Gene-specific correlation of RNA and protein levels in human cells and tissues. *Molecular Systems Biology*, *12*(10), 883.

English, J. A., Fan, Y., Föcking, M., Lopez, L. M., Hryniewiecka, M., Wynne, K.,... Cotter, D. R. (2015). Reduced protein synthesis in schizophrenia patient-derived olfactory cells. *Translational Psychiatry*, *5*, e663. https://doi.org/10.1038/tp.2015.119.

Ferrandon, D., Elphick, L., Nüsslein-Volhard, C., & St. Johnston, D. (1994). Staufen protein associates with the 3′UTR of bicoid mRNA to form particles that move in a microtubule-dependent manner. *Cell*, *79*(7), 1221–1232.

Flores, J. V., Cordero-Espinoza, L., Oeztuerk-Winder, F., Andersson-Rolf, A., Selmi, T., Blanco, S.,... Frye, M. (2017). Cytosine-5 RNA methylation regulates neural stem cell differentiation and motility. *Stem Cell Reports*, *8*(1), 112–124. https://doi.org/10.1016/j.stemcr.2016.11.014.

Fujii, K., Shi, Z., Zhulyn, O., Denans, N., & Barna, M. (2017). Pervasive translational regulation of the cell signalling circuitry underlies mammalian development. *Nature Communications*, *8*, 14443. https://doi.org/10.1038/ncomms14443.

Fukao, A., Sasano, Y., Imataka, H., Inoue, K., Sakamoto, H., Sonenberg, N.,... Fujiwara, T. (2009). The ELAV protein HuD stimulates cap-dependent translation in a Poly(A)- and eIF4A-dependent manner. *Molecular Cell*, *36*(6), 1007–1017. https://doi.org/10.1016/j.molcel.2009.11.013.

Gao, P., Postiglione, M. P., Krieger, T. G., Hernandez, L., Wang, C., Han, Z.,... Shi, S. H. (2014). Deterministic progenitor behavior and unitary production of neurons in the neocortex. *Cell*, *159*(4), 775–788. https://doi.org/10.1016/j.cell.2014.10.027.

García-Domínguez, D. J., Morello, D., Cisneros, E., Kontoyiannis, D. L., & Frade, J. M. (2011). Stabilization of Dll1 mRNA by Elavl1/HuR in neuroepithelial cells undergoing mitosis. *Molecular Biology of the Cell*, *22*(8), 1227–1239. https://doi.org/10.1091/mbc.E10-10-0808.

Gebauer, F., Preiss, T., & Hentze, M. W. (2012). From cis-regulatory elements to complex RNPs and back. *Cold Spring Harbor Perspectives in Biology*, *4*(7), a012245. https://doi.org/10.1101/cshperspect.a012245.

Geschwind, D. H., & Levitt, P. (2007). Autism spectrum disorders: Developmental disconnection syndromes. *Current Opinion in Neurobiology*, *17*(1), 103–111. https://doi.org/10.1016/j.conb.2007.01.009.

Ghazalpour, A., Bennett, B., Petyuk, V. A., Orozco, L., Hagopian, R., Mungrue, I. N., . . . Lusis, A. J. (2011). Comparative analysis of proteome and transcriptome variation in mouse. *PLoS Genetics*, *7*(6), e1001393. https://doi.org/10.1371/journal.pgen.1001393.

Ginsberg, M. R., Rubin, R. A., Falcone, T., Ting, A. H., & Natowicz, M. R. (2012). Brain transcriptional and epigenetic associations with autism. *PLoS One*, *7*(9), e44736. https://doi.org/10.1371/journal.pone.0044736.

Gkogkas, C. G., Khoutorsky, A., Ran, I., Rampakakis, E., Nevarko, T., Weatherill, D. B., . . . Sonenberg, N. (2013). Autism-related deficits via dysregulated eIF4E-dependent translational control. *Nature*, *493*(7432), 371–377. https://doi.org/10.1038/nature11628.

Groszer, M., Erickson, R., Scripture-Adams, D. D., Lesche, R., Trumpp, A., Zack, J. A., . . . Wu, H. (2001). Negative regulation of neural stem/progenitor cell proliferation by the Pten tumor suppressor gene in vivo. *Science*, *294*(5549), 2186–2189. https://doi.org/10.1126/science.1065518.

Guo, H. (2018). Specialized ribosomes and the control of translation. *Biochemical Society Transactions*, *46*(4), 855–869. https://doi.org/10.1042/BST20160426.

Hafer, N., Xu, S., Bhat, K. M., & Schedl, P. (2011). The Drosophila CPEB protein Orb2 has a novel expression pattern and is important for asymmetric cell division and nervous system function. *Genetics*, *189*(3), 907–921. https://doi.org/10.1534/genetics.110.123646.

Haider, S., & Pal, R. (2013). Integrated analysis of transcriptomic and proteomic data. *Current Genomics*, *14*(2), 91–110. https://doi.org/10.2174/1389202911314020003.

Han, J., Wang, B., Xiao, Z., Gao, Y., Zhao, Y., Zhang, J., . . . Dai, J. (2008). Mammalian target of rapamycin (mTOR) is involved in the neuronal differentiation of neural progenitors induced by insulin. *Molecular and Cellular Neuroscience*, *39*(1), 118–124. https://doi.org/10.1016/j.mcn.2008.06.003.

Hartman, N. W., Lin, T. V., Zhang, L., Paquelet, G. E., Feliciano, D. M., & Bordey, A. (2013). mTORC1 targets the translational repressor 4E-BP2, but not S6 kinase 1/2, to regulate neural stem cell self-renewal in vivo. *Cell Reports*, *5*(2), 433–444. https://doi.org/10.1016/j.celrep.2013.09.017.

Hussain, S., Sajini, A. A., Blanco, S., Dietmann, S. Lombard, P., Sugimoto, Y., . . . Frye, M. (2013). NSun2-mediated cytosine-5 methylation of vault noncoding RNA determines its processing into regulatory small RNAs. *Cell Reports*, *4*(2), 255–261. https://doi.org/10.1016/j.celrep.2013.06.029.

Hutsler, J. J., Love, T., & Zhang, H. (2007). Histological and magnetic resonance imaging assessment of cortical layering and thickness in autism spectrum disorders. *Biological Psychiatry*, *61*(4), 449–457. https://doi.org/10.1016/j.biopsych.2006.01.015.

Hwang, D. W., Jaganathan, A., Shrestha, P., Nour El-Amine, Y. J., Wang, S. H., Hammell, M., & Mills, A. A. (2018). Chromatin-mediated translational control is essential for neural cell fate specification. *Life Science Alliance*, *1*(4). https://doi.org/10.26508/lsa.201700016.

Iossifov, I., O'Roak, B. J., Sanders, S. J., Ronemus, M., Krumm, N., Levy, D., . . . Wigler, M. (2014). The contribution of de novo coding mutations to autism spectrum disorder. *Nature*, *515*(7526), 216-U136. https://doi.org/10.1038/nature13908.

Ivshina, M., Lasko, P., & Richter, J. D. (2014). Cytoplasmic polyadenylation element binding proteins in development, health, and disease. *Annual Review of Cell and Developmental Biology, 30*, 393–415. https://doi.org/10.1146/annurev-cellbio-101011-155831.

Jackson, R. J., Hellen, C. U., Pestova, T. V. (2010). The mechanism of eukaryotic translation initiation and principles of its regulation. *Nature Reviews Molecular Cell Biology, 11*(2), 113–127. https://doi.org/10.1038/nrm2838.

Jacquemont, S., Pacini, L., Jønch, A. E., Cencelli, G., Rozenberg, I., He, Y., . . . Bagni, C. (2018). Protein synthesis levels are increased in a subset of individuals with fragile X syndrome. *Human Molecular Genetics, 27*(12), 2039–2051. https://doi.org/10.1093/hmg/ddy291.

Johnstone, T. G., Bazzini, A. A., & Giraldez, A. J. (2016). Upstream ORFs are prevalent translational repressors in vertebrates. *EMBO Journal, 35*(7), 706–723. https://doi.org/10.15252/embj.201592759.

Kamenska, A., Lu, W. T., Kubacka, D., Broomhead, H., Minshall, N., Bushell, M., & Standart, N. (2014). Human 4E-T represses translation of bound mRNAs and enhances microRNA-mediated silencing. *Nucleic Acids Research, 42*(5), 3298–3313. https://doi.org/10.1093/nar/gkt1265.

Kaushik, G., & Zarbalis, K. S. (2016). Prenatal neurogenesis in autism spectrum disorders. *Frontiers in Chemistry, 4*, 12. https://doi.org/10.3389/fchem.2016.00012.

Khajuria, R. K., Munschauer, M., Ulirsch, J. C., Fiorini, C., Ludwig, L. S., McFarland, S. K., . . . Sankaran, V. G. (2018). Ribosome levels selectively regulate translation and lineage commitment in human hematopoiesis. *Cell, 173*(1), 90–103.e19. https://doi.org/10.1016/j.cell.2018.02.036.

Khoddami, V., & Cairns, B. R. (2013). Identification of direct targets and modified bases of RNA cytosine methyltransferases. *Nature Biotechnology, 31*(5), 458–464. https://doi.org/10.1038/nbt.2566.

Klauck, S. M., Felder, B., Kolb-Kokocinski, A., Schuster, C., Chiocchetti, A., Schupp, I., . . . Poustka, A. (2006). Mutations in the ribosomal protein gene RPL10 suggest a novel modulating disease mechanism for autism. *Molecular Psychiatry, 11*(12), 1073–1084. https://doi.org/10.1038/sj.mp.4001883.

Kosodo, Y., & Huttner, W. B. (2009). Basal process and cell divisions of neural progenitors in the developing brain. *Development, Growth, & Differentiation, 51*(3), 251–261. https://doi.org/10.1111/j.1440-169X.2009.01101.x.

Kosodo, Y., Toida, K., Dubreuil, V., Alexandre, P., Schenk, J., Kiyokage, E., . . . Huttner, W. B. (2008). Cytokinesis of neuroepithelial cells can divide their basal process before anaphase. *EMBO Journal, 27*(23), 3151–3163. https://doi.org/10.1038/emboj.2008.227.

Kraushar, M. L., Popovitchenko, T., Volk, N. L., & Rasin, M. R. (2016). The frontier of RNA metamorphosis and ribosome signature in neocortical development. *International Journal of Developmental Neuroscience, 55*, 131–139. https://doi.org/10.1016/j.ijdevneu.2016.02.003.

Kraushar, M. L., Thompson, K., Wijeratne, H. R., Viljetic, B., Sakers, K. Marson, J. W., . . . Rasin, M. R. (2014). Temporally defined neocortical translation and polysome assembly are determined by the RNA-binding protein Hu antigen R. *Proceedings of the National Academy of Sciences of the United States of America, 111*(36), E3815-24. https://doi.org/10.1073/pnas.1408305111.

Kraushar, M. L., Viljetic, B., H. R. Wijeratne, M. R., Thompson, K., Jiao, X., Pike, J. W., . . . Rasin, M. R. (2015). Thalamic WNT3 secretion spatiotemporally regulates the neocortical ribosome signature and mRNA translation to specify neocortical cell subtypes. *Journal of Neuroscience, 35*(31), 10911–10926. https://doi.org/10.1523/JNEUROSCI.0601-15.2015.

Kristensen, A. R., Gsponer, J., & Foster, L. J. (2013). Protein synthesis rate is the predominant regulator of protein expression during differentiation. *Molecular Systems Biology*, *9*, 689. https://doi.org/10.1038/msb.2013.47.

Kusek, G., Campbell, M., Doyle, F., Tenenbaum, S. A., Kiebler, M., & Temple, S. (2012). Asymmetric segregation of the double-stranded RNA binding protein Staufen2 during mammalian neural stem cell divisions promotes lineage progression. *Cell Stem Cell*, *11*(4), 505–516. https://doi.org/10.1016/j.stem.2012.06.006.

Laguesse, S., Creppe, C., Nedialkova, D. D., Prévot, P. P., Borgs, L., Huysseune, S., . . . Nguyen, L. (2015). A dynamic unfolded protein response contributes to the control of cortical neurogenesis. *Developmental Cell*, *35*(5), 553–567. https://doi.org/10.1016/j.devcel.2015.11.005.

Larsson, O., Tian, B., & Sonenberg, N. (2013). Toward a genome-wide landscape of translational control. *Cold Spring Harbor Perspectives in Biology*, *5*(1), a012302. https://doi.org/10.1101/cshperspect.a012302.

Lee, J. A., & Cole, G. J. (2000). Localization of transitin mRNA, a nestin-like intermediate filament family member, in chicken radial glia processes. *Journal of Comparative Neurology*, *418*(4), 473–483.

Lennox, A. L., Mao, H., & Silver, D. L. (2018). RNA on the brain: Emerging layers of post-transcriptional regulation in cerebral cortex development. *Wiley Interdisciplinary Reviews: Developmental Biology*, *7*(1). https://doi.org/10.1002/wdev.290.

Li, Y., Wang, H., Muffat, J., Cheng, A. W., Orlando, D. A., Lovén, J., . . . Jaenisch, R. (2013). Global transcriptional and translational repression in human-embryonic-stem-cell-derived Rett syndrome neurons. *Cell Stem Cell*, *13*(4), 446–458. https://doi.org/10.1016/j.stem.2013.09.001.

Liu, Y., Beyer, A., & Aebersold, R. (2016). On the dependency of cellular protein levels on mRNA abundance. *Cell*, *165*(3), 535–550. https://doi.org/10.1016/j.cell.2016.03.014.

Llorens-Bobadilla, E., Zhao, S., Baser, A., Saiz-Castro, G., Zwadlo, K., & Martin-Villalba, A. (2015). Single-cell transcriptomics reveals a population of dormant neural stem cells that become activated upon brain injury. *Cell Stem Cell*, *17*(3), 329–340. https://doi.org/10.1016/j.stem.2015.07.002.

Louros, S. R., & Osterweil, E. K. (2016). Perturbed proteostasis in autism spectrum disorders. *Journal of Neurochemistry*, *139* (6), 1081–1092. https://doi.org/10.1111/jnc.13723.

Luo, Y., Na, Z., & Slavoff, S. A. (2018). P-bodies: Composition, properties, and functions. *Biochemistry*, 57(17), 2424–2431. https://doi.org/10.1021/acs.biochem.7b01162.

Magri, L., Cominelli, M., Cambiaghi, M., Cursi, M., Leocani, L., Minicucci, F., Poliani, P. L., & Galli, R. (2013). Timing of mTOR activation affects tuberous sclerosis complex neuropathology in mouse models. *Disease Models & Mechanisms*, *6*(5), 1185–1197. https://doi.org/10.1242/dmm.012096.

Mahoney, C., Feliciano, D. M., Bordey, A., & Hartman, N. W. (2016). Switching on mTORC1 induces neurogenesis but not proliferation in neural stem cells of young mice. *Neuroscience Letters*, *614*, 112–118. https://doi.org/10.1016/j.neulet.2015.12.042.

Malinovskaya, E. M., Ershova, E. S., Golimbet, V. E., Porokhovnik, L. N., Lyapunova, N. A., Kutsev, S. I., . . . Kostyuk, S. V. (2018). Copy number of human ribosomal genes with aging: Unchanged mean, but narrowed range and decreased variance in elderly group. *Frontiers in Genetics*, *9*, 306. https://doi.org/10.3389/fgene.2018.00306.

McGeachy, A. M., & Ingolia, N. T. (2016). Starting too soon: Upstream reading frames repress downstream translation. *EMBO Journal*, *35*(7), 699–700. https://doi.org/10.15252/embj.201693946.

McGowan, H., & Pang, Z. P. (2015). Regulatory functions and pathological relevance of the MECP2 3′UTR in the central nervous system. *Cell Regeneration (London)*, *4*, 9. https://doi.org/10.1186/s13619-015-0023-x.

Meng, D., Frank, A. R., Jewell, J. L. (2018). mTOR signaling in stem and progenitor cells. *Development*, *145*(1). https://doi.org/10.1242/dev.152595.

Miyata, T., Kawaguchi, A., Okano, H., & Ogawa, M. *2001. Asymmetric inheritance of radial glial fibers by cortical neurons. Neuron*, *31*(5), 727–741.

Nedialkova, D. D., & Leidel, S. A. (2015). Optimization of codon translation rates via tRNA modifications maintains proteome integrity. *Cell*, *161*(7), 1606–1618. https://doi.org/10.1016/j.cell.2015.05.022.

Neves-Pereira, M., Müller, B., Massie, D., Williams, J. H., O'Brien, P. C., Hughes, A.,... Miedzybrodzka, Z. (2009). Deregulation of EIF4E: A novel mechanism for autism. *Journal of Medical Genetics*, *46*(11), 759–765. https://doi.org/10.1136/jmg.2009.066852.

Nicolini, C., Ahn, Y., Michalski, B., Rho, J. M., & Fahnestock, M. (2015). Decreased mTOR signaling pathway in human idiopathic autism and in rats exposed to valproic acid. *Acta Neuropathologica Communications*, 3, 3. https://doi.org/10.1186/s40478-015-0184-4.

Nishino, J., Kim, S., Zhu, Y., Zhu, H., & Morrison, S. J. (2013). A network of heterochronic genes including Imp1 regulates temporal changes in stem cell properties. *Elife*, *2*, e00924. https://doi.org/10.7554/eLife.00924.

Noctor, S. C., Flint, A. C., Weissman, T. A., Dammerman, R. S., & Kriegstein, A. R. (2001). Neurons derived from radial glial cells establish radial units in neocortex. *Nature*, *409*(6821), 714–720. https://doi.org/10.1038/35055553.

Ozgur, S., Basquin, J., Kamenska, A., Filipowicz, W., Standart, N., & Conti, E. (2015). Structure of a human 4E-T/DDX6/CNOT1 complex reveals the different interplay of DDX6-binding proteins with the CCR4-NOT complex. *Cell Reports*, *13*(4), 703–711. https://doi.org/10.1016/j.celrep.2015.09.033.

Packer, A. (2016). Neocortical neurogenesis and the etiology of autism spectrum disorder. *Neuroscience Biobehavior Review*, *64*, 185–195. https://doi.org/10.1016/j.neubiorev.2016.03.002.

Paliouras, G. N., Hamilton, L. K. Aumont, A., Joppé, S. E., Barnabé-Heider, F., & Fernandes, K. J. (2012). Mammalian target of rapamycin signaling is a key regulator of the transit-amplifying progenitor pool in the adult and aging forebrain. *Journal of Neuroscience*, *32*(43), 15012–15026. https://doi.org/10.1523/JNEUROSCI.2248-12.2012.

Pan, T. (2018). Modifications and functional genomics of human transfer RNA. *Cell, Research* *28*(4), 395–404. https://doi.org/10.1038/s41422-018-0013-y.

Paolini, N. A., Attwood, M., Sondalle, S. B., Vieira, C. M. D., van Adrichem, A. M., di Summa, F. M.,... MacInnes, A. W. (2017). A ribosomopathy reveals decoding defective ribosomes driving human dysmorphism. *American Journal of Human Genetics*, *100*(3), 506–522. https://doi.org/10.1016/j.ajhg.2017.01.034.

Parras, A., Anta, H., Santos-Galindo, M., Swarup, V., Elorza, A., Nieto-González, J. L.,... Lucas, J. J. (2018). Autism-like phenotype and risk gene mRNA deadenylation by CPEB4 mis-splicing. *Nature*, *560*(7719), 441–446. https://doi.org/10.1038/s41586-018-0423-5.

Pavitt, G. D., & Ron, D. (2012). New insights into translational regulation in the endoplasmic reticulum unfolded protein response. *Cold Spring Harbor Perspectives in Biology*, *4*(6). https://doi.org/10.1101/cshperspect.a012278.

Pilaz, L. J., Lennox, A. L., Rouanet, J. P., & Silver, D. L. (2016). Dynamic mRNA transport and local translation in radial glial progenitors of the developing brain. *Current Biology*, *26*(24), 3383–3392. https://doi.org/10.1016/j.cub.2016.10.040.

Pilaz, L. J., & Silver, D. L. (2015). Post-transcriptional regulation in corticogenesis: How RNA-binding proteins help build the brain. *Wiley Interdisciplinary Reviews: RNA*, *6*(5), 501–515. https://doi.org/10.1002/wrna.1289.

Pilaz, L. J., & Silver, D. L. (2017). Moving messages in the developing brain-emerging roles for mRNA transport and local translation in neural stem cells. *FEBS Letters*, *591*(11), 1526–1539. https://doi.org/10.1002/1873-3468.12626.

Popovitchenko, T., & Rasin, M. R. (2017). Transcriptional and post-transcriptional mechanisms of the development of neocortical lamination. *Frontiers in Neuroanatomy*, *11*, 102. https://doi.org/10.3389/fnana.2017.00102.

Porokhovnik, L. N., Passekov, V. P., Gorbachevskaya, N. L., Sorokin, A. B., Veiko, N. N., N. N. & Lyapunova, N. A. (2015). Active ribosomal genes, translational homeostasis and oxidative stress in the pathogenesis of schizophrenia and autism. *Psychiatriatric Genetics*, *25*(2), 79–87. https://doi.org/10.1097/YPG.0000000000000076.

Quan, Z., Zheng, D., & Qing, H. (2017). Regulatory roles of long non-coding RNAs in the central nervous system and associated neurodegenerative diseases. *Frontiers in Cellular Neuroscience*, *11*, 175. https://doi.org/10.3389/fncel.2017.00175.

Rodrigues, D. C., Kim, D. S., Yang, G., Zaslavsky, K. Ha, K. C., Mok, R. S., . . . Ellis, J. (2016). MECP2 is post-transcriptionally regulated during human neurodevelopment by combinatorial action of RNA-binding proteins and miRNAs. *Cell Reports*, *17*(3), 720–734. https://doi.org/10.1016/j.celrep.2016.09.049.

Roundtree, I. A., Evans, M. E., Pan, T., & He, C. (2017). Dynamic RNA modifications in gene expression regulation. *Cell*, *169*(7), 1187–1200. https://doi.org/10.1016/j.cell.2017.05.045.

Saffary, R., & Xie, Z. (2011). FMRP regulates the transition from radial glial cells to intermediate progenitor cells during neocortical development. *Journal of Neuroscience*, *31*(4), 1427–1439. https://doi.org/10.1523/JNEUROSCI.4854-10.2011.

Sanchez, C. G., Teixeira, F. K., Czech, B., Preall, J. B., Zamparini, A. L., Seifert, J. R., . . . Lehmann, R. (2016). Regulation of ribosome biogenesis and protein synthesis controls germline stem cell differentiation. *Cell Stem Cell*, *18*(2), 276–290. https://doi.org/10.1016/j.stem.2015.11.004.

Santini, E., Huynh, T. N., MacAskill, A. F., Carter, A. G., Pierre, P., Ruggero, D., . . . Klann, E. (2013). Exaggerated translation causes synaptic and behavioural aberrations associated with autism. *Nature*, *493*(7432), 411–415. https://doi.org/10.1038/nature11782.

Sartor, F., Anderson, J., McCaig, C., Miedzybrodzka, Z., & Müller, B. (2015). Mutation of genes controlling mRNA metabolism and protein synthesis predisposes to neurodevelopmental disorders. *Biochemistry Society Transactions*, *43*(6), 1259–1265. https://doi.org/10.1042/BST20150168.

Scheper, G. C., van der Knaap, M. S., & Proud, C. G. (2007). Translation matters: Protein synthesis defects in inherited disease. *Nature Reviews Genetics*, *8*(9), 711–723. https://doi.org/10.1038/nrg2142.

Shi, Z., & Barna, M. (2015). Translating the genome in time and space: Specialized ribosomes, RNA regulons, and RNA-binding proteins." *Annual Review of Cell and Developmental Biology*, *31*, 31–54. https://doi.org/10.1146/annurev-cellbio-100814-125346.

Shi, Z., Fujii, K., Kovary, K. M., Genuth, N. R., Röst, H. L., Teruel, M. N., & Barna, M. (2017). Heterogeneous ribosomes preferentially translate distinct subpools of mRNAs genome-wide. *Molecular Cell*, *67*(1), 71–83.e7. https://doi.org/10.1016/j.molcel.2017.05.021.

Shin, J., Berg, D. A., Zhu, Y., Shin, J. Y., Song, J., Bonaguidi, M. A., . . . Song, H. (2015). Single-cell RNA-seq with waterfall reveals molecular cascades underlying adult neurogenesis. *Cell Stem Cell*, *17*(3), 360–372. https://doi.org/10.1016/j.stem.2015.07.013.

Signer, R. A., Magee, J. A., Salic, A., & Morrison, S. J. (2014). Haematopoietic stem cells require a highly regulated protein synthesis rate. *Nature, 509*(7498), 49–54. https://doi.org/10.1038/nature13035.

Slavov, N., Semrau, S., Airoldi, E., Budnik, B., & van Oudenaarden, A. (2015). Differential stoichiometry among core ribosomal proteins. *Cell Reports, 13*(5), 865–873. https://doi.org/10.1016/j.celrep.2015.09.056.

Somers, J., Pöyry, T., & Willis, A. E. (2013). A perspective on mammalian upstream open reading frame function. *International Journal of Biochemistry & Cell Biology, 45*(8), 1690–1700. https://doi.org/10.1016/j.biocel.2013.04.020.

Sonenberg, N., & Hinnebusch, A. G. (2009). Regulation of translation initiation in eukaryotes: Mechanisms and biological targets. *Cell, 136*(4), 731–745. https://doi.org/10.1016/j.cell.2009.01.042.

Standart, N., & Weil, D. (2018). P-bodies: Cytosolic droplets for coordinated mRNA storage. *Trends in Genetics, 34*(8), 612–626. https://doi.org/10.1016/j.tig.2018.05.005.

Stoner, R., Chow, M. L., Boyle, M. P., Sunkin, S. M., Mouton, P. R., Roy, S., . . . Courchesne, E. (2014). Patches of disorganization in the neocortex of children with autism. *New England Journal of Medicine, 370*(13), 1209–1219. https://doi.org/10.1056/NEJMoa1307491.

Suhl, J. A., Muddashetty, R. S., Anderson, B. R., Ifrim, M. F., Visootsak, J., Bassell, G. J., Warren, S. T. (2015). A 3′ untranslated region variant in FMR1 eliminates neuronal activity-dependent translation of FMRP by disrupting binding of the RNA-binding protein HuR. *Proceedings of the National Academy of Sciences of the United States of America, 112*(47), E6553-61. https://doi.org/10.1073/pnas.1514260112.

Tahmasebi, S., Khoutorsky, A., Mathews, M. B., & Sonenberg, N. (2018). Translation deregulation in human disease. *National Review of Molecular Cell Biology*. https://doi.org/10.1038/s41580-018-0034-x.

Taverna, E., Gotz, M., & Huttner, A. B. (2014). The cell biology of neurogenesis: Toward an understanding of the development and evolution of the neocortex. *Annual Review of Cell and Developmental Biology, 30*(30), 465–502. https://doi.org/10.1146/annurev-cellbio-101011-155801.

Tebaldi, T., Zuccotti, P., Peroni, D., Köhn, M., Gasperini, L., Potrich, V., . . . Quattrone, A. (2018). HuD is a neural translation enhancer acting on mTORC1-responsive genes and counteracted by the Y3 small non-coding RNA. *Molecular Cell, 71*(2), 256–270.e10. https://doi.org/10.1016/j.molcel.2018.06.032.

Toma, K., & Hanashima, C. (2015). Switching modes in corticogenesis: Mechanisms of neuronal subtype transitions and integration in the cerebral cortex. *Frontiers in Neuroscience*, 9, 274. https://doi.org/10.3389/fnins.2015.00274.

Topol, A., English, J. A., Flaherty, E., Rajarajan, P. Hartley, B. J., Gupta, S., . . . Brennand, K. J. (2015). Increased abundance of translation machinery in stem cell-derived neural progenitor cells from four schizophrenia patients. *Translational Psychiatry*, 5, e662. https://doi.org/10.1038/tp.2015.118.

Troca-Marín, J. A., Alves-Sampaio, A., & Montesinos, M. L. (2012). Deregulated mTOR-mediated translation in intellectual disability. *Progress Neurobiology, 96*(2), 268–282. https://doi.org/10.1016/j.pneurobio.2012.01.005.

Tsai, V., Parker, W. E., Orlova, K. A., Baybis, M., Chi, A. W., Berg, B. D., . . . Crino, P. B. (2014). Fetal brain mTOR signaling activation in tuberous sclerosis complex. *Cerebral Cortex, 24*(2), 315–327. https://doi.org/10.1093/cercor/bhs310.

Tsunekawa, Y., Britto, J. M., Takahashi, M., Polleux, F., Tan, S. S., & Osumi, N. (2012). Cyclin D2 in the basal process of neural progenitors is linked to non-equivalent cell fates. *EMBO Journal*, *31*(8), 1879–1892. https://doi.org/10.1038/emboj.2012.43.

Tuoc, T. C., & Stoykova, A. (2010). Roles of the ubiquitin-proteosome system in neurogenesis. *Cell Cycle*, *9*(16), 3174–3180. https://doi.org/10.4161/cc.9.16.12551.

Tuorto, F., Liebers, R., Musch, T., Schaefer, M., Hofmann, S., Kellner, S., . . . Lyko, F. (2012). RNA cytosine methylation by Dnmt2 and NSun2 promotes tRNA stability and protein synthesis. *Nature Structural & Molecular Biology*, *19*(9), 900–905. https://doi.org/10.1038/nsmb.2357.

Tuorto, F., & Lyko, F. (2016). Genome recoding by tRNA modifications. *Open Biology*, *6*(12). https://doi.org/10.1098/rsob.160287.

Vessey, J. P., Amadei, G., Burns, S. E., Kiebler, M. A., Kaplan, D. R., & Miller, F. D. (2012). An asymmetrically localized Staufen2-dependent RNA complex regulates maintenance of mammalian neural stem cells. *Cell Stem Cell*, *11*(4), 517–528. https://doi.org/10.1016/j.stem.2012.06.010.

Wang, Y., Li, Y., Yue, M., Wang, J., Kumar, S., Wechsler-Reya, R. J., . . . Zhao, J. C. (2018). N6-methyladenosine RNA modification regulates embryonic neural stem cell self-renewal through histone modification. *Nature Neuroscience*, *21*(2), 195–206. https://doi.org/10.1038/s41593-017-0057-1.

Willsey, A. J., Sanders, S. J., Li, M., Dong, S., Tebbenkamp, A. T., Muhle, R. A., State, M. W. (2013). Coexpression networks implicate human midfetal deep cortical projection neurons in the pathogenesis of autism. *Cell*, *155*(5), 997–1007. https://doi.org/10.1016/j.cell.2013.10.020.

Yang, G., Smibert, C. A., Kaplan, D. R., & Miller, F. D. (2014). An eIF4E1/4E-T complex determines the genesis of neurons from precursors by translationally repressing a proneurogenic transcription program. *Neuron*, *84*(4), 723–739. https://doi.org/10.1016/j.neuron.2014.10.022.

Yang, X., Yang, Y., Sun, B. F., Chen, Y. S., Xu, J. W., Lai, W. Y., . . . Yang, Y. G. (2017). 5-methylcytosine promotes mRNA export—NSUN2 as the methyltransferase and ALYREF as an m. *Cell Research*, *27*(5), 606–625. https://doi.org/10.1038/cr.2017.55.

Yoon, K. J., Ringeling, F. R., Vissers, C., Jacob, F., Pokrass, M., Jimenez-Cyrus, D., . . . Song, H. (2017). Temporal control of mammalian cortical neurogenesis by m^6A methylation. *Cell*, *171*(4), 877–889.e17. https://doi.org/10.1016/j.cell.2017.09.003.

Youn, J. Y., Dunham, W. H., Hong, S. J., Knight, J. D. R., Bashkurov, M., Chen, G. I., . . . Gingras, A. C. (2018). High-density proximity mapping reveals the subcellular organization of mRNA-associated granules and bodies. *Molecular Cell*, *69*(3), 517–532.e11. https://doi.org/10.1016/j.molcel.2017.12.020.

Zahr, S. K., Yang, G., Kazan, H., Borrett, M. J., Yuzwa, S. A., Voronova, A., . . . Miller, F. D. (2018). A translational repression complex in developing mammalian neural stem cells that regulates neuronal specification. *Neuron*, *97*(3), 520–537.e6. https://doi.org/10.1016/j.neuron.2017.12.045.

Zhang, J., Kong, L., Guo, S., Bu, M., Guo, Q., Xiong, Y., . . . Chen, M. (2016). hnRNPs and ELAVL1 cooperate with uORFs to inhibit protein translation. *Nucleic Acids Research*. https://doi.org/10.1093/nar/gkw991.

Zismanov, V., Chichkov, V., Colangelo, V., Jamet, S., Wang, S., Syme, A., . . . Crist, C. (2016). Phosphorylation of eIF2α is a translational control mechanism regulating muscle stem cell quiescence and self-renewal. *Cell Stem Cell*, *18*(1), 79–90. https://doi.org/10.1016/j.stem.2015.09.020.

PART IV

NEURONAL PROTEIN SYNTHESIS AND DISEASE

CHAPTER 17

TRANSLATIONAL CONTROLS IN PAIN

SARAH LOERCH, JUNE BRYAN DE LA PEÑA, JANE SONG, JOSEPH J. PANCRAZIO, THEODORE J. PRICE, AND ZACHARY T. CAMPBELL

Introduction

The nervous system facilitates a crucial role in detection of harmful cues through a conserved process termed nociception (W. D. Tracey, Jr., 2017). It serves a critical function in the prevention of tissue damage and increases organismal fitness. Humans with congenital insensitivity to pain (CIP) often perish in childhood due to injuries or infections that fail to be recognized (Indo et al., 1996). Nociceptors are sensory neurons tasked with detection of noxious stimuli (e.g., heat, inflammatory cytokines, neurotrophic factors, capsaicin). They play a key role in both the detection and propagation of pain signals to the spinal cord that are ultimately communicated to the somatosensory cortex of the brain (Figure 17.1A). After an injury, nociceptors undergo remarkable changes in their activity (termed plasticity) that often outlive the healing process (Pace et al., 2018). Nociceptor sensitization refers to a failure of nociceptors to return to their resting state and may play a major role in the transition from acute to chronic pain (Ferrari, Bogen, & Levine, 2010). Translational control have emerged as a dominant theme in nociceptor plasticity (Khoutorsky & Price, 2018; Melemedjian & Khoutorsky, 2015). Here we provide an overview of the tremendous body of evidence in support of translation as an integral component of pain signaling. We emphasize the critical role of nociceptors given their key function in the detection and relay of pain signals.

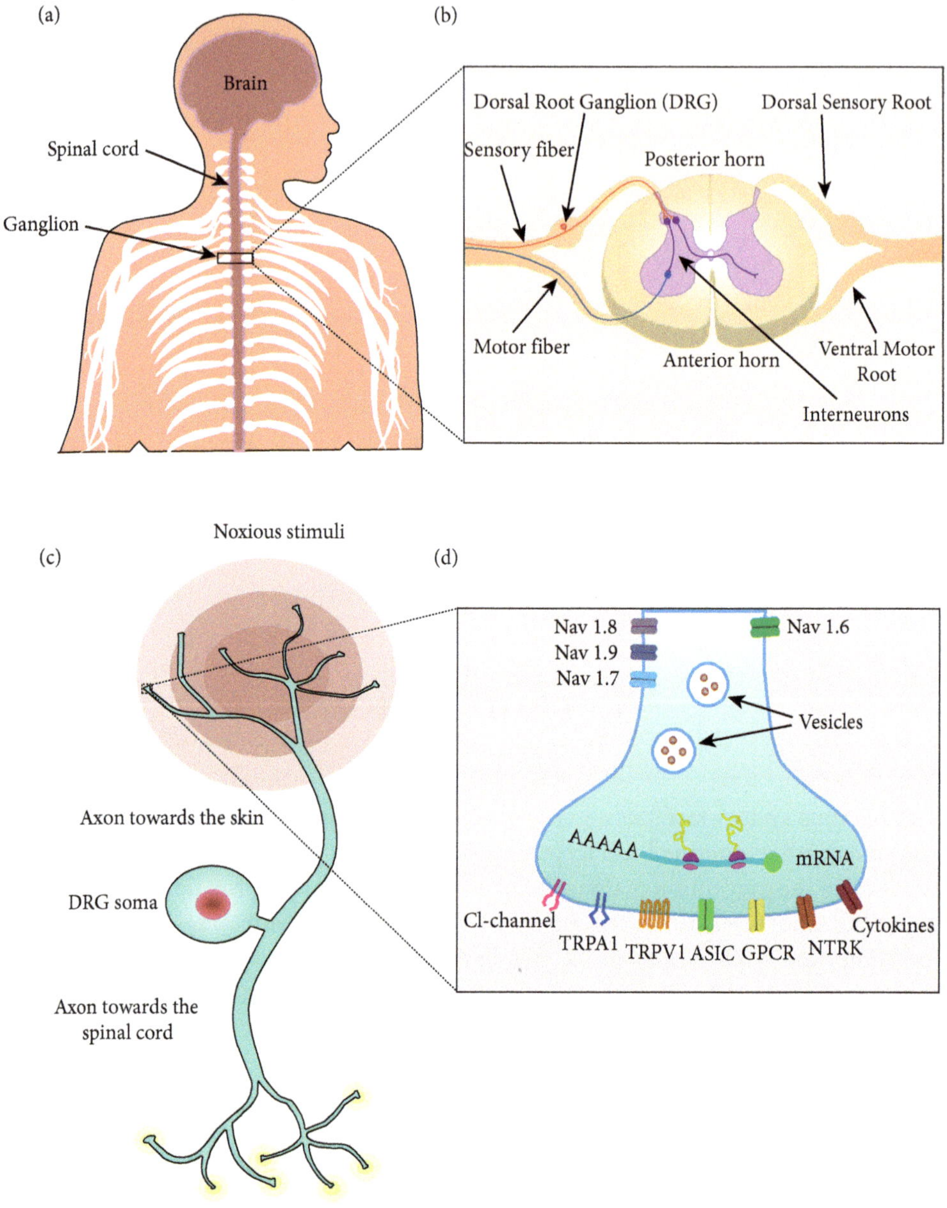

FIGURE 17.1. (A) The anatomy of the pain. (B) Nociceptors are responsible for detecting harmful stimuli. Cell bodies of nociceptors are clustered in the DRG adjacent to the spinal cord. DRG neurons have one axon with two branches: one branch (sensory fiber) innervates the skin, peripheral tissues, and internal organs. The opposing branch (sensory root) synapses with neurons in the spinal cord, which then relay the signal to the somatosensory cortex of the brain via the thalamus. (C) Nerve endings of sensory fibers can detect various noxious stimuli. (D) Receptors for various stimuli such as heat (TRPV1), environmental irritants (TRPA1), pro-inflammatory mediators (e.g., NGF), and cytokines. Synaptic vesicles can store neurotransmitters at the synapse and are controlled by voltage-gated calcium channels.

A Primer on Pain Physiology

Nociceptors are pseudounipolar neurons tasked with detection of harmful stimuli and propagation of these signals to the spinal cord (Figure 17.1B). The nucleus and many of the organelles in the nociceptor are housed in the cell body or soma. These are found in two different tissues, the dorsal root ganglia (DRG) adjacent to the spinal cord, or the trigeminal ganglia (TG) in the head. Approximately half of the neurons in the DRG and TG are nociceptors in most species. Protein and mRNA expression differ substantially between cell types giving rise to characteristic conduction velocity, diameter, and stimuli responsiveness. These differences manifest in the fibers (also known as axons) that extend from the soma (Figure 17.1C). One end innervates the skin and other peripheral organs, including most of the viscera, and is responsible for the detection of noxious and/or damaging stimuli.

Injury can result in damage to the axon and its subsequent degeneration (Davies et al., 2019). Nociceptor axons are associated with Schwann cells through structures called remak bundles, but most are not myelinated. Following injury, Schwann cells have been demonstrated to play a critical role in the clearance of debris resulting from tissue degradation and secrete molecules, including nerve growth factor (NGF), that stimulate axonal growth. This process hinges on local production of proteins in axons (Figure 17.1D). Axons projecting into the skin shed any associated Schwann cells and encounter fibroblasts and keratocytes. After an injury, in addition to inflammatory mediators secreted by immune cells (e.g., IL-6), keratinocytes release ATP, which contributes to changes in nociceptive activity and pain-associated behaviors (Moehring, Halder, Seal, & Stucky, 2018). Thus, the microenvironment surrounding the nerve fiber facilitates axonal regeneration and activity.

The most critical function accomplished by nociceptor axons is signal relay. Peripheral receptors are electrically silent in a resting state, but once threshold is reached they transmit action potentials back to the central nervous system (CNS; Barragan-Iglesias et al., 2018; Bogen, Alessandri-Haber, Chu, Gear, & Levine, 2012; Dubin & Patapoutian, 2010; Ferrari, Bogen, Chu, & Levine, 2013; Ferrari, Bogen, & Levine, 2013; Inceoglu et al., 2015; Khoutorsky et al., 2016; Melemedjian et al., 2010; Moy et al., 2017; Xu et al., 2014). These signals are received by interneurons in the dorsal horn of the spinal cord. The spinal cord transmits pain signals to the brain, where they are consciously perceived. Specific neurons act as checkpoints and determine whether a pain signal is relayed or not, thus not all signals are relayed (I. Tracey & Mantyh, 2007). Injury changes the electrophysiological and neurochemistry of neurons that detect and relay pain signals (Song, Vizcarra, Xu, Rupert, & Wong, 2003). Hypersensitivity to noxious stimuli (hyperalgesia) or innocuous stimuli (allodynia) results from neuronal plasticity that causes a lowering of pain thresholds (Figure 17.2).

A growing body of evidence suggests that translational controls are integral to sustained changes in neuronal excitability that drive persistent pain states (Barragan-Iglesias

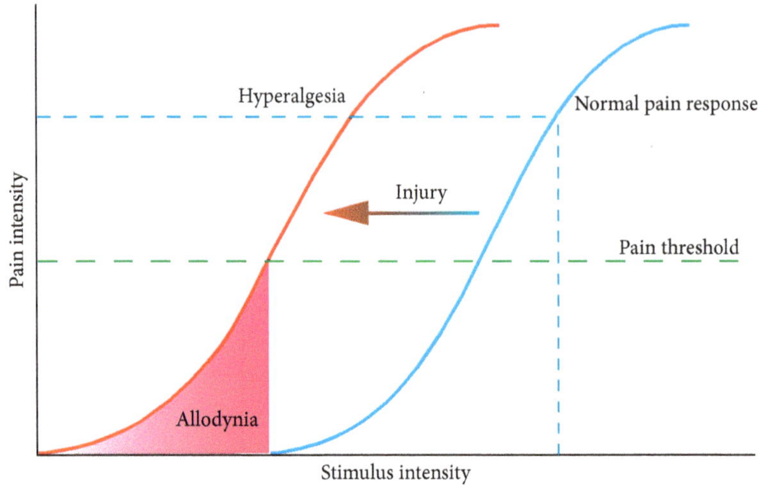

FIGURE 17.2. Injury changes pain responses. Allodynia (pain due to a stimulus that does not usually provoke pain) and hyperalgesia (increased pain from a stimulus that usually elicits pain) are commonly observed in patients after an injury. Maladaptive central changes and nociceptor sensitization contribute to the generation and maintenance of allodynia and hyperalgesia, which can ultimately lead to chronic pain (Cervero & Laird, 1996).

et al., 2019; Ferrari, Araldi, & Levine, 2015; Ferrari, Bogen, Chu, & Levine, 2013; Ferrari, Bogen, Reichling, & Levine, 2015; Geranton et al., 2009; Hunt, Winkelstein, Rutkowski, Weinstein, & DeLeo, 2001; Khoutorsky et al., 2015; Khoutorsky et al., 2016; Megat et al., 2019; Melemedjian et al., 2011; Melemedjian & Khoutorsky, 2015; Melemedjian, Khoutorsky, et al., 2013; Melemedjian, Mejia, Lepow, Zoph, & Price, 2014; Moy et al., 2017; Obara et al., 2011; Uttam, 2018). A major challenge moving forward is dissecting differential contributions of translational control to establishment as opposed to maintenance of pain states. Tremendous insights into nociceptor plasticity have resulted from studies of hyperalgesic priming. Priming refers to susceptibility to normally subthreshold noxious inputs following a noxious stimulus. The strength of this model is the ability to separate acute and prolonged pain states (Kandasamy & Price, 2015). The ever expanding pharmacopoeia for translational control applied to hyperalgesic priming will enhance our understanding of acute and chronic pain. Increasingly precise genetic and optogenetic tools are also likely to contribute key insights.

Local Translation

Neurons must modulate their function in response to a range of physiologic stimuli. A key mechanism that facilitates rapid changes in sensory fibers is local translation from polarized populations of mRNA (Jung, Gkogkas, Sonenberg, & Holt, 2014). RNA-localization is common in eukaryotes. An extreme example can be found in *Drosophila* embryos where ~70 percent of genes show distinct patterns of subcellular localization (Lecuyer et al., 2007; Tomancak et al., 2007). Most of the corresponding proteins co-localize with their transcripts, suggestive of a potential use of RNA localization to regulate sites of protein synthesis. Highly specialized cell types, including neurons, make extensive use of RNA localization. RNA-seq on neuronal processes suggests highly specific mechanisms of mRNA trafficking (Andreassi et al., 2010; Cajigas et al., 2012; Gumy et al., 2011; Minis et al., 2014; A. M. Taylor et al., 2009; Zivraj et al., 2010). During transit, translation of the mRNA is repressed often via protein factors recruited to the 3′ untranslated region (UTR; Figure 17.3A). The repertoire of RNA-binding proteins bound to mRNAs destined for local translation is controlled by a variety of signaling mechanisms (Gumy et al., 2011; T. T. Merianda et al., 2009; A. M. Taylor et al., 2009; D. E. Willis et al., 2007; D. E. Willis et al., 2011; Yudin et al., 2008; Zivraj et al., 2010). These multi-protein complexes serve critical roles in both trafficking of the mRNA and ensuring that translation is repressed until the transcript has arrived at the appropriate location within the cell. Thus, tremendous precision is achieved through cis-acting elements present in mRNA that provide all subsequent regulatory potential by trans-acting factors.

Local translation serves a key biological function. Neuronal protein synthesis can occur in the soma, synapse, or in axons. In nociceptors, axons can span vast distances (in some cases a meter or longer). Localized translation provides a means to accomplish protein biosynthesis at the site where polypeptides are required. This provides a rapid solution to the problem of generating new proteins on demand that can guide critical processes to the function of afferent fibers (such as axonal growth; Brittis, Lu, & Flanagan, 2002; Kar, Lee, & Twiss, 2018). Local translation requires instructions provided by mRNA and is executed through the combined actions of ribosomes, tRNAs, and regulatory factors. Regulatory factors play a critical role in triggering translation of the correct target at the appropriate moment when it is required. Among the best characterized examples of activity-dependent protein synthesis is local translation of the immediate early gene Arc. Arc is translated in dendrites as an integral component of learning and memory processes in the hippocampus and amygdala (Guzowski et al., 2000; Guzowski, McNaughton, Barnes, & Worley, 1999; McIntyre et al., 2005; Tzingounis & Nicoll, 2006). However, the role of Arc in peripheral neurons is unclear. While Arc is translated in the spinal cord, it appears to be dispensable for inflammatory pain (Hossaini, Jongen, Biesheuvel, Kuhl, & Holstege, 2010). What are the regulatory features present in mRNA that dictate the specificity of local translation in nociceptors? While the answer is likely transcript specific, emerging evidence suggests that analogous mechanisms to neurons in the CNS are employed, making use of untranslated regions (UTRs) to impart changes in mRNA localization (Baj, Pinhero, Vaghi, & Tongiorgi, 2016; D. E. Willis et al., 2011).

Multiple lines of evidence suggest a key role for local translation in pain. First, axons are key sites of protein synthesis particularly in nociceptors (Barragan-Iglesias et al., 2018; Kar et al., 2018; T. Merianda & Twiss, 2013; Terenzio et al., 2018). Injection of protein synthesis inhibitors into the paw blocks behavioral responses to inflammatory

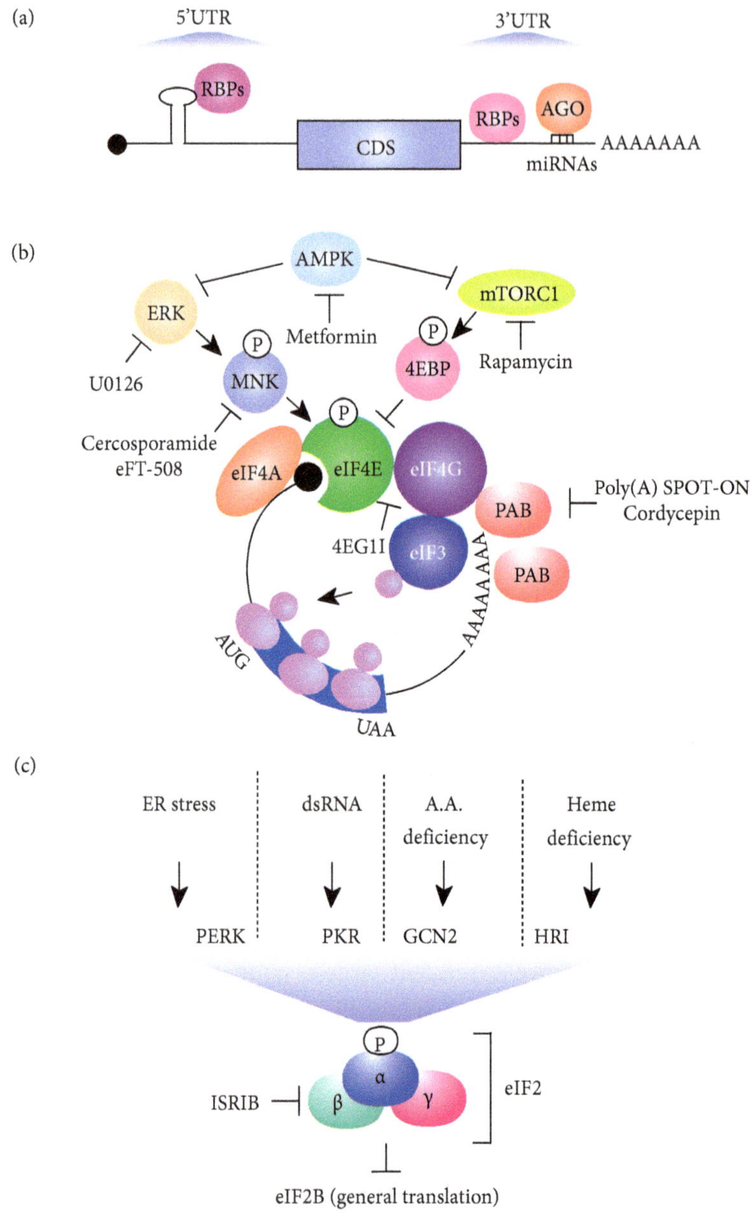

FIGURE 17.3. (A) mRNA is composed of a 5′ UTR, the coding sequence (CDS), the 3′ UTR, and the poly(A) tail. The M7G cap structure (black ball) is found on the 5′ end of the transcript. Structures in the 5′ UTR can influence translation efficiency and can also recruit RNA-binding proteins (RBPs). Similarly, the 3′ UTR contains regulatory elements that can be bound by trans-acting RNAs (e.g., miRNAs), as well as proteins. The transcript is appended with a poly(A) tail. (B) Translation initiation and key inhibitors. AMPK negatively regulates both mTOR and ERK. ERK controls MNK, which ultimately phosphorylates eIF4E. mTOR controls eIF4E availability through phosphorylation of 4EBPs. eIF4G interacts with PABP to promote initiation. eIF4G facilitates recruitment of the 40S ribosomal subunit through interactions with eIF3 (not shown). (C) Stimuli that engage the ISR are indicated, as well as downstream kinases that act on eIF2α. eIF2α-dependent translation is blocked by the small molecule ISRIB.

mediators that increase nociceptor excitability (Black et al., 2018; Melemedjian et al., 2010). Second, disruption of mRNA polyadenylation specifically in the DRG blocks hyperalgesic priming through local translation of CamKIIα (Bogen et al., 2012; Ferrari, Bogen, et al., 2013). Third, NGF increases axonal localization of a subset of mRNAs (D. Willis et al., 2005). Fourth, injection of NGF into humans promotes mechanical hypersensitivity without inflammation through a mechanism that is locally regulated (Rukwied et al., 2010). In addition, injection of NGF into an axonal branch of a single nociceptor sensitizes only that branch (Obreja et al., 2018). Fifth, proteomics of neuromas and pulse chase–labeling experiments suggest that local translation of cytoskeletal factors drives hyper-excitability after nerve damage (H. L. Huang et al., 2008). Finally, several groups have identified Nav1.8 mRNA as axonally localized after peripheral nerve injury (Hirai et al., 2017; Thakor et al., 2009). Knockdown of Nav1.8 in the sciatic nerve fiber but not the DRG blocks neuropathic pain caused by sciatic nerve entrapment (Ruangsri et al., 2011).

mRNA Structure

Eukaryotic mRNAs are resplendent with regulatory features. These include the ubiquitous 5′ 7-methylguanosine (^{m7}G) cap on the 5′ end of the transcript (Figure 17.3A). The cap is bound by the cap-binding protein eIF4E (Sonenberg, Rupprecht, Hecht, & Shatkin, 1979). Loss of the ^{m7}G renders the mRNA susceptible to rapid 5′ → 3′ degradation by exonucleases (e.g., Xrn1). mRNAs possess two UTRs that either precede the coding segment on the 5′ side or follow the stop codon on the 3′ end. The UTRs harbor regulatory information in the form of *cis*-acting structures and sequences which are bound by trans-acting regulatory factors. These include RNA-binding proteins and regulatory RNAs that act in consort with RNA-binding proteins. The 5′ UTR has distinct classes of regulatory elements that include internal ribosomal entry sequences (IRES) and upstream open reading frames (uORFs). IRES elements can overcome the need for eIF4E mediated translation initiation through recruitment of translation factors. The function of uORFs is generally to reduce protein output of the main reading frame, but they can also change the reading frame, add additional protein sequence, or encode functional peptides (Barbosa, Peixeiro, & Romao, 2013). A key property of 5′ UTRs is structural content. Secondary structure in the 5′ UTR can increase dependency on the helicase eIF4A and further refine translational output. Similar to the 5′ UTR, the 3′ UTR can encode binding sites for regulatory factors and serves as a major repository of information that can enhance or reduce translational efficiency. The 3′ UTR also provides a critical function in neurons as a source of information for specification of local translation (Aronov, Aranda, Behar, & Ginzburg, 2001; Y. S. Huang, Carson, Barbarese, & Richter, 2003; Menon et al., 2004). An additional challenge arises from dynamic changes in 3′ UTR length caused by alternative polyadenylation (APA). APA provides a mechanism to modulate poly(A) site selection and appears to be critical for localization of ion channels (e.g., $Na_v1.8$) required for nociception (Hirai et al., 2017). The final step in mRNA maturation is addition of the Poly(A) tail to the 3′ end of the mRNA (AC & M, 2008). Poly(A) tail-length is intimately linked to translational efficacy and the Poly(A)-binding protein appears to be integral to pain signaling (Barragan-Iglesias et al., 2018). Finally, targeted disruption of polyadenylation by the small molecule cordycepin reverses pain hypersensitivity (Ferrari, Bogen, et al., 2013b).

Initiation

Protein synthesis is the culmination of a complex process initiated with the birth of RNA during transcription and the emergence of nascent peptides on the ribosome. Translation can be described in a series of four subsequent steps—translation initiation, elongation, termination, and ribosome recycling. Translation initiation is the rate-limiting step and has garnered tremendous attention, as the bulk of translational control is thought to occur at this step (Hinnebusch, Ivanov, & Sonenberg, 2016). Inhibition of translation initiation in nociceptors abolishes sensitization and exemplifies the central role that translation initiation plays in pain plasticity (Melemedjian et al., 2010; Melemedjian, Tillu, et al., 2014; Moy et al., 2017). In mammals, the main initiation pathway is termed cap-dependent translation and is responsible for the initiation of most translational events under non-stress conditions (Aitken & Lorsch, 2012). However, alternative pathways exist and are essential for survival under stress conditions and viral infections (Holcik & Sonenberg, 2005). Among the best-studied examples of alternative initiation pathways are IRES. They reside in the 5′ UTR and can directly recruit the ribosome to the mRNA. While their function in pain is unclear, cellular IRES initiate translation of mRNA subsets when cap-dependent translation is compromised and could mediate translation of nociceptive factors (Komar & Hatzoglou, 2011).

Cap-Dependent Translation

Cap-dependent translation hinges on multiple complexes that recruit the ribosome to the mRNA (Aitken & Lorsch, 2012; Figure 17.3B). The eukaryotic initiation factor 4E (eIF4E) associates with the 5′ 7-methylguanosine (^{m7}G) cap of the mRNA (Sonenberg et al., 1979). eIF4E is controlled both at the level of phosphorylation at a single site and through sequestration by a protein partner, eIF4E-binding protein (4E-BP) (Pause et al., 1994; Waskiewicz, Flynn, Proud, & Cooper, 1997). eIF4E interacts with the scaffold protein eIF4G, which in turn binds the helicase eIF4A. Collectively, this tripartite complex (referred to as eIF4F) stably associates with the ^{m7}G cap. eIF4F phosphorylation globally affects mRNA translation, and in some cases alters the translation of specific subsets of mRNAs—frequently proteins that are important for cell survival (Hsieh et al., 2012).

Once assembled, eIF4F recruits the 43S pre-initiation complex (PIC) to the ^{m7}G cap. The PIC consists of the small ribosomal subunit (40S) bound to the initiation factor eIF2, initiator tRNA Met-tRNA$_i^{Met}$, and GTP. Though intrinsically active, eIF4A helicase function is further stimulated by complex formation and unwinds the 5′ UTR of the mRNA to facilitate ribosomal scanning of the 5′ UTR. Thus, the translation of many mRNAs with highly structured 5′ UTRs is eIF4A-dependent. Upon encountering the AUG start codon, the large ribosomal subunit (60S) joins the complex to form the 80S ribosome, and eIF2 is released. The joining of the large ribosomal subunit concludes successful translation initiation and transitions the ribosome into the elongation phase. eIF4G further interacts with PABP and circularizes the mRNA, possibly facilitating re-initiation after a successful round of translation (Wells, Hillner, Vale, & Sachs, 1998).

eIF4F

Several signaling cascades converge on eIF4F. Multiple lines of evidence suggest that eIF4E is central in the development of pain pathologies. While the interaction of eIF4G with eIF4E is crucial for pain amplification, as evidenced by pharmacological studies (e.g., 4EGI1; Moerke et al., 2007), a specific role of eIF4A in pain remains poorly understood. A possible reason might be that the high expression level of eIF4A and its eIF4F-independent helicase properties complicate tight regulation (Duncan & Hershey, 1983; Galicia-Vazquez, Cencic, Robert, Agenor, & Pelletier, 2012). In contrast, eIF4E has a low expression level, thus minor changes in availability by sequestration or modification can have extensive consequences on translation initiation. Two major pathways directly affect and modulate eIF4E activity: the mechanistic target of rapamycin (mTOR), and the mitogen-activated protein kinase (MAPK) pathways (Figure 17.3B; Melemedjian et al., 2010; Moy et al., 2017).

The mechanistic target of rapamycin (mTOR) signaling cascade is a dominant regulatory feature of translational control (Yanagiya et al., 2012). The mTOR catalytic subunit exists in two multimeric protein complexes, one of which is sensitive to inhibition by rapamycin (mTORC1). In neurons, the mTORC1 pathway receives input from a large variety of upstream pathways that relay external input to mTORC1, which in turn creates cellular responses (Boutouja, Stiehm, & Platta, 2019). mTORC1 upstream receptors include NMDA, Trk, and IGF-1. The downstream targets of mTOR include regulators of translation like eIF4E binding proteins (4E-BPs), p70 S6 kinase (S6K), and eEF2 kinase. The three known 4E-BP isoforms (1, 2, and 3) show a tissue-specific expression and the predominant isoform in the pain processing pathway is 4E-BP1 (Jimenez-Diaz et al., 2008; Khoutorsky et al., 2015; Melemedjian et al., 2011; Xu, Zhao, Yaster, & Tao, 2010). Phosphorylation of 4E-BPs releases eIF4E from sequestration and allows it to engage in the eIF4F complex. Inflammatory pain models using injections of the upstream activators nerve growth factor (NGF) and interleukin 6 (IL-6) revealed a rapid induction of protein synthesis in nociceptors, which is concurrent with the activation of mTORC1 as monitored by phosphorylation (Melemedjian et al., 2010). Conversely, pharmacological inhibition of mTORC1 with rapamycin-related small molecules reduces pain hypersensitivity in a wide variety of pain models (Geranton et al., 2009; Jimenez-Diaz et al., 2008; Price et al., 2007). The endogenous endothelial growth factor receptor (EGFR) ligand, Epiregulin (EREG), stimulates the mTOR pathway in DRG neurons and upregulates matrix metalloproteinase 9 (MMP-9) translation (L. J. Martin et al., 2017). MMP-9 is a regulator of inflammation and is transiently upregulated in DRG sensory neurons in models of neuropathic chronic pain (Kawasaki et al., 2008; Manicone & McGuire, 2008). Inhibitors of EGFR, used in cancer treatments, have been reported to also alleviate pain in patients with cancer-induced neuropathic pain (Kersten & Cameron, 2012; Moryl, Obbens, Ozigbo, & Kris, 2006).

While mTORC1 is a global regulator of translation, it also appears to locally alter translation in the sciatic nerve and proprioceptive DRG neurons. During neuronal injury mTOR is transiently activated and translation of its own mRNA and other survival promoting molecules is up-regulated in a 3′ UTR-dependent fashion (Terenzio et al., 2018). 3′ UTRs frequently contain localization motifs suggesting that local mRNA pools can be deposited and activated upon a stimulus, in this case injury. Local pharmacological repression of mTOR leads to reduced neuron numbers. It is not known, however, if injury-induced local translation of mTOR affects nociception plasticity.

mTORC1 is also known to specifically regulate specific subsets of transcripts. For example, mTOR regulates expression of mRNAs that contain terminal oligopyrimidine tracts in their 5′ UTRs (5′ TOP mRNAs) via 4E-BPs (Thoreen et al., 2012). A critical issue in the field is the systematic identification of mTOR targets that contribute to pain-associated behaviors. While the molecular mechanisms by which these subsets are selected remains elusive, an enticing hypothesis is that disabling cap-dependent translation favors alternative initiation pathways. While so far not investigated in nociceptors, this hypothesis is underpinned by increased IRES-dependent translation of Arc mRNA in dendrites when cap-dependent initiation is inhibited (Pinkstaff, Chappell, Mauro, Edelman, & Krushel, 2001), which is consistent with the continued translation of IRES-containing mRNAs in the presence of mTOR inhibitors (Torin-1; Thoreen et al., 2012).

S6K1 and 2 are downstream effectors of mTORC1. S6Ks act on translation elongation by phosphorylation of initiation and elongation factors like eukaryotic elongation factor 2 (eEF2; reviewed in Zoncu, Efeyan, & Sabatini, 2011). Although an important regulator of elongation, the role of S6K1/2 in pain is less clear than that of eIF4E. The investigation of S6K1 has been complicated by predominantly relying on genetic tools as small molecules targeting S6Ks lack in high specificity. In models of chronic inflammation pain, mTOR activation leads to S6K1 phosphorylation in DRG neurons but remains unaffected in neuropathic pain models (Liang et al., 2013). S6K1/2 double-knockout mice are more sensitive to mechanical stimuli with unaltered thermal sensitivity. The direct implications of S6K1/2 on elongation are masked by a negative feedback mechanism that in the long term activates the MAPK/ERK pathway. This leads to hyperexcitability of sensory neurons, allodynia, and spontaneous pain (Melemedjian, Khoutorsky, et al., 2013).

The MAPK pathway controls phosphorylation of a single residue, Ser209, in eIF4E via MAPK-interacting protein kinases (MNKs) 1 and 2 (Pyronnet et al., 1999; Waskiewicz et al., 1999). MNK1/2-mediated eIF4E phosphorylation contributes to the development of nociceptor sensitization and promotes chronic pain after injury (Moy et al., 2017). Both phosphorylation-resistant eIF4E^{S209A} mutant mice and, reciprocally, MNK knockout mice show decreased pain hypersensitivity in response to most inflammatory mediators. Inhibition of eIF4E phosphorylation also inhibits hyperalgesic priming (Melemedjian et al., 2010; Moy et al., 2017). Similar to the mTOR pathway, MNK1/2-dependent phosphorylation of eIF4E Ser209 is suspected to promote tissue-specific alternative translation of mRNA subsets. Few eIF4E-phosphorylation dependent mRNA targets have been identified so far. In the pain-processing pathway, known targets are matrix metalloproteases (MMP-2 and 9) and the key regulator of pain plasticity, *Bdnf*, in dorsal root ganglia (Moy, Khoutorsky, Asiedu, Dussor, & Price, 2018). Translation of *Bdnf* mRNA is stimulated in response to inflammation and is important for pain plasticity and hyperalgesia (Obata & Noguchi, 2006; Melemedjian, Mejia, et al., 2014; Melemedjian, Tillu, et al., 2013; Moy et al., 2018). In the DRG, eIF4E phosphorylation is required for hyperalgesic priming and promotes the translation of a specific *Bdnf* mRNA isoform (Bdnf-201), which has the longest and most structured 5′ UTR of all *Bdnf* isoforms (Moy et al., 2018). The specific translation enhancement might reflect the stimulatory role of eIF4E on the RNA helicase eIF4A, although the specific effect of eIF4E-phosphorylation is unknown. These findings highlight that local and tissue-specific eIF4E-dependent translation is a feature of pain amplification.

eIF4E in Chemotherapy-Induced Peripheral Neuropathy

While several relevant pathways for pain amplification have been identified, the translational alterations of their specific mRNA targets are mostly unknown. A ribosome profiling study identified regulators of the MAPK/ERK pathway as mRNA targets in the DRG and spinal cord dorsal horn in neuropathic pain (Uttam, 2018). Although ribosome profiling lends itself to the identification of translationally regulated mRNA expression, the cellular heterogeneity of the nervous system poses an obstacle and can confound the identification of cell type–specific changes in protein expression. A method that allows for cell type–specific analysis is translating ribosome affinity purification (TRAP; Heiman et al., 2008), which uses a tagged ribosomal protein that is specifically expressed in the desired cell type. An initial study using TRAP has described translation in nociceptors in chemotherapy (paclitaxel)-induced pain in mice (Megat et al., 2019). Sequencing of mRNA bound to tagged ribosomes and further pharmacological and mutational validation suggest that MNK1-mediated eIF4E phosphorylation increases translation of the mTORC1-activator RagA complex. In mice, pain-associated behavioral effects of paclitaxel were reversed upon injection of a MNK inhibitor called eFT508. This suggests that pharmacological disruption of cap-dependent translation may provide a means to reverse neuropathic pain states. Consistent with this notion, elimination of the sole phosphorylation site on eIF4E results in profound deficits in pain behavioral responses to inflammatory mediators (Moy et al., 2017; Moy et al., 2018). This work suggests that cap-dependent translation is integral to the persistence chemotherapy-induced neuropathic pain.

eIF2

eIF2 is another key regulator of protein translation that promotes initiation (Holcik & Sonenberg, 2005) and is a known effector in neuropathic pain (Barragan-Iglesias et al., 2019). Phosphorylation on Ser51 of the eIF2α subunit is the nexus of four pathways that collectively form the integrated stress response (ISR; Khoutorsky et al., 2016; Sidrauski, McGeachy, Ingolia, & Walter, 2015). These pathways (Figure 17.3C) are activated by viral infection (double-stranded RNA-dependent protein kinase, PKR), ER-stress (PKR-like ER kinase, PERK), amino acid deprivation (general control non-repressible 2, GCN2), oxidative stress and heme-deficiency (heme-regulated inhibitor, HRI; Lu, Han, & Chen, 2001). Ser51 phosphorylation inhibits initiation by turning eIF2 into a competitive inhibitor of its GDP exchange factor (GEF) eIF2B, rendering it inactive (Jennings, Zhou, Mohammad-Qureshi, Bennett, & Pavitt, 2013; Krishnamoorthy, Pavitt, Zhang, Dever, & Hinnebusch, 2001; Pavitt, Ramaiah, Kimball, & Hinnebusch, 1998; Yang & Hinnebusch, 1996). eIF2α phosphorylation is increased in models of diabetes-induced neuropathic pain and chronic inflammation (Barragan-Iglesias et al., 2019; Khoutorsky et al., 2016). The targetability of individual pathways and subsequently the phosphorylation state of eIF2α make it an attractive pharmacological target. For example, activation of eIF2B by the small molecule ISRIB (Tsai et al., 2018) reverts eIF2α phosphorylation via PERK and relieves both translational inhibition and diabetic pain in mice (Barragan-Iglesias et al., 2019).

eIF2α phosphorylation generally inhibits translation but stimulates translation of mRNAs containing upstream ORFs (uORFs) in the 5′ UTRs of mRNAs (Barbosa et al., 2013; Hinnebusch et al., 2016). eIF2α phosphorylation can also impact read-through of uORFs through eIF2A-dependent mechanisms (Sendoel et al., 2017). Thus, it is tempting to speculate that this transient shift from uORFs to main ORFs (mORFs) causes pain hypersensitivity, potentially by affecting the local biophysics of the cell and membrane. The molecular mechanism by which uORFs affect nociception remains to be investigated.

eIF2α in Diabetic Peripheral Neuropathy

A reactive glycolytic metabolite associated with painful diabetic pain called methylglyoxal (MGO) triggers neuropathic pain via the integrated stress response (Barragan-Iglesias et al., 2019). Intriguingly, MGO-induced pain or diabetic pain caused by ablation of insulin-producing cells (with streptozotocin) is reversed by the small molecule inhibited ISRIB that targets eIF2B. While the relevant targets are unknown, the integrated stress response has been broadly implicated in neuronal function and is likely to be key in a variety of pain states linked to increases in eIF2a phosphorylation. Indeed, hemizygous loss of eIF2a phosphorylation decreases thermal but not mechanical hypersensitivity (Khoutorsky et al., 2016).

AMP-Activated Protein Kinase

AMPK functions as a key energy sensor and has emerged as a therapeutic target for pain (Carling, Thornton, Woods, & Sanders, 2012; Price, Das, & Dussor, 2016; Price & Dussor, 2013; A. Taylor et al., 2013). Three subunits contribute to AMPK function (α, β, and γ). The γ subunit senses the AMP/ATP ratio and mediates allosteric effects on the α subunit. The catalytic domain is modulated by an upstream kinase (AMPKK). AMPK controls mTOR via two different pathways. AMPK directly inhibits mTOR activity through phosphorylation of raptor and indirectly inhibits mTOR via activation of the TSC complex. AMPK is a target for metabolic disease and cancer. AMPK agonists including metformin attenuate nascent translation and increase neuronal p-granules (Melemedjian, Mejia, et al., 2014). AMPK activators appear to attenuate allodynia caused by peripheral nerve injury and reduce the excitability of nociceptors in vitro (Melemedjian et al., 2011).

Translational Controls in the Central Nervous System

Reconsolidation Mechanisms in Pain

Reconsolidation has been coupled to protein synthesis inhibitors as a means of erasing memories, and it has clear implications for traumatic memories that can lead to pathological states (e.g., posttraumatic stress disorder). Analogous states may underlie certain nociceptive pain states. For example, mechanical hyperalgesia can be labile and susceptible to reversal by intrathecal delivery of protein synthesis inhibitors (Bonin & De Koninck, 2014). This work suggests that pain reconsolidation is probably spinally mediated and could be a useful strategy to reverse persistent pain.

Spinal Modulation

Injury can increase the excitability of nociceptors and of the spinal cord circuitry. Central sensitization refers to increases in the excitability of the spinal circuit, and it plays a major role in pain signaling. Central sensitization can amplify signals originating in the periphery (communicated by the nociceptors) destined for processing by the central nervous system. The implications are manifest in three ways: allodynia, hyperalgesia, and generalized pain to noninjured sites (secondary hyperalgesia). Central sensitization is driven in part by changes in synaptic strengthening at the dorsal horn.

A major structural model for understanding plasticity comes from understanding the key role of synaptic strength in learning and memory in the brain. Synaptic strength is modulated by opposing processes termed long-term potentiation (LTP) and long-term depression (LTD) in mammals. Learning and memory and LTP share several commonalities. Both LTP and long-term memory require protein synthesis and are blocked by mTOR inhibitors (Costa-Mattioli, Sossin, Klann, & Sonenberg, 2009; K. C. Martin, Bartsch, Bailey, & Kandel, 1999). Drugs that block LTP induction also attenuate hyperalgesia in vivo (Ruscheweyh, Wilder-Smith, Drdla, Liu, & Sandkuhler, 2011). Finally, electrical stimulation that induces LTP in rodents generates long-lasting increases in pain perception in humans (Biurrun Manresa, Kaeseler Andersen, Mouraux, & van den Broeke, 2018). Conversely, electrical devices that induce LTD show some promise in reduction of pain perception and may be useful for treating chronic pain (Rottmann, Jung, & Ellrich, 2010).

Opioid-Induced Hyperalgesia

Chronic administration of opioids can sensitize patients to acute pain through an effect called opioid induced hyperalgesia (OIH). Intriguingly, mTOR is activated in the dorsal horn of the spinal cord in a model of OIH (Xu et al., 2014). This drives an increase in nascent protein synthesis and eIF4E activity due to an increase in 4E-BP1 phosphorylation. OIH-induced mechanical hyperalgesia can be reversed by intrathecal delivery of rapamycin. While the precise site of action is unclear, as the delivery route is not specific to the spinal cord, these data suggest that neuroplasticity in the nervous system caused by opioids is controlled, at least in part, by mTOR signaling.

Conclusion

Tremendous human suffering results from poorly managed pain. Chronic pain is estimated to impact the lives of a quarter of the population in the United States (Dahlhamer et al., 2018). Existing therapies for the treatment of chronic pain include numerous opioids that interact with reward circuity in the central nervous system, contributing to their rampant misuse (Pathan & Williams, 2012). Advances in understanding the genesis of pain, particularly in the peripheral nervous system, have tremendous potential value in the identification of new therapeutic targets. Therapeutics with a peripheral site of action may provide safe and effective alternatives to opiates because they need not cross the blood brain barrier and target pain from where it originates. Translational control in peripheral sensory nociceptors has emerged as an important regulator in pain sensitization and in the development and maintenance of various chronic pain conditions. Despite the identification of upstream signaling events that mediate translation in nociceptors, we still haven't elucidated the precise mechanism by which translation leads to nociceptor hyperexcitability and synaptic plasticity. Which mRNAs are efficiently translated or repressed during a particular pain state, and what are their functions? What are commonalities among these mRNAs? Can these factors be targeted for inhibition? We hope that the answers to these key questions will provide the genesis for more effective pain management.

References

Aitken, C. E., & Lorsch, J. R. (2012). A mechanistic overview of translation initiation in eukaryotes. *Nature Structural & Molecular Biology, 19*(6), 568–576. doi:10.1038/nsmb.2303.

Andreassi, C., Zimmermann, C., Mitter, R., Fusco, S., De Vita, S., Saiardi, A., & Riccio, A. (2010). An NGF-responsive element targets myo-inositol monophosphatase-1 mRNA to sympathetic neuron axons. *Nature Neuroscience, 13*(3), 291–301. doi:10.1038/nn.2486.

Aronov, S., Aranda, G., Behar, L., & Ginzburg, I. (2001). Axonal tau mRNA localization coincides with tau protein in living neuronal cells and depends on axonal targeting signal. *Journal of Neuroscience, 21*(17), 6577–6587.

Baj, G., Pinhero, V., Vaghi, V., & Tongiorgi, E. (2016). Signaling pathways controlling activity-dependent local translation of BDNF and their localization in dendritic arbors. *Journal of Cell Science, 129*(14), 2852–2864. doi:10.1242/jcs.177626.

Barbosa, C., Peixeiro, I., & Romao, L. (2013). Gene expression regulation by upstream open reading frames and human disease. *PLoS Genetics, 9*(8), e1003529. doi:10.1371/journal.pgen.1003529.

Barragan-Iglesias, P., Kuhn, J., Vidal-Cantu, G. C., Salinas-Abarca, A. B., Granados-Soto, V., Dussor, G. O.,... Price, T. J. (2019). Activation of the integrated stress response in nociceptors drives methylglyoxal-induced pain. *Pain, 160*(1), 160–171. doi:10.1097/j.pain.0000000000001387.

Barragan-Iglesias, P., Lou, T. F., Bhat, V. D., Megat, S., Burton, M. D., Price, T. J., & Campbell, Z. T. (2018). Inhibition of Poly(A)-binding protein with a synthetic RNA mimic reduces pain sensitization in mice. *Nature Communications, 9*(1), 10. doi:10.1038/s41467-017-02449-5.

Biurrun Manresa, J., Kaeseler Andersen, O., Mouraux, A., & van den Broeke, E. N. (2018). High frequency electrical stimulation induces a long-lasting enhancement of event-related potentials but does not change the perception elicited by intra-epidermal electrical stimuli delivered to the area of increased mechanical pinprick sensitivity. *PLoS One, 13*(9), e0203365. doi:10.1371/journal.pone.0203365.

Black, B. J., Atmaramani, R., Kumaraju, R., Plagens, S., Romero-Ortega, M., Dussor, G.,... Pancrazio, J. J. (2018). Adult mouse sensory neurons on microelectrode arrays exhibit increased spontaneous and stimulus-evoked activity in the presence of interleukin-6. *Journal of Neurophysiology, 120*(3), 1374–1385. doi:10.1152/jn.00158.2018.

Bogen, O., Alessandri-Haber, N., Chu, C., Gear, R. W., & Levine, J. D. (2012). Generation of a pain memory in the primary afferent nociceptor triggered by PKC epsilon activation of CPEB. *Journal of Neuroscience, 32*(6), 2018–2026. doi:10.1523/JNEUROSCI.5138-11.2012.

Bonin, R. P., & De Koninck, Y. (2014). A spinal analog of memory reconsolidation enables reversal of hyperalgesia. *Nature Neuroscience, 17*(8), 1043–1045. doi:10.1038/nn.3758.

Boutouja, F., Stiehm, C. M., & Platta, H. W. (2019). mTOR: A cellular regulator interface in health and disease. *Cells, 8*(1). doi:10.3390/cells8010018.

Brittis, P. A., Lu, Q., & Flanagan, J. G. (2002). Axonal protein synthesis provides a mechanism for localized regulation at an intermediate target. *Cell, 110*(2), 223–235.

Cajigas, I. J., Tushev, G., Will, T. J., tom Dieck, S., Fuerst, N., & Schuman, E. M. (2012). The local transcriptome in the synaptic neuropil revealed by deep sequencing and high-resolution imaging. *Neuron, 74*(3), 453–466. doi:10.1016/j.neuron.2012.02.036.

Carling, D., Thornton, C., Woods, A., & Sanders, M. J. (2012). AMP-activated protein kinase: New regulation, new roles? *Biochemical Journal, 445*(1), 11–27. doi:10.1042/BJ20120546.

Cervero, F., & Laird, J. M. (1996). Mechanisms of touch-evoked pain (allodynia): A new model. *Pain, 68*(1), 13–23. S030439599603165X [pii].

Costa-Mattioli, M., Sossin, W. S., Klann, E., & Sonenberg, N. (2009). Translational control of long-lasting synaptic plasticity and memory. *Neuron, 61*(1), 10–26. S0896-6273(08)01089-1 [pii]. doi:10.1016/j.neuron.2008.10.055.

Dahlhamer, J., Lucas, J., Zelaya, C., Nahin, R., Mackey, S., DeBar, L.,... Helmick, C. (2018). Prevalence of chronic pain and high-impact chronic pain among adults—United States, 2016. *Morbidity and Mortality Weekly Report (MMWR), 67*(36), 1001–1006. doi:10.15585/mmwr.mm6736a2.

Davies, A. J., Kim, H. W., Gonzalez-Cano, R., Choi, J., Back, S. K., Roh, S. E.,... Oh, S. B. (2019). Natural killer cells degenerate intact sensory afferents following nerve injury. *Cell, 176*(4), 716–728 e718. doi:10.1016/j.cell.2018.12.022.

Dubin, A. E., & Patapoutian, A. (2010). Nociceptors: The sensors of the pain pathway. *Journal of Clinical Investigation, 120*(11), 3760–3772. doi:10.1172/jci42843.

Duncan, R., & Hershey, J. W. (1983). Identification and quantitation of levels of protein synthesis initiation factors in crude HeLa cell lysates by two-dimensional polyacrylamide gel electrophoresis. *Journal of Biological Chemistry, 258*(11), 7228–7235.

Ferrari, L. F., Araldi, D., & Levine, J. D. (2015). Distinct terminal and cell body mechanisms in the nociceptor mediate hyperalgesic priming. *Journal of Neuroscience, 35*(15), 6107–6116. doi:10.1523/JNEUROSCI.5085-14.2015.

Ferrari, L. F., Bogen, O., Chu, C., & Levine, J. D. (2013). Peripheral administration of translation inhibitors reverses increased hyperalgesia in a model of chronic pain in the rat. *Journal of Pain, 14*(7), 731–738. doi:10.1016/j.jpain.2013.01.779.

Ferrari, L. F., Bogen, O., & Levine, J. D. (2010). Nociceptor subpopulations involved in hyperalgesic priming. *Neuroscience, 165*(3), 896–901. S0306-4522(09)01875-2 [pii]. doi:10.1016/j.neuroscience.2009.11.029.

Ferrari, L. F., Bogen, O., & Levine, J. D. (2013). Role of nociceptor alpha CaMKII in transition from acute to chronic pain (hyperalgesic priming) in male and female rats. *Journal of Neuroscience, 33*(27), 11002–11011. doi:10.1523/JNEUROSCI.1785-13.2013.

Ferrari, L. F., Bogen, O., Reichling, D. B., & Levine, J. D. (2015). Accounting for the delay in the transition from acute to chronic pain: Axonal and nuclear mechanisms. *Journal of Neuroscience, 35*(2), 495–507. doi:10.1523/JNEUROSCI.5147-13.2015.

Galicia-Vazquez, G., Cencic, R., Robert, F., Agenor, A. Q., & Pelletier, J. (2012). A cellular response linking eIF4AI activity to eIF4AII transcription. *RNA, 18*(7), 1373–1384. doi:10.1261/rna.033209.112.

Geranton, S. M., Jimenez-Diaz, L., Torsney, C., Tochiki, K. K., Stuart, S. A., Leith, J. L., . . . Hunt, S. P. (2009). A rapamycin-sensitive signaling pathway is essential for the full expression of persistent pain states. *Journal of Neuroscience, 29*(47), 15017–15027. 29/47/15017 [pii]. doi:10.1523/JNEUROSCI.3451-09.2009.

Goldstrom, A. C., & Wickens, M. (2008). Multifunctional deadenylase complexes diversify mRNA control. *Nature Reviews Molecular Cell Biology, 9*(4), 337–344.

Gumy, L. F., Yeo, G. S., Tung, Y. C., Zivraj, K. H., Willis, D., Coppola, G., . . . Fawcett, J. W. (2011). Transcriptome analysis of embryonic and adult sensory axons reveals changes in mRNA repertoire localization. *RNA, 17*(1), 85–98. doi:10.1261/rna.2386111.

Guzowski, J. F., Lyford, G. L., Stevenson, G. D., Houston, F. P., McGaugh, J. L., Worley, P. F., & Barnes, C. A. (2000). Inhibition of activity-dependent arc protein expression in the rat hippocampus impairs the maintenance of long-term potentiation and the consolidation of long-term memory. *Journal of Neuroscience, 20*(11), 3993–4001.

Guzowski, J. F., McNaughton, B. L., Barnes, C. A., & Worley, P. F. (1999). Environment-specific expression of the immediate-early gene Arc in hippocampal neuronal ensembles. *Nature Neuroscience, 2*(12), 1120–1124.

Heiman, M., Schaefer, A., Gong, S., Peterson, J. D., Day, M., Ramsey, K. E., . . . Heintz, N. (2008). A translational profiling approach for the molecular characterization of CNS cell types. *Cell, 135*(4), 738–748. doi:10.1016/j.cell.2008.10.028.

Hinnebusch, A. G., Ivanov, I. P., & Sonenberg, N. (2016). Translational control by 5′-untranslated regions of eukaryotic mRNAs. *Science, 352*(6292), 1413–1416. doi:10.1126/science.aad9868.

Hirai, T., Mulpuri, Y., Cheng, Y., Xia, Z., Li, W., Ruangsri, S., . . . Nishimura, I. (2017). Aberrant plasticity of peripheral sensory axons in a painful neuropathy. *Scientific Reports, 7*(1), 3407. doi:10.1038/s41598-017-03390-9.

Holcik, M., & Sonenberg, N. (2005). Translational control in stress and apoptosis. *Nature Review Molecular Cell Biology, 6*(4), 318–327. nrm1618 [pii]. doi:10.1038/nrm1618.

Hossaini, M., Jongen, J. L., Biesheuvel, K., Kuhl, D., & Holstege, J. C. (2010). Nociceptive stimulation induces expression of Arc/Arg3.1 in the spinal cord with a preference for neurons containing enkephalin. *Molecular Pain, 6*, 43. doi:10.1186/1744-8069-6-43.

Hsieh, A. C., Liu, Y., Edlind, M. P., Ingolia, N. T., Janes, M. R., Sher, A., . . . Ruggero, D. (2012). The translational landscape of mTOR signalling steers cancer initiation and metastasis. *Nature, 485*(7396), 55–61. doi:10.1038/nature10912.

Huang, H. L., Cendan, C. M., Roza, C., Okuse, K., Cramer, R., Timms, J. F., & Wood, J. N. (2008). Proteomic profiling of neuromas reveals alterations in protein composition and local protein synthesis in hyper-excitable nerves. *Molecular Pain, 4*, 33. 1744-8069-4-33 [pii]. doi:10.1186/1744-8069-4-33.

Huang, Y. S., Carson, J. H., Barbarese, E., & Richter, J. D. (2003). Facilitation of dendritic mRNA transport by CPEB. *Genes & Development, 17*(5), 638–653.

Hunt, J. L., Winkelstein, B. A., Rutkowski, M. D., Weinstein, J. N., & DeLeo, J. A. (2001). Repeated injury to the lumbar nerve roots produces enhanced mechanical allodynia and persistent spinal neuroinflammation. *Spine (Phila Pa 1976), 26*(19), 2073–2079.

Inceoglu, B., Bettaieb, A., Trindade da Silva, C. A., Lee, K. S., Haj, F. G., & Hammock, B. D. (2015). Endoplasmic reticulum stress in the peripheral nervous system is a significant driver of neuropathic pain. *Proceedings of the National Academy of Sciences of the United States of America, 112*(29), 9082–9087. doi:10.1073/pnas.1510137112.

Indo, Y., Tsuruta, M., Hayashida, Y., Karim, M. A., Ohta, K., Kawano, T., . . . Matsuda, I. (1996). Mutations in the TRKA/NGF receptor gene in patients with congenital insensitivity to pain with anhidrosis. *Nature Genetics, 13*(4), 485–488. doi:10.1038/ng0896-485.

Jennings, M. D., Zhou, Y., Mohammad-Qureshi, S. S., Bennett, D., & Pavitt, G. D. (2013). eIF2B promotes eIF5 dissociation from eIF2*GDP to facilitate guanine nucleotide exchange for translation initiation. *Genes & Development, 27*(24), 2696–2707. doi:10.1101/gad.231514.113.

Jimenez-Diaz, L., Geranton, S. M., Passmore, G. M., Leith, J. L., Fisher, A. S., Berliocchi, L., . . . Hunt, S. P. (2008). Local translation in primary afferent fibers regulates nociception. *PLoS One, 3*(4), e1961. doi:10.1371/journal.pone.0001961.

Jung, H., Gkogkas, C. G., Sonenberg, N., & Holt, C. E. (2014). Remote control of gene function by local translation. *Cell, 157*(1), 26–40. doi:10.1016/j.cell.2014.03.005.

Kandasamy, R., & Price, T. J. (2015). The pharmacology of nociceptor priming. *Handbook of Experimental Pharmacology, 227*, 15–37. doi:10.1007/978-3-662-46450-2_2.

Kar, A. N., Lee, S. J., & Twiss, J. L. (2018). Expanding axonal transcriptome brings new functions for axonally synthesized proteins in health and disease. *Neuroscientist, 24*(2), 111–129. doi:10.1177/1073858417712668.

Kawasaki, Y., Xu, Z. Z., Wang, X., Park, J. Y., Zhuang, Z. Y., Tan, P. H., . . . Ji, R. R. (2008). Distinct roles of matrix metalloproteases in the early- and late-phase development of neuropathic pain. *Nature Medicine, 14*(3), 331–336. doi:10.1038/nm1723.

Kersten, C., & Cameron, M. G. (2012). Cetuximab alleviates neuropathic pain despite tumour progression. *BMJ Case Reports, 2012*. doi:10.1136/bcr.12.2011.5374.

Khoutorsky, A., Bonin, R. P., Sorge, R. E., Gkogkas, C. G., Pawlowski, S. A., Jafarnejad, S. M., . . . Sonenberg, N. (2015). Translational control of nociception via 4E-binding protein 1. *eLife, 4*. doi:10.7554/eLife.12002.

Khoutorsky, A., & Price, T. J. (2018). Translational control mechanisms in persistent pain. *Trends in Neuroscience, 41*(2), 100–114. doi:10.1016/j.tins.2017.11.006.

Khoutorsky, A., Sorge, R. E., Prager-Khoutorsky, M., Pawlowski, S. A., Longo, G., Jafarnejad, S. M., . . . Sonenberg, N. (2016). eIF2alpha phosphorylation controls thermal nociception. *Proceedings of the National Academy of Sciences of the United States of America, 113*(42), 11949–11954. doi:10.1073/pnas.1614047113.

Komar, A. A., & Hatzoglou, M. (2011). Cellular IRES-mediated translation: The war of ITAFs in pathophysiological states. *Cell Cycle, 10*(2), 229–240. doi:10.4161/cc.10.2.14472.

Krishnamoorthy, T., Pavitt, G. D., Zhang, F., Dever, T. E., & Hinnebusch, A. G. (2001). Tight binding of the phosphorylated alpha subunit of initiation factor 2 (eIF2alpha) to the regulatory subunits of guanine nucleotide exchange factor eIF2B is required for inhibition of translation initiation. *Molecular and Cellular Biology, 21*(15), 5018–5030. doi:10.1128/mcb.21.15.5018-5030.2001.

Lecuyer, E., Yoshida, H., Parthasarathy, N., Alm, C., Babak, T., Cerovina, T.,... Krause, H. M. (2007). Global analysis of mRNA localization reveals a prominent role in organizing cellular architecture and function. *Cell, 131*(1), 174–187. doi:10.1016/j.cell.2007.08.003.

Liang, L., Tao, B., Fan, L., Yaster, M., Zhang, Y., & Tao, Y. X. (2013). mTOR and its downstream pathway are activated in the dorsal root ganglion and spinal cord after peripheral inflammation, but not after nerve injury. *Brain Research, 1513*, 17–25. doi:10.1016/j.brainres.2013.04.003.

Lu, L., Han, A. P., & Chen, J. J. (2001). Translation initiation control by heme-regulated eukaryotic initiation factor 2alpha kinase in erythroid cells under cytoplasmic stresses. *Molecular and Cellular Biology, 21*(23), 7971–7980. doi:10.1128/mcb.21.23.7971-7980.2001.

Manicone, A. M., & McGuire, J. K. (2008). Matrix metalloproteinases as modulators of inflammation. *Seminars in Cell & Developmental Biology, 19*(1), 34–41. doi:10.1016/j.semcdb.2007.07.003.

Martin, K. C., Bartsch, D., Bailey, C. H., & Kandel, E. R. (1999). *Molecular mechanisms underlying learning-related long-lasting synaptic plasticity* (2nd ed.). Cambridge, MA: MIT Press.

Martin, L. J., Smith, S. B., Khoutorsky, A., Magnussen, C. A., Samoshkin, A., Sorge, R. E.,... Diatchenko, L. (2017). Epiregulin and EGFR interactions are involved in pain processing. *Journal of Clinical Investigation, 127*(9), 3353–3366. doi:10.1172/jci87406.

McIntyre, C. K., Miyashita, T., Setlow, B., Marjon, K. D., Steward, O., Guzowski, J. F., & McGaugh, J. L. (2005). Memory-influencing intra-basolateral amygdala drug infusions modulate expression of Arc protein in the hippocampus. *Proceedings of the National Academy of Sciences of the United States of America, 102*(30), 10718–10723. doi:10.1073/pnas.0504436102.

Megat, S., Ray, P. R., Moy, J. K., Lou, T. F., Barragan-Iglesias, P., Li, Y.,... Price, T. J. (2019). Nociceptor translational profiling reveals the ragulator-rag GTPase complex as a critical generator of neuropathic pain. *Journal of Neuroscience, 39*(3), 393–411. doi:10.1523/jneurosci.2661-18.2018.

Melemedjian, O. K., Asiedu, M. N., Tillu, D. V., Peebles, K. A., Yan, J., Ertz, N.,... Price, T. J. (2010). IL-6- and NGF-induced rapid control of protein synthesis and nociceptive plasticity via convergent signaling to the eIF4F complex. *Journal of Neuroscience, 30*(45), 15113–15123. 30/45/15113 [pii]. doi:10.1523/JNEUROSCI.3947-10.2010.

Melemedjian, O. K., Asiedu, M. N., Tillu, D. V., Sanoja, R., Yan, J., Lark, A.,... Price, T. J. (2011). Targeting adenosine monophosphate-activated protein kinase (AMPK) in preclinical models reveals a potential mechanism for the treatment of neuropathic pain. *Molecular Pain, 7*, 70. doi:10.1186/1744-8069-7-70.

Melemedjian, O. K., & Khoutorsky, A. (2015). Translational control of chronic pain. *Progress in Molecular Biology and Translational Science, 131*, 185–213. doi:10.1016/bs.pmbts.2014.11.006.

Melemedjian, O. K., Khoutorsky, A., Sorge, R. E., Yan, J., Asiedu, M. N., Valdez, A.,... Price, T. J. (2013). mTORC1 inhibition induces pain via IRS-1-dependent feedback activation of ERK. *Pain*. doi:10.1016/j.pain.2013.03.021.

Melemedjian, O. K., Mejia, G. L., Lepow, T. S., Zoph, O. K., & Price, T. J. (2014). Bidirectional regulation of P body formation mediated by eIF4F complex formation in sensory neurons. *Neuroscience Letters, 563*, 169–174. doi:10.1016/j.neulet.2013.09.048.

Melemedjian, O. K., Tillu, D. V., Asiedu, M. N., Mandell, E. K., Moy, J. K., Blute, V. M.,... Price, T. J. (2013). BDNF regulates atypical PKC at spinal synapses to initiate and maintain a centralized chronic pain state. *Molecular Pain, 9*, 12. doi:10.1186/1744-8069-9-12.

Melemedjian, O. K., Tillu, D. V., Moy, J. K., Asiedu, M. N., Mandell, E. K., Ghosh, S.,... Price, T. J. (2014). Local translation and retrograde axonal transport of CREB regulates IL-6-induced nociceptive plasticity. *Molecular Pain, 10*, 45. doi:10.1186/1744-8069-10-45.

Menon, K. P., Sanyal, S., Habara, Y., Sanchez, R., Wharton, R. P., Ramaswami, M., & Zinn, K. (2004). The translational repressor pumilio regulates presynaptic morphology and controls postsynaptic accumulation of translation factor eIF-4E. *Neuron, 44*(4), 663–676.

Merianda, T., & Twiss, J. (2013). Peripheral nerve axons contain machinery for co-translational secretion of axonally-generated proteins. *Neuroscience Bulletin, 29*(4), 493–500. doi:10.1007/s12264-013-1360-9.

Merianda, T. T., Lin, A. C., Lam, J. S., Vuppalanchi, D., Willis, D. E., Karin, N., . . . Twiss, J. L. (2009). A functional equivalent of endoplasmic reticulum and Golgi in axons for secretion of locally synthesized proteins. *Molecular and Cellular Neuroscience, 40*(2), 128–142. doi:10.1016/j.mcn.2008.09.008.

Minis, A., Dahary, D., Manor, O., Leshkowitz, D., Pilpel, Y., & Yaron, A. (2014). Subcellular transcriptomics-dissection of the mRNA composition in the axonal compartment of sensory neurons. *Developmental Neurobiology, 74*(3), 365–381. doi:10.1002/dneu.22140.

Moehring, F., Halder, P., Seal, R. P., & Stucky, C. L. (2018). Uncovering the cells and circuits of touch in normal and pathological settings. *Neuron, 100*(2), 349–360. doi:10.1016/j.neuron.2018.10.019.

Moerke, N. J., Aktas, H., Chen, H., Cantel, S., Reibarkh, M. Y., Fahmy, A., . . . Wagner, G. (2007). Small-molecule inhibition of the interaction between the translation initiation factors eIF4E and eIF4G. *Cell, 128*(2), 257–267.

Moryl, N., Obbens, E. A., Ozigbo, O. H., & Kris, M. G. (2006). Analgesic effect of gefitinib in the treatment of non-small cell lung cancer. *Journal of Supportive Oncology, 4*(3), 111.

Moy, J. K., Khoutorsky, A., Asiedu, M. N., Black, B. J., Kuhn, J. L., Barragan-Iglesias, P., . . . Price, T. J. (2017). The MNK-eIF4E signaling axis contributes to injury-induced nociceptive plasticity and the development of chronic pain. *Journal of Neuroscience, 37*(31), 7481–7499. doi:10.1523/jneurosci.0220-17.2017.

Moy, J. K., Khoutorsky, A., Asiedu, M. N., Dussor, G., & Price, T. J. (2018). eIF4E phosphorylation influences Bdnf mRNA translation in mouse dorsal root ganglion neurons. *Frontiers in Cellular Neuroscience, 12*, 29. doi:10.3389/fncel.2018.00029.

Obara, I., Tochiki, K. K., Geranton, S. M., Carr, F. B., Lumb, B. M., Liu, Q., & Hunt, S. P. (2011). Systemic inhibition of the mammalian target of rapamycin (mTOR) pathway reduces neuropathic pain in mice. *Pain, 152*(11), 2582–2595. doi:10.1016/j.pain.2011.07.025.

Obata, K., & Noguchi, K. (2006). BDNF in sensory neurons and chronic pain. *Neuroscience Research, 55*(1), 1–10. doi:10.1016/j.neures.2006.01.005.

Obreja, O., Rukwied, R., Nagler, L., Schmidt, M., Schmelz, M., & Namer, B. (2018). Nerve growth factor locally sensitizes nociceptors in human skin. *Pain, 159*(3), 416–426. doi:10.1097/j.pain.0000000000001108.

Pace, M. C., Passavanti, M. B., De Nardis, L., Bosco, F., Sansone, P., Pota, V., . . . Aurilio, C. (2018). Nociceptor plasticity: A closer look. *Journal of Cell Physiology, 233*(4), 2824–2838. doi:10.1002/jcp.25993.

Pathan, H., & Williams, J. (2012). Basic opioid pharmacology: an update. *British Journal of Pain, 6*(1), 11–16. doi:10.1177/2049463712438493.

Pause, A., Belsham, G. J., Gingras, A. C., Donze, O., Lin, T. A., Lawrence, J. C., Jr., & Sonenberg, N. (1994). Insulin-dependent stimulation of protein synthesis by phosphorylation of a regulator of 5′-cap function. *Nature, 371*(6500), 762–767. doi:10.1038/371762a0.

Pavitt, G. D., Ramaiah, K. V., Kimball, S. R., & Hinnebusch, A. G. (1998). eIF2 independently binds two distinct eIF2B subcomplexes that catalyze and regulate guanine-nucleotide exchange. *Genes & Development, 12*(4), 514–526.

Pinkstaff, J. K., Chappell, S. A., Mauro, V. P., Edelman, G. M., & Krushel, L. A. (2001). Internal initiation of translation of five dendritically localized neuronal mRNAs. *Proceedings of the National Academy of Sciences of the United States of America 98*(5), 2770–2775. doi:10.1073/pnas.051623398.

Price, T. J., Das, V., & Dussor, G. (2016). Adenosine monophosphate-activated protein kinase (AMPK) activators for the prevention, treatment and potential reversal of pathological pain. *Current Drug Targets, 17*(8), 908–920.

Price, T. J., & Dussor, G. (2013). AMPK: An emerging target for modification of injury-induced pain plasticity. *Neuroscience Letters, 557* (Pt A), 9–18. doi:10.1016/j.neulet.2013.06.060.

Price, T. J., Rashid, M. H., Millecamps, M., Sanoja, R., Entrena, J. M., & Cervero, F. (2007). Decreased nociceptive sensitization in mice lacking the fragile X mental retardation protein: Role of mGluR1/5 and mTOR. *Journal of Neuroscience, 27*(51), 13958–13967. 27/51/13958 [pii]. doi:10.1523/JNEUROSCI.4383-07.2007.

Pyronnet, S., Imataka, H., Gingras, A. C., Fukunaga, R., Hunter, T., & Sonenberg, N. (1999). Human eukaryotic translation initiation factor 4G (eIF4G) recruits mnk1 to phosphorylate eIF4E. *EMBO Journal, 18*(1), 270–279. doi:10.1093/emboj/18.1.270.

Rottmann, S., Jung, K., & Ellrich, J. (2010). Electrical low-frequency stimulation induces long-term depression of sensory and affective components of pain in healthy man. *European Journal of Pain, 14*(4), 359–365. doi:10.1016/j.ejpain.2009.06.001.

Ruangsri, S., Lin, A., Mulpuri, Y., Lee, K., Spigelman, I., & Nishimura, I. (2011). Relationship of axonal voltage-gated sodium channel 1.8 (NaV1.8) mRNA accumulation to sciatic nerve injury-induced painful neuropathy in rats. *Journal of Biological Chemistry, 286*(46), 39836–39847. doi:10.1074/jbc.M111.261701.

Rukwied, R., Mayer, A., Kluschina, O., Obreja, O., Schley, M., & Schmelz, M. (2010). NGF induces non-inflammatory localized and lasting mechanical and thermal hypersensitivity in human skin. *Pain, 148*(3), 407–413. S0304-3959(09)00688-5 [pii]. doi:10.1016/j.pain.2009.11.022.

Ruscheweyh, R., Wilder-Smith, O., Drdla, R., Liu, X. G., & Sandkuhler, J. (2011). Long-term potentiation in spinal nociceptive pathways as a novel target for pain therapy. *Molecular Pain, 7*, 20. doi:10.1186/1744-8069-7-20.

Sendoel, A., Dunn, J. G., Rodriguez, E. H., Naik, S., Gomez, N. C., Hurwitz, B.,... Fuchs, E. (2017). Translation from unconventional 5′ start sites drives tumour initiation. *Nature, 541*(7638), 494–499. doi:10.1038/nature21036.

Sidrauski, C., McGeachy, A. M., Ingolia, N. T., & Walter, P. (2015). The small molecule ISRIB reverses the effects of eIF2alpha phosphorylation on translation and stress granule assembly. *eLife, 4*. doi:10.7554/eLife.05033.

Sonenberg, N., Rupprecht, K. M., Hecht, S. M., & Shatkin, A. J. (1979). Eukaryotic mRNA cap binding protein: purification by affinity chromatography on sepharose-coupled m7GDP. *Proceedings of the National Academy of Sciences of the United States of America, 76*(9), 4345–4349.

Song, X. J., Vizcarra, C., Xu, D. S., Rupert, R. L., & Wong, Z. N. (2003). Hyperalgesia and neural excitability following injuries to central and peripheral branches of axons and somata of dorsal root ganglion neurons. *Journal of Neurophysiology, 89*(4), 2185–2193. doi:10.1152/jn.00802.2002.

Taylor, A., Westveld, A. H., Szkudlinska, M., Guruguri, P., Annabi, E., Patwardhan, A.,... Yassine, H. N. (2013). The use of metformin is associated with decreased lumbar radiculopathy pain. *Journal of Pain Research, 6*, 755–763. doi:10.2147/JPR.S52205.

Taylor, A. M., Berchtold, N. C., Perreau, V. M., Tu, C. H., Li Jeon, N., & Cotman, C. W. (2009). Axonal mRNA in uninjured and regenerating cortical mammalian axons. *Journal of Neuroscience, 29*(15), 4697–4707. doi:10.1523/jneurosci.6130-08.2009.

Terenzio, M., Koley, S., Samra, N., Rishal, I., Zhao, Q., Sahoo, P. K., . . . Fainzilber, M. (2018). Locally translated mTOR controls axonal local translation in nerve injury. *Science, 359*(6382), 1416–1421. doi:10.1126/science.aan1053.

Thakor, D. K., Lin, A., Matsuka, Y., Meyer, E. M., Ruangsri, S., Nishimura, I., & Spigelman, I. (2009). Increased peripheral nerve excitability and local NaV1.8 mRNA up-regulation in painful neuropathy. *Molecular Pain, 5*, 14. 1744-8069-5-14 [pii]. doi:10.1186/1744-8069-5-14.

Thoreen, C. C., Chantranupong, L., Keys, H. R., Wang, T., Gray, N. S., & Sabatini, D. M. (2012). A unifying model for mTORC1-mediated regulation of mRNA translation. *Nature, 485*(7396), 109–113. doi:10.1038/nature11083.

Tomancak, P., Berman, B. P., Beaton, A., Weiszmann, R., Kwan, E., Hartenstein, V., . . . Rubin, G. M. (2007). Global analysis of patterns of gene expression during Drosophila embryogenesis. *Genome Biology, 8*(7), R145. doi:10.1186/gb-2007-8-7-r145.

Tracey, I., & Mantyh, P. W. (2007). The cerebral signature for pain perception and its modulation. *Neuron, 55*(3), 377–391. doi:10.1016/j.neuron.2007.07.012.

Tracey, W. D., Jr. (2017). Nociception. *Current Biology, 27*(4), R129–R133. doi:10.1016/j.cub.2017.01.037.

Tsai, J. C., Miller-Vedam, L. E., Anand, A. A., Jaishankar, P., Nguyen, H. C., Renslo, A. R., . . . Walter, P. (2018). Structure of the nucleotide exchange factor eIF2B reveals mechanism of memory-enhancing molecule. *Science, 359*(6383). doi:10.1126/science.aaq0939.

Tzingounis, A. V., & Nicoll, R. A. (2006). Arc/Arg3.1: linking gene expression to synaptic plasticity and memory. *Neuron, 52*(3), 403–407. doi:10.1016/j.neuron.2006.10.016.

Uttam, S., Wong, C., Amorimb, I. S., Jafarnejad, S. M., Tansley, S. N., Yang, J., Prager-Khoutorsky, M., . . . Khoutorsky, A. (2018). Translational profiling of dorsal root ganglia and spinal cord in a mouse model of neuropathic pain. *Neurobiology of Pain, 4*, 35–44.

Waskiewicz, A. J., Flynn, A., Proud, C. G., & Cooper, J. A. (1997). Mitogen-activated protein kinases activate the serine/threonine kinases Mnk1 and Mnk2. *EMBO Journal, 16*(8), 1909–1920. doi:10.1093/emboj/16.8.1909.

Waskiewicz, A. J., Johnson, J. C., Penn, B., Mahalingam, M., Kimball, S. R., & Cooper, J. A. (1999). Phosphorylation of the cap-binding protein eukaryotic translation initiation factor 4E by protein kinase Mnk1 in vivo. *Molecular and Cellular Biology, 19*(3), 1871–1880.

Wells, S. E., Hillner, P. E., Vale, R. D., & Sachs, A. B. (1998). Circularization of mRNA by eukaryotic translation initiation factors. *Molecular Cell, 2*(1), 135–140.

Willis, D., Li, K. W., Zheng, J. Q., Chang, J. H., Smit, A. B., Kelly, T., . . . Twiss, J. L. (2005). Differential transport and local translation of cytoskeletal, injury-response, and neurodegeneration protein mRNAs in axons. *Journal of Neuroscience, 25*(4), 778–791. doi:10.1523/JNEUROSCI.4235-04.2005.

Willis, D. E., van Niekerk, E. A., Sasaki, Y., Mesngon, M., Merianda, T. T., Williams, G. G., . . . Twiss, J. L. (2007). Extracellular stimuli specifically regulate localized levels of individual neuronal mRNAs. *Journal of Cell Biology, 178*(6), 965–980. jcb.200703209 [pii]. doi:10.1083/jcb.200703209.

Willis, D. E., Xu, M., Donnelly, C. J., Tep, C., Kendall, M., Erenstheyn, M., . . . Twiss, J. L. (2011). Axonal localization of transgene mRNA in mature PNS and CNS neurons. *Journal of Neuroscience, 31*(41), 14481–14487. doi:10.1523/JNEUROSCI.2950-11.2011.

Xu, J. T., Zhao, J. Y., Zhao, X., Ligons, D., Tiwari, V., Atianjoh, F. E.,... Tao, Y. X. (2014). Opioid receptor-triggered spinal mTORC1 activation contributes to morphine tolerance and hyperalgesia. *Journal of Clinical Investigation, 124*(2), 592–603. doi:10.1172/JCI70236.

Xu, J. T., Zhao, X., Yaster, M., & Tao, Y. X. (2010). Expression and distribution of mTOR, p70S6K, 4E-BP1, and their phosphorylated counterparts in rat dorsal root ganglion and spinal cord dorsal horn. *Brain Research, 1336*, 46–57. doi:10.1016/j.brainres.2010.04.010.

Yanagiya, A., Suyama, E., Adachi, H., Svitkin, Y. V., Aza-Blanc, P., Imataka, H.,... Sonenberg, N. (2012). Translational homeostasis via the mRNA cap-binding protein, eIF4E. *Molecular Cell, 46*(6), 847–858. doi:10.1016/j.molcel.2012.04.004.

Yang, W., & Hinnebusch, A. G. (1996). Identification of a regulatory subcomplex in the guanine nucleotide exchange factor eIF2B that mediates inhibition by phosphorylated eIF2. *Molecular and Cellular Biology, 16*(11), 6603–6616. doi:10.1128/mcb.16.11.6603.

Yudin, D., Hanz, S., Yoo, S., Iavnilovitch, E., Willis, D., Gradus, T.,... Fainzilber, M. (2008). Localized regulation of axonal RanGTPase controls retrograde injury signaling in peripheral nerve. *Neuron, 59*(2), 241–252. S0896-6273(08)00492-3 [pii]. doi:10.1016/j.neuron.2008.05.029.

Zivraj, K. H., Tung, Y. C., Piper, M., Gumy, L., Fawcett, J. W., Yeo, G. S., & Holt, C. E. (2010). Subcellular profiling reveals distinct and developmentally regulated repertoire of growth cone mRNAs. *Journal of Neuroscience, 30*(46), 15464–15478. doi:10.1523/JNEUROSCI.1800-10.2010.

Zoncu, R., Efeyan, A., & Sabatini, D. M. (2011). mTOR: from growth signal integration to cancer, diabetes and ageing. *Nature Reviews Molecular Cell Biology, 12*(1), 21–35. nrm3025 [pii]. doi:10.1038/nrm3025.

CHAPTER 18

DYSREGULATED TRANSLATION IN AUTISM SPECTRUM DISORDER

EMANUELA SANTINI AND ANDERS BORGKVIST

INTRODUCTION

ACCORDING to The Diagnostic and Statistical Manual of Mental Disorders (DMS-5), autism spectrum disorder (ASD) comprises a group of neurodevelopmental disorders defined by early onset of deficits in two core behavioral domains, social interaction and repetitive, restricted behaviors (Posar & Visconti, 2017).

ASD is a complex polygenic disorder caused by the additive effect of multiple genetic variants in combination with unknown environmental factors. This genetic heterogeneity is paralleled by significant variability in ASD symptomatology, which is further complicated by the occurrence of comorbidities, such as motor deficits, intellectual disability, epilepsy, sleep abnormalities, and gastrointestinal disturbances (Abrahams & Geschwind, 2010).

The variety of symptoms and comorbidities displayed by ASD patients present a challenge, not only for the clinical diagnosis and the therapeutic options, but also for disease modelling in experimental animals as it complicates the evaluation of ASD-like core behaviors and overlap with the phenotypes observed in animal models of other neuropsychiatric disorders (Posar & Visconti, 2017). Thus, understanding how genetic variants converge on common signaling pathways resulting in pathogenic ASD-relevant behaviors is of paramount importance for the identification of disease mechanisms and their effects on relevant brain circuits. This knowledge will promote the development of novel therapeutics and the selection of the most appropriate treatments for a specific subgroup of patients.

mTOR signaling pathway is a central regulator of a variety of cellular processes, including protein synthesis. In neurons, mTOR signaling pathway is present at synapses, where it regulates synaptic plasticity and spine morphology. Although mTOR itself is not an ASD-risk gene, several of the genes encoding for signaling or effector molecules in the mTOR pathway has been associated with ASD and neurodevelopmental disorders (Borrie, Brems, Legius, & Bagni, 2017; Santini & Klann, 2014; Sossin & Costa-Mattioli, 2019; Winden, Ebrahimi-Fakhari, & Sahin, 2018).

Based on these monogenic disorders with a high prevalence of ASD and impaired mTOR signaling, it was postulated that excessive protein synthesis is one of the common molecular alterations that contribute to dysregulated synaptic and structural plasticity and behaviors associated with ASD and other neurodevelopmental disorders.

Here we focus on identified ASD-risk genes encoding for proteins involved broadly in mechanisms of translational regulation and summarize the experimental evidence in support of the hypothesis that dysregulated protein synthesis is involved in ASD. We have arbitrarily arranged these genes in subcategories based on their primary intracellular roles.

Do Protein Synthesis Levels Matter for ASD?

Although protein synthesis is one of the signaling pathways involved, at least in a subgroup of ASD, there seem to be conflicting results regarding the direction (increase vs. decrease) of dysregulated translational regulation (summarized in Table 18.1). When modeled in mice, some of the ASD associated genetic mutations (or deletions) result in increased (e.g., *eif4e*, pten) and others in decreased of global protein synthesis (e.g., *eef1a, ribosomal proteins*) and a third group in the specific translation (increase or decrease) of a smaller subset of mRNAs (e.g., *fmrp, rng105*).

Currently, it is not clear which of these effects is more critical for ASD, and all of these pathogenic variations may be equally important for ASD etiology. Thus, the question of "levels" may be reconducted to the idea that physiological neuronal functions are supported by the "right amount" of protein synthesis and that either increases or decreases result in synaptic, morphological and behavioral changes that recapitulate some aspects of the disorder (Kelleher & Bear, 2008).

Understanding the molecular mechanisms linking determined levels of protein synthesis to specific ASD phenotypes (e.g., morphological, synaptic, or behavioral) is of great interest for our understanding of the pathological changes occurring in ASD. However, more mechanistic studies are needed to fully understand the molecular details of translational control, the identity of the mRNAs dysregulated, and the specificity of these mechanisms for particular neurons and developmental time points. For instance, TSC function as a negative regulator upstream of mTOR and mouse models with genetic reduction of *tsc* (including *tsc*$^{+/-}$, see below) display ASD-like phenotypes. However, at odd with the initial prediction, it was found that protein synthesis is decreased in the

Table 18.1. Summary of the ASD-Risk Genes, Mouse Models, and Protein Synthesis Levels

ASD Risk Gene	Mouse Model	Protein Synthesis (Brain)*	Protein Synthesis (Predicted)	Reference(s)
EIF4E	*eif4e* transgenic	↑	↑	(Santini et al., 2013)
	4cbp2$^{-/-}$			(Gkogkas et al., 2013)
EEF1A	*eefla2*$^{-/-}$ (wasted mice)	Not determined	↓	(Chambers et al., 1998)
	G70S *eefla2* mutant			(Davies et al., 2017)
RPL10	–	–	↓	(Klauck et al., 2006)
UPF3B	*upf3b*$^{-/-}$	↑↓	↑↓	(L. Huang et al., 2018)
FMR1	fmrp$^{-/-}$	↑	↑	
RNG105	mgl05$^{+/-}$, *CamK2a*-Cre^{+}; *rngl05*$^{fl/fl}$	–		(Solomon et al., 2007)
		determined in cultured hippocampal neurons	↑↓	(Ohashi et al., 2016)
				(Nakayama et al., 2017)
sCELF6	*celf6*$^{-/-}$	Not determined		(L. T. Timchenko et al., 2006)
			↑↓	(N. A. Timchenko et al., 2005)
				(Iakova et al., 2004)
				(Dougherty et al., 2013)
JAKMIP1	*jakmip1*$^{-/-}$	↓	↓	(Berg et al., 2015)
RBFOX1	–	–	↑↓	(Lee et al., 2016)
				(Carreira-Rosario et al., 2016)
PTEN	*CamKII*-Crc^{+}; *pten*$^{fl/fl}$			(Kwon et al., 2006)
	pten$^{+/-}$	↑	↑	(W. C. Huang et al., 2016)
TSC1/2	*tsc2*$^{+/-}$			(Reith et al., 2013
	L7-Cre^{+}; tsc1$^{fl/fl}$,	↓		(Tsai et al., 2012)
	L7-Cre^{+}; tsc1$^{fl/+}$,	determined	↑	(Tsai et al., 2018)
	L7-Cre^{+}; tsc2$^{fl/+}$	in *tsc2*$^{+/-}$ determined		(Auerbach, Osterweil, & Bear, 2011)
DYRK1A	dyrk1a$^{+/-}$, dyrk1a$^{-/-}$	Not	↑	(Welsh et al., 1998)
				(Woods et al., 2001)

(*Continued*)

Table 18.1. (Continued)

ASD Risk Gene	Mouse Model	Protein Synthesis (Brain)*	Protein Synthesis (Predicted)	Reference(s)
SLC7A5	*Tie2-Cre*$^{+}$;*slc7a5*$^{fl/fl}$	↓	↓	(Tarlungcanu et al., 2016)
MeCP2	*mecp2*$^{-/-}$			(Li et al., 2013)
		↓	↓	(Li et al., 2013)
				(Ricciardi et al., 2011)

* Genes encoding for translational machinery proteins, RNA binding proteins, and other regulators of translation (green). Arrows indicate increases (↑) or decreases (↓) of protein synthesis; dash (–) denotes lack of models or undetermined protein synthesis levels; references refer to the mouse models, the determined, or the predicted protein synthesis levels.

hippocampus of the *tsc*$^{+/-}$ mice (Auerbach, Osterweil, & Bear, 2011), suggesting that increased mTOR activity reduces translation. We do not know the molecular mechanisms involved, but one speculation is that compensatory mechanisms acting downstream of mTOR (e.g., availability of initiation factors, phosphorylation of eIF2α, miRNA, ribosomes stalling) may ultimately determine this paradoxical effect. Moreover, it is crucial to perform a detailed evaluation of the signaling molecules involved in the control of protein synthesis in animal models of ASD. For instance, the phosphorylation state of a few signaling proteins is often utilized as a proxy for the levels of protein synthesis, which cannot fully recapitulate the complex regulation of translation in the brain and there are very few investigations about activity-dependent changes of protein synthesis in neuronal circuits important for ASD-like behaviors.

Therapeutic interventions are another aspect of the disease where an evaluation of the levels of protein synthesis may be necessary. However, given the heterogeneity of ASD, a therapeutic intervention will necessarily be tailored for the specific subgroup of patients and hopefully will target specific molecular complex within the signaling pathway regulating protein synthesis rather than the general process as such, since this may lead to a variety of unwanted side effects.

Genes Encoding for Translational Machinery Proteins

In the following subcategory we focus on identified ASD-risk genes encoding for proteins directly involved in the different phases of protein synthesis (initiation, elongation and termination).

Eukaryotic Initiation Factor 4E (EIF4E)

EIF4E encodes for the cap-binding protein eIF4E that is associated with the cap structure at the 5′ end of eukaryotic mRNAs, and it is involved in the regulation of cap-dependent translation initiation.

Briefly, translation is repressed by the eIF4E binding proteins (4E-BPs) that sequester eIF4E/mRNAs and inhibit the formation of initiation complex eIF4F. In principle, any 4E-BP-like protein that sequesters eIF4E may regulate translation. 4E-BPs inhibit translation by binding eIF4E on a large number of transcripts, while other eIF4E inhibitory proteins are tethered to specific subset of mRNAs and restrict their translation in time and/or space either by interacting directly with certain mRNA elements (i.e., 4E-T) or via association with RNA binding proteins (i.e., FMRP/CYFIP, Cup/Bruno; Richter & Sonenberg, 2005).

Upon stimulation, mTOR phosphorylates 4E-BP, which releases eIF4E from inhibition. Thus, eIF4E is free to interact with eIF4G, which in turn binds the RNA helicase eIF4A to form the initiation complex eIF4F. Formation of eIF4F recruits the ribosomes and other factors on the mRNA to initiate cap-dependent translation. eIF4G is a scaffolding protein that circularizes the mRNA by binding to poly(A) binding proteins (PABPs) located the 3′ end. Circularized mRNAs are translated more efficiently and are associated with other regulatory factors, such as initiation factor 3 and Mnk1/2 (see (Gingras, Raught, & Sonenberg, 2001) for details on translational regulation).

A genetic study on ASD patients identified a mutation in the promoter of the *EIF4E* gene and provided a direct link between ASD and the cap-binding protein eIF4E (Neves-Pereira et al., 2009). Binding studies suggested that the ASD-linked mutation enhanced the binding of the DNA polymerase to the *eIF4E* promoter, indicating that overexpression of eIF4E may be involved in ASD. Additional indirect genetic evidence suggests the involvement of *eIF4E* in ASD. For instance, genetic variants in the chromosome 4q, which contains the *eIF4E* locus, have been described in ASD (Autism Genome Project et al., 2007; Yonan et al., 2003). Furthermore, common genetic variants of *eIF4E* are correlated with the expression of particular ASD symptoms, such as repetitive, stereotyped behaviors but not intellectual disability (Waltes et al., 2014).

Animal models with increased eIF4E levels have been used to investigate the association between cap-dependent translation and ASD-like phenotypes. In these mice, overexpression of *eif4e* or deletion of *eif4ebp2*, which is the predominant 4E-BP isoform in the brain, results in a series of synaptic, morphological and behavioral phenotypes consistent with ASD in patients (Gkogkas et al., 2013; Santini et al., 2013). Moreover, several of the ASD-like impairments could be normalized by the administration of 4EGI-1, an inhibitor of the eIF4E-eIF4G interaction, suggesting that behavioral and synaptic alterations are caused by increased eIF4E-eIF4G binding and dysregulated cap-dependent translation (Gkogkas et al., 2013; Santini et al., 2013).

The identity of specific upregulated mRNA targets in these models may also contribute to the ASD-like phenotypes. For instance, mRNAs encoding for neuroligins, which are ASD risk genes and encode for scaffolding proteins, are increased in mice with overexpression of *eif4e* or deletion of *eif4ebp2* (Gkogkas et al., 2013). In addition, eIF4E regulates the translation of mRNAs with highly structured 5′ UTR (Hinnebusch, Ivanov, & Sonenberg, 2016) or mRNAs that contains 3′ interferon (IFN)-gamma-activated inhibitor of translation (GAIT) elements, which may be important for synaptogenesis and synaptic functions (Amorim, Kedia, et al., 2018). Overall, these studies provide strong evidence for a role of cap-dependent protein synthesis in ASD.

Alternatively, eIF4E may be involved in ASD because of its function in early neuronal development. eIF4E interacts with the eIF4E transporter (4E-T) in the processing body (P-bodies), which are cytoplasmic granules involved in mRNA degradation. In P-bodies, the complex eIF4e/4E-T represses the translation of pro-neurogenic mRNAs, such as mRNAs important for neuronal differentiation and specification (Yang, Smibert, Kaplan, & Miller, 2014). These distinct functions of eIF4E may work in concert to determine ASD-like phenotypes. Further work is required to parse out their relative contribution to ASD and neurodevelopmental disorders.

Eukaryotic Elongation Factor 1A (EEF1A)

EEF1A encodes for the elongation factor 1A, which is involved in the elongation phase of protein synthesis.

During the elongation phase of protein synthesis, ribosomes and other factors translate the codons on the mRNA into amino acids in the polypeptide. eEF1A is responsible for translocating the correct aminoacyl-tRNAs to the A site of the ribosome, a GTP-dependent process mediated by the GTP exchange factor eEF1B. After the addition of each amino acid to the newly synthesized polypeptide, GTP is hydrolyzed to GDP, and eEF1A is released to load new aminoacyl-tRNAs (Taha, Gildish, Gal-Ben-Ari, & Rosenblum, 2013).

eEF1A is present in two protein isoforms, eEF1A1 and eEF1A2, encoded by separate genes with mutually exclusive expression patterns during neuronal development. eEF1A1 is ubiquitously expressed while eEF1A2 is the only isoform expressed in neurons after postnatal day 21 (Nakajima et al., 2015; Pan, Ruest, Xu, & Wang, 2004).

De novo heterozygous missense mutations in *EEF1A2* has been described in patients with neurodevelopmental delays, motor impairments, intellectual disability, epilepsy and in a few cases of ASD (Helbig et al., 2016; Inui et al., 2016; W. W. Lam et al., 2016; Lopes et al., 2016; McLachlan, Sires, & Abbott, 2019; Nakajima et al., 2015; Veeramah et al., 2013) Interestingly, some of the mutations cluster around the regions of eEF1A2 that are significant for protein synthesis, indicating that dysregulated translation may be involved in the disease. For instance, the G70S mutation is located near the eEF1B binding domain, which is a site involved in the association with the aminoacyl-tRNA. In addition, the E122K mutation is located in the GTP/GDP binding domain (Nakajima et al., 2015) and causes reduction in translational fidelity in yeast (Sandbaken & Culbertson, 1988).

Homozygous mice carrying a spontaneous mutation in *the eef1a2* gene and resulting in complete ablation of the eEF1A2 protein (wasted mice) display early motor neuron degeneration, muscle reduction, smaller body size and die at postnatal day 28 (Chambers, Peters, & Abbott, 1998). This phenotype is not rescued by the expression of eEF1A2 in muscle cells, suggesting that the neuronal deletion *eef1a2* is linked to neurodegeneration (Newbery et al., 2005). This phenotype is not caused by the expression of *eef1a1* in skeletal muscles because it gradually declines during postnatal development so that by postnatal day 25, only *eef1a2* is detectable (Newbery et al., 2005). Heterozygous wasted mice grow, breed normally, and do not show neurodegeneration (Griffiths et al., 2012).

Recently, CRISP/Cas9 gene editing has been used to create a mouse line with the mutation G70S in the *eef1a2* gene, which leads to neurodevelopmental disorders in humans (Davies et al., 2017). These results showed that while the protein with the G70S mutation was expressed it was largely nonfunctional and perhaps toxic. In the future, it will be essential to generate conditional mice with restricted genetic deletion of *eef1a2* to understand the *in* vivo functions of this gene.

To the best of our knowledge, protein synthesis has not been analyzed in mice with genetic deletion or mutation in *eef1a2*. However, G70S mutation may result in a reduction of translational fidelity at the level of global protein synthesis given that ASD-associated mutations are located in eEF1A2 functional domains, and it is not known the specificity for a subset of mRNAs.

Ribosomal Proteins (RPs)

The mammalian 80S ribosome is formed by a large 60S and a small 40S ribosomal subunits, which consist of rRNA and RPs. The 40S subunit mediates the interaction between tRNAs and mRNA and regulates the selection of the correct aminoacyl-tRNAs. The 60S subunit contains the peptidyl transferase and provides the exit site for the growing polypeptide chain. Ribosomes are the effectors of translation as they decode mRNA and synthesize proteins (Shi & Barna, 2015).

Recently, mutations in the RPs have been described in patients with intellectual disability, ASD, microcephaly, and other dysmorphic features (Brooks et al., 2014; Chiocchetti et al., 2011; Klauck et al., 2006; Paolini et al., 2017).

Three missense mutations associated with ASD have been identified in the X-linked gene encoding for *RPL10*, which is a component of the large ribosomal subunit 60S (Brooks et al., 2014; Chiocchetti et al., 2011; Klauck et al., 2006). RPL10 is localized at the subunit interface and close to the catalytic peptidyl transferase center (Spahn et al., 2001), and it is necessary for the assembly of the 60S and 40S ribosomal subunits in the late phase of translation initiation (Eisinger, Dick, & Trumpower, 1997).

The ASD-linked missense mutations L206M and H213Q in *RPL10* were modeled in temperature-sensitive *rpl10* mutant yeast and produced a reduced number of assembled polysomes that overall determine a decrease in the rate of global protein synthesis (Brooks et al., 2014; Chiocchetti et al., 2011; Klauck et al., 2006).

The missense mutation, K78E, leads to a neurodevelopmental syndrome that in patients is characterized by microcephaly, seizures, growth retardation, and hypotonia.

K78E mutation significantly alters brain protein synthesis and results in microcephaly in the zebrafish model (Brooks et al., 2014; Chiocchetti et al., 2011; Klauck et al., 2006).

Finally, missense mutations in the gene encoding for RPS23 were identified in individuals showing microcephaly, hearing loss, and other dismorphic features (Paolini et al., 2017). RPS23 is a component of the small ribosomal subunit 40S and ensures translational fidelity by monitoring the complementarity between codons on the mRNAs and the anti-codons of the aminoacyl-tRNA. In EBV-immortalized lymphoblast cell lines (LCLs), ribosomes with mutant RPS23 do not affect the rate of protein synthesis but impairs accuracy in decoding mRNAs during translation (Paolini et al., 2017).

In conclusion, these studies indicate that ASD-associated mutations in RPs result in changes in translational regulation. Mutation in *RPL10* primarily reduced the mRNAs translational rates, whereas mutation in *RPS23* impairs the ability of ribosome to accurately decode mRNAs.

It remains to be determined why missense mutation in RPs results in such brain-specific effects, although RPs are expressed ubiquitously. It is becoming clear that ribosomes are not static translation complexes but are dynamically regulated throughout life. RPs show differential expression during development both in terms of time and tissue types (Huang et al., 2018), and RPs-heterogeneous ribosomes preferentially translate specific subclasses of mRNAs (Ferretti, Ghalei, Ward, Potts, & Karbstein, 2017; Shi et al., 2017). This may imply that neurons are less capable of compensating for defective RPs or that these mutations preferentially affect neuro-specific RPs-heterogeneous ribosomes. Alternatively, local translation in neurons may be more sensitive to RP disruption if ribosomes are the rate-limiting step of protein synthesis.

Up-frame Shift Protein 3B (UPF3B)

UPF3B encodes for a protein involved in nonsense-mediated mRNA decay (NMD), which is a surveillance mechanism controlling the expression of aberrant mRNAs that contain premature termination codons. NMD also affects mRNAs encoding for full-length proteins that display inefficient termination of translation (e.g., mRNAs with defective closed-loop conformation due to long 3′ UTR). Thus, NMD can be considered a post-transcriptional regulator of gene expression because, by degrading aberrant mRNAs containing premature stop codons, it limits the production of truncated proteins, which are deleterious for neurons (Kervestin & Jacobson, 2012; Lykke-Andersen & Jensen, 2015).

A recent in vitro study has highlighted a novel role for UPF3B in translation termination induced by premature termination codons. UPF3B binds to the ribosomes and the eukaryotic release factor (eRF) 3 and 1, delays translation termination, and dissociates ribosomal complexes without nascent peptide (Neu-Yilik et al., 2017). Ribosome recycling after defective translation termination is a crucial step for protein synthesis and avoids the sequestration of essential components of the translation apparatus. Alternatively, one speculation is that deficits in ribosome recycling may be crucial in neuronal local translation if the number of ribosomes is the rate-limiting factor.

Various mutations of *UPF3B* have been identified in patients displaying neurodevelopmental disorders, X-linked mental retardation and ASD (Addington et al., 2011; Laumonnier et al., 2010; Tarpey et al., 2007). These mutations result in truncated forms of UPF3B, which may disrupt the interaction with eRF3 and change the rate of translation of specific mRNAs that are required for normal neurodevelopment and plasticity. Moreover, wild-type and mutated forms of *UPF3B* are found in dendritic spines (Laumonnier et al., 2010) suggesting, that aberrant UPF3B may contribute to dysregulation of translation termination and mRNAs degradation in structures critical for neuronal plasticity.

Mice with genetic deletion of *upf3b* were generated (Huang et al., 2018) to determine the functions of NMD in vivo. These mice displayed behavioral and morphological alterations, partially overlapping the phenotypes of other mouse models of neurodevelopmental disorders. These include impaired fear learning, sleep alterations, defective sensorimotor gating, and reduced spine density and mature spines in the prefrontal cortex. Neuronal differentiation and maturation seem particularly affected by impairments in NMD (Huang et al., 2018) and most likely contribute to the ASD-like phenotype. Moreover, RNA-sequencing of the pre-frontal cortex identified a subset of mRNAs upregulated in the absence of UPF3B (L. Huang et al., 2018). Some of these mRNAs are known candidates for NMD and are involved in neuronal differentiation, maturation, and diseases. Since UPF3B is a mediator of NMD, the upregulated mRNAs are the results of defective NMD. However, a smaller subset of mRNAs is downregulated, perhaps due to an indirect effect of NMD on the translation rate of specific mRNAs (Huang et al., 2018).

Human mutation in *UPF3B* may result in a defective NMD and altered protein synthesis because of inefficient translation termination. Alternatively, defective UPF3B may shut down global translation and promote the translation of specific mRNAs. This can be explained by the known role of the UPF3B-branch of the NMD pathway on neuronal ER stress and the unfolded protein response, leading to the reduction of global translation and upregulation of stress-specific gene translation (Goetz & Wilkinson, 2017).

Genes Encoding for RNA Binding Proteins

This subcategory comprises ASD-risk genes encoding for RNA binding proteins.

Fragile X Mental Retardation 1 (FMR1)

FMR1 encodes for the fragile X mental retardation protein (FMRP), which is an RNA binding protein involved in different aspects of mRNA regulation.

FMRP has been intensively studied as a translational repressor, but the mechanism utilized to inhibit translation is still controversial. Experimental evidence indicates that both initiation and elongation phases of protein synthesis may be repressed, and the net effect may depend on the specific mRNA and neuronal stimulus.

At the initiation, FMRP interact indirectly with eIF4E through the 4E-BP-like protein CYFIP1. The binding of FMRP via CYFIP1 to eIF4E inhibits translation of specific mRNAs and blocks the interaction of eIF4E to eIF4G. The association between FMRP and CYFIP1 is an example of 4E-BP-like translational regulation tethered to specific mRNAs (Napoli et al., 2008; Zalfa et al., 2007).

During the elongation phase, FMRP associates with mRNAs bound to stalled ribosomes. Thus, FMRP functions as a molecular break by stabilizing ribosomes pausing events (Khandjian, Corbin, Woerly, & Rousseau, 1996; Stefani, Fraser, Darnell, & Darnell, 2004). Accordingly, genome-wide ribosome profiling to measure ribosomal positioning and occupancy in the cortex showed that reduction of translational pausing is prevalent in the absence of *fmrp* (Das Sharma et al., 2019). Mice with genetic deletion of *fmrp* have been used to model FXS in patients, and we have collected extensive knowledge about the behavioral, molecular, synaptic, and morphological phenotypes displayed by FXS animal models. We refer to these excellent reviews for more details on FMRP functions and FXS animal models (Banerjee, Ifrim, Valdez, Raj, & Bassell, 2018; Darnell & Klann, 2013; Zukin, Richter, & Bagni, 2009).

Here we will give a summary of the molecular mechanisms leading to dysregulated translation in FXS with emphasis on mTOR signaling and, the initiation phase of translation.

The evidence of an increased brain protein synthesis in FXS (Qin, Kang, Burlin, Jiang, & Smith, 2005) laid the foundations to the idea that dysregulated translation is involved in ASD. Moreover, the demonstration of the overactive PI3K/mTOR pathway in FXS (Gross et al., 2010; Ronesi et al., 2012; Sharma et al., 2010) provided the evidence that different genetic mutations may impinge on common molecular pathways and lead to ASD-like phenotypes. Thus, altered translation in FXS result not only from the absence of the repressor protein FMRP but also from the activation of signaling pathways (MAPK/ERK and PI3K/mTOR) regulating protein synthesis (Borrie et al., 2017; Zukin et al., 2009).

These signaling pathways converge on the formation of the eIF4F initiation complex that regulates initiation of translation and consequently, dysregulated eIF4E-sensitive mRNAs are involved in the FXS pathology. Accordingly, pharmacological reduction of eIF4E-eIF4G interactions rescues some of the ASD-like phenotypes (Santini et al., 2017) and genetic reduction (or pharmacological inhibition) of eIF4E phosphorylation normalize the molecular, synaptic, morphological and behavioral phenotypes of the FXS mouse model (Gkogkas et al., 2014). Importantly, eIF4E phosphorylation is also enhanced in FXS patients (Gross & Bassell, 2012; Hoeffer et al., 2012; Jacquemont et al., 2018). eIF4E is phosphorylated by MNK1/2 via activation of MAPK/ERK pathway (Joshi et al., 1995; Ueda, Watanabe-Fukunaga, Fukuyama, Nagata, & Fukunaga, 2004). It is important to note that the biochemical consequences of eIF4E phosphorylation and its impact on translation regulation (e.g., alteration of eIF4G and/or 4E-BP association, mRNAs specificity and efficiency) are still not well understood (Amorim, Lach, &

Gkogkas, 2018; Bramham, Jensen, & Proud, 2016; Cao et al., 2015; Gkogkas et al., 2014; Knauf, Tschopp, & Gram, 2001; McKendrick, Morley, Pain, Jagus, & Joshi, 2001; Pyronnet et al., 1999). Moreover, MNKs is recruited to eIF4G to phosphorylate eIF4E (Pyronnet et al., 1999; Shveygert, Kaiser, Bradrick, & Gromeier, 2010), providing an additional level of regulation for eIF4E phosphorylation. Finally, overexpression of *eIf4e* in mice with genetic deletion of *fmrp* results in cognitive defects in addition to ASD-like phenotypes (Huynh et al., 2015). Overall, the phenotypic alterations common in FXS and ASD patients may be associated, at least in part, with the convergent activity of eIF4E and FMRP in aberrant regulation of cap-dependent translation.

Dysregulated translation initiation does not account for the entire spectrum of FXS pathology, and critical roles are also played by the effects of FMRP deletion and mTOR alterations on different signaling pathways involved in, for instance, actin remodeling, macroautophagy and mRNA localization (Zukin et al., 2009). Further investigation on this topic is needed to parse out the contribution of individual signaling pathways to the phenotypic alteration observed in FSX and to identify the specific brain circuits affected.

RNA Granule Protein 105 (RNG105)

RNG105 encodes for an RNA-binding protein (also known as Caprin1) that is highly expressed in the brain where it regulates local translation and dendritic mRNAs localization (Shiina & Tokunaga, 2010).

In cultured hippocampal neurons, RNG105 is associated with RNA granules and inhibits the translation of bound mRNAs (Shiina, Shinkura, & Tokunaga, 2005; Solomon et al., 2007). Upon synaptic stimulation, translation of reporter mRNAs increases and is accompanied by RNG105 dissociation from RNA granules (Shiina et al., 2005). Overexpression of RNG105 leads to activation of the protein kinase R (PKR), which is one of the kinases that phosphorylate eIF2α. Phosphorylation of eIF2α reduces global protein synthesis and increases the synthesis of specific mRNAs. However, the deletion of RNG105 has no effect on global protein synthesis, suggesting that RNG105 controls the translation of a limited subset of mRNAs (Solomon et al., 2007).

In addition, it has been shown that RNG105 can bind DDX3X, a DEAD/H box helicase whose mutations are associated with ASD and X-linked mental retardation (Snijders Blok et al., 2015). It is important to note, however, that despite the localization of DDX3X on the X chromosome, most of the affected patients are heterozygote females, as complete loss of DDX3 is lethal. The association of RNG105/DDX3X with initiation factors and PABP modulates the translation of specific mRNAs (Copsey et al., 2017). It is intriguing to speculate that specific mRNAs dendritically localized by the ASD-linked RNG105 may be translated indirectly, via the coordinated activity of RNG105 and DDX3X, which is also associated with ASD.

More recently, it has been shown that genetic deletion of *rng105* results in reduced transport and the expression of proteins in dendrites of cortical cultured neurons. Thus in this study, reduced levels of dendritic RNG105 targets were not caused by translation inhibition but by reduction of mRNAs transported to the sites of local protein synthesis (Shiina, Yamaguchi, & Tokunaga, 2010).

It remains to be determined whether the association of RNG105 with a subset of mRNAs/RNA granules allows for specific RNG105 functions (local protein synthesis vs. mRNAs localization).

A whole-genome sequencing (WGS) study discovered a heterozygous nonsense mutation of the *RNG105* gene in patients with ASD (Jiang et al., 2013).

Mice heterozygous for *rng105* deletion (*rng105*$^{+/-}$) were generated to study the function of this clinically relevant mutation in vivo (Ohashi, Takao, Miyakawa, & Shiina, 2016). *rng105*$^{+/-}$ mice display ASD-like behaviors, as impaired social interaction, reduced novelty responses, and altered reversal learning. The behavioral phenotypes were accompanied by a reduced GluA1 subunit of the AMPA glutamate receptor (AMPAR) distribution in dendrites of cultured cortical neurons obtained from *rng105*$^{+/-}$ mice (Ohashi et al., 2016). However, this study did not address whether the reduction of GluR1 dendritic distribution was caused by impaired local translation or mRNA localization.

In a recent study from the same group, mRNAs localization and expression was determined in the hippocampus of conditional mice with genetic deletion of *rng105* (Nakayama et al., 2017). Overall, this study showed that deletion of *rng105* reduces the asymmetric somatodendritic localization of specific mRNAs in vivo. For instance, the dendritic localization of mRNAs encoding for AMPAR subunits and scaffolding proteins was reduced in the absence of RNG105. Mice with *rng105* conditional deletion displayed reduced hippocampal synaptic responses, impaired plasticity, and deficits in long-term memory formation (i.e., Morris Water Maze, contextual fear conditioning tests) and seizures. However, the authors did not perform any experiment with inhibitors of protein synthesis and dendritic transport, which could have ruled out a possible contribution of RNG105 as a direct translation regulator.

CUG-binding Proteins, Elav-like Family 6 (CELF6)

The CUG-binding proteins (BP), Elav-like family (CELF) are RNA binding proteins that regulate multiple aspects of gene expression including, RNA splicing, editing, deadenylation, stability, and translation. As other RBPs, CELF members bind to specific subsets of mRNAs in different cell types and conditions and are involved in several biological processes. Moreover, some members of the CELF are also regulated via phosphorylation, which further modulates their biological functions. Thus, the variety of mRNA binding properties and functions of each CELF member is incredibly wide and varies within a specific cell type and in different physiological conditions (e.g., time, stimuli; Dasgupta & Ladd, 2012).

Human CELF proteins, also known as Bruno-like (Bruno1) proteins after the homologous *Drosophila* protein Bruno (Good, Chen, Warner, & Herring, 2000), regulate translation via multiple mechanisms. For instance, in *Drosophila,* the protein Bruno binds the 3′ UTR elements of a series of mRNAs (e.g., oskar, gurken and cyclin A; Filardo &

Ephrussi, 2003; Kim-Ha, Kerr, & Macdonald, 1995; Sugimura & Lilly, 2006) and act as a repressor of translation by interacting with Cup. Cup is an eIF4E binding protein associated at the 5′ end of the mRNA. The eIF4E-Cup-Bruno complex represses translation by inhibiting the binding of eIF4E to eIF4G (Nakamura, Sato, & Hanyu-Nakamura, 2004). The association between Cup and Bruno is another example of 4E-BP-like translational regulation tethered to specific mRNAs (Richter & Sonenberg, 2005).

CELF proteins can also stimulate translation: in *Xenopus*, CELF3 binds the 3′UTR of cyclin A, which stimulates its translation (Horb & Horb, 2010). The mammalian CELF1 increases the translation of p21 by binding the G/C-rich sequence at the 5′ UTR (Iakova et al., 2004). Moreover, phosphorylation of CELF1 promotes the interaction with eIF2 and increases the scanning of the ribosomes for the start codons; thus, promoting translation (L. T. Timchenko et al., 2006; N. A. Timchenko, Wang, & Timchenko, 2005).

Recently, translational profiling of murine serotoninergic neurons and polymorphisms analysis of the corresponding human genes in the autism genetic resource exchange (AGRE) collection of ASD associated genes (Geschwind et al., 2001) implicated *CELF6* as ASD risk gene (Dougherty et. al., 2013). Moreover, a rare variant in *CELF6* generating a premature stop codon was found in one of the ASD patients (Dougherty et al., 2013).

Mice with genetic deletion of the *celf6* gene ($celf6^{-/-}$) were generated to study the functions of the *celf6* gene in vivo and to model the genetic condition relevant for ASD in humans. Mice with genetic deletion of *celf6* display impairments in ultrasonic vocalizations (USV) and cognitive inflexibility accompanied by reduction of brain serotonin levels (Dougherty et al., 2013). These ASD-like behaviors are similarly displayed by other ASD animal models (Moy et al., 2008; Scattoni, Gandhy, Ricceri, & Crawley, 2008). Additional studies are needed to determine the role of *celf6* in the regulation of serotonin levels and its contribution to ASD pathology. For instance, conditional deletions and rescue studies may confirm that the impaired behaviors observed in mice with genetic deletion of *celf6* are caused by depletion of CELF6 in serotoninergic neurons.

Unfortunately, the molecular role of CELF6 has not been identified, and the level of protein synthesis was not determined in mice with *celf6* deletion. Given the function of CELF as translational activators, we could speculate that the low levels of serotonin in the absence of CELF are caused by reduced translation of some of the enzymes involved in the synthesis of brain serotonin. This speculation necessitates further investigations so that the functional role and the mRNAs specificity of CELF6 in serotoninergic neurons can be determined.

Janus Kinase and Microtubule-Interacting Protein 1 (JAKMIP1)

JAKMIP1 encodes for an RBP highly expressed in glutamatergic neurons during brain development and synaptogenesis (Couve et al., 2004; Vidal, Valenzuela, Lujan, & Couve, 2009).

In patients with FXS and the duplication 15q-13 syndrome, the gene *JAKMIP1* is differentially expressed (Nishimura et al., 2007) suggesting that this gene may be part of the signaling pathway regulating protein synthesis, which is altered in FXS and other neurodevelopmental disorders. Accordingly, proteomic and biochemical analyses indicate that JAKMIP1 interacts with different factors involved in the regulation of protein synthesis, such as the PABP, FMRP, and ribosomes, among others (Berg et al., 2015; Kanai, Dohmae, & Hirokawa, 2004; Villace, Marion, & Ortin, 2004).

The involvement of JAKMIP1 in the regulation of protein synthesis was examined in mice with genetic deletion of *jakmip1*. Deletion of *jakmip1* leads to decreased protein synthesis, and actively translating ribosomes are depleted of JAKMIP1-binding partners such as DDX5 and PABPC1 (Berg et al., 2015). This indicates that JAKMIP1 is necessary for proper ribosome function and maintenance of physiological levels of neuronal protein synthesis. Further experiments will be necessary to clarify the specific molecular mechanism of the control of protein synthesis exerted by JAKMIP1.

Mice with genetic deletion of *jakmip1* display a series of ASD-like behaviors consistent with the phenotype observed in other animal models of ASD (e.g., repetitive behaviors, social abnormalities, altered USV). Moreover, the behavioral alterations appear early during postnatal development and are maintained throughout adulthood (Berg et al., 2015). ASD-like behaviors are accompanied by dysregulated translation of mRNAs encoding for proteins localized in the PSD compartment such as PSD-95, GluN2A, GluN2B, and Shank2.

The changes in PSD-protein composition result in a series of impairments in glutamatergic synaptic transmission present postnatally but normalized in adulthood. Since the synaptic phenotypes are present only during early postnatal development while the behavioral phenotypes persist, additional mechanisms may underlie the behavioral alterations observed in the adult. In support of this idea, it was shown that JAKMIP1 also binds mRNAs encoding for GABA receptor (Couve et al., 2004; Vidal et al., 2009). This raises the possibility that inhibitory transmission may also be impaired in mice with genetic deletion of *jakmip1* and that it may contribute to the behavioral alterations observed in adults. Thus, conditional deletion experiments restricted to certain developmental windows and cell-types are necessary to identify the mechanisms responsible for the behavioral defects caused by loss of *jakmip1*.

In addition, JAKMIP1 also associates with the microtubules motor protein kinesin-1, and it is mobile in dendrites (Vidal et al., 2012; Vidal et al., 2007) suggesting, that JAKMIP1 may also be involved in the transport and localization of mRNAs to specific neuronal sites. Thus, the neuronal phenotype observed in mice with genetic deletion of *jakmip1* may result from a reduction of global protein synthesis and altered localization of dendritic mRNAs. Further experiments will be necessary to clarify the specific molecular mechanism of the control of protein synthesis exerted by JAKMIP1.

RNA Binding Protein, Fox Homolog 1 (RBFOX1)

In mammals, there are three paralogs of *RBFOX* with different expression patterns but containing a single mRNA recognition motif (Auweter et al., 2006; Kuroyanagi, 2009). *RBFOX1* encodes for two different isoforms of RBFOX1 generated by alternative splicing and localized in the nucleus or cytoplasm, respectively (Kuroyanagi, 2009). The nuclear isoform is predominantly involved in alternative splicing while the cytoplasmic isoform regulates stability and translation of target mRNAs (Carreira-Rosario et al., 2016).

Chromosomal translocations and copy number variations of *RBFOX1* gene have been associated with ASD, intellectual disability, and epilepsy (Martin et al., 2007; Sebat et al., 2007). Moreover, *RBFOX1* has been identified as the master regulator of several ASD-related genes in a gene network analysis of *post mortem* cerebral cortex from ASD patients (Voineagu et al., 2011).

Overall, the cytoplasmic mRNA targets of RBFOX1 are enriched in synaptic- and ASD-related proteins in the brain (Lee et al., 2016; Ray et al., 2013). Thus, RBFOX1 not only control the nuclear splicing of genes related to cortical development and ASD (Fogel et al., 2012; Martin et al., 2007; Parikshak et al., 2013; Sebat et al., 2007; Voineagu et al., 2011) but it also controls the stability of mRNAs encoding for similar protein products (Lee et al., 2016; Parikshak et al., 2013; Voineagu et al., 2011). Moreover, the RBFOX1 target genes significantly overlap with FMRP target genes (Darnell et al., 2011; Lee et al., 2016) indicating that the molecular pathways downstream of these two RBP are linked and reinforce the ideas of shared pathogenic pathways between monogenic (i.e., FXS) and sporadic (i.e., RBFOX1) forms of ASD.

In neurons, cytoplasmic RBFOX1 controls translation directly by binding to (U) GCAUG sites in the 3′ UTR of mRNAs (Lee et al., 2016). The binding of RBFOX1 to mRNA 3′UTR may interfere with the binding of microRNA (miRNA) and may antagonize their activity. For instance, the 3′ UTR of the mRNA encoding for Camk2a, which is one of the RBFOX1 targets, was found to contain multiple miRNA binding sites, including the ones for miR11, miR26 and miR124. Moreover, it was recently reported that RBFOX1 antagonizes miR9 and regulates the expression of Vamp1, which in turn mediates the docking of synaptic vesicles and neurotransmitter release, specifically in inhibitory neurons of the hippocampus (Vuong et al., 2018).

Since the binding of RBFOX1 promotes translation, it has been suggested that RBFOX1 blocks AGO association and prevents miRNA induced repression/degradation (Lee et al., 2016). The potential mechanism used to promote RBFOX1 mediated translation may be to increase the stability of mRNA targets by competing with miRNA binding. Further work is needed to determine whether RBFOX1 also plays a role in mRNAs localization.

Other studies in *Drosophila* suggest that RBFOX1 can also repress the translation of specific mRNAs such as pumilio mRNA, which is involved in germ cell differentiation (Carreira-Rosario et al., 2016). Thus, additional studies are needed to understand the specific RBFOX1 regulation (repression vs. increase expression) for the different mRNA targets in neurons.

Genes Encoding for Other Regulators of Translation

This subcategory consists of ASD-risk genes encoding for regulators of mTOR signaling pathway or indirect regulator of protein synthesis.

Phosphate and Tensin Homolog (PTEN)

PTEN encodes for a phosphatase that inhibits PI3K signaling and results in the inactivation of the downstream targets AKT and mTOR (Zhou & Parada, 2012).

The *PTEN* gene is located on chromosome 10q23 and is a candidate gene for ASD, intellectual disability, and macrocephaly. Patients with loss of function mutation in *PTEN* display high predisposition to brain tumors, other types of cancers, developmental delays, and intellectual disability (Butler et al., 2005; Conti et al., 2012; Orrico et al., 2009).

Loss of function mutation has been modeled in mice carrying a genetic deletion of *pten*. From a molecular standpoint, deletion or loss of function mutation in *pten* results in constitutively active mTOR signaling pathway, and increased protein synthesis. However, only indirect measures of brain protein synthesis (i.e., levels of phosphorylation of mTOR and downstream target proteins) have been performed in these animal models.

We refer to these reviews (Borrie et al., 2017; Winden et al., 2018) for detailed information about pten animal models, and here we summarize the main findings.

Genetic deletion of *pten* in mice has dramatic effects during development, resulting in macrocephaly, anatomical brain abnormalities, seizures, and premature death (Backman et al., 2001).

Mice with conditional *pten* ablation restricted to a particular subset of neurons, for instance, in cortex and hippocampus, develop seizures, macrocephaly, and ASD-like behaviors (Kwon et al., 2006). Moreover, treatment with the mTOR inhibitor rapamycin rescues neuronal hypertrophy and normalizes ASD-like behaviors and seizures in adult conditional *pten* knock-out (Zhou et al., 2009), suggesting that the behavioral phenotypes are caused by a constitutive increase in the activity of mTOR pathway resulting from *pten* deletion. Consistently, adult pyramidal neurons in the cortex show apical dendritic growth in chronic in vivo imaging of mice with conditional *pten* deletion (Chow et al., 2009). The apical dendritic growth was specific for pyramidal neurons of layer 2/3, and it was blocked by rapamycin, suggesting that constitutive overactive mTOR signaling induces phenotypic changes in specific mature neurons.

Mice with germline haploinsufficiency of *pten* (*pten* $^{+/-}$) also exhibit similar phenotypes, including macrocephaly, neuronal hypertrophy, and ASD-like behaviors (Clipperton-Allen, Chen, & Page, 2016). However, mTOR signaling and the resulting alteration in axonal branching and connectivity of neurons in layer 5 of the prefrontal cortex are dysregulated in a critical temporal window (postnatal day 8) in mice with *pten* haploinsufficiency (W. C. Huang, Chen, & Page, 2016). Treatment with a pharmacological inhibitor of S6K1, which is an mTOR downstream target, only during this early postnatal time leads to an amelioration of the neuronal connectivity and a normalization of the social behavior analyzed later in the adulthood (W. C. Huang et al., 2016). These experiments indicate that neuronal connectivity is irreversible in adulthood and that there is a critical developmental window in which reduction of mTOR signaling has long lasting effects on connectivity and behavior mediated by this brain circuit.

The presence of a critical period of phenotypic reversibility in the animal models of *pten* is still controversial. Future studies investigating the levels of mTOR activation (and protein synthesis) induced by different *pten* deletions may confirm the existence of such a critical window in specific neurons and brain circuits. Moreover, it is also essential to consider the specific drugs used to normalize *pten* aberrant phenotypes. Indeed, while pharmacological inhibition of S6K1 impacts mTOR-mediated translation of specific mRNAs, blockade of mTOR with rapamycin will affect protein synthesis mediated by both S6K1 and 4E-BP. Furthermore, mTOR inhibition will impact a variety of other signaling pathways critical for neuronal functions and implicated in ASD as, macroautophagy.

Tubero Sclerosis Complex (TSC)

TSC1 and *TSC2* genes encode for proteins that form the TSC complex, which inhibits RHEB, a major activator of mTOR.

Loss of function mutations in *TSC1* and *TSC2* causes TSC, which is a multisystem disorder characterized by the formation of hamartomas in different organs. TSC patients display a high incidence of ASD, seizures, and intellectual disability (Curatolo & Maria, 2013). Overall, the behavioral and cognitive profiles of children with TSC and ASD are very similar, suggesting similar molecular and circuit abnormalities in the brain (Winden et al., 2018).

Loss of function mutations in the TSC complex results in increased mTOR activation and augmented protein synthesis.

Animal models of TSC have been generated with a variety of constitutional and conditional deletions of *tsc1* and *tsc2*. We refer to the following reviews for a detailed description (Borrie et al., 2017; Winden et al., 2018). Briefly, mice carrying genetic manipulation in *tsc1* and *tsc2* display similar cognitive alterations, seizure, and synaptic impairments as in ASD (Ehninger et al., 2008; Goorden, van Woerden, van der Weerd, Cheadle, & Elgersma, 2007; Nie et al., 2010). Interestingly, genetic reduction of *tsc1* (*L7*-Cre^{+}; *tsc1*$^{fl/+}$) or *tsc2* (*L7*-Cre^{+}; *tsc2*$^{fl/+}$) restricted to cerebellar Purkinje cells (PC) results in several

ASD-like phenotypes including social impairments and repetitive behaviors that are correlated with synaptic and morphological alterations (Reith et al., 2013; Tsai et al., 2012).

In these mouse models, treatment with the mTOR inhibitor rapamycin in adulthood corrects multiple behavioral and synaptic phenotypes (Nie et al., 2010; Tsai et al., 2012). Recently, it has shown that the ability of rapamycin to rescue social behaviors (but not repetitive behaviors) and other synaptic impairments are restricted to an early developmental window in mice with conditional genetic ablation of *tsc1* in PC (*L7*-Cre^{+}; *tsc1*$^{fl/fl}$; Tsai et al., 2018). These results indicate that the genetic dosage of tsc1 and possibly the level of mTOR hyperactivation is correlated to the presence of a sensitive period of intervention that will ameliorate specific ASD-like phenotypes. In this specific case, this was also correlated to the neurodegeneration of PC observed in mice with conditional genetic ablation of *tsc1* and not present in mice with conditional genetic reduction of *tsc1*.

The ability to normalize specific ASD-related phenotype in adult mice or the presence of irreversible changes occurring during a critical developmental window is a clinically relevant question that encourages additional mechanistic studies of the signaling pathways and the neuronal circuits involved. We still do not know the specific changes in molecular signaling resulting from *tsc* genetic dosage (i.e., levels of mTOR activation and protein synthesis) and whether these alterations impact more severely specific neuronal circuits. Since *tsc*-dependent behavioral phenotypes have been ameliorated with inhibition of mTOR with rapamycin, we cannot exclude the possibility that signaling pathways other than protein synthesis (i.e., macroautophagy) may be involved in the pharmacological rescue.

Since PTEN regulates mTOR signaling mainly via TSC, it is counterintuitive that some of the specific phenotypes displayed by *pten* and *tsc* animal models are normalized by different treatments with rapamycin. One speculative explanation accounting for the diversity may be that in *pten* animal models, TSC1/2 is intact, and it may act as an additional site of mTOR modulation via ERK/RSK1 and AMPK. However, it is important to note that further studies are needed to establish the level of mTOR activation in these models and how this impact protein synthesis. It will also be essential to perform these studies in specific neuronal circuits at determined time points since it seems that different neurons have specific requirements for optimal protein synthesis that ensures physiological functions.

Dual-Specificity Tyrosine-Phosphorylation-Regulated Kinase 1A (DYRK1A)

DYRK1A is a dosage-sensitive gene encoding for a protein kinase involved in neurodevelopment and tissues homeostasis (Tejedor & Hammerle, 2011)

Dysregulation in *DYRK1A* expression has been involved in multiple human pathologies, including ASD, intellectual disability, and Down Syndrome (DS; O'Roak et al., 2012; van Bon et al., 2016). Overexpression of DYRK1A kinase is present in DS mouse models and patients since *the DYRK1A* gene is localized in the DS critical region of chromosome 21 (Dowjat et al., 2007). Conversely, reduced expression *DYRK1A* caused by heterozygous disruptions and loss-of-function mutations leads to mental retardation and ASD (van Bon et al., 2016). Haploinsufficiency of DYRK1A causes microcephaly in mice and humans (Fotaki et al., 2002; Ji et al., 2015) and is associated with a wide range of morphological and functional alterations in neurons (Fotaki et al., 2002). Mice with homozygote DYRK1A deletion die in utero with growth retardation, reduced body size, and developmental delay of several organs (Fotaki et al., 2002).

DYRK1A also function as a translational regulator during neurodevelopment, and this may represent a molecular association to ASD. DYRK1A controls translation initiation by phosphorylation of ε subunit of initiation factor 2B (eIF2B) at Ser539 (Woods et al., 2001). eIF2B is a guanine nucleotide exchange factor for the initiation factor 2 (eIF2). After methionyl-tRNA/eIF2/GTP recognizes the first AUG codon, GTP is hydrolyzed, and eIF2/GDP is slowly released from the ribosome. eIF2B accelerates this reaction allowing eIF2 to associate again to GTP and participates in another round of translation initiation. DYRK1A-dependet phosphorylation of eIF2B allows a subsequent phosphorylation at Ser353 mediated by glycogen synthase kinase 3 (GSK3), which in turn inhibits eIF2B and downregulate translation (Welsh, Miller, Loughlin, Price, & Proud, 1998) This may indicate that in ASD syndromes linked to reduced *DYRK1A* expression, the ability to control eIF2B is reduced and thus, protein synthesis is dysregulated. Future studies are required to address the specific impact of *DYRK1A* gene dosage on the regulation of translation and the identity of dysregulated mRNAs.

Branched Chain Ketoacid Dehydrogenase Kinase (BCKDK) and Neutral Amino Acid Transporter 1 (SLC7A5)

Recently, whole exome sequencing analysis (WES) identified two genes encoding for proteins involved in the metabolism and transport of branched amino acid (BCAAs) into the brain that, when mutated, cause syndromic form of ASD and intellectual disability (Novarino et al., 2012; Tarlungeanu et al., 2016). Amino acids are the building blocks of proteins, contribute to different metabolic pathways and function as neurotransmitters in the brain.

The *BCKDK* gene encodes for a kinase necessary to metabolize BCAAs in the liver, muscles, and adipose tissue. Decreased levels of BCAAs were found in the plasma of patients carrying ASD associated mutation in *BCKDK* (Novarino et al., 2012).

The *SLC7A5* gene encodes for a component of an amino acid transporter localized in endothelial cells of the blood-brain barrier. ASD-associated mutation in LAT1 results in decreased levels of amino acid and BCAAs (Tarlungeanu et al., 2016).

Mice with genetic deletion of *bckdk* and mice with conditional deletion of *slc7a5* restricted to endothelial cells display similar phenotypes and behavioral impairments, such as severe psychomotor delay, seizures, microcephaly, and ASD. These altered phenotypes are also consistent with human pathology. Importantly, BCAAs-enriched diet and intraventricular injections of leucine and isoleucine, two of the amino acids that were particularly low in mice with conditional deletion of *slc7a5*, normalize the behavioral deficits (Novarino et al., 2012; Tarlungeanu et al., 2016). These experiments indicate that in BCAAs-related ASD, the behavioral impairments may be normalized by increasing the levels of amino acids in adult mice and humans.

In mice with conditional *slc7a5* deletion, reduced levels of BCAAs result in global changes of cap-dependent protein synthesis, as demonstrated by a shift in polysome profiling, indicating that fewer mRNAs are bound to actively translating ribosomes. It will be interesting to know whether mice with genetic deletion of *bckdk* display similar changes in protein synthesis.

Typically, the reduction of cap-dependent protein synthesis in a condition of amino acid deprivation is exerted by mTOR signaling. It has been shown that leucine deprivation activates Sestrin, which in turn decreases mTOR activation and consequently inhibits cap-dependent protein synthesis (Wolfson et al., 2016). However, amino acid deprivation in the brain of mice with conditional deletion of *slc7a5*, did not activate mTOR signaling pathway but instead resulted in increased levels of phosphorylated eIF2α (Tarlungeanu et al., 2016). It is possible that, in a state of amino acid deprivation, high levels of uncharged transfer RNA (tRNA) signals an insufficient amount of amino acid to produce new proteins and activates GCN2, which in turn phosphorylates eIF2α resulting in the reduction of global protein synthesis.

The molecular details that determine the phosphorylation of eIF2α rather than inactivation of mTOR await further elucidation. A possible explanation may be that mTOR activation is an acute response to amino acid deprivation while in the condition of long-term exposure to lack of amino acid, neurons will activate stress responses leading to eIF2α phosphorylation.

MicroRNA

This subcategory includes microRNA and an ASD-risk gene (*Methyl-CpG Binding Protein 2*) involved in the regulation of microRNA processing.

MicroRNA (miRNA)

miRNA are small noncoding RNA molecules (about 22 nucleotides in length) encoded within the genomes of eukaryotes either from discreet miRNA genes or as parts of introns of protein-coding genes. miRNA contributes to gene-expression regulation at the post-transcriptional level in different biological processes, such as neuronal development, neurogenesis, and plasticity (Im & Kenny, 2012; Sun & Shi, 2015).

Importantly, genome-wide miRNA expression profiling in postmortem brains from ASD patients indicates that miRNAs are dysregulated and that their targets are enriched in ASD-risk genes (Banerjee-Basu, Larsen, & Muend, 2014; Mundalil Vasu et al., 2014; Takata, Ionita-Laza, Gogos, Xu, & Karayiorgou, 2016; Wu, Parikshak, Belgard, & Geschwind, 2016). Moreover, some of the proteins involved in miRNA-mediated regulation of gene expression are also associated with ASD, for instance, missense mutations in AGO1, TNRC6, and CNOT (Crawley, Heyer, & LaSalle, 2016).

The post-transcriptional regulation mediated by miRNAs is complex and dynamically controlled by the cell type and the developmental stage. miRNAs regulate gene expression by inhibiting translation and facilitate mRNA destabilization and degradation. These inhibitory mechanisms are not mutually exclusive and may work in concert or alone, depending on the actual state of the neurons (Fabian, Sundermeier, & Sonenberg, 2010). Moreover, in certain conditions, miRNAs can also upregulate translation (Vasudevan, Tong, & Steitz, 2007, 2008). For a detailed description of the molecular mechanisms of miRNA-mediated control of translation and the possible involvement in neurodevelopmental disorders, please see (Im & Kenny, 2012; Sun & Shi, 2015; Fabian, Sundermeier, & Sonenberg, 2010; Wu, Parikshak, Belgard, & Geschwind, 2016; Lannom & Ceman, this volume Ch. 10).

Methyl-CpG Binding Protein 2 (MeCP2)

MeCP2 gene encodes for a transcriptional regulator that controls gene expression by binding specifically to methylated DNA and by recruiting regulatory proteins to modify transcription. MeCP2 is both a transcriptional repressor and activator of specific genes (Chahrour et al., 2008). Moreover, the transcriptional regulatory functions of MeCP2 are dynamically altered by phosphorylation, which determines the strength of the association to specific promoters (Tao et al., 2009).

Different types of mutations in the X-linked *MeCP2* have been associated with Rett Syndrome (RTT; Amir et al., 1999) and other developmental disorders, including cognitive disorders and ASD (C. W. Lam et al., 2000; Orrico et al., 2000).

Several animal models of RTT have been generated, including mice with genetic deletion of *mecp2* gene or carrying point mutations, partial deletions, or overexpression mimicking most of the genetic alterations identified in human patients. Overall, the animal models of RTT exhibit a broad range of behavioral, morphological and synaptic phenotypes that recapitulate the symptoms of the human disorder (Vashi & Justice, 2019)

Interestingly, reduced activity of the mTOR signaling pathway and decreased cap-dependent protein synthesis were demonstrated in human embryonic stem cell model of RTT (Li et al., 2013) and in the brain of a mouse model with deletion of *mecp2* (Li et al., 2013; Ricciardi et al., 2011).

In these models, the molecular mechanism connecting MeCP2 with mTOR signaling was not conclusively investigated and, given the localization of MeCP2 in the nucleus, it was speculated that protein synthesis dysregulation was determined by deregulation of transcriptional activity. However, the discovery that MeCP2 can also function as a repressor of miRNAs processing (Tsujimura et al., 2015), provides a novel mechanistic link between MeCP2 and mTOR signaling. Accordingly, MeCP2 binds Drosha complex, consisting of a series of proteins and RBP, and regulates the post-transcriptional processing of specific miRNAs, including miR-199a. Mature miR-199a downregulates the expression of the mTOR1 signaling inhibitors SIRT1, HIF1a, PDE4D, which in turn lead to an upregulation of mTOR activity. Thus, in RTT-like brains, MeCP2 deficiency leads to impaired miRNA processing and decreased levels of mature miR199a. Low levels of mature miR-199a cannot effectively suppress the expression of negative regulators of mTOR, and mTOR activity is attenuated (Tsujimura et al., 2015). Further complexity to this regulation may be added by the phosphorylation state of MeCP2, which we know alters its function as a transcriptional regulator and may also alter the function of miRNAs repressor or the subtypes of miRNA repressed.

Overall this indicates that the morphological and synaptic phenotypes, such as reduced cell growth and excitatory synaptic transmission, and eventually the behavioral impairments displayed in RTT models and possibly in patients, may be caused, at least in part, by an impairment of mTOR-dependent signaling and protein synthesis. The involvement of miRNA in RTT and, more generally, in developmental disorders also suggests a possible alternative therapeutic approach.

Conclusions and Future Directions

Increasing evidence suggests that ASD-linked mutations, not only in direct association with the mTOR signaling cascade but also in proteins with primary regulatory functions in other signaling pathways converge and contribute to protein synthesis dysregulation and ASD-relevant behaviors. It seems that regardless of the direction (increase vs. decrease) of translational dysregulation, altered protein synthesis leads to similar behavioral phenotypes, suggesting that physiological neuronal performances are ensured by optimal levels of protein synthesis. Despite all the knowledge acquired so far, critical questions remain open as to which mRNAs are altered in specific brain circuits, which molecular mechanisms result in dysregulated protein synthesis and which type of interventions restore protein synthesis without interfering with other cognitive functions impinging on translation.

Lastly, it is important to note that a growing number of studies are addressing whether it is possible to translate the molecular knowledge obtained from animal models to the clinical setting in patients. Increased PI3K/mTOR and ERK pathways and protein synthesis was detected in fibroblast, lymphocytes and lymphoblastoid obtained from FXS (Gross & Bassell, 2012; Hoeffer et al., 2012; Jacquemont et al., 2018) and idiopathic (Onore, Yang, Van de Water, & Ashwood, 2017; Poopal, Schroeder, Horn, Bassell, & Gross, 2016; Rosina et al., 2019) ASD patients. In addition, pharmacological inhibition of increased mTOR signaling restores protein synthesis in lymphoblastoid from FXS patients (Gross et al., 2010). Finally, dysregulated mTOR pathway was also observed in *post mortem* brain samples of patients with idiopathic ASD (Nicolini, Ahn, Michalski, Rho, & Fahnestock, 2015; Tang et al., 2014). These studies suggest that dysregulated protein synthesis is detectable in non-neuronal cells of a subgroup of patients affected by monogenic and idiopathic forms of ASD. Given that non-neuronal cells are more easily obtainable and do not require an invasive biopsy, this model may be used to test for drugs that restore protein synthesis in ASD in future clinical trials. Further work is needed to identify the subgroup of patients with ASD that display dysregulated brain protein synthesis.

The cerebral rate of protein synthesis was examined in children with developmental delay and with or without pervasive developmental disorder using L-[I-^{11}C]-leucine positron emission tomography (PET; Shandal et al., 2011). Increased protein synthesis was found in the perisylvian language zone in children with developmental delay and pervasive developmental disorder as compared to children with developmental delay but without pervasive developmental disorder (Shandal et al., 2011), indicating that protein synthesis is enhanced in a specific brain circuit of a subgroup of children with neurodevelopmental disorders. If successfully replicated, this study opens to the exciting possibility that the regional brain protein synthesis rate can be measured using noninvasive PET as a diagnostic method in children with neurodevelopmental disorders.

Acknowledgments

ES was supported by Swedish Research Council (2016-02758), The Knut and Alice Wallenberg Academy Fellowship, The Olle Engkvist Byggmästare Foundation, The Karolinska Institute Strategic Research Program in Neuroscience (StratNeuro) and The Karolinska Institute Internal Fund; AB by Swedish Research Council (2016-03129), Magnus Bergvall Foundation, Åhlens Stiftelse, The Karolinska Institute StratNeuro. We apologize to all colleagues whose work could not be adequately referenced due to space limitations.

References

Abrahams, B. S., & Geschwind, D. H. (2010). Connecting genes to brain in the autism spectrum disorders. *Archives of Neurology, 67*(4), 395–399. doi:10.1001/archneurol.2010.47

Addington, A. M., Gauthier, J., Piton, A., Hamdan, F. F., Raymond, A., Gogtay, N., . . . Rouleau, G. A. (2011). A novel frameshift mutation in UPF3B identified in brothers affected with childhood onset schizophrenia and autism spectrum disorders. *Molecular Psychiatry, 16*(3), 238–239. doi:10.1038/mp.2010.59

Amir, R. E., Van den Veyver, I. B., Wan, M., Tran, C. Q., Francke, U., & Zoghbi, H. Y. (1999). Rett syndrome is caused by mutations in X-linked MECP2, encoding methyl-CpG-binding protein 2. *Nature Genetics, 23*(2), 185–188. doi:10.1038/13810

Amorim, I. S., Kedia, S., Kouloulia, S., Simbriger, K., Gantois, I., Jafarnejad, S. M., . . . Gkogkas, C. G. (2018). Loss of eIF4E phosphorylation engenders depression-like behaviors via selective mRNA translation. *Journal of Neuroscience, 38*(8), 2118–2133. doi:10.1523/JNEUROSCI.2673-17.2018

Amorim, I. S., Lach, G., & Gkogkas, C. G. (2018). The role of the eukaryotic translation initiation factor 4E (eIF4E) in neuropsychiatric disorders. *Frontiers in Genetics, 9*, 561. doi:10.3389/fgene.2018.00561

Auerbach, B. D., Osterweil, E. K., & Bear, M. F. (2011). Mutations causing syndromic autism define an axis of synaptic pathophysiology. *Nature, 480*(7375), 63–68. doi:10.1038/nature10658

Autism Genome Project Consortium, Szatmari, P., Paterson, A. D., Zwaigenbaum, L., Roberts, W., Brian, J., . . . Meyer, K. J. (2007). Mapping autism risk loci using genetic linkage and chromosomal rearrangements. *Nature Genetics, 39*(3), 319–328. doi:10.1038/ng1985

Auweter, S. D., Fasan, R., Reymond, L., Underwood, J. G., Black, D. L., Pitsch, S., & Allain, F. H. (2006). Molecular basis of RNA recognition by the human alternative splicing factor Fox-1. *EMBO Journal, 25*(1), 163–173. doi:10.1038/sj.emboj.7600918

Backman, S. A., Stambolic, V., Suzuki, A., Haight, J., Elia, A., Pretorius, J., . . . Mak, T. W. (2001). Deletion of Pten in mouse brain causes seizures, ataxia and defects in soma size resembling Lhermitte-Duclos disease. *Nature Genetics, 29*(4), 396–403. doi:10.1038/ng782

Banerjee, A., Ifrim, M. F., Valdez, A. N., Raj, N., & Bassell, G. J. (2018). Aberrant RNA translation in fragile X syndrome: From FMRP mechanisms to emerging therapeutic strategies. *Brain Research, 1693*(Pt A), 24–36. doi:10.1016/j.brainres.2018.04.008

Banerjee-Basu, S., Larsen, E., & Muend, S. (2014). Common microRNAs target established ASD genes. *Frontiers in Neurology, 5*, 205. doi:10.3389/fneur.2014.00205

Berg, J. M., Lee, C., Chen, L., Galvan, L., Cepeda, C., Chen, J. Y., . . . Geschwind, D. H. (2015). JAKMIP1, a novel regulator of neuronal translation, modulates synaptic function and autistic-like behaviors in mouse. *Neuron, 88*(6), 1173–1191. doi:10.1016/j.neuron.2015.10.031

Borrie, S. C., Brems, H., Legius, E., & Bagni, C. (2017). Cognitive dysfunctions in intellectual disabilities: The contributions of the Ras-MAPK and PI3K-AKT-mTOR pathways. *Annual Review of Genomics and Human Genetics, 18*, 115–142. doi:10.1146/annurev-genom-091416-035332

Bramham, C. R., Jensen, K. B., & Proud, C. G. (2016). Tuning specific translation in cancer metastasis and synaptic memory: Control at the MNK-eIF4E axis. *Trends in Biochemical Sciences, 41*(10), 847–858. doi:10.1016/j.tibs.2016.07.008

Brooks, S. S., Wall, A. L., Golzio, C., Reid, D. W., Kondyles, A., Willer, J. R., . . . Davis, E. E. (2014). A novel ribosomopathy caused by dysfunction of RPL10 disrupts neurodevelopment and causes X-linked microcephaly in humans. *Genetics, 198*(2), 723–733. doi:10.1534/genetics.114.168211

Butler, M. G., Dasouki, M. J., Zhou, X. P., Talebizadeh, Z., Brown, M., Takahashi, T. N., . . . Eng, C. (2005). Subset of individuals with autism spectrum disorders and extreme macrocephaly associated with germline PTEN tumour suppressor gene mutations. *Journal of Medical Genetics, 42*(4), 318–321. doi:10.1136/jmg.2004.024646

Cao, R., Gkogkas, C. G., de Zavalia, N., Blum, I. D., Yanagiya, A., Tsukumo, Y., . . . Sonenberg, N. (2015). Light-regulated translational control of circadian behavior by eIF4E phosphorylation. *Nature Neuroscience, 18*(6), 855–862. doi:10.1038/nn.4010

Carreira-Rosario, A., Bhargava, V., Hillebrand, J., Kollipara, R. K., Ramaswami, M., & Buszczak, M. (2016). Repression of pumilio protein expression by Rbfox1 promotes germ cell differentiation. *Developmental Cell, 36*(5), 562–571. doi:10.1016/j.devcel.2016.02.010

Chahrour, M., Jung, S. Y., Shaw, C., Zhou, X., Wong, S. T., Qin, J., & Zoghbi, H. Y. (2008). MeCP2, a key contributor to neurological disease, activates and represses transcription. *Science, 320*(5880), 1224–1229. doi:10.1126/science.1153252

Chambers, D. M., Peters, J., & Abbott, C. M. (1998). The lethal mutation of the mouse wasted (wst) is a deletion that abolishes expression of a tissue-specific isoform of translation elongation factor 1alpha, encoded by the Eef1a2 gene. *Proceedings of the National Academy of Sciences of the United States of America, 95*(8), 4463–4468. doi:10.1073/pnas.95.8.4463

Chiocchetti, A., Pakalapati, G., Duketis, E., Wiemann, S., Poustka, A., Poustka, F., & Klauck, S. M. (2011). Mutation and expression analyses of the ribosomal protein gene RPL10 in an extended German sample of patients with autism spectrum disorder. *American Journal of Medical Genetics. Part A, 155A*(6), 1472–1475. doi:10.1002/ajmg.a.33977

Chow, D. K., Groszer, M., Pribadi, M., Machniki, M., Carmichael, S. T., Liu, X., & Trachtenberg, J. T. (2009). Laminar and compartmental regulation of dendritic growth in mature cortex. *Nature Neuroscience, 12*(2), 116–118. doi:10.1038/nn.2255

Clipperton-Allen, A. E., Chen, Y., & Page, D. T. (2016). Autism-relevant behaviors are minimally impacted by conditional deletion of Pten in oxytocinergic neurons. *Autism Research, 9*(12), 1248–1262. doi:10.1002/aur.1641

Conti, S., Condo, M., Posar, A., Mari, F., Resta, N., Renieri, A., . . . Parmeggiani, A. (2012). Phosphatase and tensin homolog (PTEN) gene mutations and autism: literature review and a case report of a patient with Cowden syndrome, autistic disorder, and epilepsy. *Journal of Child Neurology, 27*(3), 392–397. doi:10.1177/0883073811420296

Copsey, A. C., Cooper, S., Parker, R., Lineham, E., Lapworth, C., Jallad, D., . . . Morley, S. J. (2017). The helicase, DDX3X, interacts with poly(A)-binding protein 1 (PABP1) and caprin-1 at the leading edge of migrating fibroblasts and is required for efficient cell spreading. *Biochemical Journal, 474*(18), 3109–3120. doi:10.1042/BCJ20170354

Couve, A., Restituito, S., Brandon, J. M., Charles, K. J., Bawagan, H., Freeman, K. B., . . . Moss, S. J. (2004). Marlin-1, a novel RNA-binding protein associates with GABA receptors. *Journal of Biological Chemistry, 279*(14), 13934–13943. doi:10.1074/jbc.M311737200

Crawley, J. N., Heyer, W. D., & LaSalle, J. M. (2016). Autism and cancer share risk genes, pathways, and drug targets. *Trends in Genetics, 32*(3), 139–146. doi:10.1016/j.tig.2016.01.001

Curatolo, P., & Maria, B. L. (2013). Tuberous sclerosis. *Handbook of Clinical Neurology, 111*, 323–331. doi:10.1016/B978-0-444-52891-9.00038-5

Darnell, J. C., & Klann, E. (2013). The translation of translational control by FMRP: Therapeutic targets for FXS. *Nature Neuroscience, 16*(11), 1530–1536. doi:10.1038/nn.3379

Darnell, J. C., Van Driesche, S. J., Zhang, C., Hung, K. Y., Mele, A., Fraser, C. E., . . . Darnell, R. B. (2011). FMRP stalls ribosomal translocation on mRNAs linked to synaptic function and autism. *Cell, 146*(2), 247–261. doi:10.1016/j.cell.2011.06.013

Das Sharma, S., Metz, J. B., Li, H., Hobson, B. D., Hornstein, N., Sulzer, D., . . . Sims, P. A. (2019). Widespread alterations in translation elongation in the brain of juvenile fmr1 knockout mice. *Cell Reports, 26*(12), 3313–3322 e3315. doi:10.1016/j.celrep.2019.02.086

Dasgupta, T., & Ladd, A. N. (2012). The importance of CELF control: molecular and biological roles of the CUG-BP, Elav-like family of RNA-binding proteins. *Wiley Interdisciplinary Reviews: RNA, 3*(1), 104–121. doi:10.1002/wrna.107

Davies, F. C., Hope, J. E., McLachlan, F., Nunez, F., Doig, J., Bengani, H., . . . Abbott, C. M. (2017). Biallelic mutations in the gene encoding eEF1A2 cause seizures and sudden death in F0 mice. *Scientific Reports, 7*, 46019. doi:10.1038/srep46019

Dougherty, J. D., Maloney, S. E., Wozniak, D. F., Rieger, M. A., Sonnenblick, L., Coppola, G., . . . Heintz, N. (2013). The disruption of Celf6, a gene identified by translational profiling of serotonergic neurons, results in autism-related behaviors. *Journal of Neuroscience, 33*(7), 2732–2753. doi:10.1523/JNEUROSCI.4762-12.2013

Dowjat, W. K., Adayev, T., Kuchna, I., Nowicki, K., Palminiello, S., Hwang, Y. W., & Wegiel, J. (2007). Trisomy-driven overexpression of DYRK1A kinase in the brain of subjects with Down syndrome. *Neuroscience Letters, 413*(1), 77–81. doi:10.1016/j.neulet.2006.11.026

Ehninger, D., Han, S., Shilyansky, C., Zhou, Y., Li, W., Kwiatkowski, D. J., ... Silva, A. J. (2008). Reversal of learning deficits in a Tsc2$^{+/-}$ mouse model of tuberous sclerosis. *Nature Medicine, 14*(8), 843–848. doi:10.1038/nm1788

Eisinger, D. P., Dick, F. A., & Trumpower, B. L. (1997). Qsr1p, a 60S ribosomal subunit protein, is required for joining of 40S and 60S subunits. *Molecular and Cellular Biology, 17*(9), 5136–5145. doi:10.1128/mcb.17.9.5136

Fabian, M. R., Sundermeier, T. R., & Sonenberg, N. (2010). Understanding how miRNAs post-transcriptionally regulate gene expression. *Progress in Molecular and Subcellular Biology, 50*, 1–20. doi:10.1007/978-3-642-03103-8_1

Ferretti, M. B., Ghalei, H., Ward, E. A., Potts, E. L., & Karbstein, K. (2017). Rps26 directs mRNA-specific translation by recognition of Kozak sequence elements. *Nature Structural & Molecular Biology, 24*(9), 700–707. doi:10.1038/nsmb.3442

Filardo, P., & Ephrussi, A. (2003). Bruno regulates gurken during Drosophila oogenesis. *Mechanisms of Development, 120*(3), 289–297. Retrieved from https://www.ncbi.nlm.nih.gov/pubmed/12591598

Fogel, B. L., Wexler, E., Wahnich, A., Friedrich, T., Vijayendran, C., Gao, F., . . . Geschwind, D. H. (2012). RBFOX1 regulates both splicing and transcriptional networks in human neuronal development. *Human Molecular Genetics, 21*(19), 4171–4186. doi:10.1093/hmg/dds240

Fotaki, V., Dierssen, M., Alcantara, S., Martinez, S., Marti, E., Casas, C., . . . Arbones, M. L. (2002). Dyrk1A haploinsufficiency affects viability and causes developmental delay and abnormal brain morphology in mice. *Molecular and Cellular Biology, 22*(18), 6636–6647. doi:10.1128/mcb.22.18.6636-6647.2002

Geschwind, D. H., Sowinski, J., Lord, C., Iversen, P., Shestack, J., Jones, P., ... Committee, A. S. (2001). The autism genetic resource exchange: a resource for the study of autism and related neuropsychiatric conditions. *American Journal of Human Genetics, 69*(2), 463–466. doi:10.1086/321292

Gingras, A. C., Raught, B., & Sonenberg, N. (2001). Regulation of translation initiation by FRAP/mTOR. *Genes and Development, 15*(7), 807–826. doi:10.1101/gad.887201

Gkogkas, C. G., Khoutorsky, A., Cao, R., Jafarnejad, S. M., Prager-Khoutorsky, M., Giannakas, N., . . . Sonenberg, N. (2014). Pharmacogenetic inhibition of eIF4E-dependent Mmp9 mRNA translation reverses fragile X syndrome-like phenotypes. *Cell Reports, 9*(5), 1742–1755. doi:10.1016/j.celrep.2014.10.064

Gkogkas, C. G., Khoutorsky, A., Ran, I., Rampakakis, E., Nevarko, T., Weatherill, D. B., . . . Sonenberg, N. (2013). Autism-related deficits via dysregulated eIF4E-dependent translational control. *Nature, 493*(7432), 371–377. doi:10.1038/nature11628

Goetz, A. E., & Wilkinson, M. (2017). Stress and the nonsense-mediated RNA decay pathway. *Cellular and Molecular Life Sciences, 74*(19), 3509–3531. doi:10.1007/s00018-017-2537-6

Good, P. J., Chen, Q., Warner, S. J., & Herring, D. C. (2000). A family of human RNA-binding proteins related to the Drosophila Bruno translational regulator. *Journal of Biological Chemistry, 275*(37), 28583–28592. doi:10.1074/jbc.M003083200

Goorden, S. M., van Woerden, G. M., van der Weerd, L., Cheadle, J. P., & Elgersma, Y. (2007). Cognitive deficits in Tsc1$^{+/-}$ mice in the absence of cerebral lesions and seizures. *Annals of Neurology, 62*(6), 648–655. doi:10.1002/ana.21317

Griffiths, L. A., Doig, J., Churchhouse, A. M., Davies, F. C., Squires, C. E., Newbery, H. J., & Abbott, C. M. (2012). Haploinsufficiency for translation elongation factor eEF1A2 in aged mouse muscle and neurons is compatible with normal function. *PloS One, 7*(7), e41917. doi:10.1371/journal.pone.0041917

Gross, C., & Bassell, G. J. (2012). Excess protein synthesis in FXS patient lymphoblastoid cells can be rescued with a p110beta-selective inhibitor. *Molecular Medicine, 18*, 336–345. doi:10.2119/molmed.2011.00363

Gross, C., Nakamoto, M., Yao, X., Chan, C. B., Yim, S. Y., Ye, K., . . . Bassell, G. J. (2010). Excess phosphoinositide 3-kinase subunit synthesis and activity as a novel therapeutic target in fragile X syndrome. *Journal of Neuroscience, 30*(32), 10624–10638. doi:10.1523/JNEUROSCI.0402-10.2010

Helbig, K. L., Farwell Hagman, K. D., Shinde, D. N., Mroske, C., Powis, Z., Li, S., . . . Helbig, I. (2016). Diagnostic exome sequencing provides a molecular diagnosis for a significant proportion of patients with epilepsy. *Genetics in Medicine, 18*(9), 898–905. doi:10.1038/gim.2015.186

Hinnebusch, A. G., Ivanov, I. P., & Sonenberg, N. (2016). Translational control by 5′-untranslated regions of eukaryotic mRNAs. *Science, 352*(6292), 1413–1416. doi:10.1126/science.aad9868

Hoeffer, C. A., Sanchez, E., Hagerman, R. J., Mu, Y., Nguyen, D. V., Wong, H., . . . Tassone, F. (2012). Altered mTOR signaling and enhanced CYFIP2 expression levels in subjects with fragile X syndrome. *Genes, Brain, and Behavior, 11*(3), 332–341. doi:10.1111/j.1601-183X.2012.00768.x

Horb, L. D., & Horb, M. E. (2010). BrunoL1 regulates endoderm proliferation through translational enhancement of cyclin A2 mRNA. *Developmental Biology, 345*(2), 156–169. doi:10.1016/j.ydbio.2010.07.005

Huang, L., Shum, E. Y., Jones, S. H., Lou, C. H., Dumdie, J., Kim, H., . . . Wilkinson, M. F. (2018). A Upf3b-mutant mouse model with behavioral and neurogenesis defects. *Molecular Psychiatry, 23*(8), 1773–1786. doi:10.1038/mp.2017.173

Huang, W. C., Chen, Y., & Page, D. T. (2016). Hyperconnectivity of prefrontal cortex to amygdala projections in a mouse model of macrocephaly/autism syndrome. *Nature Communications, 7*, 13421. doi:10.1038/ncomms13421

Huynh, T. N., Shah, M., Koo, S. Y., Faraud, K. S., Santini, E., & Klann, E. (2015). eIF4E/Fmr1 double mutant mice display cognitive impairment in addition to ASD-like behaviors. *Neurobiology of Disease, 83*, 67–74. doi:10.1016/j.nbd.2015.08.016

Iakova, P., Wang, G. L., Timchenko, L., Michalak, M., Pereira-Smith, O. M., Smith, J. R., & Timchenko, N. A. (2004). Competition of CUGBP1 and calreticulin for the regulation of p21 translation determines cell fate. *EMBO Journal, 23*(2), 406–417. doi:10.1038/sj.emboj.7600052

Im, H. I., & Kenny, P. J. (2012). MicroRNAs in neuronal function and dysfunction. *Trends in Neurosciences, 35*(5), 325–334. doi:10.1016/j.tins.2012.01.004

Inui, T., Kobayashi, S., Ashikari, Y., Sato, R., Endo, W., Uematsu, M., . . . Haginoya, K. (2016). Two cases of early-onset myoclonic seizures with continuous parietal delta activity caused by EEF1A2 mutations. *Brain and Development, 38*(5), 520–524. doi:10.1016/j.braindev.2015.11.003

Jacquemont, S., Pacini, L., Jonch, A. E., Cencelli, G., Rozenberg, I., He, Y., . . . Bagni, C. (2018). Protein synthesis levels are increased in a subset of individuals with fragile X syndrome. *Human Molecular Genetics, 27*(21), 3825. doi:10.1093/hmg/ddy291

Ji, J., Lee, H., Argiropoulos, B., Dorrani, N., Mann, J., Martinez-Agosto, J. A., . . . Quintero-Rivera, F. (2015). DYRK1A haploinsufficiency causes a new recognizable syndrome with microcephaly, intellectual disability, speech impairment, and distinct facies. *European Journal of Human Genetics, 23*(11), 1473–1481. doi:10.1038/ejhg.2015.71

Jiang, Y. H., Yuen, R. K., Jin, X., Wang, M., Chen, N., Wu, X., . . . Scherer, S. W. (2013). Detection of clinically relevant genetic variants in autism spectrum disorder by whole-genome sequencing. *American Journal of Human Genetics, 93*(2), 249–263. doi:10.1016/j.ajhg.2013.06.012

Joshi, B., Cai, A. L., Keiper, B. D., Minich, W. B., Mendez, R., Beach, C. M., . . . Rhoads, R. E. (1995). Phosphorylation of eukaryotic protein synthesis initiation factor 4E at Ser-209. *Journal of Biological Chemistry, 270*(24), 14597–14603. doi:10.1074/jbc.270.24.14597

Kanai, Y., Dohmae, N., & Hirokawa, N. (2004). Kinesin transports RNA: isolation and characterization of an RNA-transporting granule. *Neuron, 43*(4), 513–525. doi:10.1016/j.neuron.2004.07.022

Kelleher, R. J., 3rd, & Bear, M. F. (2008). The autistic neuron: troubled translation? *Cell, 135*(3), 401–406. doi:10.1016/j.cell.2008.10.017

Kervestin, S., & Jacobson, A. (2012). NMD: a multifaceted response to premature translational termination. *Nature Reviews: Molecular Cell Biology, 13*(11), 700–712. doi:10.1038/nrm3454

Khandjian, E. W., Corbin, F., Woerly, S., & Rousseau, F. (1996). The fragile X mental retardation protein is associated with ribosomes. *Nature Genetics, 12*(1), 91–93. doi:10.1038/ng0196-91

Kim-Ha, J., Kerr, K., & Macdonald, P. M. (1995). Translational regulation of oskar mRNA by bruno, an ovarian RNA-binding protein, is essential. *Cell, 81*(3), 403–412. doi:10.1016/0092-8674(95)90393-3

Klauck, S. M., Felder, B., Kolb-Kokocinski, A., Schuster, C., Chiocchetti, A., Schupp, I., . . . Poustka, A. (2006). Mutations in the ribosomal protein gene RPL10 suggest a novel modulating disease mechanism for autism. *Molecular Psychiatry, 11*(12), 1073–1084. doi:10.1038/sj.mp.4001883

Knauf, U., Tschopp, C., & Gram, H. (2001). Negative regulation of protein translation by mitogen-activated protein kinase-interacting kinases 1 and 2. *Molecular and Cellular Biology, 21*(16), 5500–5511. doi:10.1128/MCB.21.16.5500-5511.2001

Kuroyanagi, H. (2009). Fox-1 family of RNA-binding proteins. *Cellular and Molecular Life Sciences, 66*(24), 3895–3907. doi:10.1007/s00018-009-0120-5

Kwon, C. H., Luikart, B. W., Powell, C. M., Zhou, J., Matheny, S. A., Zhang, W., . . . Parada, L. F. (2006). Pten regulates neuronal arborization and social interaction in mice. *Neuron, 50*(3), 377–388. doi:10.1016/j.neuron.2006.03.023

Lam, C. W., Yeung, W. L., Ko, C. H., Poon, P. M., Tong, S. F., Chan, K. Y., . . . Fok, T. F. (2000). Spectrum of mutations in the MECP2 gene in patients with infantile autism and Rett syndrome. *Journal of Medical Genetics, 37*(12), E41. doi:10.1136/jmg.37.12.e41

Lam, W. W., Millichap, J. J., Soares, D. C., Chin, R., McLellan, A., FitzPatrick, D. R., . . . Abbott, C. M. (2016). Novel de novo EEF1A2 missense mutations causing epilepsy and intellectual disability. *Molecular Genetics & Genomic Medicine, 4*(4), 465–474. doi:10.1002/mgg3.219

Laumonnier, F., Shoubridge, C., Antar, C., Nguyen, L. S., Van Esch, H., Kleefstra, T., . . . Raynaud, M. (2010). Mutations of the UPF3B gene, which encodes a protein widely expressed in neurons, are associated with nonspecific mental retardation with or without autism. *Molecular Psychiatry, 15*(7), 767–776. doi:10.1038/mp.2009.14

Lee, J. A., Damianov, A., Lin, C. H., Fontes, M., Parikshak, N. N., Anderson, E. S., . . . Martin, K. C. (2016). Cytoplasmic Rbfox1 Regulates the Expression of Synaptic and Autism-Related Genes. *Neuron, 89*(1), 113–128. doi:10.1016/j.neuron.2015.11.025

Li, Y., Wang, H., Muffat, J., Cheng, A. W., Orlando, D. A., Loven, J., . . . Jaenisch, R. (2013). Global transcriptional and translational repression in human-embryonic-stem-cell-derived Rett syndrome neurons. *Cell Stem Cell, 13*(4), 446–458. doi:10.1016/j.stem.2013.09.001

Lopes, F., Barbosa, M., Ameur, A., Soares, G., de Sa, J., Dias, A. I., . . . Maciel, P. (2016). Identification of novel genetic causes of Rett syndrome-like phenotypes. *Journal of Medical Genetics, 53*(3), 190–199. doi:10.1136/jmedgenet-2015-103568

Lykke-Andersen, S., & Jensen, T. H. (2015). Nonsense-mediated mRNA decay: an intricate machinery that shapes transcriptomes. *Nature Reviews: Molecular Cell Biology, 16*(11), 665–677. doi:10.1038/nrm4063

Martin, C. L., Duvall, J. A., Ilkin, Y., Simon, J. S., Arreaza, M. G., Wilkes, K., . . . Geschwind, D. H. (2007). Cytogenetic and molecular characterization of A2BP1/FOX1 as a candidate gene for autism. *American Journal of Medical Genetics. Part B: Neuropsychiatric Genetics, 144B*(7), 869–876. doi:10.1002/ajmg.b.30530

McKendrick, L., Morley, S. J., Pain, V. M., Jagus, R., & Joshi, B. (2001). Phosphorylation of eukaryotic initiation factor 4E (eIF4E) at Ser209 is not required for protein synthesis in vitro and in vivo. *European Journal of Biochemistry, 268*(20), 5375–5385. doi:10.1046/j.0014-2956.2001.02478.x

McLachlan, F., Sires, A. M., & Abbott, C. M. (2019). The role of translation elongation factor eEF1 subunits in neurodevelopmental disorders. *Human Mutation, 40*(2), 131–141. doi:10.1002/humu.23677

Moy, S. S., Nadler, J. J., Poe, M. D., Nonneman, R. J., Young, N. B., Koller, B. H., . . . Bodfish, J. W. (2008). Development of a mouse test for repetitive, restricted behaviors: relevance to autism. *Behavioural Brain Research, 188*(1), 178–194. doi:10.1016/j.bbr.2007.10.029

Mundalil Vasu, M., Anitha, A., Thanseem, I., Suzuki, K., Yamada, K., Takahashi, T., . . . Mori, N. (2014). Serum microRNA profiles in children with autism. *Molecular Autism, 5*, 40. doi:10.1186/2040-2392-5-40

Nakayama, K., Ohashi, R., Shinoda, Y., Yamazaki, M., Abe, M., Fujikawa, A., . . . Shiina, N. (2017). RNG105/caprin1, an RNA granule protein for dendritic mRNA localization, is essential for long-term memory formation. *Elife, 6*. doi:10.7554/eLife.29677

Nakajima, J., Okamoto, N., Tohyama, J., Kato, M., Arai, H., Funahashi, O., . . . Miyake, N. (2015). De novo EEF1A2 mutations in patients with characteristic facial features, intellectual disability, autistic behaviors and epilepsy. *Clinical Genetics, 87*(4), 356–361. doi:10.1111/cge.12394

Nakamura, A., Sato, K., & Hanyu-Nakamura, K. (2004). Drosophila cup is an eIF4E binding protein that associates with Bruno and regulates oskar mRNA translation in oogenesis. *Developmental Cell, 6*(1), 69–78. Retrieved from https://www.ncbi.nlm.nih.gov/pubmed/14723848

Napoli, I., Mercaldo, V., Boyl, P. P., Eleuteri, B., Zalfa, F., De Rubeis, S., . . . Bagni, C. (2008). The fragile X syndrome protein represses activity-dependent translation through CYFIP1, a new 4E-BP. *Cell, 134*(6), 1042–1054. doi:10.1016/j.cell.2008.07.031

Neu-Yilik, G., Raimondeau, E., Eliseev, B., Yeramala, L., Amthor, B., Deniaud, A., . . . Kulozik, A. E. (2017). Dual function of UPF3B in early and late translation termination. *EMBO Journal, 36*(20), 2968–2986. doi:10.15252/embj.201797079

Neves-Pereira, M., Muller, B., Massie, D., Williams, J. H., O'Brien, P. C., Hughes, A., . . . Miedzybrodzka, Z. (2009). Deregulation of EIF4E: a novel mechanism for autism. *Journal of Medical Genetics, 46*(11), 759–765. doi:10.1136/jmg.2009.066852

Newbery, H. J., Gillingwater, T. H., Dharmasaroja, P., Peters, J., Wharton, S. B., Thomson, D., . . . Abbott, C. M. (2005). Progressive loss of motor neuron function in wasted mice: effects of a spontaneous null mutation in the gene for the eEF1 A2 translation factor. *Journal of Neuropathology and Experimental Neurology, 64*(4), 295–303. doi:10.1093/jnen/64.4.295

Nicolini, C., Ahn, Y., Michalski, B., Rho, J. M., & Fahnestock, M. (2015). Decreased mTOR signaling pathway in human idiopathic autism and in rats exposed to valproic acid. *Acta Neuropathol Commun, 3*, 3. doi:10.1186/s40478-015-0184-4

Nie, D., Di Nardo, A., Han, J. M., Baharanyi, H., Kramvis, I., Huynh, T., . . . Sahin, M. (2010). Tsc2-Rheb signaling regulates EphA-mediated axon guidance. *Nature Neuroscience, 13*(2), 163–172. doi:10.1038/nn.2477

Nishimura, Y., Martin, C. L., Vazquez-Lopez, A., Spence, S. J., Alvarez-Retuerto, A. I., Sigman, M., . . . Geschwind, D. H. (2007). Genome-wide expression profiling of lymphoblastoid cell lines distinguishes different forms of autism and reveals shared pathways. *Human Molecular Genetics, 16*(14), 1682–1698. doi:10.1093/hmg/ddm116

Novarino, G., El-Fishawy, P., Kayserili, H., Meguid, N. A., Scott, E. M., Schroth, J., . . . Gleeson, J. G. (2012). Mutations in BCKD-kinase lead to a potentially treatable form of autism with epilepsy. *Science, 338*(6105), 394–397. doi:10.1126/science.1224631

Ohashi, R., Takao, K., Miyakawa, T., & Shiina, N. (2016). Comprehensive behavioral analysis of RNG105 (Caprin1) heterozygous mice: Reduced social interaction and attenuated response to novelty. *Scientific Reports, 6*, 20775. doi:10.1038/srep20775

Onore, C., Yang, H., Van de Water, J., & Ashwood, P. (2017). Dynamic Akt/mTOR signaling in children with autism spectrum disorder. *Frontiers in Pediatrics, 5*, 43. doi:10.3389/fped.2017.00043

O'Roak, B. J., Vives, L., Fu, W., Egertson, J. D., Stanaway, I. B., Phelps, I. G., . . . Shendure, J. (2012). Multiplex targeted sequencing identifies recurrently mutated genes in autism spectrum disorders. *Science, 338*(6114), 1619–1622. doi:10.1126/science.1227764

Orrico, A., Galli, L., Buoni, S., Orsi, A., Vonella, G., & Sorrentino, V. (2009). Novel PTEN mutations in neurodevelopmental disorders and macrocephaly. *Clinical Genetics, 75*(2), 195–198. doi:10.1111/j.1399-0004.2008.01074.x

Orrico, A., Lam, C., Galli, L., Dotti, M. T., Hayek, G., Tong, S. F., . . . Sorrentino, V. (2000). MECP2 mutation in male patients with non-specific X-linked mental retardation. *FEBS Letters, 481*(3), 285–288. doi:10.1016/s0014-5793(00)01994-3

Pan, J., Ruest, L. B., Xu, S., & Wang, E. (2004). Immuno-characterization of the switch of peptide elongation factors eEF1A-1/EF-1alpha and eEF1A-2/S1 in the central nervous system during mouse development. *Brain Research: Developmental Brain Research, 149*(1), 1–8. doi:10.1016/j.devbrainres.2003.10.011

Paolini, N. A., Attwood, M., Sondalle, S. B., Vieira, C., van Adrichem, A. M., di Summa, F. M., . . . MacInnes, A. W. (2017). A Ribosomopathy Reveals Decoding Defective Ribosomes Driving Human Dysmorphism. *American Journal of Human Genetics, 100*(3), 506–522. doi:10.1016/j.ajhg.2017.01.034

Parikshak, N. N., Luo, R., Zhang, A., Won, H., Lowe, J. K., Chandran, V., . . . Geschwind, D. H. (2013). Integrative functional genomic analyses implicate specific molecular pathways and circuits in autism. *Cell, 155*(5), 1008–1021. doi:10.1016/j.cell.2013.10.031

Poopal, A. C., Schroeder, L. M., Horn, P. S., Bassell, G. J., & Gross, C. (2016). Increased expression of the PI3K catalytic subunit p110delta underlies elevated S6 phosphorylation and protein synthesis in an individual with autism from a multiplex family. *Molecular Autism, 7*, 3. doi:10.1186/s13229-015-0066-4

Posar, A., & Visconti, P. (2017). Autism spectrum disorders: the troubles with the Diagnostic and Statistical Manual of Mental Disorders 5th Edition. *Journal of Pediatric Neurosciences, 12*(1), 114–115. doi:10.4103/jpn.JPN_147_16

Pyronnet, S., Imataka, H., Gingras, A. C., Fukunaga, R., Hunter, T., & Sonenberg, N. (1999). Human eukaryotic translation initiation factor 4G (eIF4G) recruits mnk1 to phosphorylate eIF4E. *EMBO Journal, 18*(1), 270–279. doi:10.1093/emboj/18.1.270

Qin, M., Kang, J., Burlin, T. V., Jiang, C., & Smith, C. B. (2005). Postadolescent changes in regional cerebral protein synthesis: an in vivo study in the FMR1 null mouse. *Journal of Neuroscience, 25*(20), 5087–5095. doi:10.1523/JNEUROSCI.0093-05.2005

Ray, D., Kazan, H., Cook, K. B., Weirauch, M. T., Najafabadi, H. S., Li, X., . . . Hughes, T. R. (2013). A compendium of RNA-binding motifs for decoding gene regulation. *Nature, 499*(7457), 172–177. doi:10.1038/nature12311

Reith, R. M., McKenna, J., Wu, H., Hashmi, S. S., Cho, S. H., Dash, P. K., & Gambello, M. J. (2013). Loss of Tsc2 in Purkinje cells is associated with autistic-like behavior in a mouse model of tuberous sclerosis complex. *Neurobiology of Disease, 51*, 93–103. doi:10.1016/j.nbd.2012.10.014

Ricciardi, S., Boggio, E. M., Grosso, S., Lonetti, G., Forlani, G., Stefanelli, G., . . . Broccoli, V. (2011). Reduced AKT/mTOR signaling and protein synthesis dysregulation in a Rett syndrome animal model. *Human Molecular Genetics, 20*(6), 1182–1196. doi:10.1093/hmg/ddq563

Richter, J. D., & Sonenberg, N. (2005). Regulation of cap-dependent translation by eIF4E inhibitory proteins. *Nature, 433*(7025), 477–480. doi:10.1038/nature03205

Ronesi, J. A., Collins, K. A., Hays, S. A., Tsai, N. P., Guo, W., Birnbaum, S. G., . . . Huber, K. M. (2012). Disrupted Homer scaffolds mediate abnormal mGluR5 function in a mouse model of fragile X syndrome. *Nature Neuroscience, 15*(3), 431–440, S431. doi:10.1038/nn.3033

Rosina, E., Battan, B., Siracusano, M., Di Criscio, L., Hollis, F., Pacini, L., . . . Bagni, C. (2019). Disruption of mTOR and MAPK pathways correlates with severity in idiopathic autism. *Translational Psychiatry, 9*(1), 50. doi:10.1038/s41398-018-0335-z

Sandbaken, M. G., & Culbertson, M. R. (1988). Mutations in elongation factor EF-1 alpha affect the frequency of frameshifting and amino acid misincorporation in Saccharomyces cerevisiae. *Genetics, 120*(4), 923–934. Retrieved from https://www.ncbi.nlm.nih.gov/pubmed/3066688

Santini, E., Huynh, T. N., Longo, F., Koo, S. Y., Mojica, E., D'Andrea, L., . . . Klann, E. (2017). Reducing eIF4E-eIF4G interactions restores the balance between protein synthesis and actin dynamics in fragile X syndrome model mice. *Science Signaling, 10*(504). doi:10.1126/scisignal.aan0665

Santini, E., Huynh, T. N., MacAskill, A. F., Carter, A. G., Pierre, P., Ruggero, D., . . . Klann, E. (2013). Exaggerated translation causes synaptic and behavioural aberrations associated with autism. *Nature, 493*(7432), 411–415. doi:10.1038/nature11782

Santini, E., & Klann, E. (2014). Reciprocal signaling between translational control pathways and synaptic proteins in autism spectrum disorders. *Science Signaling, 7*(349), re10. doi:10.1126/scisignal.2005832

Scattoni, M. L., Gandhy, S. U., Ricceri, L., & Crawley, J. N. (2008). Unusual repertoire of vocalizations in the BTBR T+tf/J mouse model of autism. *PloS One, 3*(8), e3067. doi:10.1371/journal.pone.0003067

Sebat, J., Lakshmi, B., Malhotra, D., Troge, J., Lese-Martin, C., Walsh, T., . . . Wigler, M. (2007). Strong association of de novo copy number mutations with autism. *Science, 316*(5823), 445–449. doi:10.1126/science.1138659

Shandal, V., Sundaram, S. K., Chugani, D. C., Kumar, A., Behen, M. E., & Chugani, H. T. (2011). Abnormal brain protein synthesis in language areas of children with pervasive developmental disorder: a L-[1-11C]-leucine PET study. *Journal of Child Neurology, 26*(11), 1347–1354. doi:10.1177/0883073811405200

Sharma, A., Hoeffer, C. A., Takayasu, Y., Miyawaki, T., McBride, S. M., Klann, E., & Zukin, R. S. (2010). Dysregulation of mTOR signaling in fragile X syndrome. *Journal of Neuroscience, 30*(2), 694–702. doi:10.1523/JNEUROSCI.3696-09.2010

Shi, Z., & Barna, M. (2015). Translating the genome in time and space: specialized ribosomes, RNA regulons, and RNA-binding proteins. *Annual Review of Cell and Developmental Biology, 31*, 31–54. doi:10.1146/annurev-cellbio-100814-125346

Shi, Z., Fujii, K., Kovary, K. M., Genuth, N. R., Rost, H. L., Teruel, M. N., & Barna, M. (2017). Heterogeneous ribosomes preferentially translate distinct subpools of mRNAs genome-wide. *Molecular Cell, 67*(1), 71–83 e77. doi:10.1016/j.molcel.2017.05.021

Shiina, N., Shinkura, K., & Tokunaga, M. (2005). A novel RNA-binding protein in neuronal RNA granules: regulatory machinery for local translation. *Journal of Neuroscience, 25*(17), 4420–4434. doi:10.1523/JNEUROSCI.0382-05.2005

Shiina, N., & Tokunaga, M. (2010). RNA granule protein 140 (RNG140), a paralog of RNG105 localized to distinct RNA granules in neuronal dendrites in the adult vertebrate brain. *Journal of Biological Chemistry, 285*(31), 24260–24269. doi:10.1074/jbc.M110.108944

Shiina, N., Yamaguchi, K., & Tokunaga, M. (2010). RNG105 deficiency impairs the dendritic localization of mRNAs for Na+/K+ ATPase subunit isoforms and leads to the degeneration of neuronal networks. *Journal of Neuroscience, 30*(38), 12816–12830. doi:10.1523/JNEUROSCI.6386-09.2010

Shveygert, M., Kaiser, C., Bradrick, S. S., & Gromeier, M. (2010). Regulation of eukaryotic initiation factor 4E (eIF4E) phosphorylation by mitogen-activated protein kinase occurs through modulation of Mnk1-eIF4G interaction. *Molecular and Cellular Biology, 30*(21), 5160–5167. doi:10.1128/MCB.00448-10

Snijders Blok, L., Madsen, E., Juusola, J., Gilissen, C., Baralle, D., Reijnders, M. R., . . . Kleefstra, T. (2015). Mutations in DDX3X are a common cause of unexplained intellectual disability with gender-specific effects on Wnt signaling. *American Journal of Human Genetics, 97*(2), 343–352. doi:10.1016/j.ajhg.2015.07.004

Solomon, S., Xu, Y., Wang, B., David, M. D., Schubert, P., Kennedy, D., & Schrader, J. W. (2007). Distinct structural features of caprin-1 mediate its interaction with G3BP-1 and its induction of phosphorylation of eukaryotic translation initiation factor 2alpha, entry to cytoplasmic stress granules, and selective interaction with a subset of mRNAs. *Molecular and Cellular Biology, 27*(6), 2324–2342. doi:10.1128/MCB.02300-06

Sossin, W. S., & Costa-Mattioli, M. (2019). Translational control in the brain in health and disease. *Cold Spring Harbor Perspectives in Biology, 11*(8). doi:10.1101/cshperspect.a032912

Spahn, C. M., Beckmann, R., Eswar, N., Penczek, P. A., Sali, A., Blobel, G., & Frank, J. (2001). Structure of the 80S ribosome from Saccharomyces cerevisiae—tRNA-ribosome and subunit-subunit interactions. *Cell, 107*(3), 373–386. doi:10.1016/s0092-8674(01)00539-6

Stefani, G., Fraser, C. E., Darnell, J. C., & Darnell, R. B. (2004). Fragile X mental retardation protein is associated with translating polyribosomes in neuronal cells. *Journal of Neuroscience, 24*(33), 7272–7276. doi:10.1523/JNEUROSCI.2306-04.2004

Sugimura, I., & Lilly, M. A. (2006). Bruno inhibits the expression of mitotic cyclins during the prophase I meiotic arrest of Drosophila oocytes. *Developmental Cell, 10*(1), 127–135. doi:10.1016/j.devcel.2005.10.018

Sun, E., & Shi, Y. (2015). MicroRNAs: Small molecules with big roles in neurodevelopment and diseases. *Experimental Neurology, 268*, 46–53. doi:10.1016/j.expneurol.2014.08.005

Taha, E., Gildish, I., Gal-Ben-Ari, S., & Rosenblum, K. (2013). The role of eEF2 pathway in learning and synaptic plasticity. *Neurobiology of Learning and Memory, 105*, 100–106. doi:10.1016/j.nlm.2013.04.015

Takata, A., Ionita-Laza, I., Gogos, J. A., Xu, B., & Karayiorgou, M. (2016). De Novo synonymous mutations in regulatory elements contribute to the genetic etiology of autism and schizophrenia. *Neuron, 89*(5), 940–947. doi:10.1016/j.neuron.2016.02.024

Tang, G., Gudsnuk, K., Kuo, S. H., Cotrina, M. L., Rosoklija, G., Sosunov, A., . . . Sulzer, D. (2014). Loss of mTOR-dependent macroautophagy causes autistic-like synaptic pruning deficits. *Neuron, 83*(5), 1131–1143. doi:10.1016/j.neuron.2014.07.040

Tao, J., Hu, K., Chang, Q., Wu, H., Sherman, N. E., Martinowich, K., . . . Sun, Y. E. (2009). Phosphorylation of MeCP2 at Serine 80 regulates its chromatin association and neurological function. *Proceedings of the National Academy of Sciences of the United States of America, 106*(12), 4882–4887. doi:10.1073/pnas.0811648106

Tarlungeanu, D. C., Deliu, E., Dotter, C. P., Kara, M., Janiesch, P. C., Scalise, M., . . . Novarino, G. (2016). Impaired amino acid transport at the blood brain barrier is a cause of autism spectrum disorder. *Cell, 167*(6), 1481–1494 e1418. doi:10.1016/j.cell.2016.11.013

Tarpey, P. S., Raymond, F. L., Nguyen, L. S., Rodriguez, J., Hackett, A., Vandeleur, L., . . . Gecz, J. (2007). Mutations in UPF3B, a member of the nonsense-mediated mRNA decay complex, cause syndromic and nonsyndromic mental retardation. *Nature Genetics, 39*(9), 1127–1133. doi:10.1038/ng2100

Tejedor, F. J., & Hammerle, B. (2011). MNB/DYRK1A as a multiple regulator of neuronal development. *FEBS J, 278*(2), 223–235. doi:10.1111/j.1742-4658.2010.07954.x

Timchenko, L. T., Salisbury, E., Wang, G. L., Nguyen, H., Albrecht, J. H., Hershey, J. W., & Timchenko, N. A. (2006). Age-specific CUGBP1-eIF2 complex increases translation of CCAAT/enhancer-binding protein beta in old liver. *Journal of Biological Chemistry, 281*(43), 32806–32819. doi:10.1074/jbc.M605701200

Timchenko, N. A., Wang, G. L., & Timchenko, L. T. (2005). RNA CUG-binding protein 1 increases translation of 20-kDa isoform of CCAAT/enhancer-binding protein beta by interacting with the alpha and beta subunits of eukaryotic initiation translation factor 2. *Journal of Biological Chemistry, 280*(21), 20549–20557. doi:10.1074/jbc.M409563200

Tsai, P. T., Hull, C., Chu, Y., Greene-Colozzi, E., Sadowski, A. R., Leech, J. M., . . . Sahin, M. (2012). Autistic-like behaviour and cerebellar dysfunction in Purkinje cell Tsc1 mutant mice. *Nature, 488*(7413), 647–651. doi:10.1038/nature11310

Tsai, P. T., Rudolph, S., Guo, C., Ellegood, J., Gibson, J. M., Schaeffer, S. M., . . . Sahin, M. (2018). Sensitive periods for cerebellar-mediated autistic-like behaviors. *Cell Reports, 25*(2), 357–367 e354. doi:10.1016/j.celrep.2018.09.039

Tsujimura, K., Irie, K., Nakashima, H., Egashira, Y., Fukao, Y., Fujiwara, M., . . . Nakashima, K. (2015). miR-199a Links MeCP2 with mTOR signaling and its dysregulation leads to Rett Syndrome phenotypes. *Cell Reports, 12*(11), 1887–1901. doi:10.1016/j.celrep.2015.08.028

Ueda, T., Watanabe-Fukunaga, R., Fukuyama, H., Nagata, S., & Fukunaga, R. (2004). Mnk2 and Mnk1 are essential for constitutive and inducible phosphorylation of eukaryotic initiation factor 4E but not for cell growth or development. *Molecular and Cellular Biology, 24*(15), 6539–6549. doi:10.1128/MCB.24.15.6539-6549.2004

van Bon, B. W., Coe, B. P., Bernier, R., Green, C., Gerdts, J., Witherspoon, K., . . . Eichler, E. E. (2016). Disruptive de novo mutations of DYRK1A lead to a syndromic form of autism and ID. *Molecular Psychiatry, 21*(1), 126–132. doi:10.1038/mp.2015.5

Vashi, N., & Justice, M. J. (2019). Treating Rett syndrome: from mouse models to human therapies. *Mammalian Genome, 30*(5-6), 90–110. doi:10.1007/s00335-019-09793-5

Vasudevan, S., Tong, Y., & Steitz, J. A. (2007). Switching from repression to activation: microRNAs can up-regulate translation. *Science, 318*(5858), 1931–1934. doi:10.1126/science.1149460

Vasudevan, S., Tong, Y., & Steitz, J. A. (2008). Cell-cycle control of microRNA-mediated translation regulation. *Cell Cycle, 7*(11), 1545–1549. doi:10.4161/cc.7.11.6018

Veeramah, K. R., Johnstone, L., Karafet, T. M., Wolf, D., Sprissler, R., Salogiannis, J., . . . Hammer, M. F. (2013). Exome sequencing reveals new causal mutations in children with epileptic encephalopathies. *Epilepsia, 54*(7), 1270–1281. doi:10.1111/epi.12201

Vidal, R. L., Fuentes, P., Valenzuela, J. I., Alvarado-Diaz, C. P., Ramirez, O. A., Kukuljan, M., & Couve, A. (2012). RNA interference of Marlin-1/Jakmip1 results in abnormal morphogenesis and migration of cortical pyramidal neurons. *Molecular and Cellular Neurosciences, 51*(1-2), 1–11. doi:10.1016/j.mcn.2012.07.007

Vidal, R. L., Ramirez, O. A., Sandoval, L., Koenig-Robert, R., Hartel, S., & Couve, A. (2007). Marlin-1 and conventional kinesin link GABAB receptors to the cytoskeleton and regulate receptor transport. *Molecular and Cellular Neurosciences, 35*(3), 501–512. doi:10.1016/j.mcn.2007.04.008

Vidal, R. L., Valenzuela, J. I., Lujan, R., & Couve, A. (2009). Cellular and subcellular localization of Marlin-1 in the brain. *BMC Neuroscience, 10*, 37. doi:10.1186/1471-2202-10-37

Villace, P., Marion, R. M., & Ortin, J. (2004). The composition of Staufen-containing RNA granules from human cells indicates their role in the regulated transport and translation of messenger RNAs. *Nucleic Acids Research, 32*(8), 2411–2420. doi:10.1093/nar/gkh552

Voineagu, I., Wang, X., Johnston, P., Lowe, J. K., Tian, Y., Horvath, S., . . . Geschwind, D. H. (2011). Transcriptomic analysis of autistic brain reveals convergent molecular pathology. *Nature, 474*(7351), 380–384. doi:10.1038/nature10110

Vuong, C. K., Wei, W., Lee, J. A., Lin, C. H., Damianov, A., de la Torre-Ubieta, L., . . . Black, D. L. (2018). Rbfox1 regulates synaptic transmission through the inhibitory neuron-specific vSNARE Vamp1. *Neuron, 98*(1), 127–141 e127. doi:10.1016/j.neuron.2018.03.008

Waltes, R., Gfesser, J., Haslinger, D., Schneider-Momm, K., Biscaldi, M., Voran, A., . . . Chiocchetti, A. G. (2014). Common EIF4E variants modulate risk for autism spectrum disorders in the high-functioning range. *Journal of Neural Transmission (Vienna), 121*(9), 1107–1116. doi:10.1007/s00702-014-1230-2

Welsh, G. I., Miller, C. M., Loughlin, A. J., Price, N. T., & Proud, C. G. (1998). Regulation of eukaryotic initiation factor eIF2B: glycogen synthase kinase-3 phosphorylates a conserved serine which undergoes dephosphorylation in response to insulin. *FEBS Letters, 421*(2), 125–130. doi:10.1016/s0014-5793(97)01548-2

Winden, K. D., Ebrahimi-Fakhari, D., & Sahin, M. (2018). Abnormal mTOR activation in autism. *Annual Review of Neuroscience, 41*, 1–23. doi:10.1146/annurev-neuro-080317-061747

Wolfson, R. L., Chantranupong, L., Saxton, R. A., Shen, K., Scaria, S. M., Cantor, J. R., & Sabatini, D. M. (2016). Sestrin2 is a leucine sensor for the mTORC1 pathway. *Science, 351*(6268), 43–48. doi:10.1126/science.aab2674

Woods, Y. L., Cohen, P., Becker, W., Jakes, R., Goedert, M., Wang, X., & Proud, C. G. (2001). The kinase DYRK phosphorylates protein-synthesis initiation factor eIF2Bepsilon at Ser539 and the microtubule-associated protein tau at Thr212: potential role for DYRK as a glycogen synthase kinase 3-priming kinase. *Biochemical Journal, 355*(Pt 3), 609–615. doi:10.1042/bj3550609

Wu, Y. E., Parikshak, N. N., Belgard, T. G., & Geschwind, D. H. (2016). Genome-wide, integrative analysis implicates microRNA dysregulation in autism spectrum disorder. *Nature Neuroscience, 19*(11), 1463–1476. doi:10.1038/nn.4373

Yang, G., Smibert, C. A., Kaplan, D. R., & Miller, F. D. (2014). An eIF4E1/4E-T complex determines the genesis of neurons from precursors by translationally repressing a proneurogenic transcription program. *Neuron, 84*(4), 723–739. doi:10.1016/j.neuron.2014.10.022

Yonan, A. L., Alarcon, M., Cheng, R., Magnusson, P. K., Spence, S. J., Palmer, A. A., . . . Gilliam, T. C. (2003). A genomewide screen of 345 families for autism-susceptibility loci. *American Journal of Human Genetics, 73*(4), 886–897. doi:10.1086/378778

Zalfa, F., Eleuteri, B., Dickson, K. S., Mercaldo, V., De Rubeis, S., di Penta, A., . . . Bagni, C. (2007). A new function for the fragile X mental retardation protein in regulation of PSD-95 mRNA stability. *Nature Neuroscience, 10*(5), 578–587. doi:10.1038/nn1893

Zhou, J., Blundell, J., Ogawa, S., Kwon, C. H., Zhang, W., Sinton, C., . . . Parada, L. F. (2009). Pharmacological inhibition of mTORC1 suppresses anatomical, cellular, and behavioral abnormalities in neural-specific Pten knock-out mice. *Journal of Neuroscience, 29*(6), 1773–1783. doi:10.1523/JNEUROSCI.5685-08.2009

Zhou, J., & Parada, L. F. (2012). PTEN signaling in autism spectrum disorders. *Current Opinion in Neurobiology, 22*(5), 873–879. doi:10.1016/j.conb.2012.05.004

Zukin, R. S., Richter, J. D., & Bagni, C. (2009). Signals, synapses, and synthesis: how new proteins control plasticity. *Front Neural Circuits, 3*, 14. doi:10.3389/neuro.04.014.2009

CHAPTER 19

NEURONAL MRNA TRANSLATION IN ADDICTION

EMMA PUIGHERMANAL AND
EMMANUEL VALJENT

Introduction

Drug addiction involves persistent neurobiological alterations, leading to dysregulated reward signals and increased cravings, and ultimately to habits compromising life decision making (Koob & Volkow, 2016). A critical step toward identifying neural mechanisms underlying these complex behavioral alterations was the development of rodent models that mimic distinct features of drug-related behavior (Box 1) (Belin-Rauscent et al., 2016). Over the past decades, considerable progress has been made in the identification of the neuronal substrates by which addictive drugs can hijack synaptic plasticity mechanisms in key reward circuits, including the ventral tegmental area, the nucleus accumbens, the extended amygdala, and the prefrontal cortex (Luscher & Malenka, 2011). In addition to transcriptional and epigenetic modifications, accumulating evidence indicates that altered translational control might be a key process triggering long-lasting changes and participating in the emergence of drug-adaptive behaviors.

Translation is a complex process as its initiation, elongation, and termination steps are controlled by coordinated interactions between ribosomes and translational factors. Initiation refers to the recruitment of the ribosome and the tRNA carrying the first amino acid residue at the mRNA start codon. Initiation is the major rate-limiting step and is therefore tightly regulated by several eukaryotic initiation factors (eIFs; Buffington et al., 2014). Elongation is another checkpoint of translation, during which the polypeptide chain is formed by the sequential addition of amino acids. This step requires two eukaryotic elongation factors (eEF1 and eEF2) whose functions are also

Box 19.1. Approaches Used to Study Addiction-Rrelative Behaviors

TECHNIQUE/PARADIGM	DESCRIPTION

Conditioned Place Preference (CPP)

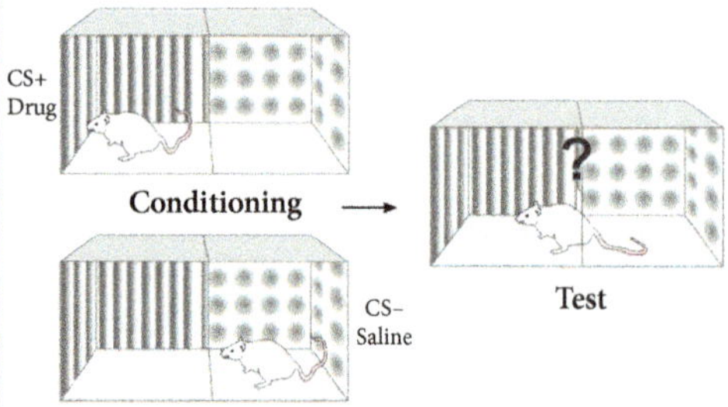

Animals are conditioned to recevie a drug in one experimental chamber and its vehicle in the other. Chambers differ in contextual cues (e.g. walls have distinct visual patterns and floors different textures). A conditioned place preference (CPP) is found if animals spend significantly more time in the drug-paired compartment versus the vehicle-paired compartment in a subsequent test, performed under drug-free conditions. Typically, drugs of abuse such as cocaine elicit rewarding responses and produce CPP.

Drug self-administration

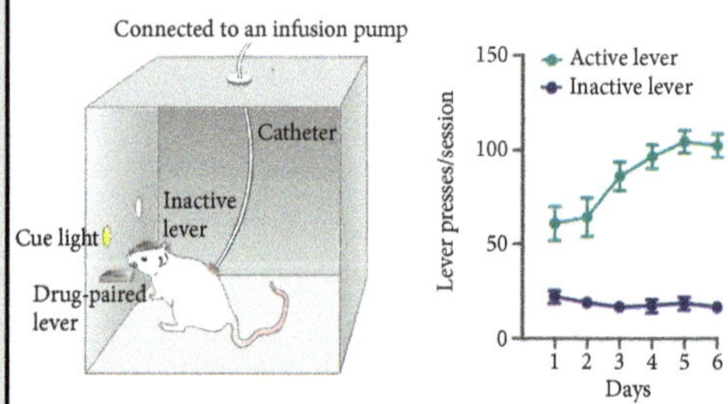

This approach is based on the fact that drugs can function as reinforcers. Thus, during daily sessions in an operant cage, animals perform a response—such as pressing a lever or nose poking (active lever/hole)—that delivers a dose of a drug (usually through an intravenous catheter connected to an infusion syringe) that reinforces the behavior. This paradigm can be used to assess many addiction-related behaviors, including acquisition of drug intake, motivation to self-administer a drug, extinction and relapse to drug seeking.

Drug-induced locomotor sensitization

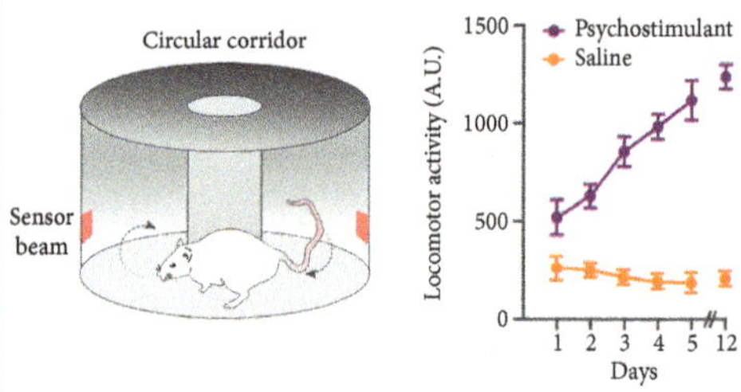

Repeated administration of psychostimulants, such as cocaine or d-amphetamine, in rodents induces a progressive increase in drug-induced locomotor activity, which is higher compared to that elicited by a single administration. This hyperlocomotion is long-lasting and can be observed after a challenge injection of the drug (e.g. day 12) following a drug-free period. This model represents a useful tool to study the bases of long-term behavioral plasticity.

regulated (Dever & Green, 2012). Finally, in addition to their well-characterized roles in stabilization and mRNA transport, RNA-binding proteins (RBPs) and microRNAs also play an important role in translational control.

In this chapter, we review recent reports that addictive drugs regulate translational control. We focus mainly on mTORC1 signaling and the role of the initiation factor eIF2.

We also discuss how these translation control mechanisms can contribute to long-lasting neuronal circuit adaptations and drug-adaptive behavior.

Role of Protein Synthesis in Drug-Altered Behavior and Neuronal Plasticity

The first indications that protein synthesis is involved in drug-related behavior came from Karler and colleagues, who reported that inhibiting protein synthesis prevented sensitized behavioral responses to cocaine and amphetamine (Karler et al., 1993). Since then, several studies have shown that most, if not all, long-lasting behavioral alterations of addictive drugs require protein synthesis. Systemic or intra-cerebro-ventricular administrations of, the translation inhibitors, anisomycin or cycloheximide attenuate locomotor sensitization to cocaine and morphine (Bernardi et al., 2007; Luo et al., 2011; Valjent et al., 2010) and different phases of conditioned place preference induced by morphine, cocaine, nicotine, and methamphetamine (Fan et al., 2010; Kuo et al., 2007; Milekic et al., 2006; Robinson & Franklin, 2007; Taubenfeld et al., 2010; Valjent et al., 2006; Xue et al., 2017; Yu et al., 2013b). Blockade of protein synthesis also impairs the acquisition of cocaine self-administration (Mierzejewski et al., 2006) and alcohol-, cocaine-, and nicotine-seeking induced by exposure to cues previously associated with drug intake (Dunbar &Taylor, 2016; von der Goltz et al., 2009; Xue et al., 2017). Finally, the local infusion of these inhibitors was instrumental to identify those brain areas in which protein synthesis was essential to control specific components of drug-related behavior (Figure 19.1).

Addiction-related behaviors in rodent models are thought to result in part from persistent synaptic adaptations in reward circuits, known as drug-evoked plasticity (for review see (Luscher, 2016). Thus, enhancement of excitatory synaptic transmission in mesolimbic dopamine neurons occurs in response to addictive drugs. This synaptic modification, which is characterized by a transient increase in AMPAR/NMDAR ratio, is sensitive to protein synthesis inhibitors (Argilli et al., 2008). Similarly, blocking protein synthesis prevents the accumulation of calcium-permeable AMPARs in medium-sized spiny neurons of the nucleus accumbens during protracted withdrawal from cocaine self-administration (Scheyer et al., 2014).

Addictive drugs also trigger persistent morphological changes in the size of cell bodies and in spine morphology in various neuronal types including mesolimbic dopamine cells, nucleus accumbens medium-sized spiny neurons, and prefrontal cortex pyramidal neurons (Russo et al., 2010). These structural remodelings are likely underpinned by de novo protein synthesis. Indeed, protein synthesis inhibitors attenuated the increase in spine density, in the nucleus accumbens and in the lateral amygdala, of animals that underwent conditioned place preference to cocaine and methamphetamine, respectively

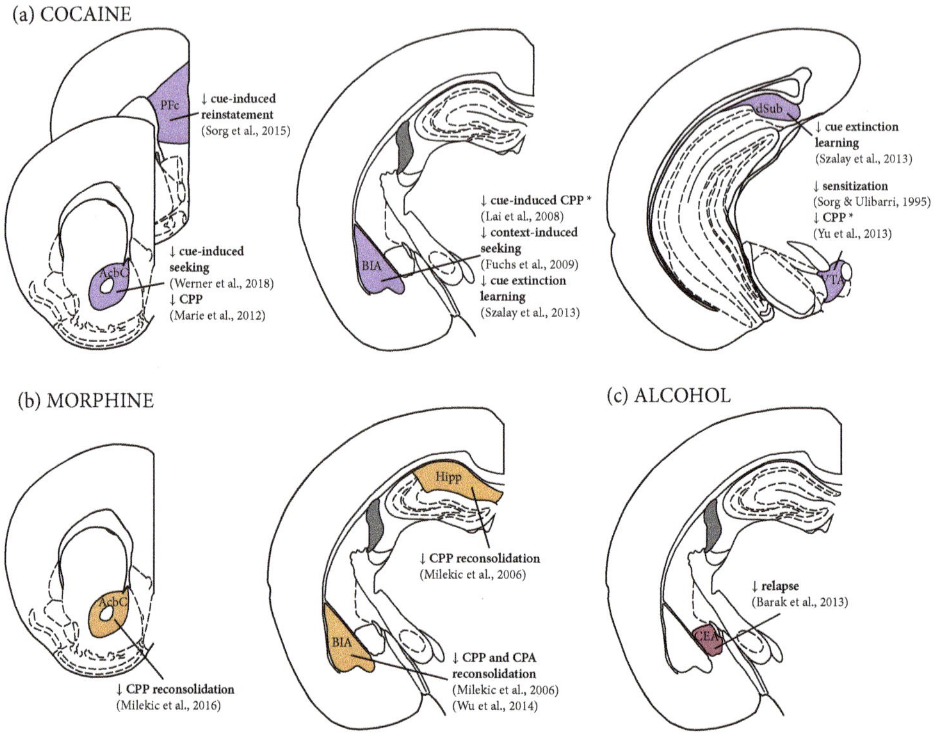

FIGURE 19.1. Key brain areas where protein synthesis inhibitors control specific components of drug-adaptive behavior. Shaded areas represent brain structures in which infusion of anisomycin or cycloheximide (*) reduce specific (a) cocaine-, (b) morphine-, and (c) alcohol-altered behavior. (PFc, prefrontal cortex; AcbC, nucleus accumbens core; VTA, ventral tegmental area; CEA, central amygdala; BlA, basolateral amygdala; Hipp, hippocampus; dSub, dorsal subiculum; CPP, conditioned place preference.)

(Marie et al., 2012; Yu et al., 2016). Despite this evidence, the translation-regulating mechanisms underpinning the long-lasting neural plasticity induced by addictive drugs have only just begun to be elucidated.

Regulation and Role of mTORC1 Pathway in Drug Addiction

The role of the mammalian or mechanistic target of rapamycin complex 1 (mTORC1) pathway in protein-synthesis-dependent synaptic plasticity and memory formation has been the focus of considerable study over the last decade (Santini et al., 2014). The mTORC1 complex is regulated by the convergence of distinct signaling cascades that depend on numerous receptors that respond to tyrosine phosphorylation, ions, and

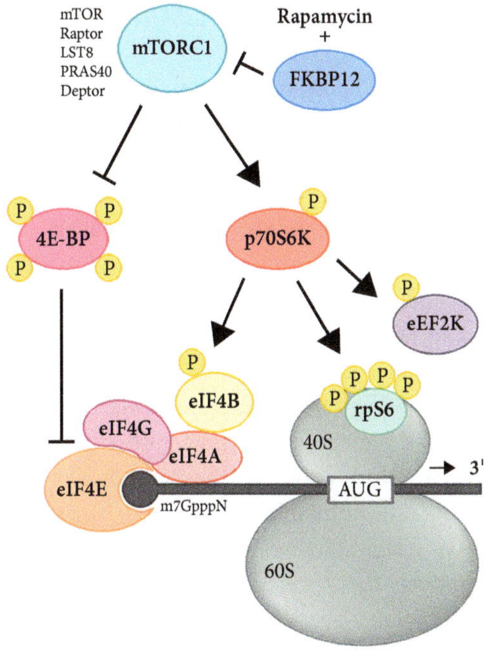

FIGURE 19.2. Translation initiation control by the mTORC1 signaling pathway. mTORC1 mediates cap-dependent translation via the phosphorylation of its primary downstream substrates: 4E-BPs and p70S6Ks. 4E-BPs compete with the scaffolding protein eIF4G for a common binding site on eIF4E. mTORC1-mediated phosphorylation of 4E-BPs prevents their binding to eIF4E, which allows eIF4E to assemble with eIF4G and eIF4A to form the eIF4F complex and bind the mRNA cap structure (m7GpppN). mTORC1 also controls translation through the phosphorylation of p70S6Ks, which in turn phosphorylates eIF4B (a cofactor of eIF4A), the ribosomal protein S6, and eEF2K (a kinase of eEF2 involved in translation elongation). Rapamycin binds to FKBP12 and forms a complex that directly binds and inhibits mTORC1.

other second messengers (Figure 19.2). mTORC1 positively controls the initiation of translation by regulating the formation of the eIF4F complex (composed of eIF4G, eIF4A, and eIF4E) that recognizes the 5′-m7G cap-structure of mRNAs. Indeed, activated mTORC1 controls crucial core components of the translation machinery: the p70 ribosomal S6 kinases 1 and 2 (p70S6K1 and p70S6K2) and the eIF4E-binding proteins (4E-BPs; Santini et al., 2014). The 4E-BPs prevent the formation of the eIF4F complex by interacting with the cap-binding protein eIF4E. mTORC1-mediated phosphorylation of 4E-BP releases eIF4E, thereby allowing its assembly with eIF4G and eIF4A, which is crucial to recruit the ribosome during translation initiation. mTORC1 also mediates eIF4G and p70S6K phosphorylation. Activated p70S6K can in turn phosphorylate eIF4B leading to increased eIF4A catalytic activity. Activated p70S6K also regulates the ribosomal protein of the 40S ribosomal subunit, rpS6, whose phosphorylation state has been implicated in the translation of certain mRNAs (Biever et al., 2015b; Puighermanal et al., 2017). The regulation of mTORC1 pathway by addictive drugs and its role in drug-adaptive behaviors are dealt with in two separate sections: (a) regulation of mTORC1 and its downstream effectors and (b) role of mTOR in drug-induced behaviors and neuronal plasticity.

The regulation of the mTORC1 pathway by addictive drugs has been one of the most studied molecular mechanisms involved in translation control. Indeed, with the notable exception of amphetamine, all addictive drugs studied regulate the phosphorylation state of mTORC1 and/or its direct downstream substrates (p70S6K and 4E-BP) in specific brain areas (Table 19.1). Interestingly, phosphorylation of these factors is also altered in animals that learn to self-administer drugs, and in animals exposed to context or cues previously associated with drug intake (Table 19.1). The regulation of rpS6 phosphorylation (Ser235/6 and Ser240/4) in response to addictive drugs or during drug-related behaviors has been widely studied and reviewed recently (Biever et al., 2015b). In contrast, little is known regarding the modulation of eIF4B or eIF4G. To date only one study described an increased phosphorylation of both translation initiation factors, in the hippocampus after THC administration (Puighermanal et al., 2009). Finally, addictive drugs also regulate the translation initiation factor eIF4E. Indeed, acute administration of d-amphetamine or THC increases eIF4E phosphorylation in the striatum and the hippocampus, respectively (Biever et al., 2015a; Puighermanal et al., 2009; Sutton & Caron, 2015). Phosphorylation is also enhanced in the nucleus accumbens after repeated exposure to cocaine or cues induced cocaine-seeking (Sutton & Caron, 2015; Werner et al., 2018). Although eIF4E phosphorylation has been widely associated with cap-dependent translation, recent studies have identified a small subset of phospho-eIF4E-sensitive mRNAs in distinct brain areas (Amorim et al., 2018; Cao et al., 2015; Gkogkas et al., 2014). Future studies are required to elucidate the mRNAs whose translation depends on eIF4E phosphorylation in the context of drug addiction. Overall, addictive drugs, including cocaine, methamphetamine, morphine, alcohol, THC, and nicotine, hijack mTORC1 signaling and/or translational machinery factors in multiple addiction-related brain areas.

Several pharmacological and genetic studies support a functional role for mTORC1 in drug-related behaviors. For example, systemic administration of the mTORC1 inhibitor rapamycin prevents both the induction and expression of cocaine or nicotine-induced locomotor sensitization in rats (Gao et al., 2014; Wu et al., 2011). Unlike in rats, only the expression of locomotor sensitization to cocaine is altered in mice (Bailey et al., 2012). mTORC1 inhibition also reduces cocaine-seeking (James et al., 2016) as well as the expression and the reconsolidation of conditioned place preference induced by morphine, cocaine, and alcohol (Bailey et al., 2012; Lin et al., 2014; Neasta et al., 2010). Finally, alcohol-related behaviors are attenuated in animals treated with rapamycin (Barak et al., 2013; Beckley et al., 2016; Lin et al., 2014; Neasta et al., 2010). Supporting the results obtained by using these preclinical models, a small randomized double-blind clinical trial showed that systemic rapamycin administration suppresses cue-induced drug craving in abstinent heroin addicts (Shi et al., 2009).

A causal link between mTORC1 activation in specific brain areas and drug-related behavior has been established by local infusions of mTORC1 inhibitors. Specifically, microinjections of rapamycin into the nucleus accumbens attenuated many

Table 19.1. Regulation of mTORC1 (S2488), p70S6K (T389), and 4E-BP1 (T37/46) by drugs of abuse and drug-related behaviors in vivo

Drugs of Abuse (doses, mg/kg)		Species	p-mTOR	p-p70S6K	p4E-BP1	Brain Areas	References
Cocaine	5	Mouse	ND	=	=	VTA	Huang et al. (2016)
	15		=[1]	ND	ND	Acb	Sutton and Caron (2015)
	10	Rat	↑	ND	ND	Acb	Luo et al. (2016)
	15*	Mouse	=[2]	↑	↑	Acb	Sutton and Caron (2015)
			ND	=	ND	Acb	Bailey et al. (2012)
	20*		↑	ND	ND	Acb	Cahill et al. (2016)
Amphetamine	5	Mouse	=	ND	ND	Striatum	Rapanelli et al. (2014)
	10	Mouse	=	=[3]	=	Striatum	Biever et al. (2015a)
	10*		ND	=	=	Striatum	Biever et al. (2017)
Methamphetamine	2*	Rat	ND	↑(NA)	ND	Acb	Narita et al. (2005)
Nicotine	0.35*	Rat	ND	↑	↑	BIA	Gao et al. (2014)
Morphine	25* (pellet)	Mouse	=	↑	ND	VTA	Mazei-Robison et al. (2011)
	75* (pellet)	Rat	=	=	↑[4]	VTA/Acb	
Alcohol	2.5 (g/kg)	Mouse	ND	↑	↑	Acb	Neasta et al. (2010)
THC	10	Mouse	↑	↑[5]	↑	Hipp	Puighermanal et al. (2009)
			ND	↑	ND	Striatum/Amy/PFc	Puighermanal et al. (2013)
Drug-Related Behaviors							
Retrieval of cocaine-related memories		Mouse	↓ (NA)	↓ (NA)	ND	Acb/Hipp	Shi et al. (2014)
			↑	↑	ND	Acb	Wang et al. (2010)

(*Continued*)

Table 19.1. (Continued)

Drugs of Abuse (doses, mg/kg)	Species	p-mTOR	p-p70S6K	p4E-BP1	Brain Areas	References
Alcohol Binge 20% (4h/d)		ND	↑	↑	Acb	Neasta et al. (2010)
Single Alcohol Drinking		ND	↑	↑	Acb	Neasta et al. (2010)
Cue-Induced Cocaine-Seeking	Rat	= [6]	=	= [7]	Acb	Werner et al. (2018)
Cocaine SA		↑	ND	ND	Acb	Cahill et al. (2016)
Withdrawal from Cocaine SA		↑	↑	ND	Acb	James et al. (2014)
Morphine-induced CPP		↑	↑	ND	Hipp	Cui et al. (2010)
Retrieval of Alcohol-Related Memories		ND	↑	↑	CEA/OFc/mPFc	Barak et al. (2013)

Notes: * repeated administration. NA, not available; ND, not determined

[1] p-mTOR S2481 (↑), T2446 (=)

[2] p-mTOR S2481 and T2446 (↑)

[3] p70S6K T421/424 (=)

[4] 4E-BP1 T37/46 (=) in the Acb

[5] p70S6K T421/424 (↑)

[6] p-mTOR S2481 (=)

[7] p-4E-BP1 S65 (↑)

cocaine-related behaviors, including hyperlocomotion, drug-seeking, and cue-induced reinstatement (Cahill et al., 2016; James et al., 2014; Wang et al., 2010; Werner et al., 2018), as well as methamphetamine locomotor sensitization (Narita et al., 2005). Likewise, nicotine-induced locomotor sensitization is prevented by mTORC1 inhibition in the basolateral amygdala (Gao et al., 2014). Finally, binge drinking and alcohol consumption are reduced when rapamycin is infused into the nucleus accumbens, whereas alcohol-seeking behavior requires mTORC1 activation in the central amygdala (Barak et al., 2013; Neasta et al., 2010).

The generation of cell type- and brain area-specific conditional knockout mice of the different components of the mTOR signaling pathway has uncovered its role in specific circuits. For example, selective knockdown of mTOR in the ventral tegmental area was sufficient to impair cocaine-induced conditioned place preference, as well as to prevent the increase in AMPAR/NMDAR ratio and reduction of GABAergic inhibition in dopaminergic neurons induced by cocaine (X. Liu et al., 2018). On the other hand, disruption of mTOR or its mTORC1 partner raptor in D1R-expressing neurons attenuated the enhanced locomotion induced by cocaine (Sutton & Caron, 2015). These results suggest a crucial role of mTOR in mediating cocaine-reward properties and synaptic plasticity. In addition to mTORC1, mTOR can bind other partners including rictor to form complex 2 (mTORC2), which has been involved in structural modifications via cytoskeletal rearrangements. Increasing evidence has demonstrated a role of mTORC2 in drug-related behaviors and structural modifications. Indeed, inactivation of rictor in the ventral tegmental area abolishes morphine-induced conditioned place preference (Mazei-Robison et al., 2011). Strikingly, the hyperlocomotor response to amphetamine is enhanced in mice with reduced mTORC2 function in the dorsal striatum (Dadalko et al., 2015). Lastly, mTORC2 is also a key player in drug-induced synaptic remodeling, through its action on cytoskeleton rearrangements. Notably, inactivation of rictor in the ventral tegmental area blocks the decrease in dopaminergic neuron soma size induced by morphine (Mazei-Robison et al., 2011). Additionally, rictor inactivation prevents mushroom spines from increasing in size and density after consuming alcohol (Laguesse et al., 2018). Altogether, compelling data demonstrate a crucial role of mTOR (either in complex 1 or 2) in drug-related behaviors and neuronal plasticity.

Regulation and Role of the Translational Initiation Factor eIF2α in Drug Addiction

Translation initiation begins with the formation of the 43S preinitiation complex, which comprises the 40S ribosome subunit, some eukaryotic initiation factors, and the ternary complex formed by the interaction of eIF2-GTP and the initiator methionine-charged tRNA. Subsequently, the 43S preinitiation complex binds the mRNA and the eIF4F complex to collectively form the 48S complex that then scans along the mRNA for the start codon. Upon AUG recognition, eIF2 hydrolyzes GTP to GDP and dissociates from the mRNA, permitting the binding of the 60S ribosomal subunit and elongation of the polypeptide chain. eIF2 remains bound to GDP until eIF2B exchanges GDP for GTP on eIF2 allowing therefore another round of initiation. This step is tightly controlled by the state of phosphorylation of eIF2α, which is regulated by the activity of four kinases (HRI, PKR, PERK, and GCN2) and two phosphatase complexes (PP1/GADD34 and PP1/CReP). Phosphorylation of eIF2α reduces the activity of the guanine nucleotide exchange factor of eIF2, eIF2B, thereby reducing ternary complex formation. This ultimately decreases the global protein synthesis and causes the enhancement of the translation of a small number of mRNAs with upstream open reading frames in their 5′UTRs (Buffington et al., 2014).

The regulation of eIF2α phosphorylation by addictive drugs has only recently begun to be assessed. Jian and colleagues were the first to report a rapid and transient eIF2α dephosphorylation in the basolateral amygdala after cocaine- and morphine-related memories are retrieved (Jian et al., 2014). Since then, decreased eIF2α phosphorylation was found in the ventral tegmental area after a single administration of alcohol, nicotine, or methamphetamine (Huang et al., 2016). This decreased phosphorylation was also observed in mice that were acutely or repeatedly given cocaine, and in the nucleus accumbens of rats exposed to cues promoting cocaine-seeking (Placzek et al., 2016a, 2016b; Werner et al., 2018). However, depending on the brain areas analyzed, decreased eIF2α phosphorylation is not systematically observed in response to a single injection of addictive drugs. Indeed, eIF2α phosphorylation is unchanged in the mouse striatum of mice receiving a single injection of cocaine or amphetamine (Biever et al., 2017; Huang et al., 2016). The regimen of drug administration also represents an important factor since mice that are repeatedly given amphetamine do have increased eIF2α phosphorylation in the striatum (Biever et al., 2017). Finally, there are differences between species, as, unlike in mice, acute amphetamine increases eIF2α phosphorylation in the rat striatum and prefrontal cortex (Xue et al., 2016).

The role of eIF2α phosphorylation in drug-related behaviors has been considerably elucidated with the help of a selective inhibitor of eIF2α phosphatases, Sal003. Infusion of Sal003 into intra-basolateral amygdala immediately after morphine or cocaine memory retrieval, impaired the reconsolidation of drug-induced conditioned place preference (Jian et al., 2014). Sal003 infusion into the nucleus accumbens also reduces cocaine-seeking (Werner et al., 2018). eIF2α phosphorylation in the ventral tegmental area appears to be necessary and sufficient to regulate the sensitivity to cocaine, as preventing eIF2α dephosphorylation in this brain area abolishes the increased AMPAR/NMDAR ratio and conditioned place preference induced by cocaine in adolescent mice. Conversely, mice given ISRIB, a pharmacological compound that reverses the effects of eIF2α phosphorylation (Sidrauski et al., 2013) or Eif2s1$^{S/A}$ heterozygous knock-in mice display an enhanced conditioned place preference to a low dose of cocaine (Huang et al., 2016; Placzek et al., 2016a, 2016b).

As mentioned earlier, enhanced eIF2α phosphorylation favor by an indirect mechanism the translation of mRNAs containing upstream open reading frames in their 5′ UTR (Buffington et al., 2014). Activating transcription factor 4 (ATF4) and oligophrenin 1 (OPHN1) have been identified as specific targets through which eIF2α regulates drug-adaptive behaviors. ATF4 is rapidly downregulated in the basolateral amygdala after retrieval of cocaine- and morphine-related memories and this transient downregulation is necessary for the reconsolidation of conditioned place preference to morphine (Jian et al., 2014). OPHN1, on the other hand, appears to be the main conduit through which, eIF2α regulates cocaine sensitivity in the ventral tegmental area. Indeed, reduced OPHN1 levels in this brain structure enhance the rewarding properties of cocaine (Huang et al., 2016). However, the modulation of eIF2α phosphorylation is not systematically associated with altered ATF4 and OPHN1 expression. For example, there were no changes in eIF2α phosphorylation in the striatum of mice repeatedly given amphetamine or in the nucleus accumbens of rats exposed to cues inducing cocaine-seeking (Biever et al., 2017; Werner et al., 2018). Instead, recent work suggests that the translational regulation of selective uORF-bearing mRNAs might depend on the brain areas and the cell type where eIF2α phosphorylation takes place (Biever et al., 2017; Buffington et al., 2014). Altogether, these data illustrated how the translational factors, that play a key role in memory and synaptic plasticity such as eIF2α, can be modulated by addictive drugs and lead to persistent synaptic rearrangements and drug-adaptive behaviors (Costa-Mattioli et al., 2005, 2007, 2009).

RNA-binding Proteins and Addictive Drug Action

RNA-binding proteins (RBPs) participate in post-transcriptional control of gene expression. In addition to their major role in regulating splicing, polyadenylation, stabilization, and mRNA transport, RBPs also critically regulate mRNA translation, mostly as repressors (Hentze et al., 2018). Despite the identification of more than 500 RBPs, only a few are known to be regulated by addictive drugs and to play a role in psychostimulant-related behaviors.

The Fragile X mental retardation protein, FMRP, encoded by the *fmr1* gene, is involved in mRNA transport and the control of translation (De Rubeis et al., 2012). To date, only one study has documented the ability of drugs of abuse to regulate FMRP levels. In it, acute cocaine administration increases FMRP levels in the brain (Tiruchinapalli et al., 2008). Neither its expression nor its phosphorylation level changed in synaptoneurosomes from the nucleus accumbens after protracted abstinence from cocaine (Werner et al., 2018). Several studies however indicate that FMRP plays a role in behavioral response to addictive psychostimulant drugs. For example, acute amphetamine-induced psychomotor effects, such as locomotion and stereotypies, are attenuated in *Fmr1*-deficient mice (Fulks et al., 2010; Ventura et al., 2004). Cocaine-induced hyperlocomotion is reduced (Fish et al., 2013), but this effect is accompanied by an enhancement of stereotypies (Smith et al., 2014). *Fmr1* knockout mice also have impaired sensitized locomotor responses and conditioned place preference to cocaine. However, cocaine response is normal in mice with a targeted deletion of FMRP in dopaminergic neurons, suggesting that functional FMRP in this cell type is not critical to mediate cocaine-adaptive behaviors (Smith et al., 2014). Interestingly, nucleus accumbens-specific knockdown of *fmr1* gene is sufficient to reduce behavioral sensitization but not cocaine reward (Smith et al., 2014).

One of the mechanisms by which FMRP has been proposed to repress translation is through its CYtoplasmic FMRP-Interacting Proteins, 1 and 2 (Napoli et al., 2008). Mutant CYFIP2 mice have reduced acute and sensitized responses to cocaine and methamphetamine. However the mechanistic role of CYFIP proteins in addiction remains to be determined (Kumar et al., 2013). Determining whether drugs of abuse interfere with the ability of FMRP to properly control the transport and translation of some of its target mRNAs in specific brain areas should help to better understand how FMRP participates in the long-lasting modifications induced by addictive drugs.

The embryonic lethal abnormal vision (ELAV)-like proteins/Hu family belongs to a class of RBPs that are enriched in neurons (Darnell, 2013). The four members (ELAVL1-4 also known as HuR, HuB, HuC, and HuD) preferentially bind to AU-rich motifs in 3′UTRs and are involved in mRNA transport, stability, and regulation of translation (Darnell, 2013). In rat brain, cocaine upregulates both HuR and HuD (Tiruchinapalli et al., 2008) and these proteins also bind mRNAs that are upregulated by cocaine, such as *homer* (Brakeman et al., 1997) and *gap43* (Tiruchinapalli et al., 2008). The ELAV/GAP-43 protein:RNA pairing is also increased in hippocampus during prolonged withdrawal from cocaine self-administration (Pascale et al., 2016). Recently, increased expression of HuD and two of its previously identified targets, *Bdnf* and *Camk2a*, was observed in the nucleus accumbens after conditioned place preference to cocaine. A causal link between enhanced HuD expression and the rewarding properties of cocaine was further provided by the use of HuD-overexpressing mice, which displayed an enhanced conditioned place preference induced by a low dose of cocaine (Oliver et al., 2018).

Zipcode-binding protein-1 (ZBP1) belongs to a group of RBPs that regulates neuronal transport and dendritic localization of mRNAs, including β-actin (Huttelmaier et al., 2005). Because ZBP1 is barely expressed in adulthood, a transgenic mouse line overexpressing ZBP1 selectively in forebrain neurons was generated to test the role of this RBP in cocaine-related behaviors (Lapidus et al., 2012). Strikingly, cocaine-induced conditioned place preference is abolished in this mouse line, a phenotype most likely due to the abnormal enhanced expression of ZBP1 target mRNAs as a result of ectopic ZBP1 expression (Lapidus et al., 2012).

Toward the Identification of Drug-Induced mRNA Translation in Genetically Identified Cell Populations

Although polysome profiling and puromycin incorporation are powerful methods to assess whether addictive drugs alter global translational rates, pioneering techniques have been used recently to isolate mRNA and identify translated mRNAs in specific cell types within specific brain areas (Box 2). The first approach, Translating Ribosome Affinity Purification was developed in the Heintz lab (Heiman et al., 2008). TRAP uses a bacterial artificial chromosome in transgenic mice to express an EGFP-tagged ribosomal protein in any cell type of interest, for example, D1R- or D2R-expressing neurons. Tagged ribosomes from these two cell types were immunoprecipitated with anti-GFP-coated magnetic beads and their bound mRNAs were extracted for microarrays analysis. The comparison of the translatome profiles between saline- and cocaine-treated mice revealed hundreds of differentially expressed genes between D1R- and D2R-expressing neuronal populations. Interestingly, repeated cocaine exposure selectively upregulated mRNAs associated with $GABA_A$ receptors in D1R-expressing cells. This may help to account for the specific electrophysiological properties of this cell population (Heiman et al., 2008). TRAP and RNA-seq were also used to uncover chronic nicotine-induced upregulation of ribosome-associated mRNAs in discrete populations of α5 nicotinic acetylcholine receptor-positive cells of the interpeduncular nucleus. Among the differentially expressed genes, neuronal nitric oxide synthase and somatostatin genes were found to be key players in nicotine reward (Ables et al., 2017).

Another approach of interest is RiboTag, which is also based on the cell type-specific immunoprecipitation of tagged ribosomes (Sanz et al., 2009). In this method a HA-tagged ribosomal protein is expressed in a Cre-dependent manner (Box 2). This technique was key to uncover that cocaine exposure differentially regulates the expression of the early growth response 3 (Egr3) and one of its targets, the peroxisome proliferator-activated receptor gamma coactivator-1α (PGC-1α), in distinct neuronal populations in the nucleus accumbens. While cocaine enhanced *Egr3* and *PGC-1α* ribosome-associated mRNAs in D1R-expressing cells, it reduced them in D2R-positive cells (Chandra et al., 2015, 2017a). Consequently, Egr3 and PGC-1α overexpression in D1R-expressing neurons enhanced conditioned place preference and locomotor sensitization to cocaine, whereas overexpressing them in D2R neurons had the opposite phenotype (Chandra et al., 2015, 2017a). A bidirectional regulation of translation of dynamin-related protein-1 (Drp1) mRNA has also been reported in the striatum after repeated cocaine administration. This GTPase that regulates mitochondrial fission is upregulated in D1R-expressing cells, but downregulated in D2R-containing neurons. This enhanced expression is

Box 19.2. Methods Used to Study the Modulation of mRNA Translation by Drugs of Abuse

APPROACH	APPLICATIONS	REFERENCES
Polysome profiling		
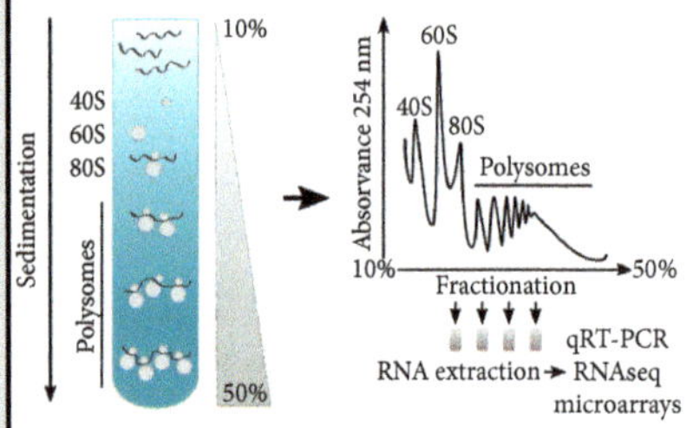	Assessment of global mRNA translation *For instance, if a treatment increases overall translational rates, the polysome peaks are lower whereas the monosome peak is higher compared to control.* Isolation of mRNA pools from monosomes and polysomes (light and heavy fractions) followed by qRT-PCR of targeted genes or genome-wide analysis.	Biever et al. 2015 Biever et al, 2017
Puromycin incorporation		
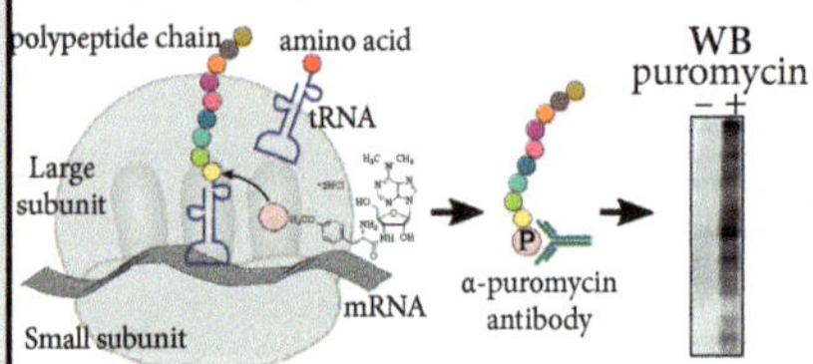	Assessment of global mRNA translation *Puromycin (P) binds to the nascent peptide chain and can be detected by Western Blot (WB) or immunofluorescence using specific antibodies. If a treatment alters global translation, puromycin staining changes.*	Biever et al, 2015 Biever et al, 2017
Tagged-ribosome immunoprecipitation		
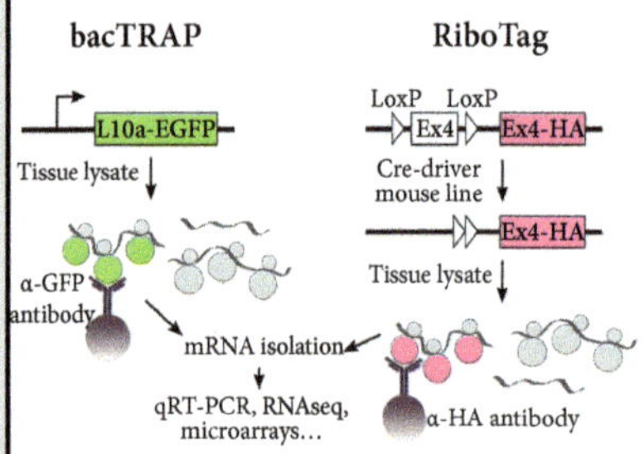	Cell type-specific translational profile *The ribosomal proteins Rpl10a (bacTRAP) and Rpl22 (RiboTag) are tagged with EGFP and HA epitopes, respectively, in genetically-defined cell populations. Using specific antibodies, tagged ribosomes can be immunoprecipitated and their bound mRNAs isolated. Expression of the transcripts can then be assessed by qRT-PCR or genome-wide analysis.*	Heiman et al, 2008 Chandra et al, 2015 Chandra et al, 2017 Ables et al, 2017 Liu et al, 2018

functionally relevant since selective knockdown of *Drp1* in D1R-positive cells blocked cocaine-seeking behavior. Conversely, overexpression of the fission-promoting Drp1(S637A) mutant enhanced cocaine-seeking after prolonged abstinence (Chandra et al., 2017b). Finally, using the same methodology, the ten-eleven translocation family member TET1 mRNA was found to be translationally upregulated in pyramidal neurons of the dorsal hippocampus during memory retrieval, where it plays a crucial role in cocaine-associated memory reconsolidation (C. Liu et al., 2018). It is worth mentioning that many of these studies assessed the levels of mRNAs bound to ribosomes, but that does not necessary imply that they are translationally regulated. Indeed, association with ribosomes is not always associated with mRNAs being translated into proteins since stalled polyribosomes on repressed mRNAs have been reported (Graber et al., 2013). Combination of transcriptional and/or translational effects can however be easily assessed by referring the abundance of each ribosome-bound mRNA to its abundance in the input fraction that contains all the mRNAs of the cell.

Perspectives

There is now compelling evidence that repeated exposure to addictive drugs triggers de novo protein synthesis leading to long-lasting changes in circuit function and behavior. Therefore, identifying novel regulated mRNAs and how and where their translation occurred might help to better understand the action of addictive drugs.

Most past studies assessing translational control by addictive drugs have focused on initiation factors. However, elongation is also an important step of translation and future studies should tackle this issue. For example, the phosphorylation of the eukaryotic elongation factor 2 (eEF2) decreases the rate of peptide chain elongation and promotes the translation of a subset of mRNAs involved in synaptic plasticity (Park et al., 2008; Scheetz et al., 2000; Verpelli et al., 2010). In this regard, repeated amphetamine exposure increases phospho-eEF2, which parallels a decrease in global translation and an upregulation of the activity-regulated cytoskeleton-associated protein (Arc/Arg3.1) in D1R-expressing cells (Biever et al., 2017).

Most research into mTOR signaling in the context of drug addiction has focused on mTORC1-dependent protein synthesis. However, mTOR orchestrates numerous cellular processes and whether other physiological roles of mTORC1 are involved in drug-related plasticity and/or behavior remains to be addressed. For example, mTORC1 inhibition induces the formation of autophagic vacuoles in prejunctional dopaminergic axons, which evokes dopamine release and decreases synaptic vesicle numbers (Hernandez et al., 2012). Given that addictive drugs trigger dopamine release, it will be interesting to examine any possible mTOR-dependent effect in autophagy. The contribution of mTORC2 in the adaptive responses associated with drugs of abuse is also poorly understood and should be investigated more carefully in the future.

Mechanisms of translational control by addictive drugs have been studied in different brain areas such as the ventral tegmental area, striatum, and amygdala, which all contain intermingled heterogeneous cell types. Of note, the signaling pathways responsible for these mechanisms can produce different outcomes depending on the cell population. For example, eIF2α phosphorylation promotes translation of *Atf4* in glutamatergic neurons, whereas it inhibits translation of *Ifnγ* in GABAergic cells (Buffington et al., 2014). Therefore, subsequent studies, assessing the cell type where addictive drugs modulate signaling cascades controlling translation, are required. Glial cells require particular attention as most studies have focused on neurons. On the other hand, the development of approaches such as the bacTRAP or RiboTag has significantly contributed to uncover the specific cell types in which translation of some mRNAs is regulated by addictive drugs (Ables et al., 2017; Chandra et al., 2015; Heiman et al., 2008; C. Liu et al., 2018). However, more thorough studies of the subset of mRNAs that is translated in specific cell populations and of their association with addiction-related processes are necessary to better understand the actions of drugs of abuse. The TRAP and RiboTag analytic techniques will be key to determining the mRNAs that are translated in a specific circuit during an addiction-related behavioral task. These approaches will be particularly powerful when combined with Cre-dependent viral vectors and transgenic mouse lines with promoters of immediate early genes, such as c-fos and Arc (Guenthner et al., 2013).

Neuronal function depends on highly localized molecular signaling and the precise distribution of mRNAs, and their translational regulators, at particular subcellular compartments needs to be addressed in more detail. Local translation has been reported in both dendrites (Holt and Schuman, 2013) and axons (Shigeoka et al., 2016). However, the precise subcellular localization of the mRNAs that are dysregulated by drugs of abuse is still unknown. This knowledge would allow a level of selectivity, both in terms of which mRNAs, but also as to which different neuronal sub-regions should be targeted for therapeutic benefit. Overall, the elucidation of the complex mechanisms of translational control as well as the mRNAs that are translated in response to drugs of abuse at both the cellular and subcellular level should provide novel targets for treatments of addictive processes.

References

Ables, J. L., Gorlich, A., Antolin-Fontes, B., Wang, C., Lipford, S. M., Riad, M. H., . . . Ibanez-Tallon, I. (2017). Retrograde inhibition by a specific subset of interpeduncular alpha5 nicotinic neurons regulates nicotine preference. *Proceedings of the National Academy of Sciences of the United States of America*, *114*(49), 13012–13017.

Amorim, I. S., Kedia, S., Kouloulia, S., Simbriger, K., Gantois, I., Jafarnejad, S.M., . . . Gkogkas, C. G. (2018). Loss of eIF4E phosphorylation engenders depression-like behaviors via selective mRNA translation. *Journal of Neuroscience*, *38*(8), 2118–2133.

Argilli, E., Sibley, D. R., Malenka, R. C., England, P. M., & Bonci, A. (2008). Mechanism and time course of cocaine-induced long-term potentiation in the ventral tegmental area. *Journal of Neuroscience*, *28*(37), 9092–9100.

Bailey, J., Ma, D., & Szumlinski, K. K. (2012). Rapamycin attenuates the expression of cocaine-induced place preference and behavioral sensitization. *Addiction Biology*, *17*(2), 248–258.

Barak, S., Liu, F., Ben Hamida, S., Yowell, Q. V., Neasta, J., Kharazia, V., . . . Ron D. (2013). Disruption of alcohol-related memories by mTORC1 inhibition prevents relapse. *Nature Neuroscience*, *16*(8), 1111–1117.

Beckley, J. T., Laguesse, S., Phamluong, K., Morisot, N., Wegner, S. A., Ron, D. (2016). The first alcohol drink triggers mTORC1-dependent synaptic plasticity in nucleus accumbens dopamine D1 receptor neurons. *Journal of Neuroscience*, *36*(3), 701–713.

Belin-Rauscent, A., Fouyssac, M., Bonci, A., & Belin D. (2016). How preclinical models evolved to resemble the diagnostic criteria of drug addiction. *Biological Psychiatry*, *79*(1), 39–46.

Bernardi, R. E., Lattal, K. M., & Berger, S. P. (2007). Anisomycin disrupts a contextual memory following reactivation in a cocaine-induced locomotor activity paradigm. *Behavioral Neuroscience*, *121*(1), 156–163.

Biever, A., Puighermanal, E., Nishi, A., David, A., Panciatici, C., Longueville, S., . . . Valjent, E. (2015a). PKA-dependent phosphorylation of ribosomal protein S6 does not correlate with translation efficiency in striatonigral and striatopallidal medium-sized spiny neurons. *Journal of Neuroscience, 35*(10), 4113–4130.

Biever, A., Valjent, E., & Puighermanal, E. (2015b). Ribosomal protein S6 phosphorylation in the nervous system: From regulation to function. *Frontiers in Molecular Neuroscience, 8*, 75.

Biever, A., Boubaker-Vitre, J., Cutando, L., Gracia-Rubio, I., Costa-Mattioli, M., Puighermanal, E., & Valjent, E. (2017). Repeated exposure to D-amphetamine decreases global protein synthesis and regulates the translation of a subset of mRNAs in the striatum. *Frontiers in Molecular Neuroscience, 9*, 165.

Brakeman, P. R., Lanahan, A. A., O'Brien, R., Roche, K., Barnes, C. A., Huganir, R, L., & Worley, P. F. (1997). Homer: A protein that selectively binds metabotropic glutamate receptors. *Nature, 386*(6622), 284–288.

Buffington, S. A., Huang, W., & Costa-Mattioli, M. (2014). Translational control in synaptic plasticity and cognitive dysfunction. *Annual Review of Neuroscience, 37*, 17–38.

Cahill, M. E., Bagot, R. C., Gancarz, A. M., Walker D. M., Sun, H., Wang, Z. J., . . . Nestler, E. J. (2016). Bidirectional synaptic structural plasticity after chronic cocaine administration occurs through Rap1 small GTPase signaling. *Neuron, 89*(3), 566–582.

Cao, R., Gkogkas, C. G., de Zavalia, N., Blum, I. D., Yanagiya, A., Tsukumo, Y., . . . Sonenberg, N. (2015). Light-regulated translational control of circadian behavior by eIF4E phosphorylation. *Nature Neuroscience, 18*(6), 855–862.

Chandra, R., Engeln, M., Francis, T. C., Konkalmatt, P., Patel, D., & Lobo, M. K. (2017a). A role for peroxisome proliferator-activated receptor gamma coactivator-1alpha in nucleus accumbens neuron subtypes in cocaine action. *Biological Psychiatry, 81*(7), 564–572.

Chandra, R., Engeln, M., Schiefer, C., Patton, M. H., Martin, J. A., Werner, C.T., . . . Lobo, M. K. (2017b). Drp1 mitochondrial fission in D1 neurons mediates behavioral and cellular plasticity during early cocaine abstinence. *Neuron, 96*(6), 1327–1341 e6.

Chandra, R., Francis, T. C., Konkalmatt, P., Amgalan, A., Gancarz, A. M., Dietz, D. M., & Lobo, M. K. 2015. Opposing role for Egr3 in nucleus accumbens cell subtypes in cocaine action. *Journal of Neuroscience, 35*(20), 7927–7937.

Costa-Mattioli, M., Gobert, D., Harding, H., Herdy, B., Azzi, M., Bruno, M., Sonenberg, N. (2005). Translational control of hippocampal synaptic plasticity and memory by the eIF2alpha kinase GCN2. *Nature, 436*(7054), 1166–1173.

Costa-Mattioli, M., Gobert, D., Stern, E., Gamache, K., Colina, R., Cuello, C., . . . Sonenberg, N. (2007). eIF2alpha phosphorylation bidirectionally regulates the switch from short- to long-term synaptic plasticity and memory. *Cell, 129*(1), 195–206.

Costa-Mattioli, M., Sossin, W. S., Klann, E., & Sonenberg, N. (2009). Translational control of long-lasting synaptic plasticity and memory. *Neuron, 61*(1), 10–26.

Cui, Y., Zhang, X. Q., Xin, W. J., Jing, J., & Liu, X. G. (2010). Activation of phosphatidylinositol 3-kinase/Akt-mammalian target of Rapamycin signaling pathway in the hippocampus is essential for the acquisition of morphine-induced place preference in rats. *Neuroscience, 171*(1), 134–143.

Dadalko, O. I., Siuta, M., Poe, A., Erreger, K., Matthies, H. J., Niswender, K., & Galli, A. (2015). mTORC2/rictor signaling disrupts dopamine-dependent behaviors via defects in striatal dopamine neurotransmission. *Journal of Neuroscience, 35*(23), 8843–8854.

Darnell, R. B. (2013). RNA protein interaction in neurons. *Annual Review of Neuroscience, 36*, 243–270.

De Rubeis, S., Fernandez, E., Buzzi, A., Di Marino, D., & Bagni, C. (2012). Molecular and cellular aspects of mental retardation in the Fragile X syndrome: From gene mutation/s to spine dysmorphogenesis. *Advances in Experimental Medicine and Biology, 970*, 517–551.

Dever, T E., & Green, R. (2012). The elongation, termination, and recycling phases of translation in eukaryotes. *Cold Spring Harbor Perspectives in Biology, 4*(7), a013706.

Dunbar, A. B., & Taylor, J. R. (2016). Inhibition of protein synthesis but not beta-adrenergic receptors blocks reconsolidation of a cocaine-associated cue memory. *Learning and Memory, 23*(8), 391–398.

Fan, H. Y., Cherng, C. G., Yang, F. Y., Cheng, L. Y., Tsai, C. J., Lin, L. C., & Yu, L. (2010). Systemic treatment with protein synthesis inhibitors attenuates the expression of cocaine memory. *Behavioural Brain Research, 208*(2), 522–527.

Fish, E. W., Krouse, M. C., Stringfield, S. J., Diberto, J. F., Robinson, J. E., & Malanga, C. J. (2013). Changes in sensitivity of reward and motor behavior to dopaminergic, glutamatergic, and cholinergic drugs in a mouse model of fragile X syndrome. *PLoS One, 8*(10), e77896.

Fuchs, R. A., Bell, G. H., Ramirez, D. R., Eaddy, J. L, Su, Z. I. (2009). Basolateral amygdala involvement in memory reconsolidation processes that facilitate drug context-induced cocaine seeking. *European Journal of Neuroscience*, 30(5), 889–900.

Fulks, J. L., O'Bryhim, B. E., Wenzel, S. K., Fowler, S. C., Vorontsova, E., Pinkston, J. W., . . . Johnson, M. A. (2010). Dopamine release and uptake impairments and behavioral alterations observed in mice that model fragile x mental retardation syndrome. *ACS Chemical Neuroscience, 1*(10), 679–690.

Gao, Y., Peng, S., Wen, Q., Zheng, C., Lin, J., Tan, Y., . . . Li, Y. (2014). The mammalian target of rapamycin pathway in the basolateral amygdala is critical for nicotine-induced behavioural sensitization. *International Journal of Neuropsychopharmacology, 17*(11), 1881–1894.

Gkogkas, C G., Khoutorsky, A., Cao, R., Jafarnejad, S. M., Prager-Khoutorsky, M., Giannakas, N., . . . Sonenberg, N. (2014). Pharmacogenetic inhibition of eIF4E-dependent Mmp9 mRNA translation reverses fragile X syndrome-like phenotypes. *Cell Reports, 9*(5), 1742–1755.

Graber, T. E., Hebert-Seropian, S., Khoutorsky, A., David, A., Yewdell, J. W., Lacaille, J. C., & Sossin, W. S. (2013). Reactivation of stalled polyribosomes in synaptic plasticity. *Proceedings of the National Academy of Sciences of the United States of America, 110*(40), 16205–16210.

Guenthner, C. J., Miyamichi, K., Yang, H. H., Heller, H. C., & Luo L. (2013). Permanent genetic access to transiently active neurons via TRAP: Targeted recombination in active populations. *Neuron, 78*(5), 773–784.

Heiman, M., Schaefer, A., Gong, S., Peterson, J. D., Day, M., Ramsey, K. E., . . . Heintz, N. (2008). A translational profiling approach for the molecular characterization of CNS cell types. *Cell, 135*(4), 738–748.

Hentze, M. W., Castello, A., Schwarzl, T., & Preiss, T. (2018). A brave new world of RNA-binding proteins. *Nature Reviews: Molecular and Cell Biology, 19*(5), 327–341.

Hernandez, D., Torres, C. A., Setlik, W., Cebrian, C., Mosharov, E. V., Tang, G., . . . Sulzer, D. (2012). Regulation of presynaptic neurotransmission by macroautophagy. *Neuron, 74*(2), 277–284.

Holt, C. E., & Schuman, E. M. (2013). The central dogma decentralized: New perspectives on RNA function and local translation in neurons. *Neuron, 80*(3), 648–657.

Huang, W., Placzek, A. N., Viana Di Prisco, G., Khatiwada, S., Sidrauski, C., Krnjevic, K., . . . Costa-Mattioli, M. (2016). Translational control by eIF2alpha phosphorylation regulates vulnerability to the synaptic and behavioral effects of cocaine. *Elife, 5*, e12052.

Huttelmaier, S., Zenklusen, D., Lederer, M., Dictenberg, J., Lorenz, M., Meng X.,... Singer, R. H. (2005). Spatial regulation of beta-actin translation by Src-dependent phosphorylation of ZBP1. *Nature, 438*(7067), 512–515.

James, M. H., Quinn, R. K., Ong, L. K., Levi, E. M., Charnley, J. L., Smith, D. W.,... Dayas C. V. (2014). mTORC1 inhibition in the nucleus accumbens "protects" against the expression of drug seeking and "relapse" and is associated with reductions in GluA1 AMPAR and CAMKIIalpha levels. *Neuropsychopharmacology, 39*(7), 1694–1702.

James, M. H., Quinn, R. K., Ong, L. K., Levi, E. M., Smith, D. W., Dickson, P. W., & Dayas, C. V. (2016). Rapamycin reduces motivated responding for cocaine and alters GluA1 expression in the ventral but not dorsal striatum. *European Journal of Pharmacology, 784*, 147–154.

Jian, M., Luo, Y. X., Xue, Y. X., Han, Y., Shi, H. S., Liu, J. F.,... Lu, L. (2014). eIF2alpha dephosphorylation in basolateral amygdala mediates reconsolidation of drug memory. *Journal of Neuroscience, 34*(30), 10010–10021.

Karler, R., Finnegan, K. T., & Calder, L. D. (1993). Blockade of behavioral sensitization to cocaine and amphetamine by inhibitors of protein synthesis. *Brain Research, 603*(1), 19–24.

Koob, G. F., Volkow, N. D. (2016). Neurobiology of addiction: A neurocircuitry analysis. *Lancet Psychiatry, 3*(8), 760–773.

Kumar, V., Kim, K., Joseph, C., Kourrich, S., Yoo, S. H., Huang, H. C.,... Takahashi, J. S. (2013). C57BL/6N mutation in cytoplasmic FMRP interacting protein 2 regulates cocaine response. *Science, 342*(6165), 1508–1512.

Kuo, Y. M., Liang, K. C., Chen, H H., Cherng, C. G., Lee, H. T., Lin, Y.,... Yu, L. (2007). Cocaine-but not methamphetamine-associated memory requires de novo protein synthesis. *Neurobiology Learning Memory, 87*(1), 93–100.

Laguesse, S., Morisot, N., Phamluong, K., Sakhai, S. A., & Ron, D. (2018). mTORC2 in the dorsomedial striatum of mice contributes to alcohol-dependent F-actin polymerization, structural modifications, and consumption. *Neuropsychopharmacology, 43*(7), 1539–1547.

Lai, Y. T., Fan, H. Y., Cherng, C. G., Chiang, C. Y., Kao, G. S., & Yu, L. (2008). Activation of amygdaloid PKC pathway is necessary for conditioned cues-provoked cocaine memory performance. *Neurobiology of Learning and Memory, 90*(1), 164–170.

Lapidus, K. A., Nwokafor, C., Scott, D., Baroni, T. E., Tenenbaum, S. A., Hiroi, N.,... Czaplinski, K. (2012). Transgenic expression of ZBP1 in neurons suppresses cocaine-associated conditioning. *Learning and Memory, 19*(2), 35–42.

Lin, J., Liu, L., Wen, Q., Zheng, C., Gao, Y., Peng, S.,... Li, Y. (2014). Rapamycin prevents drug seeking via disrupting reconsolidation of reward memory in rats. *International Journal of Neuropsychopharmacology, 17*(1), 127–136.

Liu, X., Li, Y., Yu, L., Vickstrom, C. R., & Liu, Q. S. (2018). VTA mTOR signaling regulates dopamine dynamics, cocaine-induced synaptic alterations, and reward. *Neuropsychopharmacology, 43*(5), 1066–1077.

Liu, C., Sun, X., Wang, Z., Le, Q., Liu, P., Jiang, C.,... Ma, L. (2018). Retrieval-induced upregulation of Tet3 in pyramidal neurons of the dorsal hippocampus mediates cocaine-associated memory reconsolidation. *International Journal of Neuropsychopharmacology, 21*(3), 255–266.

Luo, Y. X., Han, H., Shao, J., Gao, Y., Yin, X., Zhu, W. L.,... Shi, H. S. (2016). mTOR signalling in the nucleus accumbens shell is critical for augmented effect of TFF3 on behavioural response to cocaine. *Science Reports, 6*, 278–295.

Luo, J., Jing, L., Qin, W. J., Zhang, M., Lawrence, A. J., Chen, F., & Liang, J. H. 2011. Transcription and protein synthesis inhibitors reduce the induction of behavioural sensitization to a single morphine exposure and regulate Hsp70 expression in the mouse nucleus accumbens. *International Journal of Neuropsychopharmacology, 14*(1), 107–121.

Luscher, C. (2016). The emergence of a circuit model for addiction. *Annual Review of Neuroscience, 39*, 257–276.

Luscher, C., Malenka, R. C. (2011). Drug-evoked synaptic plasticity in addiction: From molecular changes to circuit remodeling. *Neuron, 69*(4), 650–663.

Marie, N., Canestrelli, C., Noble, F. (2012). Transfer of neuroplasticity from nucleus accumbens core to shell is required for cocaine reward. *PLoS One, 7*(1), e30241.

Mazei-Robison, M. S., Koo, J. W., Friedman, A. K., Lansink, C. S., Robison, A. J., Vinish, M.,... Nestler, E. J (2011). Role for mTOR signaling and neuronal activity in morphine-induced adaptations in ventral tegmental area dopamine neurons. *Neuron, 72*(6), 977–990.

Mierzejewski, P., Siemiatkowski, M., Radwanska, K., Szyndler, J., Bienkowski, P., Stefanski, R.,... Kostowski W. (2006). Cycloheximide impairs acquisition but not extinction of cocaine self-administration. *Neuropharmacology, 51*(2), 367–373.

Milekic, M. H., Brown, S. D., Castellini, C., Alberini, C. M. (2006). Persistent disruption of an established morphine conditioned place preference. *Journal of Neuroscience, 26*(11), 3010–3020.

Napoli, I., Mercaldo, V., Boyl, P. P., Eleuteri, B., Zalfa, F., De Rubeis, S.,....Bagni, C. (2008). The fragile X syndrome protein represses activity-dependent translation through CYFIP1, a new 4E-BP. *Cell, 134*(6), 1042–1054.

Narita, M., Akai, H., Kita, T., Nagumo, Y., Sunagawa, N., Hara, C.,... Suzuki, T. (2005). Involvement of mitogen-stimulated p70-S6 kinase in the development of sensitization to the methamphetamine-induced rewarding effect in rats. *Neuroscience, 132*(3), 553–560.

Neasta, J., Ben Hamida, S., Yowell, Q., Carnicella, S., & Ron, D. (2010). Role for mammalian target of rapamycin complex 1 signaling in neuroadaptations underlying alcohol-related disorders. *Proceedings of the National Academy of Sciences of the United States of America, 107*(46), 20093–20098.

Oliver, R. J., Brigman, J. L., Bolognani, F., Allan, A. M., Neisewander, J. L., & Perrone-Bizzozero, N. I. (2018). Neuronal RNA-binding protein HuD regulates addiction-related gene expression and behavior. *Genes Brain Behavior, 17*(4), e12454.

Park, S., Park, J. M., Kim, S., Kim, J. A., Shepherd, J. D., Smith-Hicks, C. L.,... Worley, P. F. (2008). Elongation factor 2 and fragile X mental retardation protein control the dynamic translation of Arc/Arg3.1 essential for mGluR-LTD. *Neuron, 59*(1), 70–83.

Pascale, A., Osera, C., Moro, F., Di Clemente, A., Giannotti, G., Caffino, L.,... Cervo, L. (2016). Abstinence from cocaine-self-administration activates the nELAV/GAP-43 pathway in the hippocampus: A stress-related effect? *Hippocampus, 26*(6), 700–704.

Placzek, A. N., Molfese, D. L., Khatiwada, S., Viana Di Prisco, G., Huang, W.,... Costa-Mattioli, M. (2016a). Translational control of nicotine-evoked synaptic potentiation in mice and neuronal responses in human smokers by eIF2alpha. *Elife*, 5:e12056.

Placzek, A. N., Prisco, G. V., Khatiwada, S., Sgritta, M., Huang, W., Krnjevic, K.,... Costa-Mattioli, M. (2016b). eIF2alpha-mediated translational control regulates the persistence of cocaine-induced LTP in midbrain dopamine neurons. *Elife, 5*, e17517.

Puighermanal, E., Biever, A., Pascoli, V., Melser, S., Pratlong, M., Cutando, L.,... Valjent, E. (2017). Ribosomal protein S6 phosphorylation is involved in novelty-induced locomotion, synaptic plasticity and mRNA translation. *Frontiers in Molecular Neuroscience, 10*, 419.

Puighermanal, E., Busquets-Garcia, A., Gomis-Gonzalez, M., Marsicano, G., Maldonado, R., & Ozaita, A. (2013). Dissociation of the pharmacological effects of THC by mTOR blockade. *Neuropsychopharmacology, 38*(7), 1334–1343.

Puighermanal, E., Marsicano, G., Busquets-Garcia, A., Lutz, B., Maldonado, R., & Ozaita, A. (2009). Cannabinoid modulation of hippocampal long-term memory is mediated by mTOR signaling. *Nature Neuroscience, 12*(9), 1152–1158.

Rapanelli, M., Frick, L. R., Pogorelov, V., Ota, K. T., Abbasi, E., Ohtsu, H., & Pittenger, C. (2014). Dysregulated intracellular signaling in the striatum in a pathophysiologically grounded model of Tourette syndrome. *European Neuropsychopharmacology, 24*(12), 1896–1906.

Robinson, M. J., & Franklin, K. B. (2007). Effects of anisomycin on consolidation and reconsolidation of a morphine-conditioned place preference. *Behavioral Brain Research, 178*(1), 146–153.

Russo, S. J., Dietz, D. M., Dumitriu, D., Morrison, J.H., Malenka, R. C., & Nestler E. J. (2010). The addicted synapse: Mechanisms of synaptic and structural plasticity in nucleus accumbens. *Trends in Neuroscience, 33*(6), 267–276.

Santini, E., Huynh, T. N., & Klann, E. (2014). Mechanisms of translation control underlying long-lasting synaptic plasticity and the consolidation of long-term memory. *Progress in Molecular Biology and Translational Science, 122*, 131–167.

Sanz, E., Yang, L., Su, T., Morris, D. R., McKnight, G. S., & Amieux, P. S. (2009). Cell-type-specific isolation of ribosome-associated mRNA from complex tissues. *Proceedings of the National Academy of Sciences of the United States of America, 106*(33), 13939–13944.

Scheetz, A. J., Nairn, A. C., & Constantine-Paton, M. (2000). NMDA receptor-mediated control of protein synthesis at developing synapses. *Nature Neuroscience, 3*(3), 211–216.

Scheyer, A. F., Wolf, M. E., & Tseng, K. Y. (2014). A protein synthesis-dependent mechanism sustains calcium-permeable AMPA receptor transmission in nucleus accumbens synapses during withdrawal from cocaine self-administration. *Journal of Neuroscience, 34*(8), 3095–3100.

Shi, J., Jun, W., Zhao, L. Y., Xue, Y. X., Zhang, X. Y., Kosten, T. R., & Lu, L. (2009). Effect of rapamycin on cue-induced drug craving in abstinent heroin addicts. *European Journal of Pharmacology, 615*(1–3), 108–112.

Shi, X., Miller, J. S., Harper, L. J, Poole, R. L., Gould, T. J., & Unterwald, E. M. (2014). Reactivation of cocaine reward memory engages the Akt/GSK3/mTOR signaling pathway and can be disrupted by GSK3 inhibition. *Psychopharmacology (Berlin), 231*(16), 3109–3118.

Shigeoka, T., Jung, H., Jung, J., Turner-Bridger, B., Ohk, J., Lin, J. Q.,...Holt C, E. (2016). Dynamic axonal translation in developing and mature visual circuits. *Cell, 166*(1), 181–192.

Sidrauski, C., Acosta-Alvear, D., Khoutorsky, A., Vedantham, P., Hearn, B. R., Li, H.,...Walter, P. (2013). Pharmacological brake-release of mRNA translation enhances cognitive memory. Elife, 2, e00498.

Smith, L. N., Jedynak, J. P., Fontenot, M. R., Hale, C. F., Dietz, K. C., Taniguchi, M.,...Cowan, C. W. (2014). Fragile X mental retardation protein regulates synaptic and behavioral plasticity to repeated cocaine administration. *Neuron, 82*(3), 645–658.

Sorg, B. A., Todd, R. P., Slaker, M., & Churchill, L. (2015). Anisomycin in the medial prefrontal cortex reduces reconsolidation of cocaine-associated memories in the rat self-administration model. *Neuropharmacology, 92*, 25–33.

Sorg, B. A., & Ulibarri, C. (1995). Application of a protein synthesis inhibitor into the ventral tegmental area, but not the nucleus accumbens, prevents behavioral sensitization to cocaine. *Synapse, 20*(3), 217–224.

Sutton, L. P., Caron, M. G. (2015). Essential role of D1R in the regulation of mTOR complex1 signaling induced by cocaine. *Neuropharmacology, 99*, 610–619.

Szalay, J. J., Jordan, C. J., & Kantak, K. M. (2013). Neural regulation of the time course for cocaine-cue extinction consolidation in rats. *European Journal of Neuroscience, 37*(2), 269–277.

Taubenfeld, S. M., Muravieva, E. V., Garcia-Osta, A., & Alberini, C. M. (2010). Disrupting the memory of places induced by drugs of abuse weakens motivational withdrawal in a context-dependent manner. *Proceedings of the National Academy of Sciences of the United States of America, 107*(27), 12345–12350.

Tiruchinapalli, D. M., Caron, M. G., & Keene, J. D. (2008). Activity-dependent expression of ELAV/Hu RBPs and neuronal mRNAs in seizure and cocaine brain. *Journal of Neurochemistry, 107*(6), 1529–1543.

Valjent, E., Bertran-Gonzalez, J., Aubier, B., Greengard, P., Herve, D., & Girault, J. A. (2010). Mechanisms of locomotor sensitization to drugs of abuse in a two-injection protocol. *Neuropsychopharmacology, 35*(2), 401–415.

Valjent, E., Corbille, A. G., Bertran-Gonzalez, J., Herve, D., & Girault, J. A. (2006). Inhibition of ERK pathway or protein synthesis during reexposure to drugs of abuse erases previously learned place preference. *Proceedings of the National Academy of Sciences of the United States of America, 103*(8), 2932–2937.

Ventura, R., Pascucci, T., Catania, M. V., Musumeci, S. A., & Puglisi-Allegra, S. (2004). Object recognition impairment in Fmr1 knockout mice is reversed by amphetamine: Involvement of dopamine in the medial prefrontal cortex. *Behavioral Pharmacology, 15*(5–6):433–442.

Verpelli, C., Piccoli, G., Zibetti, C., Zanchi, A., Gardoni, F., Huang, K., . . . Sala, C. (2010). Synaptic activity controls dendritic spine morphology by modulating eEF2-dependent BDNF synthesis. *Journal of Neuroscience, 30*(17), 5830–5842.

von der Goltz, C., Vengeliene, V., Bilbao, A., Perreau-Lenz, S., Pawlak, C. R., Kiefer, F., & Spanagel, R. (2009). Cue-induced alcohol-seeking behaviour is reduced by disrupting the reconsolidation of alcohol-related memories. *Psychopharmacology (Berlin), 205*(3), 389–397.

Wang, X., Luo, Y. X., He, Y. Y., Li, F. Q., Shi, H. S., Xue, L. F., . . . Lu, L. (2010). Nucleus accumbens core mammalian target of rapamycin signaling pathway is critical for cue-induced reinstatement of cocaine seeking in rats. *Journal of Neuroscience, 30*(38), 12632–12641.

Werner, C. T., Stefanik, M. T., Milovanovic, M., Caccamise, A., & Wolf, M. E. (2018). Protein translation in the nucleus accumbens is dysregulated during cocaine withdrawal and required for expression of incubation of cocaine craving. *Journal of Neuroscience, 38*(11), 2683–2697.

Wu, J., McCallum, S. E., Glick, S. D., & Huang, Y. (2011). Inhibition of the mammalian target of rapamycin pathway by rapamycin blocks cocaine-induced locomotor sensitization. *Neuroscience, 172*, 104–109.

Wu, Y., Li, Y., Yang, X., Sui, N. (2014). Differential effect of beta-adrenergic receptor antagonism in basolateral amygdala on reconsolidation of aversive and appetitive memories associated with morphine in rats. *Addiction Biology, 19*(1), 5–15.

Xue, Y. X., Chen, Y. Y., Zhang, L. B., Zhang, L. Q., Huang, G. D., Sun, S. C., . . . Lu, L. (2017). Selective inhibition of amygdala neuronal ensembles encoding nicotine-associated memories inhibits nicotine preference and relapse. *Biological Psychiatry, 82*(11), 781–793.

Xue, B., Fitzgerald, C. A., Jin, D. Z., Mao, L. M., & Wang, J. Q. (2016). Amphetamine elevates phosphorylation of eukaryotic initiation factor 2alpha (eIF2alpha) in the rat forebrain via activating dopamine D1 and D2 receptors. *Brain Research, 1646*, 459–466.

Yu, Y. J., Chang, C. H., & Gean, P. W. (2013b). AMPA receptor endocytosis in the amygdala is involved in the disrupted reconsolidation of methamphetamine-associated contextual memory. *Neurobiology of Learning and Memory*, *103*, 72–81.

Yu, Y. J., Huang, C. H., Chang, C. H., & Gean, P. W. (2016). Involvement of protein phosphatases in the destabilization of methamphetamine-associated contextual memory. *Learning and Memory*, *23*(9), 486–493.

Yu, F., Zhong, P., Liu, X., Sun, D., Gao, H.Q., & Liu, Q. S. (2013a). Metabotropic glutamate receptor I (mGluR1) antagonism impairs cocaine-induced conditioned place preference via inhibition of protein synthesis. *Neuropsychopharmacology*, *38*(7), 1308–1321.

CHAPTER 20

DYSREGULATED PROTEIN SYNTHESIS IN MAJOR DEPRESSIVE DISORDER

CHELCIE F. HEANEY AND
KIMBERLY F. RAAB-GRAHAM

Introduction

Major depressive disorder (MDD) is a heterogeneous disorder, wherein the symptoms and treatment efficacy are highly variable among individuals. Current treatments range from typical antidepressants, such as selective serotonin reuptake inhibitors (SSRIs), to electroconvulsive therapy. While these therapies have shown promise in some cases, only 35–50% of patients undergoing treatment experience relief from depressive symptoms, and the rate of relapse is high (Murrough, 2012; Rush et al., 2011; Undurraga & Baldessarini, 2012). Moreover, the onset of efficacy of typical antidepressants is delayed, thus increasing the risk of suicide (Hieronymus, Nilsson, & Eriksson, 2016). Within the last decade, glutamate N-methyl-D-aspartate receptor (NMDAR) antagonists, such as ketamine, have emerged as viable treatments for MDD due to their rapid onset of symptom relief, thus bridging the gap between the administration of typical antidepressants and the onset of their efficacy (Trullas et al., 1991). In addition, NMDAR antagonists show promise in *treatment-resistant* patients, for whom typical antidepressants are ineffective (Murrough, 2012). While ketamine appears to be the Holy Grail to treat

MDD, its abuse potential is high and it can induce psychosis at high doses (Y. Liu et al., 2016). Thus, the discovery of a safe pharmaceutical that elicits the same rapid-acting antidepressant effects of ketamine is necessary.

MDD Is a Disease of the Synapse

Patients with MDD exhibit alterations in neural tissue volume and connectivity within the frontal cortex and hippocampus, regions and circuits responsible for cognition, emotional regulation, and stress-responsiveness (Koolschijn et al., 2009; Bearden et al., 2009; Genzel et al., 2015). Specifically, these regions exhibit decreased volume and activity (Drevets, Price, & Furey 2008; Murrough et al., 2011), and patients with MDD also exhibit fewer synapses (Kang et al., 2012) and decreased synaptic proteins (Feyissa et al., 2009) within these regions. Preclinical rodent models of depression display similar neural alterations, including significant loss of dendrites, dendritic spines, and synaptic excitation (Liston et al., 2006; Radley et al., 2008; Yuen et al., 2012).

Together, these data suggest that depression is a disease of synapses, leading to a breakdown of communication between neurons. Thus, drugs that increase synaptic efficacy and new synapse formation are likely to also ameliorate MDD-associated symptoms. Research regarding the molecular basis of learning and memory provides insight into mechanisms that strengthen synapses and induce structural plasticity; namely, the necessity of local dendritic protein synthesis and repression (for review, see Graber, McCamphill, & Sossin, 2013). These studies provide a new understanding of and insight into the molecular bases and synaptic dysregulations underlying MDD.

MDD Is a mTORopathy

mTOR (mammalian target of rapamycin), a serine/threonine kinase, affects several signaling pathways, including those regulating cell growth, autophagy, and mRNA translation. mTOR is composed of two complexes, complex 1 (C1) and complex 2 (C2). mTORC1 activity regulates local protein synthesis in dendrites, and this specific role is required for long-term increases in synaptic strength, structural plasticity, and cognition (Graber, McCamphill, & Sossin, 2013), all domains that are affected in patients with MDD (Millan et al., 2012; Drevets, Price, & Furey, 2008; Murrough et al., 2011; Kang et al., 2012). For our purposes, we will focus solely on mTORC1.

mTORC1 has two primary targets that affect translation, S6 kinase (S6K1) and eukaryotic initiation factor 4E-binding protein (4E-BP1; see Figure 20.1, green). S6K1 is activated by phosphorylation, which in turn leads to increased translation. S6K1 regulates ribosomal protein S6, eukaryotic translation initiation factor 4B (eIF4B), and eukaryotic elongation factor 2 kinase (eEF2K; see Figure 20.1, dark blue). Conversely, 4E-BP1 phosphorylation by mTORC1 prevents it from binding to eIF4E, which enables cap-dependent translation (Hay & Sonenberg, 2004; Laplante & Sabatini, 2009), and animals with disrupted phosphorylation of eIF4E exhibit depressive-like behaviors and dysregulated serotonin (Aguilar-Valles et al., 2018). Previous studies demonstrate that serotonin increases the phosphorylation of 4EBP and activates S6K in an mTORC1-dependent manner in *Aplysia* neurons (Khan, Pepio, & Sossin, 2001; Carroll, Dyer, & Sossin 2006). Together, these data suggest that serotonin, eIF4E, and 4E-BP1 may participate in a feedback loop (see Figure 20.1, dark cyan). Not surprisingly, downregulation of the mTORC1 pathway—including mTOR and its downstream targets, p-S6K1, eIF4B, and p-eIF4B—as well as dysregulated serotonergic mechanisms are observed in patients with MDD (Jernigan et al., 2011; Ressler & Nemeroff, 2000). Thus, mTORC1 itself is a promising candidate to target as a means to treat depression.

Nearly 20 years after the discovery of the antidepressant effects of NMDAR antagonists (Trullas et al., 1991), low, subanesthetic doses of ketamine, which rapidly decrease MDD symptoms (Berman et al., 2000; Diazgranados et al., 2010), were discovered to activate mTOR and mTORC1-dependent protein synthesis of several synaptic proteins within 30 minutes of administration (N. Li et al., 2010). Ketamine's proposed antidepressant mechanism of action helps elucidate the importance of protein translation in treating MDD. Within 30 minutes of exposure, ketamine decreases the phosphorylation of eEF2 (effectively promoting translation) within the hippocampus, and a corresponding increase in brain derived neurotrophic factor (BDNF) expression is also observed (Autry et al., 2011). When animals are pretreated with a protein synthesis inhibitor, ketamine fails to produce an antidepressant-like behavioral effect, suggesting that eEF2-driven translation of BDNF is crucial to the antidepressant-like effects of ketamine (Autry et al., 2011). Interestingly, mTORC1-S6K activity leads to phosphorylation of eEF2K, inhibiting its action of repressing elongation through eEF2 (McCamphill, Ferguson, & Sossin, 2017). These data strongly point toward a role for mTOR relieving repression of elongation of BDNF mRNA. Transgenic animals with altered MAPK-extracellular regulated kinase (ERK)1/2-eIF4E phosphorylation respond similarly to control animals in a number of behavioral domains when treated with ketamine, demonstrating that the mechanism of action for ketamine is MAPK-ERK1/2-independent (Aguilar-Valles et al., 2018). Ketamine also strengthens synaptic connections and induces synaptic complexity within the prefrontal cortex (PFC) of an animal model. Specifically, at a dose that produced antidepressant-like behavioral effects, ketamine increased dendritic tufting, maturation, and excitatory responsiveness of PFC neurons. These crucial findings demonstrate that activation of the mTORC1 protein synthesis pathway underlies the antidepressant efficacy of ketamine, which further emphasizes the critical role of mTORC1-regulated protein synthesis in treating MDD, thus possibly rectifying the mTORC1-related disruptions observed in MDD.

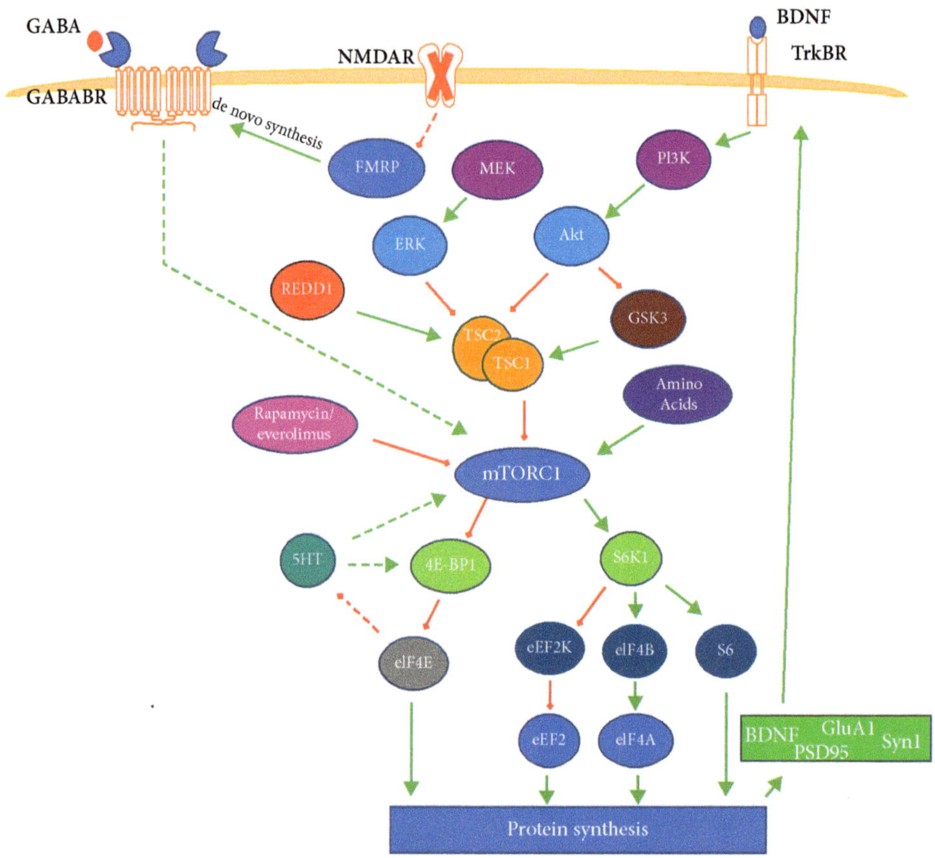

FIGURE 20.1. Several pathways influence the activity of the mTORC1 signaling pathway. mTORC1 affects protein synthesis via the inhibition or activation of several pathways. mTORC1 is activated by amino acids but can be inhibited by the compound rapamycin or the tuberous sclerosis complex (TSC). TSC is activated by REDD1 and GSK3. ERK and Akt inhibit TSC. Akt also inhibits GSK3. Akt is activated by PI3K via BDNF-activated TrkB receptors. Here, NMDAR antagonism is being depicted as inhibiting FMRP expression, which enables the translation of GABABR mRNA and de novo protein synthesis of GABABRs. New GABABRs are coupled to L-type calcium channels, which leads to mTORC1 activity. mTORC1 has two primary targets that affect translation, S6 kinase (S6K1 and downstream ribosomal protein S6) and eukaryotic initiation factor 4E-binding protein (4E-BP1), regulating translation initiation (eIF4E and eIF4B) and elongation (eEF2K/eEF2). Serotonin (5HT) can activate S6K in an mTORC1-dependent manner and phosphorylate 4EBP; affecting 4E-BP1 phosphorylation of eIF4E dysregulates 5HT. The mTORC1 pathway is implicated in promoting the synthesis of the synaptic proteins PSD95, Synapsin1 (Syn1), and GluA1, as well as BDNF, which can then feedforward to activate mTORC1 via TrkB receptors.

Red lines with diamond ends indicate inhibition; green lines with arrow heads indicate activation. Dashed lines indicate pathways that have not yet been fully determined.

Preclinical Models Provide Further Evidence for mTORC1-Regulated Protein Synthesis as an Effective Treatment of MDD

Preclinical models of depression are extremely valuable, as they allow scientists to probe for (1) the underlying neurobiological substrates of depression and (2) the efficacy of antidepressant medications at the molecular, cellular, and behavioral level. Animal models of depression typically involve exposure to chronic mild stress (CMS), repeated social defeat, or chronic administration of stress hormones. Stressed animals exhibit neuronal abnormalities such as altered dendritic morphology, decreased spine number, and decreased evoked excitatory responsiveness, similar to human patients with MDD (N. Li et al.,; Radley et al., 2006; Radley et al., 2008; R.-J. Liu & Aghajanian, 2008). Further, stress decreases the phosphorylation and activity of the mTORC1 signaling pathway with a corresponding reduction in the expression of synaptic proteins regulated by mTORC1, including PSD95 (postsynaptic density protein 95), Synapsin1, and GluR1 (α-amino-3-hydroxy-5-methyl-4-isoxazolepropionic acid [AMPA] glutamate receptor subunit 1), similar to what is seen in postmortem tissue from patients with MDD (Feyissa et al., 2009). Importantly, stress-induced reduced protein expression and dendritic/synaptic abnormalities are reversed with the administration of ketamine (N. Li et al., 2011). Moreover, when mTORC1 signaling is blocked with rapamycin (see Figure 20.1, pink), the positive effects of ketamine at the molecular, cellular, and behavioral level are blocked.

Despite the many data collected by many laboratories, ketamine's ability to activate mTORC1 was puzzling. These data seemed paradoxical because NMDAR activation is usually required for mTORC1-dependent signaling, synaptic strengthening, and increased spine density (Raab-Graham & Niere, 2017). NMDARs are tetrameric ligand- and voltage-gated cation channels and consist of several subunit types: GluN1, GluN2 (GluN2A-D), and GluN3 (GluN3A–B). Functional receptors require two GluN1 and two GluN2 or GluN3 subunits, with different configurations affecting region specificity, kinetics, open probability, and deactivation time (Paoletti, Bellone, & Zhou, 2013). Interestingly, GluN2A- and GluN2B-containing receptors are most prominent within the hippocampus and PFC (Paoletti, Bellone, & Zhou, 2013). NMDARs are considered "coincidence detectors," since both conditions of ligand-binding and membrane depolarization must occur to displace a magnesium ion that blocks the ion channel pore so that calcium may enter the cell via the receptor. This calcium then activates a number of signaling cascades, including initiating the mTORC1 translational pathway and synaptic strengthening. Although both GluN2A- and GluN2B-containing receptors generate large calcium currents when activated, GluN2B-mediated currents are longer lasting (Bloodgood, Sabatini, & Van Dongen 2009). In developing synapses,

NMDAR-mediated calcium influx activates calcium-dependent eEF2K, which then phosphorylates eEF2 (Scheetz, Nairn, & Constantine-Paton, 2000). This phosphorylation may slow the local rate of protein translation, and GluN2B-mediated calcium may keep eEF2K active longer than GluN2A-mediated conductance, thus leading to greater inhibition of global translation. Elongation, then, instead of initiation, consequently would become the rate-limiting step in protein synthesis.

Ketamine acts to block the channel pore and prevents calcium from entering the cell. The best evidence, however, that the activation of mTORC1 signaling mediates the antidepressant-like effects of ketamine was provided by studies that demonstrated that not all NMDAR antagonists exhibit antidepressant-like effects. For example, memantine, a partial NMDAR antagonist that also blocks the channel pore and which is used to treat Alzheimer's disease, does not promote antidepressant-like behavioral effects (Zhang et al., 2017; Gideons, Kavalali, & Monteggia, 2014). Memantine appears to more effectively affect GluN2A-containing NMDARs, whereas ketamine is more effective at GluN2B-containing NMDARs (Glasgow et al., 2017), thus suggesting that ketamine acts to decrease the GluN2B-mediated eEF2 phosphorylation, whereas memantine might increase or not alter this effect. Transgenic animals with cortical GluN2B subunits removed exhibit decreased behavioral despair in the tail suspension test (TST) similar to ketamine-treated animals, and these transgenic animals are not affected by ketamine treatment (Miller et al., 2014). Further, other GluN2B-specific antagonists are effective at treating both animal models of MDD and human patients with MDD (Maeng et al., 2008; Preskorn et al., 2008). What is more, memantine does not increase phosphorylation of mTORC1 (Zhang et al., 2017), nor does it affect the phosphorylation of eEF2K like ketamine, thus rendering memantine unable to promote activity of the mTORC1 signaling pathway and protein synthesis (Zhang et al., 2017; Gideons, Kavalali, & Monteggia, 2014). Thus, crucial differences between memantine and ketamine include subunit specificity and the ability to initiate mTORC1-dependent protein synthesis. Together, these data provide strong support for the hypothesis that the antidepressant-like effects of NMDAR antagonists rely on activity of the mTORC1 signaling pathway.

If this hypothesis is correct, then one can predict how altering genes downstream of mTORC1 signaling will affect animal models of MDD. For instance, knocking out one such gene would correspond to dendritic and spine deficits, as well as depressive-like behaviors. Indeed, similar to patients with MDD (Jernigan et al., 2011), genetically knocking down S6K1 in the medial PFC of animals generates depressive-like behaviors in the absence of stress (Dwyer et al., 2015). Reduced S6K1 activity (and thus, decreased activity of the mTORC1 signaling pathway) also blocks the antidepressant-like effects of ketamine on the behavior of these animals. Conversely, overexpression of S6K1 (and, thus, increased activity of the mTORC1 signaling pathway) produces antidepressant-like behaviors and is protective against stress-induced behavioral changes. Also, increased S6K1 activity increases neuronal complexity via neuronal branch number compared to controls. Collectively, these data strongly support the hypothesis that mTORC1 signaling pathway and, thus, mTORC1-dependent protein synthesis is disrupted in MDD. Therefore, endogenous *upstream* targets that regulate mTORC1 may also be good candidates for pharmacological treatment of MDD.

One such potential target is REDD1 (regulated in development and DNA damage responses 1), an upstream inhibitor of mTORC1 (see Figure 20.1, red). REDD1 is increased in the dorsolateral PFC of patients with MDD, and CMS increases mRNA and protein expression of REDD1 in the PFC of rats (Ota et al., 2014). These increases are also associated with decreased expression of p-S6K, p-4EBP1, and p-Akt. Further, REDD1 knockout (KO) mice display a stress resilient phenotype wherein stress exposure does not decrease sucrose preference in these animals, nor does stress affect p-S6K or p-4EBP expression in the PFC (Ota et al., 2014). In addition, while stress exposure promotes decreased spine density, this effect is not seen in REDD1 KO mice. In contrast, REDD1 overexpression decreases p-S6K and p-mTOR in the PFC and leads to decreased spine density (Ota et al., 2014). Also, overexpression of REDD1 leads to a pro-depressive phenotype even in the absence of stress. Thus, inhibiting REDD1, and disinhibiting the mTORC1 signaling pathway, may help treat depressive-like symptoms.

Importantly, there are reports of NMDAR antagonists improving performance in behavioral tests of despair, motivation, and self-care, without prior stress. A single administration of ketamine (N. Li et al., 2010) or Ro 25-6981, a GluN2B-specific antagonist (Workman, Niere, & Raab-Graham, 2013), decreases immobility in the forced swim test (FST) and TST, indicating less behavioral despair. Workman et al. (2015) also demonstrated that a single administration of Ro 25-6981 increases grooming frequency in the splash test, a behavior that is a measurement of self-care. Importantly, co-administration of rapamycin, an mTORC1 inhibitor, blocks these behavioral effects. These drugs improve performance in tasks without chronic administration or without prior exposure to stress, which may be a result of rapidly boosting proteins that mediate antidepressant-like effects through mTORC1 activation.

Does Inhibition of mTORC1-Regulated Protein Synthesis Promote MDD?

Rapamycin and rapamycin analogs, like everolimus, are FDA approved treatments for disorders where mTORC1 signaling is overactive (see Figure 20.1, pink). However, the hypothesis that the mTORC1 signaling pathway is disrupted in MDD predicts that chronic administration of mTORC1 inhibitors may cause depressive-like behaviors. A recent study by Russo et al. (2016) addresses this particular prediction. The authors treated mice with synthetic glucocorticoids (a common method used to activate and increase endogenous stress mechanisms) with or without everolimus in their drinking water and examined performance in measures of cognition, despair, and anxiety. While mice treated with everolimus exhibited enhanced cognitive performance, and rescued the cognitive impairment from the glucocorticoid treatment, the mice also exhibited increased anxiety- and depressive-like behaviors. These data suggest that caution should be exercised with the use of mTOR inhibitors in neurological disorders such as autism spectrum disorder and Alzheimer's disease, two disorders with comorbid depression (Starkstein et al., 2005; Simonoff et al., 2008).

Where in the Brain Should mTORC1 Be Activated to Treat MDD?

mTOR is ubiquitously expressed throughout the body and overactive mTORC1 is implicated in many diseases, including cancer, autism spectrum disorder, and Alzheimer's disease (for review, see Siddiqui & Sonenberg, 2015 and Raab-Graham & Niere, 2017). Thus, a potential challenge is to design pharmaceuticals that adjust mTORC1 activity within an optimal range without causing somatic problems. A necessary first step is to determine the brain regions that exhibit decreased mTORC1 activity due to MDD. Preclinical animal models of depression are extremely useful because they exhibit alterations to mTORC1 activity within brain regions that parallel those affected in human patients with MDD, as detailed earlier. However, not all of the data agree on which regions are affected. For example, Chandran et al. (2013) examined whether CMS induces changes to the mTORC1 signaling pathway within the frontal cortex, hippocampus, amygdala, and dorsal raphe. The authors found that only the amygdala showed stress-related effects. In addition to decreased expression of phosphorylated mTOR, p-70S6K, and S6, stress-exposed animals also displayed decreased ERK1/2, and Akt/ protein kinase B (Akt; see Figure 20.1, light blue). Moreover, the authors measured p-GluR1, as an indicator for surface-expressed and thus active glutamate receptors, and observed a parallel reduction in the stressed animals. These data imply that the stress-induced reduction in AMPAR signaling and synaptic efficacy are specific to the amygdala (Chandran et al., 2013). In contrast, animals exposed to the FST exhibit less hippocampal p-mTOR and BDNF compared to animals treated with ketamine, which also decreased immobility in the FST (Yang, Hu, et al., 2013). Consistent with these data, other studies indicate that the PFC and hippocampus are sensitive to the effects of NMDAR antagonist treatment. For instance, in a study that examined proteomic changes between non-stressed rats treated with ketamine, ketamine treatment increased hippocampal expression of mTOR in treated animals compared to controls (Wesseling et al., 2015). Similar to ketamine, treatment with Ro 25-6981 also increases the phosphorylation of mTOR, mTOR-dependent synthesis of BDNF, and GluA1 in the PFC in parallel with reducing behavioral despair, as measured by the FST (Workman, Niere, & Raab-Graham, 2013).

Niere et al. (2016) showed that the largest effect of mTORC1 activity is remodeling protein composition at the synapse. One difference between the studies discussed in the previous paragraph is which neuronal fractions were assayed. Chandran et al. (2013) assayed total cellular lysates, rather than synaptic fractions, of the brain regions they examined. Assessment of the total cell lysate may not be sensitive enough to detect changes to mTORC1 activity apparent at the synapse. These data argue that targeting mTORC1 in specific brain regions, and perhaps only at synapses, may provide the most effective results.

It should also be noted that the target mRNAs translated upon mTORC1 activation may vary across brain regions. For example, chronic inhibition of calcineurin, a calcium and calmodulin-dependent serine/threonine phosphatase, in the PFC promotes depressive-like behaviors in animals and downregulates the mTORC1 signaling pathway (Yu et al., 2013). Ketamine treatment increases the expression of the regulatory subunit of calcineurin, Ppp3cb, in the frontal cortex of rats; however, this treatment also downregulates Ppp3cb in the hippocampus (Wesseling et al., 2015). Further, others have demonstrated that Ppp3cb expression increases in the cortex with mTOR activity (Niere et al., 2016). These data add to the complexity of understanding MDD therapies because, even if mTORC1 is activated both in the PFC and hippocampus, mRNAs may be differentially regulated in MDD and by ketamine.

Targeting Protein Synthesis Pathways via Traditional Antidepressants and Other Treatments

Other drugs with antidepressant-like effects may produce their effects via the mTORC1 signaling pathway and mTORC1-dependent protein synthesis. However, as in the case of typical antidepressants, little is known about their mechanisms of action. Thus, in order to determine whether several typical antidepressants utilize the mTORC1 signaling pathway, Park et al. (2014) examined the effect of chronic treatment on primary rat hippocampal neurons. The authors found that, similar to ketamine, chronic exposure to escitalopram (SSRI), paroxetine (SSRI), and tranylcypromine (MAO inhibitor) increases the phosphorylation of several mTORC1 pathway markers, including mTOR, 4EBP1, and p70S6K. Further, phosphorylation of upstream activators of the mTOR pathway, Akt and ERK, also increased with chronic treatment of these antidepressants. The ability for the antidepressants to increase expression of p-mTOR was blocked with PI3K (phosphoinositide 3 kinase), MEK (mitogen-activated protein kinase kinase, also known as MAP2K), and mTOR inhibitors (see Figure 20.1, light purple). Similar to ketamine, these antidepressants also increase dendritic outgrowth and branching in an mTORC1-dependent manner. Increases in the synaptic markers PSD95 and synaptophysin were also found to be mTORC1-dependent.

In agreement with these studies, proteomic studies conducted by Shen et al. (2017) sought to determine the molecular mechanisms associated with venlafaxine (SNRI) treatment. The authors administered venlafaxine to mice and then compared differences in hippocampal metabolites between treated and control animals. The authors identified 27 significantly different metabolites via gas chromatography-mass spectrometry (GC-MS), for which they then developed a molecular interaction network. Of the pathways identified, the MAPK-ERK1/2 and PI3K-Akt pathways were highly correlated with the differentially expressed metabolites. Finally, the authors verified via western blotting that targets within each pathway were differentially affected between the control and treated groups. They found significant increases in hippocampal ERK1/2, p-Akt, CREB, BDNF, p-C-Raf, and p-MEK in the venlafaxine-treated animals (Shen et al., 2017).

Several other typical antidepressants, however, exhibit different effects. The antidepressants fluoxetine (SSRI), sertraline (SSRI), and imipramine (tricyclic) increase the phosphorylation of Akt and ERK without affecting mTORC1 or its downstream targets in primary rat hippocampal neurons (Park et al. 2014), agreeing with previous research showing no effect on the phosphorylation of mTOR, 4E-BP1, or p-S6K in the PFC (N. Li et al., 2010). It should be noted that these antidepressants also increase dendritic complexity and synaptic markers, but these results are unaffected by mTORC1 inhibition (Park et al., 2014). Also, there may be discrete regional effects of these antidepressants. For example, chronic in vivo fluoxetine administration (21 days) increases phosphorylation of both eIF4E and eEF2 within the dentate gyrus. eEF2 phosphorylation was also increased within the hippocampus and PFC (Dagestad et al., 2006). While the conclusions from Park et al. (2014) are limited because they only investigated the effects of these antidepressants on the mTORC1 pathway in cultured hippocampal neurons, the authors suggest that these three antidepressants may activate CREB and BDNF via the ERK pathway and eEF2 in an mTORC1-independent manner to increase synaptic markers and dendritic complexity. However, transgenic animals with altered MAPK-ERK1/2-eIF4E phosphorylation do not exhibit a different response to fluoxetine treatment compared to controls, thus suggesting that fluoxetine does not affect eIF4E through the MAPK-ERK1/2 pathway (Aguilar-Valles et al., 2018). As mentioned earlier, because ketamine has a high abuse potential, it is not prudent to utilize this drug to treat a disorder that is marked with high comorbid substance use disorders (Kessler et al., 2003). Thus, researchers are continuously investigating other drugs that may similarly affect the mTORC1 signaling pathway and produce antidepressant-like effects. Dwyer et al. (2012) demonstrate that the mGluR2/3 antagonist LY 341495 also increases mTORC1 signaling targets and synaptic proteins in PFC synaptoneurosomes similar to ketamine. Further, LY 341495 decreases immobility in the FST in an mTORC1-dependent manner. Similar increases to synaptic proteins within hippocampal synaptoneurosomes are reported; however, not all of the same effects seen in the PFC are evident in this hippocampal preparation (Dwyer, Lepack, &Duman, 2012). These differences could reflect either region-specific effects of the stress or drug treatment.

Another avenue researchers are exploring is augmenting the effects of low, sub-effective doses of ketamine with other drugs that lack abuse potential. Co-administration of GSK-3β (a serine/threonine kinase upstream of mTORC1) inhibitors with sub-effective doses of ketamine decreases immobility in the FST, similar to the effects of higher ketamine doses (R.-J. Liu et al., 2013; see Figure 20.1, dark red). In addition, lithium administered with ketamine increases phosphorylation of PFC synaptosomal GSK-3β, mTORC1 signaling pathway markers (mTOR, p-S6K), as well as Akt and ERK, with some changes lasting up to 24 hours post-administration. Lithium and ketamine co-administration also increases synaptic responses and dendritic complexity (R.-J. Liu et al., 2013).

Interestingly, while most accept that ketamine drives its antidepressant-like effect via NMDAR antagonism, some data suggest that ketamine's active metabolite, (2S,6S;2R,6R)-hydroxynorketamine (HNK), mediates the drug's antidepressant properties (Kavalali & Monteggia, 2018). Zanos et al. (2016) demonstrate that administration of HNK promotes antidepressant-like behavioral effects in mice in an NMDAR inhibition-independent manner. The authors also report that neither ketamine nor HNK administration affected phosphorylation of mTOR within hippocampal or PFC synaptosomes, unlike previous reports detailed earlier. Further, they demonstrated that both ketamine and HNK decrease hippocampal phosphorylation of eEF2, but increase expression of BDNF, GluA1, and GluA2. Again, in contrast to others' findings, increased expression of these proteins occurred only in hippocampal synaptosomes and not PFC, and only at 24 hours post-administration (Zanos et al., 2016). Importantly, the authors suggest that HNK does not have the adverse side effects of ketamine. However, independent groups suggest that HNK can block NMDARs and have been unable to replicate the antidepressant-like effects of HNK on behavior (Suzuki et al., 2017; Yang et al., 2017; Shirayama & Hashimoto, 2018).

The mTORC1 pathway is also activated by amino acids (see Figure 20.1, dark purple). Baranyi et al. (2016) investigated whether patients with MDD exhibited any changes to branched chain amino acids (BCAA) in their plasma. They found that compared to controls, patients with MDD have significantly lower levels of the BCAAs valine, leucine, and isoleucine, which would lead to decreased activation of the mTORC1 signaling pathway and protein synthesis. Further, the authors report a negative correlation between the concentration of these BCAAs and scores on self-report and observer-based psychiatric assessments. These data corroborate previous research that identified an increase in amino acids after chronic antidepressant treatment in animals (Webhofer et al., 2011).

Exercise, which is often touted as a non-pharmaceutical treatment of MDD, also rescues deficits in the mTORC1 signaling pathway after stress. Fang et al. (2013) demonstrate that immobilization stress reduces BDNF, phosphorylation of TrkB, Akt, GSK-3β, mTOR, and p-70S6K within the hippocampus. Additionally, the authors found decreased expression of the synaptic markers synaptophysin, PSD95, b-neurexin, and neuroligin in response to stress. However, treadmill exercise reversed the effect of stress on these markers, providing some evidence for the effectiveness of exercise as an MDD treatment.

Collectively, these data argue that, regardless of the upstream signal, the key unifying target for relief of MDD symptoms is the activation of protein synthesis pathways that increase synaptic proteins and synaptic strength.

Targeting mRNA Translational Repression Factors to Mitigate MDD

RNA-binding proteins (RBPs) and microRNAs (miRs), small noncoding RNAs, scan the transcriptome to silence mRNA translation of *specific* transcripts (Raab-Graham & Niere, 2017). Thus, manipulating levels of translational repressors may be an additional avenue to alleviate depression. The following sections describe recent reports indicating the roles of translational repressors in MDD and how they may be targeted to reverse depressive-like symptoms.

Establishing the Function of the RBP Fragile X Mental Retardation Protein in MDD

Fragile X mental retardation protein (FMRP) is one of the most well characterized RBPs. FMR1 premutation carriers, people without fragile X syndrome (FXS) but who are likely to have children with FXS, have higher rates of MDD and depressed mood (Johnston et al., 2001; Roberts et al., 2009). These data suggest that insufficient levels of FMRP can lead to MDD. Consistent with this finding, patients with MDD express reduced levels of FMRP in the lateral cerebellum (Fatemi et al., 2013). Recent work characterizing the antidepressant pathway of NMDAR antagonists demonstrates that de novo protein synthesis of the metabotropic inhibitory neurotransmitter receptor gamma-aminobutyric acid B (GABABR) drives the activation of mTORC1 necessary for antidepressant effects (Workman, Niere, & Raab-Graham, 2013; Workman et al., 2015; see Figure 20.1). GABABR mRNA is a target of FMRP (Darnell et al., 2011) and *Fmr1* KO mice exhibit increased basal levels of GABABR protein in hippocampal synaptosomes (Wolfe et al., 2016). NMDAR antagonists promote the synthesis of new GABABRs and their surface expression (Workman, Niere, & Raab-Graham, 2013; Workman et al., 2015; Wolfe et al., 2016), however, this antidepressant-mediated effect is absent in cultured primary hippocampal neurons from *Fmr1* KO mice (Wolfe et al., 2016). Further, the antidepressant-like effects associated with NMDAR antagonists require GABABR activation (Workman, Niere, & Raab-Graham, 2013; Workman et al., 2015) and is absent in *Fmr1* KO mice (Wolfe et al., 2016). Thus, understanding how NMDAR antagonism promotes the translation of FMRP-targeted mRNAs may lead to new avenues of MDD treatments.

Targeting the miR Landscape and Downstream Protein Synthesis in the Brain to Treat MDD

Many drugs alter the expression of miRs, usually resulting in the inverse expression of their mRNA targets. Importantly, many miRs are specific to the brain, and specifically targeting them avoids the problem of somatic side effects associated with mTOR activity. One of the best examples of this interaction is demonstrated by Lopez et al. (2014). First, the authors demonstrated that the expression of miR-1202 is decreased in the ventrolateral PFC in patients with MDD compared to controls. Next, they showed that miR-1202 is primate-specific and enriched in the brain compared to other tissues. Further, the mRNA of a predicted target of miR-1202, metabotropic glutamate receptor 4 (GRM4), was demonstrated to be negatively correlated with the expression of miR-1202 within the PFC. The authors then investigated the effect of antidepressants on the expression of miR-1202 and GRM4 in patients with MDD. Regardless of antidepressant use, miR-1202 was decreased in patients with MDD compared to controls; however, within the MDD group, miR-1202 was increased in patients being treated with antidepressants. Further, GRM4 was upregulated in brain tissue from patients with MDD who were not being treated with antidepressants. Importantly, however, antidepressant use mitigated differences in GRM4 expression between controls and MDD patients. The authors went on to demonstrate a bidirectional interaction between miR-1202 and GRM4, and found that only antidepressants that have direct effects on serotonin or serotonin transporters regulate miR-1202 expression. Finally, they demonstrated that miR-1202 expression was decreased in MDD patients who were treatment naïve compared to controls. Importantly, in patients who were treated and whose depression was alleviated, miR-1202 expression was increased. They further found a negative correlation between miR-1202 expression and severity of MDD symptoms. All together, these data strongly support targeting miR-1202 as a treatment for MDD.

Higuchi et al. (2016) demonstrated that chronic ultra-mild stress (similar to chronic mild stress, but without food or water deprivation, and no nociceptive events) decreased expression of miR-124, which was rescued with chronic treatment of imipramine. Hippocampal overexpression of miR-124 increased resilience to stress, whereas inhibition led to greater reactivity to mild stress. To further demonstrate the protective role of miR-124, overexpression of miR-124 prevented stress-induced increases in GSK-3β mRNA and protein. In contrast, Bahi et al. (2014) found that socially defeated rats express increased hippocampal miR-124 and downregulated BDNF, a direct target of miR-124. These authors also found that overexpression of miR-124 in the hippocampus increased depressive-like behaviors, whereas inhibition or infusion of BDNF decreased these behaviors. Such findings emphasize the heterogeneity of MDD and suggest that finding specific biomarkers that would indicate effectiveness of specific antidepressant treatment is desperately needed.

Detection of Peripheral Biomarkers

Biomarkers for MDD are needed, but *how* and *what* to assess in patients with MDD has remained elusive. However, in a letter to the editor, Yang et al. (2013) describe peripheral changes to mTORC1 signaling targets after administering ketamine to three patients with MDD. Over the course of 2 hours, p-mTOR, p-GSK-3β, and p-eEF2 increased above baseline, whereas scores on self-reported and observer-based measures of depressive symptoms steadily decreased. Interestingly, Li et al. (2013) found a significant increase in GSK-3β mRNA in blood from patients with MDD prior to antidepressant treatment, compared to controls. This increase was brought to control baseline levels with 8 weeks of escitalopram (SSRI) treatment. These results are unexpected and promising. The phosphorylation state of these proteins is a good readout for mTORC1 activity and suggests that increased activity of mTORC1 in the brain can be also observed in the blood. In contrast, mRNA levels do not always correlate with protein levels, perhaps explaining why Li et al. (2013) detected less GSK-3β mRNA in the blood while Yang et al. (2013) detected increased levels of the active form of the protein. Nonetheless, these data are promising and thus provide potential markers to screen for antidepressant efficacy in humans.

Importantly, miRs that are dysregulated in the brain can often be detected peripherally. miR-132 represses the expression of BDNF and MeCP2, an important enzyme that methylates DNA and silences transcription (Su et al., 2015; Zimmermann et al., 2015). Thus, changes in miR-132 in the brain are likely to have short-term (via BDNF mRNA repression) and long-term effects (MeCP2) on depression. A recent study examined the relationship between the expression of BDNF, MeCP2, and miR-132 in the blood of patients with MDD (Su et al., 2015). They found an increase in miR-132 and a corresponding decreased BDNF and MeCP2, as would be predicted. To further verify the relationship between BDNF, MeCP2, and miR-132, Su et al. (2015) examined changes to these targets by using the chronic unpredictable stress animal model. They found that stress increased miR-132 expression in the hippocampus but decreased total protein expression of MeCP2 and BDNF, similar to what was observed in the blood from patients with MDD. Capitalizing on an animal model, they overexpressed miR-132, which further decreased MeCP2 and BDNF expression. In contrast, when they downregulated miR-132, there was a corresponding increase in MeCP2 and BDNF.

More recently, Lopez et al. (2017) examined the effects of treating MDD with desvenlafaxine (SNRI) on peripheral miR-1202 and neural activity associated with a Go/NoGo task. They compared performance and miR-1202 expression prior to antidepressant treatment and after 8 weeks of treatment. The researchers found that after treatment, peripheral miR-1202 levels were correlated with altered network connectivity in task-based and resting-state activity, suggesting that desvenlafaxine may modulate network activity through the glutamatergic system.

Assessment of peripheral miRs may help design personalized treatment for depression. For example, Issler et al. (2014) found that overexpression of miR-135a in serotonergic neurons promotes stress resiliency and altered serotonin metabolism, but also prevents stress-related reductions to serotonin. Further, they found that miR-135a is decreased peripherally, as well as in the dorsal raphe nucleus in patients with MDD and in suicide victims. While SSRI treatment did not affect peripheral miR-135, the authors demonstrate that cognitive behavioral therapy increased miR-135 expression. The authors did not report efficacy of either treatment in decreasing symptoms associated with

MDD. However, these studies do raise the question of whether peripheral detection of miRs can predict which treatment course will be most effective for individuals with MDD. Together, these data suggest that miRs are useful peripheral biomarkers for MDD, and that further validation of miRs that are dysregulated in MDD will help researchers identify potential treatments for MDD. In addition, identifying miRs and their targets that are affected in MDD will help researchers further understand the underlying etiology of MDD.

Conclusion

Together, the data presented here demonstrate that effective antidepressant treatments influence protein synthesis, either by directly targeting protein synthesis pathways (such as the mTORC1 signaling pathway) or by targeting translation repressors (such as RNA binding proteins or microRNAs). What is more, these treatments aim to increase synaptic functioning and appear to combat the loss of neural volume and disrupted circuitry. One remaining issue, however, is the delayed onset of efficacy of traditional antidepressants. Researchers are currently studying the mechanisms behind ketamine's rapid effects in order to develop new therapeutics that are both fast-acting and non-addictive.

Future Directions

1. Developing effective targets for treatment that avoid abuse potential and negative side effects by specifically targeting brain regions and mapping how antidepressants work in specific regions and on which transcripts.
2. Because MDD is heterogenous, developing personalized therapeutics with programs like the Connectivity Map (CMAP) may help optimize treatment based on mRNA profiles rather than general chemical transduction.
3. Using membrane potential as a guide for treatment, since many ion channels are expressed only in the brain. Pharmaceutical companies have designed many drugs that target specific ion channels; could these drugs be repurposed?
4. Some populations with comorbid MDD may be unresponsive to current treatments. For example, people with FXS lack FMRP, which is required for the effects of rapid antidepressants with NMDAR antagonists. The neurobiology of MDD in these populations need to be studied independently.
5. MDD that appears during adolescence may arise differently than in adults. This population may also need to be studied independently in order to determine how the etiology of MDD is different, and whether different drug treatment strategies are needed.

6. People with MDD are more likely to also develop a substance abuse disorder. Is this comorbidity due to people with MDD self-medicating with alcohol and/or illicit drugs, or does MDD, the disorder itself, affect neurobiology in such a way to make people more susceptible to addiction? Can mechanisms that underlie antidepressant efficacy provide insight into treating substance abuse disorders?

References

Aguilar-Valles, A., Haji, N., De Gregorio, D., Matta-Camacho, E., Eslamizade, M. J., Popic, J., Sharma, V., et al. (2018). Translational control of depression-like behavior via phosphorylation of eukaryotic translation initiation factor 4E. *Nature Communications*, *9*(1), 1–14. Article No. 2459. https://doi.org/10.1038/s41467-018-04883-5.

Autry, A. E., Adachi, M., Nosyreva, E., Na, E. S., Los, M. F., Cheng, P. F., . . . Monteggia, L. M. (2011). NMDA receptor blockade at rest triggers rapid behavioural antidepressant responses. *Nature*, *475*(7354), 91–95. https://doi.org/10.1038/nature10130.

Bahi, A., Chandrasekar, V., & Dreyer, J.-L. (2014). Selective lentiviral-mediated suppression of microRNA124a in the hippocampus evokes antidepressants-like effects in rats. *Psychoneuroendocrinology*, *46*(August), 78–87. https://doi.org/10.1016/j.psyneuen.2014.04.009.

Baranyi, A., Amouzadeh-Ghadikolai, O., von Lewinski, D., Rothenhäusler, H.-B., Theokas, S., Robier, C., . . . Meinitzer, A. (2016). Branched-chain amino acids as new biomarkers of major depression: A novel neurobiology of mood disorder. *PLoS One*, *11*(8), e0160542. https://doi.org/10.1371/journal.pone.0160542.

Bearden, C. E., Thompson, P. M., Avedissian, C., Klunder, A. D., Nicoletti, M., Dierschke, N., . . . Soares, J. C. (2009). Altered hippocampal morphology in unmedicated patients with major depressive illness. *ASN Neuro*, *1*(4), AN20090026. https://doi.org/10.1042/AN20090026.

Berman, R. M., Cappiello, A., Anand, A., Oren, D. A., Heninger, G. R., Charney, D. S., & Krystal, J. H. (2000). Antidepressant effects of ketamine in depressed patients. *Biological Psychiatry*, *47*(4), 351–354. https://doi.org/10.1016/S0006-3223(99)00230-9.

Bloodgood, B. L., Bernardo, L., Sabatini, & Van Dongen, A. M. (2009). NMDA receptor-mediated calcium transients in dendritic spines. In Antonius M. Van Dongen (Ed.), *Biology of the NMDAR Receptor*. Boca Raton, FL: CRC Press/Taylor & Francis.

Carroll, M., Dyer, J., & Sossin, W. S. (2006). Serotonin increases phosphorylation of synaptic 4EBP through TOR, but eukaryotic initiation factor 4E levels do not limit somatic cap-dependent translation in Aplysia neurons. *Molecular and Cellular Biology*, *26*(22), 8586–8598. https://doi.org/10.1128/MCB.00955-06.

Chandran, A., Iyo, A. H., Jernigan, C. S., Legutko, B., Austin, M. C., & Karolewicz, B. (2013). Reduced phosphorylation of the mTOR signaling pathway components in the amygdala of rats exposed to chronic stress. *Progress in Neuro-Psychopharmacology and Biological Psychiatry*, *40*(January), 240–245. https://doi.org/10.1016/j.pnpbp.2012.08.001.

Dagestad, G., Kuipers, S. D., Messaoudi, E., & Bramham, C. R. (2006). Chronic fluoxetine induces region-specific changes in translation factor eIF4E and eEF2 activity in the rat brain. *European Journal of Neuroscience*, *23*(10), 2814–2818. https://doi.org/10.1111/j.1460–9568.2006.04817.x.

Darnell, J. C., Van Driesche, S. J., Zhang, C., Hung, K. Y. S., Mele, A., Fraser, C. E., Stone, E. F., et al. (2011). FMRP stalls ribosomal translocation on mRNAs linked to synaptic function and autism. *Cell*, *146*(2), 247–261. https://doi.org/10.1016/j.cell.2011.06.013.

Diazgranados, N., Lobna, A. I., Brutsche, N. E, Ameli, R., Henter, I. D., Luckenbaugh, D. A., Machado-Vieira, R., & Zarate, C. A. (2010). Rapid resolution of suicidal ideation after a single infusion of an N-Methyl-D-Aspartate antagonist in patients with treatment-resistant major depressive disorder. *Journal of Clinical Psychiatry*, *71*(12), 1605–1611. https://doi.org/10.4088/JCP.09m05327blu.

Drevets, W. C., Price, J. L., & Furey, M. L. (2008). Brain structural and functional abnormalities in mood disorders: implications for neurocircuitry models of depression. *Brain Structure & Function*, *213*(1–2), 93–118. https://doi.org/10.1007/s00429-008-0189-x.

Dwyer, J. M., Lepack, A. E., & Duman, R. S. (2012). mTOR activation is required for the antidepressant effects of mGluR2/3 blockade. *International Journal of Neuropsychopharmacology*, *15*(4), 429–434. https://doi.org/10.1017/S1461145711001702.

Dwyer, J. M., Maldonado-Avilés, J. G., Lepack, A. E. DiLeone, R. J., & Duman, R. S. (2015). "Ribosomal Protein S6 Kinase 1 signaling in prefrontal cortex controls depressive behavior." *Proceedings of the National Academy of Sciences*, *112*(19), 6188–6193. https://doi.org/10.1073/pnas.1505289112.

Fang, Z. H., Lee, C. H., Seo, M. K. Cho, H., Lee, J. G., Lee, J. B., Park, S. W., & Kim, Y. H. (2013). Effect of treadmill exercise on the BDNF-mediated pathway in the hippocampus of stressed rats. *Neuroscience Research*, *76*(4), 187–194. https://doi.org/10.1016/j.neures.2013.04.005.

Fatemi, S. H., Folsom, T. D., Rooney, R. J., & Thuras, P. D. (2013). mRNA and protein expression for novel GABAA receptors Θ and P2 are altered in schizophrenia and mood disorders: Relevance to FMRP-mGluR5 signaling pathway. *Translational Psychiatry*, *3*(6), e271. https://doi.org/10.1038/tp.2013.46.

Feyissa, A. M., Chandran, A., Stockmeier, C. A., & Karolewicz, B. (2009). Reduced levels of NR2A and NR2B subunits of NMDA receptor and PSD-95 in the prefrontal cortex in major depression. *Progress in Neuro-Psychopharmacology and Biological Psychiatry*, *33*(1), 70–75. https://doi.org/10.1016/j.pnpbp.2008.10.005.

Genzel, L., Dresler, M., Cornu, M., Jäger, E., Konrad, B., Adamczyk, M., Friess, E.,... Goya-Maldonado, R. (2015). Medial prefrontal-hippocampal connectivity and motor memory consolidation in depression and schizophrenia. *Biological Psychiatry*, *77*(2), 177–186. https://doi.org/10.1016/j.biopsych.2014.06.004.

Gideons, E. S., Kavalali, E. T., & Monteggia, L. M. (2014). Mechanisms underlying differential effectiveness of memantine and ketamine in rapid antidepressant responses. *Proceedings of the National Academy of Sciences of the United States of America*, *111*(23), 8649–8654. https://doi.org/10.1073/pnas.1323920111.

Glasgow, N. G., Povysheva, N. V., Azofeifa, A. M., & Johnson, J. W. (2017). Memantine and ketamine differentially alter NMDA receptor desensitization. *Journal of Neuroscience*, *37*(40), 9686–9704. https://doi.org/10.1523/JNEUROSCI.1173-17.2017.

Graber, T. E., McCamphill, P. K., & Sossin, W. S. (2013). A recollection of mTOR signaling in learning and memory. *Learning & Memory*, *20*(10), 518–530. https://doi.org/10.1101/lm.027664.112.

Hay, N., & Sonenberg, N. (2004). Upstream and downstream of mTOR. *Genes & Development*, *18*(16), 1926–1945. https://doi.org/10.1101/gad.1212704.

Hieronymus, F., Nilsson, S., & Eriksson, E. (2016). A mega-analysis of fixed-dose trials reveals dose-dependency and a rapid onset of action for the antidepressant effect of three selective serotonin reuptake inhibitors. *Translational Psychiatry*, *6*(6), e834–34. https://doi.org/10.1038/tp.2016.104.

Higuchi, F., Uchida, S., Yamagata, H., Abe-Higuchi, N., Hobara, T., Hara, K., Kobayashi, A., et al. (2016). Hippocampal microRNA-124 enhances chronic stress resilience in mice. *Journal of Neuroscience, 36*(27), 7253–7267. https://doi.org/10.1523/JNEUROSCI.0319-16.2016.

Issler, O., Haramati, S., Paul, E. D., Maeno, H., Navon, I., Zwang, R., Gil, S., et al. (2014). MicroRNA 135 is essential for chronic stress resiliency, antidepressant efficacy, and intact serotonergic activity. *Neuron, 83*(2), 344–360. https://doi.org/10.1016/j.neuron.2014.05.042.

Jernigan, C. S., Goswami, D. B., Austin, M. C., Iyo, A. H., Chandran, A., Stockmeier, C. A., & Karolewicz, B. (2011). The mTOR signaling pathway in the prefrontal cortex is compromised in major depressive disorder. *Progress in Neuro-Psychopharmacology and Biological Psychiatry, 35*(7), 1774–1779. https://doi.org/10.1016/j.pnpbp.2011.05.010.

Johnston, C., Eliez, S., Dyer-Friedman, J., Hessl, D., Glaser, B., Blasey, C., Taylor, A., & Reiss, A. (2001). Neurobehavioral phenotype in carriers of the Fragile X premutation. *American Journal of Medical Genetics Part B: Neuropsychiatric Genetics, 103*(4), 314–319.

Kang, H.-J., Voleti, B., Hajszan, T., Rajkowska, G., Stockmeier, C. A., Licznerski, P., Lepack, A., et al. (2012). Decreased expression of synapse-related genes and loss of synapses in major depressive disorder. *Nature Medicine, 18*(9), 1413–1417. https://doi.org/10.1038/nm.2886.

Kavalali, E. T., & Monteggia, L. M. (2018). The ketamine metabolite 2R,6R-hydroxynorketamine blocks NMDA receptors and impacts downstream signaling linked to antidepressant effects. *Neuropsychopharmacology, 43*(1), 221–222. https://doi.org/10.1038/npp.2017.210.

Kessler, R. C., Berglund, P., Demler, O., Jin, R., Koretz, D., Merikangas, K. R., . . . Wang, S. (2003). The epidemiology of major depressive disorder: Results from the National Comorbidity Survey Replication (NCS-R). *Journal of the American Medical Association, 289*(23), 3095–3105. https://doi.org/10.1001/jama.289.23.3095.

Khan, A., Pepio, A. M., & Sossin, W. S. (2001). Serotonin activates S6 kinase in a rapamycin-sensitive manner in Aplysia Synaptosomes. *Journal of Neuroscience, 21*(2), 382–391. https://doi.org/10.1523/JNEUROSCI.21-02-00382.2001.

Koolschijn, P. C. M. P., van Haren, N. E. M., Lensvelt-Mulders, G., Hulshoff Pol, H. E., & Kahn, R. S. (2009). Brain volume abnormalities in major depressive disorder: A meta-analysis of magnetic resonance imaging studies. *Human Brain Mapping, 30*(11), 3719–3735. https://doi.org/10.1002/hbm.20801.

Laplante, M., & Sabatini, D. M. (2009). mTOR signaling at a glance. *Journal of Cell Science, 122*(Pt 20), 3589–3594. https://doi.org/10.1242/jcs.051011.

Li, N., Lee, B., Liu, R.-J., Banasr, M., Dwyer, J. M., Iwata, M., Li, X. Y., . . . Duman, R. S. (2010). mTOR-dependent synapse formation underlies the rapid antidepressant effects of NMDA antagonists. *Science, 329*(5994), 959–964. https://doi.org/10.1126/science.1190287.

Li, N., Liu, R. J., Dwyer, J. M., Banasr, M., Lee, B., Son, H.,& . . . Duman, R. S. (2011). Glutamate N-Methyl-D-Aspartate receptor antagonists rapidly reverse behavioral and synaptic deficits caused by chronic stress exposure. *Biological Psychiatry, 69*(8), 754–761. https://doi.org/10.1016/j.biopsych.2010.12.015.

Li, S.-X., Liu, L.-J., Xu, L.-Z., Gao, L., Wang, X.-F., Zhang, J.-T., & Lu, L. (2013). Diurnal alterations in circadian genes and peptides in major depressive disorder before and after escitalopram treatment. *Psychoneuroendocrinology, 38*(11), 2789–2799. https://doi.org/10.1016/j.psyneuen.2013.07.009.

Liston, C., Miller, M. M., Goldwater, D. S., Radley, J. J., Rocher, A. B., Hof, P. R, . . . Bruce S McEwen. 2006. Stress-induced alterations in prefrontal cortical dendritic morphology predict selective impairments in perceptual attentional set-shifting. *Journal of Neuroscience, 26*(30), 7870–7874. https://doi.org/10.1523/JNEUROSCI.1184-06.2006.

Liu, R.-J., & Aghajanian, G. K. (2008). Stress blunts serotonin- and hypocretin-evoked EPSCs in prefrontal cortex: Role of corticosterone-mediated apical dendritic atrophy. *Proceedings of the National Academy of Sciences of the United States of America, 105*(1), 359–364. https://doi.org/10.1073/pnas.0706679105.

Liu, R.-J., Fuchikami, M., Dwyer, J. M., Lepack, A. E., Duman, R. S., & Aghajanian, G. K. (2013). GSK-3 inhibition potentiates the synaptogenic and antidepressant-like effects of subthreshold doses of ketamine. *Neuropsychopharmacology, 38*(11), 2268–2277. https://doi.org/10.1038/npp.2013.128.

Liu, Y., Lin, D., Wu, B., & Zhou, W. (2016). Ketamine abuse potential and use disorder. *Brain Research Bulletin, 126*(Pt 1), 68–73. https://doi.org/10.1016/j.brainresbull.2016.05.016.

Lopez, J. P., Lim, R., Cruceanu, C., Crapper, L., Fasano, C., Labonte, B., Maussion, G., et al. 2014. miR-1202 is a primate-specific and brain-enriched microRNA involved in major depression and antidepressant treatment. *Nature Medicine, 20*(7), 764–768. https://doi.org/10.1038/nm.3582.

Lopez, J. P., Pereira, F., Richard-Devantoy, S., Berlim, M., Chachamovich, E., Fiori, L. M., . . . Jollant, F. (2017). Co-variation of peripheral levels of mir-1202 and brain activity and connectivity during antidepressant treatment. *Neuropsychopharmacology, 42*(10), 2043–2051. https://doi.org/10.1038/npp.2017.9.

Maeng, S., Zarate, C. A., Du, J., Schloesser, R. J., McCammon, J., Chen, G., & Manji, H. K. (2008). Cellular mechanisms underlying the antidepressant effects of ketamine: role of alpha-amino-3-hydroxy-5-methylisoxazole-4-propionic acid receptors. *Biological Psychiatry, 63*(4), 349–352. https://doi.org/10.1016/j.biopsych.2007.05.028.

McCamphill, P. K., Ferguson, L., & Sossin, W. S. (2017). A decrease in eukaryotic elongation factor 2 phosphorylation is required for local translation of sensorin and long-term facilitation in Aplysia." *Journal of Neurochemistry, 142*(2), 246–259. https://doi.org/10.1111/jnc.14030.

Millan, M. J., Agid, Y., Brüne, M., Bullmore, E. T., Carter, C. S., Clayton, N. S., Connor, R. et al. (2012). Cognitive dysfunction in psychiatric disorders: Characteristics, causes and the quest for improved therapy. *Nature Reviews Drug Discovery, 11*(2), 141–168. https://doi.org/10.1038/nrd3628.

Miller, O. H., Yang, L., Wang, C.-C., Hargroder, E. A., Zhang, Y., Delpire, E., & Hall, B. J. (2014). GluN2B-containing NMDA receptors regulate depression-like behavior and are critical for the rapid antidepressant actions of ketamine. *eLife*, 3(October), 513. https://doi.org/10.7554/eLife.03581.

Murrough, J. W. (2012). Ketamine as a novel antidepressant: From synapse to behavior. *Clinical Pharmacology & Therapeutics, 91*(2), 303–309. https://doi.org/10.1038/clpt.2011.244.

Murrough, J. W., Iacoviello, B., Neumeister, A., Charney, D. S., & Iosifescu, D. V. (2011). Cognitive dysfunction in depression: Neurocircuitry and new therapeutic strategies. *Neurobiology of Learning and Memory, 96*(4), 553–563. https://doi.org/10.1016/j.nlm.2011.06.006.

Niere, F., Namjoshi, S., Song, E., Dilly, G. A., Schoenhard, G., Zemelman, B. V., . . . Raab-Graham, K. F. (2016). Analysis of proteins that rapidly change upon mechanistic/mammalian target of rapamycin complex 1 (mTORC1) repression identifies Parkinson protein 7 (PARK7) as a novel protein aberrantly expressed in tuberous sclerosis complex (TSC). *Molecular & Cellular Proteomics, 15*(2), 426–444. https://doi.org/10.1074/mcp.M115.055079.

Ota, K. T., Liu, R.-J., Voleti, B., Maldonado-Avilés, J. G., Duric, V., Iwata, M., Dutheil, S., et al. 2014. REDD1 is essential for stress-induced synaptic loss and depressive behavior. *Nature Medicine, 20*(5), 531–535. https://doi.org/10.1038/nm.3513.

Paoletti, P., Bellone, C., & Zhou, Q. (2013). NMDA receptor subunit diversity: Impact on receptor properties, synaptic plasticity and disease. *Nature Reviews Neuroscience, 14*(6), 383–400. https://doi.org/10.1038/nrn3504.

Park, S. W., Lee, J. G., Seo, M. K., Lee, C. H., Cho, H. Y., Lee, B. J., Seol, W., & Kim, Y. H. (2014). Differential effects of antidepressant drugs on mTOR Signalling in rat hippocampal neurons. *International Journal of Neuropsychopharmacology, 17*(11), 1831–1846. https://doi.org/10.1017/S1461145714000534.

Preskorn, S. H., Baker, B., Kolluri, S., Menniti, F. S., Krams, M., & Landen, J. W. (2008). An innovative design to establish proof of concept of the antidepressant effects of the NR2B subunit selective N-Methyl-D-Aspartate antagonist, CP-101,606, in patients with treatment-refractory major depressive disorder. *Journal of Clinical Psychopharmacology, 28*(6), 631–637. https://doi.org/10.1097/JCP.0b013e31818a6cea.

Raab-Graham, K. F., & Niere, F. (2017). mTOR referees memory and disease through mRNA repression and competition. *FEBS Letters, 591*(11), 1540–1554. https://doi.org/10.1002/1873-3468.12675.

Radley, J. J., Rocher, A. B., Miller, M., Janssen, W. G. M., Liston, C., Hof, P. R., . . . Morrison, J. H. (2006). Repeated stress induces dendritic spine loss in the rat medial prefrontal cortex. *Cerebral Cortex, 16*(3), 313–320. https://doi.org/10.1093/cercor/bhi104.

Radley, J. J., Rocher, A. B., Rodriguez, A., Ehlenberger, D. B., Dammann, M., McEwen, B. S., . . . Hof, P. R. (2008). Repeated stress alters dendritic spine morphology in the rat medial prefrontal cortex. *Journal of Comparative Neurology, 507*(1), 1141–50. https://doi.org/10.1002/cne.21588.

Ressler, K. J., & Nemeroff, C. B. (2000). Role of serotonergic and noradrenergic systems in the pathophysiology of depression and anxiety disorders. *Depression and Anxiety, 12*(Suppl. 1), 2–19. https://doi.org/10.1002/1520-6394(2000)12:1+<2::AID-DA2>3.0.CO;2-4.

Roberts, J. E., Bailey, D. B., Mankowski, J., Ford, A., Sideris, J., Weisenfeld, L. A., . . . Golden, R. N. (2009). Mood and anxiety disorders in females with the FMR1 premutation. *American Journal of Medical Genetics Part B: Neuropsychiatric Genetics*, 150B(1), 130–139. https://doi.org/10.1002/ajmg.b.30786.

Rush, A. J., Trivedi, M. H., Stewart, J. W., Nierenberg, A. A., Fava, M., Kurian, B. T., Warden, D., et al. (2011). Combining medications to enhance depression outcomes (co-med): acute and long-term outcomes of a single-blind randomized study. *American Journal of Psychiatry, 168*(7), 689–701. https://doi.org/10.1176/appi.ajp.2011.10111645.

Russo, E., Leo, A., Crupi, R., Aiello, R., Lippiello, P., Spiga, R., et al. (2016). Everolimus improves memory and learning while worsening depressive- and anxiety-like behavior in an animal model of depression. *Journal of Psychiatric Research, 78*(July), 1–10. https://doi.org/10.1016/j.jpsychires.2016.03.008.

Scheetz, A. J., Nairn, A. C., & Constantine-Paton, M. (2000). NMDA receptor-mediated control of protein synthesis at developing synapses. *Nature Neuroscience, 3*(3), 211–216. https://doi.org/10.1038/72915.

Shen, P., Hu, Q., Dong, M., Bai, S., Liang, Z., Chen, Z., Li, P., et al. (2017). Venlafaxine exerts antidepressant effects possibly by activating MAPK-ERK1/2 and P13K-AKT pathways in the hippocampus. *Behavioural Brain Research, 335*(August), 63–70. https://doi.org/10.1016/j.bbr.2017.08.011.

Shirayama, Y., & Hashimoto, K. (2018). Lack of antidepressant effects of (2R,6R)-hydroxynorketamine in a rat learned helplessness model: Comparison with (R)-ketamine. *International Journal of Neuropsychopharmacology, 21*(1), 84–88. https://doi.org/10.1093/ijnp/pyx108.

Siddiqui, N., & Sonenberg, N. (2015). Signalling to eIF4E in Cancer. *Biochemical Society Transactions, 43*(5), 763–772. https://doi.org/10.1042/BST20150126.

Simonoff, E., Pickles, A., Charman, T., Chandler, S., Loucas, T., & Baird, G. (2008). Psychiatric disorders in children with autism spectrum disorders: Prevalence, comorbidity, and associated factors in a population-derived sample. *Journal of the American Academy of Child and Adolescent Psychiatry, 47*(8), 921–929. https://doi.org/10.1097/CHI.0b013e318179964f.

Starkstein, S. E., Jorge, R., Mizrahi, R., & Robinson, R. G. (2005). The construct of minor and major depression in Alzheimer's disease. *American Journal of Psychiatry, 162*(11), 2086–2093. https://doi.org/10.1176/appi.ajp.162.11.2086.

Su, M., Hong, J., Zhao, Y., Liu, S., & Xue, X. (2015). MeCP2 controls hippocampal brain-derived neurotrophic factor expression via homeostatic interactions with microRNA-132 in rats with depression. *Molecular Medicine Reports, 12*(4), 5399–5406. https://doi.org/10.3892/mmr.2015.4104.

Suzuki, K., Nosyreva, E., Hunt, K. W., Kavalali, E. T., & Monteggia, L. M. (2017). Effects of a ketamine metabolite on synaptic NMDAR function. *Nature, 546*(7659), E1–E3. https://doi.org/10.1038/nature22084.

Trullas, R., Folio, T., Young, A., Miller, R., Boje, K., & Skolnick, P. (1991). 1-Aminocyclopropanecarboxylates exhibit antidepressant and anxiolytic actions in animal models. *European Journal of Pharmacology, 203*(3), 379–385. https://doi.org/10.1016/0014-2999(91)90894-V.

Undurraga, J., & Baldessarini, R. J. (2012). Randomized, placebo-controlled trials of antidepressants for acute major depression: Thirty-year meta-analytic review. *Neuropsychopharmacology, 37*(4), 851–864. https://doi.org/10.1038/npp.2011.306.

Webhofer, C., Gormanns, P., Tolstikov, V., Zieglgänsberger, W., Sillaber, I., Holsboer, F., & Turck, C. W. (2011). Metabolite profiling of antidepressant drug action reveals novel drug targets beyond monoamine elevation. *Translational Psychiatry, 1*(12), e58. https://doi.org/10.1038/tp.2011.56.

Wesseling, H., Rahmoune, H., Tricklebank, M., Guest, P. C., & Bahn, S. (2015). A targeted multiplexed proteomic investigation identifies ketamine-induced changes in immune markers in rat serum and expression changes in protein kinases/phosphatases in rat brain. *Journal of Proteome Research, 14*(1), 411–421. https://doi.org/10.1021/pr5009493.

Wolfe, S. A., Workman, E. R., Heaney, C. F., Niere, F., Namjoshi, S., Cacheaux, L. P., Farris, S. P., et al. (2016). FMRP regulates an ethanol-dependent shift in GABABR function and expression with rapid antidepressant properties. *Nature Communications, 7*(September), 12867. https://doi.org/10.1038/ncomms12867.

Workman, E. R., Haddick, P. C. G., Bush, K., Dilly, G. A., Niere, F., Zemelman, B. V., Raab-Graham, K. F. (2015). Rapid antidepressants stimulate the decoupling of GABAB receptors from GIRK/Kir3 channels through increased protein stability of 14-3-3η. *Molecular Psychiatry, 20*(3), 298–310. https://doi.org/10.1038/mp.2014.165.

Workman, E. R., Niere, F., & Raab-Graham, K. F. (2013). mTORC1-dependent protein synthesis underlying rapid antidepressant effect requires GABABR signaling. *Neuropharmacology, 73*, 192–203. https://doi.org/10.1016/j.neuropharm.2013.05.037.

Yang, C., Hu, Y.-M., Zhou, Z.-Q., Zhang, G.-F., & Yang, J.-J. (2013). Acute administration of ketamine in rats increases hippocampal BDNF and mTOR levels during forced swimming test. *Upsala Journal of Medical Sciences, 118*(1), 3–8. https://doi.org/10.3109/03009734.2012.724118.

Yang, C., Qu, Y., Abe, M., Nozawa, D., Chaki, S., & Hashimoto, K. (2017). (R)-Ketamine shows greater potency and longer lasting antidepressant effects than its metabolite

(2R,6R)-hydroxynorketamine. *Biological Psychiatry*, *82*(5), e43–e44. https://doi.org/10.1016/j.biopsych.2016.12.020.

Yang, C., Zhou, Z.-Q., Gao, Z.-Q. Shi, J.-Y., & Yang, J.-J. (2013). Acute increases in plasma mammalian target of rapamycin, glycogen synthase kinase-3β, and eukaryotic elongation factor 2 phosphorylation after ketamine treatment in three depressed patients. *Biological Psychiatry*, *73*(12), e35–e36. https://doi.org/10.1016/j.biopsych.2012.07.022.

Yu, J.-J., Zhang, Y., Wang, Y., Wen, Z.-Y., Liu, X.-H., Qin, J., & Yang, J.-L. (2013). Inhibition of calcineurin in the prefrontal cortex induced depressive-like behavior through mTOR signaling pathway. *Psychopharmacology*, *225*(2), 361–72. https://doi.org/10.1007/s00213-012-2823-9.

Yuen, E. Y., Wei, J., Liu, W., Zhong, P., Li, X., & Yan, Z. (2012). Repeated stress causes cognitive impairment by suppressing glutamate receptor expression and function in prefrontal cortex. *Neuron*, *73*(5), 962–977. https://doi.org/10.1016/j.neuron.2011.12.033.

Zanos, P., Moaddel, R., Morris, P. J., Georgiou, P., Fischell, J., Elmer,G. I., Alkondon, M., et al. 2016. NMDAR inhibition-independent antidepressant actions of ketamine metabolites. *Nature*, *533*(7604), 481–486. https://doi.org/10.1038/nature17998.

Zhang, K., Yamaki, V. N., Wei, Z., Zheng, Y., & Cai, X. (2017). Differential regulation of GluA1 expression by ketamine and memantine. *Behavioural Brain Research*, *316*(January), 152–159. https://doi.org/10.1016/j.bbr.2016.09.002.

Zimmermann, C. A., Hoffmann, A., Raabe, F., & Spengler, D. (2015). Role of Mecp2 in experience-dependent epigenetic programming. *Genes*, *6*(1), 60–86. https://doi.org/10.3390/genes6010060.

CHAPTER 21

DYSREGULATION OF NEURONAL PROTEIN SYNTHESIS IN ALZHEIMER'S DISEASE

TAO MA

INTRODUCTION

ALZHEIMER'S disease (AD) is the most common form of dementia syndromes with aging as the greatest known risk factor (Alzheimer's Association, 2019; Querfurth & LaFerla, 2010). Concomitant with the rapid growth of the aging population worldwide, the incidence of AD is escalating significantly, becoming a global threat to public health and a pressing "epidemic" for the 21st century Unfortunately, no interventions have been discovered to either slow the progress of AD or cure the disease, and recent clinical trials have not succeeded in identifying disease-modifying strategies (Dyer, Renner, & Bachmann, 2006; Holtzman, Goate, Kelly, & Sperling, 2011; Holtzman, Morris, & Goate, 2011; Roberson & Mucke, 2006). Thus, there is an urgent need to develop novel therapeutics targeting AD pathophysiology based on solid mechanistic studies.

The hallmarks of AD brain pathology include cerebral plaques and neurofibrillary tangles, which consist of aggregated β-amyloid peptide (Aβ) and abnormally hyperphosphorylated tau proteins, respectively (Querfurth & LaFerla, 2010). Of note, while there is no doubt about the importance of Aβ and tau phosphorylation (p-tau) as biomarkers or pathological diagnostic criteria for AD, the exact roles of Aβ and tau phosphorylation in AD etiology are still under debate (Herrup, 2015; Ma & Klann, 2012; Tse & Herrup, 2017). Substantial evidence indicates synaptic dysfunction as a key event in AD pathophysiology, and accordingly AD has been called a disease of "synaptic failure" (Ma & Klann, 2012; Selkoe, 2002; Walsh & Selkoe, 2004). Understanding the molecular and cellular mechanisms underlying AD-associated synaptic dysfunction/failure will help identify novel diagnostic biomarkers and perhaps, more importantly, therapeutic targets for AD (Ma & Klann, 2012; Rowan, Klyubin, Wang, & Anwyl, 2005; Teich et al., 2015). Substantial evidence over the past few decades has established that long-lasting forms of synaptic plasticity (persistent change in neuronal circuits) and memory require de novo protein synthesis (or mRNA translation). Recent studies indicate an important role of mRNA translation impairments in AD-associated synaptic failure and cognitive defects (Beckelman et al., 2019; Jan et al., 2017; Ma et al., 2013). For this chapter, I will focus on several signaling pathways controlling protein synthesis and their roles (when dysregulated) in AD-associated dementia syndrome.

Role of eIF2α Phosphorylation in AD

Protein synthesis takes place in three phases: initiation, elongation, and termination. Each phase requires the action of multiple translational factors to facilitate the process. Translation is highly controlled at the initiation phase. During the initiation phase, eukaryotic initiation factor 2 (eIF2) binds GTP and Met–tRNA$_i^{Met}$ to form a ternary complex to interact with the small ribosomal subunit. The exchange of GTP and GDP-bound form of the complex is catalyzed by another translational factor: eukaryotic initiation factor 2B (eIF2B). Phosphorylation of eIF2 on the α subunit (eIF2α) at the Ser51 site inhibits the catalytic function of eIF2B and consequently blocks the GDP/GTP exchange. Therefore, general protein synthesis is inhibited when there is increased phosphorylation of eIF2α (Klann & Dever, 2004; Richter, Bassell, & Klann, 2015; Trinh et al., 2014). On the other hand, eIF2α phosphorylation increases translation of a selective subset of mRNAs with upstream open reading frames (uORFs) in their 5′ untranslated region (UTR). One of the best known examples of such mRNAs encodes activating transcription factor 4 (ATF4), a repressor of long-term synaptic plasticity and memory formation (Costa-Mattioli, Sossin, Klann, & Sonenberg, 2009; Klann, Antion, Banko, & Hou, 2004; Trinh et al., 2014; R. Wek, 2018).

There are four known eIF2α kinases: double-stranded-RNA-dependent protein kinase (PKR); heme-regulated inhibitor kinase (HRI); general control non-derepressible-2 (GCN2), and PKR-like ER (endoplasmic reticulum) kinase (PERK; R.C. Wek, Jiang, & Anthony, 2006). As indicated in their names, the classic view of the eIF2α kinases considers that each kinase is activated by a specific type of stimulus or cellular stress (R. C. Wek & Cavener, 2007). However, it is oversimplified to conclude that one type of stressing stimulus activates only one eIF2α kinase. For example, during oxidative stress conditions, multiple eIF2α kinases, particularly PERK and GCN2, are recruited simultaneously or sequentially to cope with cellular homeostasis (Jiang et al., 2004; Zhan, Narasimhan, & Wek, 2004). Notably, each of the four kinases phosphorylates eIF2α on the same site Ser51 (Trinh & Klann, 2013).

Multiple studies indicate elevated eIF2α phosphorylation in neurodegenerative diseases and point to PERK suppression as a potential therapeutic strategy for cognitive impairments associated with these diseases (Ma & Klann, 2014; Moreno et al., 2012; Radford, Moreno, Verity, Halliday, & Mallucci, 2015). Accumulation of misfolded proteins represents one of the key brain pathologies of neurodegenerative diseases. These misfolded proteins cause significant cellular stress and induce activation of the signaling pathways associated with unfolded protein response (UPR). The UPR is composed of three key effectors: activating transcription factor 6 (ATF6), inositol-requiring enzyme 1 (IRE1), and PERK (Ronald C. Wek & Cavener, 2007). Dysregulations of signaling pathways associated with all three effectors (ATF6, IRE1, and PERK) have been indicated in AD (Cornejo & Hetz, 2013; Gerakis & Hetz, 2018; Hetz & Saxena, 2017). In this review I focus the discussion on PERK signaling, from the view of protein synthesis regulation (Figure 21.1). Activation of PERK leads to increased eIF2α phosphorylation and inhibition of general protein synthesis. Suppression of general protein synthesis temporarily can be beneficial, since it helps conserve energy resources while enhancing translation of mRNAs related to anti-stress response (e.g., ATF4), thus preparing cells to cope with stress (Ma & Klann, 2014; Paschen, Proud, & Mies, 2007). During the pathogenesis of certain neurodegenerative diseases (e.g., AD), however, activation of PERK and eIF2α phosphorylation persist as a result from severe cellular stress and UPR, causing prolonged repression of protein synthesis. Prolonged elevation of eIF2α phosphorylation and inhibition of mRNA translation are detrimental because long-lasting forms of synaptic plasticity and memory formation/consolidation are dependent on de novo

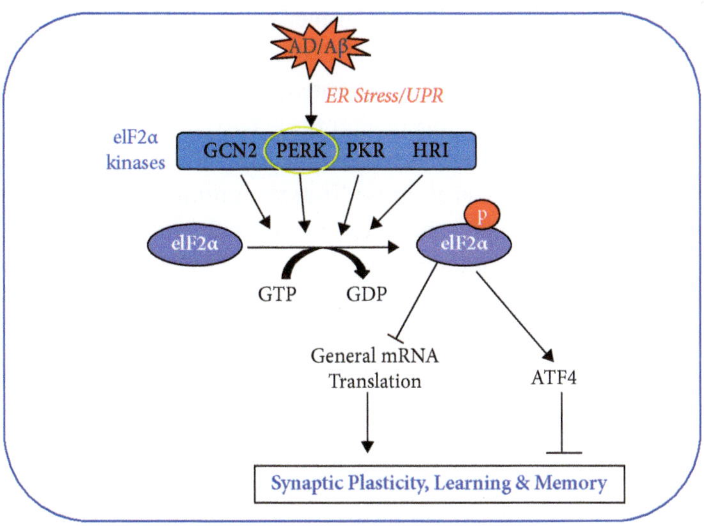

FIGURE 21.1. Schematic model depicting potential links of eIF2α phosphorylation and eIF2α kinases to AD. Briefly, AD-associated pathological processes induce ER stress and unfolded protein response (UPR), leading to activation of eIF2α kinase PERK. Phosphorylation of eIF2α by PERK results in inhibition of general protein synthesis (mRNA translation), and protein synthesis is required for memory formation and long-lasting forms of synaptic plasticity. Arrows denote activation and blunted lines indicate inhibition.

protein synthesis (Alberini, 2008; Costa-Mattioli et al., 2009; Kelleher, Govindarajan, & Tonegawa, 2004).

AD is characterized by ER stress and aberrant signaling associated with UPR. In agreement, abnormal hyper-phosphorylation of eIF2α has been identified in brains of AD patients and in AD model mice (Chang, Wong, Ng, & Hugon, 2002; Kim et al., 2010; Ma et al., 2013; O'Connor et al., 2008). To study whether elevated eIF2α phosphorylation contributes to AD pathophysiology, investigators reduce brain eIF2α phosphorylation in a mouse model of AD by genetically suppressing eIF2α kinase PERK in neurons of the forebrain and hippocampus (Ma et al., 2013). They found that spatial memory impairments displayed in aged AD model mice were significantly improved by genetic reduction of PERK. Furthermore, AD-associated defects in hippocampal long-term potentiation (LTP), one of the most intensively studied forms of synaptic plasticity that is considered as a cellular model for memory, were also improved with PERK suppression. In addition, they demonstrated that application of the general protein synthesis inhibitor anisomycin was able to reverse the LTP improvement associated with PERK suppression, indicating that the beneficial effects of reducing eIF2α phosphorylation on AD-associated synaptic failure are dependent on protein synthesis (Ma et al., 2013). Long-term depression (LTD) is another important form of synaptic plasticity that is known to be impaired in AD (S. Li et al., 2009). Studies suggest that metabotropic glutamate receptor 5 (mGluR5) may mediate the neurotoxic effects of Aβ (Hu et al., 2014; Um et al., 2013). It was later demonstrated that mGluR-LTD failure in AD model mice can also be alleviated by suppressing PERK activity genetically or pharmacologically using small molecule PERK antagonist GSK2606414 (W. Yang, Zhou, Cavener, Klann, & Ma, 2016).

As a novel and selective PERK inhibitor, GSK2606414 holds promise as a potential treatment for neurodegenerative diseases. Moreno and colleagues treated prion disease model mice with GSK2606414 before or after the occurrence of clinical symptoms, and found marked improvement in both behavior deficiency and brain pathology associated with the prion disease (Moreno et al., 2013). The same group also showed that treatment of GSK2606414 is effective in preventing tau-mediated neuropathology and behavioral defects in a mouse model of frontotemporal dementia (FTD; Radford et al., 2015).

There are several caveats in proposing PERK inhibition as a therapeutic strategy for AD or other neurodegenerative diseases. Regulation of PERK is critical for normal pancreatic function, maintenance of glucose homeostasis, and development of skeletal system (P. Zhang et al., 2002). For example, a mutation in the gene encoding PERK (*EIF2AK3*) is linked to Wolcott-Rallison syndrome in humans, which is characterized by neonatal diabetes mellitus (Sene´e et al., 2004). Consistently, it was reported that oral treatment of GSK2606414 in mice caused hyperglycemia and weight loss (Moreno et al., 2013). These non-neuronal harmful side effects have to be considered when PERK inhibitors are applied systematically. Moreover, it was reported that brain-specific suppression of PERK in mice leads to significant impairments in cognition including severe behavioral inflexibility (Trinh et al., 2012), which could be related to the roles of PERK in handling cellular stress. Studies also show that interruption of eIF2α phosphorylation homeostasis in mice leads to impaired behavioral and neuronal plasticity (Costa-Mattioli et al., 2007). Therefore, in order for PERK inhibition to be a viable therapeutic approach, it is critical to optimize the dose and duration of PERK inhibitor treatment to "normalize" eIF2α phosphorylation, thus keeping a balance between de novo protein synthesis (for learning and memory) and maintaining essential UPR response under stressing conditions.

Dysregulation of mTORC1 Signaling and AD

mTOR (mammalian target of rapamycin, also known as mechanistic target of rapamycin) is an evolutionarily conserved protein kinase that plays a critical role in cell growth and protein synthesis (Albert & Hall, 2015; Laplante & Sabatini, 2012). Depending on the binding proteins associated with mTOR and sensitivity to rapamycin, mTOR assembles into two complexes, mTORC1 (mammalian target of rapamycin complex 1) and mTORC2 (mammalian target of rapamycin complex 2; Albert & Hall, 2015; Hoeffer & Klann, 2010; Q. Yang & Guan, 2007). Briefly, mTORC1 contains raptor, an essential and non-enzymatic subunit of the complex that is required for rapamycin's inhibitory effect, and mLST8, which binds to the kinase domain of mTOR. The interactions between mTOR, raptor, and mLST8 are thought to determine the access of mTOR to its downstream targets (Q. Yang & Guan, 2007). Most studies reported so far on AD focus on regulation of mTORC1 only.

One known upstream regulator of mTORC1 involves PI3K signaling pathway through the tuberous sclerosis complex, consisting of TSC1 (hamartin) and TSC2 (tuberin). The 3′-phosphoinositides PI(3,4)P2 and PI(3,4,5)P3, produced by PI3K, bind to the pleckstrin homology domain of the Ser/Thr kinase AKT (PKB) and recruit it to the membrane. There, AKT can be phosphorylated at Thr308 in its activation loop by phosphoinositide-dependent-kinase 1 (PDK1), which also has a pleckstrin homology domain. In addition to its phosphorylation at Thr308, AKT can be phosphorylated on the hydrophobic motif site Ser473 by a kinase initially referred as "PDK2" but of unknown identity. This kinase was later demonstrated to be the long-sought mTORC2. While the activity of AKT can be potentiated significantly by phosphorylation at Ser473, Thr308 phosphorylation is thought to be both necessary and sufficient for TSC2 phosphorylation (Fruman et al., 2017; Manning & Cantley, 2007; Q. Yang & Guan, 2007). Besides AKT, multiple signaling molecules are known to regulate mTORC1 activity directly or via TSC2 including p90 ribosomal S6 kinase (RSK), MAPK/ERK, AMP-activated protein kinase (AMPK), and glycogen synthase kinase-3 (GSK3). Compared to AKT and ERK which are positive regulators of mTORC1, both AMPK and GSK3 are negative regulators of mTORC1 (i.e., activation of AMPK or GSK3 results in suppression of mTORC1 signaling; Choo, Roux, & Blenis, 2006). Probably the best known downstream targets of mTORC1 are S6K1 (or p70S6K) and a repressor protein of eukaryotic initiation factor 4E (eIF4E) named 4EBP (eIF4E binding protein) (Hoeffer & Klann, 2010). Translation is highly regulated at the initiation step during which a ribosome is recruited to the 5′ end of an mRNA, which in all nuclear-transcribed mRNAs possesses the cap structure m^7GpppN (where "m" represents a methyl group and "N" refers to any nucleotide). The cap is specifically recognized and bound by eIF4E, which is a subunit of a complex termed eIF4F that contains two other proteins: eIF4G (a scaffolding protein) and eIF4A (an RNA helicase). Following its binding to the 5′cap, eIF4F (attached to the 40S subunit through an interaction between eIF4G and eIF3) is thought to melt the secondary structure of the 5′-UTR, thereby facilitating scanning to the start codon, where the 60S subunit joins and translation commences. The initiation factor eIF4B, which enhances the helicase activity of eIF4A, also contributes to removing the secondary structure of the transcript (Gingras et al., 1999). S6K1, on the other hand, plays a critical role in regulation of cell size and glucose homeostasis (Meyuhas & Dreazen, 2009). S6K1 has eight known phosphorylation sites distributed in three functional domains—catalytic, linker, and autoinhibitory. At least five of the phosphorylation sites—Thr229, Thr389, Ser404, Ser411, and Ser421—are sensitive to rapamycin, among which Thr389 is reported to be phosphorylated by mTORC1 and is often used as a readout for mTORC1 signaling regulation (Fumagalli & Thomas, 2000; Lehman, Calvo, & Gomez-Cambronero, 2003; Loreni, Thomas, & Amaldi, 2000; Y. Zhang et al., 2001). Moreover, mTOR signaling mediates translation of a specific class of mRNAs characterized by presence of the terminal oligopyrimidine (TOP) at their 5′ end. Interestingly, many of the TOP mRNAs encode components of the "translational machinery" including ribosomal proteins and elongation factors (e.g., eEF1A). Thus, activation of the mTOR pathway is considered to increase translational capacity (Hay & Sonenberg, 2004; Meyuhas, 2000; Panayiotis Tsokas, Ma, Iyengar, Landau, & Blitzer, 2007).

Numerous studies have demonstrated a key role of mTORC1 in neuronal function. The activation of mTORC1 signaling and the consequent boost of translational capacity and cap-dependent translation initiation are crucial for memory consolidation and long-lasting forms of synaptic plasticity. A plethora of studies using genetically modified mice or pharmacological agents (e.g., rapamycin) demonstrated that integral mTORC1 signaling is required for normal learning, memory and synaptic plasticity (Graber, McCamphill, & Sossin, 2013; Hoeffer & Klann, 2010).

Multiple lines of evidence indicate a role of mTORC1 signaling dysregulation and AD pathophysiology; although controversy arises regarding how exactly mTORC1 signaling is dysregulated in AD. Downregulation of the mTORC1 signaling was reported in APP/PS1 AD model mice. Consistently, the mTORC1 inhibitor rapamycin exacerbates Aβ neurotoxicity (Lafay-Chebassier et al., 2005; Lafay-Chebassier et al., 2006). In Tg2576 AD model mice, it was revealed that brain mTORC1 signaling, assessed by S6K1 activity, is decreased at young (3-month-old) and middle-aged (9-month-old) mice, but unaltered in old mice (over 20-month- old; Ma et al., 2010). In agreement, levels of elongation factor eEF1A, one of the "TOP" mRNA encoded proteins controlled by mTORC1/S6K1 signaling that is indicated in long-lasting synaptic plasticity, are also reduced in AD (Beckelman et al., 2016; P. Tsokas et al., 2005). In contrast, elevated mTORC1 signaling is reported in the 3XTg-AD model mice (Caccamo, Majumder, Richardson, Strong, & Oddo, 2010). Moreover, a study on the PDAPP (known as J20) AD model mice reported no change in mTORC1 signaling (Spilman et al., 2010). mTORC1 functions as a "hub" to integrate many signaling cascades in response to diverse stimuli such as stress, energy status alteration, and inflammation (Hoeffer & Klann, 2010; Reiling & Sabatini, 2006). It is likely that certain differences in generation of AD models may contribute to the discrepancies in regulation of mTORC1 signaling. For example, mutations in tau (in 3xTg-AD model mice) may induce unique regulation of mTORC1 pathway, or alter the response of the mTORC1 signaling to Aβ. In line with such inconsistency on mTORC1 dysregulation in AD, there is also debate on whether mTORC1 should be inhibited or activated for treatment of AD-associated cognition deficits and synaptic failure (Ma & Klann, 2012; Talboom, Velazquez, & Oddo, 2015).

Aging is one of the best known risk factors for AD. Several studies in late 2000 on the association between the mTORC1 signaling and mammalian aging garnered a lot of attention from the AD research field. One study in mice showed that genetic deletion of S6K1, an established downstream effector of mTORC1, resulted in increased life span and resistance to multiple age-related pathologies including immune and motor dysfunction, and insulin insensitivity (Selman et al., 2009). In agreement, another study published in the same year showed that aged mice fed with rapamycin live longer than control groups (Harrison et al., 2009). Worth mentioning is that neither studies addressed whether aging-related decline in synaptic plasticity and cognition are improved by inhibiting mTORC1 signaling. A later study reported that while rapamycin may extend life span, it has limited beneficial effects on aging phenotypes. In another words, the longevity effects of rapamycin feeding might be due to its drug effects unrelated to aging such as cancer limiting effects (Neff et al., 2013). In two lines of AD model mice, treatment with rapamycin for over two months was able to rescue AD-associated cognitive deficits (Caccamo et al., 2010; Spilman et al., 2010). In agreement, genetic suppression of mTOR or S6K1 prevented cognitive impairments and brain pathology in AD model mice (Caccamo et al., 2015; Caccamo, Pinto, Messina, Branca, & Oddo, 2014). On the other hand, it was reported that hippocampal synaptic plasticity impairments caused by Aβ application are reversed by up-regulating mTORC1 signaling via pharmacological methods or genetic manipulation such as suppression of TSC2 or FK506-binding protein 12 (FKBP12), both of which are negative regulators of mTORC1 (Hoeffer et al., 2008; Ma et al., 2010). Several reasons may explain the seemingly conflicting findings regarding the relationship between mTORC1 signaling regulation and AD pathophysiology. First, in the aforementioned studies using rapamycin, mTORC1 signaling was manipulated differently (e.g., different dose and duration of treatment). For example, while mTORC2 usually is considered rapamycin-insensitive, prolonged treatment of rapamycin can lead to inhibition of mTORC2 and consequently interference of AKT signaling, which functions as an upstream regulator of mTORC1 (Sarbassov et al., 2006). Moreover, several feedback loops exist in the mTORC1 signaling pathway such as the reciprocal effects between S6K1 and AMPK, a central molecular energy sensor functioning to maintain cellular energy homeostasis, dysregulation of which is indicated in AD (Hardie, 2014;

Ma et al., 2014; Selman et al., 2009). Thus, the effects of mTORC1 manipulation on AD might be attributed to influence from feedback or compensatory regulations (Selman et al., 2009). The complexity of mTORC1 signaling regulation in AD can also be demonstrated by studies on roles of insulin in neuronal diseases. Insulin activates the PI3K-PDK-AKT cascade, which is upstream of mTORC1 (Arnold et al., 2018; Piper, Selman, McElwee, & Partridge, 2008). Insulin-PI3K-PDK-AKT signaling is impaired by Aβ treatment (De Felice et al., 2009; Lee, Kumar, Fu, Rosen, & Querfurth, 2009; Magrané et al., 2005; Townsend, Mehta, & Selkoe, 2007). Also, insulin treatment can improve cognitive function in AD patients (Craft et al., 2012; Reger et al., 2008). On the other hand, suppression of insulin signaling can protect a transgenic mouse model of AD from cognitive decline (Cohen et al., 2009). To summarize, the role of mTORC1 signaling in AD remain unclear, and further investigation is necessary to determine whether rapamycin can be considered as a feasible treatment for AD.

Role of eEF2K/eEF2 Signaling in AD

As discussed earlier, eIF2α phosphorylation and mTORC1 signaling play vital roles in controlling the initiation phase of mRNA translation. While much attention has been devoted to the initiation process for translational control, accumulating evidence indicates that control at the elongation phase is critical in modulation of protein synthesis during cellular responses to deficiency of nutrients and energy. In fact, most (> 95%) of the energy and amino acids used in protein synthesis are consumed during the elongation phase (Browne & Proud, 2002; Kenney, Moore, Wang, & Proud, 2014). Elongation is primarily regulated through the eukaryotic elongation factor 2 kinase (eEF2K), the only known kinase for eEF2. Phosphorylation of eEF2 on Thr56 by eEF2K prevents it from binding to the ribosome, thus disrupting peptide growth and general protein synthesis (Kenney et al., 2014). Selective eEF2K inhibitors, such as NH125, AG-484954 and JAN-384, have been developed, but have not yet been tested in vivo for effects on AD-related abnormalities (Arora et al., 2003; Arora et al., 2004; Chen et al., 2011; Kenney et al., 2016). Previous studies, mostly from non-neuronal systems, have identified multiple signaling molecules as upstream regulators in the eEF2K-eEF2 signaling network, including mTORC1 and AMPK (Hardie, 2004, 2014; Horman et al., 2002). In brief, the mTORC1 inhibits activity of eEF2K via phosphorylation, either directly or indirectly through its downstream effector S6K1. Thus, activation of mTORC1 leads to dephosphorylation of eEF2 and consequently promotion of translation elongation, which, in concert with mTORC1-controlled translation initiation, allows the stimulation of general protein synthesis. In contrast, AMPK stimulates activity of eEF2K, either via phosphorylating eEF2K (on a different site from mTORC1), or by inhibiting mTORC1 activity, leading to eEF2 phosphorylation and thus suppression of general protein synthesis (Kenney et al., 2014). AMPK is a central molecular sensor critical for maintenance of cellular energy homeostasis, and AMPK activity is important in maintaining long-lasting forms of synaptic plasticity (Potter et al., 2010). AMPK is activated in response to low-energy state, and AD pathogenesis has been linked to abnormalities in neuronal energy metabolism (Lin & Beal, 2006; Ma et al., 2012; Ma & Klann, 2012; Massaad, Washington, Pautler, & Klann, 2009). Consistently, over-activation of AMPK has been revealed in brain tissue of AD mouse models and human AD patients (Ma et al., 2014; Vingtdeux, Davies, Dickson, & Marambaud, 2011). Most recently it was reported that repression of AMPK activity either pharmacologically or genetically results in mitigation of AD-related impairments of hippocampal LTP, a well-studied synaptic model for learning and memory (Ma et al., 2014). Furthermore, inhibition of LTP by application of amyloid beta (Aβ) is also linked to reducing kinase activity of AKT/PKB, a canonical upstream regulator for both mTORC1 (stimulation) and AMPK (inhibition; Figure 21.2). It was reported that the reduction of AKT activity in AD results from increased caspase-3 activity associated with reactive oxygen species (ROS), which is implicated in many neurodegenerative diseases (Dumont et al., 2009; Jo et al., 2011).

Regulation of protein synthesis via elongation is particularly important in cellular environments where the translational capacity is low (e.g., neuronal dendrites), thus both initiation and elongation processes need to be upregulated to fulfill the substantial requirements of new protein synthesis associated with synaptic plasticity (Sutton & Schuman, 2006). For example, it was demonstrated that treatment of neurons with brain-derived neurotrophic factor (BDNF), a key player in neuronal plasticity, reduced

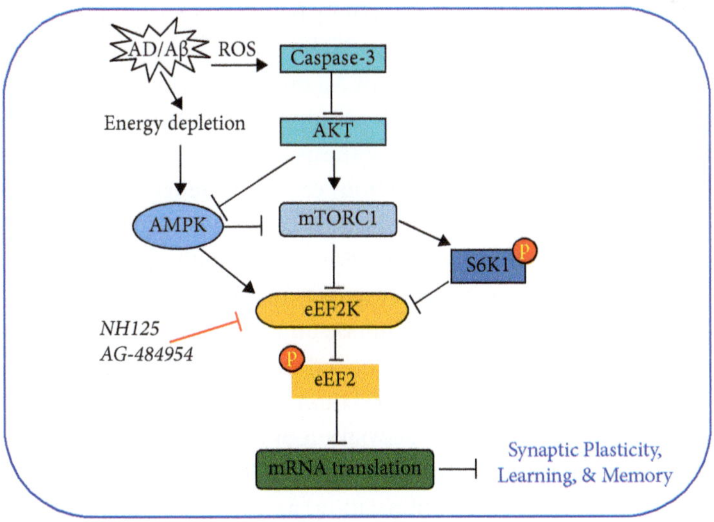

FIGURE 21.2. Schematic model of eEF2K/eEF2 signaling and its potential link to AD. Phosphorylation of eEF2 by eEF2K results in repression of general protein synthesis (mRNA translation). AMP-activated protein kinase (AMPK) activates eEF2K and thus increases eEF2 phosphorylation. In contrast, mammalian target of rapamycin complex 1 (mTORC1) inhibits eEF2K activity and decreases eEF2 phosphorylation, either directly or via its downstream target S6K1. NH125 and AG-484594 (or A-484594) are both selective inhibitors of eEF2K.

eEF2 phosphorylation, leading to promotion of translation elongation and general protein synthesis (Takei et al., 2009). Moreover, it was recently reported that eEF2K-eEF2 signaling controls synthesis of microtubule-related proteins, which play an important role in regulating dendritic spine morphology that is modified during synaptic plasticity, learning and memory (Kenney et al., 2016). In transgenic *eEF2K* homozygous knockout mice, late, protein synthesis-dependent LTP is enhanced (Park et al., 2008). As for LTD, another form of synaptic plasticity, genetic deletion of *eEF2K* results in defects in mGluR-LTD, but does not alter LTD that is N-methyl-D-aspartate (NMDA) receptor-dependent (Park et al., 2008). Interestingly, while phosphorylation of eEF2 (particularly via PKA activation) decreases general protein synthesis, it can increase translation of certain mRNAs including those for Arc and αCaMKII, which may play important roles in mGluR-LTD (Park et al., 2008; Taha, Gildish, Gal-Ben-Ari, & Rosenblum, 2013). Conversely, LTP is impaired in transgenic mice in which eEF2K is overexpressed (Im et al., 2009). Consistently, transgenic mice with eEF2K overexpression display impaired long-term contextual fear memory, but normal long-term cued fear memory. In addition, long-term hippocampus-dependent spatial memory, as assessed by the Morris water maze test, is also impaired in transgenic mice overexpressing eEF2K (Im et al., 2009).

Recent studies demonstrated abnormal hyper-phosphorylation of eEF2 (Thr56 site) in brain tissues of multiple lines of AD model mice, and in *post mortem* brain tissues from AD patients (Jan et al., 2017; X. Li, Alafuzoff, Soininen, Winblad, & Pei, 2005; Ma et al., 2014). Furthermore, hippocampal LTP failure induced by Aβ is rescued by the eEF2K inhibitor NH125 (Ma et al., 2014). In agreement, studies conducted in primary cultured neurons show that repression of eEF2K activity using compound A-484594 (a selective eEF2K inhibitor structurally distinct from NH125) or a genetic knockdown approach (shRNAs) alleviate the neurotoxic effects of Aβ on dendrite formation and cell viability (Jan et al., 2017). In a recent report, eEF2K is suppressed with genetic approaches in two separate lines of AD model mice. The hypothesis to be tested is whether suppression of eEF2 phosphorylation (via eEF2K inhibition) can alleviate AD-associated pathophysiology. First, it was found that AD-associated de novo protein synthesis deficits in hippocampi are improved by eEF2K suppression. Using multiple behavioral assays, the investigators demonstrated that cognitive defects in aged AD model mice are alleviated with repression of eEF2K. Consistently, hippocampal LTP impairments displayed in AD model mice are also mitigated by eEF2K knockdown. Interestingly, AD brain pathology including Aβ plaques and tau phosphorylation is unaltered with eEF2K inhibition. Moreover, reducing eEF2K results in alleviation of AD-related deficits in postsynaptic density formation, dendritic spine morphology, and dendritic polyribosome (clusters of ribosomes involved in active mRNA translation) assembly (Beckelman et al., 2019).

Conclusion

AD is the most common form of dementia syndromes without any cure currently available. Elucidation of the molecular mechanisms underlying AD etiology would provide important insights for the development of novel therapeutic targets and diagnostic biomarkers. Emerging evidence indicate an important role of protein synthesis dysregulation in AD pathophysiology. Recent studies suggest that boosting overall protein synthesis capacity via manipulation of multiple signaling pathways controlling mRNA translation can confer beneficial effects to AD-associated cognitive impairments and synaptic failure, thus representing a potential therapeutic strategy for AD and related dementia syndromes.

Future Perspectives

1. Both hypoactive and hyperactive protein synthesis are associated with impairments of cognitive function and synaptic plasticity. While boosting general protein synthesis could be a promising therapeutic strategy for AD, it is critical to fine-tune the treatment paradigm to avoid significant side effects due to interruption of mRNA translation homeostasis.
2. AD is a multifactorial disease that may involve dysregulation of multiple body systems. It would be interesting to determine whether better outcomes arise from targeting protein synthesis at the level of whole body or brain only. Along the same line, it would be intriguing to investigate the effects of manipulating protein synthesis "locally" (i.e., at dendrites or synapses), or in non-neuronal cells such as astrocytes and microglia.
3. In addition to general protein synthesis, it would be important to elucidate whether specifically targeting mRNA translation of plasticity-related proteins would be an effective therapeutic avenue. Findings from comprehensive, large-scale proteomics studies are expected to contribute significantly to the field.
4. Many studies using genetic approaches discussed in this chapter are elegantly designed and performed. Meanwhile, it is important to design and develop small molecule agents targeting protein synthesis regulation with minimal off-target effects to help advance the studies to human trials.
5. There is also an urgent need to develop diagnostic biomarkers for AD, especially at its early stage. Future studies, particularly in humans, are necessary to determine the correlation between dysregulations of signaling molecules discussed in this chapter (eIF2α, eEF2, etc.) and AD.
6. It is attractive to apply the strategy of targeting protein synthesis dysregulation to other AD-related dementia (ADRD).

Acknowledgments

I thank Antoine Almonte for comments and help in editing the manuscript. T. M. is supported by National Institutes of Health grants R01 AG055581 and R01 AG056622; the Alzheimer's Association grant NIRG-15-362799; the BrightFocus Foundation grant A2017457S; and Wake Forest University School of Medicine. I apologize for those authors whose studies were not discussed in this manuscript due to space limitation.

References

Alberini, C. M. (2008). The role of protein synthesis during the labile phases of memory: Revisiting the skepticism. *Neurobiology of Learning and Memory, 89*(3), 234–246.

Albert, V., & Hall, M. N. (2015). mTOR signaling in cellular and organismal energetics. *Current Opinion in Cell Biology, 33*, 55–66.

Alzheimer's Association. (2019). 2019 Alzheimer's disease facts and figures. *Alzheimer's Dementia, 15*(2019), 321–387.

Arnold, S. E., Arvanitakis, Z., Macauley-Rambach, S. L., Koenig, A. M., Wang, H-H.-Y., Ahima, R. S., . . . Nathan, D. M. (2018). Brain insulin resistance in type 2 diabetes and Alzheimer disease: Concepts and conundrums. *Nature Reviews Neurology, 14*(3), 168–181.

Arora, S., Yang, J.-M.-, Kinzy, T. G., Utsumi, R., Okamoto, T., Kitayama, T., . . . Hait, W. N. (2003). Identification and characterization of an inhibitor of eukaryotic elongation factor 2 kinase against human cancer cell lines. *Cancer Research, 63*(20), 6894–6899.

Arora, S., Yang, J.-M., Utsumi, R., Okamoto, T., Kitayama, T., & Hait, W. N. (2004). P-glycoprotein mediates resistance to histidine kinase inhibitors. *Molecular Pharmacology, 66*(3), 460–467.

Beckelman, B. C., Day, S., Zhou, X., Donohue, M., Gouras, G. K., Klann, E., . . . Ma, T. (2016). Dysregulation of elongation factor 1A expression is correlated with synaptic plasticity impairments in Alzheimer's disease. *Journal of Alzheimers Disease, 54*(2), 669–678.

Beckelman, B. C., Yang, W., Kasica, N. P., Zimmermann, H. R., Zhou, X., Keene, C. D., . . . Ma, T. (2019). Genetic reduction of eEF2 kinase alleviates pathophysiology in Alzheimer's disease model mice. *Journal of Clinical Investigation, 129*(2), 820–833.

Browne, G. J., & Proud, C. G. (2002). Regulation of peptide-chain elongation in mammalian cells. *European Journal of Biochemistry, 269*(22), 5360–5368.

Caccamo, A., Branca, C., Talboom, J. S., Shaw, D. M., Turner, D., Ma, L., . . . Oddo, S. (2015). Reducing ribosomal protein s6 kinase 1 expression improves spatial memory and synaptic plasticity in a mouse model of Alzheimer's disease. *Journal of Neuroscience, 35*(41), 14042–14056.

Caccamo, A., Majumder, S., Richardson, A., Strong, R., & Oddo, S. (2010). Molecular interplay between mammalian target of rapamycin (mTOR), amyloid-beta, and Tau: Effects on cognitive impairments. *Journal of Biological Chemistry, 285*(17), 13107–13120.

Caccamo, A., Pinto, V. De, Messina, A., Branca, C., & Oddo, S. (2014). Genetic reduction of mammalian target of rapamycin ameliorates Alzheimer's disease-like cognitive and pathological deficits by restoring hippocampal gene expression signature. *Journal of Neuroscience, 34*(23), 7988–7998.

Chang, R. C., Wong, A. K., Ng, H. K., & Hugon, J. (2002). Phosphorylation of eukaryotic initiation factor-2alpha (eIF2alpha) is associated with neuronal degeneration in Alzheimer's disease. *Neuroreport, 13*(18), 2429–2432.

Chen, Z., Gopalakrishnan, S. M., Bui, M.-H., Soni, N. B., Warrior, U., Johnson, E. F., . . . Glaser, K. B. (2011). 1-Benzyl-3-cetyl-2-methylimidazolium iodide (NH125) induces phosphorylation of eukaryotic elongation factor-2 (eEF2): A cautionary note on the anticancer mechanism of an eEF2 kinase inhibitor. *Journal of Biological Chemistry, 286*(51), 43951–43958.

Choo, A. Y., Roux, P. P., & Blenis, J. (2006). Mind the GAP: Wnt steps onto the mTORC1 train. *Cell, 126*, 834–836.

Cohen, E., Paulsson, J. F., Blinder, P., Burstyn-Cohen, T., Du, D., Estepa, G., . . . Dillin, A. (2009). Reduced IGF-1 signaling delays age-associated proteotoxicity in mice. *Cell, 139*(6), 1157–1169.

Cornejo, V. H., & Hetz, C. (2013). The unfolded protein response in Alzheimer's disease. *Seminars in Immunopathology, 35*(3), 277–292.

Costa-Mattioli, M., Gobert, D., Stern, E., Gamache, K., Colina, R., Cuello, C., . . . Sonenberg, N. (2007). eIF2alpha phosphorylation bidirectionally regulates the switch from short- to long-term synaptic plasticity and memory. *Cell, 129*(1), 195–206.

Costa-Mattioli, M., Sossin, W. S., Klann, E., & Sonenberg, N. (2009). Translational control of long-lasting synaptic plasticity and memory. *Neuron, 61*(1), 10–26.

Craft, S., Baker, L. D., Montine, T. J., Minoshima, S., Watson, G. S., Claxton, A., . . . Gerton, B. (2012). Intranasal insulin therapy for Alzheimer disease and amnestic mild cognitive impairment: A pilot clinical trial. *Archives of Neurology, 69*(1), 29–38.

De Felice, F. G., Vieira, M. N. N., Bomfim, T. R., Decker, H., Velasco, P. T., Lambert, M. P., . . . Klein, W. L. (2009). Protection of synapses against Alzheimer's-linked toxins: Insulin signaling prevents the pathogenic binding of Abeta oligomers. *Proceedings of the National Academy of Sciences USA, 106*(6), 1971–1976.

Dumont, M., Wille, E., Stack, C., Calingasan, N. Y., Beal, M. F., & Lin, M. T. (2009). Reduction of oxidative stress, amyloid deposition, and memory deficit by manganese superoxide dismutase overexpression in a transgenic mouse model of Alzheimer's disease. *FASEB Journal, 23*(8), 2459–2466.

Dyer, M. R., Renner, W. A., & Bachmann, M. F. (2006). A second vaccine revolution for the new epidemics of the 21st century. *Drug Discovery Today, 11*(21–22), 1028–1033.

Fruman, D. A., Chiu, H., Hopkins, B. D., Bagrodia, S., Cantley, L. C., & Abraham, R. T. (2017). The PI3K pathway in human disease. *Cell, 170*(4), 605–635.

Fumagalli, S., & Thomas, G. (2000). S6 phosphorylation and signal transduction. In N. Sonenberg, W. B. Hershey, & M. B. Mathews (Eds.), *Translational Control of Gene Expression* (pp. 695–717). Cold Spring Harbor, NY: Cold Spring Harbor Laboratory Press.

Gerakis, Y., & Hetz, C. (2018). Emerging roles of ER stress in the etiology and pathogenesis of Alzheimer's disease. *FEBS Journal, 285*(6), 995–1011.

Gingras, A. C., Gygi, S. P., Raught, B., Polakiewicz, R. D., Abraham, R. T., Hoekstra, M. F., . . . Sonenberg, N. (1999). Regulation of 4E-BP1 phosphorylation: a novel two-step mechanism. *Genes & Development, 13*(11), 1422–1437.

Graber, T. E., McCamphill, P. K., & Sossin, W. S. (2013). A recollection of mTOR signaling in learning and memory. *Learning & Memory, 20*(10), 518–530.

Hardie, D. G. (2004). The AMP-activated protein kinase pathway—new players upstream and downstream. *Journal of Cell Science, 117*(23), 5479–5487.

Hardie, D. G. (2014). AMPK—sensing energy while talking to other signaling pathways. *Cell Metabolism, 20*(6), 939–952.

Harrison, D. E., Strong, R., Sharp, Z. D., Nelson, J. F., Astle, C. M., Flurkey, K., . . . Miller, R. A. (2009). Rapamycin fed late in life extends lifespan in genetically heterogeneous mice. *Nature, 460*(7253), 392–395.

Hay, N., & Sonenberg, N. (2004). Upstream and downstream of mTOR. *Genes & Development, 18*(16), 1926–1945.

Herrup, K. (2015). The case for rejecting the amyloid cascade hypothesis. *Nature Neuroscience, 18*(6), 794–799.

Hetz, C., & Saxena, S. (2017). ER stress and the unfolded protein response in neurodegeneration. *Nature Reviews Neurology, 13*(8), 477–491.

Hoeffer, C. A., & Klann, E. (2010). mTOR signaling: At the crossroads of plasticity, memory and disease. *Trends in Neuroscience, 33*(2), 67–75.

Hoeffer, C. A., Tang, W., Wong, H., Santillan, A., Patterson, R. J., Martinez, L. A., ... Klann, E. (2008). Removal of FKBP12 enhances mTOR-Raptor interactions, LTP, memory, and perseverative/repetitive behavior. *Neuron, 60*(5), 832–845.

Holtzman, D. M., Goate, A., Kelly, J., & Sperling, R. (2011). Mapping the road forward in Alzheimer's disease. *Science Translation Medicine, 3*(114), 114ps148.

Holtzman, D. M., Morris, J. C., & Goate, A. M. (2011). Alzheimer's disease: The challenge of the second century. *Science Translation Medicine, 3*(77), 77sr71.

Horman, S., Browne, G. J., Krause, U., Patel, J. V., Vertommen, D., Bertrand, L., ... Rider, M. H. (2002). Activation of AMP-activated protein kinase leads to the phosphorylation of elongation factor 2 and an inhibition of protein synthesis. *Current Biology, 12*(16), 1419–1423.

Hu, N.-W., Nicoll, A. J., Zhang, D., Mably, A. J., O'Malley, T., Purro, S. A., Terry, C., ... Rowan, M. J. (2014). mGlu5 receptors and cellular prion protein mediate amyloid-β-facilitated synaptic long-term depression in vivo. *Nature Communications, 5*, 3374.

Im, H.-I., Nakajima, A., Gong, B., Xiong, X., Mamiya, T., Gershon, E. S., ... Tang, Y-P. (2009). Post-training dephosphorylation of eEF-2 promotes protein synthesis for memory consolidation. *PLoS One, 4*(10), e7424.

Jan, A., Jansonius, B., Delaidelli, A., Somasekharan, S. P., Bhanshali, F., Vandal, M., ... Sorensen, P. H. (2017). eEF2K inhibition blocks Aβ42 neurotoxicity by promoting an NRF2 antioxidant response. *Acta Neuropathologica, 133*(1), 101–119.

Jiang, H.-Y., Wek, S. A., McGrath, B. C., Lu, D., Hai, T., Harding, H. P., ... Wek, R. C. (2004). Activating transcription factor 3 is integral to the eukaryotic initiation factor 2 kinase stress response. *Molecular and Cellular Biology, 24*(3), 1365–1377.

Jo, J., Whitcomb, D. J., Olsen, K. M., Kerrigan, T. L., Lo, S.-C., Bru-Mercier, G., ... Cho, K. (2011). Aβ(1-42) inhibition of LTP is mediated by a signaling pathway involving caspase-3, Akt1 and GSK-3β. *Nature Neuroscience, 14*(5), 545–547.

Kelleher, R. J., 3rd, Govindarajan, A., & Tonegawa, S. (2004). Translational regulatory mechanisms in persistent forms of synaptic plasticity. *Neuron, 44*(1), 59–73.

Kenney, J. W., Genheden, M., Moon, K.-M., Wang, X., Foster, L. J., & Proud, C. G. (2016). Eukaryotic elongation factor 2 kinase regulates the synthesis of microtubule-related proteins in neurons. *Journal of Neurochemistry, 136*(2), 276–284.

Kenney, J. W., Moore, C. E., Wang, X., & Proud, C. G. (2014). Eukaryotic elongation factor 2 kinase, an unusual enzyme with multiple roles. *Advances in Biological Regulation, 55*, 15–27.

Kim, S. M., Yoon, S. Y., Choi, J. E., Park, J. S., Choi, J. M., Nguyen, T., & Kim, D. H. (2010). Activation of eukaryotic initiation factor-2 α-kinases in okadaic acid-treated neurons. *Neuroscience, 169*(4), 1831–1839.

Klann, E., Antion, M. D., Banko, J. L., & Hou, L. (2004). Synaptic plasticity and translation initiation. *Learning and Memory, 11*(4), 365–372.

Klann, E., & Dever, T. E. (2004). Biochemical mechanisms for translational regulation in synaptic plasticity. *Nature Reviews Neuroscience, 5*(12), 931–942.

Lafay-Chebassier, C., Paccalin, M., Page, G., Barc-Pain, S., Perault-Pochat, M. C., Gil, R., ... Hugon, J. (2005). mTOR/p70S6k signaling alteration by Abeta exposure as well as in APP-PS1 transgenic models and in patients with Alzheimer's disease. *Journal of Neurochemistry, 94*(1), 215–225.

Lafay-Chebassier, C., Pérault-Pochat, M. C., Page, G., Bilan, A. R., Damjanac, M., Pain, S., ... Hugon, J. (2006). The immunosuppressant rapamycin exacerbates neurotoxicity of Abeta peptide. *Journal of Neuroscience Research, 84*(6), 1323–1334.

Laplante, M., & Sabatini, D. M. (2012). mTOR signaling in growth control and disease. *Cell, 149*(2), 274–293.

Lee, H.-K., Kumar, P., Fu, Q., Rosen, K. M., & Querfurth, H. W. (2009). The insulin/Akt signaling pathway is targeted by intracellular beta-amyloid. *Molecular Biology of the Cell, 20*(5), 1533–1544.

Lehman, J. A., Calvo, V., & Gomez-Cambronero, J. (2003). Mechanism of ribosomal p70S6 kinase activation by granulocyte macrophage colony-stimulating factor in neutrophils: cooperation of a MEK-related, THR421/SER424 kinase and a rapamycin-sensitive, m-TOR-related THR389 kinase. *Journal of Biological Chemistry, 278*(30), 28130–28138.

Li, S., Hong, S., Shepardson, N. E., Walsh, D. M., Shankar, G. M., & Selkoe, D. (2009). Soluble oligomers of amyloid Beta protein facilitate hippocampal long-term depression by disrupting neuronal glutamate uptake. *Neuron, 62*(6), 788–801.

Li, X., Alafuzoff, I., Soininen, H., Winblad, B., & Pei, J.-J. (2005). Levels of mTOR and its downstream targets 4E-BP1, eEF2, and eEF2 kinase in relationships with tau in Alzheimer's disease brain. *FEBS Journal, 272*(16), 4211–4220.

Lin, M. T., & Beal, M. F. (2006). Mitochondrial dysfunction and oxidative stress in neurodegenerative diseases. *Nature, 443*(7113), 787–795.

Loreni, F., Thomas, G., & Amaldi, F. (2000). Transcription inhibitors stimulate translation of 5' TOP mRNAs through activation of S6 kinase and the mTOR/FRAP signalling pathway. *European Journal of Biochemistry, 267*(22), 6594–6601.

Ma, T., Chen, Y., Vingtdeux, V., Zhao, H., Viollet, B., Marambaud, P., & Klann, E. (2014). Inhibition of AMP-activated protein kinase signaling alleviates impairments in hippocampal synaptic plasticity induced by amyloid β. *Journal of Neuroscience, 34*(36), 12230–12238.

Ma, T., Du, X., Pick, J. E., Sui, G., Brownlee, M., & Klann, E. (2012). Glucagon-like peptide-1 cleavage product GLP-1 (9-36) amide rescues synaptic plasticity and memory deficits in Alzheimer's disease model mice. *Journal of Neuroscience, 32*(40), 13701–13708.

Ma, T., Hoeffer, C. A., Capetillo-Zarate, E., Yu, F., Wong, H., Lin, M. T.,... Gouras, G. K. (2010). Dysregulation of the mTOR pathway mediates impairment of synaptic plasticity in a mouse model of Alzheimer's disease. *PLoS One, 5*(9), e12845.

Ma, T., & Klann, E. (2012). Amyloid b: Linking synaptic plasticity failure to memory disruption in Alzheimer's disease. *Journal of Neurochemistry, 120 Suppl. 1*, 140–148.

Ma, T., & Klann, E. (2014). PERK: A novel therapeutic target for neurodegenerative diseases? *Alzheimers Research & Therapy, 6*(3), 30.

Ma, T., Trinh, M. A., Wexler, A. J., Bourbon, C., Gatti, E., Pierre, P.,... Klann, E. (2013). Suppression of eIF2α kinases alleviates Alzheimer's disease-related plasticity and memory deficits. *Nature Neuroscience, 16*(9), 1299–1305.

Magrané, J., Rosen, K. M., Smith, R. C., Walsh, K., Gouras, G. K., & Querfurth, H. W. (2005). Intraneuronal beta-amyloid expression downregulates the Akt survival pathway and blunts the stress response. *Journal of Neuroscience, 25*(47), 10960–10969.

Manning, B. D., & Cantley, L. C. (2007). AKT/PKB signaling: navigating downstream. *Cell, 129*, 1261–1274.

Massaad, C. A., Washington, T. M., Pautler, R. G., & Klann, E. (2009). Overexpression of SOD-2 reduces hippocampal superoxide and prevents memory deficits in a mouse model of Alzheimer's disease. *Proceedings of the National Academy of Sciences USA, 106*(32), 13576–13581.

Meyuhas, O. (2000). Synthesis of the translational apparatus is regulated at the translational level. *European Journal of Biochemistry, 267*(21), 6321–6330.

Meyuhas, O., & Dreazen, A. (2009). Ribosomal protein S6 kinase from TOP mRNAs to cell size. *Progress in Molecular Biology and Translational Science, 90*, 109–153.

Moreno, J. A., Halliday, M., Molloy, C., Radford, H., Verity, N., Axten, J. M., . . . Mallucci, G. R. (2013). Oral treatment targeting the unfolded protein response prevents neurodegeneration and clinical disease in prion-infected mice. *Science Translational Medicine, 5*(206), 206ra138.

Moreno, J. A., Radford, H., Peretti, D., Steinert, J. R., Verity, N., Martin, M. G., . . . Mallucci, G. R. (2012). Sustained translational repression by eIF2α-P mediates prion neurodegeneration. *Nature, 485*(7399), 507–511.

Neff, F., Flores-Dominguez, D., Ryan, D. P., Horsch, M., Schröder, S., Adler, T., . . . Ehninger, D. (2013). Rapamycin extends murine lifespan but has limited effects on aging. *Journal of Clinical Investigation, 123*(8), 3272–3291. doi: 10.1172/JCI67674

O'Connor, T., Sadleir, K. R., Maus, E., Velliquette, R. A., Zhao, J., Cole, S. L., . . . Vassar, Rt. (2008). Phosphorylation of the translation initiation factor eIF2alpha increases BACE1 levels and promotes amyloidogenesis. *Neuron, 60*(6), 988–1009.

Park, S., Park, J. M., Kim, S., Kim, J.-A., Shepherd, J. D., Smith-Hicks, C. L., . . . Worley, P. F. (2008). Elongation factor 2 and fragile X mental retardation protein control the dynamic translation of Arc/Arg3.1 essential for mGluR-LTD. *Neuron, 59*(1), 70–83.

Paschen, W., Proud, C. G., & Mies, G. (2007). Shut-down of translation, a global neuronal stress response: Mechanisms and pathological relevance. *Current Pharmaceutical Design, 13*(18), 1887–1902.

Piper, M. D. W., Selman, C., McElwee, J. J., & Partridge, L. (2008). Separating cause from effect: How does insulin/IGF signalling control lifespan in worms, flies and mice? *Journal of Internal Medicine, 263*(2), 179–191.

Potter, W. B., O'Riordan, K. J., Barnett, D., Osting, S. M. K., Wagoner, M., Burger, C., & Roopra, A. (2010). Metabolic regulation of neuronal plasticity by the energy sensor AMPK. *PLoS One, 5*(2), e8996.

Querfurth, H. W., & LaFerla, F. M. (2010). Alzheimer's disease. *New England Journal of Medicine, 362*(4), 329–344.

Radford, H., Moreno, J. A., Verity, N., Halliday, M., & Mallucci, G. R. (2015). PERK inhibition prevents tau-mediated neurodegeneration in a mouse model of frontotemporal dementia. *Acta Neuropathologica, 130*(5), 633–642.

Reger, M. A., Watson, G. S., Green, P. S., Wilkinson, C. W., Baker, L. D., Cholerton, B., . . . Craft, S. (2008). Intranasal insulin improves cognition and modulates beta-amyloid in early AD. *Neurology, 70*(6), 440–448.

Reiling, J. H., & Sabatini, D. M. (2006). Stress and mTORture signaling. *Oncogene, 25*(48), 6373–6383.

Richter, J. D., Bassell, G. J., & Klann, E. (2015). Dysregulation and restoration of translational homeostasis in fragile X syndrome. *Nature Reviews Neuroscience, 16*(10), 595–605.

Roberson, E. D., & Mucke, L. (2006). 100 years and counting: Prospects for defeating Alzheimer's disease. *Science, 314*(5800), 781–784.

Rowan, M. J., Klyubin, I., Wang, Q., & Anwyl, R. (2005). Synaptic plasticity disruption by amyloid beta protein: Modulation by potential Alzheimer's disease modifying therapies. *Biochemical Society Transactions, 33*(4), 563–567.

Sarbassov, D. D., Ali, S. M., Sengupta, S., Sheen, J.-H., Hsu, P. P., Bagley, A. F., . . . Sabatini, D. M. (2006). Prolonged rapamycin treatment inhibits mTORC2 assembly and Akt/PKB. *Molecular Cell, 22*(2), 159–168.

Selkoe, D. J. (2002). Alzheimer's disease is a synaptic failure. *Science, 298*(5594), 789–791.

Selman, C., Tullet, J. M. A., Wieser, D., Irvine, E., Lingard, S. J., Choudhury, Agharul I., . . . Withers, D. J. (2009). Ribosomal protein S6 kinase 1 signaling regulates mammalian life span. *Science, 326*(5949), 140–144.

Sene'e, V., Vattem, K. M., Dele'pine, M., Rainbow, L. A., Haton, C., Lecoq, A.,... Julier, C. (2004). Wolcott-Rallison syndrome: Clinical, genetic, and functional study of EIF2AK3 mutations and suggestion of genetic heterogeneity. *Diabetes, 53*(7), 1876–1883.

Spilman, P., Podlutskaya, N., Hart, M. J., Debnath, J., Gorostiza, O., Bredesen, D.,... Galvan, V. (2010). Inhibition of mTOR by rapamycin abolishes cognitive deficits and reduces amyloid-beta levels in a mouse model of Alzheimer's disease. *PLoS One, 5*(4), e9979.

Sutton, M. A., & Schuman, E. M. (2006). Dendritic protein synthesis, synaptic plasticity, and memory. *Cell, 127*, 49–58.

Taha, E., Gildish, I., Gal-Ben-Ari, S., & Rosenblum, K. (2013). The role of eEF2 pathway in learning and synaptic plasticity. *Neurobiology of Learning and Memory, 105*, 100–106.

Takei, N., Kawamura, M., Ishizuka, Y., Kakiya, N., Inamura, N., Namba, H., & Nawa, H. (2009). Brain-derived neurotrophic factor enhances the basal rate of protein synthesis by increasing active eukaryotic elongation factor 2 levels and promoting translation elongation in cortical neurons. *Journal of Biological Chemistry, 284*(39), 26340–26348.

Talboom, J. S., Velazquez, R., & Oddo, S. (2015). The mammalian target of rapamycin at the crossroad between cognitive aging and Alzheimer's disease. *NPJ Aging and the Mechanisms of Disease, 1*, 15008.

Teich, A. F., Nicholls, R. E., Puzzo, D., Fiorito, J., Purgatorio, R., Fa', M., & Arancio, O. (2015). Synaptic therapy in Alzheimer's disease: A CREB-centric approach. *Neurotherapeutics, 12*(1), 29–41.

Townsend, M., Mehta, T., & Selkoe, D. J. (2007). Soluble Abeta inhibits specific signal transduction cascades common to the insulin receptor pathway. *Journal of Biological Chemistry, 282*(46), 33305–33312.

Trinh, M. A., Kaphzan, H., Wek, R. C., Pierre, P., Cavener, D. R., & Klann, E. (2012). Brain-specific disruption of the eIF2α kinase PERK decreases ATF4 expression and impairs behavioral flexibility. *Cell Reports, 1*(6), 676–688.

Trinh, M. A., & Klann, E. (2013). Translational control by eIF2α kinases in long-lasting synaptic plasticity and long-term memory. *Neurobiology of Learning and Memory, 105*, 93–99.

Trinh, M. A., Ma, T., Kaphzan, H., Bhattacharya, A., Antion, M. D., Cavener, D. R.,... Klann, E. (2014). The eIF2α kinase PERK limits the expression of hippocampal metabotropic glutamate receptor-dependent long-term depression. *Learning and Memory, 21*(5), 298–304.

Tse, K.-H., & Herrup, K. (2017). Re-imagining Alzheimer's disease: The diminishing importance of amyloid and a glimpse of what lies ahead. *Journal of Neurochemistry, 143*(4), 432–444.

Tsokas, P., Grace, E. A., Chan, P., Ma, T., Sealfon, S. C., Iyengar, R.,... Blitzer, R. D. (2005). Local protein synthesis mediates a rapid increase in dendritic elongation factor 1A after induction of late long-term potentiation. *Journal of Neuroscience, 25*(24), 5833–5843.

Tsokas, P., Ma, T., Iyengar, R., Landau, E. M., & Blitzer, R. D. (2007). Mitogen-activated protein kinase upregulates the dendritic translation machinery in long-term potentiation by controlling the mammalian target of rapamycin pathway. *Journal Neuroscience, 27*, 5885–5894.

Um, J.-W., Kaufman, A. C., Kostylev, M., Heiss, J. K., Stagi, M., Takahashi, H.,... Strittmatter, S. M. (2013). Metabotropic glutamate receptor 5 is a coreceptor for Alzheimer aβ oligomer bound to cellular prion protein. *Neuron, 79*(5), 887–902.

Vingtdeux, V., Davies, P., Dickson, D. W., & Marambaud, P. (2011). AMPK is abnormally activated in tangle- and pre-tangle-bearing neurons in Alzheimer's disease and other tauopathies. *Acta Neuropathologica, 121*(3), 337–349.

Walsh, D. M., & Selkoe, D. J. (2004). Deciphering the molecular basis of memory failure in Alzheimer's disease. *Neuron, 44*(1), 181–193.

Wek, R. C. (2018). Role of eIF2α kinases in translational control and adaptation to cellular stress. *Cold Spring Harbor Perspectives in Biology, 10*(7), pii: a032870.

Wek, R. C., & Cavener, D. R. (2007). Translational control and the unfolded protein response. *Antioxidants & Redox Signaling, 9*(12), 2357–2371.

Wek, R. C., Jiang, H.-Y., & Anthony, T. G. (2006). Coping with stress: eIF2 kinases and translational control. *Biochemical Society Transactions, 34*(Pt 1), 7–11.

Yang, Q., & Guan, K.-L. (2007). Expanding mTOR signaling. *Cell Research, 17*, 666–681.

Yang, W., Zhou, X., Cavener, H. R., Zimmermann D. R., Klann, E., & Ma, T. (2016). Repression of the eIF2α kinase PERK alleviates mGluR-LTD impairments in a mouse model of Alzheimer's disease. *Neurobiology of Aging, 41*, 19–24.

Zhan, K., Narasimhan, J., & Wek, R. C. (2004). Differential activation of eIF2 kinases in response to cellular stresses in *Schizosaccharomyces pombe*. *Genetics, 168*(4), 1867–1875.

Zhang, P., McGrath, B., Li, S., Frank, A., Zambito, F., Reinert, J., . . . Cavener, D. R. (2002). The PERK eukaryotic initiation factor 2 alpha kinase is required for the development of the skeletal system, postnatal growth, and the function and viability of the pancreas. *Molecular and Cellular Biology, 22*(11), 3864–3874.

Zhang, Y., Dong, Z., Nomura, M., Zhong, S., Chen, N., & Bode, A. M. (2001). Signal transduction pathways involved in phosphorylation and activation of p70S6K following exposure to UVA irradiation. *Journal of Biological Chemistry, 276*(24), 20913–20923.

CHAPTER 22

RNA-BINDING PROTEINS AND TRANSLATION IN NEURODEGENERATIVE DISEASE

KENT E. DUNCAN

INTRODUCTION

MAJOR advances in human genetics coupled with increasing appreciation of the importance of post-transcriptional regulatory mechanisms have revealed key roles for RNA-binding proteins (RBPs) in neurodegenerative diseases. In parallel, altered translation and balance between protein synthesis and turnover have also emerged as major molecular control points in the cellular pathology of several neurodegenerative diseases. As key mediators of mRNA-specific translational control, RBPs seem strong candidates to orchestrate translation in neurodegenerative diseases. But do they?

This review explores potential connections between RBPs and translational control in specific neurodegenerative diseases. Its fundamental goal is to provide a critical evaluation of the evidence that altered translation of specific mRNAs mediated by RNA-binding proteins may play an important role in specific neurodegenerative diseases. Specific suggestions for how this evidence could be made stronger through future experimentation are presented in context. Broader implications for RNA biology in neurodegenerative disease are also covered, without shying away from unorthodox hypotheses for disease etiology.

A major focus is on amyotrophic lateral sclerosis (ALS) and frontotemporal dementia (FTD) since human genetic studies strongly support a causal role for RNA-binding proteins in these neurodegenerative diseases. Particular emphasis in this section is placed on TDP-43 in ALS/FTD, since several studies support translational effects of overexpression of WT and/or patient mutant alleles of TDP-43 on specific mRNA targets that could conceivably be relevant to disease. Both these studies and their potential implications for disease are described in some detail. Evidence for altered translation in fused in sarcoma (FUS) disease models is also covered, as is the potential for translational regulation by other RBPs implicated in ALS/FTD.

Subsequently, neurodegenerative diseases are presented where evidence for altered translation is substantial, but the contribution of RNA-binding proteins is less clear from human genetics. These include Huntington's disease, where a role for MID1 and associated proteins in modulating translation driven by extended CAG repeats was first suggested by biochemical and cell-based studies. The emerging topic of repeat-associated non-AUG-initiated (RAN) translation and its modulation by RBPs is covered in the context of Fragile X tremor/ataxia syndrome and ALS/FTD caused by *C9ORF72* repeat expansions. Connections of key Parkinson's proteins, PINK1, PARKIN, and LRRK2 to RBP-mediated translation are reviewed, as is the potential role of atypical RBP DJ-1.

Because the article focuses on protein-mRNA interactions in translational control of protein synthesis, several topics which might conceivably fit the theme are not covered. Canonical translational control via signaling pathways, such as mechanisms that target 4E-binding proteins or eIF2α phosphorylation are only covered inasmuch as they relate to effects on RNA-binding proteins. Neurodegeneration caused by mutations in amino acyl tRNA synthetases or other effects on tRNA biology are not covered. MicroRNA function in neurodegenerative disease is also only covered briefly and specifically in the context of direct connections to RNA-binding proteins implicated in disease. Comprehensive reviews on these topics are available for interested readers (e.g. Juzwik et al., 2019; Kapur, Monaghan, & Ackerman, 2017; Kirchner & Ignatova, 2015; Shen et al., 2018), including several chapters from this volume (see the chapters in this volume including chapters 2, 3, 10, 11 and 23)

Throughout the article, emphasis is placed on how likely specific proteins and mechanisms are to be relevant to disease in patients. Evidence from human genetic and in vivo studies in animal models takes primacy, but studies using mainly cell-based or cell-free assays are also discussed. The extent to which such "bottom up" approaches will reveal mechanisms that actually operate in vivo in neurodegenerative disease is not yet clear. However, the hope is that they will motivate in vivo experiments designed to determine whether the mechanisms defined in culture are truly relevant to disease. Ideally, iterative combinations of in vitro and in vivo approaches, coupled with rigorous human genetic analysis, will provide significant new insight into disease mechanisms and reveal new risk factors and targets for therapeutic intervention.

RNA-Binding Proteins and Translational Control in Amyotrophic Lateral Sclerosis and Frontotemporal Dementia

Amyotrophic lateral sclerosis (ALS) is a progressive, adult-onset neurodegenerative disorder affecting both upper and lower motor neurons (UMNs and LMNs) (Rowland & Shneider, 2001). Disease typically involves focal onset, followed by gradual spreading and a steady decline in motor function leading to complete ataxia and paralysis, with most patients dying within 2-5 years. The two approved therapies only modestly slow disease progression, highlighting an urgent need to develop better therapeutics, which would be aided by better understanding of the molecular mechanisms that drive disease (Jaiswal, 2019).

Identifying ALS genes has defined key molecular pathways in disease and enabled development of cell and animal models (Cook & Petrucelli, 2019). ALS can be sporadic (sALS) or familial (fALS), accounting for ~90% and ~10% of ALS, respectively Mutations in superoxide dismutase 1 (*SOD1*) were the first identified genetic cause of fALS and were used to generate mouse models that develop motor symptoms strikingly similar to human ALS (Joyce, Fratta, Fisher, & Acevedo-Arozena, 2011). Lack of success with therapeutic development based on these models has led some to suggest that *SOD1*-driven ALS might be a special form of disease (Turner et al., 2013).

RNA-Binding Protein Aggregates and Mutants in ALS and FTD

The ALS field was revolutionized when TAR DNA-binding protein-43 (TDP-43) was identified in ubiquitinated aggregates in brain and spinal cord of most ALS patients, (Arai et al., 2006; Neumann et al., 2006). "TDP-43 pathology" is now a hallmark of both ALS and Frontotemporal dementia (FTD), a heterogeneous disorder caused by cortical neurodegeneration affecting language and behavior (Rademakers, Neumann, & Mackenzie, 2012).

TDP-43 mutations are found in fALS, sALS, and FTD patients (Kabashi et al., 2008; Sreedharan et al., 2008; Therrien, Dion, & Rouleau, 2016), accounting for 4.2% of fALS and ~1% sALS in Europeans (reviewed in Mejzini et al., 2019). Despite its name, TDP-43 has two RNA-recognition motifs (RRMs) and seems primarily to be an RBP. Patient mutations cluster in its glycine-rich domain. A different set of ALS/FTD patients have mutations in another RBP, fused in sarcoma (FUS), and FUS protein aggregates in affected neurons. Thus, the four most commonly mutated genes in ALS include two encoding RNA-binding proteins which are also mutated in FTD (reviewed in Baradaran-Heravi, Van Broeckhoven, & van der Zee, 2020).

The most common genetic cause of ALS and FTD in Europeans is a hexanucleotide repeat expansion in *C9ORF72* (DeJesus-Hernandez et al., 2011; Turner et al., 2017). Repeat expansion leads to aberrant mRNAs with toxic properties (see below), further implicating aberrant RNA metabolism in ALS and FTD. Significant overlap in both mutated genes and pathological protein aggregate markers in affected neurons has led to the suggestion of an ALS/FTD "disease spectrum" (Ling, Polymenidou, & Cleveland, 2013; Rademakers et al., 2012) with altered RNA metabolism underlying neurodegeneration.

TDP-43 Binds to Many Cellular RNAs and Affects Multiple Aspects of RNA Regulation

TDP-43 aggregates are observed in degenerating MNs upon autopsy for ~97% of ALS patients. While this observation does not resolve when aggregates appear in disease or whether they are themselves toxic, it certainly makes TDP-43 of particular interest. Many studies have addressed TDP-43's normal functions and numerous cell and animal models of TDP-43-driven ALS/FTD have been developed. Models are often based on expression of patient mutant alleles of TDP-43, however merely overexpressing WT TDP-43 above a certain level can also trigger disease (Taylor, Brown, & Cleveland, 2016). In both fly and rodent models, TDP-43 expression within MNs appears sufficient for significant disease symptoms (Estes et al., 2011; Huang, Tong, Bi, Zhou, & Xia, 2012). This does not preclude contributions of other cells, but suggests an important role for TDP-43 deregulation within neurons that ultimately degenerate.

TDP-43 directly interacts with thousands of RNAs, often by binding UG repeats (Bhardwaj, Myers, Buratti, & Baralle, 2013; Polymenidou et al., 2011; Tollervey et al., 2011). Functional studies implicate TDP-43 in many post-transcriptional aspects of gene expression, including mRNA splicing, stability, transport, and translation (Cohen, Lee, & Trojanowski, 2011), as well as microRNA processing and long-noncoding RNA (lncRNA) expression (Ling et al., 2013; Polymenidou et al., 2012). Surprisingly, the characteristic aggregates observed at end-stage with patient autopsy material do not appear to be necessary for toxicity in numerous animal and cellular TDP-43 ALS models (Arnold et al., 2013; Taylor et al., 2016), but TDP-43's RNA-binding activity is necessary (Voigt et al., 2010).

Patient neurons with cytoplasmic TDP-43 aggregates frequently also show nuclear depletion (Dammer et al., 2012; Dormann & Haass, 2011; Giordana et al., 2010), but this is not required for motor neuron dysfunction in mouse models of ALS (Arnold et al., 2013). Other observations support TDP-43 gain-of-function in disease. For example, fALS patient mutations in TDP-43 are dominant and preserve most wild-type (WT) protein functions. Moreover, simply overexpressing WT TDP-43 at a high enough level is toxic in cells and leads to motor symptoms in animals (reviewed in Cohen et al., 2011)

Finally, since ~95% of TDP-43 is nuclear at steady-state, any accumulation in the cytoplasm- as observed in diseased cells- is effectively a gain of function there. But what is TDP-43 doing in the cytoplasm? Obviously, it can aggregate, as observed in patient neurons at end-stage of disease. However, since this does not seem to be necessary for initiation or progression of motor symptoms in animal models, it seems likely that mislocalized TDP-43 in disease affects cytoplasmic RNA metabolism even in the absence of aggregation, at least in early stages. TDP-43 CLIP-Seq experiments also indicated 3′UTR interactions (Polymenidou et al., 2011; Tollervey et al., 2011), suggesting a role in regulating cytoplasmic mRNA fate (Gebauer & Hentze, 2004; Kuersten & Goodwin, 2003). In principle, 3′UTR-bound TDP-43 could affect the stability, localization, or translation of these mRNAs.

TDP-43 Associates with Ribosome-mRNA Complexes and Can Affect General Translation

Proteomics revealed two major groups of TDP-43 interaction partners: nuclear splicing factors and cytoplasmic translation/ribosome components (Freibaum, Chitta, High, & Taylor, 2010; Sephton et al., 2011). TDP-43 knockdown globally increased translation in mammalian cell lines indirectly (Fiesel, Weber, Supper, Zell, & Kahle, 2012). In SH-SY5Y cells, oxidative stress was shown to lead to TDP-43 association with stalled ribosomes and stress granules (Higashi et al., 2013). Another study found TDP-43 in polysome fractions without stress and that TDP-43 lacking its nuclear localization signal reduced general translation, apparently due to interaction with the 40S ribosomal subunit component RACK1 (Russo et al., 2017). TDP-43 has also been found in proteomics of large RNA granules containing mainly polysomes (Elvira et al., 2006). However, overexpression of WT or a patient mutant TDP-43 in a mouse motor-neuronal cell line did not affect general translation (Neelagandan et al., 2019).

Evidence from Mammalian Cell Lines and Primary Neuronal Cultures that mRNA-Specific Translational Repression Is a Normal Function of TDP-43

Studies focused on TDP-43's normal function in dendrites revealed translational repression. TDP-43 co-localized with the post-synaptic marker, PSD-95 and mRNP granule components. Inducing neuronal activity with KCl, led to increased TDP-43 association with dendritic granules and expanded their size (I. F. Wang, Wu, Chang, & Shen, 2008). This study is frequently cited for demonstrating that TDP-43 is a translational repressor and/or regulates local translation. However, the only translation assay used Rabbit Reticulocyte Lysates (RRL) with GST-TDP-43 added at a > 1000X molar excess over a luciferase reporter with no mRNA stability control. It seems premature to conclude from this that TDP-43 has a physiological function as a translation repressor in neurons or that this is relevant to disease. A separate study in cultured-neuron dendrites also found that TDP-43 responds to neuronal activity and that ALS-patient mutants showed significantly reduced mobility in comparison to the WT protein, but translation was not evaluated. (Liu-Yesucevitz et al., 2014).

Later studies provided more compelling evidence for translational repression of specific mRNAs as a normal function of TDP-43. Depleting TDP-43 in primary cultures of mouse hippocampal neurons led to increased RAC1 protein levels without mRNA-level changes and a cycloheximide chase experiment supported an effect on protein synthesis, rather than stability (Majumder et al., 2012). A follow-up study (Majumder, Chu, Chatterjee, Swamy, & Shen, 2016), explored TDP-43 translational repression of RAC1 and other mRNAs in much greater detail, revealing TDP-43 interaction with UG/GU repeats in the *RAC1* mRNA 3′UTR and the fragile X mental retardation protein (FMRP) augmenting repression. The mechanism seems to involve interaction with CYFIP, an eIF4E-binding protein implicated in translational repression by FMRP (Napoli et al., 2008). Multiple studies suggest a key role for FMRP in translational repression of dendritically-localized mRNAs to support normal synaptic development and function (reviewed in Fernandez, Rajan, & Bagni, 2013; see the chapter in this volume by Taesun Eom, Muslimov, Iacoangeli, and Tiedge, Chapter 6). Imaging assays suggested a local effect of TDP-43 depletion on translation of these mRNAs, an idea strongly supported by a recent live-imaging reporter study (Chu, Majumder, Chatterjee, Huang, & Shen, 2019).

Altogether, TDP-43 and FMRP appear to cooperate in translational regulation of a common set of mRNA targets to promote aspects of neuronal cell biology that are crucial for normal neuronal development. However, because these results are based on loss of TDP-43 function it is not immediately clear how they relate to TDP-43's role in causing neurodegeneration.

Translational Repression of Futsch/MAP1B and Hsc70-4/HSPA8 mRNAs by TDP-43 in Flies

Using established fly models of ALS overexpressing either WT or mutant human TDP-43 (Estes et al., 2011), TDP-43 was shown (mainly with polysome fractionation assays) to repress translation of two mRNAs: *Futsch* (Coyne et al., 2014) and *Hsc 70-4* (Coyne et al., 2017). Protein levels for Futsch were reduced at the larval NMJ and increased in the motor neuron soma and the authors presented evidence for an mRNA localization defect (Coyne et al., 2014), reminiscent of effects of patient mutations on mRNA trafficking in axons of cultured rodent neurons (Alami et al., 2014). Hsc70-4 protein levels were also selectively reduced at synapses and this correlated with effects on the synaptic vesicle cycle and altered neurotransmission (Coyne et al., 2017). Importantly, genetic rescue experiments in fly ALS models supported the notion that deregulation of translation of Futsch and Hsc70-4 contributes to TDP-43's disease-promoting role. Moreover, both studies also provided evidence for deregulation of the orthologous proteins in disease in mammals, although it was not clear these effects are due to altered translation.

TDP-43 Enhances Translation of Specific mRNAs In Motor-Neuronal Cells and Primary Cortical Neurons

The only published genome-wide analysis of TDP-43 on translation revealed a novel function as a translational enhancer for specific mRNAs (Neelagandan et al., 2019). The impact of low-level overexpression of WT or patient mutant human TDP-43 on translation of specific mRNAs was assessed by ribosome profiling (Ingolia, Brar, Rouskin, McGeachy, & Weissman, 2012). This method was applied to both a cholinergic motor neuron cell line (Salazar-Grueso, Kim, & Kim, 1991), as well as cultured mouse cortical neurons, and revealed relatively few mRNAs with altered ribosome density, but without corresponding effects on mRNA levels in response to mutant TDP-43 expression.

Increased ribosome density on several mRNAs was validated by polysome profiling and endogenous TDP-43 interacted directly with these mRNAs. Modest increases in encoded protein levels were also observed by quantitative immunostaining. 5′ and 3′UTR luciferase reporters provided additional evidence that increased ribosome density reflects enhanced translation, rather than ribosome stalling. One target, DENND4A, was selectively regulated by the mutant protein through a specific region in its 3′UTR, although the basis for this allele-specific effect was not determined. Collectively, these experiments defined a new function for TDP-43 as a translational enhancer in cultured motor neuron-like cells. Moreover, among the few identified targets were mRNAs encoding proteins that had previously been implicated in neurodegenerative disease by several other studies, suggesting a potential link between translational enhancement of these mRNAs by TDP-43 and ALS.

TDP-43 Enhances Translation of CAMTA1 and DENND4A "Master Regulators" of Neurodegenerative Transcriptional Programs

Two mRNA targets for translational enhancement by TDP-43 in MN1 cells encoded CAMTA1 and DENND4A (Neelagandan et al., 2019), which had previously been implicated in cognitive performance and neurodegenerative disease by other studies. In humans, CAMTA1 variants are implicated in neurobehavioral abnormalities (Shinawi, Coorg, Shimony, Grange, & Al-Kateb, 2015), episodic memory performance (Huentelman et al., 2007), and ALS patient survival (Fogh et al., 2016). CAMTA1 knockout mice develop neurodegeneration and ataxia (Long et al., 2014). While it is not obvious that translational up-regulation would phenocopy the knockout, overexpression can sometimes lead to loss-of-function phenotypes, particularly with transcription factors (Prelich, 2012).

CAMTA1 and DENND4A were independently identified in two separate studies that searched for "master regulators" (MRs) of neurodegenerative transcription in a mouse in vivo model of Parkinson's disease (Brichta et al., 2015) and an in vitro model of ALS involving co-culturing of motor neurons and astrocytes (Ikiz et al., 2015). CAMTA1 and DENND4A were among the three MRs that overlapped. In the ALS co-culture study, knocking down DENND4A markedly increased motor neuron survival, suggesting that it is a motor neuron "death driver." This provides a straightforward potential explanation for how translational upregulation of *DENND4A* mRNA by mutant TDP-43 could contribute to disease.

It remains to be seen whether altered translation of *CAMTA1* and *DENND4A* mRNAs occurs in MNs in vivo and affects the timing of disease onset or progression. Nevertheless, these results reveal a different mode of action for TPD-43 from previously described repression. They also suggest testable hypotheses for the future, in particular that alterations in the translational status of neurodegeneration transcriptional master regulators are an early event in TDP-43-driven ALS, predisposing motor neurons to unleash the degenerative transcriptional response at a lower "stressor threshold" (Saxena & Caroni, 2011).

FUS Knockdown and ALS/FTD Patient Mutants Affect General Translation in Cultured Cells

Over 50 mutations in the RBP FUS are found in a subset of patients with familial, as well as sporadic fALS/FTD (reviewed in Baradaran-Heravi et al., 2020). FUS has an N-terminal low-complexity domain, RGG and RRM RNA-binding domains, and a PY-NLS recognized by the transportin-1/karyopherin-$\beta 2$ import receptor (Lee et al., 2006). Superficially, FUS and TDP-43 are similar (reviewed in Nussbacher, Batra, Lagier-Tourenne, & Yeo, 2015), although direct comparisons of genome-wide RNA-Seq data suggest important functional differences (Kapeli et al., 2016). Like TDP-43, FUS is mainly nuclear at steady-state in healthy cells (Andersson et al., 2008), but shuttles between nucleus and cytoplasm (Lagier-Tourenne, Polymenidou, & Cleveland, 2010). Patient mutations are frequently found in the PY-NLS, (Da Cruz & Cleveland, 2011). In cells, they lead to decreased interaction with the import receptor, increased FUS levels in the cytoplasm (Dormann et al., 2010), and aggregation via liquid-liquid phase separation (LLPS) (L. Guo et al., 2018; Hofweber et al., 2018; Qamar et al., 2018; Yoshizawa et al., 2018). While phase separation and aggregation might be the whole story, increased FUS localization in the cytoplasm and processes of CNS neurons and glia in disease could obviously also affect the translation, localization, and/or stability of any of the many mRNAs that it appears able to bind to in these cellular compartments.

Potential effects of FUS on translation have been examined in a few cases, typically with loss-of-function approaches. FUS knockdown in NIH/3T3 cells led to reduced levels of Kank2 protein, without affecting mRNA levels, and pulse labeling newly synthesized proteins with the methionine analogue azidohomoalanine (AHA) (Dieterich et al., 2007) supported this being a translational effect (Yasuda et al., 2013). *Kank2* mRNA and FUS protein were found in "APC RNA granules," which contribute to local translation in cell protrusions. Patient mutant FUS did not affect translation of Kank2, suggesting that this FUS function is not compromised in ALS/FTD.

Several additional loss-of-function studies have shown that depleting or deleting FUS in cultured hippocampal neurons alters dendritic spine morphology and density (Hicks et al., 2000; Udagawa et al., 2015) and one of these also showed that FUS accumulated in spines in response to mGLUR5 receptor activation (Fujii et al., 2005). It was suggested this could be due to FUS affecting dendritic mRNA localization and/or translation, although direct translation assays were not performed and patient mutants with these phenotypes were not described, making potential disease relevance hard to assess.

One study found that FUS mutations led to reduced translation in both cultured cell lines and human patient fibroblasts (Kamelgarn et al., 2018). Mutant FUS also affected nonsense-mediated decay (NMD), perhaps because NMD is intimately coupled to translation (Karousis & Muhlemann, 2019). Future work will be needed to understand the mechanistic basis for the impact of mutant FUS on translation and NMD.

Reduced Levels of Newly Synthesized Proteins in Axons of FUS ALS/FTD Mouse Models

New humanized mouse models of mutant FUS-driven ALS/FTD highlight altered neuronal translational control as a pathological mechanism (Lopez-Erauskin et al., 2018). Direct comparison of WT and mutant FUS lines revealed that only mutants developed age-dependent symptoms. Strikingly, this did not involve either nuclear depletion of FUS protein or aggregation, but activation of stress chaperones and the integrated stress response via eIF2α phosphorylation, with FUS mutant proteins and phospho-eIF2α accumulating in mutant axons. Puromycin labeling and detection in situ revealed decreased levels of puromycin-containing proteins in FUS mutant axons of cultured neurons and LMNs in vivo, suggesting decreased local protein synthesis. However, this result might also potentially reflect defective axonal protein transport, which is known to be strongly perturbed in cultured LMNs differentiated from patient-derived induced pluripotent stem cells (W. Guo, Naujock, et al., 2017).

FUS Interacts with miRNA Machinery to Promote Gene Silencing and Patient Mutations Compromise This Function

FUS can influence miRNA regulation of mRNA targets by interacting with Argonaute effector proteins (T. Zhang et al., 2018). Although FUS and TDP-43 have both been implicated previously in miRNA biogenesis (Kawahara & Mieda-Sato, 2012; King et al., 2014; Morlando et al., 2012), this study revealed that FUS can function directly together with mature miRISC complexes to influence directly its gene silencing activity toward specific FUS mRNA targets. This conclusion rests on a solid series of biochemical and functional miRNA reporter assays in mammalian cell lines and FUS KO mouse embryonic fibroblasts, backed in vivo by genetics in *C. elegans*. The loss-of-function FUS phenotype in miRNA-mediated repression was not complemented by the ALS-patient frameshift mutation R495X, which leads to a truncated protein. In contrast, the patient mutation R521C functioned as well as WT FUS. Given that the effect of FUS on miRNA function appears to be due to loss of function and is patient allele-specific, it is not clear at present how generally this function would be compromised in disease.

In Zhang et al. 2018 FUS promoted mRNA decay. However, miRNAs can also regulate translation and mRNA-interacting RBPs have been proposed to influence which effect occurs (Behm-Ansmant, Rehwinkel, & Izaurralde, 2006). Thus, FUS might conceivably repress translation of other mRNA targets relevant to disease via its interaction with AGO proteins and miRISC. Future work will be needed to determine whether altered FUS-AGO interaction also affects translation and is an important contributor to disease in vivo in mammalian ALS/FTD models.

Does FUS Have Specific Translational mRNA Targets in the CNS in Disease?

In sum, several studies directly support effects of FUS on translation or link it to pathways regulating translation in an mRNA-specific manner. As an RBP, FUS clearly has the potential to regulate translation of specific mRNAs, perhaps in a local manner in neuronal axons or dendrites. Nevertheless, published studies to date have not reported effects of mutant FUS alleles on translation of specific mRNAs. Genome-wide studies of translation, like the one described above for TDP-43 (Neelagandan et al., 2019), could readily be performed with neurons derived from FUS models of ALS/FTD described above or in vitro-differentiated LMNs from patient iPSCs (e.g. as in W. Guo, Naujock, et al., 2017). It would be particularly interesting to perform this separately for the axons and somato-dendritic compartment with cultured neurons or spinal cords in vivo. Integrating such a global analysis of translation in FUS mutant neurons with data from direct binding assays could help to identify mRNA-specific effects on translation due to direct mutant FUS-mRNA interactions in the axon. Ribosome profiling of non-neuronal cells has identified several mRNAs that are unusually resistant or hypersensitive to eIF2α phosphorylation (Andreev et al., 2015), but how this translational landscape might look for a motor neuronal axon is completely unknown. It is especially interesting to consider the possibility that mutant FUS interactions with specific mRNAs might determine how they respond to eIF2α phosphorylation in the axon.

Other RBPs Implicated in ALS/FTD: hnRNPA2B1 and A1, MATR3, TIA1, and Ataxin-2

Genetic and functional studies implicate additional RBPs in ALS/FTD (reviewed in Baradaran-Heravi et al., 2020). These include heterogeneous nuclear ribonucleoprotein (hnRNP) hnRNPA1 and hnRNPA2/B1 (Kim et al., 2013), as well as Matrin-3 (MATR3) (Johnson et al., 2014), T-cell-restricted intracellular antigen 1 (TIA1) (Mackenzie et al., 2017), and Ataxin-2 (Elden et al., 2010).

hnRNPA2/B1 and hnRNPA1 Are Implicated in Translation in Neurons and Other Cell Types

hnRNPA2/B1 (two protein isoforms encoded by the *hnRNPA2B1* gene) and hnRNPA1 have classical RRMs, RGG domains, and glycine rich low-complexity/prion-like domains where mutations are mainly found (Kim et al., 2013). They are ubiquitous shuttling RBPs and splicing regulators (Huelga et al., 2012) and there is evidence they directly affect translation in non-ALS/FTD contexts. For example, hnRNPA1 has been reported to couple mRNA export and translation (Roy et al., 2014) to modulate translation of mRNAs with key roles in cancer (Roy, Huang, Seckl, & Pardo, 2017), and to affect use of internal ribosome entry sites in the cytoplasm (Cammas et al., 2007; Lewis et al., 2007).

hnRNPB1 is not well characterized, but several studies support direct translational control of specific mRNAs by hnRNPA2. In non-neuronal cultured cells, it cooperated with other hnRNP proteins and upstream open-reading frames for translational control of the *AXIIR* mRNA (J. Zhang et al., 2017). Moreover, hnRNPA2 has been implicated in translational control in the nervous system through 3′UTR-mediated regulation. In cultured oligodendrocytes, hnRNPA2 contributes to localization and local translation of myelin basic protein (*MBP*) mRNA in glial processes (White et al., 2008). A combined study using cell culture and in vitro translation revealed that translational regulation of MBP mRNA by hnRNPA2 involves physical interaction with specific A2 elements which are also found in other mRNAs (Kosturko et al., 2006). However, a more recent study concluded that hnRNPA2 only controls mRNA localization and that other proteins regulate translation (Torvund-Jensen, Steengaard, Reimer, Fihl, & Laursen, 2014). A2 elements contribute to dendritic targeting of locally translated mRNAs, including *CamkIIα* and *Arg3.1/ARC*, but whether they impact translation of these mRNAs has not been determined (Gao, Tatavarty, Korza, Levin, & Carson, 2008). hnRNPA2 also plays a key role in neuronal transport of the non-coding BC RNAs, that have been shown to regulate dendritic translation.

Collectively, these observations suggest hnRNPA1 and hnRNPA2 could potentially contribute to ALS/FTD by affecting translation of specific mRNAs through direct binding or via interactions with non-coding regulatory RNAs. Evidence for axonal "dying back" pathology (Gillingwater & Wishart, 2013) motivates exploring an in vivo role for MBP translational control by hnRNPA2 in oligodendrocytes for axonal maintenance in ALS/FTD. A potential broader role for hRNPA2 in regulation of mRNA localization and local translation in neurons should also be considered (reviewed in Sinnamon & Czaplinski, 2011).

MATR3 Colocalizes with SGs and Is Implicated in mRNA Export and HIV1 mRNA Translation

MATR3 is another mainly nuclear, shuttling RBP with two classical RRMs, two zinc finger motifs, an NLS and low-complexity domains. Patient mutations are found outside of the RNA-binding domains and the NLS in regions with no defined functional domains (Johnson et al., 2014). Previous studies of MATR3 function implicate it in multiple aspects of RNA regulation, including nuclear pre-mRNA splicing (Coelho et al., 2015), as well as the export and translation of HIV1 mRNA (Yedavalli & Jeang, 2011). Whether MATR3 affects translation of cellular mRNAs is unknown. However, proteomic and functional studies of patient mutant proteins in cell culture, including in murine motor neuronal NSC34 cells, have revealed interactions with the mRNA export machinery that affect cellular mRNA export (Boehringer et al., 2017), as well as SG components as a major interacting group (Iradi et al., 2018). MATR3 was detected in cytoplasmic TDP-43 aggregates in sALS patient spinal motor neurons (Tada et al., 2018), suggesting a potential cytoplasmic function in disease related to interaction with TDP-43.

TIA1 Affects Translation of Specific mRNAs and SG Formation in Non-neuronal Cells, whereas ALS/FTD Mutations Promote LLPS and Affect SG Dynamics

TIA1 has three canonical RRMs, as well as a C-terminal low complexity domain and well-characterized roles in both the nucleus and cytoplasm in post-transcriptional regulation, particularly in immune cells and inflammatory responses (Stoecklin & Anderson, 2006). In the cytoplasm, TIA1 is a key component of SGs, influencing their formation and dynamics (N. Kedersha et al., 2000; Panas, Ivanov, & Anderson, 2016). Indeed, TIA1 appears to be a key mediator of mRNA partitioning to SGs in response to translational inhibition resulting from eIF2α phosphorylation by stress-induced kinases (N. L. Kedersha, Gupta, Li, Miller, & Anderson, 1999). In non-neuronal cells, TIA1 has been shown to regulate translation of specific mRNAs, including those with terminal oligopyrimidine (TOP) motifs in their 5′ UTR (Damgaard & Lykke-Andersen, 2011) and several with AU-rich elements (AREs) in their 3′ UTRs (Kawai et al., 2006; Lopez de Silanes et al., 2005; Piecyk et al., 2000). Although AREs are often associated with rapid mRNA decay, the impact of AREs on mRNA stability or translation depends on both mRNA context and the specific RBP bound (reviewed in Garcia-Maurino et al., 2017) and evidence is emerging that the same RBP can use distinct molecular mechanisms for mRNA decay and translational control (reviewed in Otsuka, Fukao, Funakami, Duncan, & Fujiwara, 2019).

TIA1 was implicated genetically in ALS/FTD when novel rare variants in the protein coding sequence were identified in patients by whole exome sequencing with an ALS/FTD family and follow-up association (Mackenzie et al., 2017). Several novel rare variants in the protein coding sequence were identified, all in the low-complexity domain (LCD), which is important for TIA1's role in SG formation and prion-like aggregation in cells (Gilks et al., 2004). TIA-1's LCD also mediates LLPS (Y. Lin, Protter, Rosen, & Parker, 2015), as observed for LCDs in other RBPs, including FUS (see above). Consistent with a potential pathological effect of the rare variants found in patients, purified recombinant proteins bearing patient variant alleles of TIA1 displayed more rapid LLPS in cell-free assays and expressing them in HeLa cells reduced SG dynamics and the mobility and solubility of SG-recruited TDP-43.

A causal role for TIA1 variants in disease seems not to be completely resolved (Baradaran-Heravi et al., 2018; Gu et al., 2018; van der Spek et al., 2018; K. Zhang et al., 2018). However, even if mutations in the TIA1 coding sequence do not ultimately prove to be a significant cause of ALS, this would not preclude an important downstream role for TIA1 in disease pathology. The patient mutations might provide evidence of effects on TIA1 that can be triggered in disease by other routes when the protein is WT (e.g. altered regulation of TIA1 function via neuronal stress signaling pathways activated in disease). In fact, a similar line of thinking is also required for TDP-43: pathology, without mutations, is present in almost all ALS and modest overexpression of WT TDP-43 appears sufficient to trigger disease in multiple models.

How might altered TIA1 function connect to translation in ALS/FTD? One speculative possibility is that it might control polysome disassembly and partitioning of specific mRNAs to SGs in vulnerable neurons, as has been observed in non-neuronal cells. In any case, it will be interesting to examine how translational control by TIA-1 relates to LLPS, SG regulation, mRNA turnover, and interplay with other RBPs.

Ataxin-2 Associates with 3′ UTRs and PABP to Regulate mRNA Stability/Translation and Intermediate-length Polyglutamine Repeat Expansions in Ataxin-2 Are a Risk-factor for ALS

Despite lacking typical mRNA-binding domains, PAR-CLIP sequencing studies demonstrate that Ataxin-2 interacts with U-rich regions in mRNA 3′UTRs, enhancing stability and protein output (Yokoshi et al., 2014). This unusual RBP has been implicated genetically and functionally in two distinct neurodegenerative diseases. Expansion of the CAG codon polyglutamine (polyQ) repeat in Ataxin-2 to > 34 causes spinocerebellar ataxia (SCA)2 (reviewed in Orr & Zoghbi, 2007). However, a landmark paper that began with a genetic screen for modifiers of TDP-43 toxicity in *S. cerevisiae*, found that Ataxin-2 is a modifier of TDP-43 toxicity in multiple ALS/FTD models and that intermediate length expansions (polyQ 27-33) in Ataxin-2 are a risk factor for ALS (Elden et al., 2010). Targeting Ataxin-2 with antisense oligonucleotides in the CNS also improved survival and motor function in a mouse model of TDP-43-driven ALS (L. A. Becker et al., 2017), suggesting a potential therapeutic strategy. Although translational control of individual mRNAs by Ataxin-2 was not examined directly in these studies, interaction with the cytoplasmic poly(A) binding protein (PABP1, a.k.a. PABPC1) provided circumstantial evidence, since PABP1 is a key translational regulator in animals (reviewed in Smith, Blee, & Gray, 2014).

Several other studies support a role for Ataxin-2 in translational regulation. Its *C. elegans* homolog regulates translation in the germline (Ciosk, DePalma, & Priess, 2004). In *Drosophila,* Ataxin-2 physically assembled with polysomes, directly interacted with PABP1 (Satterfield & Pallanck, 2006), and was shown to activate PERIOD at the translational level to control circadian biology (Lim & Allada, 2013; Y. Zhang, Ling, Yuan, Dubruille, & Emery, 2013). In a mouse model of SCA2, Ataxin-2 with long repeats specifically led to reduced levels of *RGS8* mRNA and a stronger reduction in levels of RGS8 protein, suggesting a combined effect on *RGS8* mRNA stability and translation (Dansithong et al., 2015). Long polyQ Ataxin-2 also reduced the RGS8 protein levels generated in a coupled in vitro transcription/translation system, although whether this was due to an effect on mRNA levels was not examined. The authors proposed that reduced levels of RGS8, a regulator of G-protein signaling, might contribute to SCA2 pathogenesis by prolonging mGluR1-stimulated Ca^{2+} release, thereby increasing intracellular Ca^{2+} levels. Whether Ataxin-2's disease modifying effects in ALS/FTD involve effects on translation of RGS8 or other mRNAs remains to be seen. However, extant data strongly suggest this protein will function in the cytoplasm through direct binding to PABP to influence the fate of specific mRNAs in affected CNS neurons of ALS/FTD patients and disease models.

Translational Control and Non-canonical RBPs in Huntington's disease

Neurodegeneration in Huntington's disease (HD) strongly affects the midbrain striatum and later the cortex, with movement, cognitive, and psychiatric symptoms appearing in middle age that progressively worsen (reviewed in McColgan & Tabrizi, 2018). A clear genetic cause underlies HD: expansion of CAG repeats in the gene encoding the protein

Huntingtin (Huntington's Disease Collaborative Research Group, 1993). A repeated CAG stretch is present in all alleles, but tends to be relatively short in the general population, with < 27 considered normal. Whereas repeat lengths of 36–39 are incompletely penetrant, the presence of more than 39 repeats essentially predicts disease.

Several mechanisms for HD pathology have been proposed (reviewed in Bates et al., 2015). The prevailing theory is that the expanded repeats in Huntingtin protein cause disease. However, it is still debated whether this involves loss of a normal function of Huntingtin protein or a toxic gain-of-function. A dominant-negative mode is supported by cell culture findings that Huntingtin protein can auto-dimerize and the protein with expanded polyglutamine tracts is more prone to aggregation, but it remains unclear how exactly expanded repeats cause disease in patient striatum.

Appealing as protein-based concepts for HD pathogenesis may be, numerous drug development efforts based on blocking or reversing Huntingtin aggregation have not yet yielded a successful therapy (Wyant, Ridder, & Dayalu, 2017). While this certainly does not mean the underlying assumption is wrong, it has naturally led to interest in other potential pathogenic mechanisms in HD. One obvious alternative is an RNA-based mechanism in which the repeat expansion in the mRNA encoding HTT is itself toxic (reviewed in McColgan & Tabrizi, 2018). This can involve sequestration of RBPs by the repeat-containing mRNA, as first observed for splicing factors in myotonic dystrophy (Miller et al., 2000), but other mechanisms have also been described. In HD, this has motivated efforts to identify proteins that interact specifically with the expanded CAG repeat mRNA and determine how they might contribute to disease.

The MID1 Complex Binds Expanded CAG Repeats and Stimulates Translation in Cultured Cells

Several papers using biochemical and cell culture assays suggest a potential role for translational control mediated by the MID1 complex in association with expanded CAG-repeat mRNA in HD pathobiology. MID1 is a protein found by biotinylated RNA-affinity purification to interact selectively with the HTT-derived mRNA when it has expanded CAG repeats. Moreover, it also appears to upregulate translation of the encoded protein. Co-immunoprecipitation data suggests that MID1 binds to the expanded CAG repeat RNA as part of a complex with three other proteins: PP2Ac phosphatase, mTOR, and S6K signaling kinases. The model in the case of HD is that MID1 directly interacts with the mRNA and brings these components together for mRNA-specific translational control. MID1 upregulates mTOR by promoting degradation of the phosphatase that negatively regulates it, PP2Ac (Krauss et al., 2013). Presumably, this leads to locally enhanced phosphorylation of 4E-BPs and Rps6, thereby activating translation of the MID-associated mRNA, in this case *HTT* mRNA with an expanded CAG repeat region. Multiple lines of evidence point to a repeat-induced increase in translation.

Despite being in an "mRNP complex" (Aranda-Orgilles et al., 2008), MID1 is not a canonical RBP and has been primarily characterized as a ubiquitin ligase mutated in patients with the neurodevelopmental disorder, Opitz syndrome (Trockenbacher et al., 2001). It is still not clear how the MID1 complex interacts with RNA. A specific sequence (MIDAS) for translational activation by MID1 was identified in the firefly luciferase coding sequence and many cellular mRNAs (Aranda-Orgilles et al., 2011). However, both the extent to which MID1 generally modulates translation through MIDAS elements, as well as their potential involvement in neurodegenerative disease remain unclear.

MID1 has also been suggested to be important for translational control in other CAG repeat expansion diseases. The MID1 complex was found to interact with CAG repeat mRNAs derived from *ATXN2*, *ATXN3*, and *ATXN7* and enhance translational output (Griesche et al., 2016). However, translation from CAA repeats was not affected.

Evidence that the MID1 Complex Is Relevant to HD in Vivo: Metformin Rescues Circuit and Behavioral Defects in an HD Mouse Model

Intriguing as it is, evidence for translational effects via MID1 in Huntington's described above comes exclusively from in vitro translation and cell lines; there is no genetic evidence from either humans or animal models supporting a role for MID1 in HD pathobiology in vivo. Mutations in *MID1* cause the neurodevelopmental disorder Opitz syndrome (Trockenbacher et al., 2001), linking MID1 function to human brain development, but not obviously to HD. However, there is a growing interest in the idea that the seeds of HD- and perhaps other neurodegenerative diseases- may be planted during development, i.e. that altered brain development ultimately leads to neurodegeneration later in life (Kerschbamer & Biagioli, 2015). In fact, a recent study exploring this idea in a mouse model of HD provides pharmacological evidence that MID1 may indeed play an important role in modifying HD phenotypes in vivo (Arnoux et al., 2018). This paper used systems neuroscience approaches to monitor altered cortical network interactions in a mouse model of HD at ages that are "very far from disease onset (VFDO)." This revealed early effects on network activity long before any evidence of aggregation or other molecular pathologies. Both network function defects and behavioral abnormalities were fully reversed by supplying the FDA-approved diabetes drug metformin in drinking water. On a molecular level, metformin treatment led to a selective reduction in levels of HTT mutant protein, without affecting levels of the wild type HTT protein, making it the only described compound with this selective activity. In cultured cells, the effect of metformin on HTT protein levels was occluded by MID1 knockdown, which led to reduced HTT protein levels that were not further reduced by metformin treatment. Taken together, this multidisciplinary study highlights altered cortical development as an early aspect of PD disease etiology and further implicates MID1 as a potentially relevant target of metformin in treating these HD symptoms in the mouse model.

While metformin clearly ameliorates network and neurological phenotypes in HD mice, less clear is whether these in vivo effects are entirely due to direct targeting of translational control of CAG repeat mRNAs bound by MID1. Arnoux et al., 2018 and earlier in vitro work in cancer cells (Demir, Koehler, Schneider, Schweiger, & Klocker, 2014) supports a functional role for MID1 in metformin efficacy. Nevertheless, metformin does not act directly on the MID1 complex, but leads to PP2A activation by disrupting it (Kickstein et al., 2010), thereby impacting many different cellular functions regulated by mTORC1 signaling. For example, other studies have highlighted effects of metformin on mTORC1 signaling that generally affect translational control via 4E-BPs (e.g. Bhat et al., 2017; Gantois et al., 2017). This suggests that achieving selective effects on mutant HTT protein levels with metformin may require careful dosing. There is also evidence that metformin can potentially target other signaling pathways, particularly AMPK in HD models (Jin et al., 2016; Vazquez-Manrique et al., 2016), although AMPK seems not to be critical for metformin function in a number of contexts (Gantois, Popic, Khoutorsky, & Sonenberg, 2019).

Even if metformin is inhibiting PP2A exclusively through the MID1 complex in vivo, it is still not clear this relates to MID1's proposed role as a translational regulator that interacts with expanded-repeat mRNAs. To determine whether the MID1 translational control mechanisms described using biochemistry and cell culture assays are relevant to Huntington's disease, two things seem crucial: (1) stronger validation in animal models of HD that MID1 is a modifier of disease onset or progression; (2) in vivo demonstration that the MID1 complex binds to and regulates translation of expanded-repeat *HTT* mRNA in the brain of HD models. Ideally this would be demonstrated in relevant cell types whose function is compromised in HD. If these things can be demonstrated, translational activation by the MID1 complex might be a potentially interesting target to explore for therapeutic targeting in HD and perhaps other neurodegenerative diseases. Regardless of the specific mechanism, the general strategy of repurposing metformin or developing similar molecules to target disease with a neurodevelopmental component seems promising. Metformin treatment reversed neurodevelopmental phenotypes in mouse models of Fragile X syndrome (Gantois et al., 2017) and is therefore being explored for treatment of FXS and other neurological disorders (reviewed in Gantois et al., 2019). It will be interesting to see whether these effects also involve MID1 and its described capacity for RNA-binding and translational control of specific mRNAs.

Does MID1 Have a Role in Other Neurodegenerative Diseases?

Although MID1 is mainly implicated in CAG repeat mRNA regulation based on cell culture studies, it has been suggested to potentially be involved in other neurodegenerative diseases. For example, one study indicating that metformin targets the MID1 complex examined its effects on Tau phosphorylation in primary neurons, suggesting a possible link to Alzheimer's disease (Kickstein et al., 2010). A potential connection between MID1 and TDP-43 (and therefore ALS/FTD) has also been reported (Neelagandan et al., 2019). The mRNA encoding MID1-interacting protein 1/MIG12 was one of very few found to be translationally deregulated by TDP-43 expression in *both* primary cortical neurons and cultured MN1 cells. MIG12 binds MID1 (Berti, Fontanella, Ferrentino, & Meroni, 2004) and altering MID1 levels abrogated the enhancement of MIG12 protein levels driven by TDP-43 overexpression, suggesting that MID1 can modulate the impact of enhanced translation of its binding partner at the protein level. It will be important to determine whether this regulation also operates in vivo and contributes to disease in animal models of TDP-43-driven ALS/FTD. If so, future work should explore the underlying mechanism and whether it relates to previously described functions of MID1 in translation and the response to metformin.

RBP Regulation of Repeat-associated Non-AUG (RAN) Translation

Non-Huntington's neurological disorders caused by repeat expansions include spinal cerebellar ataxias (SCAs) and fragile X tremor ataxia syndrome (FXTAS) (Orr & Zoghbi, 2007). Moreover, a significant proportion of ALS/FTD is caused by expanded GGGGCC (G4C2) hexanucleotide repeats at the *C9ORF72* locus ("C9-ALS/FTD," DeJesus-Hernandez et al., 2011; Gijselinck et al., 2012; Renton et al., 2011). Most of these expanded repeats are outside the CDS and therefore cause disease via RNA-based mechanisms. This can involve sequestration of specific RNA-binding proteins away from their normal targets (Mohan, Goodwin, & Swanson, 2014). Alternatively, the repeat expansions themselves can be translated in the absence of a typical AUG start codon. Repeat-associated non-AUG (RAN) translation, was originally described for expanded CAG repeats in *SCA8* (Zu et al., 2011). Translation of repeats in different reading frames can generate dipeptide repeat proteins whose composition depends on the repeat and the reading frame. As first demonstrated for C9-ALS/FTD, these "DPR proteins" are detectable in CNS of patients and animal disease models (Ash et al., 2013; Gendron et al., 2013; Mori et al., 2013; Zu et al., 2013). While the contribution of individual DPR proteins in specific diseases is still debated, interfering with their production is typically neuroprotective and molecular mechanisms underlying their production are beginning to be elucidated (reviewed in Green, Linsalata, & Todd, 2016; Zu, Pattamatta, & Ranum, 2018).

RBPs Modulate RAN Translation in FXTAS and C9-ALS/FTD

Directed genetic screens for modifiers of RAN translation have revealed modulatory roles for RBPs. Using reverse genetics in flies and cultured mammalian cells, one study identified several translation factors and RNA helicases affecting RAN translation from

CGG repeats (Linsalata et al., 2019). The only classical RBP examined in this study was *Drosophila* SF2 (human SRSF1), known primarily as a splicing factor, but also implicated in translation (reviewed in Das & Krainer, 2014). SF2 knockdown strongly suppressed RAN translation, but the basis for this effect was not analyzed further in this study. SRSF1 was independently found to interact with expanded G4C2 repeat RNA from the C9ORF72 locus and promote its splicing-independent nuclear export and thereby RAN translation and toxic DPR production (Hautbergue et al., 2017). This suggests a conserved role for this RBP in promoting RAN translation from different repeats, although it is not clear that the mechanisms are the same.

Knocking down the RNA helicase belle/DDX3X also strongly suppressed FXTAS CGG-repeat RAN translation in flies (Linsalata et al., 2019), but had the opposite effect on RAN translation from the G4C2 repeat expansion found in C9 ALS/FTD (Cheng et al., 2019), highlighting context-specific modulation. Although RNA helicases are not classical RBPs, DDX3X is a bona fide component of the mRNA interactome (Castello et al., 2012) and displays affinity for specific mRNA features, particularly G quadruplex structures (Herdy et al., 2018).

Ability to modulate RAN translation make DDX3X a potential therapeutic target in repeat disorders and add to interest in targeting this molecule. DDX3X is already a target in drug discovery efforts for cancer (Riva & Maga, 2019) and was recently shown to be a key mutated gene causing autosomal-dominant neurodevelopmental disorders in girls (Lennox et al., 2020). Additional targets and potential lead compounds may emerge from cell-based screens for small-molecule modulators of RAN translation (Green et al., 2019).

One important question for the future will be to understand how much overlap there is among modulators of RAN translation driven by different repeat sequences. From the therapeutic perspective, it will also be important to determine whether common proteins, such as DDX3X, can be targeted in multiple diseases or whether more disease-specific molecular targets will prove superior for particular indications.

Translational Control by RBPs in Parkinson's Disease

Parkinson's disease (PD) features a characteristic, debilitating movement disorder and non-motor symptoms (Poewe et al., 2017). Pathological hallmarks of PD are intracellular α-synuclein aggregates ("Lewy bodies") and dopaminergic (DA) neuron loss in the substantia nigra region of midbrain striatum. Few PD cases are familial, but the responsible genes have enabled disease modeling and provided clues into pathobiology (Dawson, Ko, & Dawson, 2010). Multiple studies implicate translation in PD pathogenesis (reviewed in Lu, Gehrke, & Wu, 2014). While canonical RBPs are not implicated in PD by human genetics, functional studies in PD models suggest they control translation in pathobiological contexts. There is also evidence for translational control by non-canonical RBPs in PD.

Local Translation of Mitochondrial mRNAs Regulated by PINK1 and Parkin

Mutations in Pten-induced kinase 1 (*PINK1/PARK6*) or Parkin (*PARK2*) cause early-onset autosomal recessive PD, probably due to defective mitochondrial maintenance (reviewed in Pickrell & Youle, 2015). Pink1 localizes to the mitochondrial outer membrane (Valente et al., 2004) where it functions together with the ubiquitin ligase, Parkin, to maintain mitochondrial function and thereby dopaminergic neuron survival (Clark et al., 2006; Park et al., 2006; Yang et al., 2006). Neither protein has classical RNA-binding domains, but both are implicated in translational control in PD models. Genetic and pharmacological approaches revealed that activating eIF4E-binding proteins (4E-BPs) suppresses pathological Pink1 and parkin phenotypes in flies and human parkin mutant dopaminergic neurons. (Liu & Lu, 2010; Tain et al., 2009).

Pink1 and Parkin were also shown to promote mitochondrial localization and local translation of specific mRNAs on the mitochondrial outer membrane, apparently to sustain levels of specific mitochondrial oxidative phosphorylation components (Gehrke et al., 2015). Translational activation involved displacement of the repressive RBPs, Pumilio and Glorund/hnRNP-F. While these results reveal new roles for Parkin and PINK1 in regulating mRNA-specific translation of mitochondrial components, the importance of this specific mechanism in PD is not clear. Both proteins can influence mitochondrial function via other mechanisms and efficient post-translational import of mitochondrial components (reviewed in T. Becker, Song, & Pfanner, 2019) raises issues about how important their local translation would be. It will be interesting to see whether this mechanism is also found in vivo in mammalian PINK1/Parkin models and whether it proves relevant in other forms of PD.

The idea of local translation associated with organelles appears to extend to other neurodegenerative diseases. Recent work highlights local translation in association with late endosomes as being important in axonal mitochondria maintenance and disrupted by mutants that lead to axonal dysfunction in Charcot Marie Tooth neuropathy (Cioni et al., 2019). Although a role for PINK1 or Parkin in this specific pathway was not described, it is interesting to consider a potential role for these proteins in localized translation in axonal mitochondria, as well as possible crosstalk between local translation on endosomes and mitochondria. While PD is usually viewed as a disease driven fundamentally by dopaminergic neuron death, there are increasing indications that PD may involve problems with synaptic homeostasis prior to neuronal loss (reviewed in Soukup, Vanhauwaert, & Verstreken, 2018). Viewed through this lens, localized translational control by PINK1 and Parkin in association with mitochondria might conceivably reflect a broader role for organelle-associated local translation in synaptic maintenance that is just beginning to be appreciated (reviewed in Pushpalatha & Besse, 2019).

Patient Kinase-Activating Mutations in LRRK2 Activate Translation and Promote Neurodegeneration by Phosphorylating Specific Regulatory Proteins, including RPS15

Gain-of-function mutations in leucine-rich repeat kinase 2 (LRRK2, PARK8) cause late-onset, autosomal-dominant PD, as well as some cases of sporadic PD (Paisan-Ruiz et al., 2004; Zimprich et al., 2004). LRRK2 toxicity in PD involves increased phosphorylation of vesicular trafficking components, but it also upregulates translation via its kinase activity in a pathological context (reviewed in Martin, Kim, Dawson, & Dawson, 2014). In addition to protein-protein interactions with translation initiation factors, LRRK2 PD models based on the kinase-activating G2019S mutation revealed effects on dopaminergic neuron maintenance in *Drosophila* via mTOR pathway and eIF4E-binding proteins (4E-BPs) (Imai et al., 2008) and promoting neurodegeneration in both fly and human cell culture PD models by phosphorylating ribosomal protein S15 (Martin, Kim, Lee, et al., 2014). Both studies also showed enhanced translation by mutant LRRK2, which was interpreted as a general effect. However, it is not immediately obvious why dopaminergic neurons would be especially sensitive to increased general translation. A better understanding of which proteins are being synthesized at a higher level could elucidate how increased translation via RPS15 phosphorylation might lead to selective neuronal vulnerability. It might seem obvious that RPS15's contribution to pathogenesis relates to its role on the ribosome during translation, but it could involve an off-ribosome, "moonlighting" role of RPS15 (W. Wang et al., 2015) and/or an effect on ribosome biogenesis (reviewed in Bohnsack & Bohnsack, 2019; Hetman & Slomnicki, 2019).

Another apparent link to translational control in *Drosophila* involves physical and functional interactions of pathogenic LRRK2 with the miRNA system, in particular Argonaute proteins and specific miRNA-mRNA target pairs (Gehrke, Imai, Sokol, & Lu, 2010). However, the importance of mutant LRKK2 interaction with miRNA regulatory components as a disease-driving mechanism in mammalian PD models remains to be demonstrated.

Mutations in DJ-1 and Effects on Translation of Specific mRNAs as a Possible Cause of PD

Mutations in *DJ-1 (PARK 7)* cause rare, early-onset autosomal recessive PD (Bonifati et al., 2003). DJ-1 has been implicated in many cellular processes, particularly the response to oxidative stress (Reviewed in Repici & Giorgini, 2019), but which functions are relevant to PD remains unclear. In addition to transcriptional regulation by DJ-1, there is also evidence for post-transcriptional functions. DJ-1 was identified as a component of an "RNA-binding complex" before it was implicated in PD (Hod, Pentyala, Whyard, & El-Maghrabi, 1999) and is homologous to a bacterial protein that binds ribosomes and regulates translation (Kthiri et al., 2010).

Mammalian DJ-1 was reported to bind to specific mRNAs and potentially regulate their translation (van der Brug et al., 2008). Importantly, DJ-1 mutations from PD patients showed reduced mRNA target binding, suggesting a potential pathological role involving failure to properly bind and control translation of specific mRNAs bound by DJ-1. Oxidation of DJ-1 has been observed in brains of sporadic PD patients (Bandopadhyay et al., 2004; Choi et al., 2006) and DJ-1's oxidation state appeared to affect its RNA-binding capability. A follow up study validated interactions of DJ-1with several targets in non-diseased human brains and also detected changes in protein levels for four DJ-bound mRNAs in patient brains (Blackinton et al., 2009). mRNA levels were not changed, consistent with altered post-transcriptional regulation of these targets in PD brains, potentially at the translational level.

While the relevance of DJ-1 RNA-binding in PD remains unclear (Cookson, 2017), a recent paper shows interaction and colocalization of DJ-1 with SGs in mammalian cells upon induction of osmotic or oxidative stress and also identified specific mRNAs that associate with DJ-1 and co-localize to SGs under these conditions (Repici et al., 2019). Inducing NMDA excitotoxicity of primary neurons or treating dopaminergic neurons with toxins to mimic PD also reportedly led to DJ-1 association with SGs. While this study did not directly assess translation, it provides additional evidence for mRNA interaction by DJ-1 and the potential for it to mediate post-transcriptional regulation in the context of neurodegenerative disease. Because DJ-1 has numerous functions with no connection to RNA-binding or translational control, it will be crucial to obtain evidence from in vivo disease models to support the notion that these specific activities are disease-relevant.

Some Important Issues for the Future

Several future directions specific to individual diseases and RBPs were mentioned above. Here, the review outlines larger issues to resolve and frontiers with potential for major impact on the broader field in the years to come.

Do Other RBPs Lacking Typical RNA-Binding Motifs Contribute to Neurodegenerative Disease?

Several examples discussed above suggest that RBPs need not contain classical RNA-binding motifs. Indeed, mRNA "interactome capture" proteomics suggests that many proteins lacking these motifs can bind mRNAs directly in living cells (Baltz et al., 2012; Castello et al., 2012). Time will tell how many of these interactions have a discernible function, but it is easy to imagine that some of these non-canonical RBPs might control neuronal protein synthesis and thereby contribute to disease mechanisms.

Published RNA interactome-capture studies to date have been performed in non-neuronal cell lines. Formulating stronger hypotheses about novel RBPs in neurodegenerative disease would require use of a more physiologically relevant starting material, such as aging brain from animal disease models. These experiments might support roles of DJ-1 or MID1 as unconventional RBPs in PD and HD, respectively, but seem equally likely to reveal other novel RBPs. It is interesting to consider that, as originally suggested for DJ-1 (van der Brug et al., 2008), unappreciated roles for proteins in RNA regulation might help to explain apparent diverse genetic causes of some diseases. Perhaps many more RBPs have already been implicated in neurodegenerative disease by genetics; we just don't realize yet these proteins are RBPs!

What Is the Relationship between RBP Aggregates, Stress Granules, Phase Transitions, and Translation in ALS/FTD and Other Diseases?

SGs were suggested early on to play key roles in ALS pathogenesis (Li, King, Shorter, & Gitler, 2013; Ramaswami, Taylor, & Parker, 2013) and interest in their potential contributions to neurodegeneration remains high (reviewed in Wolozin & Ivanov, 2019). For diseases where RBPs are key components of pathological aggregates, it seems simple to envisage that stress leads first to accumulation of an RBP in the cytoplasm in SGs, followed by large-scale aggregation in a "feed-forward" mechanism. More recently, the realization that RBPs with "prion-like" LCDs will undergo LLPS and that this can lead to aggregation has refined the model. Nevertheless, the key unresolved issues remain: what is the actual relationship between liquid-liquid phase transitions, stress granules, and the pathological RBP aggregates found in patient neurons and glia on autopsy? Is altered translation caused by these phenomena or causing them? And which effects actually drive disease onset and progression?

Several recent studies have proposed that mutant RBPs may first undergo LLPS in association with SGs and then form larger aggregates (reviewed in Mikhaleva & Lemke, 2018). However, a detailed study of TDP-43 behavior in non-cycling SH-SY5Y cells suggests that cytoplasmic LLPS can lead directly to aggregation and nuclear clearing, without SGs being involved (Gasset-Rosa et al., 2019). Another study using an optogenetic approach in HEK293 cells also suggests that SGs do not promote aggregation and that RNA binding by TDP-43 (as would be expected in SGs) actually *protects* against phase transitions (Mann et al., 2019).

If SGs are not directly relevant to LLPS or aggregation then what is their role in disease? Unfortunately, we still know much more about SG components and the requirements for their formation than what function they actually serve (Protter &

Parker, 2016). While it is generally agreed that mRNAs in SGs are translationally silent, an important issue is what fraction of any mRNA or RBP in the cytoplasm is actually present in SGs. When this has actually been examined for TDP-43, only a small fraction appears to be associated, even after severe stress (Gasset-Rosa et al., 2019; Higashi et al., 2013). Obviously, if only a small fraction of an RBP and the cellular mRNAs it regulates are in SGs, it is hard to understand how they can play a major role in mediating translational inhibition. Future experiments using kinetic and quantitative approaches will be crucial to resolve these functional relationships and develop a better understanding of the role SGs play in disease.

Another important issue to resolve is whether the contribution of LLPS to disease is solely to promote aggregation or whether there is also a direct connection between LLPS and translation in disease, independent of aggregation. There is direct evidence that inducing LLPS of FUS can impact on axonal translation (Qamar et al., 2018). However, this was observed with ex vivo preparations of developing *Xenopus* embryonic retinal axons. While this pioneering system to study local translation in developing axonal growth cones has led to many important insights (Lin & Holt, 2007), its limitations as a model for age-dependent neurodegenerative disease must be acknowledged. Whether patient mutant alleles expressed at physiologically relevant concentrations would show similar behavior in adult motor neurons remains to be seen.

The finding that LLPS can promote RBP aggregation, which is the pathological hallmark in patient neurons on autopsy, seems to suggest a straightforward pathobiological mechanism. Less straightforward is how to reconcile pathological LLPS-driven aggregation with numerous mouse models of ALS/FTD driven by TDP-43 or FUS mutants that show neither nuclear clearing nor visible aggregates (Arnold et al., 2013; Lopez-Erauskin et al., 2018). One possible solution could be that disease has two major stages, distinguished by fundamentally different pathology: symptoms can be initiated without aggregation, but full progression with cell death depends on aggregation. This might fit with the idea that many neurodegenerative diseases are largely "synaptopathies" in the earliest phases (Gillingwater & Wishart, 2013). However, it also raises the question of what role cell death actually plays in the disease course.

The alternative possibility is that aggregation may not need to produce visible aggregates to be toxic. According to this view, even though the animal models lack detectable aggregates, proteins are actually aggregating constantly, but are simply cleared at a sufficient rate to prevent the aggregates from becoming large and easily detectable by standard light microscopy. The situation might actually be similar within patient neurons during earlier stages of disease. Since we only have information from end-stage, there is no way to know when in disease large, visible aggregates form in patient neurons and whether aggregate size correlates with toxicity. These issues are not new in the neurodegeneration field, but as RBPs and LLPS become a focus, it will be important to consider these details when forming models of disease pathobiology. One thing that might help to resolve this would be to develop the capability to monitor and manipulate both protein synthesis and decay rates, as well as LLPS and "micro-aggregation," longitudinally in vivo in affected neurons of disease models.

How Do eIF2α Phosphorylation and RBP Regulation of Translation Intersect in Disease?

In addition to ALS/FTD, there is evidence that eIF2α phosphorylation also plays important roles in other neurodegenerative diseases, including HD (Leitman et al., 2014; Reijonen, Putkonen, Norremolle, Lindholm, & Korhonen, 2008), PD (Colla et al., 2012; Ryu et al., 2002; Silva et al., 2005), Alzheimer's disease (Ohno, 2014), and infectious prion-driven neurodegeneration (Moreno et al., 2012). In AD, there is even precedent for local activation of ATF4 translation in axons as a mechanism for retrograde signaling to the nucleus (Baleriola et al., 2014), similar to retrograde signaling mechanisms in acute axonal injury paradigms (reviewed in Koley, Rozenbaum, Fainzilber, & Terenzio, 2019).

If translational control by eIF2α phosphorylation contributes to multiple neurodegenerative diseases, it would make sense to examine whether common RBPs may also be involved in modulating the translational response at the level of individual mRNAs in these diseases. It seems impossible to imagine they do not, yet potential interplay between translation control by eIF2α phosphorylation and RBPs remains a largely unexplored topic, not just in the context of neurodegeneration. Although ATF4 receives the most attention as a uORF-regulated mRNA, at least half of annotated human mRNAs have at least one AUG-initiated uORF in their 5′UTR (Calvo, Pagliarini, & Mootha, 2009) and other mRNAs whose translation responds paradoxically to eIF2α phosphorylation are beginning to emerge (Andreev et al., 2015). There is precedent from in vitro translation assays for RBP binding to 5′UTRs making uORFs regulatory (Medenbach, Seiler, & Hentze, 2011) and interplay between uORF and RBP regulation has also been described in cultured cells (J. Zhang et al., 2017). While it is still not clear how extensively this mechanism is used to regulate protein synthesis in vivo, it is easy to imagine that it could play an important role in determining the impact of eIF2α phosphorylation on protein production from specific mRNAs.

eIF2α phosphorylation may represent a common, downstream point of convergence and therefore a potentially attractive target for therapeutic development across multiple neurodegenerative diseases. With SG induction emerging as potentially important in ALS/FTD, the contribution of RBPs and SGs to other neurodegenerative diseases merits further attention. Gaining a better understanding of the interplay between eIF2α phosphorylation and RBP function in mRNA-specific translational control in vulnerable neurons, particularly in axons, seems likely to be a fruitful avenue for the future.

It is important to keep in mind that eIF2α phosphorylation in the brain does not just get activated under stressful conditions. Rather, it plays a crucial role in regulating synaptic plasticity and memory consolidation, with mice typically showing increased memory performance in many different behavioral paradigms upon inhibition of eIF2α phosphorylation (Costa-Mattioli et al., 2005; Costa-Mattioli et al., 2007; Trinh & Klann, 2013). Moreover, ISRIB, an experimental drug targeting this pathway, leads to memory enhancement in mice (Sidrauski et al., 2013) and reverses cognitive effects in a mouse model of Down syndrome (Zhu et al., 2019). ISRIB directly binds to eIF2B, activates its guanine nucleotide exchange (GEF) function (Sidrauski, Tsai, et al., 2015; Tsai et al., 2018; Zyryanova et al., 2018), and thereby eliminates the translational inhibition, ISR activation, and SG formation that normally accompany increased eIF2α phosphorylation (Sidrauski, McGeachy, Ingolia, & Walter, 2015). In biochemical and cell-based assays, ISRIB also rescued the effects of partial loss-of-function mutations in eIF2B subunits that cause axonal demyelination in children with vanishing white matter (VWM) disease (Wong et al., 2018). Moreover, chronic treatment with 2BAct, an ISRIB-derivative with a superior solubility and pharmacokinetic profile, fully rescued demyelination and all associated neurological deficits in a mouse model of VWM disease based on a patient mutation in eIF2B (Wong et al., 2019).

Current preclinical studies with ISRIB and molecules with a related mechanism of action seem to reveal only benefits from interfering with eIF2α phosphorylation's downstream effects in the CNS. However, it seems like there must be a downside to chronically inhibiting or bypassing eIF2α phosphorylation in the brain or systemically e.g. due to interference with the normal immune response to viral infection. Indeed, ISRIB is just one of many other compounds targeting this pathway, most of which have not successfully translated to clinical use (Perez-Arancibia, Rivas, & Hetz, 2018). A recent study found that ISRIB no longer inhibits the ISR when levels of eIF2α phosphorylation exceed a threshold level (Rabouw et al., 2019). The authors present this as a benefit and suggest this may be one reason ISRIB is tolerated in vivo so well, but it also implies that ISRIB-like molecules might not be effective in therapeutic contexts where high eIF2α phosphorylation levels are relevant. Thus, the potential therapeutic value of ISRIB-like compounds for neurodegenerative diseases may hinge on how high eIF2α phosphorylation levels actually are in vulnerable CNS cell populations. Potential differences in levels of eIF2B across these cell populations will probably also be relevant. Understanding the impact of threshold effects, as well as what distinguishes a positive or negative response to eIF2α phosphorylation in different brain regions and diseases, seems fundamental if we want to target eIF2α regulation and the ISR to treat neurodegenerative disease (Bellato & Hajj, 2016; Bond, Lopez-Lloreda, Gannon, Akay-Espinoza, & Jordan-Sciutto, 2020; Halliday & Mallucci, 2015). It will be extremely interesting to see whether neurodegenerative diseases driven by RBPs will also respond to ISRIB-like molecules and whether RBPs will modify responses to these drugs in specific CNS cell types.

Does Pathogen Exposure Play a Role in Triggering Age-Related Neurodegenerative Diseases?

In addition to being a common aspect of multiple "classical" neurodegenerative diseases, eIF2α phosphorylation and associated effects on translation are also seen in neurodegenerative diseases caused by infectious agents, including prion-driven neurodegeneration (Moreno et al., 2012). While this is typically viewed as a common end reached by different routes, it might actually reflect a common role for pathogen-induced stress or stress related to re-activation of a pathogen-related "molecular memory." The idea that prior infection or previous exposure to infectious agents might play a role in development of classical genetically-caused and sporadic neurodegenerative diseases is not new, but still remains largely unexplored. Recent work highlights roles in AD, where infection with Herpes Simplex Virus 1 (Eimer et al., 2018; Readhead et al., 2018) and potentially other pathogens are increasingly implicated (Ashraf et al., 2019), as well as intestinal infection in PD (Matheoud et al., 2019). In ALS, there is evidence from human patient data and animal models that activation of endogenous retroviruses may play a role in disease (Douville & Nath, 2017; Krug et al., 2017; Manghera, Ferguson-Parry, & Douville, 2016) and might even contribute to non-cell-autonomous pathological spreading (Chang & Dubnau, 2019). How this connects with translation is not completely clear, but RBPs including TDP-43 can interact with these RNA elements. Moreover, dsRNAs produced during viral infection or reactivation of retroviruses and transposable elements can activate PKR, leading to eIF2α phosphorylation with its associated effects on translation (Balachandran & Barber, 2007) and SG formation (McCormick & Khaperskyy, 2017). Further investigation in this direction would seem especially timely, as the potential long-term consequences of SARS-CoV-2 infection and COVID-19 on the nervous system of former patients become increasingly relevant to a significant fraction of the world's population (Baig, Khaleeq, Ali, & Syeda, 2020; Troyer, Kohn, & Hong, 2020).

How Can We Best Model What Actually Happens to Translation in Vulnerable Patient Neurons?

Understanding how translation contributes to neurodegenerative disease hinges on accurate disease models. Standard cell lines (e.g. HeLa/HEK) express very few neuronal proteins and it seems unlikely that specific mRNA targets or pathways of RBPs in these cells will be the same in neuronal populations affected by disease. Even "neuronal cell lines" (e.g. SH-SY5Y) and primary neuronal cultures (typically from embryos) may not model conditions in an aged human CNS very accurately. Patient-derived human neuronal cultures from induced pluripotent stem cells (iPSC) or trans-differentiation technologies offer an interesting alternative, though potential effects on translation in these cells seem largely unexplored. Despite the obvious appeal of "disease-in-a-dish" approaches, there are numerous challenges for accurate modeling of age-related neurodegenerative diseases (reviewed for ALS in Guo, Fumagalli, Prior, & Van Den Bosch, 2017; Sances et al., 2016).

For the time being, it seems like animal models will remain unavoidable for physiological disease modeling. Despite the power of invertebrate and aquatic models for forward genetics and in vivo compound screening, the much more similar anatomical and physiological features seem to make rodent models essential (Van Damme, Robberecht, & Van Den Bosch, 2017). Humanized mouse knock-in approaches (for example Lopez-Erauskin et al., 2018) may reduce pleiotropic effects of over- or mis-expression.

Determining the contribution of translational regulation by RNA-binding proteins to disease using these models will depend on several factors. Above all, it seems essential to develop robust technologies to monitor translation of specific mRNAs accurately over time in vivo within vulnerable neuronal populations. It will also be important to be able to monitor the impact on protein levels in these cells. Having such approaches would enable validation of mechanisms that have been identified using cell culture or in vitro translation, but could also be useful for screening the impact on translation directly in cells of interest in vivo with readouts such as RNA-seq or proteomics. A complete appreciation for altered translation by RNA-binding proteins in driving neurodegenerative disease will also require significant cooperation between the scientific communities studying translation mechanism, local translation, and in vivo neurobiology. Historically these are three separate scientific "tribes" with fundamentally different mindsets, but increased interaction seems certain to lead to a better understanding of how RNA and translation-based mechanisms contribute to neurodegenerative disease. This, in turn, holds promise not only for a better fundamental understanding of translational control in the aging brain, but also for expanding the extremely limited and largely palliative therapeutic options that exist currently for neurodegenerative diseases.

References

Alami, N. H., Smith, R. B., Carrasco, M. A., Williams, L. A., Winborn, C. S., Han, S. S. W., et al. (2014). Axonal transport of TDP-43 mRNA granules is impaired by ALS-causing mutations. *Neuron, 81*(3), 536–543.

Andersson, M. K., Stahlberg, A., Arvidsson, Y., Olofsson, A., Semb, H., Stenman, G., et al. (2008). The multifunctional FUS, EWS and TAF15 proto-oncoproteins show cell type-specific expression patterns and involvement in cell spreading and stress response. *BMC Cell Biology, 9*, 37.

Andreev, D. E., O'Connor, P. B., Fahey, C., Kenny, E. M., Terenin, I. M., Dmitriev, S. E., et al. (2015). Translation of 5′ leaders is pervasive in genes resistant to eIF2 repression. *Elife, 4*, e03971.

Arai, T., Hasegawa, M., Akiyama, H., Ikeda, K., Nonaka, T., Mori, H., et al. (2006). TDP-43 is a component of ubiquitin-positive tau-negative inclusions in frontotemporal lobar degeneration and amyotrophic lateral sclerosis. *Biochemical and Biophysical Research Commununications, 351*(3), 602–611.

Aranda-Orgilles, B., Rutschow, D., Zeller, R., Karagiannidis, A. I., Kohler, A., Chen, C., et al. (2011). Protein phosphatase 2A (PP2A)-specific ubiquitin ligase MID1 is a sequence-dependent regulator of translation efficiency controlling 3-phosphoinositide-dependent protein kinase-1 (PDPK-1). *Journal of Biological Chemistry, 286*(46), 39945–39957.

Aranda-Orgilles, B., Trockenbacher, A., Winter, J., Aigner, J., Kohler, A., Jastrzebska, E., et al. (2008). The Opitz syndrome gene product MID1 assembles a microtubule-associated ribonucleoprotein complex. *Human Genetics, 123*(2), 163–176.

Arnold, E. S., Ling, S. C., Huelga, S. C., Lagier-Tourenne, C., Polymenidou, M., Ditsworth, D., et al. (2013). ALS-linked TDP-43 mutations produce aberrant RNA splicing and adult-onset motor neuron disease without aggregation or loss of nuclear TDP-43. *Proceedings of the National Academy of Sciences of the United States of America, 110*(8), E736–E745.

Arnoux, I., Willam, M., Griesche, N., Krummeich, J., Watari, H., Offermann, N., et al. (2018). Metformin reverses early cortical network dysfunction and behavior changes in Huntington's disease. *Elife, 7*.

Ash, P. E., Bieniek, K. F., Gendron, T. F., Caulfield, T., Lin, W. L., Dejesus-Hernandez, M., et al. (2013). Unconventional translation of C9ORF72 GGGGCC expansion generates insoluble polypeptides specific to c9FTD/ALS. *Neuron, 77*(4), 639–646.

Ashraf, G. M., Tarasov, V. V., Makhmutovsmall a, C. A., Chubarev, V. N., Avila-Rodriguez, M., Bachurin, S. O., et al. (2019). The possibility of an infectious etiology of Alzheimer disease. *Molecular Neurobiology, 56*(6), 4479–4491.

Baig, A. M., Khaleeq, A., Ali, U., & Syeda, H. (2020). Evidence of the COVID-19 virus targeting the CNS: Tissue distribution, host-virus interaction, and proposed neurotropic mechanisms. *ACS Chemical Neuroscience, 11*(7), 995–998.

Balachandran, S., & Barber, G. N. (2007). PKR in innate immunity, cancer, and viral oncolysis. *Methods in Molecular Biology, 383*, 277–301.

Baleriola, J., Walker, C. A., Jean, Y. Y., Crary, J. F., Troy, C. M., Nagy, P. L., et al. (2014). Axonally synthesized ATF4 transmits a neurodegenerative signal across brain regions. *Cell, 158*(5), 1159–1172.

Baltz, A. G., Munschauer, M., Schwanhausser, B., Vasile, A., Murakawa, Y., Schueler, M., et al. (2012). The mRNA-bound proteome and its global occupancy profile on protein-coding transcripts. *Molecular Cell, 46*(5), 674–690.

Bandopadhyay, R., Kingsbury, A. E., Cookson, M. R., Reid, A. R., Evans, I. M., Hope, A. D., et al. (2004). The expression of DJ-1 (PARK7) in normal human CNS and idiopathic Parkinson's disease. *Brain, 127*(Pt 2), 420–430.

Baradaran-Heravi, Y., Dillen, L., Nguyen, H. P., Van Mossevelde, S., Baets, J., De Jonghe, P., et al. (2018). No supportive evidence for TIA1 gene mutations in a European cohort of ALS-FTD spectrum patients. *Neurobiology of Aging, 69*, 293 e299–293 e211.

Baradaran-Heravi, Y., Van Broeckhoven, C., & van der Zee, J. (2020). Stress granule mediated protein aggregation and underlying gene defects in the FTD-ALS spectrum. *Neurobiology of Disease, 134*, 104639.

Bates, G. P., Dorsey, R., Gusella, J. F., Hayden, M. R., Kay, C., Leavitt, B. R., et al. (2015). Huntington disease. *Nature Revies Disase Primers, 1*, 15005.

Becker, L. A., Huang, B., Bieri, G., Ma, R., Knowles, D. A., Jafar-Nejad, P., et al. (2017). Therapeutic reduction of ataxin-2 extends lifespan and reduces pathology in TDP-43 mice. *Nature, 544*(7650), 367–371.

Becker, T., Song, J., & Pfanner, N. (2019). Versatility of Preprotein Transfer from the Cytosol to Mitochondria. *Trends in Cell Biology, 29*(7), 534–548.

Behm-Ansmant, I., Rehwinkel, J., & Izaurralde, E. (2006). MicroRNAs silence gene expression by repressing protein expression and/or by promoting mRNA decay. *Cold Spring Harbor Symposia on Quantitative Biology, 71*, 523–530.

Bellato, H. M., & Hajj, G. N. (2016). Translational control by eIF2alpha in neurons: Beyond the stress response. *Cytoskeleton (Hoboken), 73*(10), 551–565.

Berti, C., Fontanella, B., Ferrentino, R., & Meroni, G. (2004). Mig12, a novel Opitz syndrome gene product partner, is expressed in the embryonic ventral midline and co-operates with Mid1 to bundle and stabilize microtubules. *BMC Cell Biology, 5*, 9.

Bhardwaj, A., Myers, M. P., Buratti, E., & Baralle, F. E. (2013). Characterizing TDP-43 interaction with its RNA targets. *Nucleic Acids Research, 41*(9), 5062–5074.

Bhat, M., Yanagiya, A., Graber, T., Razumilava, N., Bronk, S., Zammit, D., et al. (2017). Metformin requires 4E-BPs to induce apoptosis and repress translation of Mcl-1 in hepatocellular carcinoma cells. *Oncotarget, 8*(31), 50542–50556.

Blackinton, J., Kumaran, R., van der Brug, M. P., Ahmad, R., Olson, L., Galter, D., et al. (2009). Post-transcriptional regulation of mRNA associated with DJ-1 in sporadic Parkinson disease. *Neuroscience Letters, 452*(1), 8–11.

Boehringer, A., Garcia-Mansfield, K., Singh, G., Bakkar, N., Pirrotte, P., & Bowser, R. (2017). ALS Associated Mutations in Matrin 3 Alter Protein-Protein Interactions and Impede mRNA Nuclear Export. *Science Reports, 7*(1), 14529.

Bohnsack, K. E., & Bohnsack, M. T. (2019). Uncovering the assembly pathway of human ribosomes and its emerging links to disease. *EMBO Journal, 38*(13), e100278.

Bond, S., Lopez-Lloreda, C., Gannon, P. J., Akay-Espinoza, C., & Jordan-Sciutto, K. L. (2020). The integrated stress response and phosphorylated eukaryotic initiation factor 2alpha in neurodegeneration. *Journal of Neuropathology & Experimental Neurology, 79*(2), 123–143.

Bonifati, V., Rizzu, P., van Baren, M. J., Schaap, O., Breedveld, G. J., Krieger, E., et al. (2003). Mutations in the DJ-1 gene associated with autosomal recessive early-onset parkinsonism. *Science, 299*(5604), 256–259.

Brichta, L., Shin, W., Jackson-Lewis, V., Blesa, J., Yap, E. L., Walker, Z., et al. (2015). Identification of neurodegenerative factors using translatome-regulatory network analysis. *Nature Neuroscience, 18*(9), 1325–1333.

Calvo, S. E., Pagliarini, D. J., & Mootha, V. K. (2009). Upstream open reading frames cause widespread reduction of protein expression and are polymorphic among humans. *Proceedings of the National Academy of Science of the United States of America, 106*(18), 7507–7512.

Cammas, A., Pileur, F., Bonnal, S., Lewis, S. M., Leveque, N., Holcik, M., et al. (2007). Cytoplasmic relocalization of heterogeneous nuclear ribonucleoprotein A1 controls translation initiation of specific mRNAs. *Molecular Biology of the Cell, 18*(12), 5048–5059.

Castello, A., Fischer, B., Eichelbaum, K., Horos, R., Beckmann, B. M., Strein, C., et al. (2012). Insights into RNA biology from an atlas of mammalian mRNA-binding proteins. *Cell, 149*(6), 1393–1406.

Chang, Y. H., & Dubnau, J. (2019). The gypsy endogenous retrovirus drives non-cell-autonomous propagation in a Drosophila TDP-43 model of neurodegeneration. *Current Biology, 29*(19), 3135–3152 e3134.

Cheng, W., Wang, S., Zhang, Z., Morgens, D. W., Hayes, L. R., Lee, S., et al. (2019). CRISPR-Cas9 Screens Identify the RNA Helicase DDX3X as a Repressor of C9ORF72 (GGGGCC)n Repeat-Associated Non-AUG Translation. *Neuron, 104*(5), 885–898 e888.

Choi, J., Sullards, M. C., Olzmann, J. A., Rees, H. D., Weintraub, S. T., Bostwick, D. E., et al. (2006). Oxidative damage of DJ-1 is linked to sporadic Parkinson and Alzheimer diseases. *Journal of Biological Chemistry, 281*(16), 10816–10824.

Chu, J. F., Majumder, P., Chatterjee, B., Huang, S. L., & Shen, C. J. (2019). TDP-43 regulates coupled dendritic mRNA transport-translation processes in co-operation with FMRP and Staufen1. *Cell Reports, 29*(10), 3118–3133 e3116.

Cioni, J. M., Lin, J. Q., Holtermann, A. V., Koppers, M., Jakobs, M. A. H., Azizi, A., et al. (2019). Late Endosomes Act as mRNA Translation Platforms and Sustain Mitochondria in Axons. *Cell, 176*(1–2), 56–72 e15.

Ciosk, R., DePalma, M., & Priess, J. R. (2004). ATX-2, the C. elegans ortholog of ataxin 2, functions in translational regulation in the germline. *Development, 131*(19), 4831–4841.

Clark, I. E., Dodson, M. W., Jiang, C., Cao, J. H., Huh, J. R., Seol, J. H., et al. (2006). Drosophila pink1 is required for mitochondrial function and interacts genetically with parkin. *Nature, 441*(7097), 1162–1166.

Coelho, M. B., Attig, J., Bellora, N., Konig, J., Hallegger, M., Kayikci, M., et al. (2015). Nuclear matrix protein Matrin3 regulates alternative splicing and forms overlapping regulatory networks with PTB. *EMBO Journal, 34*(5), 653–668.

Cohen, T. J., Lee, V. M., & Trojanowski, J. Q. (2011). TDP-43 functions and pathogenic mechanisms implicated in TDP-43 proteinopathies. *Trends in Molecular Medicine, 17*(11), 659–667.

Colla, E., Coune, P., Liu, Y., Pletnikova, O., Troncoso, J. C., Iwatsubo, T., et al. (2012). Endoplasmic reticulum stress is important for the manifestations of alpha-synucleinopathy in vivo. *Journal of Neuroscience, 32*(10), 3306–3320.

Cook, C., & Petrucelli, L. (2019). Genetic convergence brings clarity to the enigmatic red line in ALS. *Neuron, 101*(6), 1057–1069.

Cookson, M. R. (2017). RNA-binding proteins implicated in neurodegenerative diseases. *Wiley Interdisciplinary Reviews: RNA, 8*(1).

Costa-Mattioli, M., Gobert, D., Harding, H., Herdy, B., Azzi, M., Bruno, M., et al. (2005). Translational control of hippocampal synaptic plasticity and memory by the eIF2alpha kinase GCN2. *Nature, 436*(7054), 1166–1173.

Costa-Mattioli, M., Gobert, D., Stern, E., Gamache, K., Colina, R., Cuello, C., et al. (2007). eIF2alpha phosphorylation bidirectionally regulates the switch from short- to long-term synaptic plasticity and memory. *Cell, 129*(1), 195–206.

Coyne, A. N., Lorenzini, I., Chou, C. C., Torvund, M., Rogers, R. S., Starr, A., et al. (2017). Post-transcriptional inhibition of Hsc70-4/HSPA8 expression leads to synaptic vesicle cycling defects in multiple models of ALS. *Cell Reports, 21*(1), 110–125.

Coyne, A. N., Siddegowda, B. B., Estes, P. S., Johannesmeyer, J., Kovalik, T., Daniel, S. G., et al. (2014). Futsch/MAP1B mRNA is a translational target of TDP-43 and is neuroprotective in a Drosophila model of amyotrophic lateral sclerosis. *Journal of Neuroscience, 34*(48), 15962–15974.

Da Cruz, S., & Cleveland, D. W. (2011). Understanding the role of TDP-43 and FUS/TLS in ALS and beyond. *Current Opinion in Neurobiology, 21*(6), 904–919.

Damgaard, C. K., & Lykke-Andersen, J. (2011). Translational coregulation of 5′TOP mRNAs by TIA-1 and TIAR. *Genes & Development, 25*(19), 2057–2068.

Dammer, E. B., Fallini, C., Gozal, Y. M., Duong, D. M., Rossoll, W., Xu, P., et al. (2012). Coaggregation of RNA-binding proteins in a model of TDP-43 proteinopathy with selective RGG motif methylation and a role for RRM1 ubiquitination. *PLoS One, 7*(6), e38658.

Dansithong, W., Paul, S., Figueroa, K. P., Rinehart, M. D., Wiest, S., Pflieger, L. T., et al. (2015). Ataxin-2 regulates RGS8 translation in a new BAC-SCA2 transgenic mouse model. *PLoS Genetics, 11*(4), e1005182.

Das, S., & Krainer, A. R. (2014). Emerging functions of SRSF1, splicing factor and oncoprotein, in RNA metabolism and cancer. *Molecular Cancer Research, 12*(9), 1195–1204.

Dawson, T. M., Ko, H. S., & Dawson, V. L. (2010). Genetic animal models of Parkinson's disease. *Neuron, 66*(5), 646–661.

DeJesus-Hernandez, M., Mackenzie, I. R., Boeve, B. F., Boxer, A. L., Baker, M., Rutherford, N. J., et al. (2011). Expanded GGGGCC hexanucleotide repeat in noncoding region of C9ORF72 causes chromosome 9p-linked FTD and ALS. *Neuron, 72*(2), 245–256.

Demir, U., Koehler, A., Schneider, R., Schweiger, S., & Klocker, H. (2014). Metformin anti-tumor effect via disruption of the MID1 translational regulator complex and AR downregulation in prostate cancer cells. *BMC Cancer, 14*, 52.

Dieterich, D. C., Lee, J. J., Link, A. J., Graumann, J., Tirrell, D. A., & Schuman, E. M. (2007). Labeling, detection and identification of newly synthesized proteomes with bioorthogonal non-canonical amino-acid tagging. *Nature Protocols, 2*(3), 532–540.

Dormann, D., & Haass, C. (2011). TDP-43 and FUS: a nuclear affair. *Trends in Neuroscience, 34*(7), 339–348.

Dormann, D., Rodde, R., Edbauer, D., Bentmann, E., Fischer, I., Hruscha, A., et al. (2010). ALS-associated fused in sarcoma (FUS) mutations disrupt Transportin-mediated nuclear import. *EMBO Journal, 29*(16), 2841–2857.

Douville, R. N., & Nath, A. (2017). Human endogenous retrovirus-K and TDP-43 expression bridges ALS and HIV neuropathology. *Frontiers in Microbiology, 8*, 1986.

Eimer, W. A., Vijaya Kumar, D. K., Navalpur Shanmugam, N. K., Rodriguez, A. S., Mitchell, T., Washicosky, K. J., et al. (2018). Alzheimer's disease-associated beta-amyloid is rapidly seeded by herpesviridae to protect against brain infection. *Neuron, 100*(6), 1527–1532.

Elden, A. C., Kim, H. J., Hart, M. P., Chen-Plotkin, A. S., Johnson, B. S., Fang, X., et al. (2010). Ataxin-2 intermediate-length polyglutamine expansions are associated with increased risk for ALS. *Nature, 466*(7310), 1069–1075.

Elvira, G., Wasiak, S., Blandford, V., Tong, X. K., Serrano, A., Fan, X., et al. (2006). Characterization of an RNA granule from developing brain. *Molecular & Cellular Proteomics, 5*(4), 635–651.

Estes, P. S., Boehringer, A., Zwick, R., Tang, J. E., Grigsby, B., & Zarnescu, D. C. (2011). Wild-type and A315T mutant TDP-43 exert differential neurotoxicity in a Drosophila model of ALS. *Human Molecular Genetics, 20*(12), 2308–2321.

Fernandez, E., Rajan, N., & Bagni, C. (2013). The FMRP regulon: from targets to disease convergence. *Frontiers in Neuroscience, 7*, 191.

Fiesel, F. C., Weber, S. S., Supper, J., Zell, A., & Kahle, P. J. (2012). TDP-43 regulates global translational yield by splicing of exon junction complex component SKAR. *Nucleic Acids Res., 40*(6), 2668–2682.

Fogh, I., Lin, K., Tiloca, C., Rooney, J., Gellera, C., Diekstra, F. P., et al. (2016). Association of a locus in the CAMTA1 gene with survival in patients with sporadic amyotrophic lateral sclerosis. *JAMA Neurology, 73*(7), 812–820.

Freibaum, B. D., Chitta, R. K., High, A. A., & Taylor, J. P. (2010). Global analysis of TDP-43 interacting proteins reveals strong association with RNA splicing and translation machinery. *Journal of Proteome Research, 9*(2), 1104–1120.

Fujii, R., Okabe, S., Urushido, T., Inoue, K., Yoshimura, A., Tachibana, T., et al. (2005). The RNA binding protein TLS is translocated to dendritic spines by mGluR5 activation and regulates spine morphology. *Current Biology, 15*(6), 587–593.

Gantois, I., Khoutorsky, A., Popic, J., Aguilar-Valles, A., Freemantle, E., Cao, R., et al. (2017). Metformin ameliorates core deficits in a mouse model of fragile X syndrome. *Nature Medicine, 23*(6), 674–677.

Gantois, I., Popic, J., Khoutorsky, A., & Sonenberg, N. (2019). Metformin for treatment of fragile X syndrome and other neurological disorders. *Annual Review of Medicine, 70*, 167–181.

Gao, Y., Tatavarty, V., Korza, G., Levin, M. K., & Carson, J. H. (2008). Multiplexed dendritic targeting of alpha calcium calmodulin-dependent protein kinase II, neurogranin, and activity-regulated cytoskeleton-associated protein RNAs by the A2 pathway. *Molecular Biology of Cell, 19*(5), 2311–2327.

Garcia-Maurino, S. M., Rivero-Rodriguez, F., Velazquez-Cruz, A., Hernandez-Vellisca, M., Diaz-Quintana, A., De la Rosa, M. A., et al. (2017). RNA Binding Protein Regulation and Cross-Talk in the Control of AU-rich mRNA Fate. *Frontiers in Molecular Biosciences, 4*, 71.

Gasset-Rosa, F., Lu, S., Yu, H., Chen, C., Melamed, Z., Guo, L., et al. (2019). Cytoplasmic TDP-43 de-mixing independent of stress granules drives inhibition of nuclear import, loss of nuclear TDP-43, and cell death. *Neuron, 102*(2), 339–357 e337.

Gebauer, F., & Hentze, M. W. (2004). Molecular mechanisms of translational control. *Nature Reviews Molecular Cell Biology, 5*(10), 827–835.

Gehrke, S., Imai, Y., Sokol, N., & Lu, B. (2010). Pathogenic LRRK2 negatively regulates microRNA-mediated translational repression. *Nature, 466*(7306), 637–641.

Gehrke, S., Wu, Z., Klinkenberg, M., Sun, Y., Auburger, G., Guo, S., et al. (2015). PINK1 and Parkin control localized translation of respiratory chain component mRNAs on mitochondria outer membrane. *Cell Metabolism, 21*(1), 95–108.

Gendron, T. F., Bieniek, K. F., Zhang, Y. J., Jansen-West, K., Ash, P. E., Caulfield, T., et al. (2013). Antisense transcripts of the expanded C9ORF72 hexanucleotide repeat form nuclear RNA foci and undergo repeat-associated non-ATG translation in c9FTD/ALS. *Acta Neuropathologica, 126*(6), 829–844.

Gijselinck, I., Van Langenhove, T., van der Zee, J., Sleegers, K., Philtjens, S., Kleinberger, G., et al. (2012). A C9orf72 promoter repeat expansion in a Flanders-Belgian cohort with disorders of the frontotemporal lobar degeneration-amyotrophic lateral sclerosis spectrum: A gene identification study. *Lancet Neurology, 11*(1), 54–65.

Gilks, N., Kedersha, N., Ayodele, M., Shen, L., Stoecklin, G., Dember, L. M., et al. (2004). Stress granule assembly is mediated by prion-like aggregation of TIA-1. *Molecular Biology of the Cell, 15*(12), 5383–5398.

Gillingwater, T. H., & Wishart, T. M. (2013). Mechanisms underlying synaptic vulnerability and degeneration in neurodegenerative disease. *Neuropathology and Applied Neurobiology, 39*(4), 320–334.

Giordana, M. T., Piccinini, M., Grifoni, S., De Marco, G., Vercellino, M., Magistrello, M., et al. (2010). TDP-43 redistribution is an early event in sporadic amyotrophic lateral sclerosis. *Brain Pathology, 20*(2), 351–360.

Green, K. M., Linsalata, A. E., & Todd, P. K. (2016). RAN translation-What makes it run? *Brain Research, 1647*, 30–42.

Green, K. M., Sheth, U. J., Flores, B. N., Wright, S. E., Sutter, A. B., Kearse, M. G., et al. (2019). High-throughput screening yields several small-molecule inhibitors of repeat-associated non-AUG translation. *Journal of Biological Chemistry, 294*(49), 18624–18638.

Griesche, N., Schilling, J., Weber, S., Rohm, M., Pesch, V., Matthes, F., et al. (2016). Regulation of mRNA Translation by MID1: A Common Mechanism of Expanded CAG Repeat RNAs. *Frontiers in Cellular Neuroscience, 10*, 226.

Gu, X., Chen, Y., Wei, Q., Cao, B., Ou, R., Yuan, X., et al. (2018). Mutation screening of the TIA1 gene in Chinese patients with amyotrophic lateral sclerosis/frontotemporal dementia. *Neurobiology of Aging, 68*, 161 e161–161 e163.

Guo, L., Kim, H. J., Wang, H., Monaghan, J., Freyermuth, F., Sung, J. C., et al. (2018). Nuclear-Import Receptors Reverse Aberrant Phase Transitions of RNA-binding proteins with prion-like domains. *Cell, 173*(3), 677–692 e620.

Guo, W., Fumagalli, L., Prior, R., & Van Den Bosch, L. (2017). Current advances and limitations in modeling ALS/FTD in a dish using induced pluripotent stem cells. *Frontiers in Neuroscience, 11*, 671.

Guo, W., Naujock, M., Fumagalli, L., Vandoorne, T., Baatsen, P., Boon, R., et al. (2017). HDAC6 inhibition reverses axonal transport defects in motor neurons derived from FUS-ALS patients. *Nature Communications, 8*(1), 861.

Halliday, M., & Mallucci, G. R. (2015). Review: Modulating the unfolded protein response to prevent neurodegeneration and enhance memory. *Neuropathology and Applied Neurobiology, 41*(4), 414–427.

Hautbergue, G. M., Castelli, L. M., Ferraiuolo, L., Sanchez-Martinez, A., Cooper-Knock, J., Higginbottom, A., et al. (2017). SRSF1-dependent nuclear export inhibition of C9ORF72 repeat transcripts prevents neurodegeneration and associated motor deficits. *Nature Communications, 8*, 16063.

Herdy, B., Mayer, C., Varshney, D., Marsico, G., Murat, P., Taylor, C., et al. (2018). Analysis of NRAS RNA G-quadruplex binding proteins reveals DDX3X as a novel interactor of cellular G-quadruplex containing transcripts. *Nucleic Acids Research, 46*(21), 11592–11604.

Hetman, M., & Slomnicki, L. P. (2019). Ribosomal biogenesis as an emerging target of neurodevelopmental pathologies. *Journal of Neurochemistry, 148*(3), 325–347.

Hicks, G. G., Singh, N., Nashabi, A., Mai, S., Bozek, G., Klewes, L., et al. (2000). FUS deficiency in mice results in defective B-lymphocyte development and activation, high levels of chromosomal instability and perinatal death. *Nature Genetics, 24*(2), 175–179.

Higashi, S., Kabuta, T., Nagai, Y., Tsuchiya, Y., Akiyama, H., & Wada, K. (2013). TDP-43 associates with stalled ribosomes and contributes to cell survival during cellular stress. *Journal of Neurochemistry, 126*(2), 288–300.

Hod, Y., Pentyala, S. N., Whyard, T. C., & El-Maghrabi, M. R. (1999). Identification and characterization of a novel protein that regulates RNA-protein interaction. *Journal of Cellular Biochemistry, 72*(3), 435–444.

Hofweber, M., Hutten, S., Bourgeois, B., Spreitzer, E., Niedner-Boblenz, A., Schifferer, M., et al. (2018). Phase separation of FUS is suppressed by its nuclear import receptor and arginine methylation. *Cell, 173*(3), 706–719 e713.

Huang, C., Tong, J., Bi, F., Zhou, H., & Xia, X. G. (2012). Mutant TDP-43 in motor neurons promotes the onset and progression of ALS in rats. *Journal of Clinical Investigation, 122*(1), 107–118.

Huelga, S. C., Vu, A. Q., Arnold, J. D., Liang, T. Y., Liu, P. P., Yan, B. Y., et al. (2012). Integrative genome-wide analysis reveals cooperative regulation of alternative splicing by hnRNP proteins. *Cell Reports, 1*(2), 167–178.

Huentelman, M. J., Papassotiropoulos, A., Craig, D. W., Hoerndli, F. J., Pearson, J. V., Huynh, K. D., et al. (2007). Calmodulin-binding transcription activator 1 (CAMTA1) alleles predispose human episodic memory performance. *Human Molecular Genetics, 16*(12), 1469–1477.

The Huntington's Disease Collaborative Research Group. (1993). "A novel gene containing a trinucleotide repeat that is expanded and unstable on Huntington's disease chromosomes." *Cell, 72*(6), 971–983.

Ikiz, B., Alvarez, M. J., Re, D. B., Le Verche, V., Politi, K., Lotti, F., et al. (2015). The regulatory machinery of neurodegeneration in in vitro models of amyotrophic lateral sclerosis. *Cell Reports, 12*(2), 335–345.

Imai, Y., Gehrke, S., Wang, H. Q., Takahashi, R., Hasegawa, K., Oota, E., et al. (2008). Phosphorylation of 4E-BP by LRRK2 affects the maintenance of dopaminergic neurons in Drosophila. *EMBO Journal, 27*(18), 2432–2443.

Ingolia, N. T., Brar, G. A., Rouskin, S., McGeachy, A. M., & Weissman, J. S. (2012). The ribosome profiling strategy for monitoring translation in vivo by deep sequencing of ribosome-protected mRNA fragments. *Nature Protocols, 7*(8), 1534–1550.

Iradi, M. C. G., Triplett, J. C., Thomas, J. D., Davila, R., Crown, A. M., Brown, H., et al. (2018). Characterization of gene regulation and protein interaction networks for Matrin 3 encoding mutations linked to amyotrophic lateral sclerosis and myopathy. *Science Reports, 8*(1), 4049.

Jaiswal, M. K. (2019). Riluzole and edaravone: A tale of two amyotrophic lateral sclerosis drugs. *Medicinal Research Reviews, 39*(2), 733–748.

Jin, J., Gu, H., Anders, N. M., Ren, T., Jiang, M., Tao, M., et al. (2016). Metformin protects cells from mutant huntingtin toxicity through activation of AMPK and modulation of mitochondrial dynamics. *NeuroMolecular Medicine, 18*(4), 581–592.

Johnson, J. O., Pioro, E. P., Boehringer, A., Chia, R., Feit, H., Renton, A. E., et al. (2014). Mutations in the Matrin 3 gene cause familial amyotrophic lateral sclerosis. *Nature Neuroscience, 17*(5), 664–666.

Joyce, P. I., Fratta, P., Fisher, E. M., & Acevedo-Arozena, A. (2011). SOD1 and TDP-43 animal models of amyotrophic lateral sclerosis: Recent advances in understanding disease toward the development of clinical treatments. *Mammalian Genome, 22*(7–8), 420–448.

Juzwik, C. A., S, S. D., Zhang, Y., Paradis-Isler, N., Sylvester, A., Amar-Zifkin, A., et al. (2019). microRNA dysregulation in neurodegenerative diseases: A systematic review. *Progress in Neurobiology, 182*, 101664.

Kabashi, E., Valdmanis, P. N., Dion, P., Spiegelman, D., McConkey, B. J., Vande Velde, C., et al. (2008). TARDBP mutations in individuals with sporadic and familial amyotrophic lateral sclerosis. *Nature Genetics, 40*(5), 572–574.

Kamelgarn, M., Chen, J., Kuang, L., Jin, H., Kasarskis, E. J., & Zhu, H. (2018). ALS mutations of FUS suppress protein translation and disrupt the regulation of nonsense-mediated decay. *Proceedings of the National Academy of Science of the United States of America, 115*(51), E11904–E11913.

Kapeli, K., Pratt, G. A., Vu, A. Q., Hutt, K. R., Martinez, F. J., Sundararaman, B., et al. (2016). Distinct and shared functions of ALS-associated proteins TDP-43, FUS and TAF15 revealed by multisystem analyses. *Nature Communications, 7*, 12143.

Kapur, M., Monaghan, C. E., & Ackerman, S. L. (2017). Regulation of mRNA translation in neurons-a matter of life and death. *Neuron, 96*(3), 616–637.

Karousis, E. D., & Muhlemann, O. (2019). Nonsense-mediated mRNA decay begins where translation ends. *Cold Spring Harbor Perspectives in Biology, 11*(2).

Kawahara, Y., & Mieda-Sato, A. (2012). TDP-43 promotes microRNA biogenesis as a component of the Drosha and Dicer complexes. *Proceedings of the National Academy of Sciences of the United States of America, 109*(9), 3347–3352.

Kawai, T., Lal, A., Yang, X., Galban, S., Mazan-Mamczarz, K., & Gorospe, M. (2006). Translational control of cytochrome c by RNA-binding proteins TIA-1 and HuR. *Molecular and Cellular Biology, 26*(8), 3295–3307.

Kedersha, N., Cho, M. R., Li, W., Yacono, P. W., Chen, S., Gilks, N., et al. (2000). Dynamic shuttling of TIA-1 accompanies the recruitment of mRNA to mammalian stress granules. *Journal of Cell Biology, 151*(6), 1257–1268.

Kedersha, N. L., Gupta, M., Li, W., Miller, I., & Anderson, P. (1999). RNA-binding proteins TIA-1 and TIAR link the phosphorylation of eIF-2 alpha to the assembly of mammalian stress granules. *Journal of Cell Biology, 147*(7), 1431–1442.

Kerschbamer, E., & Biagioli, M. (2015). Huntington's disease as neurodevelopmental disorder: Altered chromatin regulation, coding, and non-coding RNA transcription. *Frontiers in Neuroscience, 9*, 509.

Kickstein, E., Krauss, S., Thornhill, P., Rutschow, D., Zeller, R., Sharkey, J., et al. (2010). Biguanide metformin acts on tau phosphorylation via mTOR/protein phosphatase 2A

(PP2A) signaling. *Proceedings of the National Academy of Sciences of the United States of America, 107*(50), 21830–21835.

Kim, H. J., Kim, N. C., Wang, Y. D., Scarborough, E. A., Moore, J., Diaz, Z., et al. (2013). Mutations in prion-like domains in hnRNPA2B1 and hnRNPA1 cause multisystem proteinopathy and ALS. *Nature, 495*(7442), 467–473.

King, I. N., Yartseva, V., Salas, D., Kumar, A., Heidersbach, A., Ando, D. M., et al. (2014). The RNA-binding protein TDP-43 selectively disrupts microRNA-1/206 incorporation into the RNA-induced silencing complex. *Journal of Biological Chemistry, 289*(20), 14263–14271.

Kirchner, S., & Ignatova, Z. (2015). Emerging roles of tRNA in adaptive translation, signalling dynamics and disease. *Nature Reviews Genetics, 16*(2), 98–112.

Koley, S., Rozenbaum, M., Fainzilber, M., & Terenzio, M. (2019). Translating regeneration: Local protein synthesis in the neuronal injury response. *Neuroscience Research, 139*, 26–36.

Kosturko, L. D., Maggipinto, M. J., Korza, G., Lee, J. W., Carson, J. H., & Barbarese, E. (2006). Heterogeneous nuclear ribonucleoprotein (hnRNP) E1 binds to hnRNP A2 and inhibits translation of A2 response element mRNAs. *Molecular Biology of the Cell, 17*(8), 3521–3533.

Krauss, S., Griesche, N., Jastrzebska, E., Chen, C., Rutschow, D., Achmuller, C., et al. (2013). Translation of HTT mRNA with expanded CAG repeats is regulated by the MID1-PP2A protein complex. *Nature Communications, 4*, 1511.

Krug, L., Chatterjee, N., Borges-Monroy, R., Hearn, S., Liao, W. W., Morrill, K., et al. (2017). Retrotransposon activation contributes to neurodegeneration in a Drosophila TDP-43 model of ALS. *PLoS Genetics, 13*(3), e1006635.

Kthiri, F., Gautier, V., Le, H. T., Prere, M. F., Fayet, O., Malki, A., et al. (2010). Translational defects in a mutant deficient in YajL, the bacterial homolog of the parkinsonism-associated protein DJ-1. *Journal of Bacteriology, 192*(23), 6302–6306.

Kuersten, S., & Goodwin, E. B. (2003). The power of the 3′ UTR: translational control and development. *Nat Rev Genet, 4*(8), 626–637.

Lagier-Tourenne, C., Polymenidou, M., & Cleveland, D. W. (2010). TDP-43 and FUS/TLS: Emerging roles in RNA processing and neurodegeneration. *Human Molecular Genetics, 19*(R1), R46–R64.

Lee, B. J., Cansizoglu, A. E., Suel, K. E., Louis, T. H., Zhang, Z., & Chook, Y. M. (2006). Rules for nuclear localization sequence recognition by karyopherin beta 2. *Cell, 126*(3), 543–558.

Leitman, J., Barak, B., Benyair, R., Shenkman, M., Ashery, U., Hartl, F. U., et al. (2014). ER stress-induced eIF2-alpha phosphorylation underlies sensitivity of striatal neurons to pathogenic huntingtin. *PLoS One, 9*(3), e90803.

Lennox, A. L., Hoye, M. L., Jiang, R., Johnson-Kerner, B. L., Suit, L. A., Venkataramanan, S., et al. (2020). Pathogenic DDX3X mutations impair RNA metabolism and neurogenesis during fetal cortical development. *Neuron, 106*(3), 404–420 e408.

Lewis, S. M., Veyrier, A., Hosszu Ungureanu, N., Bonnal, S., Vagner, S., & Holcik, M. (2007). Subcellular relocalization of a trans-acting factor regulates XIAP IRES-dependent translation. *Molecular Biology of the Cell, 18*(4), 1302–1311.

Li, Y. R., King, O. D., Shorter, J., & Gitler, A. D. (2013). Stress granules as crucibles of ALS pathogenesis. *Journal of Cell Biology, 201*(3), 361–372.

Lim, C., & Allada, R. (2013). ATAXIN-2 activates PERIOD translation to sustain circadian rhythms in Drosophila. *Science, 340*(6134), 875–879.

Lin, A. C., & Holt, C. E. (2007). Local translation and directional steering in axons. *EMBO Journal, 26*(16), 3729–3736.

Lin, Y., Protter, D. S., Rosen, M. K., & Parker, R. (2015). Formation and maturation of phase-separated liquid droplets by RNA-binding proteins. *Molecular Cell, 60*(2), 208–219.

Ling, S. C., Polymenidou, M., & Cleveland, D. W. (2013). Converging mechanisms in ALS and FTD: Disrupted RNA and protein homeostasis. *Neuron, 79*(3), 416–438.

Linsalata, A. E., He, F., Malik, A. M., Glineburg, M. R., Green, K. M., Natla, S., et al. (2019). DDX3X and specific initiation factors modulate FMR1 repeat-associated non-AUG-initiated translation. *EMBO Report, 20*(9), e47498.

Liu, S., & Lu, B. (2010). Reduction of protein translation and activation of autophagy protect against PINK1 pathogenesis in Drosophila melanogaster. *PLoS Genetics, 6*(12), e1001237.

Liu-Yesucevitz, L., Lin, A. Y., Ebata, A., Boon, J. Y., Reid, W., Xu, Y. F., et al. (2014). ALS-linked mutations enlarge TDP-43-enriched neuronal RNA granules in the dendritic arbor. *Journal of Neuroscience, 34*(12), 4167–4174.

Long, C., Grueter, C. E., Song, K., Qin, S., Qi, X., Kong, Y. M., et al. (2014). Ataxia and Purkinje cell degeneration in mice lacking the CAMTA1 transcription factor. *Proceedings of the National Academy of Sciences of the United States of America, 111*(31), 11521–11526.

Lopez-Erauskin, J., Tadokoro, T., Baughn, M. W., Myers, B., McAlonis-Downes, M., Chillon-Marinas, C., et al. (2018). ALS/FTD-Linked Mutation in FUS Suppresses Intra-axonal Protein Synthesis and Drives Disease Without Nuclear Loss-of-Function of FUS. *Neuron, 100*(4), 816–830 e817.

Lopez de Silanes, I., Galban, S., Martindale, J. L., Yang, X., Mazan-Mamczarz, K., Indig, F. E., et al. (2005). Identification and functional outcome of mRNAs associated with RNA-binding protein TIA-1. *Molecular and Cellular Biology, 25*(21), 9520–9531.

Lu, B., Gehrke, S., & Wu, Z. (2014). RNA metabolism in the pathogenesis of Parkinsons disease. *Brain Research, 1584*, 105–115.

Mackenzie, I. R., Nicholson, A. M., Sarkar, M., Messing, J., Purice, M. D., Pottier, C., et al. (2017). TIA1 Mutations in amyotrophic lateral sclerosis and frontotemporal dementia promote phase separation and alter stress granule dynamics. *Neuron, 95*(4), 808–816 e809.

Majumder, P., Chen, Y. T., Bose, J. K., Wu, C. C., Cheng, W. C., Cheng, S. J., et al. (2012). TDP-43 regulates the mammalian spinogenesis through translational repression of Rac1. *Acta Neuropathologica, 124*(2), 231–245.

Majumder, P., Chu, J. F., Chatterjee, B., Swamy, K. B., & Shen, C. J. (2016). Co-regulation of mRNA translation by TDP-43 and Fragile X Syndrome protein FMRP. *Acta Neuropathologica, 132*(5), 721–738.

Manghera, M., Ferguson-Parry, J., & Douville, R. N. (2016). TDP-43 regulates endogenous retrovirus-K viral protein accumulation. *Neurobiology of Disease, 94*, 226–236.

Mann, J. R., Gleixner, A. M., Mauna, J. C., Gomes, E., DeChellis-Marks, M. R., Needham, P. G., et al. (2019). RNA binding antagonizes neurotoxic phase transitions of TDP-43. *Neuron, 102*(2), 321–338 e328.

Martin, I., Kim, J. W., Dawson, V. L., & Dawson, T. M. (2014). LRRK2 pathobiology in Parkinson's disease. *Journal of Neurochemistry, 131*(5), 554–565.

Martin, I., Kim, J. W., Lee, B. D., Kang, H. C., Xu, J. C., Jia, H., et al. (2014). Ribosomal protein s15 phosphorylation mediates LRRK2 neurodegeneration in Parkinson's disease. *Cell, 157*(2), 472–485.

Matheoud, D., Cannon, T., Voisin, A., Penttinen, A. M., Ramet, L., Fahmy, A. M., et al. (2019). Intestinal infection triggers Parkinson's disease-like symptoms in Pink1(-/-) mice. *Nature, 571*(7766), 565–569.

McColgan, P., & Tabrizi, S. J. (2018). Huntington's disease: a clinical review. *European Journal of Neurology, 25*(1), 24–34.

McCormick, C., & Khaperskyy, D. A. (2017). Translation inhibition and stress granules in the antiviral immune response. *Nature Reviews Immunology, 17*(10), 647–660.

Medenbach, J., Seiler, M., & Hentze, M. W. (2011). Translational control via protein-regulated upstream open reading frames. *Cell, 145*(6), 902–913.

Mejzini, R., Flynn, L. L., Pitout, I. L., Fletcher, S., Wilton, S. D., & Akkari, P. A. (2019). ALS genetics, mechanisms, and therapeutics: Where are we now? *Frontiers in Neuroscience, 13*, 1310.

Mikhaleva, S., & Lemke, E. A. (2018). Beyond the transport function of import receptors: What's all the FUS about? *Cell, 173*(3), 549–553.

Miller, J. W., Urbinati, C. R., Teng-Umnuay, P., Stenberg, M. G., Byrne, B. J., Thornton, C. A., et al. (2000). Recruitment of human muscleblind proteins to (CUG)(n) expansions associated with myotonic dystrophy. *EMBO Journal, 19*(17), 4439–4448.

Mohan, A., Goodwin, M., & Swanson, M. S. (2014). RNA-protein interactions in unstable microsatellite diseases. *Brain Research, 1584*, 3–14.

Moreno, J. A., Radford, H., Peretti, D., Steinert, J. R., Verity, N., Martin, M. G., et al. (2012). Sustained translational repression by eIF2alpha-P mediates prion neurodegeneration. *Nature, 485*(7399), 507–511.

Mori, K., Weng, S. M., Arzberger, T., May, S., Rentzsch, K., Kremmer, E., et al. (2013). The C9orf72 GGGGCC repeat is translated into aggregating dipeptide-repeat proteins in FTLD/ALS. *Science, 339*(6125), 1335–1338.

Morlando, M., Dini Modigliani, S., Torrelli, G., Rosa, A., Di Carlo, V., Caffarelli, E., et al. (2012). FUS stimulates microRNA biogenesis by facilitating co-transcriptional Drosha recruitment. *EMBO Journal, 31*(24), 4502–4510.

Napoli, I., Mercaldo, V., Boyl, P. P., Eleuteri, B., Zalfa, F., De Rubeis, S., et al. (2008). The fragile X syndrome protein represses activity-dependent translation through CYFIP1, a new 4E-BP. *Cell, 134*(6), 1042–1054.

Neelagandan, N., Gonnella, G., Dang, S., Janiesch, P. C., Miller, K. K., Kuchler, K., et al. (2019). TDP-43 enhances translation of specific mRNAs linked to neurodegenerative disease. *Nucleic Acids Research, 47*(1), 341–361.

Neumann, M., Sampathu, D. M., Kwong, L. K., Truax, A. C., Micsenyi, M. C., Chou, T. T., et al. (2006). Ubiquitinated TDP-43 in frontotemporal lobar degeneration and amyotrophic lateral sclerosis. *Science, 314*(5796), 130–133.

Nussbacher, J. K., Batra, R., Lagier-Tourenne, C., & Yeo, G. W. (2015). RNA-binding proteins in neurodegeneration: Seq and you shall receive. *Trends in Neuroscience, 38*(4), 226–236.

Ohno, M. (2014). Roles of eIF2alpha kinases in the pathogenesis of Alzheimer's disease. *Frontiers in Molecular Neuroscience, 7*, 22.

Orr, H. T., & Zoghbi, H. Y. (2007). Trinucleotide repeat disorders. *Annual Review of Neuroscience, 30*, 575–621.

Otsuka, H., Fukao, A., Funakami, Y., Duncan, K. E., & Fujiwara, T. (2019). Emerging Evidence of Translational Control by AU-Rich Element-Binding Proteins. *Frontiers in Genetics, 10*, 332.

Paisan-Ruiz, C., Jain, S., Evans, E. W., Gilks, W. P., Simon, J., van der Brug, M., et al. (2004). Cloning of the gene containing mutations that cause PARK8-linked Parkinson's disease. *Neuron, 44*(4), 595–600.

Panas, M. D., Ivanov, P., & Anderson, P. (2016). Mechanistic insights into mammalian stress granule dynamics. *Journal of Cell Biology, 215*(3), 313–323.

Park, J., Lee, S. B., Lee, S., Kim, Y., Song, S., Kim, S., et al. (2006). Mitochondrial dysfunction in Drosophila PINK1 mutants is complemented by parkin. *Nature, 441*(7097), 1157–1161.

Perez-Arancibia, R., Rivas, A., & Hetz, C. (2018). (off)Targeting UPR signaling: the race toward intervening ER proteostasis. *Expert Opinion on Therapeutic Targets, 22*(2), 97–100.

Pickrell, A. M., & Youle, R. J. (2015). The roles of PINK1, parkin, and mitochondrial fidelity in Parkinson's disease. *Neuron, 85*(2), 257–273.

Piecyk, M., Wax, S., Beck, A. R., Kedersha, N., Gupta, M., Maritim, B., et al. (2000). TIA-1 is a translational silencer that selectively regulates the expression of TNF-alpha. *EMBO Journal, 19*(15), 4154–4163.

Poewe, W., Seppi, K., Tanner, C. M., Halliday, G. M., Brundin, P., Volkmann, J., et al. (2017). Parkinson disease. *Nature Reviews Disease Primers, 3*, 17013.

Polymenidou, M., Lagier-Tourenne, C., Hutt, K. R., Bennett, C. F., Cleveland, D. W., & Yeo, G. W. (2012). Misregulated RNA processing in amyotrophic lateral sclerosis. *Brain Research, 1462*, 3–15.

Polymenidou, M., Lagier-Tourenne, C., Hutt, K. R., Huelga, S. C., Moran, J., Liang, T. Y., et al. (2011). Long pre-mRNA depletion and RNA missplicing contribute to neuronal vulnerability from loss of TDP-43. *Nature Neuroscience, 14*(4), 459–468.

Prelich, G. (2012). Gene overexpression: uses, mechanisms, and interpretation. *Genetics, 190*(3), 841–854.

Protter, D. S. W., & Parker, R. (2016). Principles and properties of stress granules. *Trends in Cell Biology, 26*(9), 668–679.

Pushpalatha, K. V., & Besse, F. (2019). Local translation in axons: When membraneless RNP granules meet membrane-bound organelles. *Frontiers in Molecular Biosciences, 6*, 129.

Qamar, S., Wang, G., Randle, S. J., Ruggeri, F. S., Varela, J. A., Lin, J. Q., et al. (2018). FUS Phase Separation Is Modulated by a Molecular Chaperone and Methylation of Arginine Cation-pi Interactions. *Cell, 173*(3), 720–734 e715.

Rabouw, H. H., Langereis, M. A., Anand, A. A., Visser, L. J., de Groot, R. J., Walter, P., et al. (2019). Small molecule ISRIB suppresses the integrated stress response within a defined window of activation. *Proceedings of the National Academy of Sciences of the United States of America, 116*(6), 2097–2102.

Rademakers, R., Neumann, M., & Mackenzie, I. R. (2012). Advances in understanding the molecular basis of frontotemporal dementia. *Nature Reviews Neurology, 8*(8), 423–434.

Ramaswami, M., Taylor, J. P., & Parker, R. (2013). Altered ribostasis: RNA-protein granules in degenerative disorders. *Cell, 154*(4), 727–736.

Readhead, B., Haure-Mirande, J. V., Funk, C. C., Richards, M. A., Shannon, P., Haroutunian, V., et al. (2018). Multiscale analysis of independent Alzheimer's cohorts finds disruption of molecular, genetic, and clinical networks by human herpesvirus. *Neuron, 99*(1), 64–82 e67.

Reijonen, S., Putkonen, N., Norremolle, A., Lindholm, D., & Korhonen, L. (2008). Inhibition of endoplasmic reticulum stress counteracts neuronal cell death and protein aggregation caused by N-terminal mutant huntingtin proteins. *Experimental Cell Research, 314*(5), 950–960.

Renton, A. E., Majounie, E., Waite, A., Simon-Sanchez, J., Rollinson, S., Gibbs, J. R., et al. (2011). A hexanucleotide repeat expansion in C9ORF72 is the cause of chromosome 9p21-linked ALS-FTD. *Neuron, 72*(2), 257–268.

Repici, M., & Giorgini, F. (2019). DJ-1 in Parkinson's Disease: Clinical insights and therapeutic perspectives. *Journal of Clinical Medicine, 8*(9).

Repici, M., Hassanjani, M., Maddison, D. C., Garcao, P., Cimini, S., Patel, B., et al. (2019). The Parkinson's disease-linked protein DJ-1 associates with cytoplasmic mRNP granules during stress and neurodegeneration. *Molecular Neurobiology, 56*(1), 61–77.

Riva, V., & Maga, G. (2019). From the magic bullet to the magic target: Exploiting the diverse roles of DDX3X in viral infections and tumorigenesis. *Future Medicinal Chemistry, 11*(11), 1357–1381.

Rowland, L. P., & Shneider, N. A. (2001). Amyotrophic lateral sclerosis. *New England Journal of Medicine, 344*(22), 1688–1700.

Roy, R., Durie, D., Li, H., Liu, B. Q., Skehel, J. M., Mauri, F., et al. (2014). hnRNPA1 couples nuclear export and translation of specific mRNAs downstream of FGF-2/S6K2 signalling. *Nucleic Acids Research, 42*(20), 12483–12497.

Roy, R., Huang, Y., Seckl, M. J., & Pardo, O. E. (2017). Emerging roles of hnRNPA1 in modulating malignant transformation. *Wiley Interdisciplinary Reviews: RNA, 8*(6).

Russo, A., Scardigli, R., La Regina, F., Murray, M. E., Romano, N., Dickson, D. W., et al. (2017). Increased cytoplasmic TDP-43 reduces global protein synthesis by interacting with RACK1 on polyribosomes. *Human Molecular Genetics, 26*(8), 1407–1418.

Ryu, E. J., Harding, H. P., Angelastro, J. M., Vitolo, O. V., Ron, D., & Greene, L. A. (2002). Endoplasmic reticulum stress and the unfolded protein response in cellular models of Parkinson's disease. *Journal of Neuroscience, 22*(24), 10690–10698.

Salazar-Grueso, E. F., Kim, S., & Kim, H. (1991). Embryonic mouse spinal cord motor neuron hybrid cells. *NeuroReport, 2*(9), 505–508.

Sances, S., Bruijn, L. I., Chandran, S., Eggan, K., Ho, R., Klim, J. R., et al. (2016). Modeling ALS with motor neurons derived from human induced pluripotent stem cells. *Nature Neuroscience, 19*(4), 542–553.

Satterfield, T. F., & Pallanck, L. J. (2006). Ataxin-2 and its Drosophila homolog, ATX2, physically assemble with polyribosomes. *Human Molecular Genetics, 15*(16), 2523–2532.

Saxena, S., & Caroni, P. (2011). Selective neuronal vulnerability in neurodegenerative diseases: From stressor thresholds to degeneration. *Neuron, 71*(1), 35–48.

Sephton, C. F., Cenik, C., Kucukural, A., Dammer, E. B., Cenik, B., Han, Y., et al. (2011). Identification of neuronal RNA targets of TDP-43-containing ribonucleoprotein complexes. *Journal of Biological Chemistry, 286*(2), 1204–1215.

Shen, Y., Yu, X., Zhu, L., Li, T., Yan, Z., & Guo, J. (2018). Transfer RNA-derived fragments and tRNA halves: biogenesis, biological functions and their roles in diseases. *Journal of Molecular Medicine (Berlin), 96*(11), 1167–1176.

Shinawi, M., Coorg, R., Shimony, J. S., Grange, D. K., & Al-Kateb, H. (2015). Intragenic CAMTA1 deletions are associated with a spectrum of neurobehavioral phenotypes. *Clinical Genetics, 87*(5), 478–482.

Sidrauski, C., Acosta-Alvear, D., Khoutorsky, A., Vedantham, P., Hearn, B. R., Li, H., et al. (2013). Pharmacological brake-release of mRNA translation enhances cognitive memory. *Elife, 2*, e00498.

Sidrauski, C., McGeachy, A. M., Ingolia, N. T., & Walter, P. (2015). The small molecule ISRIB reverses the effects of eIF2alpha phosphorylation on translation and stress granule assembly. *Elife, 4*.

Sidrauski, C., Tsai, J. C., Kampmann, M., Hearn, B. R., Vedantham, P., Jaishankar, P., et al. (2015). Pharmacological dimerization and activation of the exchange factor eIF2B antagonizes the integrated stress response. *Elife, 4*, e07314.

Silva, R. M., Ries, V., Oo, T. F., Yarygina, O., Jackson-Lewis, V., Ryu, E. J., et al. (2005). CHOP/GADD153 is a mediator of apoptotic death in substantia nigra dopamine neurons in an in vivo neurotoxin model of parkinsonism. *Journal of Neurochemistry, 95*(4), 974–986.

Sinnamon, J. R., & Czaplinski, K. (2011). mRNA trafficking and local translation: the Yin and Yang of regulating mRNA localization in neurons. *Acta Biochimica Biophysica Sinica (Shanghai), 43*(9), 663–670.

Smith, R. W., Blee, T. K., & Gray, N. K. (2014). Poly(A)-binding proteins are required for diverse biological processes in metazoans. *Biochemical Society Transactions, 42*(4), 1229–1237.

Soukup, S. F., Vanhauwaert, R., & Verstreken, P. (2018). Parkinson's disease: convergence on synaptic homeostasis. *EMBO Journal, 37*(18).

Sreedharan, J., Blair, I. P., Tripathi, V. B., Hu, X., Vance, C., Rogelj, B., et al. (2008). TDP-43 mutations in familial and sporadic amyotrophic lateral sclerosis. *Science, 319*(5870), 1668–1672.

Stoecklin, G., & Anderson, P. (2006). Posttranscriptional mechanisms regulating the inflammatory response. *Advances in Immunology, 89*, 1–37.

Tada, M., Doi, H., Koyano, S., Kubota, S., Fukai, R., Hashiguchi, S., et al. (2018). Matrin 3 Is a Component of Neuronal Cytoplasmic Inclusions of Motor Neurons in Sporadic Amyotrophic Lateral Sclerosis. *American Journal of Pathology, 188*(2), 507–514.

Tain, L. S., Mortiboys, H., Tao, R. N., Ziviani, E., Bandmann, O., & Whitworth, A. J. (2009). Rapamycin activation of 4E-BP prevents parkinsonian dopaminergic neuron loss. *Nature Neuroscience, 12*(9), 1129–1135.

Taylor, J. P., Brown, R. H., Jr., & Cleveland, D. W. (2016). Decoding ALS: from genes to mechanism. *Nature, 539*(7628), 197–206.

Therrien, M., Dion, P. A., & Rouleau, G. A. (2016). ALS: Recent developments from genetics studies. *Current Neurology and Neuroscience Reports, 16*(6), 59.

Tollervey, J. R., Curk, T., Rogelj, B., Briese, M., Cereda, M., Kayikci, M., et al. (2011). Characterizing the RNA targets and position-dependent splicing regulation by TDP-43. *Nature Neuroscience, 14*(4), 452–458.

Torvund-Jensen, J., Steengaard, J., Reimer, L., Fihl, L. B., & Laursen, L. S. (2014). Transport and translation of MBP mRNA is regulated differently by distinct hnRNP proteins. *Journal of Cell Science, 127*(Pt 7), 1550–1564.

Trinh, M. A., & Klann, E. (2013). Translational control by eIF2alpha kinases in long-lasting synaptic plasticity and long-term memory. *Neurobiology of Learning and Memory, 105*, 93–99.

Trockenbacher, A., Suckow, V., Foerster, J., Winter, J., Krauss, S., Ropers, H. H., et al. (2001). MID1, mutated in Opitz syndrome, encodes an ubiquitin ligase that targets phosphatase 2A for degradation. *Nature Genetics, 29*(3), 287–294.

Troyer, E. A., Kohn, J. N., & Hong, S. (2020). Are we facing a crashing wave of neuropsychiatric sequelae of COVID-19? Neuropsychiatric symptoms and potential immunologic mechanisms. *Brain, Behavior, and Immunology 87*, 34–39.

Tsai, J. C., Miller-Vedam, L. E., Anand, A. A., Jaishankar, P., Nguyen, H. C., Renslo, A. R., et al. (2018). Structure of the nucleotide exchange factor eIF2B reveals mechanism of memory-enhancing molecule. *Science, 359*(6383).

Turner, M. R., Al-Chalabi, A., Chio, A., Hardiman, O., Kiernan, M. C., Rohrer, J. D., et al. (2017). Genetic screening in sporadic ALS and FTD. *Journal of Neurology, Neurosurgery, and Psychiatry, 88*(12), 1042–1044.

Turner, M. R., Hardiman, O., Benatar, M., Brooks, B. R., Chio, A., de Carvalho, M., et al. (2013). Controversies and priorities in amyotrophic lateral sclerosis. *Lancet Neurology, 12*(3), 310–322.

Udagawa, T., Fujioka, Y., Tanaka, M., Honda, D., Yokoi, S., Riku, Y., et al. (2015). FUS regulates AMPA receptor function and FTLD/ALS-associated behaviour via GluA1 mRNA stabilization. *Nature Communications, 6*, 7098.

Valente, E. M., Abou-Sleiman, P. M., Caputo, V., Muqit, M. M., Harvey, K., Gispert, S., et al. (2004). Hereditary early-onset Parkinson's disease caused by mutations in PINK1. *Science, 304*(5674), 1158–1160.

Van Damme, P., Robberecht, W., & Van Den Bosch, L. (2017). Modelling amyotrophic lateral sclerosis: progress and possibilities. *Disease Models & Mechanisms, 10*(5), 537–549.

van der Brug, M. P., Blackinton, J., Chandran, J., Hao, L. Y., Lal, A., Mazan-Mamczarz, K., et al. (2008). RNA binding activity of the recessive parkinsonism protein DJ-1 supports involvement in multiple cellular pathways. *Proceedings of the National Academy of Sciences of the United States of America, 105*(29), 10244–10249.

van der Spek, R. A., van Rheenen, W., Pulit, S. L., Kenna, K. P., Ticozzi, N., Kooyman, M., et al. (2018). Reconsidering the causality of TIA1 mutations in ALS. *Amyotrophic Lateral Sclerosis and Frontotemporal Degeneration, 19*(1–2), 1–3.

Vazquez-Manrique, R. P., Farina, F., Cambon, K., Dolores Sequedo, M., Parker, A. J., Millan, J. M., et al. (2016). AMPK activation protects from neuronal dysfunction and vulnerability across nematode, cellular and mouse models of Huntington's disease. *Human Molecular Genetics, 25*(6), 1043–1058.

Voigt, A., Herholz, D., Fiesel, F. C., Kaur, K., Muller, D., Karsten, P., et al. (2010). TDP-43-mediated neuron loss in vivo requires RNA-binding activity. *PLoS One, 5*(8), e12247.

Wang, I. F., Wu, L. S., Chang, H. Y., & Shen, C. K. (2008). TDP-43, the signature protein of FTLD-U, is a neuronal activity-responsive factor. *Journal of Neurochemistry, 105*(3), 797–806.

Wang, W., Nag, S., Zhang, X., Wang, M. H., Wang, H., Zhou, J., et al. (2015). Ribosomal proteins and human diseases: pathogenesis, molecular mechanisms, and therapeutic implications. *Medicinal Research Reviews, 35*(2), 225–285.

White, R., Gonsior, C., Kramer-Albers, E. M., Stohr, N., Huttelmaier, S., & Trotter, J. (2008). Activation of oligodendroglial Fyn kinase enhances translation of mRNAs transported in hnRNP A2-dependent RNA granules. *Journal of Cell Biology, 181*(4), 579–586.

Wolozin, B., & Ivanov, P. (2019). Stress granules and neurodegeneration. *Nature Reviews Neuroscience, 20*(11), 649–666.

Wong, Y. L., LeBon, L., Basso, A. M., Kohlhaas, K. L., Nikkel, A. L., Robb, H. M., et al. (2019). eIF2B activator prevents neurological defects caused by a chronic integrated stress response. *Elife, 8*.

Wong, Y. L., LeBon, L., Edalji, R., Lim, H. B., Sun, C., & Sidrauski, C. (2018). The small molecule ISRIB rescues the stability and activity of vanishing white matter disease eIF2B mutant complexes. *Elife, 7*.

Wyant, K. J., Ridder, A. J., & Dayalu, P. (2017). Huntington's disease: Update on treatments. *Current Neurology and Neuroscience Reports, 17*(4), 33.

Yang, Y., Gehrke, S., Imai, Y., Huang, Z., Ouyang, Y., Wang, J. W., et al. (2006). Mitochondrial pathology and muscle and dopaminergic neuron degeneration caused by inactivation of Drosophila Pink1 is rescued by Parkin. *Proceedings of the National Academy of Sciences of the United States of America, 103*(28), 10793–10798.

Yasuda, K., Zhang, H., Loiselle, D., Haystead, T., Macara, I. G., & Mili, S. (2013). The RNA-binding protein FUS directs translation of localized mRNAs in APC-RNP granules. *Journal of Cell Biology, 203*(5), 737–746.

Yedavalli, V. S., & Jeang, K. T. (2011). Matrin 3 is a co-factor for HIV-1 Rev in regulating post-transcriptional viral gene expression. *Retrovirology, 8*, 61.

Yokoshi, M., Li, Q., Yamamoto, M., Okada, H., Suzuki, Y., & Kawahara, Y. (2014). Direct binding of Ataxin-2 to distinct elements in 3′ UTRs promotes mRNA stability and protein expression. *Molecular Cell, 55*(2), 186–198.

Yoshizawa, T., Ali, R., Jiou, J., Fung, H. Y. J., Burke, K. A., Kim, S. J., et al. (2018). Nuclear import receptor inhibits phase separation of FUS through binding to multiple sites. *Cell, 173*(3), 693–705 e622.

Zhang, J., Kong, L., Guo, S., Bu, M., Guo, Q., Xiong, Y., et al. (2017). hnRNPs and ELAVL1 cooperate with uORFs to inhibit protein translation. *Nucleic Acids Research, 45*(5), 2849–2864.

Zhang, K., Liu, Q., Shen, D., Tai, H., Fu, H., Liu, S., et al. (2018). Genetic analysis of TIA1 gene in Chinese patients with amyotrophic lateral sclerosis. *Neurobiology of Aging, 67*, 201 e209–201 e210.

Zhang, T., Wu, Y. C., Mullane, P., Ji, Y. J., Liu, H., He, L., et al. (2018). FUS regulates activity of microrna-mediated gene silencing. *Molecular Cell, 69*(5), 787–801 e788.

Zhang, Y., Ling, J., Yuan, C., Dubruille, R., & Emery, P. (2013). A role for Drosophila ATX2 in activation of PER translation and circadian behavior. *Science, 340*(6134), 879–882.

Zhu, P. J., Khatiwada, S., Cui, Y., Reineke, L. C., Dooling, S. W., Kim, J. J., et al. (2019). Activation of the ISR mediates the behavioral and neurophysiological abnormalities in Down syndrome. *Science, 366*(6467), 843–849.

Zimprich, A., Biskup, S., Leitner, P., Lichtner, P., Farrer, M., Lincoln, S., et al. (2004). Mutations in LRRK2 cause autosomal-dominant parkinsonism with pleomorphic pathology. *Neuron, 44*(4), 601–607.

Zu, T., Gibbens, B., Doty, N. S., Gomes-Pereira, M., Huguet, A., Stone, M. D., et al. (2011). Non-ATG-initiated translation directed by microsatellite expansions. *Proceedings of the National Academy of Sciences of the United States of America, 108*(1), 260–265.

Zu, T., Liu, Y., Banez-Coronel, M., Reid, T., Pletnikova, O., Lewis, J., et al. (2013). RAN proteins and RNA foci from antisense transcripts in C9ORF72 ALS and frontotemporal dementia. *Proceedings of the National Academy of Sciences of the United States of America, 110*(51), E4968–E4977.

Zu, T., Pattamatta, A., & Ranum, L. P. W. (2018). Repeat-associated non-ATG translation in neurological diseases. *Cold Spring Harbor Perspectives in Biology, 10*(12).

Zyryanova, A. F., Weis, F., Faille, A., Alard, A. A., Crespillo-Casado, A., Sekine, Y., et al. (2018). Binding of ISRIB reveals a regulatory site in the nucleotide exchange factor eIF2B. *Science, 359*(6383), 1533–1536.

CHAPTER 23

ROLE OF EUKARYOTIC INITIATION FACTOR eIF2B IN VANISHING WHITE MATTER DISEASE

TRUUS E. M. ABBINK, LISANNE E. WISSE, XUEMIN WANG, AND CHRISTOPHER G. PROUD

Abbreviations

AGP	adenosine diphosphate-glucose pyrophosphorylase
AHA	azidohomoalanine
ATF4	activating transcription factor 4
eIF	eukaryotic initiation factor
GCN2	general control nonderepressible 2
GDF	GDI-dissociation factor
GDI	GDP-dissociation inhibitor
GDP	guanosine diphosphate
GEF	guanine nucleotide-exchange factor
GFAPδ	glial fibrillary acidic protein δ
GTP	guanosine triphosphate
ISR	integrated stress response
ISRIB	ISR inhibitor
MBP	myelin basic protein
MOG	myelin oligodendrocyte glycoprotein
MRI	magnetic resonance imaging
Met-tRNAMeti	initiator methionyl-transfer RNA
NF	Asn-Phe

OPC	oligodendrocyte precursor cell
PERK	protein kinase RNA-like endoplasmic reticulum kinase
uORF	upstream open reading frame
VWM	vanishing white matter
WT	wild-type

Vanishing White Matter

Vanishing white matter (VWM; also termed childhood ataxia with diffuse central nervous system hypomyelination, CACH) is one of the most prevalent inherited childhood white matter disorders (van der Knaap, Breiter, Naidu, Hart, & Valk, 1999; van der Knaap et al., 1977). The diagnosis of VWM is based on magnetic resonance imaging (MRI) and genetic testing. The MRI pattern characteristic for VWM shows diffuse abnormality of the cerebral white matter. When the disease progresses, increasing amounts of white matter are replaced by fluid (van der Knaap et al., 1977). The brain consists of gray and white matter. The gray matter contains the neurons, while the white matter contains axons myelinated by oligodendrocytes. The myelin is high in fat and makes myelinated structures of the brain white. In VWM the myelin is not maintained properly and the white matter literally vanishes. Currently, there is no curative treatment for VWM.

The clinical severity of the disease is quite variable and correlates mostly with the age of onset (Fogli et al., 2004a; van der Knaap et al., 1998). Three groups have been discriminated (Fogli & Boespflug-Tanguy, 2006; van der Knaap, Pronk, & Scheper, 2006). The classical phenotype is associated with an early childhood onset at an age between 2 and 6 years (van der Knaap et al., 2006). A severe phenotype is observed in patients with a disease onset before the age of 2 years (antenatal or infantile onset) (van der Knaap et al., 2006). The third group comprises patients with a relatively mild phenotype, characterized by a late childhood or adult onset (van der Knaap et al., 2006).

The classical phenotype is most frequently observed. Patients display chronic as well as episodic neurological deterioration (van der Knaap et al., 1977, 1998). They develop ataxia and spasticity and become wheelchair-dependent (van der Knaap et al., 1977, 1998; van der Lei et al., 2010; Hanefeld et al., 1993). Eventually VWM patients die prematurely, usually a few years after diagnosis. Episodic deterioration is provoked by minor head trauma, infections with fever, or acute fright (van der Knaap et al., 2006; Vermeulen et al., 2005). These episodes may end in an unexplained coma, which in some cases results in death (van der Knaap et al., 2006). Alternatively, the episodes result in slow and partial recovery. In general, VWM patients have no or only mild epilepsy (van der Knaap et al., 2006). Patients receive interventions that prevent episodic deterioration (van der Knaap et al., 2006).

The brain is the most affected organ in all patients (van der Knaap et al., 2006). Severe cases of VWM with antenatal onset are affected also in other organs, including lens, liver, kidney, pancreas, and ovaries (Fogli et al., 2003; van der Knaap et al., 2003). In addition ovarian dysfunction has been observed in adolescent and adult female patients (Fogli et al., 2003; van der Knaap et al., 2003).

Postmortem neuropathological examination shows an abnormal morphology of astrocytes and oligodendrocytes in the white matter of the central nervous system. White matter oligodendrocytes look foamy, and white matter astrocytes have thick and blunted processes (Bugiani et al., 2011; Van Haren, van der Voorn, Peterson, van der Knaap, & Powers, 2004). The morphology of astrocytes in gray matter is normal (Bugiani et al., 2011). Interestingly, increased numbers of white matter astrocytes are positive for nestin, a marker for immature astrocytes. Reduced numbers of astrocytes express S100β, a marker for mature astrocytes. Combined these observations indicate that increased numbers of astrocytes are immature (Bugiani et al., 2011; Middeldorp et al., 2010; Raponi et al., 2007). In addition, high numbers of oligodendrocyte precursor cells (OPCs) are found in affected white matter structures (Bugiani et al., 2011; Van Haren et al., 2004). In general, astrocytes are important for blood–brain barrier formation and maintenance, uptake of neurotransmitters, regulation of oxygen, and glucose availability for neurons (Lundgaard, Osório, Kress, Sanggaard, & Nedergaard, 2014; Garden & Campbell 2016). Astrocytes have also been described to metabolically support oligodendrocytes (Hirrlinger & Nave, 2014; Kiray, Lindsay, Hosseinzadeh, & Barnett, 2016). Upon injury, astrocytes form scar tissue to help prevent further damage and maintain brain/tissue homeostasis in a process called "reactive astrogliosis". Oligodendrocytes are the cells that produce the extensive myelin sheets around the axons of neurons. Functional myelin is essential for the normal conduction of action potentials in the brain (Garden & Campbell, 2016). It has been postulated that, in VWM, white matter astrocytes and oligodendrocytes do not mature properly and therefore cannot execute their normal function.

Mutations in eIF2B Cause VWM

VWM is a genetic disease with a recessive mode of inheritance. Due to a founder effect, the incidence of VWM in the Netherlands is relatively high. Together with the characteristic MRI pattern, this founder effect facilitated research into the underlying genetic cause. The first gene associated with VWM was found in a genetic linkage study of a group of Dutch families (Leegwater et al., 1999). This study localized the gene on chromosome 3q27 (Leegwater et al., 1999). Sequence analyses identified a T91A mutation in the gene *EIF2B5* in VWM patients that segregated with disease (Leegwater et al., 2001). The mutation was found in a homozygous state. Some patients were compound heterozygous for a second mutation in *EIF2B5* (Leegwater et al., 2001). A second group of patients from another geographical region with relatively high incidence was investigated. Sequence analysis of this group identified the E213G mutation in *EIF2B2* (Leegwater et al., 2001). Patients without mutations in *EIF2B2* or *EIF2B5* are homozygous or compound heterozygous for mutations in the remaining *EIF2B1*, *EIF2B3*, or *EIF2B4* genes (van der Knaap et al., 2002). Taken together, these findings lead to the conclusion that VWM is caused by mutations in any of the five genes encoding the subunits of eIF2B (α, β, γ, δ, and ε, in order of increasing size).

Founder mutations for VWM exist not only in the Dutch population; others have been described, such as in the Cree population in North America (R195H in eIF2Bε) and in China (I346T in eIF2Bγ) (Fogli et al., 2002; Zhang et al., 2015). In other countries patients with sporadic mutations in *EIF2B1–5* are found in all five subunits (Pronk, van Kollenburg, Scheper, & van der Knaap, 2006). At the moment there are more than 100 different mutations known to be involved in VWM (Pronk et al., 2006; Bugiani, Boor, Powers, Scheper, & van der Knaap, 2010). Almost two thirds of the patients have mutations in *EIF2B5*, and only a few have mutations in *EIF2B1* (Pronk et al., 2006).

The majority of mutations are of a missense character. Frameshift or nonsense mutations are always compound heterozygous with a missense mutation, and this missense mutation is associated with a mild phenotype (Pronk et al., 2006; Bugiani et al., 2010). The catalytic domain in eIF2Bε is relatively spared for pathogenic mutations (Pronk et al., 2006). The absence of homozygous frameshift or nonsense mutations in VWM patients likely reflects the function of eIF2B, which is essential for survival. The eIF2B protein is highly conserved in eukaryotes, and loss of either the eIF2Bβ, eIF2Bγ, eIF2Bδ, or eIF2Bε subunit makes yeast nonviable, indicating that—as expected for a translation factor—eIF2B knockout is not compatible with life (Hinnebusch, 1996). The genotype–phenotype correlation in VWM is difficult to determine due to the phenotypic variability between sibling patients (Bugiani et al., 2010). However, some mutations, such as R113H in *EIF2B5* in a homozygous state, always correlate with a relatively late onset and slow disease progression. Moreover, this mutation is not conserved between mammals: for example, mice and rats have a histidine at the equivalent position. This lack of conservation may explain the relatively mild phenotype associated with the R113H mutation (van der Lei et al., 2010; Bugiani et al., 2010). The R195H mutation in *EIF2B5* is associated with a severe phenotype (Black et al., 1988). An overview of the mutations referred to in this chapter is given in Table 23.1.

eIF2B Function and Regulation

Regulation of mRNA translation plays a key role in cell and organismal physiology. This encompasses both the overall regulation of the rate of protein synthesis, to match the cell's needs or resources, and the modulation of the translation of specific mRNAs and thus the production of specific proteins or protein isoforms. Since the translation of mRNAs for certain transcription factors is regulated, control of mRNA translation can affect the expression of many genes, including ones whose mRNAs are not themselves subject to translational control.

The regulatory network involving the translation initiation factors eIF2 and eIF2B (eukaryotic initiation factor 2 and 2B) provides an excellent and well-studied example of just this type of control system (Pavitt, 2005). eIF2 brings the initiator methionyl-tRNA (Met-tRNAMeti) to the ribosome to recognize the start codon in mRNAs and thus get translation underway (Figure 23.1). In order to bind Met-tRNAMeti, eIF2 must itself be associated with guanosine triphosphate (GTP). eIF2·GTP ·Met-tRNAMeti is known

Table 23.1. Overview of eIF2B Mutations, the Effects on eIF2B Structure and Activity and Disease Severity

Mutation	Effect on eIF2B structure	Effect on eIF2B activity[a]	Disease severity
eIF2Bα V183F	Loss of eIF2Bα dimerization (Wortham et al., 2014; Wortham & Proud, 2015)	↓ (Wortham & Proud, 2015)	Mild (Ohlenbusch et al., 2005)
eIF2Bβ E213G	No effect (Li et al., 2004; Richardson, Mohammad, & Pavitt, 2004)	↓↓ (Li et al., 2004)	Classical (Fogli et al., 2004a)
eIF2Bβ P291S	Loss of complex integrity (Liu et al., 2011)		
eIF2Bβ V316D	Loss of complex integrity (Li et al., 2004; Liu et al., 2011)	↓↓↓ (Li et al., 2004)	
eIF2Bγ R225Q	No effect (Liu et al., 2011)	± (Liu et al., 2011)	Classical (Liu et al., 2011)
eIF2Bγ I346T			Classical (Wu et al., 2009)
eIF2Bδ R357W	Loss of complex integrity (Liu et al., 2011)	↓↓↓ (Liu et al., 2011)	Classical (Fogli et al., 2004a)
eIF2Bδ A391D	No effect (Liu et al., 2011)	± (Liu et al., 2011)	Severe (van der Knaap et al., 2003)
eIF2Bδ R483W	Loss of complex integrity (Liu et al., 2011)	↓↓↓ (Liu et al., 2011)	Severe (van der Knaap et al., 2003)
eIF2Bε T91A	Reduced binding to complex (Li et al., 2004)	↓↓ (Li et al., 2004)	Classical (Fogli et al., 2004a; van der Lei et al., 2010)
eIF2Bε R113H	No effect (Wang, Wortham, Liu, & Proud, 2012)	↓↓ (Li et al., 2004)	Mild (van der Lei et al., 2010)
eIF2Bε R136H		↓ (Geva et al., 2010)	Classical (Kantor et al., 2005)
eIF2Bε R195H		↓↓↓ (Li et al., 2004)	Severe (Fogli et al., 2004a)
eIF2Bε Y495C	No effect (Liu et al., 2011)	± (Liu et al., 2011)	Severe (van der Knaap et al., 2003)

[a] eIF2B activity (relative to control) ↓70–90%; ↓↓50–70%; ↓↓↓<50%; ±indicates no significant difference from controls.

as a "ternary complex." The GTP is hydrolyzed during translation initiation, and eIF2 leaves the ribosome as inactive eIF2·guanosine diphosphate (GDP). Since GDP dissociates only slowly from eIF2, an extra factor, eIF2B, is needed to maintain the required rate of translation initiation. eIF2B acts as a GDP-dissociation stimulating factor to promote release of GDP from eIF2 to allow its replacement by GTP and regenerate active eIF2·GTP (Williams, Price, Loughlin, & Proud, 2001) (Figure 23.1). eIF2B is generally referred to as a guanine nucleotide-exchange factor (GEF). In yeast, it has also been shown to promote the dissociation of a further factor, eIF5, from eIF2 to allow nucleotide exchange. eIF5 normally impairs the release of GDP from eIF2 and therefore has GDP-dissociation inhibitor (GDI) activity (Jennings et al., 2016; Jennings & Pavitt, 2010). eIF2B thus acts as a GDI-dissociation factor, or GDF (Jennings & Pavitt, 2010). eIF5 actually has dual roles, acting both as a GDI and a GTPase-activator protein for eIF2·GTP (Paulin, Campbell, O'Brien, Loughlin, & Proud, 2001) (Figure 23.1). This allows proper control of the initiation stage of protein synthesis under normal conditions and when eIF2 is phosphorylated in response to a range of stress conditions (Figure 23.2). It has not yet been tested whether eIF2B from other species, including mammals, also displays these additional roles; but given the high degree of similarity between eIF2, eIF2B, and eIF5 from diverse species, this seems likely.

The eIF2/eIF2B system is subject to several types of regulation. The first to be discovered and the best understood involves the phosphorylation of eIF2 (Pavitt, 2005) (Figure 23.2). eIF2 comprises three different subunits, α–γ. eIF2α undergoes phosphorylation on Ser51, an event which can be catalyzed by any one of four different kinases in mammals (Figure 23.2). When phosphorylated in this way, eIF2 becomes a competitive inhibitor of its own GEF, eIF2B, thus slowing down the production of active eIF2·GTP (Rowlands, Panniers, & Henshaw, 1988). Overall protein synthesis is thus inhibited. Importantly, this mechanism can actually increase translation of the coding regions of some mRNAs.

This increase is because those mRNAs contain short "open reading frames" upstream (uORF) of the main coding region. When eIF2·GTP·Met-tRNAMeti levels are high, at least some of these so-called uORFs are translated, preventing ribosomes from reaching the start of the main coding region. On the other hand, when eIF2B is inhibited and eIF2·GTP·Met-tRNAMeti levels are low, translation of the uORFs is impaired and the main coding region is translated more frequently (Pavitt, 2005). Thus, production of the corresponding proteins is actually enhanced when eIF2B activity is impaired, for example, when eIF2 is phosphorylated. In mammals, the most widely studied example of such an mRNA encodes activating transcription factor 4 (ATF4). This provides a clear example of the ability of translational control to regulate multiple genes; in the case of ATF4, the effects are widened by the fact that some genes controlled by ATF4 are themselves transcription factors—for instance, CHOP. Analyses of postmortem brain tissue from VWM patients shows increased expression of ATF4 and CHOP in white matter glia (van der Voorn et al., 2005; van Kollenburg et al., 2006).

Recent studies on yeast by Pavitt and colleagues have revealed several unusual features of the eIF2B/eIF2 GEF system (Jennings, Kershaw, Adomavicius, & Pavitt, 2017). For example, eIF2B has similar affinity for eIF2 whether bound to GDP or GTP; one

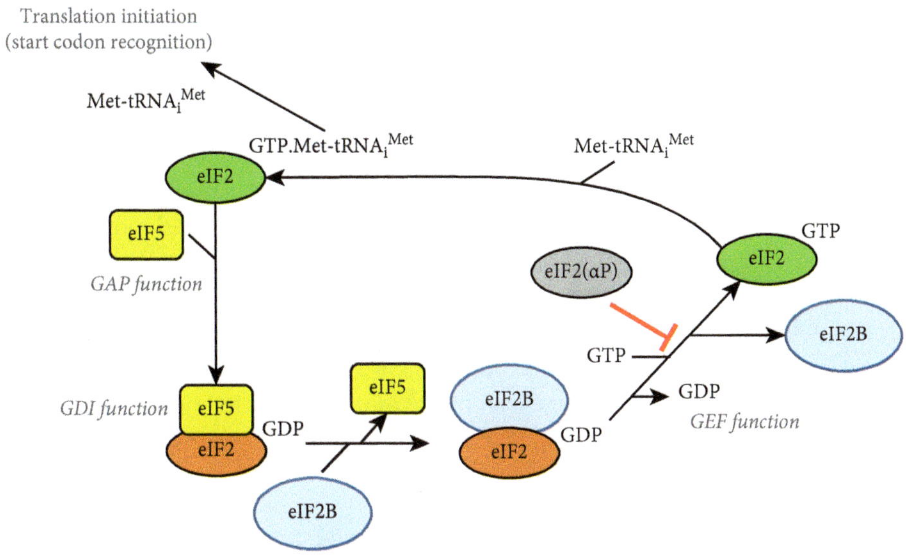

FIGURE 23.1. Role of eIF2B. eIF2B promotes the dissociation of GDP from eIF2, allowing GTP to bind to form active eIF2.GTP complexes. This figure also illustrates the function of eIF2, to recruit – in its GTP-bound form – the initiator methionyl-tRNA to the 40S ribosomal subunit and the dual role of eIF5 as a GAP for eIF2-bound GTP and a GDI for the resulting eIF2-GDP complex. Red and green indicate, respectively, the inactive and active forms of eIF2.

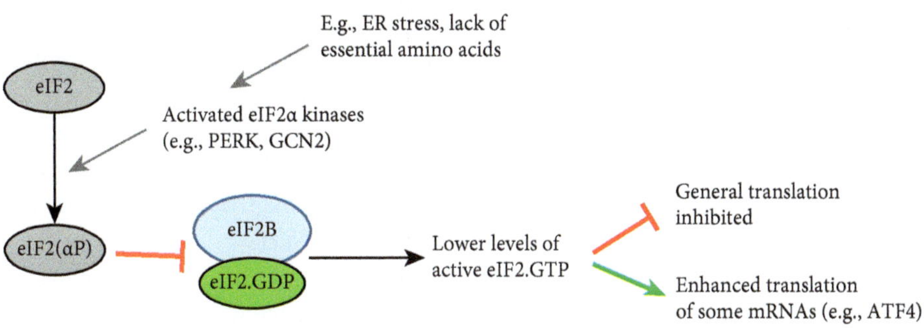

FIGURE 23.2. Regulation of eIF2B by phosphorylation of eIF2. Under stress conditions, such as endoplasmic reticulum stress or lack of essential amino acids, the eIF2α kinases PERK and GCN2, respectively, are activated. The resulting phosphorylation of eIF2 on its α subunit creates eIF2(αP) which competitively inhibits eIF2B, so that eIF2 builds up in its inactive GDP-bound form. This slows down general translation initiation but actually enhances the translation of a small subset of mRNAs which have upstream open reading frames (uORFs) in their 5'-UTRs. One such mRNA encodes ATF4, activating transcription factor 4, which promotes expression of a range of genes involved in coping with cell stress.

implication of this is that eIF2B and Met-tRNAMeti compete with each other for binding to eIF2·GTP. Formation of ternary complexes depends on Met-tRNAMeti "winning" this competition. In contrast, by binding eIF2, eIF5 can bind and stabilize ternary complexes. When eIF2 is phosphorylated on its α subunit, its increased affinity for eIF2B means that binding of eIF2 to eIF5 or Met-tRNAMeti is outcompeted by eIF2B, which will impair eIF2's ability to bind Met-tRNAMeti and form ternary complexes. This provides an additional "brake" mechanism to slow down translation initiation (Hronova & Valasek, 2017). Again, it will be important to investigate whether this also applies to the mammalian factors.

It is possible that some VWM mutations affect this delicate balance of interaction between eIF2B and eIF2; indeed, some mutations are already known to do so (Li, Wang, van der Knaap, & Proud, 2004), but effects of VWM mutations on these "new" functions of eIF2B have yet to be tested.

eIF2B Structure in Relation to Function

eIF2B is a multimeric protein; it has five nonidentical types of subunit, α–ε (Cigan, Bushman, Boal, & Hinnebusch, 1993). Two αβγδε heteropentamers associate to form a heterodecamer (Gordiyenko et al., 2014; Wortham, Martinez, Gordiyenko, Robinson, & Proud, 2014) (Figure 23.3). However, the data for the yeast and human complexes show considerable disparities. In particular, the data for the human complex indicate an important role for α–α dimers in stabilizing the decameric complex (Wortham et al., 2014), while in contrast this subunit appears to be dispensable for formation of the yeast (*Saccharomyces cerevisiae*) protein complex (Gordiyenko et al., 2014). Subsequent determination of the crystal structure of eIF2B from a quite distinct yeast species, *Schizosaccharomyces pombe*, revealed an α–α dimer (Kashiwagi, Ito, & Yokoyama, 2017; Kashiwagi et al., 2016), similar to the situation for human eIF2B.

eIF2Bε contains the catalytic GEF domain toward its C terminus (Gomez, Mohammad, & Pavitt, 2002) (Figure 23.4). Its N-terminal region shows sequence similarity to eIF2Bγ, with which it associates to form a "catalytic subcomplex," and to nucleotidyl transferases and acyltransferases (Pavitt, Ramaiah, Kimball, & Hinnebusch, 1998). The three-dimensional structures of these subunits resemble that of adenosine diphosphate-glucose pyrophosphorylase (AGP). The similarity of eIF2B subunits to these types of proteins may have significance for the function, or perhaps more likely the regulation, of the eIF2B complex; but this remains to be investigated. However, the structure of the potential nucleotide binding pocket in eIF2Bγ differs from that in AGP and is likely unable to bind ligand. eIF2Bε is extended at its C terminus compared to eIF2Bγ and contains additional regions that include a HEAT domain and an Asn-Phe (NF) pair that is important for GEF activity (Gomez et al., 2002; Gomez & Pavitt, 2000). This region of eIF2Bε extends out from the main 3D structure of the decameric complex.

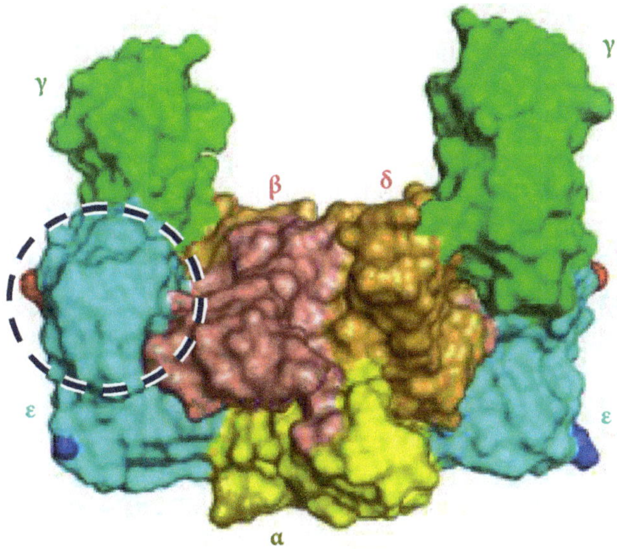

FIGURE 23.3. Three-dimensional structure of eIF2B. Overall structure of eIF2B. The β, γ, δ and ε subunits are shown, respectively, in pink, green, orange, and cyan. The two α subunits are shown in yellow and olive. The "NF" motifs (which augment eIF2B's GEF function) and the visible C termini of the ε subunits are shown in red and blue, respectively. The substrate, eIF2, docks into the "central cavity" with GEF function being associated with the distal face (dotted circle), with which the GTP-binding γ-subunit of eIF2 associates.

A adapted with permission from Kashiwagi, Ito, & Yokoyama (2017).

The α, β, and δ subunits also show mutual homology and similarity to ribose-1,5-bisphosphate isomerase. At least in yeast, they can form a stable "regulatory complex" (Pavitt et al., 1998) (Figure 23.4). This term reflects the facts that mutations in these subunits can make eIF2B insensitive to phosphorylation of eIF2 and that eIF2Bα is absolutely required for this control. The actual roles of the noncatalytic subunits of eIF2B are not fully understood. Some are reported to bind nucleotides, which could play roles in the catalytic activity or the control of eIF2B (Dholakia, Francis, Haley, & Wahba, 1989; Kuhle, Eulig, & Ficner, 2015), while others (e.g., eIF2Bγ [Jennings & Pavitt, 2014; Jennings, Zhou, Mohammad-Qureshi, Bennett, & Pavitt, 2013]) likely contribute to eIF2B's GDF function.

The effects of VWM mutations on eIF2B's GEF activity have been tested in a number of studies. Two approaches have been used. In the first, wild-type (WT) or mutated eIF2B subunits were overexpressed in human cell lines, and their activity and certain other properties were then studied (Li et al., 2004; Liu et al., 2011; Matsukawa et al., 2011; Wortham & Proud, 2015). In the second, GEF activity was measured in lysates of cells from VWM patients (Fogli et al., 2004b; Horzinski et al., 2009; Liu et al., 2011). In many cases, eIF2B GEF activity was found to be decreased in both types of test systems. However, in some cases, it was unchanged or actually higher than for WT eIF2B. For example, the first kind of study showed that the A391D mutation in eIF2Bδ and the Y495C change in eIF2Bε had little or no effect on GEF activity or the integrity of eIF2B α–ε pentamers, even though they are associated with very severe disease (Liu et al., 2011). This result was confirmed with experiments that assessed GEF activity of endogenous eIF2B in lysates from lymphoblasts from patients homozygous for these mutations. The authors of this study concluded that such mutations may affect (at that time, unknown) additional functions of eIF2B. The exciting discovery that eIF2B exerts GDF function raises the possibility that some VWM mutations may affect this property of eIF2B. A391 lies at the β/δ interface (Kashiwagi et al., 2017); mutating it to Asp does not impair the overall interaction between the subunits (Liu et al., 2011) but might affect their geometry, thereby affecting the function of eIF2B. It should be noted that another study, where the A391 mutation was "knocked in" to the chromosomal copy of the *EIF2B4* gene in Chinese hamster ovary cells, reported diminished GEF activity (Sekine et al., 2016). These authors introduced discriminatory silent mutations (introduced to check the efficacy of their CRISPR/Cas9 genome editing procedure), which in a human setting are predicted to affect mRNA splicing. It is possible that levels of eIF2Bδ mRNA and protein were reduced. This could well contribute to the observed phenotype but was not investigated.

As noted, the initial studies of the effects of VWM mutations in *EIF2B2*, *EIF2B3*, and *EIF2B5* on the function of eIF2B indicated that all the mutations tested reduced GEF activity, albeit to differing extents (Li et al., 2004; Liu et al., 2011; Matsukawa et al., 2011). Some mutations also affected the association between eIF2B subunits or increased association with phosphorylated eIF2. Two implications of this were that (1) restoring eIF2B GEF activity might provide potential therapies and (2) measuring GEF activity in lysates of cells from potential VWM patients could provide a diagnostic tool (Fogli et al., 2004b; Horzinski et al., 2009). The subsequent discovery that some mutations do not cause

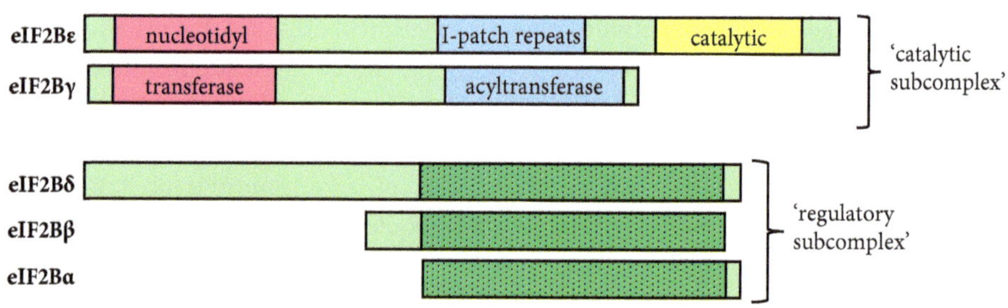

FIGURE 23.4. Layout of individual eIF2B subunits (domains) and the catalytic and regulatory subcomplexes of eIF2B. Homologous regions are shown in matching colours, and are labelled to show the nucleotidyltransferase and acyltransferase regions of eIF2Bγ and eIF2Bε. The latter contain the so-called I-patch repeats. The catalytic (GEF) domain of eIF2Bε is shown, as are the mutually homologous C-terminal regions of eIF2Bα, βand δ. eIF2Bγ and eIF2Bε together form the catalytic subcomplex, which displays some GEF activity; full activity of mammalian eIF2B requires all five types of subunit. The other three subunits form the regulatory subcomplex.

reduced GEF activity (or may even increase it [Liu et al., 2011; Wortham & Proud, 2015]) promoted a revision of the earlier conclusions for the diagnosis or treatment of VWM.

Given that the effects of VWM mutations on the GEF activity of eIF2B differ so markedly (changes are not even evident for some mutations), a key question is, do all VWM mutations affect the formation of ternary complexes (which is the key parameter as far as translation initiation is concerned)? As eIF2B exerts other effects on eIF2 in addition to acting as its GEF (i.e., acting as a GDI to displace eIF5 and competing with Met-tRNAMeti), it is indeed possible that even mutations that do not reduce GEF activity still impair ternary complex levels by affecting other properties of eIF2B. It is important to devise ways to test this possibility, especially in cells and tissues.

Data from structural studies (Figure 23.3), first on the β and δ subunits of eIF2B (Kuhle et al., 2015) and then on the entire complex (Kashiwagi et al., 2016, 2017) have provided crucial insights into both the mechanism of action and regulation of eIF2B and the potential basis for the effects of a number of VWM mutations. Here, we will focus primarily on the implications of these studies for our understanding of VWM disease.

The crystal structure of the eIF2B complex confirms it is a heterodecamer (Kashiwagi et al., 2016, 2017) and shows that the $\alpha_2\beta_2\delta_2$ hexamer is flanked by the two γε dimeric catalytic subcomplexes (Figure 23.3). Many VWM mutations are located at the subunit interfaces, suggesting that VWM frequently involves defects in the proper assembly of the decameric complex and/or correct orientation of its component subunits.

Cross-linking data suggest that eIF2 (specifically its GTP-binding γ subunit) binds to the distal surface of the complex, where eIF2Bγ and eIF2Bε are located (Kashiwagi et al., 2016, 2017). Some cross-links are affected by phosphorylation of eIF2, notably those involving the NF-containing region of eIF2Bε. The NF domain is important for catalysis of nucleotide exchange, and a number of VWM mutations are located in this region.

In contrast, other cross-links which do not involve the NF motif, are not affected by the phosphorylation status of eIF2. Indeed, eIF2 also binds to a second site in eIF2B, the so-called central cavity within the $\alpha_2\beta_2\delta_2$ hexamer. This is true for both phosphorylated and nonphosphorylated eIF2, but phosphorylated eIF2 interacts with higher affinity. Docking studies (Kashiwagi et al., 2016) indicate that eIF2 cannot readily bind at the same time both to the central cavity and to the NF domain. This provides a potential explanation for the inhibitory effect of eIF2 phosphorylation on the GEF function of eIF2B; because phosphorylated eIF2 binds better to the central cavity, it is less able to access the NF domain, which is important for catalysis of the GEF reaction (Figure 23.3). Interestingly, however, few VWM mutations map to the central cavity region. This indicates that effects on the nonproductive interaction between this feature and eIF2 are not a major cause of the defects caused by VWM mutations (Kashiwagi et al., 2016, 2017). Unexpectedly, the cavity is not positively charged, as one might have expected given its greater affinity for phosphorylated eIF2. The basis of the higher affinity of P-eIF2 for the cavity, and thus the ability of P-eIF2 to inhibit eIF2B function, therefore remains obscure. The high affinity of P-eIF2 for eIF2B means that, in cells where eIF2 levels substantially exceed those of eIF2B, even low levels of P-eIF2 cause substantial inhibition of eIF2B activity by preventing eIF2B from acting on nonphosphorylated eIF2.

However, several VWM mutations map to the same regions of eIF2B as the cross-links that are insensitive to eIF2 phosphorylation. This may indicate that such mutations affect the GDF function of eIF2B, which is also not affected by phosphorylation of eIF2. Mutational studies show that distinct residues in eIF2B (e.g., in eIF2Bγ) are involved in its GEF and GDF functions (Jennings & Pavitt, 2014).

The new structural models help us rationalize the effects of a number of VWM mutations. As Kuhle et al. (2015) point out, many lie close to the interface between the β and δ subunits in their mutual dimer. Furthermore, the correct association between β and δ is needed to create the binding site for the ε subunit (Kashiwagi et al., 2016). Indeed, other VWM mutations are at exposed residues in a hydrophobic surface of eIF2Bδ and are likely involved in protein–protein interactions (Bogorad et al., 2014).

Some mutations that affect residues at or close to the β/δ interface have been tested and shown to impair binding between the subunits (e.g., R357W in eIF2Bδ and V316D and P291S in eIF2Bβ [Li et al., 2004; Liu et al., 2011]). However, another mutation at the same interface did not impair association between subunits (A391D [Liu et al., 2011]). The V183F mutation in eIF2Bα lies at the α–α dimerization interface and impairs the integrity of the whole complex (Wortham & Proud, 2015). R225Q in eIF2Bγ lies on the distal surface of the complex, which interacts with eIF2; and, consistent with this, eIF2 binds less well to complexes containing this mutation (Liu et al., 2011).

Thus, functional consequences of mutations at residues involved in the interactions between subunits in GEF activity and GDF function all seem likely to play roles in the pathogenesis of VWM. However, some VWM mutations are in residues that are exposed on the surface of the protein that are not associated with known functions of eIF2B (Kashiwagi et al., 2016). This raises the possibility that these mutations affect roles of eIF2B that are not yet understood.

Mouse Models for VWM

In 2010, the first transgenic mouse model for VWM was described (Geva et al., 2010). This mouse model is homozygous for the R132H (R136H in human) mutation in eIF2Bε, which is associated with a classical phenotype in patients (Kantor et al., 2005). The mutation affected eIF2 GDP release in brain lysates by approximately 20% (Kantor et al., 2005). The effects of this mutation on the mouse phenotype are subtle and do not include a reduction in life span. The mice show some motor problems on rotarod and open field tests. Their body weight and body fat content are reduced. Brain analyses indicated a high proportion of small-caliber axons, reduced myelin, an increased number of OPCs, but reduced numbers of astrocytes at 3 weeks of age in the internal capsule white matter structure (Geva et al., 2010). At an age of 4 months, the OPC and astrocyte numbers had normalized, although the mice still performed less well at the rotarod than WT mice. The authors concluded that the R132H mutation negatively affected normal brain white matter development. Myelin recovery in the eIF2Bε[R132H] mice was impaired after cuprizone-induced demyelination as well, indicating that remyelination was also affected by the mutation. Follow-up studies with the R132H mouse model showed time-dependent changes in expression of mRNAs with various functions (Marom, Ulitsky, Cabilly, Shamir, & Elroy-Stein, 2011). This genome-wide analysis was not informative on the mechanism by which that R132H mutation altered the levels of expression of these mRNAs.

A recent proteomic study on cerebrum brain samples of the eIF2Bε[R132H] mice revealed a difference in expression of mitochondrial proteins as well as proteins important for protein synthesis and degradation via he proteasome (Gat-Viks, Geiger, Barbi, Raini, & Elroy-Stein, 2015). Indeed, a subtle reduction in 20S proteasome activity was reported in brain lysates from 21-day-old eIF2Bε[R132H] mice. At this age, the white matter may not have become fully myelinated (Baumann & Pham-Dinh, 2001). It would be informative to investigate if the reduction in 20S proteasome activity is due to the developmental difference or more directly to differences in eIF2Bε activity. The 20S proteasome function would have to be measured at different ages and compared to the gene expression differences at these ages.

Most recently, Raini et al. (2017) reported that astrocytes from the R132H VWM knock-in mouse model show an imbalance in mitochondrial proteins involved in oxidative phosphorylation, impaired mitochondrial function (i.e., decreased oxygen consumption), and (presumably as a compensatory effect) enhanced mitochondrial biogenesis. Since almost all mitochondrial proteins are actually made in the cytoplasm, these effects may be a consequence of dysregulation of protein synthesis due to altered eIF2B function.

Another approach for generating a mouse model for VWM has been reported, in which eIF2B activity was inhibited in oligodendrocytes by activation of the eIF2α kinase protein kinase RNA-like endoplasmic reticulum kinase (PERK) (Lin et al., 2008, 2014). Mice were made in which PERK was activated in oligodendrocytes in a cell-autonomous manner. PERK was activated in young and adult animals; especially oligodendrocytes in young mice appeared vulnerable to PERK activation. When activated at postnatal day 10, PERK in oligodendrocytes induced a tremoring phenotype as early as 4 days; and the mice died approximately 14 days later (Lin et al., 2014). When PERK was activated at postnatal day 28, tremor and death were not observed for a further 2 weeks. The young mice showed modest hypomyelination in the brain, whereas the adult mice did not. The astrocytes were not investigated in this model, so at the moment it is unclear if these cells functioned normally or not. It is difficult to interpret the results of the study: the data show that activation of PERK signaling is detrimental for young, but not adult, oligodendrocytes. It remains unclear whether eIF2B inhibition is the sole factor for determining tremor and hypomyelination. Also, the approach for activating PERK in the oligodendrocytes did not activate the stress sensors ATF6 and IRE1α in the endoplasmic reticulum, which are usually activated all together in response to endoplasmic reticulum stress (Walter & Ron, 2011). The chaperone protein BIP was not induced by

PERK activation (Lin et al., 2008, 2014). Increased BIP expression is a result of ATF6 activation and constitutes an important negative feedback loop, which restores the stress sensors PERK, ATF6, and IRE1a to normal levels (Walter & Ron, 2011). As such, the model is perhaps best considered as a PERK "out-of-control" model rather than a model for VWM, in which diverse and often subtle mutations in eIF2B result in white matter pathology in young and adult patients.

Recently, additional transgenic knock-in mouse models have been described for VWM: the *Eif2b4*R484W/R484W (*2b4ho*) and the *Eif2b5*R191H/R191H (*2b5ho*) lines (Dooves et al., 2016a). The *2b4ho* mice are homozygous for R484W in eIF2Bδ and the *2b5ho* mice, for R191H in eIF2Bε. The equivalent mutations in human eIF2Bδ and eIF2Bε (R483W and R195H, respectively) are associated with a severe phenotype (Fogli et al., 2004a; van der Knaap et al., 2003). A variety of analyses show that the mouse models are representative of the human disease. For example, the *2b4ho* as well as the *2b5ho* mice develop ataxia as visualized by a footprint assay with a slight variation in age at onset (7 and 5 months, respectively) (Dooves et al., 2016a). Both mouse models have a reduced life span, and again the humane endpoint varies between the two genotypes (19 months for *2b4ho* and 8 months for *2b5ho* mice) (Dooves et al., 2016a). Crossbreeding of the *2b4ho* and *2b5ho* mice resulted in more severely affected genotypes: the average survival of double homozygous animals (*2b4ho2b5ho*) was 3 weeks, and that of double transgenic animals heterozygous for either mutation and homozygous for the other (*2b42b5he/ho*) was 4 months.

The white matter in all these mutant mice shows perturbed myelination and progressive myelin vacuolation (Dooves et al., 2016a). Myelin vacuolation was reported to correlate with disease severity, genotype, and disease stage. The progressive nature of the vacuolation indicates deficient myelin formation, maturation, and maintenance. Although white matter damage was pronounced, formation of scar tissue (reactive gliosis) was not observed. This suggests that astrocytes in the white matter are unable to fulfill their normal reactive function to tissue damage. As in patients, increased numbers of astrocytes double-positive for nestin and glial fibrillary acidic protein δ (GFAPδ) were found in white matter of *2b5ho* mice, without increased GFAPα expression, indicating an immature state (Dooves et al., 2016a). These double-positive astrocytes have an abnormal morphology with thick, coarse processes (Dooves et al., 2016a). The astrocyte morphology in the gray matter appears normal. The number of mature oligodendrocytes was also decreased, and the number of OPCs was increased in *2b4ho2b5ho* and *2b42b5he/ho* mice (Dooves et al., 2016a).

A co-culture assay of astrocytes and OPCs was developed in which OPC maturation was measured by counting cells that express mature markers (myelin basic protein [MBP] and myelin oligodendrocyte glycoprotein [MOG]). Co-cultures with mutant (*2b4ho* or *2b5ho*) cells yielded reduced MBP- and MOG-positive cells when compared to WT co-cultures, indicative of impaired OPC maturation in mutant cells (Dooves et al., 2016a). This block in OPC maturation was not observed when mutant OPCs were co-cultured with WT astrocytes, indicating that the OPC maturation program itself is not directly affected by the mutation in eIF2B. Also, co-cultures with *2b4ho* astrocytes and WT OPCs showed that *2b4ho* astrocytes reduced the number of MBP- and MOG-positive cells (Dooves et al., 2016a). This finding strongly suggests that mutant astrocytes are compromised in supporting OPC maturation and substantially affected by the mutation in eIF2B.

The block in OPC maturation in vitro appears to be due to one or more inhibitory factors secreted by *2b4ho* astrocytes as cell-free, conditioned medium from *2b4ho* astrocyte cultures blocked OPC maturation in co-cultures of WT astrocytes and WT OPCs (Dooves et al., 2016a). This observation suggests that mutant astrocytes produce a factor that inhibits OPC maturation rather than being deficient in secreting one or more factors that promote OPC maturation. The secreted inhibitory factor(s) has not yet been identified, and further investigation is obviously needed to identify this factor.

Future Perspective

VWM is a devastating disease for which there is so far no cure or even treatment. Recent research into the structure and functions of eIF2B as well as studies using VWM disease models and patient materials have provided significant new insights into some aspects of eIF2B in relation to the disease.

A major remaining question concerns why mutations in genes for a protein which is crucial for protein synthesis in all cells should cause a neurological disease. Some patients do show defects in other tissues, but their disease is still primarily neurological. At this stage, one can only speculate about the possible reasons: perhaps the eIF2/eIF2B ratio or the levels of eIF5 or Met-tRNA are particularly low (or high) in glial cells, perhaps eIF2B exerts a novel and cell-specific unknown function in glial cells, or eIF2 and levels of ternary complexes affect the translation of a glia-specific mRNA(s) important for glial function. It is probably not due to a high rate of protein synthesis in glial cells since other cells, such as fast dividing cells (e.g., immune cells or cells in the jejunum) or pancreatic cells, producing many secreted factors with high protein synthetic rates are not affected.

For mechanistically unknown reasons, astrocytes appear to be most affected by the pathogenic mutations in eIF2B. Targets that are potentially interesting for therapy development may be involved in astrocyte proteasome or mitochondrial function (Gat-Viks et al., 2015; Raini et al., 2017). Targeting proteasome or mitochondria, however, have the risk of being too general and affecting other cell types as well; thus, drugs lacking astrocyte specificity may ultimately result in unwanted side effects. Alternatively, it may be of interest to identify the factor(s) secreted by mutant astrocytes that affects OPC maturation in co-culture test systems. Perhaps these factors are specifically secreted by astrocytes, providing a potentially elegant target for cell-specific treatment. Alternatively, diseased astrocytes could be replaced by healthy astrocytes lacking eIF2B mutations (Dooves, van der Knaap, & Heine, 2016b). As the inhibitory factor may persist in the extracellular matrix of the affected brain (Dooves et al., 2016b), its removal is likely to be necessary for successful cell replacement therapy.

Identifying the Secreted Factors that Inhibit OPC Maturation Is Not an Easy Task

Currently we do not know what the relevant factors secreted from astrocytes are. We may assume that they are proteins as eIF2B mutations likely affect translation of specific mRNAs. However, we cannot rule out that the factors are of a different nature. Their abnormal synthesis and secretion may, for instance, be the result of abnormal expression of the enzymes involved in their metabolism. Insight into the mRNA and protein expression profile of WT and mutant astrocytes is therefore needed first, and using an open screen ("omics") approach is likely to be most insightful. Polysomal profiling or proteomics screening of mouse brain tissue or astrocyte cultures from WT and mutant mice would be insightful. Obtaining astrocyte-specific protein expression profiling can be done by creating mutant mouse models in which the ribosomes are tagged with, for instance, a hemagglutinin tag (Sanz et al., 2009).

Subsequently, the secreted factors in astrocyte-conditioned medium can be further investigated to identify the OPC maturation inhibitory factor. To first gain insight into the nature of the inhibitory factor, the astrocyte-conditioned medium can be subjected to proteinase K treatment. If MBP and MOG expression is restored upon protein K treatment, the factor is likely a polypeptide. Proteomics may identify the protein composition of the inhibitory medium. Secretome composition of other cell types has successfully been determined with azidohomoalanine (AHA) labeling, which allows purification of the secreted proteins from culture medium factor (e.g., bovine serum albumin and growth factors in the medium) (Eichelbaum, Winter, Berriel Diaz, Herzig, & Krijgsveld, 2012). Proteins synthesized by WT and mutant astrocytes can be labeled with AHA in combination with pulsed stable isotope labeling of new proteins to strengthen quantification (Howden et al., 2013; Huo et al., 2012). Of course, it should first be determined whether the labeling technique does not interfere with the OPC inhibition function. The labeled samples can be fractionated on the basis of molecular weight, as has been reported for other test systems (Coyle, 1995; Polazzi et al., 2009; Sugino et al., 2016). The fractions that retain the inhibitory capacity can be subjected to mass spectrometry. If the factor is not a protein but is, for example, a metabolite, exosome, or fatty acid derivative, a metabolomics, rather than a proteomic, type of screening may be warranted. Eventually compounds can be searched for that interfere with the synthesis or activity of the inhibitory factor. One should keep in mind that successful targeting of the inhibitory factor may only provide relief of the abnormal OPC maturation. Other affected astrocytic functions, for example, reactive astrogliosis, may not be repaired by this treatment. VWM may therefore not be completely solved by such a specific therapy.

Discovery of a Small Molecule that Enhances eIF2B Activity

An approach that possibly targets all disease aspects lies in direct targeting of eIF2B enzyme activity. Recently, the small molecule integrated stress response (ISR) inhibitor was identified in a phenotypic, cell-based screen that scored absence of ATF4 induction after ISR activation by thapsigargin and tunicamycin (Sidrauski et al., 2013). In a recent genetic screen, mutations in eIF2Bδ were identified that made cells unresponsive to ISR inhibitor (ISRIB), indicating that eIF2Bδ is the target of ISRIB (Sekine et al., 2015). ISRIB enhanced eIF2B GEF activity in vitro and stabilized the eIF2B complex in biochemical analyses (Sidrauski et al., 2015). ISRIB prevents and resolves an already activated stress response. ISRIB has been tested in mice: it possesses good pharmacological properties and crosses the blood–brain barrier, although insolubility of the substance may be a problem to reach an efficacious dose.

A recent study assessed ISRIB's impact in prion-infected mice (Halliday et al., 2015). In this neurodegenerative disease model, eIF2B in the brain is inhibited by infectious prion particles. ISRIB prevented prion-related disease symptoms and neurodegeneration. ISRIB did not reduce the prion accumulation or eIF2 phosphorylation; however, prion-related disease symptoms were greatly diminished, and ATF4 expression in brain was reduced. This result is expected considering ISRIB's action on eIF2B (downstream of eIF2 phosphorylation and upstream of ATF4 induction). Prolonged treatment resulted in weight loss in the treated mice (Halliday et al., 2015). It is not clear if this weight loss was due to the ongoing prion infection, to ISRIB, or to their combination. Overall, it is reasonable to test ISRIB in VWM mice. One caveat is the possibility that (some) mutations in eIF2B that cause VWM may interfere with ISRIB's action. Furthermore, given the very different biochemical consequences of VWM mutations on eIF2B function, ISRIB may not prevent the consequences of all such mutations. If all mutations do affect ternary complex levels, trying to restore such levels may represent a possible therapeutic strategy (although it not clear how to achieve this in vivo). Also, if mutations in eIF2B interfere in a yet undiscovered astrocyte- or oligodendrocyte-specific function that is unrelated to the GEF activity, ISRIB may not be useful. Further research is needed to investigate ISRIB's therapeutic potential in VWM. In the case that ISRIB is not suitable for therapy in patients, other ISR modulators may be of interest for therapy development (Smith & Mallucci, 2016; Sundaram, Lee, & Shenolikar, 2017).

Acknowledgments

Work on eIF2B in C.G.P.'s lab was supported by the UK Biotechnology & Biological Sciences Research Council.

References

Baumann, N., & Pham-Dinh, D. (2001). Biology of oligodendrocyte and myelin in the mammalian central nervous system. *Physiological Reviews*, *81*(2), 871–927.

Black, D. N., Booth, F., Watters, G. V., Andeermann, E., Dumont, C., Halliday, W. C., . . . O'gorman, G. (1988). Leukoencephalopathy among native Indian infants in northern Quebec and Manitoba. *Annals of Neurology*, *24*(4), 490–496.

Bogorad, A. M., Xia, B., Sandor, D. G., Mamonov, A. B., Cafarella, T. R., Jehle, S., . . . Marintchev, A. (2014). Insights into the architecture of the eIF2Balpha/beta/delta regulatory subcomplex. *Biochemistry, 53*(21), 3432–3445.

Bugiani, M., Boor, I., Powers, J. M., Scheper, G. C., & van der Knaap, M. S. (2010). Leukoencephalopathy with vanishing white matter: A review. *Journal of Neuropathology and Experimental Neurology, 69*(10), 987–996.

Bugiani, M., Boor, I., van Kollenburg, B., Postma, N., Polder, E., van Berkel, C., . . . van der Knaap, M. S. (2011). Defective glial maturation in vanishing white matter disease. *Journal of Neuropathology and Experimental Neurology, 70*(1), 69–82.

Cigan, A. M., Bushman, J. L., Boal, T. R., & Hinnebusch, A. G. (1993). A protein complex of translational regulators of GCN4 mRNA is the guanine nucleotide-exchange factor for translation initiation factor 2 in yeast. *Proceedings of the National Academy of Sciences of the United States of America, 90*(11), 5350–5354.

Coyle, D. E. (1995). Adaptation of C6 glioma cells to serum-free conditions leads to the expression of a mixed astrocyte–oligodendrocyte phenotype and increased production of neurite-promoting activity. *Journal of Neuroscience Research, 41*(3), 374–385.

Dholakia, J. N., Francis, B. R., Haley, B. E., & Wahba, A. J. (1989). Photoaffinity labeling of the rabbit reticulocyte guanine nucleotide exchange factor and eukaryotic initiation factor 2 with 8-azidopurine nucleotides. Identification of GTP- and ATP-binding domains. *Journal of Biological Chemistry, 264*, 20638–20642.

Dooves, S., Bugiani, M., Postman, N. L., Polder, E., Land, N., Horan, S. T., . . . van der Knaap, M. S. (2016a). Astrocytes are central in the pathomechanisms of vanishing white matter. *Journal of Clinical Investigation, 126*(4), 1512–1524.

Dooves, S., van der Knaap, M. S., & Heine, V. M. (2016b). Stem cell therapy for white matter disorders: don't forget the microenvironment! *Journal of Inherited Metabolic Disease, 39*(4), 513–518.

Eichelbaum, K., Winter, M., Berriel Diaz, M., Herzig, S., & Krijgsveld, J. (2012). Selective enrichment of newly synthesized proteins for quantitative secretome analysis. *Nature Biotechnology, 30*(10), 984–990.

Fogli, A., & Boespflug-Tanguy, O. (2006). The large spectrum of eIF2B-related diseases. *Biochemical Society Transactions, 34*(Pt 1), 22–29.

Fogli, A., Rodriguez, D., Eymard-Pierre, E., Bouhour, F., Labauge, P., Meaney, B.F., . . . Boespflug-Tanguy, O. (2003). Ovarian failure related to eukaryotic initiation factor 2B mutations. *American Journal of Human Genetics, 72*(6), 1544–1550.

Fogli, A., Schiffmann, R., Bertini, E., Ughetto, S., Combes, P., Eymard-Pierre, E., . . . Boespflug-Tanguy, O. (2004a). The effect of genotype on the natural history of eIF2B-related leukodystrophies. *Neurology, 62*(9), 1509–1517.

Fogli, A., Schiffmann, R., Hugendubler, L., Combes, P., Bertini, E., Rodriguez, D., . . . Boespflug-Tanguy, O. (2004b). Decreased guanine nucleotide exchange factor activity in eIF2B-mutated patients. *European Journal of Human Genetics, 12*, 561–566.

Fogli, A., Wong, K., Eymard-Pierre, E., Wenger, J., Bouffard, J. P., Goldin, E., . . . Schiffmann, R. (2002). Cree leukoencephalopathy and CACH/VWM disease are allelic at the EIF2B5 locus. *Annals of Neurology, 52*(4), 506–510.

Garden, G. A., & Campbell, B. M. (2016). Glial biomarkers in human central nervous system disease. *Glia, 64*(10), 1755–1771.

Gat-Viks, I., Geiger, T., Barbi, M., Raini, G., & Elroy-Stein, O. (2015). Proteomics-level analysis of myelin formation and regeneration in a mouse model for vanishing white matter disease. *Journal of Neurochemistry, 134*(3), 513–526.

Geva, M., Cabilly, Y., Assaf, Y., Mindroul, N., Marom, L., Raini, G., . . . Elroy-Stein, O. (2010). A mouse model for eukaryotic translation initiation factor 2B-leucodystrophy reveals abnormal development of brain white matter. *Brain, 133*, 2448–2461.

Gomez, E., Mohammad, S. S., & Pavitt, G. D. (2002). Characterization of the minimal catalytic domain within eIF2B: The guanine-nucleotide exchange factor for translation initiation. *EMBO Journal, 21*(19), 5292–5301.

Gomez, E., & Pavitt, G. D. (2000). Identification of domains and residues within translation initiation factor eIF2Bepsilon required for guanine nucleotide-exchange reveals a novel activation function promoted by eIF2B complex formation. *Molecular and Cellular Biology, 20*, 3965–3976.

Gordiyenko, Y., Schmidt, C., Jennings, M. D., Matak-Vinkovic, D., Pavitt, G. D., & Robinson, C. V. (2014). eIF2B is a decameric guanine nucleotide exchange factor with a $\gamma 2\varepsilon 2$ tetrameric core. *Nature Communications, 5*, 3902.

Halliday, M., Radford, H., Sekine, Y., Moreno, J., Verity, N., le Quesne, J., . . . Mallucci, G. R. (2015). Partial restoration of protein synthesis rates by the small molecule ISRIB prevents neurodegeneration without pancreatic toxicity. *Cell Death & Disease, 6*, e1672.

Hanefeld, F., Holzbach, U., Kruse, B., Wilichowski, E., Christen, H. J., & Frahm, J. (1993). Diffuse white matter disease in three children: An encephalopathy with unique features on magnetic resonance imaging and proton magnetic resonance spectroscopy. *Neuropediatrics, 24*(5), 244–248.

Hinnebusch, A. G. (1996). Translational control of GCN4: Gene-specific regulation by phosphorylation of eIF2. In J. W. B. Hershey, M. B. Matthews, & N. Sonenberg (Eds.), *Translational Control* (pp. 199–244). CSH Monographs 30. Cold Spring Harbor, NY: Cold Spring Harbor Laboratories.

Hirrlinger, J., & Nave, K. A. (2014). Adapting brain metabolism to myelination and long-range signal transduction. *Glia, 62*(11), 1749–1761.

Horzinski, L., Huyghe, A., Cardoso, M. C., Gonthier, C., Ouchchane, L., Schiffmann, R., . . . Fogli, A. (2009). Eukaryotic initiation factor 2B (eIF2B) GEF activity as a diagnostic tool for EIF2B-related disorders. *PLoS One, 4*(12), e8318.

Howden, A. J. M., Geoghegan, V., Katsch, K., Efstathiou, G., Bhushan, B., Boutureira, O., . . . Acuto, O. (2013). QuaNCAT: Quantitating proteome dynamics in primary cells. *Nature Methods, 10*(4), 343–346.

Hronova, V., & Valasek, L. S. (2017). An emergency brake for protein synthesis. *Elife, 6, e27085.*

Huo, Y., Iadevaia, V., Yao, Z., Kelly, I., Cosulich, S., Guichard, S., . . . Proud, C. G. (2012). Stable isotope-labelling analysis of the impact of inhibition of the mammalian target of rapamycin on protein synthesis. *Biochemical Journal, 444*(1), 141–151.

Jennings, M. D., Kershaw, C. J., Adomavicius, T., & Pavitt, G. D. (2017). Fail-safe control of translation initiation by dissociation of eIF2alpha phosphorylated ternary complexes. *Elife, 6*, e24542.

Jennings, M. D., Kershaw, C. J., White, C., Hoyle, D., Richardson, J. P., Costello, J. L., . . . Pavitt, G. D. (2016). eIF2beta is critical for eIF5-mediated GDP-dissociation inhibitor activity and translational control. *Nucleic Acids Research, 44*(20), 9698–9709.

Jennings, M. D., & Pavitt, G. D. (2010). eIF5 has GDI activity necessary for translational control by eIF2 phosphorylation. *Nature, 465*(7296), 378–381.

Jennings, M. D., & Pavitt, G. D. (2014). A new function and complexity for protein translation initiation factor eIF2B. *Cell Cycle, 13*(17), 2660–2665.

Jennings, M. D., Zhou, Y., Mohammad-Qureshi, S. S., Bennett, D., & Pavitt, G. D. (2013). eIF2B promotes eIF5 dissociation from eIF2*GDP to facilitate guanine nucleotide exchange for translation initiation. *Genes and Development, 27*(24), 2696–2707.

Kantor, L., Harding, H. P., Ron, D., Shiffmann, R., Kaneski, C. R., Kimball, S. R., & Elroy-Stein, O. (2005). Heightened stress response in primary fibroblasts expressing mutant eIF2B genes from CACH/VWM leukodystrophy patients. *Human Genetics, 118*(1), 99–106.

Kashiwagi, K., Ito, T., & Yokoyama, S. (2017). Crystal structure of eIF2B and insights into eIF2–eIF2B interactions. *FEBS Journal, 284*, 868–874.

Kashiwagi, K., Takahashi, M., Nishimoto, M., Hiyama, T. B., Higo, T., Umehara, T., . . . Yokoyama, S. (2016). Crystal structure of eukaryotic translation initiation factor 2B. *Nature, 531*(7592), 122–125.

Kiray, H., Lindsay, S. L., Hosseinzadeh, S., & Barnett, S. C. (2016). The multifaceted role of astrocytes in regulating myelination. *Experimental Neurology, 283*(Pt. B), 541–549.

Kuhle, B., Eulig, N. K., & Ficner, R. (2015). Architecture of the eIF2B regulatory subcomplex and its implications for the regulation of guanine nucleotide exchange on eIF2. *Nucleic Acids Research, 43*, 9994–10014.

Leegwater, P. A., Könst, A. A., Kuyt, B., Sandkuijl, L. A., Naidu, S., Oudejans, C. B., . . . van der Knapp, M. S. (1999). The gene for leukoencephalopathy with vanishing white matter is located on chromosome 3q27. *American Journal of Human Genetics, 65*(3), 728–734.

Leegwater, P. A., Vermeulen, G., Könst, A. A., Naidu, S., Mulders, J., Visser, A., . . . van der Knapp, M. S. (2001). Subunits of the translation initiation factor eIF2B are mutant in leukoencephalopathy with vanishing white matter. *Nature Genetics, 29*(4), 383–388.

Li, W., Wang, X., van der Knaap, M. S., & Proud, C. G. (2004). Mutations linked to leukoencephalopathy with vanishing white matter impair the function of the eukaryotic initiation factor 2B complex in diverse ways. *Molecular and Cellular Biology, 24*(8), 3295–3306.

Lin, W., Kunkler, P. E., Harding, H. P., Ron, D., Kraig, R. P., & Popko, B. (2008). Enhanced integrated stress response promotes myelinating oligodendrocyte survival in response to interferon-gamma. *American Journal of Pathology, 173*(5), 1508–1517.

Lin, Y., Pang, X., Huang, G., Jamison, S., Fang, J., Harding, H. P., . . . Lin, W. (2014). Impaired eukaryotic translation initiation factor 2B activity specifically in oligodendrocytes reproduces the pathology of vanishing white matter disease in mice. *Journal of Neuroscience, 34*(36), 12182–12191.

Liu, A. R., van der Lei, H. D., Wang, X., Wortham, N. C., Tang, H., van Berkel, C. G., . . . Proud, C. G. (2011). Severity of vanishing white matter disease does not correlate with deficits in eIF2B activity or the integrity of eIF2B complexes. *Human Mutation, 32*, 1036–1045.

Liu, R., van der Lei, H. D., Wang, X., Wortham, N. C., Tang, H., van Berkel, C. G., . . . Proud, C. G. (2011). Severity of vanishing white matter disease does not correlate with deficits in eIF2B activity or the integrity of eIF2B complexes. *Human Mutation, 32*(9), 1036–1045.

Lundgaard, I., Osório, M. J., Kress, B. T., Sanggaard, S., & Nedergaard, M. (2014). White matter astrocytes in health and disease. *Neuroscience, 276*, 161–173.

Marom, L., Ulitsky, I., Cabilly, Y., Shamir, R., & Elroy-Stein, O. (2011). A point mutation in translation initiation factor eIF2B leads to function—and time-specific changes in brain gene expression. *PLoS One, 6*(10), e26992.

Matsukawa, T., Wang, X., Liu, R., Wortham, N. C., Onuki, Y., Kubota, A., . . . Tsuji, S. (2011). Adult-onset leukoencephalopathies with vanishing white matter with novel missense mutations in EIF2B2, EIF2B3, and EIF2B5. *Neurogenetics, 12*, 259–261.

Middeldorp, J., Boer, K., Sluijs, J. A., De Fillippis, L., Encha-Razavi, F., Vescovi, A. L., . . . Hol, E. M. (2010). GFAPdelta in radial glia and subventricular zone progenitors in the developing human cortex. *Development, 137*(2), 313–321.

Ohlenbusch, A., Henneke, M., Brockmann, K., Goerg, M., Hanefeld, F., Kohlschütter, A., & Gärtner, J. (2005). Identification of ten novel mutations in patients with eIF2B-related disorders. *Human Mutation, 25*(4), 411.

Paulin, F. E., Campbell, L. E., O'Brien, K., Loughlin, J., & Proud, C. G. (2001). Eukaryotic translation initiation factor 5 (eIF5) acts as a classical GTPase-activator protein. *Current Biology, 11*(1), 55–59.

Pavitt, G. D. (2005). eIF2B, a mediator of general and gene-specific translational control. *Biochemical Society Transactions, 33*(Pt. 6), 1487–1492.

Pavitt, G. D., Ramaiah, K. V., Kimball, S. R., & Hinnebusch, A. G. (1998). eIF2 independently binds two distinct eIF2B subcomplexes that catalyse and regulate guanine-nucleotide exchange. *Genes & Development, 12*, 514–526.

Polazzi, E., Altamira, L. E., Eleuteri, S., Barbaro, R., Casadio, C., Contestabile, A., & Monti, B. (2009). Neuroprotection of microglial conditioned medium on 6-hydroxydopamine-induced neuronal death: Role of transforming growth factor beta-2. *Journal of Neurochemistry, 110*(2), 545–556.

Pronk, J.C., van Kollenburg, B., Scheper, G. C., & van der Knaap, M. S. (2006). Vanishing white matter disease: A review with focus on its genetics. *Mental Retardation and Developmental Disabilities Research Reviews, 12*(2), 123–128.

Raini, G., Sharet, R., Herrero, M., Atzmon, A., Shenoy, A., Geiger, T., & Elroy-Stein, O. (2017). Mutant eIF2B leads to impaired mitochondrial oxidative phosphorylation in vanishing white matter disease. *Journal of Neurochemistry, 141*, 694–707.

Raponi, E., Agenes, F., Delphin, C., Assard, N., Baudier, J., Legraverend, C., & Deloulme, J. C. (2007). S100B expression defines a state in which GFAP-expressing cells lose their neural stem cell potential and acquire a more mature developmental stage. *Glia, 55*(2), 165–177.

Richardson, J. P., Mohammad, S. S., & Pavitt, G. D. (2004). Mutations causing childhood ataxia with central nervous system hypomyelination reduce eukaryotic initiation factor 2B complex formation and activity. *Molecular and Cellular Biology, 24*(6), 2352–2363.

Rowlands, A. G., Panniers, R., & Henshaw, E. C. (1988). The catalytic mechanism of guanine nucleotide exchange factor action and competitive inhibition by phosphorylated eukaryotic initiation factor 2. *Journal of Biological Chemistry, 263*(12), 5526–5533.

Sanz, E., Yang, L., Su, T., Morris, D. R., McKnight, G. S., & Amieux, P. S. (2009). Cell-type-specific isolation of ribosome-associated mRNA from complex tissues. *Proceedings of the National Academy of Sciences of the United States of America, 106*(33), 13939–13944.

Sekine, Y., Zyryanova, A., Crespillo-Casado, A., Amin-Wetzel, N., Harding, H. P., & Ron, D. (2016). Paradoxical sensitivity to an integrated stress response blocking mutation in vanishing white matter cells. *PLoS One, 11*(11), e0166278.

Sekine, Y., Zyryanova, A., Crespillo-Casado, A., Fischer, P. M., Harding, H. P., & Ron, D. (2015). Stress responses. Mutations in a translation initiation factor identify the target of a memory-enhancing compound. *Science, 348*(6238), 1027–1030.

Sidrauski, C., Acosta-Alvear, D., Khoutorsky, A., Vedantham, P., Hearn, B. R., Li, H., . . . Walter, P. (2013). Pharmacological brake-release of mRNA translation enhances cognitive memory. *Elife, 2*, e00498.

Sidrauski, C., Tsai, J. C., Kampmann, M., Hearn, B. R., Vedantham, P., Jaishankar, P., . . . Walter, P. (2015). Pharmacological dimerization and activation of the exchange factor eIF2B antagonizes the integrated stress response. *Elife, 4*, e07314.

Smith, H. L., & Mallucci, G. R. (2016). The unfolded protein response: Mechanisms and therapy of neurodegeneration. *Brain, 139*(Pt. 8), 2113–2121.

Sugino, I. K., Sun, Q., Springer, C., Cheewatrakoolpong, N., Liu, T., Li, H., & Zarbin, M. A. (2016). Two bioactive molecular weight fractions of a conditioned medium enhance RPE cell survival on age-related macular degeneration and aged Bruch's Membrane. *Translational Vision Science & Technology*, *5*(1), 8.

Sundaram, J. R., Lee, I. C., & Shenolikar, S. (2017). Translating protein phosphatase research into treatments for neurodegenerative diseases. *Biochemical Society Transactions*, *45*, 101–112.

van der Knaap, M. S., Barth, P. G., Gabreëls, F. J., Franzoni, E., Begeer, J. H., Stroink, H., . . . Valk, J. (1977). A new leukoencephalopathy with vanishing white matter. *Neurology*, *48*(4), 845–855.

van der Knaap, M. S., Breiter, S. N., Naidu, S., Hart, A. A., & Valk, J. (1999). Defining and categorizing leukoencephalopathies of unknown origin: MR imaging approach. *Radiology*, *213*(1), 121–133.

van der Knaap, M. S., Kamphorst, W., Barth, P. G., Kraaijeeveld, C. L., Gut, E., & Valk, J. (1998). Phenotypic variation in leukoencephalopathy with vanishing white matter. *Neurology*, *51*(2), 540–547.

van der Knaap, M. S., Leegwater, P. A., Könst, A. A., Visser, A., Naidu, S., Oudejans, C. B., . . . Pronk, J. C. (2002). Mutations in each of the five subunits of translation initiation factor eIF2B can cause leukoencephalopathy with vanishing white matter. *Annals of Neurology*, *51*(2), 264–270.

van der Knaap, M. S., Pronk, J. C., & Scheper, G. C. (2006). Vanishing white matter disease. *Lancet Neurology*, *5*(5), 413–423.

van der Knaap, M. S., van Berkel, C. G., Herms, J., van Coster, R., Baethmann, M., Naidu, S., . . . Pronk, J. C. (2003). eIF2B-related disorders: Antenatal onset and involvement of multiple organs. *American Journal of Human Genetics*, *73*(5), 1199–1207.

van der Lei, H. D., van Berkel, C. G., van Wieringen, W. N., Brenner, C., Feigenbaum, A., Mercimek-Mahmutoglu, S., . . . van der Knaap, M. S. (2010). Genotype–phenotype correlation in vanishing white matter disease. *Neurology*, *75*(17), 1555–1559.

van der Voorn, J. P., van Kollenburg, B., Bertrand, G., Van Haren, K., Scheper, G. C., Powers, J. M., & van der Knaap, M. S. (2005). The unfolded protein response in vanishing white matter disease. *Journal of Neuropathology and Experimental Neurology*, *64*(9), 770–775.

Van Haren, K., van der Voorn, J. P., Peterson, D. R., van der Knaap, M. S., & Powers, J. M. (2004). The life and death of oligodendrocytes in vanishing white matter disease. *Journal of Neuropathology and Experimental Neurology*, *63*(6), 618–630.

van Kollenburg, B., van Dijk, J., Garbern, J., Thomas, A. A., Scheper, G. C., Powers, J. M., & van der Knaap, M. S. (2006). Glia-specific activation of all pathways of the unfolded protein response in vanishing white matter disease. *Journal of Neuropathology and Experimental Neurology*, *65*(7), 707–715.

Vermeulen, G., Seidl, R., Mercimek-Mahmutoglu, S., Rotteveel, J. J., Scheper, G. C., & van der Knaap, M. S. (2005). Fright is a provoking factor in vanishing white matter disease. *Annals of Neurology*, *57*(4), 560–563.

Walter, P., & Ron, D. (2011). The unfolded protein response: From stress pathway to homeostatic regulation. *Science*, *334*(6059), 1081–1086.

Wang, X. M., Wortham, N. C., Liu, R., & Proud, C. G. (2012) Identification of residues that underpin interactions within the eukaryotic initiation factor (eIF2) 2B complex. *Journal of Biological Chemistry*, *287*(11), 8263–8274.

Williams, D. D., Price, N. T., Loughlin, A. J., & Proud, C. G. (2001). Characterisation of the mammalian initiation factor eIF2B complex as a GDP-dissociation stimulator protein. *Journal of Biological Chemistry*, *276*(27), 24697–24703.

Wortham, N. C., & Proud, C. G. (2015). Biochemical effects of mutations in the gene encoding the alpha subunit of eukaryotic initiation factor (eIF) 2B associated with vanishing white matter disease. *BMC Medical Genetics, 16*, 64.

Wortham, N. C., Martinez, M., Gordiyenko, Y., Robinson, C. V., & Proud, C. G. (2014). Analysis of the subunit organization of the eIF2B complex reveals new insights into its structure and regulation. *FASEB Journal, 28*(5), 2225–2237.

Wu, Y., Pan, Y., Du, L., Wang, J., Gu, Q., Gao, Z., . . . Jiang, Y. (2009). Identification of novel EIF2B mutations in Chinese patients with vanishing white matter disease. *Journal of Human Genetics, 54*(2), 74–77.

Zhang, H. H., Dai, L., Chen, N., Zang, L., Leng, X., Du, L., . . . Wu, Y. (2015). Fifteen novel EIF2B1-5 mutations identified in Chinese children with leukoencephalopathy with vanishing white matter and a long term follow-up. *PLoS One, 10*(3), e0118001.

Index

Note: Tables and figures are indicated by an italic "*t*" and "*f*", respectively, following the page number.

For the benefit of digital users, indexed terms that span two pages (e.g., 52–53) may, on occasion, appear on only one of those pages

A-443654 91–2
A-484594 542
abt-5 332
acetylation
  of histone 286–7, 287*f*
  in LTP maintenance 286
Actb mRNA
  localization of 361–3
  transport of 363
  ZBP1 and 364, 369–71
Actg mRNA, localization of 361–3
activating transcription factor 4 (ATF4)
  addiction and 497
  eIF2 phosphorylation and 46–7, 600
  GCN2-ATF4 signaling pathway 6*f*, 12
  ISR inhibition and 51–2, 610–11
  mutations and 54–5
activating transcription factor 6 (ATF6)
  AD and 534–6
  eIF2B and 607–8
activity regulated cytoskeletal protein (ARC) 136–7, 280–2, 282*f*
AD. *See* Alzheimer's disease
$ADAR_3$. *See* adenosine deaminase
ADAT. *See* adenosine deaminase tRNA specific
addiction. *See also* substance abuse
  brain areas and 490*f*
  eIF2α in 495–7
  eIF4E and 492
  elongation phase and 501
  mTORC1 in 490–5, 491*f*, 493*t*
  mTOR in 495, 501
  neurobiological alterations in 487
  protein synthesis in 489–90
  RNA-binding proteins 497–9
  study of 488
    methods for 499–501
  synaptic plasticity and 487
adenomatous polyposis coli (APC) protein 373–4
adenosine deaminase ($ADAR_3$) 155
adenosine deaminase tRNA specific (ADAT) 156
adenosine monophosphate-activated protein kinase (AMPK)
  AD and 540–1
  aging and 72
  eEF2K regulation by 65, 66*f*
  pain and 439
AG-484954 540–1
AGC kinase superfamily 81–2
aging
  AD and 533, 539–40
  BiP and 140
  eEF2 and eEF2K in 72
  S6K1 and 92
AGO. *See* Argonaute
AGO1. *See* Argonaute 1
AGO2. *See* Argonaute 2
AHA. *See* azidohomoalanine
Akt. *See* protein kinase B
alcohol
  brain areas and 490*f*
  eIF2α and 496
  mTORC1 signaling and 32, 493*t*
  rapamycin and 492–5

aldicarb 330–1, 339
alg-1 344
AlkB homolog 5, histone H2A dioxygenase (ALKBH5) 167–8
allodynia 428*f*, 429
ALM neuron 328–9
alr-1 328–9
ALS. *See* amyotrophic lateral sclerosis
alternative polyadenylation (APA) 327*f*
  pain and 433–4
  regulation of 335–6
alternative splicing by RBPs 324–36, 327*f*
Aly/REF export factor (ALYREF) 163
Alzheimer's disease (AD)
  aging and 533, 539–40
  eEF2 and eEF2K in 72–3
  eEF2K/eEF2 signaling in 540–2, 541*f*
  eIF2α in 534–7, 535*f*
  eIF4E phosphorylation and 7–8
  hallmarks of 533–4
  IRES and 138–9
  ISR and 52–3
  mTORC1 signaling dysregulation and 537–40
  mTOR inhibitors and 517–18
  PERK inhibition in 53–4
  prevalence of 533
  rapamycin and 539–40
  S6K1 and 92
  synaptic tagging and capture and 276–8
  treatment for 536
α-amino-3-hydroxy-5-methyl-4-isoxazolepropionic acid receptors (AMPARs)
  ASD and 464
  in E-LTP and L-LTP 271
  with GluA2 subunits 271–2
  LTD and 302
  regulation of 333–4
  In synaptic tagging and capture 274–5
(2R)-amino-5-phosphonovaleric acid (APV) 303–4
amino acids, mTORC1 and 521
AMPARs. *See* α-amino-3-hydroxy-5-methyl-4-isoxazolepropionic acid receptors
amphetamines
  eIF2α and 496
  FMRP and 497–8
  mTORC1 signaling and 32, 493*t*
  mTORC2 signaling and 495
AMPK. *See* adenosine monophosphate-activated protein kinase
β-amyloid peptide 138–9, 533–4, 540–1
amyloid precursor protein (APP)
  IRES and 138–9
  ISR and 52–3
amyotrophic lateral sclerosis (ALS)
  familial 553
  FUS and 224
    knockdown 558–9
    miRNA machinery and 559–60
    mRNA translational targets 560–1
    reduced levels of synthesized proteins in 559
  ISR and 53–4
  RBPs in 553–64
    aggregates and mutants 553–4
    Ataxin-2 561, 563–4
    hnRNPA2/B1 and A1 561–2
    MATR3 561–2
    RAN translation modulation in 568–9
    TIA1 561–3
  RNA-binding protein de-regulation in 323–4
  sporadic 553
  TDP-43
    CAMTA1 and DENND4A translation enhancement 557–8
    Futsch/MAP1B and Hsc70-4/HSPA8 mRNA repression 556–7
    motor-neuronal cell translation enhancement 557
    mRNA-specific translational repression 555–6
    ribosome-mRNA complexes and 555
    RNA regulation and 554–5
Angelman Syndrome, S6K1 and 86–7, 91–2
anisomycin 31–2, 73, 489
ANKRD52-PPP6C phosphatase 220
antidepressant treatments
  efficacy of 515

for MDD 511–12
mTORC1 and 33
protein synthesis pathway targeting with 519–22
S6K1 and 89
anti-psychotic treatments, 4E-BPs 33
antisense oligonucleotide (ASO) 164–5
APA. *See* alternative polyadenylation
APC protein. *See* adenomatous polyposis coli protein
Aplysia model
CPEB in 199
synaptic plasticity and memory role in 203–4
eEF2/eEF2K pathway in 68–9
IRES in 137
local translation in 194–5
stimulus-dependent 243–4
APM-3 339
apoptosis
IRES and 134–5, 141–3
N-myc and 142–3
APP. *See* amyloid precursor protein
APV. *See* (2R)-amino-5-phosphonovaleric acid
ARC. *See* activity regulated cytoskeletal protein
Arc/Arg3.1 *See* activity regulated cytoskeletal protein
AREs. *See* AU rich elements
arginine-rich motifs (ARMs) 106
Argonaute (AGO)
P bodies, RISC particles and 250–1
RISC particles and 367–8
Argonaute 1 (AGO1), ASD and 473
Argonaute 2 (AGO2)
DND1 and 224
FMRP and 221–2
importance of 219
miRNA biogenesis and 218
MOV10 and 222–3, 223*f*
RBPs interaction with 220–1
regulation of 220
ARMs. *See* arginine-rich motifs
ASD. *See* autism spectrum disorders
Ash1 240
ASO. *See* antisense oligonucleotide
associative plasticity
LTP and LTD interactions 280, 281*f*
metaplasticity and 276–9, 278*f*, 279*f*
synaptic tagging and capture and 272–4, 273*f*
astrocytes, in VWMD 597, 609–10
AT 13148 92
Ataxin-2 561, 563–4
ATF4. *See* activating transcription factor 4
ATF6. *See* activating transcription factor 6
A-to-I editing enzyme 155
inosine formation 156, 173
microRNAs 163–4
AU rich elements (AREs) 241
HuD stabilization of 366
Aurora A kinase 197–8, 205–6
autism spectrum disorders (ASD)
BC RNA and 111–12
causes of 453
definition of 453
4E-BP2 in 28–9
gene risks for 454, 455*t*
initiation mutations and 412
MeCP2 473–4
microRNA and 472–3
mTOR and 410–11, 453–6
inhibitors 517–18
protein synthesis levels in 454–6, 455*t*
regulator genes 468–72
BCKDK 471–2
DYRK1A 470–1
PTEN 468–9
SLC7A5 471–2
TSC 469–70
ribosomal proteins and 411–12
RNA-binding proteins and 323–4, 461–8
CELF6 464–5
FMR1 461–3
JAKMIP1 465–7
RBFOX1 467–8
RNG105 463–4
S6K1 and 91–2
stalled polysomes and 257
symptoms and comorbidities of 453
therapeutic interventions for 456

autism spectrum disorders (ASD) (*Continued*)
translational machinery proteins and 456–61
eEF1A 458–9
eIF4E 412, 457–8
RPs 459–60
UPF3B 460–1
autophagy, 4E-BP2 and 27–8
AVE neuron 333–4
axonal localization motif 362–3
axons 359
elongation of, 4E-BP2 in 28
mRNA localization in 360–1
regulation of 361–5, 361*f*
stability regulation of 365–9
storage of 372–4
regeneration of
m6A and 172
miRNA-dependent regulation of 344
mTORC1 in 375
piRNA-mediated regulation 344
RNA granule-associated genes 343–4
RTCB-1 342–3
UNC-75/CELF in 327*f*, 329–31, 331*f*
stress granules and 372–3
translation in 360
direct regulation of 374–9
HuD and 366
mechanisms of 369
NMD and 366–7
RBPs and 369–72, 370*f*
regulation of 360–79, 361*f*
UPF1 and 367
axotomy 378–9
azidohomoalanine (AHA) 610

Bbs2 408
BC. *See* brain cytoplasmic
BCAA. *See* branched chain amino acids
BCKDK. *See* branched chain ketoacid dehydrogenase kinase
Bdnf 437
BDNF. *See* brain-derived neurotrophic factor
BDU neuron 328–9
behavioral tagging (BT) 279–80
beta-actin 240–1
BiP
function of 140
IRES and 140
BIP, eIF2B and 607–8
blmp-1 339–40
Bowen-Conradi syndrome 159
BP. *See* CUG-binding proteins
brain cytoplasmic (BC) RNA 107
cognitive impairment and 111–12, 113*f*
eIF4B phosphorylation and 109
localization of regulatory 117–18
mGluRs and 111
regulatory 108
retroposition and 118
as translation repressors 107–8
brain-derived neurotrophic factor (BDNF)
eEF2 phosphorylation and 69
eIF2 phosphorylation and 48, 541–2
eIF4E expression and phosphorylation 7–8
LIMK1 and 225
in local translation 194–5
in dendrites 363–4
stimulus-dependent 243–4
MDD and 512–15, 514*f*, 524
MNKs activation of 7
NMDAR and 271
S6K1 and 87
TrkB and 137–8
Val633Met variation in 94
branched chain amino acids (BCAA), mTORC1 and 521
branched chain ketoacid dehydrogenase kinase (BCKDK) 471–2
Brn1 405–6
bruceantin 251
Bruno-like proteins 464–5
b-snoRNAs 160–1
BT. *See* behavioral tagging

C9ORF72 554
Ca^{2+}/calmodulin-dependent protein kinase II (CaMKII)
abundance of 240–1
CPEB1 and 200–1
eEF2K pathways and 70–1

FMRP and 107
in inverse synaptic tagging 280–2, 282*f*
local translation and 243–4
in LTM stabilization 195–6
metaplasticity and 276, 278*f*
NMDAR and 271
nociception and 431–3
PSD and 274–5
SP and 282
Ca^{2+}/calmodulin-dependent protein kinase IV (CaMKIV), in CRE-dependent transcription 270–1
CACH. *See* childhood ataxia with diffuse central nervous system hypomyelination
CAD cells. *See* Catecholaminergic neuronal tumor cells
Cajal bodies 160
calcineurin 519
CaMKII. *See* Ca^{2+}/calmodulin-dependent protein kinase II
CaMKIV. *See* Ca^{2+}/calmodulin-dependent protein kinase IV
cAMP
eEF2K regulation by 65, 66*f*
synaptic stimulation and 270–1
cAMP response element binding protein (CREB)
activation of 270–1
in hippocampus 270–1
local translation of 114–15
NMDAR and 271
phosphorylation of 270–1
CAMTA1 translation enhancement 557–8
cancer
CRD-BP and 224–5
eIF4B phosphorylation and 110–11
4E-BP1 phosphorylation and 24–5
4E-HP in 9
IRES and 134–5
Malat1 and 165
miRNAs in 9
N-myc and 143
nuclear eIF4E and 10–11
XIAP IRES translation for 141–2
cap-binding complex (CBC) 161–2
cap-dependent initiation 3–4, 132, 133*f*
cap-dependent translation 3–4, 299
4E-BPs in 5, 6*f*, 34, 374–5
LRRK2 and 309–10
for LTP and LTM 13
in NSCs 401–2
oxygen levels and 9
pain and 434–5
regulation of 299–300
requirements for 374–5
cap-independent initiation 134–5
CAPRIN-1. *See* cytoplasmic activation/proliferation-associated protein 1
CAR-1 343–4
5-carbamoylmethyl (ncm^5) modification 402
Casein kinase 2, G3BP1 phosphorylation 373
Catecholaminergic neuronal tumor (CAD) cells 363–4
CBC. *See* cap-binding complex
CBP. *See* CREB-binding protein
Ccnd2 408–10
Cdk5. *See* cyclin dependent kinase 5
CELF. *See* CUG-binding and ELAV factor
CELF6 464–5
cell proliferation, IRES and 134–5
cellular IRESs 136
central nervous system (CNS). *See also* brain
homeostatic plasticity in 302–3
IRES translation in 136–8
m6A in 166–71
neural stem cells and 397–8
RNA modifications in 154–5
translational controls in 439–41
opioid-induced hyperalgesia 440–1
reconsolidation mechanisms in pain 439–40
spinal modulation 440
cerebral ischemic injury, XIAP and 142
CGI-99 245–6
ChAT. *See* choline acetyltransferase
chemotherapy-induced peripheral neuropathy, eIF4E in 437–8
childhood ataxia with diffuse central nervous system hypomyelination (CACH) 596. *See also* vanishing white matter disease
choline acetyltransferase (ChAT) 201–2

CIP. *See* congenital insensitivity to pain
circadian rhythms, 4E-BPs control of 29–30
circular RNAs (circRNAs) 165
cis-acting targeting elements 115–17, 116*f*
CK2delta 335–6
cleavage and polyadenylation specificity factors (CPSF) 197
CLIP. *See* crosslinking immunoprecipitations
CNOT 473
CNQX. *See* 6-cyano-7-nitroquinoxaline-2,3-dione
CNS. *See* central nervous system
cocaine
  ATF4 and 497
  brain areas and 490*f*
  eEF2 and eEF2K in 73
  ELAV and 498
  FMRP and 497–8
  mTOR and 495
  mTORC1 signaling and 32, 493*t*
  OPHN1 and 497
  rapamycin and 492–5
  RiboTag and 439
  TRAP and 439
  ZBP1 and 498
coding region determinant-binding protein (CRD-BP), in miRNA regulation 224–5
cognitive deficits. *See also* Alzheimer's disease
  ADAT and 156
  BC RNA and 111–12, 113*f*
  DYRK1A in 470–1
  FXS and 221
  ISR and 44, 52–3, 53*f*
  NSXLID 157
  PTEN and 468
  RBFOX1 and 467
  ribosomal proteins and 411–12
  stalled polysomes and 257
coincidence detectors 515–16
conditioned place preference (CPP) 488–9
conditioned taste aversion (CTA)
  BT and 279–80
  eEF2 phosphorylation 67–8
  S6K1 and 89
congenital insensitivity to pain (CIP) 427
cortical gamma oscillations, BC RNA and 111–12
cortical malformations, mTORC1 and 27–8
Couch potato 326–7
CPE. *See* cytoplasmic polyadenylation element
CPEB. *See* cytoplasmic element binding protein; cytoplasmic polyadenylation element binding protein
CPEB1. *See* cytoplasmic polyadenylation element binding protein 1
CPEB2. *See* cytoplasmic polyadenylation element binding protein 2
CPEB3. *See* cytoplasmic polyadenylation element binding protein 3
CPEB4. *See* cytoplasmic polyadenylation element binding protein 4
CPP. *See* Conditioned Place Preference
CPSF. *See* cleavage and polyadenylation specificity factors
CRAC. *See* cross-linking and analysis of cDNAs
CRD-BP. *See* coding region determinant-binding protein
CREB. *See* cAMP response element binding protein
CREB-binding protein (CBP) 285–6
cross-linking and analysis of cDNAs (CRAC) 404
crosslinking immunoprecipitations (CLIP) 363
cross-tagging 280, 281*f*
CSNK1A1 220
CTA. *See* conditioned taste aversion
CUG-binding and ELAV factor (CELF) 327*f*, 329–31, 331*f*
  ASD and 464–5
CUG-binding proteins (BP) 464
Cux1 405–6
6-cyano-7-nitroquinoxaline-2,3-dione (CNQX) 303–4
cyclin dependent kinase 5 (Cdk5), S6K1 and 87
cycloheximide 67
  addiction and 489
  LRRK2 and 309–10
CYFIP. *See* cytoplasmic FMR1 interacting protein

CYFIP2. *See* cytoplasmic FMR1 interacting protein 2
cytoplasmic activation/proliferation-associated protein 1 (CAPRIN-1), neuronal RNA granules and 245–6
cytoplasmic FMR1 interacting protein (CYFIP) 248–9
  addiction and 498
  ASD and 462
cytoplasmic FMR1 interacting protein 2 (CYFIP2) 173
cytoplasmic polyadenylation
  local translation and 196–9
  RBPs and 371–2
cytoplasmic polyadenylation element (CPE)
  CPEB binding to 371–2
  cytoplasmic polyadenylation with 197–8
cytoplasmic polyadenylation element binding protein (CPEB)
  cytoplasmic polyadenylation with 197–8
  family of 199
  prion-like properties of 204–6
  synaptic plasticity and memory role of 199–203
    invertebrate 203–6
cytoplasmic polyadenylation element binding protein 1 (CPEB1)
  aggregation of 205–6
  functions of 199
  synaptic plasticity and memory role of 200
cytoplasmic polyadenylation element binding protein 2 (CPEB2)
  functions of 199
  synaptic plasticity and memory role of 201
cytoplasmic polyadenylation element binding protein 3 (CPEB3)
  aggregation of 205–6
  functions of 199
  synaptic plasticity and memory role of 202
cytoplasmic polyadenylation element binding protein 4 (CPEB4)
  functions of 199
  synaptic plasticity and memory role of 203
cytosine-5 methylation (m^5C) 402

daf-2 333
DCC. *See* deleted in colon cancer protein
DD. *See* Dorsal type D
DDX3X 463, 569
DDX6 343–4
dead end 1 (DND1), in miRNA regulation 224
deamidation
  of 4E-BP2 25–6
  inosine and 162–3
deleted in colon cancer protein (DCC) 373–4
dendrin 114–15, 136–7, 240–1
dendritic targeting elements (DTEs) 115, 116*f*, 117–18
DENND4A translation enhancement 557–8
dentate gyrus (DG)
  LTP consolidation in 7
  ribosomes in 240–1
dephosphorylation
  of AGO2 220
  of eEF2 68
  of eIF2 46*f*, 47
  of eIF2α 196–7, 496
  of eIF4B 109–11, 110*f*
  of FMRP 109, 110*f*, 228
  translational regulation with 108–11
depressive disorders. *See also* major depressive disorder
  lncRNAs and circRNAs in 165
  mTORC1 and 33
  S6K1 and 516–17
desvenlafaxine 524
DG. *See* dentate gyrus
DGCR8 218–19
DGCR14 334–5
DHPG. *See* 3,5-dihydroxyphenylglycine
diabetic peripheral neuropathy, eIF2α in 439
Dicer, in axons 368
DiGeorge syndrome 334
3,5-dihydroxyphenylglycine (DHPG) 51
direct neurogenesis 397–8
DJ-1, PD and 571–2
dlk-1 334–6
DNA methyltransferases (DNMTs) 287–8
DND1. *See* dead end 1
DNMTs. *See* DNA methyltransferases
dorsal root ganglia (DRG) 429
  mTORC1 in 435–6
  polyadenylation in 431–3
Dorsal type D (DD) 338–9

double-stranded RNA-dependent protein kinase (PKR)
  AD and 52–3, 534
  ASD and 463
  Down syndrome and 52–3
  eIF2 phosphorylation 45–6, 46*f*, 49–50
  eIF2α phosphorylation 299, 377–8
  inhibition of 52
  LTM and 48–9, 51–2
  pain and 438
Down syndrome (DS)
  DYRK1A in 470–1
  ISR and 52–3, 53*f*
DRG. *See* dorsal root ganglia
DROSHA 218–19
Drosha complex 473–4
Drosophila model
  Bruno protein 464–5
  CPEB in
    orthologs of 199
    synaptic plasticity and memory role in 204
  FMRP orthologs in 221–2
  homeostatic enhancement in 306
  local translation in 430–1
  LRRK2 in 309–10
drug addiction. *See* addiction
drug-induced locomotor sensitization 488–9
drug self-administration 488
DS. *See* Down syndrome
DTEs. *See* dendritic targeting elements
dual-specificity tyrosine-phosphorylation-regulated kinase 1A (DYRK1A) 470–1
dyn-1 330
dynamin 302
DYRK1A. *See* dual-specificity tyrosine-phosphorylation-regulated kinase 1A
DYRK1A 470–1
dysautonomia 156–7
dyskeratosis congenita 134–5

E2F protein 3 (E2F3) 224
early long-term depression (E-LTD), cross-tagging 280, 281*f*
early long-term potentiation (E-LTP)
  CPEB1 and 201
  cross-tagging 280, 281*f*
  decay of 271
  ISR and 50–1
  mTORC1 in 31
  presynaptic mechanisms in 270
  protein synthesis and 301
  S6K1 and 84–5
  in synaptic tagging and capture 272–5, 273*f*, 274*f*
EC-CA2 synapses. *See* entorhinal cortical CA2 synapses
eEF1A. *See* eukaryotic elongation factor 1A
eEF2. *See* eukaryotic elongation factor 2
eEF2K. *See* eukaryotic elongation factor 2 kinase
eFT508 437–8
EGFR. *See* endothelial growth factor receptor
egg-laying hormone (ELH) 137
eIF2. *See* eukaryotic initiation factor 2
eIF2B. *See* eukaryotic initiation factor 2B
eIF2α. *See* eukaryotic initiation factor 2α
eIF3. *See* eukaryotic translation initiation factor 3
eIF3E. *See* eukaryotic translation initiation factor 3E
eIF4A. *See* eukaryotic translation initiation factor 4A
eIF4B. *See* eukaryotic translation initiation factor 4B
elf4e 457–8, 462–3
  eIF4E sensitivity element (4ESE) 10–11
  eIF4E transporter protein (4E-T)
    ASD and 458
    4E-HP and 9
    in NSCs 405
    nuclear eIF4E and 11
    translation repression by 6*f*, 10
eIF4E. *See* eukaryotic translation initiation factor 4E
eif4ebp2 See eukaryotic translation initiation factor 4E-binding protein 2
eIF4F. *See* eukaryotic translation initiation factor 4F
eIF4G. *See* eukaryotic translation initiation factor 4G
80S ribosome
  ASD and 459
  formation of 105
  in IRES 134–5

EIF4A3. *See* eukaryotic initiation factor 4A3
EJC. *See* exon junction complex
ELAV. *See* embryonic lethal abnormal visual system
Elavl4. *See* embryonic lethal abnormal vision-like 4
electroconvulsive therapy 511–12
elongation phase 487–8
  addiction and 501
  ASD and 458–9
  in biogenesis 66
  eEF2 in 64
  NMDAR and 304
  regulation of 64, 540–1
Elongator complex (ELP) 156–7, 402
E-LTD. *See* early long-term depression
E-LTP. *See* early long-term potentiation
embryonic lethal abnormal vision-like 4 (Elavl4/HuD)
  addiction and 498
  in mRNA axonal localization 363–4, 366
  in neurogenesis 404
embryonic lethal abnormal visual system (ELAV)
  addiction and 498
  MEC-8 and 328
  in mRNA localization 241
  neuronal RNA granules and 245–6
emetine 251–2
endophilin 302
endoplasmic reticulum (ER)
  mRNA recruitment to 226–7
  stress granules and 246–8
  stress of
    AD and 536
    Elongator complex and 156–7
    4E-BP2 and 27–8
    PERK and 402
endothelial growth factor receptor (EGFR) 435–6
entorhinal cortical CA2 (EC-CA2) synapses 282
epigenetics, in LTP maintenance 285–8
  acetylation 286
  CREB 285
  DNA methylation 287–8
  HDAC inhibitors 286–7
  histone acetyltransferases 285–6
epilepsy, *RBFOX1* and 467
Epiregulin 435–6
epitranscriptomics 153–4
  in CNS 154–5
ER. *See* endoplasmic reticulum
eRF. *See* eukaryotic release factor
ERK1/2. *See* extracellular signal regulated kinase
escitalopram 519, 523–4
ESS-2 334–5
eukaryotic elongation factor 1A (eEF1A), ASD and 458–9
eukaryotic elongation factor 2 (eEF2)
  AD and 540–2, 541*f*
  addiction and 501
  dephosphorylation of 68
  eEF2K and 64–5
  in elongation phase 64
  memory consolidation with 65, 67–70
  in neurological diseases 70–3, 71*f*
  phosphorylation of 64–6, 540–2
    BDNF and 69
    in fear conditioning training 68
    in gustatory cortex 67–8
    ketamine and 70–1
    mTORC1 pathway 68–9
    serotonin 68–9
    stalled polysomes and 256
  proteostasis phosphorylation and 66–7
  regulation of 65
  S6K1 and 84
  synaptic plasticity with 65, 67–70
eukaryotic elongation factor 2 kinase (eEF2K)
  AD and 540–2, 541*f*
  in elongation phase 64–5
  memory formation with 67–70
  mTORC1 pathway and 68–9
  in neurological diseases 70–3, 71*f*
  NMDAR and 515–16
  proteostasis and 67
  regulation of 65, 66*f*
  S6K1 and 84
  synaptic plasticity with 67–70
  in synaptic transmission 69–70

eukaryotic initiation factor 2 (eIF2)
  dephosphorylation of 46f, 47
  in initiation phase 63–4
  ISR and 43
    translation initiation 44–7, 45f, 46f
  Met-tRNA$_i^{Met}$ and 599–600, 601f
  in NSCs 401–2
  pain and 434–5, 438–9
  phosphorylation of 46f, 45–6, 46–7, 438, 600, 601f
    LTM and 48
    LTP and 48
    Sal003 and 49–50
  VWM and 602
eukaryotic initiation factor 2B (eIF2B)
  ATF6 and 607–8
  DYRK1A and 471
  enhancement of 610–11
  function and regulation of 598–602, 601f
  in initiation phase 63–4
  ISR and translation initiation 44–5, 45f, 46f
  mutations in 54, 597–8, 599t, 602–6, 609
  PERK and 607–8
  structures of 45–6
    function and 602–6, 603f, 604f
  subunits of 602–3, 603f
  VWMD and 54, 597–8, 599t
eukaryotic initiation factor 2α (eIF2α)
  in AD 534–7, 535f
  in addiction 495–7
  dephosphorylation of 196–7, 496
  in diabetic peripheral neuropathy 439
  pain and 438–9
  phosphorylation of 63–4, 299, 377–9, 438, 463, 496, 534, 536
    RBP regulation and 575–6
eukaryotic initiation factor 4A3 (EIF4A3) 366–7
eukaryotic release factor (eRF) 460
eukaryotic translation initiation factor 3 (eIF3), as m6A readers 168–9
eukaryotic translation initiation factor 3E (eIF3E), in translation initiation 132, 133f
eukaryotic translation initiation factor 4A (eIF4A)
  BC RNAs and 108
  in cap-dependent translation 299
  as RNA helicases 106–7
  in translation initiation 106, 132, 133f
eukaryotic translation initiation factor 4B (eIF4B)
  BC RNAs and 108
  phosphorylation of 109–11, 110f
  S6K1 and 84
  in translation initiation 106
eukaryotic translation initiation factor 4E (eIF4E)
  addiction and 492
  Alzheimer's disease and 7–8
  APP and 138–9
  ASD and 412, 457–8
  in cap-dependent translation 299, 374–5
  in chemotherapy-induced peripheral neuropathy 437–8
  fragile X syndrome and 7–8
  imaging of activity of 13
  IRES and 136–7
  Maskin and 197–8
  MDD and 512–15, 514f
  in NSCs 401
  in nucleus 10–11, 13
    regulation of 6f, 11–12
  pain and 434–7
  in RAN translation 6f, 12
  regulation of
    eIF4E-binding proteins in 5, 6f, 24
    eIF4G interaction 3–4, 8–9, 401, 435
    by 4E-T 6f, 10
    GCN2-ATF4 signaling pathway 6f, 12
    by microRNAs and 4EHP 6f, 9–10
    phosphorylation 5–8, 6f
  TOR and 306–7
  translation initiation with 3, 23, 63–4, 132, 133f
eukaryotic translation initiation factor 4E-binding protein 1 (4E-BP1)
  brain distribution of 25
  circadian rhythms and 30
  gene regulation by 26–7
  MDD and 512–15, 514f
  mTORC1 and 24
  phosphorylation of 24–5

eukaryotic translation initiation factor 4E-binding protein 2 (4E-BP2) 26–32, 457–8
- in autism and synaptic transmission 28–9
- autophagy and 27–8
- in axon elongation 28
- brain distribution of 25
- deamidation of 25–6
- in dentate gyrus 7
- ER stress and 27–8
- gene regulation by 26–7
- in memory formation 31–2
- phosphorylation of 25

eukaryotic translation initiation factor 4E-binding protein 3 (4E-BP3), brain distribution of 25

eukaryotic translation initiation factor 4E-binding proteins (4E-BPs)
- AD and 537–8
- anti-psychotic treatments with 33
- ASD and 457–8
- biochemical considerations of 23–4
- in brain development 27–34
- brain distribution of 25
- in cap-dependent translation 374–5
- circadian rhythm control with 29–30
- eIF4E regulation by 5, 6*f*, 24, 401, 457–8
- homeostatic plasticity and 307–8
- in memory formation 30–2
- mTORC1 and 490–5, 491*f*
- in neurodegenerative diseases 33–4
- in neurodevelopmental disorders 27–34
- pain and 435–6
- phosphorylation of 24–5, 63–4, 196–7, 299–300, 375
- in psychiatric disorders 32–3
- S6K1 and 86–7
- in synaptic plasticity 30–2

eukaryotic translation initiation factor 4F (eIF4F), protease 2A and 134–5

eukaryotic translation initiation factor 4G (eIF4G)
- eIF4E interaction with 3–4, 8–9, 401, 435
- protease 2A and 134–5
- in translation initiation 106, 132, 133*f*

everolimus 517–18

exc-7 328, 330

exercise, for MDD 521

exon junction complex (EJC), mRNA transport particles and 248–9

extracellular signal regulated kinase (ERK1/2)
- CREB activation 270–1
- pain and 436–7
- S6K1 and 83–5
- synaptic cooperation and competition and 275–6, 277*f*

familial amyotrophic lateral sclerosis (fALS) 553. *See also* amyotrophic lateral sclerosis

Far Upstream Element-Binding Protein 2 (FUBP2) 366

fat mass and obesity associated demethylase (FTO) 167–8, 170–2

fear conditioning training
- eEF2 phosphorylation in 68
- m6A and 172
- S6K1 and 89–90

fibroblast growth factor 2 (FGF-2), IRES and 138

5hmC. *See* 5-hydroxymethylcytosine

5mC. *See* 5-methylcytosine

FK506-binding protein 12 (FKBP12), AD and 539–40

flp-1 332

fluoxetine 520

FMRP. *See* fragile X mental retardation protein

forskolin 48

40S ribosome
- ASD and 459
- in initiation 298
- in IRES 133*f*, 134–5
- ternary complex and 298–9

43S preinitiation complex (PIC)
- formation of 105–6, 298
- recruitment of 132, 133*f*, 434–5
- ternary complex and 298–9

48S initiation complex, formation of 105

4E-BP1. *See* eukaryotic translation initiation factor 4E-binding protein 1

4E-BP2. *See* eukaryotic translation initiation factor 4E-binding protein 2
4E-BP3. *See* eukaryotic translation initiation factor 4E-binding protein 3
4E-BPs. *See* eukaryotic translation initiation factor 4E-binding proteins
4EGI-1 8
4E-homologous protein (4EHP), eIF4E regulation by 6*f*, 9–10
4ESE. *See* eIF4E sensitivity element
4E-T. *See* eIF4E transporter protein
fragile X mental retardation protein (FMRP) 462–3
  activity-dependent regulation with 227–8
  addiction and 497–8
  ASD and 461–3
  Ccnd2 and 409–10
  dephosphorylation of 228
  Fmr1 409–11, 461–3, 497–8
  functional absence of 107
  FXS and 107, 155, 170, 255
  interactions with 363
  m6A and 170
  in major depressive disorder 522
  miR-132 and miR-125b 225
  miR-181d 225–6
  miRNA regulation by 221–2
  MOV10 and 222–3, 223*f*
  mRNA localization with 227
  neuronal RNA granules and 245–6
  P bodies, RISC particles and 250–1
  phosphorylation of 109, 110*f*, 228
  RNA 107
  S6K1 and 84–5, 90–1
  stalled polysomes and 255, 462
  as translational suppressor 369–71
fragile X syndrome (FXS)
  BC RNA and 112
  dysregulated translation in 462–3
  eIF4E phosphorylation and 7–8
  FMRP absence and 107, 155, 170, 255
  4EGI-1 and 8
  JAKMIP1 in 466
  mTOR and 410–11
  RAN translation and 12
  S6K1 and 86–7, 90–1
fragile X tremor ataxia syndrome (FXTAS) 568
  RBP modulation of RAN translation in 568–9
frontotemporal dementia (FTD)
  FUS
    knockdown 558–9
    miRNA machinery and 559–60
    mRNA translational targets 560–1
    reduced levels of synthesized proteins in 559
  PERK inhibition in 53–4
  RBPs in 553–64
    aggregates and mutants 553–4
    Ataxin-2 561, 563–4
    hnRNPA2/B1 and A1 561–2
    MATR3 561–2
    RAN translation modulation in 568–9
    TIA1 561–3
  TDP-43
    CAMTA1 and DENND4A translation enhancement 557–8
    Futsch/MAP1B and Hsc70-4/HSPA8 mRNA repression 556–7
    motor-neuronal cell translation enhancement 557
    mRNA-specific translational repression 555–6
    ribosome-mRNA complexes and 555
    RNA regulation and 554–5
  treatment for 536
FS-115 91–2
FTD. *See* frontotemporal dementia
FTO. *See* fat mass and obesity associated demethylase
FUBP2. *See* Far Upstream Element-Binding Protein 2
functional specificity, alternative splicing by RBPs and 324–36, 327*f*
fused in sarcoma (FUS)
  in ALS 323–4, 553–4
  altered translation in 552
  in FTD 553–4
  gene silencing and 559–60
  knockdown 558–9
  in miRNA regulation 221–2, 224

mRNA translational targets of 560–1
reduced levels of synthesized proteins and 559
Futsch/MAP1B mRNA repression 556–7
FXS. *See* fragile X syndrome
FXTAS. *See* fragile X tremor ataxia syndrome

G3BP1. *See* Ras GTPase-activating protein binding protein 1
G3BP2. *See* Ras GTPase-activating protein binding protein 2
G3BPs. *See* RAS GTPase-activating protein binding proteins
G9a/GLP 285–6
GABABR. *See* gamma-aminobutyric acid B
GABAergic synapses 338*f*, 339
Galloway-Mowat syndrome 157
gamma-aminobutyric acid B (GABABR) 522
Gap43 mRNA 364, 366
GAR-2 333
GC. *See* gustatory cortex
GCN2. *See* general control nonspecific 2
GCN2-ATF4 signaling pathway 6*f*, 12
GDI-dissociation factor (GDF) 599–600, 603
GDP exchange factor (GEF), pain and 438
GEF. *See* guanine nucleotide-exchange factor
Gemin3 332
gene expression
control of 105
LTP and 270–1
miRNA regulation of 218
NSC differentiation and 398
RNA modifications and 154–5
general control nonspecific 2 (GCN2)
AD and 534
eIF2 phosphorylation 45–6, 46*f*, 48
eIF2α phosphorylation 299, 377–8
pain and 438
gene regulation, by 4E-BP1 and 4E-BP2 26–7
gene silencing, FUS and 559–60
germline development 2 poly (A) polymerase (Gld2) 197–8
GFAPδ. *See* glial fibrillary acidic protein δ
Gld2. *See* germline development 2 poly (A) polymerase
glial fibrillary acidic protein δ (GFAPδ) 608
GLO-1 339
GLO-4 339
GluA1 subunits
ASD and 464
calcium and synthesis of 304
TTX and 303–4
GluA2 subunits
AMPARs with 271–2
LTD and 302
TTX and 303–4
GluN2B-specific antagonist 517
GluR1 515
glutamate
eEF2 regulation by 65
local dendritic protein synthesis and 283–4
TrkB and 137–8
glutamate receptor 4 (GRM4) 522–3
glycine-tryptophan repeat (GW) protein, P bodies, RISC particles and 250–1
glycogen synthase kinase 3β (GSK-3 β)
4E-BP1 and 24–5
ketamine with 520–1
S6K1 and 83–4, 86–7
GRLD-1 333–4
GRM4. *See* glutamate receptor 4
growth hormone, CPEB1 and 200
GRP78. *See* BiP
GSK-3 β. *See* glycogen synthase kinase 3β
GSK2606414 536
guanabenz 53
guanine nucleotide-exchange factor (GEF) 599–600, 603–4
gustatory cortex (GC), eEF2 phosphorylation in 67–8
GW protein. *See* glycine-tryptophan repeat protein

H3K9me3 341–2
H89 2HCl 92
haloperidol
4E-BP and 33
S6K1 and 86–7
hamartin 537–8
HATs. *See* histone acetyltransferases
hbl-1 338–9
HD. *See* Huntington's Disease
HDAC inhibitors. *See* histone deacetylase inhibitors

Hebbian plasticity 297–8
  protein translation and 300–2
    LTD and 301–2
    LTP and 300–1
hematopoietic stem cells 398–9
heme-regulated inhibitor (HRI)
  AD and 534
  eIF2 phosphorylation 45–6, 46*f*
  eIF2α phosphorylation 299, 377–8
heterogeneous nuclear ribonucleoprotein A1 (hnRNP A1) 561–2
heterogeneous nuclear ribonucleoprotein A2 (hnRNP A2)
  ALS/FTD and 561–2
  dendritic RNA 115–17
  ID4 and 119
  mRNA transport particles and 248–9
heterogeneous nuclear ribonucleoprotein B1 (hnRNP B1) 561–2
heterogeneous nuclear ribonucleoprotein C1 (hnRNP C1), XIAP and 142
heterogeneous nuclear ribonucleoprotein K (hnRNP K)
  FMRP and 221
  MOR and 141
heterogeneous nuclear ribonucleoprotein Q (hnRNP Q), neuronal RNA granules and 245–6
heterogeneous nuclear ribonucleoprotein R (hnRNP R), RBP transport of 363
heterogeneous nuclear ribonucleoprotein U (hnRNP U), neuronal RNA granules and 245–6
hippocampal subfield CA1
  emetine and 283
  LTD at 301–2
hippocampal subfield CA2 282–3
histone acetyltransferases (HATs), in long-term memory 285–6
histone deacetylase (HDAC) inhibitors, in long-term memory 286–7, 287*f*
histone methyltransferases (HMTs) 285–6
HIV1 mRNA 562
HMTs. *See* histone methyltransferases
HNK. *See* (2S,6S;2R,6R)-hydroxynorketamine
hnRNP A1. *See* heterogeneous nuclear ribonucleoprotein A1
hnRNP A2. *See* heterogeneous nuclear ribonucleoprotein A2
hnRNP B1. *See* heterogeneous nuclear ribonucleoprotein B1
hnRNP C1. *See* heterogeneous nuclear ribonucleoprotein C1
hnRNP K. *See* heterogeneous nuclear ribonucleoprotein K
hnRNP Q. *See* heterogeneous nuclear ribonucleoprotein Q
hnRNP U. *See* heterogeneous nuclear ribonucleoprotein U
homeostasis. *See also* proteostasis; synaptic homeostasis
  energy 91–2
  glucose 89
  ISR and 47
  metaplasticity and 276–8
  of NSCs 399
  SAM 160
homeostatic plasticity 297–8
  at neuromuscular junction 305–8
    postsynaptic 4E-BP 307–8
    postsynaptic TOR 306–7
  protein translation and 302–5
  in vertebrate CNS
    postsynaptic compensation 303–5
    presynaptic compensation 305
homoharringtonine 251
HOTAIR. *See* HOX transcript antisense RNA
HOXA9 11–12
HOX transcript antisense RNA (HOTAIR) 165
HRI. *See* heme-regulated inhibitor
Hsc70-4/HSPA8 mRNA repression 556–7
5-HT. *See* serotonin
H-Twist 142–3
HuD. *See* embryonic lethal abnormal vision-like 4
human antigen R (HuR)
  addiction and 498
  in miRNA regulation 224
  neuronal subtype identity and 406
Huntington's Disease (HD)
  ISR in 54
  mechanisms for 565

MID1 complex in 565–6
metformin and 566–7
RBPs in 564–8
therapy for 565
HuR. *See* human antigen R
5-hydroxymethylcytosine (5hmC) 163
(2S,6S;2R,6R)-hydroxynorketamine (HNK) 521
hyperalgesia 428*f*, 429
opioid-induced 440–1

I. *See* inosine
IAP family. *See* Inhibitor of Apoptosis family
ID element 4 subtype (ID4) 119
IGF2BPs. *See* IGF2 mRNA-binding proteins
IGF2 mRNA-binding proteins (IGF2BPs) 168–9
imipramine 520
importin 8 6*f*, 11
Importin-β1 363–4
indirect neurogenesis 397–8
induced pluripotent stem cells (iPSCs), neurodevelopmental disorders and 411
Inhibitor of Apoptosis (IAP) family 141–2
Inhibitory post-synaptic currents (IPSC) 338–9
initiation phase. *See also* internal ribosome entry site mechanism
ASD and 457–8
cap-dependent 3–4, 132, 133*f*
cap-independent 134–5
eIF4E and 63–4
eIF4G with 8–9
internal 134–5
ISR, eIF2 and 44–7, 45*f*, 46*f*
mechanism of 105–6, 487–8
miRNA in 9
mTORC1 and 490–2, 491*f*
pain and 434
rate-limiting 132
regulation in 105
RNA helicases in 106–7
injury 429
inosine (I)
A-to-I editing and 156
in miRNA 163–4
in mRNA 162–6
regulation of 173
inositol 1,4,5-triphosphate receptor type 1 114–15
inositol-requiring enzyme 1 (IRE1), AD and 534–6
insulin
AD and 539–40
S6K1 and 92
integrated stress response (ISR)
cell type and mechanism of action of 51–2
in cognitive disorders 44, 52–3, 53*f*
eIF2 and translation initiation 44–7, 45*f*, 46*f*
inactivation of
genetic and pharmacological 48–9
LTP and 48
LTM formation and 44, 47–50
activation 49–50
mechanism of 43
mutations in components of 54–5
in neurodegenerative disorders 52–4, 53*f*
synaptic plasticity and 50–1
translational reprogramming by 44–7
intermediate progenitor cell (IPC) 397–8
mTORC1 in 401
protein synthesis rate in 399
intermediate term facilitation (ITF) 252
internal initiation 134–5
internal ribosome entry site (IRES) mechanism
as alternative 133*f*, 134
in axons 369
cellular 136
in CNS 136–8
neurodegenerative disease with
Alzheimer's disease 138–9
Parkinson disease 139–41
neuronal apoptosis control by 141–3
opioid stimulation of 141
pain and 433–4
translation initiation mode 134–6
inverse synaptic tagging 280–2, 282*f*
IPC. *See* intermediate progenitor cell
IPSC. *See* Inhibitory post-synaptic currents
iPSCs. *See* induced pluripotent stem cells
IRE1. *See* inositol-requiring enzyme 1

IRES mechanism. *See* internal ribosome entry site mechanism
IRES Trans-Acting Factors (ITAFs) 134–5
  cellular need for 136
  hnRNP C1 as 142
  hnRNP K as 141
ISR. *See* integrated stress response
ISR inhibitor (ISRIB) 48–9, 53–4
  addiction and 496
  eIF2B and 610–11
  pain and 438
ITAFs. *See* IRES Trans-Acting Factors
ITF. *See* intermediate term facilitation

JAKMIP1. *See* Janus kinase and microtubule-interacting protein 1
JAN-384 540–1
Janus kinase and microtubule-interacting protein 1 (JAKMIP1) 465–7

K78E mutation 459–60
keratinocytes 429
ketamine 33
  antidepressants with 520–1
  eEF2 phosphorylation and 70–1
  for MDD 511–13, 515, 519
  mechanism of 521
  mTORC1 and 515–17
  S6K1 and 89
KH-type splicing regulatory protein (KHSRP) 366–8
KIN-20 335–6
kinesin-1 466
kinesin5a (KIF5a) 245–6

late long-term depression (L-LTD), cross-tagging 280, 281*f*
late long-term potentiation (L-LTP)
  cross-tagging 280, 281*f*
  decay of 271
  degradation and synthesis balance for 284–5
  4EGI-1 and 8
  ISR and 50–1
  local dendritic protein synthesis 283–4
  mTORC1 in 31
  PKMζ in 271, 273
  protein synthesis and 4, 269, 301
  S6K1 and 84–5
  in synaptic tagging and capture 272–5, 273*f*, 274*f*
learning
  novel taste 67–8
  protein synthesis in 217
  synaptic plasticity and 269
  translation and 82
lep-2 341
lep-5 341
let-7 338–41, 344
leucine-rich pentatricopeptide repeat protein (LRPPRC) 10–11
leucine-rich repeat kinase 2 (LRRK2) 24–5
  cap-dependent translation and 309–10
  in PD 571
Lewy bodies, IRES and 139–40
Lim-domain containing protein kinase 1 (LIMK1) 225
lin-4 339–40
lin-28 339–41
lin-41 344
linc-60 341–2
linc-73 341–2
lincRNA. *See* long intergenic coding RNA
LINE-1 222
liquid-liquid phase separation (LLPS)
  neuronal RNA granules and 246–8
  of RBPs 369–71, 573–5
  stress granules and 249–50
lithium, ketamine with 520–1
LLPS. *See* liquid-liquid phase separation
L-LTD. *See* late long-term depression
L-LTP. *See* late long-term potentiation
lncRNA. *See* long noncoding RNA
local translation
  activity-dependent regulation of 227–8
  control of 239
  cytoplasmic polyadenylation and 196–9
  dendritic 283–4
  function of 431
  mEPSC and 304
  of mRNA 114–15, 131–2
    control of 226–7
  pain and 430–3, 432*f*

prevalence of 430–1
principles of 240
serotonin and 243–4
at synapse 107
persistence of change and 195–6
synaptic plasticity and 194–5
stimulus-dependent 243–4
long intergenic coding RNA (lincRNA), in locomotion and vesicle trafficking 341–2
long noncoding RNA (lncRNA)
modification of 165–6
in neuronal RNAs 155
in sexual dimorphic neuronal gene expression 340–1
long-term depression (LTD). *See also* early long-term depression; late long-term depression
AD and 536
cross-tagging 280, 281*f*
4E-BP2 in 31–2
Hebbian plasticity and 301–2
induction of 300
ISR and 44
memory and 193
mGluR-LTD and 243–4
process of 297–8
protein synthesis and 301–2
S6K1 and 85
synaptic plasticity and 50–1
long-term facilitation (LTF)
eEF2/eEF2K pathway and 68–9
memory and 193
protein synthesis inhibitors and 194–5
long-term memory (LTM) 43–4
behavioral tagging 279–80
cap-dependent translation for 13
DNA methylation in 287–8
4EGI-1 and 8
HATs in 285–6
HDAC inhibitors 286–7
ISR and formation of 44, 47–50
activation of 49–50
genetic and pharmacological inhibition 48
maintenance of 4
miR-134 and 225
persistence and 195–6
PKR and 48–9, 51–2
stabilization of 195–6
synapse-based model of 194–5
synaptic cooperation and competition and 275–6, 277*f*
translation in 4
long-term potentiation (LTP). *See also* early long-term potentiation; late long-term potentiation
AD and 536, 540–1
aging and 72
cap-dependent translation for 13
cross-tagging 280, 281*f*
decay of 271–2
DNA methylation and 287–8
epigenetics in 285–8
acetylation 286
CREB 285
DNA methylation 287–8
HDAC inhibitors 286–7
histone acetyltransferases 285–6
gene expression and 270–1
Hebbian plasticity and 300–1
induction of 300
ISR and 44
inactivation 48
maintenance of 282–3
mGluRs in 276–8
miR-134 and 225
MNKs in 7
post-synaptic changes and 270
process of 297–8
protein synthesis for 4, 270–2
reversal of 271
S6K1 and 84–5
synaptic plasticity and 50, 269
temporally distinct phases of 50–1
LRPPRC. *See* leucine-rich pentatricopeptide repeat protein
LRRK2. *See* leucine-rich repeat kinase 2
LSM14 343–4
LTD. *See* long-term depression
LTF. *See* long-term facilitation
LTM. *See* long-term memory
LTP. *See* long-term potentiation
LY2584702 92

m1A. *See* N1-methyladenosine
m^{5}C. *See* cytosine-5 methylation
m6A. *See* N6-methyladenosine
m7GpppN cap structure 3, 23, 132, 133*f*, 161–2, 299, 374–5, 433–4
macrocephaly
  pten and 468–9
  PTEN and 468
major depressive disorder (MDD)
  BCCAs and 521
  biomarkers for 523–5
  exercise for 521
  FMRP in 522
  miRNAs and 522–3
  mRNA translational repression factors for 522–3
  mTOR and 512–15, 514*f*
  mTORC1-regulated protein synthesis
    brain region of 518–19
    promotion of 517–18
    as treatment 515–17
  symptoms of 511–12
  synapse and 512
MAK kinase. *See* mitogen-activated kinase
Malat1. *See* metastasis associated lung adenocarcinoma transcript 1
mammalian/mechanistic target of rapamycin (mTOR)
  activation of 375
  addiction and 495, 501
  aging and 72
  ASD and 410–11, 453–6, 472
  cap-dependent translation 299–300
  eIF4E control by 5, 6*f*
  homeostatic plasticity and 305
    4E-BP and 307–8
    at NMJ 306–7
  local dendritic protein synthesis and 283–4
  MDD and 512–15, 514*f*
  MeCP2 and 473–4
  mRNA inhibition by 26
  neurodevelopmental disorders and 410–11
  pain and 435–6
  phosphorylation of 196–7
  PTEN and 470
  regulation of 375–7, 376*f*, 490–2
  substance abuse and 32
  TSC complex and 469
mammalian/mechanistic target of rapamycin complex 1 (mTORC1)
  AD and dysregulation of 537–40
  in addiction 490–5, 491*f*, 493*t*
  amino acids and 521
  antidepressants and 519–20
  in axon regeneration 375
  circadian rhythms and 30
  cortical malformations and 27–8
  depressive disorders and 33
  eEF2 and eEF2K phosphorylation 68–9, 540–1
  eEF2K regulation by 65, 66*f*
  4E-BPs and 24, 401
  initiation phase and 490–2, 491*f*
  ketamine and 515–17
  MDD and 512–15, 514*f*
    brain region of 518–19
    promotion of 517–18
    as treatment 515–17
  in memory formation 30–1
  mRNA translation and 26
  neuronal function and 538–9
  in NSCs 401–2
  pain and 435–6
  in postnatal development 28
  Raptor and 24
  REDD1 and 517
  S6K1 and 82–5, 90–2
  serotonin and 243–4
  substance abuse and 32
mammalian/mechanistic target of rapamycin complex 2 (mTORC2), in addiction 495
mammalian site 20-like kinase (MST1) 5–7
MAP1B. *See* microtubule-associated protein 1B
MAP2. *See* microtubule-associated protein 2
MAPK. *See* mitogen-activated protein kinase
MAP kinase-interacting kinases (MNKs)
  in eIF4E regulation 5–8, 6*f*
  in long-term potentiation 7
  pain and 437

Maskin 197–8
MAT2A. *See* methionine adenosyltransferase 2A
maternal and embryonic leucine zipper kinase (MELK) 110–11
MATR3. *See* Matrin-3
Matrin-3 (MATR3) 561–2
matrix metalloproteinase 2 (MMP-2) 437
matrix metalloproteinase 9 (MMP-9) 435–7
MBL-1 328–9
MBP. *See* myelin basic protein
mcm^5 modification. *See* 5-methoxycarbonylmethyl modification
mcm5s2U. *See* 5-methoxycarbonylmethyl thiouridine
MDD. *See* major depressive disorder
mec-2 328
MEC-8, mechanosensation with 326–9
MeCP2. *See* methyl-CpG binding protein 2
MeCP2 473–4
MEF-2 337
MEHMO. *See* mental impairment, epilepsy, hypogonadism, microcephaly, and obesity
MEL-46 333
MELK. *See* maternal and embryonic leucine zipper kinase
memantine 516
memory function. *See also* long-term memory
  aging and 72
  CPEB role in 199–203
    CPEB1 200
    CPEB2 201
    CPEB3 202
    CPEB4 203
    invertebrate 203–6
    persistence of 204–6
  eEF2 and eEF2K 65, 67–70
  forms of 43–4
  4E-BPs in 30–2
  NMDAR in 7
  PABP in 198–9
  phosphorylation and 196–7. *See also* long-term memory
  protein synthesis in 217, 270–2
  synaptic plasticity and 269
  translation and 82
mental impairment, epilepsy, hypogonadism, microcephaly, and obesity (MEHMO) 54–5
mEPSC. *See* miniature excitatory postsynaptic current
messenger RNA (mRNA). *See also* translation
  axonal localization motifs 362–3
  axon localization of 360–1
    regulation of 361–5, 361*f*
    stability regulation of 365–9
    storage of 372–4
  circadian rhythms and 30
  circularization of 198–9
  cytoplasmic polyadenylation of 196–9
  eIF2 phosphorylation and 46–7, 46*f*
  FMRP and 221
  4E-T and 10
  half-lives of 365–6
  IRES and 134–5
  local translation of 114–15, 131–2
    control of 226–7
  of MEC-8 327–8
  in memory formation 30–1
  miRNA and 217–18, 219*f*
  MNKs and translation of 5–7
  modification of 132, 161–2
    5mC and 5hmC 163
    inosine 162–6
    m6A 162, 166–71
  mTORC1 translation of 26
  mTOR inhibition of 26
  neuronal process abundance of 241
  neuron specificity of 241
  nuclear eIF4E and 10–11
  pain and 433–4
  post-transcriptional regulation of 241–3, 242–243*f*
  RBPs in regulation of 220–5, 241
    in NSCs 404–5
  recognition of 132
  ribosome translation of 3–4
  structure of 433–4
  TDP-43 repression of 555–6

messenger RNA (mRNA). (*Continued*)
translation of
in brain 4, 12–13, 23
gene expression control and 105
phases of 3, 63–4, 105, 298, 374–5, 534
regulation of 196–9
schizophrenia and 410–11
in transport, quantity of 252–3
metabotropic glutamate receptor-mediated LTD (mGluR-LTD)
inverse synaptic tagging 280–2, 282*f*
ISR and 51–2
local translation and 243–4
Raptor and 31–2
S6K1 and 85
metabotropic glutamate receptors (mGluRs)
BC RNA and 111
LTD and 301–2
in LTP 276–8
metalloprotein 9 (MMP-9) 7–8
metaplasticity phenomenon
L-LTP decay and 271
plasticity and synaptic competition with 276–9, 278*f*, 279*f*
metastasis associated lung adenocarcinoma transcript 1 (Malat1) 165
metformin 566–7
methamphetamines. *See* amphetamines
methionine adenosyltransferase 2A (MAT2A) 160
methionyl tRNA (Met-tRNA$_i^{Met}$)
eIF2 and 599–600, 601*f*
inhibition of 299
in NSCs 401–2
pain and 434–5
ternary complex and 298–9
in translation 298
5-methoxycarbonylmethyl (mcm^5) modification 402
5-methoxycarbonylmethyl thiouridine (mcm5s2U) 156
methylation
of DNA 287–8
of tRNAs 156
methyl-CpG binding protein 2 (MeCP2) 473–4, 524
5-methylcytosine (5mC) 155, 163
methylglyoxal (MGO) 439
METTL3 166, 172–3
METTL14 166, 172–3
METTL16. 160
Met-tRNA$_i^{Met}$. *See* methionyl tRNA
mGluR-LTD. *See* metabotropic glutamate receptor-mediated LTD
mGluRs. *See* metabotropic glutamate receptors
MGO. *See* methylglyoxal
microcephaly
ADAT and 156
ribosomal proteins and 411–12
microRNAs (miRNA)
activity-dependent regulation of 227–8
ASD and 472–3
axon regeneration regulation 344
biogenesis of 217–18, 219*f*
factors in 218–20
in cancer 9
eIF4E regulation by 6*f*, 9–10
fused in sarcoma and 559–60
gene expression by 218
heterochronic pathway 339–40
MDD and 522–3
MeCP2 and 473–4
modification of 163–5
mRNA stability and 365–6
in nervous system development 225–6
in neuronal RNAs 155
RBPs in regulation of 220–5
FMRP 221–2
MOV10 222–5, 223*f*
RISC particles and 250–1
in RNA degradation 367–8
in synaptic plasticity 9–10
microtubule-associated protein 1B (MAP1B) 302, 369–71
microtubule-associated protein 2 (MAP2)
dendrite localization of 361–2
IRES and 136–7
local translation and 114–15
MICU-1 343–4
MID1 complex
HD and 565–6
metformin and 566–7
TDP-43 and 323–4
miniature excitatory postsynaptic current (mEPSC)

eEF2K function 69
4E-BP2 and 29
homeostatic plasticity and 303–5
local translation and 304
mir-1 337–8, 338*f*
mir-2 333
miR-16 367–8
miR-19a 367–8
mir-51 338*f*, 339
mir-84 338–9, 338*f*
miR-124 523
miR-125b 225
miR-132 225, 524
miR-134 225
miR-135 524–5
miR-135a 524–5
miR-181d 225–6
miR-199a 473–4
miR-338 367–8
miR-1202 524
mir-5550 341–2
miRISC. *See* miRNA-induced silencing complex
miRNA. *See* microRNAs
miRNA-induced silencing complex (miRISC) 9
miRNA Recognition Element (MRE) 218
mitogen-activated kinase (MAK kinase), S6K1 and 83–4
mitogen-activated protein kinase (MAPK)
CREB activation 270–1
eEF2K regulation by 65, 66*f*
pain and 435–7
S6K1 and 83–4
MKRN2 341
MMP-2. *See* matrix metalloproteinase 2
MMP-9. *See* matrix metalloproteinase 9; metalloprotein 9
MNKs. *See* MAP kinase-interacting kinases
MOG. *See* myelin oligodendrocyte glycoprotein
MOR. *See* mu-opioid receptor
morphine. *See also* opioids
ATF4 and 497
brain areas and 490*f*
mTORC1 signaling and 493*t*
rapamycin and 492–5
Morris Water Maze (MWM)
4E-BP1 and 31
ISR and 68
S6K1 and 89
motor neuron
developmental synapse remodeling of 338–9, 338*f*
S6K1 and 539–40
SMN-1/SMN regulation of function of 327*f*, 331–3
MOV10 219–20, 222–5, 223*f*
MRE. *See* miRNA Recognition Element
mRNA. *See* messenger RNA
mRNA co-immunoprecipitations (RIP) 363
mRNA transport particles
proteins of 248–9
shared components of 246–8, 247*f*
MST1. *See* mammalian site 20-like kinase
mTOR. *See* mammalian/mechanistic target of rapamycin
mTORC1. *See* mammalian/mechanistic target of rapamycin complex 1
mTORC2. *See* mammalian/mechanistic target of rapamycin complex 2
Munc19-1 369–71
mu-opioid receptor (MOR), IRES and 141
mutations
ASD and 412, 458
ATF4 and 54–5
BCKDK 471–2
CELF6 464–5
in eEF1A 458
in eef1a2 458–9
in eIF2B 54, 597–8, 599*t*, 602–6, 609
FMR1 461–3
in ISR components 54–5
K78E 459–60
MeCP2 473–4
PTEN 468–9
R132H 606–7
RBFOX1 and 467–8
in ribosomal proteins 459
RNG105 463–4
RPS23 460
SLC7A5 471–2
in *SOD1* 553
TDP-43 553–4
UPF3B 460–1

MWM. *See* Morris Water Maze
myasthenia gravis 305–6
Myc family
  IRES and 143
  NSC protein synthesis rate and 399–400
myelin basic protein (MBP)
  local translation of 240–1
  OPC maturation 608–10
myelin oligodendrocyte glycoprotein (MOG) 608–10
myelin vacuolation 608
myeloid cell leukemia 1 protein 110–11

N1-methyladenosine (m1A)
  FTO and 167–8, 171–2
  modifications with 173–4
N6-methyladenosine (m6A)
  brain abundance of 155
  in CNS from development to regeneration 166–71
  erasers 167–8
  modifications with 162
    lncRNA 165
    miRNA 164
    regulation of 172–3
  readers/anti-readers 168–9
  in regulation of local translation in neurons 169–71
  writer complex 166–7
N6-threonyl-carbamoyl-adenosine (t6A) 157
NAB-1 328–9
nAChR subunits. *See* nicotinic acetylcholine receptor subunits
Nanos1 405
Ncl. *See* Nucleolin protein
ncm^5 modification. *See* 5-carbamoylmethyl modification
ncRNA. *See* non-coding RNA
NDD. *See* neurodevelopmental disorders
NECs. *See* neuroepithelial cells
nerve growth factor (NGF)
  eIF4E expression and phosphorylation 7–8
  FTO and 170
  nociceptors and 431–3
  Schwann cells and 429
Netrin 339–40
Netrin 1 371–2
Neurabin 87
neural stem cell (NSC)
  adult 399, 400*f*
  differentiation of 397–8
  4E-T in 405
  gene expression and 398
  global protein synthesis in 398–400
    cap-dependent translation, initiation factors and mTORC1 401–2
    rate of 398–9
    ribosome biogenesis 403
    tRNA and RNA modifications 402–3
  HuD and 404
  lineage progression of 399–400, 400*f*
  mTORC1 and self-renewal of 28, 401
  target-specific translational control of neurogenesis in 403–10
    asymmetric division and spatial control 408–10, 409*f*
    RBP-mediated 404–5
    for temporal genesis of neuronal subtypes 405–6, 407*f*
    uORFs 406–8
neurexin (*nrx-1*) 329–30, 337, 338*f*
neurodegenerative disorders. *See also* Alzheimer's disease; Parkinson disease; *specific disorders*
  eIF2α phosphorylation and RBPs in 575–6
  4E-BPs in 33–4
  with IRES 138–41
  ISR and 44, 52–4, 53*f*
  MID1 complex in 567–8
  pathogen exposure and age-related 576–7
  RAN translation in 12
  RBPs in 551
    noncanonical 572–3
  S6K1 in 90–2
  translation modeling in 577–8
neurodevelopmental disorders (NDD). *See also specific disorders*
  4E-BPs in 27–34
  iPSCs and 411
  mTORC1 and 27–8
  neurogenesis in 410–13
    cis-regulatory element alterations 413
    dysregulation of 410–11

RBPs 412–13
translational machinery perturbation 411–12
stalled polysomes and 257
UPF3B mutations and 461
neuroepithelial cells (NECs) 397–400
protein synthesis rate of 399–400
neurofibrillary tangles of hyperphosphorylated tau protein 139, 533–4
neurogenesis
adult NSCs during 399
direct 397–8
Elavl4 and 404
4E-T in 10
indirect 397–8
miRNA role in 225–6
in neurodevelopmental disorders 410–13
cis-regulatory element alterations 413
dysregulation of 410–11
RBPs 412–13
translational machinery perturbation 411–12
target-specific translational control of 403–10
asymmetric division and spatial control 408–10, 409*f*
RBP-mediated 404–5
for temporal genesis of neuronal subtypes 405–6, 407*f*
uORFs 406–8
transcriptome in 323–4
translational control in 400–3
cap-dependent translation, initiation factors and mTORC1 401–2
ribosome biogenesis 403
tRNA and RNA modifications 402–3
UNC-75/CELF in 327*f*, 329–31, 331*f*
neurogranin 115, 136–7
local translation and 114–15
Neuroguidin 197–8
neuroligins 457–8
genes for 26–7, 29
nlg-1 337, 338*f*
neurological disease
eEF2 and eEF2K in 70–3, 71*f*
Elongator complex and 156–7
S6K1 in 90–3
neuromuscular junction (NMJ), homeostatic plasticity at 305–8
LRRK2 and 309–10
postsynaptic 4E-BP 307–8
postsynaptic TOR 306–7
neuronal ankyrin isoform 335–6
neuronal RNA granules
ASD and 463
components of 245–6
discovery of 244–5
FMRP and 255
formation of 226
liquid-liquid phase separation and 246–8
P bodies compared with 250–1
RISC particles compared with 250–1
shared components of 246–8, 247*f*
stalled polysomes in 251–2
stress granules compared with 249–50
neurons 359
neuropathy, peripheral
chemotherapy-induced, eIF4E in 437–8
diabetic, eIF2α in 439
neuropeptide Y receptor 5 71–2
neuropsychiatric disorders. *See* psychiatric disorders
neurotrophin receptors, cellular functions and 137–8
neutral amino acid transporter 1 (SLC7A5) 471–2
NGF. *See* nerve growth factor
nicotine
eIF2α and 496
mTORC1 signaling and 493*t*
rapamycin and 492–5
nicotinic acetylcholine receptor (nAChR) subunits 337
nlg-1. *See* neuroligins
NMD. *See* nonsense-mediated decay
N-methyl-D-aspartate receptors (NMDAR)
AD and 541–2
antagonists of 517, 522
eEF2 phosphorylation 65
eIF4E phosphorylation and 7
local dendritic protein synthesis and 283–4
LTD and 301–2
in LTP decay 271–2
LTP-stimulation and 48

N-methyl-D-aspartate receptors (NMDAR) (*Continued*)
  MDD and 511–12
  metaplasticity phenomenon and 271
  mTORC1 and 515–16
  protein translation elongation and 304
NMJ. *See* neuromuscular junction
N-myc, IRES and 142–3
nociception 427. *See also* pain
nociceptors 427, 428*f*
  physiology of 429–30, 430*f*
  plasticity of 8, 427, 429–30
no-go decay, stalled polysome prevention of 256–7
non-coding RNA (ncRNA) 217–18
  axon localization of 368–9
  in synapse formation and function 336–42
    heterochronic miRNA pathway 339–40
    lincRNA locomotion and vesicle trafficking 341–2
    lncRNA in sexual dimorphic neuronal gene expression 340–1
    mir-1 337–8, 338*f*
    mir-51 338*f*, 339
    mir-84 338–9, 338*f*
nonsense-mediated decay (NMD)
  ASD and 460–1
  functions of 366–7
  smg-2 for 329–30
  stalled polysome prevention of 256–7
non-syndromic X-linked intellectual disability (NSXLID) 157
NORT. *See* novel object recognition test
novel object recognition test (NORT), BC RNA and 112
novel taste learning, eEF2 phosphorylation and 67–8
NOXA. *See* phorbol-12-myristate-13-acetate-induced protein 1
Nrn1 mRNA 366
nrx-1. *See* neurexin
NSC. *See* neural stem cell
NSXLID. *See* non-syndromic X-linked intellectual disability
Nucleolin protein (Ncl) 363–4
nucleus, eIF4E in 10–11, 13
  regulation of 6*f*, 11–12

OIH.o *See* opioid-induced hyperalgesia
oligodendrocyte precursor cells (OPCs) 597, 608–10
oligophrenin-1 (OPHN1) 52
  addiction and 497
OPCs. *See* oligodendrocyte precursor cells
OPHN1. *See* oligophrenin-1
opioid-induced hyperalgesia (OIH) 440–1
opioids, IRES stimulation by 141
Orb1 199, 204
Orb2 199, 204–6
oxygen, cap-dependent translation and 9
oxytocin 71–2

p70 riboosomal S6 kinase (S6K)
  phosphorylation of 299–300
  pten and 469
p70 ribosomal S6 kinase 1 (S6K1)
  AD and 537–40
  in behavior 88–90
  depression and 516–17
  historical timeline of research on 81–3
  MDD and 512–15, 514*f*
  mTORC1 and 490–2, 491*f*
  natural variation in 94
  in neurological disease 90–3
  pain and 436–7
  phosphorylation of 84
  reagents for research on 93–4
  signaling and synaptic plasticity 83–6
  steady state *versus* activity-driven translation 94–5
  structural plasticity of neurons and 86–8, 88*f*
  structure of 83
p70 ribosomal S6 kinase 2 (S6K2)
  in behavior 89
  historical timeline of research on 81–3
  mTORC1 and 490–2, 491*f*
  pain and 436–7
p300 285–6
p300/CBP associated factor (PCAF) 285–6
PABP. *See* poly(A)-binding protein
PAI1. *See* plasminogen activator inhibitor 1
pain 427, 428*f*
  AMPK and 439
  cap-dependent translation and 434–5

CNS translational controls
opioid-induced hyperalgesia 440–1
reconsolidation mechanisms 439–40
spinal modulation 440
eIF2 and 434–5, 438–9
eIF2α and 438–9
eIF4E and 434–7
in chemotherapy-induced peripheral neuropathy 437–8
initiation and 434
local translation and 430–3, 432*f*
mRNA structure and 433–4
physiology of 429–30, 430*f*
plasticity of 8, 427, 434
PAP. *See* poly (A) polymerase
parasympathetic nervous system, CPEB2 in 201–2
Parkin 570–1
Parkinson disease (PD)
IRES and 139–41
LRRK2 and 309–10
PERK inhibition in 53–4
RBP translational control in 569–72
DJ-1 571–2
LRRK2 571
PINK1 and Parkin 570–1
S6K1 and 92
tau IRES and 139
parkinsonism, 4E-BPs in 33–4
PARN. *See* poly (A) ribonuclease
paroxetine 519
P bodies
neuronal RNA granules compared with 250–1
shared components of 246–8, 247*f*
PCAF. *See* p300/CBP associated factor
PD. *See* Parkinson disease
PDCD4 84
PDK1. *See* phosphoinositide-dependent-kinase 1
pentylenetetrazole (PTZ) 339
peripheral neuropathy
chemotherapy-induced, eIF4E in 437–8
diabetic, eIF2α in 439
PERK. *See* PKR-like ER kinase
perturbed myelination 608
PF-4708671 82, 84–7, 91–2
phorbol-12-myristate-13-acetate-induced protein 1 (NOXA) 142–3
phosphate and tensin homolog (PTEN) 399, 468–70
phosphate-inositol 3 kinase (PI3K) 375–7, 376*f*
phosphatidyl inositol 3 kinase (PI-3K), S6K1 and 83–4
phosphoinositide-dependent-kinase 1 (PDK1)
AD and 537–8
S6K1 and 83–4
phosphorylation
of AGO2 220
of CPEB1 201
of CREB 270–1
of eEF2 64–6, 540–2
BDNF and 69
in fear conditioning training 68
in gustatory cortex 67–8
ketamine and 70–1
mTORC1 pathway 68–9
serotonin 68–9
stalled polysomes and 256
of eIF2 43, 45–7, 46*f*, 438, 600, 601*f*
LTM and 48
LTP and 48
Sal003 and 49–50
of eIF2α 63–4, 299, 377–9, 438, 463, 496, 534, 536
RBP regulation and 575–6
of eIF4B 109–11, 110*f*
of eIF4E 5–8, 6*f*
of FMRP 109, 110*f*, 228
of 4E-BPs 24–5, 63–4, 196–7, 299–300, 375
G3BP1 373
memory function and 196–7
of mTOR 196–7
of S6K 299–300
of S6K1 84
synaptic changes and 195–6
translational regulation with 108–11
of ZBP1 369–71
PI3K. *See* phosphate-inositol 3 kinase
PI-3K. *See* phosphatidyl inositol 3 kinase
PIC. *See* 43S preinitiation complex
PINK1. *See* Pten-induced kinase 1

piwi-interacting RNA (piRNA) 176
axon localization of 368–9
axon regeneration regulation 344
PKA. *See* protein kinase A
PKC. *See* protein kinase C
PKMζ. *See* protein kinase M-zeta
PKR. *See* double-stranded RNA-dependent protein kinase
PKR-like ER kinase (PERK)
AD and 534–7, 535*f*
eIF2B and 607–8
eIF2 phosphorylation 45–6, 46*f*
eIF2α phosphorylation 299, 377–8
LTM and 48–9
neurodegeneration and 53–4
in NSCs 402
pain and 438
TBI and 53
plasminogen activator inhibitor 1 (PAI1) 245–6
plasticity. *See also* associative plasticity; Hebbian plasticity; homeostatic plasticity; metaplasticity phenomenon; synaptic plasticity
of hippocampal subfield CA2 282–3
nociceptor 8, 427
pain 8, 427, 434
plasticity-related products (PRPs)
BT and 279–80
cross-tagging 280
L-LTP and 269
metaplasticity and 276, 278*f*
proteasomal degradation and 284–5
PSD and 274–5
synaptic cooperation and competition and 275–6, 277*f*
in synaptic tagging and capture 272–5, 273*f*, 274*f*
PML. *See* promyelocytic leukemia protein
PMSE. *See* polyhydramnios, megacephaly, symptomatic epilepsy
poliovirus 134–5
poly(A)-binding protein (PABP)
in mRNA circularization 198–9, 457
in translation initiation 106, 132, 133*f*
polyadenylation. *See also* cytoplasmic polyadenylation
alternative 327*f*
pain and 433–4
regulation of 335–6
in DRG 431–3
poly(A) tail 132, 133*f*, 196–9
localized modulation of 371–2
polyhydramnios, megacephaly, symptomatic epilepsy (PMSE) 91–2
poly (A) polymerase (PAP) 197
polypyrimidine tract-binding protein (PTB)
IRES and 137–8
in miRNA regulation 224
poly (A) ribonuclease (PARN) 197–8
polysomes. *See* stalled polysomes
postnatal development, mTORC1 in 28
postsynaptic compensation, homeostatic plasticity and 303–5
postsynaptic density (PSD), synaptic tagging and 274–5
post-transcriptional regulation
of mRNA 241–3, 242–243*f*
RNA modifications and 154–5
PP1. *See* protein phosphatase 1
PP2A. *See* protein-phosphatase 2A
Prader-Willi syndrome 160–1
PRDE-1 344
PRDM1 339–40
precursor-miRNA (pre-miRNA) 218
localization of 368
pre-miRNA. *See* precursor-miRNA
presenilin 1 (PSEN1), ISR and 52–3
presynaptic compensation, homeostatic plasticity and 305
prg-1 344
PRH. *See* proline-rich homeodomain
primary transcript of miRNA gene (pri-miRNA) 218
prion disease, ISR and 53–4
prions 204–5
progesterone 197–8
progressive myelin vacuolation 608
proline-rich homeodomain (PRH) 11–12
promyelocytic leukemia protein (PML) 11–12
protease 2A 134–5
proteasomal degradation, PRPs synthesis and 284–5

protein kinase A (PKA), synaptic stimulation and 270–1
protein kinase B (Akt)
G3BP1 phosphorylation 373
mTOR regulation 375–7, 376*f*
S6K1 and 82–4
protein kinase C (PKC), NMDAR and 271
protein kinase M-zeta (PKMζ)
AMPARs and 271–2
cross-tagging 280
in L-LTM 271
local translation of 114–15
in LTM stabilization 195–6
metaplasticity and 276, 278*f*
SP and 282
in synaptic tagging and capture 273, 273*f*, 274*f*
protein kinase R. *See* double-stranded RNA-dependent protein kinase
protein phosphatase 1 (PP1) 47
protein-phosphatase 2A (PP2A)
FMRP and 109, 110*f*
S6K1 and 85
protein synthesis in NSCs
control of 400–3
cap-dependent translation 401
eIF2 401–2
eIF4E 401
mTORC1 401
ribosome biogenesis 403
tRNA and RNA modifications 402–3
in neurodevelopmental disorders 410–13
cis-regulatory element alterations 413
dysregulation of 410–11
RBPs 412–13
translational machinery perturbation 411–12
rate of 398–9
adult 399
lineage progression 399–400
target-specific translational control of neurogenesis in 403–10
asymmetric division and spatial control 408–10, 409*f*
RBP-mediated 404–5
for temporal genesis of neuronal subtypes 405–6, 407*f*
uORFs 406–8
proteostasis
eEF2 phosphorylation and 66–7
ISR and 43–4
synapse organization and 193–4
Prox1 408
PRPs. *See* plasticity-related products
PSD. *See* postsynaptic density
PSD-95 112, 228, 555–6
PSEN1. *See* presenilin 1
pseudouridine 153–4, 174
pseudouridylate synthase (PUS3) 157
psychiatric disorders
eEF2 and eEF2K in 70–2
4E-BPs in 32–3
PTB. *See* polypyrimidine tract-binding protein
PTEN. *See* phosphate and tensin homolog
Pten-induced kinase 1 (PINK1) 570–1
PTZ. *See* pentylenetetrazole
PUM1. *See* Pumilio
Pum2 405–6
Pumilio (PUM1), in miRNA regulation 224
pur-alpha 245–6, 257
pur-beta 245–6
puromycin 4, 217, 244–5, 251–2
PUS3. *See* pseudouridylate synthase

queuosine 176

RAB-3 337, 339–40
RACK1 555
radial glial cells (RGCs) 397–8
protein synthesis rate of 399–400
RagA complex 437–8
RAIP motif 24
Ran binding protein 1 (*Ranbp1*) 363–4
RAN translation. *See* repeat-associated non-AUG translation
rapamycin
AD and 539–40
addiction and 492–5
aging and 539–40
antidepressant effects of 33
APP and 138–9

rapamycin (*Continued*)
LRRK2 and 309–10
for MDD 517–18
pten and 468
S6K1 and 91–2
serotonin and 243–4
substance abuse and 32
tsc and 470
Raphin1 54
rapid tRNA decay pathway (RTD) 158
Raptor
LTD and 31–2
mTORC1 and 24
RAPTOR protein 375
Ras GTPase-activating protein binding protein 1 (G3BP1) 168–9
phosphorylation of 373
Ras GTPase-activating protein binding protein 2 (G3BP2) 168–9
RAS GTPase-activating protein binding proteins (G3BPs)
neuronal RNA granules and 245–6
stress granules and 249–50
RA signaling. *See* retinoic acid signaling
RBFOX1. *See* RNA binding protein, Fox homolog
RBM15B. *See* RNA-binding motif protein 15B
RBPMS. *See* RNA-binding proteins with multiple splicing
RBPs. *See* RNA-binding proteins
rde-1 328
reconsolidation mechanisms in pain 439–40
regulated in development and DNA damage response 1 (REDD1) 33, 517
Rent1 366–7
repeat-associated non-AUG (RAN) translation
eIF4E in 6*f*, 12
RBPs and 552, 568–9
in FXTAS 568–9
retinoic acid (RA) signaling, GluA1 synthesis and 304
retroposition 118
retroposons 118
Rett syndrome 411, 473–4
RGCs. *See* radial glial cells
ribonucleases (RNase) 365–6
ribonucleic acids (RNAs). *See also* brain cytoplasmic RNA; dendritic RNAs; microRNAs; mRNA; transfer RNAs; *specific types*
adducts 176
classes of 155–6
FMRP 107
miRNA degradation of 367–8
modification of 153
5mC and 5hmC 163
in CNS 154–5
inducible and erasable 171–4
inosine 162–6, 173
lncRNA 165–6
m1A 173–4
m6A 162, 172–3
miRNAs 163–5
mRNAs 161–2
pseudouridine 174
rRNAs 159
tRNAs 156–9
UsnRNA and snoRNA 159–61
P bodies and 250–1
TDP-43 and 554–5
ribonucleoparticle (RNP) complex 11–12. *See also* neuronal RNA granules
dendritic RNAs 115–16
formation of 363
in mRNA localization control 226–7
nuclear eIF4E and 10–11
ribopuromycylation (RPM) 251–2
ribosomal proteins (RPs), ASD and 459–60
ribosomal RNA (rRNA) 155–6
modifications to 159
snoRNAs and 161
ribosomal S6 kinase B2. *See* p70 ribosomal S6 kinase 1
ribosomes. *See also* 80S ribosome; 40S ribosome
biogenesis of 403
in dentate gyrus 240–1
FMRP stalling of 228
in IRES 134–5
local accumulation of 66
mRNA translation by 3–4

recycling of 132, 460
in RNA granules 244–5
RiboTag, addiction and 439, 500
Rictor 87, 375, 495
RIP. *See* mRNA co-immunoprecipitations
RISC accessory proteins 219–20
RISC particles. *See* RNA-induced silencing complexes particles
RNA-binding motif protein 15B (RBM15B) 333–4
RNA-binding proteins (RBPs)
addiction and 497–9
in ALS and FTD 553–64
aggregates and mutants 553–4, 573–5
Ataxin-2 561, 563–4
CAMTA1 and DENND4A translation enhancement 557–8
FUS knockdown 558–9
Futsch/MAP1B and Hsc70-4/HSPA8 mRNA repression 556–7
hnRNPA2/B1 and A1 561–2
MATR3 561–2
miRNA machinery and 559–60
motor-neuronal cell translation enhancement 557
mRNA-specific translational repression 555–6
mRNA translational targets 560–1
reduced levels of synthesized proteins in 559
ribosome-mRNA complexes and 555
RNA regulation and 554–5
TIA1 561–3
alternative splicing by 324–36, 327*f*
ASD and 323–4, 461–8
CELF6 464–5
FMR1 461–3
JAKMIP1 465–7
RBFOX1 467–8
RNG105 463–4
axonal mRNA translation with 369–72, 370*f*
of *C. elegans* 325*t*
in dendritic RNA transport 115
eIF2α phosphorylation and 575–6
Fox homolog (RBFOX1) 328, 467–8
in HD 564–8
liquid phase separation of 369–71
major depressive disorder and 522
MID1 complex 565–6
applications of 567–8
metformin 566–7
in miRNA regulation 220–5, 241–3, 242–243*f*, 398
FMRP 221–2
MOV10 222–5, 223*f*
mRNA stability and 365–6
mRNA transport particles and 248–9
NDD and 324
in neurodegenerative diseases 551
non-canonical 572–3
in neurodevelopmental disorders 412–13
in neuronal RNA granules 226
PD translational control by 569–72
DJ-1 571–2
LRRK2 571
PINK1 and Parkin 570–1
RAN translation and 568–9
in FXTAS 568–9
RNP with 363
sequence motif recognition by 362–3
target mRNA translation and 364, 365*f*
translational activation and repression in NSCs 404–5
RNA-binding proteins with multiple splicing (RBPMS), mechanosensation and 326–9
RNA epigenetics 153–4
decoding 174–6
RNA granule-associated genes 343–4
RNA granule protein 105 (RNG105) 463–4
RNA granules. *See* neuronal RNA granules
RNA helicases, in initiation phase 106–7, 132, 133*f*
RNA-induced silencing complexes (RISC) particles
formation of 367–8
neuronal RNA granules compared with 250–1
RNA ligase, RTCB-1 342–3
RNA recognition motif (RRM) 106
in RBM15B 333–4
in RBPMS 326–7
in TDP-43 553–4

RNA regulators 118
RNA regulon 364
RNAs. *See* ribonucleic acids
RNase. *See* ribonucleases
RNA transcription, translation, and transport factor (RTRAF) 245–6
RNA transport 114
  activity-dependent 118–19
  cis-acting targeting elements and trans-acting transport factors 115–17, 116*f*
RNA viruses, cap-independent initiation of translation with 134–5
RNG105. *See* RNA granule protein 105
RNP complex. *See* ribonucleoparticle complex
Ro 25-6981 517
RPL10 459
RPM. *See* ribopuromycylation
RPM-1 334–5
RPs. *See* ribosomal proteins
RPS23 mutations 460
RRM. *See* RNA recognition motif
rRNA. *See* ribosomal RNA
RTCB-1 342–3
RTD. *See* rapid tRNA decay pathway
RTRAF. *See* RNA transcription, translation, and transport factor
Ryanodine receptors (RyR) 276

S6K. *See* p70 ribosomal S6 kinase
S6K1. *See* p70 ribosomal S6 kinase 1
S6K2. *See* p70 ribosomal S6 kinase 2
S6 kinase A group 81–2
S6 kinase B group 81–2
SAD-1 328–9
S-adenosylmethionine (SAM), m6A and 160
Sal003 49–50, 496
sALS. *See* sporadic amyotrophic lateral sclerosis
SAM. *See* S-adenosylmethionine
SCAs. *See* spinal cerebellar ataxias
Schaffer collateral-CA2 (SC-CA2) synapses 282
schizophrenia
  4E-BPs and 33
  lncRNAs and circRNAs in 165
  mRNA translation and 410–11
Schwann cells 429
sciatic nerve crush 378
selective serotonin reuptake inhibitors (SSRIs) 511–12, 519
Sema3A 379
SERBP1. *See* serpine1 mRNA binding protein
Serine/threonine-protein kinase (SMG1) 366–7
serotonin (5-HT)
  CPEB and 203–4
  eEF2 phosphorylation and 68–9
  local translation and 243–4
  rapamycin and 243–4
  snoRNA-mediated RNA modification and 160–1
  synaptic changes and 195–6
serpine1 mRNA binding protein (SERBP1) 245–6
sertraline 520
Sestrin 472
sexual dimorphic neuronal gene expression, lncRNA in 340–1
SG. *See* stress granules
short-term facilitation, protein synthesis inhibitors and 194–5
short-term memory 43–4
SH-SY5Y neuronal cell lines 137–8, 555
Signal transducer and activator of transcription 3 (*Stat3α*) 363–4
single nucleotide polymorphisms (SNP), in tau IRES 139
SKAR 84
SKF-38393 279–80
SLC7A5. *See* neutral amino acid transporter 1
small nuclear RNA (snRNA) 155–6
small nucleolar RNA (snoRNA) 155–6
  modifications to 159–61
Smaug2 405
SMG1. *See* Serine/threonine-protein kinase
smg-2 329–30
SMN. *See* survival motor neuron
SNB-1 339
SNCA. *See* α-synuclein
snoRNA. *See* small nucleolar RNA
SNP. *See* single nucleotide polymorphisms
snRNA. *See* small nuclear RNA
social approach test, S6K1 and 89
social interactions, BC RNA and 112

SOD1. *See* superoxide dismutase 1
SOR. *See* spatial object recognition
Sox2 406–7
SP. *See* Substance P
spatial object recognition (SOR), BT and 279–80
Spen paralog and ortholog C-terminal domain (SPOC) 333–4
spinal cerebellar ataxias (SCAs) 568
spinal cord, injury and 429
spinal modulation of pain 440
spinal muscular atrophy 331–3
spinogenesis, S6K1 in 87
spinophilin 87
spliceosome 159–60
SPOC. *See* Spen paralog and ortholog C-terminal domain
sporadic amyotrophic lateral sclerosis (sALS) 553. *See also* amyotrophic lateral sclerosis
SSRIs. *See* selective serotonin reuptake inhibitors
SSUP-72 335–6
stalled polysomes
  eEF2 phosphorylation 256
  FMRP and 255, 462
  mechanism of 253–7, 254*f*
  neurodevelopmental disorders and 257
  in neuronal RNA granules 251–2
  nonsense-mediated and no-go decay prevention 256–7
  puromycin and 244–5
  synaptic plasticity and 252
  UPF1–Staufen 2 interactions 255–6
Stat3a. *See* Signal transducer and activator of transcription 3
Staufen 115, 245–6
Staufen 2 particle
  mRNA transport particles and 248–9
  stalled polysomes and 255–6
Staufen-mediated decay 367
staurosporine 142
STC hypothesis. *See* synaptic tagging and capture
STEP. *See* striatal-enriched protein tyrosine phosphatase
STET. *See* strong tetanization
stimulus dependent local translation, synaptic plasticity and 243–4
STRADA deletion 91–2
stress granules (SG)
  ALS/FTD and 573–5
  axonal translation and 369
  axons and 372–3
  neuronal RNA granules compared with 249–50
  shared components of 246–8, 247*f*
striatal-enriched protein tyrosine phosphatase (STEP) 107, 302
strong tetanization (STET) 272–4, 274*f*, 277*f*, 278*f*
structural plasticity of neurons, S6K1 and 86–8, 88*f*
substance abuse
  eEF2 and eEF2K in 73
  4E-BPs in 32
Substance P (SP) 282
sucrose sedimentation, neuronal RNA granules and 244–5
SunTag method 251–2
superoxide dismutase 1 (*SOD1*), mutations in 553
survival motor neuron (SMN) 327*f*, 331–3
syd-2 334–5
syd-9 330
sydn-1 335–6
symplekin 197–8
synapse
  AD and 533–4
  adaptability of 297–8
  cooperation and competition in 275–6, 277*f*
  cross-tagging 280, 281*f*
  eEF2/eEF2K pathway and 69
  formation of 114–15
    cryptic splicing of DKL-1 during 334
    ncRNA role in 336–42
  local translation at 107
    persistence of change and 195–6
  m6A at 169–70, 172–3
  Malata1 at 165
  MDD and 512
  protein synthesis at 193–9
  tripartite 95, 169–70
Synapsin1 515

synaptic capture hypothesis 194–5
synaptic competition 275–6, 277*f*
  metaplasticity and 276–9, 278*f*, 279*f*
synaptic cooperation 275–6, 277*f*
synaptic failure 533–4
synaptic homeostasis
  disease implications of 308–10
  LRRK2 and 309–10
  maintenance of 193–4
  mir-1 in 337–8, 338*f*
  postsynaptic receptor function and 302–3
synaptic plasticity. *See also* memory function
  addiction and 487
  CPEB role in 199–203
    CPEB1 200
    CPEB2 201
    CPEB3 202
    CPEB4 203
    invertebrate 203–6
  eEF2 and eEF2K in 65, 67–70, 256
  FMRP and loss of 221
  forms of 269
  4E-BPs in 30–2
  intermediate term facilitation 252
  ISR and 50–2
  local translation and 194–5
    stimulus-dependent 243–4
  LTP and 50, 269
  memory storage and 193
  miR-134 and 225
  miRNAs in 9–10
  NMDAR in 7
  pre- and post-synaptic activity and 297–8
  protein synthesis for 107
  S6K1 and 83–6
  stalled polysomes and 252, 256
  translation in 4
synaptic tagging 194–5
  inverse 280–2, 282*f*
synaptic tagging and capture (STC)
  cross-tagging 280, 281*f*
  hypothesis 272–5, 273*f*, 274*f*
  inverse synaptic tagging 280–2, 282*f*
  metaplasticity 276–9, 278*f*, 279*f*
  synaptic cooperation and competition 275–6, 277*f*
synaptic transmission
  eEF2K in 69–70
  4E-BP2 in 28–9
  ISR and 44
    inactivation of 48
  UNC-75/CELF in 327*f*, 329–31, 331*f*
synaptotagmin binding cytoplasmic RNA interacting protein (SYNCRIP)
  neurodevelopmental disorders and 257
  neuronal RNA granules and 245–6
α-synuclein (SNCA), IRES and 139–40

t6A. *See* N6-threonyl-carbamoyl-adenosine
TACs. *See* transit amplifying cells
TAR DNA-binding protein-43 (TDP-43)
  ALS and 323–4, 553
  CAMTA1 translation enhancement 557–8
  DENND4A translation enhancement 557–8
  FTD and 553
  Futsch/MAP1B mRNA repression 556–7
  Hsc70-4/HSPA8 mRNA repression 556–7
  MID1 complex and 567–8
  motor-neuronal cell translation enhancement 557
  mRNA-specific translational repression 555–6
  mutations 553–4
  overexpression of 552
  primary cortical neuron translation enhancement 557
  ribosome-mRNA complexes and 555
  RNA regulation and 554–5
target of rapamycin (TOR). *See* mammalian/mechanistic target of rapamycin
taste recognition task. *See* conditioned taste aversion; novel object recognition test
tau protein
  axon localization of 361–2
  IRES and 139
TBI. *See* traumatic brain injury
T-cell-restricted intracellular antigen-1 protein (TIA1)
  ALS and 323–4, 561–3
  nerve regeneration and 373

tdc-1 332
TDP-43. *See* TAR DNA-binding protein-43
ten-eleven translocator family (TET) 163, 499–501
ternary complex 298–9
TET. *See* ten-eleven translocator family
tetrodotoxin (TTX) 138, 303–4
texture recognition, BC RNA and 112
TG. *See* trigeminal ganglia
thapsigargin 610–11
THC, mTORC1 signaling and 493*t*
theta-burst stimulation, LTP and 300–1
TIA1. *See* T-cell-restricted intracellular antigen-1 protein
tiar-2 343
Tle4 405–6
TNRC6. *See* tri-nucleotide-repeat containing proteins
tom-1 329–30
TOR. *See* mammalian/mechanistic target of rapamycin
Torin-1 26
TOR (target of rapamycin) signaling motif (TOS) 24
Tp53inp2 368–9
trans-acting transport factors 115–17, 116*f*
transcription
  rate changes of 364
  translation and 131
transfer RNAs (tRNAs) 155–6
  modification of 155–9
  in NSCs 402–3
transit amplifying cells (TACs) 28
Transitin 408–9
translating ribosome affinity purification (TRAP) 437–8
  addiction and 439, 500
translation. *See also* cap-dependent translation; elongation phase; initiation phase; local translation; neuronal translation
  in axons 360
    direct regulation of 374–9
    HuD and 366
    mechanisms of 369
    NMD and 366–7
    RBPs and 369–72, 370*f*
    regulation of 360–79, 361*f*
    UPF1 and 367
  in brain 4, 12–13, 23
  circadian rhythms and 30
  dephosphorylation regulation of 108–11
  dysregulation 111–14, 113*f*
    FXS and 462–3
  4E-T repression of 6*f*, 10
  gene expression control and 105
  general control of 132–4, 133*f*
  Hebbian plasticity and 300–2
    LTD and 301–2
    LTP and 300–1
  ISR and reprogramming of 44–7
  in learning 82
  in memory formation 4, 82
  miRNA in 9
  pain and 429–30
  phosphorylation regulation of 108–11
  regulation of 131, 196–9
    ternary complex 298–9
  repeat-associated non-AUG 6*f*, 12
  steps of 3, 63–4, 105, 298, 374–5, 487–8, 534
  at synapse 193–9
  transcription and 131
transport granule 363
transport particle 363
tranylcypromine 519
TRAP. *See* translating ribosome affinity purification
traumatic brain injury (TBI), ISR and 53
trazodone 53–4
trichostatin A (TSA) 286–7
trigeminal ganglia (TG) 429
Trim32 408
tri-nucleotide-repeat containing proteins (TNRC6) 9, 473
tripartite synapse 95, 169–70
TrkB, IRES and 137–8
tRNA methyltransferase 10 (TRMT10) 158
tRNAs. *See* transfer RNAs
TSA. *See* trichostatin A
TSC. *See* tuberous sclerosis complex
TSC 469–70

tsc1 469–70
tsc2 469–70
TTX. *See* tetrodotoxin
tuberin 537–8
tuberous sclerosis complex (TSC) 308–9, 375–7, 376*f*
  AD and 537–40
  ASD and 454–6, 469–70
  mTOR and 410–11
tunicamycin 610–11
twk-39 330

U2AF 331–2
ubiquitin/proteasome system (UPS) 156–7
unc-16 329–30
unc-26 332–3
unc-29 337
unc-32 329–30
unc-44 335–6
UNC-49 339
unc-52 326–9
unc-55 338–9, 341–2
unc-57 332–3
unc-63 337
unc-64 330–1
unc-75 327*f*, 329–31, 331*f*
unc-104 341–2
unfolded protein response (UPR) 156–7, 378–9, 402, 534–6
uORFs. *See* upstream open reading frames
UPF1. *See* Upstream frameshift protein 1
upf3b 461
up-frame shift protein 3B (UPF3B), ASD and 460–1
UPR. *See* unfolded protein response
UPS. *See* ubiquitin/proteasome system
Upstream frameshift protein 1 (UPF1)
  neuronal RNA granules 245–6
  NMD and 366–7
  stalled polysomes and 255–6
  Staufen-mediated decay 367
upstream open reading frames (uORFs) 406–8
  eIF2 600
  pain and 433–4, 438
uridine modifications in tRNA
  Elongator complex and 156–7
  pseudouridylate synthase and 157
uridylate-rich small nuclear RNAs (UsnRNAs) 155–6
  modifications to 159–61
  snoRNAs and 161

vanishing white matter disease (VWMD)
  astrocytes in 597
  clinical severity of 596
  diagnosis of 596
  eIF2 and 602
  eIF2B mutations and 54, 597–8, 599*t*
  ISRIB and 611
  mouse models for 597–8
  neurological focus of 609
  organ impacts of 596–7
  phenotype of 596
VD. *See* Ventral type D
venlafaxine 519–20
Ventral type D (VD) 338–9
Vimentin 372
VP16-CREB 270–1
VWMD. *See* vanishing white matter disease

weak tetanization (WTET) 273–4, 274*f*, 277*f*, 278*f*
Williams-Beuren syndrome 159
WT1 associated protein (WTAP) 172
WTET. *See* weak tetanization

XBP-1 342, 378
XIAP, IRES and 141–2
X inactive specific transcript (XIST) 165
X-linked mental retardation 463

yeast, m6A modification in 171
Y-maze, S6K1 and 89
YT521-B homology family 168–9
YTHDC1 168–9
YTHDC2 168–9
YTHDF1 168–70
YTHDF2 168–70
YTHDF3 168–70
$YX_4L\phi$ motif 23–5

ZBP. *See* zipcode binding protein
ZBP1. *See* zipcode binding protein 1
zipcode binding protein (ZBP)
  neuronal RNA granules and 245–6
  in RNA delivery 115–16, 116*f*
zipcode binding protein 1 (ZBP1)
  addiction and 498
  competition for 364
  mRNA localization with 227
  mRNA transport particles and 248–9
  phosphorylation of 369–71
  RBPs and 369–71
zip-code motif 362–3